中国科学院科学出版基金资助出版

国外数学名著系列（影印版） 33

Galois Theory of Linear Differential Equations

线性微分方程的伽罗瓦理论

Marius van der Put　Michael F. Singer

科学出版社
北京

图字：01-2006-7398

Marius van der Put，Michael F. Singer：Galois Theory of Linear Differential Equations

图书在版编目（CIP）数据

线性微分方程的伽罗瓦理论＝Galois Theory of Linear Differential Equations/（荷）范德帕特（van der Put，M.）等著.—影印版.—北京：科学出版社，2007

（国外数学名著系列）

ISBN 978-7-03-018301-9

Ⅰ. 线… Ⅱ. 范… Ⅲ. 线性方程：微分方程-英文 Ⅳ. O175

中国版本图书馆CIP数据核字（2006）第154744号

责任编辑：范庆奎/责任印刷：徐晓晨/封面设计：黄华斌

科学出版社 出版

北京东黄城根北街16号

邮政编码：100717

http://www.sciencep.com

北京凌奇印刷有限责任公司 印刷

科学出版社发行 各地新华书店经销

*

2007年1月第 一 版 开本：B5（720×1000）

2016年4月第二次印刷 印张：28 3/4

字数：538 000

POD定价： 188.00元

（如有印装质量问题，我社负责调换）

《国外数学名著系列》(影印版) 序

要使我国的数学事业更好地发展起来，需要数学家淡泊名利并付出更艰苦地努力。另一方面，我们也要从客观上为数学家创造更有利的发展数学事业的外部环境，这主要是加强对数学事业的支持与投资力度，使数学家有较好的工作与生活条件，其中也包括改善与加强数学的出版工作。

从出版方面来讲，除了较好较快地出版我们自己的成果外，引进国外的先进出版物无疑也是十分重要与必不可少的。从数学来说，施普林格(Springer)出版社至今仍然是世界上最具权威的出版社。科学出版社影印一批他们出版的好的新书，使我国广大数学家能以较低的价格购买，特别是在边远地区工作的数学家能普遍见到这些书，无疑是对推动我国数学的科研与教学十分有益的事。

这次科学出版社购买了版权，一次影印了 23 本施普林格出版社出版的数学书，就是一件好事，也是值得继续做下去的事情。大体上分一下，这 23 本书中，包括基础数学书 5 本，应用数学书 6 本与计算数学书 12 本，其中有些书也具有交叉性质。 这些书都是很新的，2000 年以后出版的占绝大部分，共计 16 本，其余的也是 1990 年以后出版的。这些书可以使读者较快地了解数学某方面的前沿，例如基础数学中的数论、代数与拓扑三本，都是由该领域大数学家编著的“数学百科全书”的分册。对从事这方面研究的数学家了解该领域的前沿与全貌很有帮助。按照学科的特点，基础数学类的书以“经典”为主，应用和计算数学类的书以“前沿”为主。这些书的作者多数是国际知名的大数学家，例如《拓扑学》一书的作者诺维科夫是俄罗斯科学院的院士，曾获“菲尔兹奖”和“沃尔夫数学奖”。这些大数学家的著作无疑将会对我国的科研人员起到非常好的指导作用。

当然，23 本书只能涵盖数学的一部分，所以，这项工作还应该继续做下去。更进一步，有些读者面较广的好书还应该翻译成中文出版，使之有更大的读者群。

总之，我对科学出版社影印施普林格出版社的部分数学著作这一举措表示热烈的支持，并盼望这一工作取得更大的成绩。

王　元

2005 年 12 月 3 日

Preface

This book is an introduction to the algebraic, algorithmic, and analytic aspects of the Galois theory of homogeneous linear differential equations. Although the Galois theory has its origins in the 19th Century and was put on a firm footing by Kolchin in the middle of the 20th Century, it has experienced a burst of activity in the last 30 years. In this book we present many of the recent results and new approaches to this classical field. We have attempted to make this subject accessible to anyone with a background in algebra and analysis at the level of a first-year graduate student. Our hope is that this book will prepare and entice the reader to delve further.

Here we will describe the contents of this book. Various researchers are responsible for the results described here. We will not attempt to give proper attributions here but refer the reader to each of the individual chapters for appropriate bibliographic references.

The Galois theory of linear differential equations (which we shall refer to simply as differential Galois theory) is the analog for linear differential equations of the classical Galois theory for polynomial equations. The natural analog of a field in our context is the notion of a differential field. This is a field k together with a derivation $\partial : k \to k$, that is, an additive map that satisfies $\partial(ab) = \partial(a)b + a\partial(b)$ for all $a, b \in k$ (we will usually denote ∂a for $a \in k$ as a'). Except for those in Chap. 13, all differential fields will be of characteristic zero. A linear differential equation is an equation of the form $\partial Y = AY$ where A is an $n \times n$ matrix with entries in k, although sometimes we shall also consider scalar linear differential equations $L(y) = \partial^n y + a_{n-1}\partial^{n-1} y + \cdots + a_0 y = 0$ (these objects are in general equivalent, as we show in Chap. 2). One has the notion of a "splitting field," the Picard-Vessiot extension, which contains "all" solutions of $L(y) = 0$ and in this case has the additional structure of being a differential field. The differential Galois group is the group of field automorphisms of the Picard-Vessiot field fixing the base field and commuting with the derivation. Although defined abstractly, this group can be easily represented as a group of matrices and has the structure of a linear algebraic group, that is, it is a group of invertible matrices defined by the vanishing of a set of polynomials in the entries of these matrices. There is a Galois correspondence identifying differential subfields with linear algebraic subgroups of the Galois group. Corresponding to the notion of solvability by radicals for polynomial equations is the notion of solvability in terms of integrals, exponentials, and algebraics, that is, solvable in terms of liouvillian functions, and one can characterize this in terms of the differential Galois group as well.

Chapter 1 presents these basic facts. The main tools come from the elementary algebraic geometry of varieties over fields that are not necessarily algebraically closed and the theory of linear algebraic groups. In Appendix A we develop the results necessary for the Picard-Vessiot theory.

In Chap. 2, we introduce the ring $k[\partial]$ of differential operators over a differential field k, that is, the (in general, noncommutative) ring of polynomials in the symbol ∂ where multiplication is defined by $\partial a = a' + a\partial$ for all $a \in k$. For any differential equation $\partial Y = AY$ over k one can define a corresponding $k[\partial]$-module in much the same way that one can associate an $F[X]$-module with any linear transformation of a vector space over a field F. If $\partial Y = A_1 Y$ and $\partial Y = A_2 Y$ are differential equations over k and M_1 and M_2 are their associated $k[\partial]$-modules, then $M_1 \simeq M_2$ as $k[\partial]$-modules if and only if there is an invertible matrix Z with entries in k such that $Z^{-1}(\partial - A_1)Z = \partial - A_2$, i.e., $A_2 = Z^{-1}A_1Z - Z^{-1}Z'$. We say two equations are *equivalent over k* if such a relation holds. We show that equivalent equations have the same Galois groups and so can define the Galois group of a $k[\partial]$-module. This chapter is devoted to further studying the elementary properties of modules over $k[\partial]$ and their relationship to linear differential equations. Furthermore, the tannakian equivalence between differential modules and representations of the differential Galois group is presented.

In Chap. 3, we study differential equations over the field of fractions $k = \mathbf{C}((z))$ of the ring of formal power series $\mathbf{C}[[z]]$ over the field of complex numbers, provided with the usual differentiation $\frac{d}{dz}$. The main result is to classify $k[\partial]$-modules over this ring or, equivalently, show that any differential equation $\partial Y = AY$ can be put in a normal form over an algebraic extension of k (an analog of the Jordan Normal Form of complex matrices). In particular, we show that any equation $\partial Y = AY$ is equivalent (over a field of the form $\mathbf{C}((t))$, $t^m = z$ for some integer $m > 0$) to an equation $\partial Y = BY$ where B is a block diagonal matrix where each block B_i is of the form $B_i = q_i I + C_i$ where $q_i \in t^{-1}\mathbf{C}[t^{-1}]$ and C_i is a constant matrix. We give a proof (and formal meaning) of the classical fact that any such equation has a solution matrix of the form $Z = Hz^L e^Q$, where H is an invertible matrix with entries in $\mathbf{C}((t))$, L is a constant matrix (i.e., with coefficients in $\mathbf{C}$) z^L means $e^{\log(z)L}$, and Q is a diagonal matrix whose entries are polynomials in t^{-1} without a constant term. A differential equation of this type is called *quasisplit* (because of its block form over a finite extension of $\mathbf{C}((z))$). Using this, we are able to explicitly give a universal Picard-Vessiot extension containing solutions for all such equations. We also show that the Galois group of the above equation $\partial Y = AY$ over $\mathbf{C}((z))$ is the smallest linear algebraic group containing a certain commutative group of diagonalizable matrices (the *exponential torus*) and one more element (the *formal monodromy*) and these can be explicitly calculated from its normal form. In this chapter we also begin the study of differential equations over $\mathbf{C}(\{z\})$, the field of fractions of the ring of convergent power series $\mathbf{C}\{z\}$. If A has entries in $\mathbf{C}(\{z\})$, we show that the equation $\partial Y = AY$ is equivalent over $\mathbf{C}((z))$ to a unique (up to equivalence over $\mathbf{C}(\{z\})$) equation with entries in $\mathbf{C}(\{z\})$, that is *quasisplit*. This latter fact is key to understanding the analytic behavior of solutions of these equations and will be used repeatedly in succeeding chapters. In Chaps. 2 and 3 we also use the language of tannakian categories to describe some of these results. This theory is explained in Appendix B. This appendix also contains a proof of the general result that the category of $k[\partial]$-modules for a differential field k forms a tannakian category and explains how one can deduce from this the fact that the Galois groups of the associated equations are linear algebraic groups. In general, we shall use tannakian categories throughout the book to deduce facts about categories of special $k[\partial]$-modules, i.e., deduce facts about the Galois groups of restricted classes of differential equations.

In Chap. 4, we consider the "direct" problem, which is to calculate explicitly for a given differential equation or differential module its Picard-Vessiot ring and its differential Galois group. A complete answer for a given differential equation should, in principle, provide all the algebraic information about the differential equation. Of course this can only be achieved for special base fields k, such as $\overline{\mathbf{Q}}(z)$, $\partial z = 1$ (where $\overline{\mathbf{Q}}$ is the algebraic closure of the field of rational numbers). The direct problem requires factoring many differential operators L over k. A right-hand factor $\partial - u$ of L (over k or over an algebraic extension of k) corresponds to a special solution f of $L(f) = 0$, which can be rational, exponential, or liouvillian. Some of the ideas involved here were already present in Beke's classical work on factoring differential equations. The "inverse" problem, namely to construct a differential equation over k with a prescribed differential Galois group G and action of G on the solution space, is treated for a *connected* linear algebraic group in Chap. 11. In the opposite case that G is a finite group (and with base field $\overline{\mathbf{Q}}(z)$) an effective algorithm is presented together with examples for equations of order 2 and 3. We note that some of the algorithms presented in this chapter are efficient and others are only the theoretical basis for an efficient algorithm.

Starting with Chap. 5, we turn to questions that are, in general, of a more analytic nature. Let $\partial Y = AY$ be a differential equation where A has entries in $\mathbf{C}(z)$, where $\mathbf{C}$ is the field of complex numbers and $\partial z = 1$. A point $c \in \mathbf{C}$ is said to be a *singular point* of the equation $\partial Y = AY$ if some entry of A is not analytic at c (this notion can be extended to the point at infinity on the Riemann sphere $\mathbf{P}$ as well). At any point p on the manifold $\mathbf{P}\backslash\{$the singular points$\}$, standard existence theorems imply that there exists an invertible matrix Z of functions, analytic in a neighborhood of p, such that $\partial Z = A\,Z$. Furthermore, one can analytically continue such a matrix of functions along any closed path γ, yielding a new matrix Z_γ which must be of the form $Z_\gamma = ZA_\gamma$ for some $A_\gamma \in \mathrm{GL}_n(\mathbf{C})$. The map $\gamma \mapsto A_\gamma$ induces a homomorphism, called the monodromy homomorphism, from the fundamental group $\pi_1(\mathbf{P}\backslash\{$the singular points$\}, c)$ into $\mathrm{GL}_n(\mathbf{C})$. As explained in Chap. 5, when all the singular points of $\partial Y = AY$ are regular singular points (that is, all solutions have at most polynomial growth in sectors at the singular point), the smallest linear algebraic group containing the image of this homomorphism is the Galois group of the equation. In Chaps. 5 and 6 we consider the inverse problem: Given points $\{p_0, \ldots, p_n\} \subset \mathbf{P}^1$ and a representation $\pi_1(\mathbf{P}\backslash\{p_1, \ldots, p_n\}, p_0) \to \mathrm{GL}_n(\mathbf{C})$, does there exist a differential equation with regular singular points having this monodromy representation? This is one form of Hilbert's 21st Problem and we describe its positive solution. We discuss refined versions of this problem that demand the existence of an equation of a more restricted form, as well as the existence of scalar linear differential equations having prescribed monodromy. Chapter 5 gives an elementary introduction to this problem concluding with an outline of the solution depending on basic facts concerning sheaves and vector bundles. In Appendix C, we give an exposition of the necessary results from sheaf theory needed in this and later sections. Chapter 6 contains deeper results concerning Hilbert's 21st problem and uses the machinery of connections on vector bundles, material that is developed in Appendix C and this chapter.

In Chap. 7, we study the analytic meaning of the formal description of solutions of a differential equation that we gave in Chap. 3. Let $w \in \mathbf{C}(\{z\})^n$ and let A be a matrix with entries in $\mathbf{C}(\{z\})$. We begin this chapter by giving analytic meaning to formal solutions $\hat{v} \in \mathbf{C}((z))^n$ of equations of the form $(\partial - A)\hat{v} = w$. We consider open sectors $S = S(a, b, \rho) = \{z \mid z \neq 0, \arg(z) \in (a, b) \text{ and } |z| < \rho(\arg(z))\}$, where $\rho(x)$ is a continuous positive function of a real variable and $a \leq b$ are real numbers and functions f analytic in S and define what it means for a formal series $\sum a_i z^i \in \mathbf{C}((z))$ to be the asymptotic expansion of f in S. We show that for any formal solution $\hat{v} \in \mathbf{C}((z))^n$ of $(\partial - A)\hat{v} = w$ and any sector $S = S(a, b, \rho)$ *with* $|a - b|$ *sufficiently small and suitable* ρ, there is a vector of functions v analytic in S satisfying $(\partial - A)v = w$ such that each entry of v has the corresponding entry in $\hat{v}$ as its asymptotic expansion. The vector v is referred to as an *asymptotic lift* of $\hat{v}$. In general, there will be many asymptotic lifts of $\hat{v}$ and the rest of the chapter is devoted to describing conditions that guarantee uniqueness. This leads us to the study of *Gevrey functions* and *Gevrey asymptotics*. Roughly stated, the main result, the multisummation theorem, allows us to associate, in a functorial way, to any formal solution $\hat{v}$ of $(\partial - A)\hat{v} = w$ and all but a finite number (mod 2π) of directions d, a unique asymptotic lift in an open sector $S(d - \epsilon, d + \epsilon, \rho)$ for suitable ϵ and ρ. The exceptional values of d are called the *singular directions* and are related to the so-called *Stokes phenomenon*. They play a crucial role in the succeeding chapters where we give an analytic description of the Galois group as well as a classification of meromorphic differential equations. Sheaves and their cohomology are the natural way to take analytic results valid in small neighborhoods and describe their extension to larger domains and we use these tools in this chapter. The necessary facts are described in Appendix C.

In Chap. 8 we give an analytic description of the differential Galois group of a differential equation $\partial Y = AY$ over $\mathbf{C}(\{z\})$ where A has entries in $\mathbf{C}(\{z\})$. In Chap. 3, we show that any such equation is equivalent to a unique *quasisplit* equation $\partial Y = BY$

with the entries of B in $\mathbf{C}(\{z\})$ as well, that is there exists an invertible matrix $\hat{F}$ with entries in $\mathbf{C}((z))$ such that $\hat{F}^{-1}(\partial - A)\hat{F} = \partial - B$. The Galois groups of $\partial Y = BY$ over $\mathbf{C}(\{z\})$ and $\mathbf{C}((z))$ coincide and are generated (as linear algebraic groups) by the associated exponential torus and formal monodromy. The differential Galois group G' over $\mathbf{C}(\{z\})$ of $\partial Y = BY$ is a subgroup of the differential Galois group of $\partial Y = AY$ over $\mathbf{C}(\{z\})$. To see what else is needed to generate this latter differential Galois group we note that the matrix $\hat{F}$ also satisfies a differential equation $\hat{F}' = A\hat{F} - \hat{F}B$ over $\mathbf{C}(\{z\})$ and so the results of Chap. 7 can be applied to $\hat{F}$. Asymptotic lifts of $\hat{F}$ can be used to yield isomorphisms of solution spaces of $\partial Y = AY$ in overlapping sectors, and using this we describe how, for each singular direction d of $\hat{F}' = A\hat{F} - \hat{F}B$, one can define an element St_d (called the *Stokes map in the direction* d) of the Galois group G of $\partial Y = AY$ over $\mathbf{C}(\{z\})$. Furthermore, it is shown that G is the smallest linear algebraic group containing the Stokes maps $\{St_d\}$ and G'. Various other properties of the Stokes maps are described in this chapter.

In Chap. 9, we consider the meromorphic classification of differential equations over $\mathbf{C}(\{z\})$. If one fixes a quasisplit equation $\partial Y = BY$, one can consider pairs $(\partial - A, \hat{F})$, where A has entries in $\mathbf{C}(\{z\})$, $\hat{F} \in \mathrm{GL}_n(\mathbf{C}((z)))$ and $\hat{F}^{-1}(\partial - A)\hat{F} = \partial - B$. Two pairs $(\partial - A_1, \hat{F}_1)$ and $(\partial - A_2, \hat{F}_2)$ are called *equivalent* if there is a $G \in \mathrm{GL}_n(\mathbf{C}(\{z\}))$ such that $G(\partial - A_1)G^{-1} = \partial - A_2$ and $\hat{F}_2 = \hat{F}_1 G$. In this chapter, it is shown that the set E of equivalence classes of these pairs is in bijective correspondence with the first cohomology set of a certain sheaf of nonabelian groups on the unit circle, the *Stokes sheaf*. We describe how one can, furthermore, characterize those sets of matrices that can occur as Stokes maps for some equivalence class. This allows us to give the above cohomology set the structure of an affine space. These results will be further used in Chaps. 10 and 11 to characterize those groups that occur as differential Galois groups over $\mathbf{C}(\{z\})$.

In Chap. 10, we consider certain differential fields k and certain classes of differential equations over k and explicitly describe the *universal Picard-Vessiot ring* and its group of differential automorphisms over k, the *universal differential Galois group*, for these classes. For the special case $k = \mathbf{C}((z))$ this universal Picard-Vessiot ring is described in Chap. 3. Roughly speaking, a universal Picard-Vessiot ring is the smallest ring such that any differential equation $\partial Y = AY$ (with A an $n \times n$ matrix) in the given class has a set of n independent solutions with entries from this ring. The group of differential automorphisms over k will be an affine group scheme, and for any equation in the given class its Galois group will be a quotient of this group scheme. The necessary information concerning affine group schemes is presented in Appendix B. In Chap. 10, we calculate the universal Picard-Vessiot extension for the class of regular differential equations over $\mathbf{C}((z))$, the class of arbitrary differential equations over $\mathbf{C}((z))$ and the class of meromorphic differential equations over $\mathbf{C}(\{z\})$.

In Chap. 11, we consider the problem of, given a differential field k, determining which linear algebraic groups can occur as differential Galois groups for linear differential equations over k. In terms of the previous chapter, this is the, a priori, easier problem of determining the linear algebraic groups that are quotients of the universal Galois group. We begin by characterizing those groups that are differential Galois groups over $\mathbf{C}((z))$. We then give an analytic proof of the fact that any linear algebraic group occurs as a differential Galois group of a differential equation $\partial Y = AY$ over $\mathbf{C}(z)$, and describe the minimal number and type of singularities of such an equation that are necessary to realize a given group. We end by discussing an algebraic (and constructive) proof of this result for connected linear algebraic groups and give explicit details when the group is semisimple.

In Chap. 12, we consider the problem of finding a fine moduli space for the equivalence classes E of differential equations considered in Chap. 9. In that chapter, we describe how E has a natural structure as an affine space. Nonetheless, it can be shown that there does not exist a universal family of equations parameterized by E. To remedy this situation, we show the classical result that for any meromorphic differential equation $\partial Y = AY$, there is a differential equation $\partial Y = BY$ where B has coefficients in $\mathbf{C}(z)$ (i.e., a differential equation on the Riemann sphere) having singular points at 0 and ∞ such that the singular point at infinity is regular and such that the equation is equivalent to the original equation when both are considered as differential equations over $\mathbf{C}(\{z\})$. Furthermore, this latter equation can be identified with a (meromorphic) connection on a free vector bundle over the Riemann sphere. In this chapter we show that, loosely speaking, there exists a fine moduli space for connections on a fixed free vector bundle over the Riemann sphere having a regular singularity at infinity and an irregular singularity at the origin together with an extra piece of data (corresponding to fixing the formal structure of the singularity at the origin).

In Chap. 13, the differential field K has characteristic $p > 0$. A perfect field (i.e., $K = K^p$) of characteristic $p > 0$ has only the zero derivation. Thus we have to assume that $K \neq K^p$. In fact, we will consider fields K such that $[K : K^p] = p$. A nonzero derivation on K is then unique up to a multiplicative factor. This seems to be a good analog of the most important differential fields $\mathbf{C}(z)$, $\mathbf{C}(\{z\})$, $\mathbf{C}((z))$ in characteristic zero. Linear differential equations over a differential field of characteristic $p > 0$ have attracted, for various reasons, a lot of attention. Some references are [90, 139, 152, 153, 162, 205, 217, 227, 229, 8, 226]. One reason is Grothendieck's conjecture on p-curvatures, which states that the differential Galois group of a linear differential equation in characteristic zero is finite if and only if the p-curvature of the reduction of the equation modulo p is zero for almost all p. N. Katz has extended this conjecture to one that states that the Lie algebra of the differential Galois group of a linear differential equation in characteristic zero is determined by the collection of its p-curvatures (for almost all p). In this chapter we will classify a differential module over K essentially by the Jordan normal form of its p-curvature. Algorithmic considerations make this procedure effective. A glimpse at order two equations gives an indication of how this classification could be used for linear differential equations in characteristic 0. A more or less obvious observation is that these linear differential equations in positive characteristic behave very differently from what might be expected from the characteristic zero case. A different class of differential equations in positive characteristic, namely the iterative differential equations, is introduced. The chapter ends with a survey on iterative differential modules.

Appendix A contains the tools from the theory of affine varieties and linear algebraic groups that are needed, particularly in Chap. 1. Appendix B contains a description of the formalism of tannakian categories that are used throughout the book. Appendix C describes the results from the theory of sheaves and sheaf cohomology that are used in the analytic sections of the book. Finally, Appendix D discusses systems of linear partial differential equations and the extent to which the results of this book are known to generalize to this situation.

Conspicuously missing from this book are discussions of the arithmetic theory of linear differential equations as well as the Galois theory of nonlinear differential equations. A few references are [162, 197, 199, 222, 223, 293, 294, 295, 296]. We have also not described the recent applications of differential Galois theory to Hamiltonian mechanics for which we refer to [11] and [213]. For an extended historical treatment of linear differential equations and group theory in the 19th century see [113].

Notation and Terminology. We shall use the letters **C**, **N**, **Q**, **R**, and **Z** to denote the complex numbers, the non-negative integers, the rational numbers, the real numbers, and the integers, respectively. Authors of any book concerning functions of a complex variable are confronted with the problem of how to use the terms analytic and holomorphic. We consider these terms synonymous and use them interchangeably but with an eye to avoiding such infelicities as "analytic differential" and "holomorphic continuation".

Acknowledgments. We have benefited from conversations with and the comments of many colleagues. Among those we especially wish to thank are A. Bolibruch, B.L.J. Braaksma, O. Gabber, M. van Hoeij, M. Loday-Richaud, B. Malgrange, C. Mitschi, J.-P. Ramis, F. Ulmer, and several anonymous referees.

The second author was partially supported by National Science Foundation Grants CCR-9731507 and CCR-0096842 during the preparation of this book.

Table of Contents

Algebraic Theory

1 Picard-Vessiot Theory

In this chapter we give the basic algebraic results from the differential Galois theory of linear differential equations. Other presentations of some or all of this material can be found in the classics of Kaplansky [151] and Kolchin [162] (and Kolchin's original papers that have been collected in [25]) as well as the recent book of Magid [183] and the papers [231] and [173].

1.1 Differential Rings and Fields

The study of polynomial equations leads naturally to the notions of rings and fields. For studying differential equations, the natural analogs are *differential rings* and *differential fields*, which we now define. All the rings considered in this chapter are assumed to be commutative, to have a unit element and to contain **Q**, the field of the rational numbers.

Definition 1.1 A *derivation* on a ring R is a map $\partial : R \to R$ having the properties that for all $a, b \in R$,

$$\begin{aligned} \partial(a+b) &= \partial(a) + \partial(b), \quad \text{and} \\ \partial(ab) &= \partial(a)b + a\partial(b). \end{aligned}$$

A ring R equipped with a derivation is called a *differential ring* and a field equipped with a derivation is called a *differential field*. We say a differential ring $S \supset R$ is a *differential extension* of the differential ring R or a *differential ring over* R if the derivation of S restricts on R to the derivation of R. □

Very often we will denote the derivation of a differential ring by $a \mapsto a'$. Furthermore, a derivation on a ring will also be called a differentiation.

Examples 1.2 The following are differential rings.
1. Any ring R with trivial derivation, i.e., $\partial = 0$.
2. Let R be a differential ring with derivation $a \mapsto a'$. One defines the *ring of differential polynomials in* $y_1, \ldots, y_n$ *over* R, denoted by $R\{\{y_1, \ldots, y_n\}\}$,

in the following way. For each $i = 1, \ldots, n$, let $y_i^{(j)}$, $j \in \mathbf{N}$ be an infinite set of distinct indeterminates. For convenience we will write y_i for $y_i^{(0)}$, y_i' for $y_i^{(1)}$ and y_i'' for $y_i^{(2)}$. We define $R\{\{y_1, \ldots, y_n\}\}$ to be the polynomial ring $R[y_1, y_1', y_1'', \ldots, y_2, y_2', y_2'', \ldots, y_n, y_n', y_n'', \ldots]$. We extend the derivation of R to a derivation on $R\{\{y_1, \ldots, y_n\}\}$ by setting $(y_i^{(j)})' = y_i^{(j+1)}$. □

Continuing with Example 1.2.2, let S be a differential ring over R and let $u_1, \ldots, u_n \in S$. The prescription $\phi : y_i^{(j)} \mapsto u_i^{(j)}$ for all i, j, defines an R-linear *differential homomorphism* from $R\{\{y_1, \ldots, y_n\}\}$ to S, that is ϕ is an R-linear homomorphism such that $\phi(v') = (\phi(v))'$ for all $v \in R\{\{y_1, \ldots, y_n\}\}$. This formalizes the notion of evaluating differential polynomials at values u_i. We will write $P(u_1, \ldots, u_n)$ for the image of P under ϕ. When $n = 1$ we shall usually denote the ring of differential polynomials as $R\{\{y\}\}$. For $P \in R\{\{y\}\}$, we say that P *has order* n if n is the smallest integer such that P belongs to the polynomial ring $R[y, y', \ldots, y^{(n)}]$.

Examples 1.3 The following are differential fields. Let C denote a field.
1. $C(z)$, with derivation $f \mapsto f' = \frac{df}{dz}$.
2. The field of formal Laurent series $C((z))$ with derivation $f \mapsto f' = \frac{df}{dz}$.
3. The field of convergent Laurent series $\mathbf{C}(\{z\})$ with derivation $f \mapsto f' = \frac{df}{dz}$.
4. The field of all meromorphic functions on any open connected subset of the extended complex plane $\mathbf{C} \cup \{\infty\}$, with derivation $f \mapsto f' = \frac{df}{dz}$.
5. $\mathbf{C}(z, e^z)$ with derivation $f \mapsto f' = \frac{df}{dz}$. □

The following defines an important property of elements of a differential ring.

Definition 1.4 Let R be a differential ring. An element $c \in R$ is called a *constant* if $c' = 0$. □

In Exercise 1.5.1, the reader is asked to show that the set of constants in a ring forms a ring and in a field forms a field. The ring of constants in Examples 1.2.1 and 1.2.2 is R. In Examples 1.3.1 and 1.3.2, the field of constants is C. In the other examples the field of constants is $\mathbf{C}$. For the last example this follows from the embedding of $\mathbf{C}(z, e^z)$ in the field of the meromorphic functions on $\mathbf{C}$.

The following exercises give many properties of these concepts.

Exercises 1.5
1. *Constructions with rings and derivations*
Let R be any differential ring with derivation ∂.
(a) Let $t, n \in R$ and assume that n is invertible. Prove the formula $\partial(\frac{t}{n}) = \frac{\partial(t)n - t\partial(n)}{n^2}$.
(b) Let $I \subset R$ be an ideal. Prove that ∂ induces a derivation on R/I if and only if $\partial(I) \subset I$.

(c) Let the ideal $I \subset R$ be generated by $\{a_j\}_{j \in J}$. Prove that $\partial(I) \subset I$ if $\partial(a_j) \in I$ for all $j \in J$.
(d) Let $S \subset R$ be a multiplicative subset, i.e., $0 \notin S$ and for any two elements $s_1, s_2 \in S$ one has $s_1 s_2 \in S$. We recall that the *localization of R with respect to S* is the ring RS^{-1}, defined as the set of equivalence classes of pairs (r, s) with $r \in R,\ s \in S$. The equivalence relation is given by $(r_1, s_1) \sim (r_2, s_2)$ if there is an $s_3 \in S$ with $s_3(r_1 s_2 - r_2 s_1) = 0$. The symbol $\frac{r}{s}$ denotes the equivalence class of the pair (r, s). Prove that there exists a unique derivation ∂ on RS^{-1} such that the canonical map $R \to RS^{-1}$ commutes with ∂. Hint: Use the fact that $tr = 0$ implies $t^2\partial(r) = 0$.
(e) Consider the polynomial ring $R[X_1, \ldots, X_n]$ and a multiplicative subset $S \subset R[X_1, \ldots, X_n]$. Let $a_1, \ldots, a_n \in R[X_1, \ldots, X_n]S^{-1}$ be given. Prove that there exists a unique derivation ∂ on $R[X_1, \ldots, X_n]S^{-1}$ such that the canonical map $R \to R[X_1, \ldots, X_n]S^{-1}$ commutes with ∂ and $\partial(X_i) = a_i$ for all i.
(We note that the assumption $\mathbf{Q} \subset R$ is not used in this exercise.)

2. *Constants*
Let R be any differential with derivation ∂.
(a) Prove that the set of constants C of R is a subring containing 1.
(b) Prove that C is a field if R is a field.
Assume that $K \supset R$ is an extension of differential fields.
(c) Assume that $c \in K$ is algebraic over the constants C of R. Prove that $\partial(c) = 0$. Hint: Let $P(X)$ be the minimal monic polynomial of c over C. Differentiate the expression $P(c) = 0$ and use the fact that $\mathbf{Q} \subset R$.
(d) Show that $c \in K$, $\partial(c) = 0$ and c is algebraic over R, implies that c is algebraic over the field of constants C of R. Hint: Let $P(X)$ be the minimal monic polynomial of c over R. Differentiate the expression $P(c) = 0$ and use $\mathbf{Q} \subset R$.

3. *Derivations on field extensions*
Let F be a field (of characteristic 0) and let ∂ be a derivation on F. Prove the following statements.
(a) Let $F \subset F(X)$ be a transcendental extension of F. Choose an $a \in F(X)$. There is a unique derivation $\tilde{\partial}$ of $F(X)$, extending ∂, such that $\tilde{\partial}(X) = a$.
(b) Let $F \subset \tilde{F}$ be a finite extension, then ∂ has a unique extension to a derivation of $\tilde{F}$. Hint: $\tilde{F} = F(a)$, where a satisfies some irreducible polynomial over F. Use part (1) of these exercises and $\mathbf{Q} \subset F$.
(c) Prove that ∂ has a unique extension to any field $\tilde{F}$ that is algebraic over F (and, in particular, to the algebraic closure of F).
(d) Show that (b) and (c) are, in general, false if F has characteristic $p > 0$. Hint: Let F_p be the field with p elements and consider the field extension $\mathbf{F}_p(x^p) \subset \mathbf{F}_p(x)$, where x is transcendental over $\mathbf{F}_p$.
(e) Let F be a perfect field of characteristic $p > 0$ (i.e., $F^p =: \{a^p |\ a \in F\}$ is equal to F). Show that the only derivation on F is the zero derivation.

(f) Suppose that F is a field of characteristic $p > 0$ such that $[F : F^p] = p$. Give a construction of all derivations on F. Hint: Compare with the beginning of Sect. 13.1.

4. *Lie algebras of derivations*
A *Lie algebra* over a field C is a C-vector space V equipped with a map $[\ ,\ ] : V \times V \to V$ that satisfies the rules:
(i) The map $(v, w) \mapsto [v, w]$ is linear in each factor.
(ii) $[[u, v], w] + [[v, w], u] + [[w, u], v] = 0$ for all $u, v, w \in V$ (Jacobi identity).
(iii) $[u, u] = 0$ for all $u \in V$.
The antisymmetry $[u, v] = -[v, u]$ follows from

$$0 = [u + v, u + v] = [u, u] + [u, v] + [v, u] + [v, v] = [u, v] + [v, u].$$

The standard example of a Lie algebra over C is $\mathrm{M}_n(C)$, the vector space of all $n \times n$ matrices over C, with $[A, B] := AB - BA$. Another example is the Lie algebra $\mathfrak{sl}_n \subset \mathrm{M}_n(C)$ consisting of the matrices with trace 0. The brackets of $\mathfrak{sl}_n$ are again defined by $[A, B] = AB - BA$. The notions "homomorphism of Lie algebras" and "Lie subalgebra" are obvious. We will say more on Lie algebras when they occur in connection with the other themes of this text.
(a) Let F be any field and let $C \subset F$ be a subfield. Let $\mathrm{Der}(F/C)$ denote the set of all derivations ∂ of F such that ∂ is the zero map on C. Prove that $\mathrm{Der}(F/C)$ is a vector space over F. Prove that for any two elements $\partial_1, \partial_2 \in \mathrm{Der}(F/C)$, the map $\partial_1\partial_2 - \partial_2\partial_1$ is again in $\mathrm{Der}(F/C)$. Conclude that $\mathrm{Der}(F/C)$ is a Lie algebra over C.
(b) Assume now that the field C has characteristic 0 and that F/C is a finitely generated field extension. One can show that there is an intermediate field $M = C(z_1, \ldots, z_d)$ with M/C purely transcendental and F/M finite. Prove, with the help of Exercise 1.5.3, that the dimension of the F-vector space $\mathrm{Der}(F/C)$ is equal to d. □

1.2 Linear Differential Equations

Let k be a differential field with field of constants C. Linear differential equations over k can be presented in various forms. The somewhat abstract setting is that of a differential module.

Definition 1.6 A *differential module* (M, ∂) (or simply M) of dimension n is a k-vector space, as dimension n equipped with an additive map $\partial : M \to M$ that has the property: $\partial(fm) = f'm + f\partial m$ for all $f \in k$ and $m \in M$. □

A differential module of dimension one has thus the form $M = Ke$ and the map ∂ is completely determined by the $a \in k$ given by $\partial e = ae$. Indeed, $\partial(fe) = (f' + fa)e$ for all $f \in k$. More generally, let $e_1, \ldots, e_n$ be a basis of M over k, then ∂ is completely determined by the elements ∂e_i, $i = 1, \ldots, n$. Define the matrix $A = (a_{i,j}) \in \mathrm{M}_n(k)$ by the condition $\partial e_i = -\sum_j a_{j,i}e_j$. The minus sign is introduced for historical reasons and is of no importance. Then for any element $m = \sum_{i=1}^n f_i e_i \in M$ the element ∂m has the form $\sum_{i=1}^n f_i' e_i - \sum_{i=1}^n (\sum_j a_{i,j} f_j) e_i$. The equation $\partial m = 0$ then has the translation $(y_1', \ldots, y_n')^T = A(y_1, \ldots, y_n)^T$. This brings us to a second possibility for expressing linear differential equations. First, we give some notations.

The differentiation on k is extended to vectors in k^n and to matrices in $\mathrm{M}_n(k)$ by component-wise differentiation. Thus for $y = (y_1, \ldots, y_n)^T \in k^n$ and $A = (a_{i,j}) \in \mathrm{M}_n(k)$ one writes $y' = (y_1', \ldots, y_n')^T$ and $A' = (a_{i,j}')$. We note that there are obvious rules like $(AB)' = A'B + AB'$, $(A^{-1})' = -A^{-1}A'A^{-1}$, and $(Ay)' = A'y + Ay'$, where A, B are matrices and y is a vector. A linear differential equation in matrix form or a *matrix differential equation* over k of dimension n reads $y' = Ay$, where $A \in \mathrm{M}_n(k)$ and $y \in k^n$.

As we have seen, a choice of a basis of the differential module M over k translates M into a matrix differential equation $y' = Ay$. If one chooses another basis of M over k, then y is replaced by Bf for some $B \in \mathrm{GL}_n(k)$. The matrix differential equation for this new basis reads $f' = \tilde{A}f$, where $\tilde{A} = B^{-1}AB - B^{-1}B'$. Two matrix differential equations given by matrices A and $\tilde{A}$ are called *equivalent* if there is a $B \in \mathrm{GL}_n(k)$ such that $\tilde{A} = B^{-1}AB - B^{-1}B'$. Thus two matrix differential equations are equivalent if they are obtained from the same differential module. It is, furthermore, clear that any matrix differential equation $y' = Ay$ comes from a differential module, namely $M = k^n$ with standard basis $e_1, \ldots, e_n$ and ∂ given by the formula $\partial e_i = -\sum_j a_{j,i}e_j$. In this chapter we will mainly work with matrix differential equations.

Lemma 1.7 *Consider the matrix equation $y' = Ay$ over k of dimension n. Let $v_1, \ldots, v_r \in k^n$ be solutions, i.e., $v_i' = Av_i$ for all i. If the vectors $v_1, \ldots, v_r \in V$ are linearly dependent over k then they are linearly dependent over C.*

Proof. The lemma is proved by induction on r. The case $r = 1$ is trivial. The induction step is proved as follows. Let $r > 1$ and let the $v_1, \ldots, v_r$ be linearly dependent over k. We may assume that any proper subset of $\{v_1, \ldots, v_r\}$ is linearly independent over k. Then there is a unique relation $v_1 = \sum_{i=2}^r a_i v_i$ with all $a_i \in k$. Now

$$0 = v_1' - Av_1 = \sum_{i=2}^r a_i' v_i + \sum_{i=2}^r a_i(v_i' - Av_i) = \sum_{i=2}^r a_i' v_i.$$

Thus all $a_i' = 0$ and so all $a_i \in C$. □

Lemma 1.8 *Consider the matrix equation $y' = Ay$ over k of dimension n. The solution space V of $y' = Ay$ in k is defined as $\{v \in k^n \mid v' = Av\}$. Then V is a vector space over C of dimension $\leq n$.*

Proof. It is clear that V is a vector space over C. The lemma follows from Lemma 1.7 since any $n+1$ vectors in V are linearly dependent over k. □

Suppose that the solution space $V \subset k^n$ of the equation $y' = Ay$ of dimension n satisfies $\dim_C V = n$. Let $v_1, \dots, v_n$ denote a basis of V. Let $B \in \mathrm{GL}_n(k)$ be the matrix with columns $v_1, \dots, v_n$. Then $B' = AB$. This brings us to the following definition.

Definition 1.9 Let R be a differential ring, containing the differential field k and having C as its set of constants. Let $A \in \mathrm{M}_n(k)$. An invertible matrix $F \in \mathrm{GL}_n(R)$ is called a *fundamental matrix for the equation* $y' = Ay$ if $F' = AF$ holds. □

Suppose that $F, \tilde{F} \in \mathrm{GL}_n(R)$ are both fundamental matrices. Define M by $\tilde{F} = FM$. Then

$$A\tilde{F} = \tilde{F}' = F'M + FM' = AFM + FM' \text{ and thus } M' = 0.$$

We conclude that $M \in \mathrm{GL}_n(C)$. In other words, the set of all fundamental matrices (inside $\mathrm{GL}_n(R)$) for $y' = Ay$ is equal to $F \cdot \mathrm{GL}_n(C)$.

Here is a third possibility for formulating differential equations. A (linear) *scalar differential equation* over the field k is an equation of the form

$$L(y) = b \text{ where } b \in k, \text{ and}$$
$$L(y) := y^{(n)} + a_{n-1}y^{(n-1)} + \cdots + a_1 y' + a_0 y \text{ with all } a_i \in k.$$

A solution of such an equation in a differential extension $R \supset k$ is an element $f \in R$ such that $f^{(n)} + a_{n-1}f^{(n-1)} + \cdots + a_1 f' + a_0 f = b$. The equation is called *homogeneous of order n* if $b = 0$. Otherwise, the equation is called *inhomogeneous of order n*.

There is a standard way of producing a matrix differential equation $y' = A_L y$ from a homogeneous scalar linear differential equation $L(y) = y^{(n)} + a_{n-1}y^{(n-1)} + \cdots + a_1 y' + a_0 y = 0$. The *companion matrix* A_L of L is the following

$$A_L = \begin{pmatrix} 0 & 1 & 0 & 0 & \dots & 0 \\ 0 & 0 & 1 & 0 & \dots & 0 \\ \vdots & \vdots & \vdots & \vdots & \dots & \vdots \\ 0 & 0 & 0 & 0 & \dots & 1 \\ -a_0 & -a_1 & \dots & \dots & \dots & -a_{n-1} \end{pmatrix}.$$

One easily verifies that this companion matrix has the following property. For any extension of differential rings $R \supset k$, the map $y \mapsto Y := (y, y', \dots, y^{(n-1)})^T$ is an isomorphism of the solution space $\{y \in R \mid L(y) = 0\}$ of L onto the solution space of $\{Y \in R^n \mid Y' = AY\}$ of the matrix differential equation $Y' = AY$. In other words, one can view a scalar differential equation as a special case of a matrix differential equation. Lemma 1.8 translates for homogeneous scalar equations.

Lemma 1.10 *Consider an nth order homogeneous scalar equation $L(y) = 0$ over k. The solution space V of $L(y) = 0$ in k is defined as $\{v \in k|\ L(v) = 0\}$. Then V is a vector space over C of dimension $\leq n$.*

In Sect. 2.1 it will be shown that, under the assumption that k contains a non-constant element, any differential module M of dimension n over k contains a *cyclic vector* e. The latter means that $e, \partial e, \ldots, \partial^{n-1}e$ forms a basis of M over k. The $n+1$ elements $e, \partial e, \ldots, \partial^n e$ are linearly dependent over k. Thus there is a unique relation $\partial^n e + b_{n-1}\partial^{n-1} + \cdots + b_1\partial e + b_0 e = 0$ with all $b_i \in k$. The transpose of the matrix of ∂ on the basis $e, \partial e, \ldots, \partial^{n-1}e$ is a companion matrix. This suffices to prove the assertion that any matrix differential equation is equivalent to a matrix equation $Y' = A_L Y$ for a scalar equation $Ly = 0$. In what follows we will use the three ways to formulate linear differential equations.

In analogy to matrix equations we say that a set of n solutions $\{y_1, \ldots, y_n\}$ (say in a differential extension $R \supset k$ having C as constants) of an order n equation $L(y) = 0$, linearly independent over the constants C, is a *fundamental set of solutions of* $L(y) = 0$. This clearly means that the solution space of L has dimension n over C and that $y_1, \ldots, y_n$ is a basis of that space.

Lemma 1.7 also has a translation. We introduce the classical wronskians.

Definition 1.11 Let R be a differential field and let $y_1, \ldots, y_n \in R$. The *wronskian matrix* of $y_1, \ldots, y_n$ is the $n \times n$ matrix

$$W(y_1, \ldots, y_n) = \begin{pmatrix} y_1 & y_2 & \cdots & y_n \\ y_1' & y_2' & \cdots & y_n' \\ \vdots & \vdots & \cdots & \vdots \\ y_1^{(n-1)} & y_2^{(n-1)} & \cdots & y_n^{(n-1)} \end{pmatrix}.$$

The *wronskian*, $wr(y_1, \ldots, y_n)$ of $y_1, \ldots, y_n$ is $\det(W(y_1, \ldots, y_n))$. □

Lemma 1.12 *Elements $y_1, \ldots, y_n \in k$ are linearly dependent over C if and only if $wr(y_1, \ldots, y_n) = 0$.*

Proof. There is a monic scalar differential equation $L(y) = 0$ of order n over k such that $L(y_i) = 0$ for $i = 1, \ldots, n$. One constructs L by induction. Put $L_1(y) = y' - \frac{y_1'}{y_1}y$, where the term $\frac{y_1'}{y_1}$ is interpreted as 0 if $y_1 = 0$. Suppose that $L_m(y)$ has been constructed such that $L_m(y_i) = 0$ for $i = 1, \ldots, m$. Define now $L_{m+1}(y) = L_m(y)' - \frac{L_m(y_{m+1})'}{L_m(y_{m+1})}L_m(y)$, where the term $\frac{L_m(y_{m+1})'}{L_m(y_{m+1})}$ is interpreted as 0 if $L_m(y_{m+1}) = 0$. Then $L_{m+1}(y_i) = 0$ for $i = 1, \ldots, m+1$. Then $L = L_n$ has the required property. The columns of the wronskian matrix are solutions of the associated companion matrix differential equation $Y' = A_L Y$. Now apply Lemma 1.7. □

Corollary 1.13 *Let $k_1 \subset k_2$ be differential fields with fields of constants $C_1 \subset C_2$. The elements $y_1, \dots, y_n \in k_1$ are linearly independent over C_1 if and only if they are linearly independent over C_2.*

Proof. The elements $y_1, \dots, y_n \in k_1$ are linearly dependent over C_2 if and only if $wr(y_1, \dots, y_n) = 0$. Another application of Lemma 1.12 implies that the same equivalence holds over C_1. □

We now come to our first problem. Suppose that the solution space of $y' = Ay$ over k is too small, i.e., its dimension is strictly less than n or, equivalently, there is no fundamental matrix in $\mathrm{GL}_n(k)$. How can we produce enough solutions in a larger differential ring or differential field? This is the subject of Sect. 1.3, Picard-Vessiot extensions. A second related problem is to make the solutions as explicit as possible.

The situation is somewhat analogous to the case of an ordinary polynomial equation $P(X) = 0$ over a field K. Suppose that P is a separable polynomial of degree n. Then one can construct a splitting field $L \supset K$ that contains precisely n solutions $\{\alpha_1, \dots, \alpha_n\}$. Explicit information on the α_i can be obtained from the action of the Galois group on $\{\alpha_1, \dots, \alpha_n\}$.

Exercises 1.14

1. *Homogeneous versus inhomogeneous equations*
Let k be a differential field and $L(y) = b$, with $b \neq 0$, an nth order inhomogeneous linear differential equation over k. Let

$$L_h(y) = b(\frac{1}{b}L(y))' .$$

(a) Show that any solution in k of $L(y) = b$ is a solution of $L_h(y) = 0$.
(b) Show that for any solution v of $L_h(y) = 0$ there is a constant c such that v is a solution of $L(y) = cb$.
This construction allows one to reduce questions concerning nth order inhomogeneous equations to $(n + 1)$st order homogeneous equations.

2. *Some order one equations over $C((z))$*
Let C be an algebraically closed field of characteristic 0. The differential field $K = C((z))$ is defined by $' = \frac{d}{dz}$. Let $a \in K,\ a \neq 0$.
(a) When does $y' = a$ have a solution in K?
(b) When does $y' = a$ have a solution in $\bar{K}$, the algebraic closure of K? We note that every finite algebraic extension of K has the form $C((z^{1/n}))$.
(c) When does $y' = ay$ have a nonzero solution in K?
(d) When does $y' = ay$ have a nonzero solution in $\bar{K}$?

3. *Some order one equations over $C(z)$*
C denotes an algebraically closed field of characteristic 0. Let $K = C(z)$ be the differential field with derivation $' = \frac{d}{dz}$. Let $a \in K$ and let

$$a = \sum_{i=1}^{N} \sum_{j=1}^{n_i} \frac{c_{ij}}{(z-\alpha_i)^j} + p(z)$$

be the partial fraction decomposition of a with $c_{ij} \in C$, N a non-negative integer, the n_i positive integers, and p a polynomial. Prove the following statements.
(a) $y' = a$ has a solution in K if and only if each c_{i1} is zero.
(b) $y' = ay$ has a solution $y \in K$, $y \neq 0$ if and only if each c_{i1} is an integer, each $c_{ij} = 0$ for $j > 1$, and $p = 0$.
(c) $y' = ay$ has a solution $y \neq 0$ that is algebraic over K if and only if each c_{i1} is a rational number, each $c_{ij} = 0$ for $j > 1$, and $p = 0$.
The above can be restated in terms of differential forms:
(a') $y' = a$ has a solution in K if and only if the residue of $a\,dz$ at every point $z = c$ with $c \in C$ is zero.
(b') $y' = ay$ has a solution in K^* if and only if $a\,dz$ has at most poles of order 1 on $C \cup \{\infty\}$ and its residues are integers.
(c') $y' = ay$ has a solution $y \neq 0$ that is algebraic over K if and only if $a\,dz$ has at most poles of order 1 at $C \cup \{\infty\}$ and its residues are rational numbers.

4. *Regular matrix equations over* $C((z))$
$C[[z]]$ will denote the ring of all formal power series with coefficients in the field C. We note that $C((z))$ is the field of fractions of $C[[z]]$ (cf. Exercise 1.3.2).
(a) Prove that a matrix differential equation $y' = Ay$ with $A \in \mathrm{M}_n(C[[z]])$ has a unique fundamental matrix B of the form $1 + \sum_{n>0} B_n z^n$ where 1 denotes the identity matrix and with all $B_n \in \mathrm{M}_n(C)$.
(b) A matrix equation $Y' = AY$ over $C((z))$ is called *regular* if the equation is equivalent to an equation $v' = \tilde{A}v$ with $\tilde{A} \in \mathrm{M}_n(C[[z]])$. Prove that an equation $Y' = AY$ is regular if and only if there is a fundamental matrix with coefficients in $C((z))$.

5. *Wronskians*
Let k be a differential field, $Y' = AY$ a matrix differential equation over k, and $L(y) = y^{(n)} + a_{n-1}y^{(n-1)} + \cdots + a_0 y = 0$ a homogeneous scalar differential equation over k.
(a) If Z is a fundamental matrix for $y' = Ay$, show that $(\det Z)' = \mathrm{tr}A \cdot (\det Z)$, where tr denotes the trace. Hint: Let $z_1, \dots, z_n$ denote the columns of Z. Then $z_i' = Az_i$. Observe that $\det(z_1, \dots, z_n)' = \sum_{i=1}^{n} \det(z_1, \dots, z_i', \dots, z_n)$. Consider the trace of A w.r.t. the basis $z_1, \dots, z_n$.
(b) Let $\{y_1, \dots, y_n\} \subset k$ be a fundamental set of solutions of $L(y) = 0$. Show that $w = wr(y_1, \dots, y_n)$ satisfies $w' = -a_{n-1}w$. Hint: Use the companion matrix of L.

6. *A Result of Ritt*
Let k be a differential field with field of constants C and assume $k \neq C$. Let P be a nonzero element of $k\{\{y_1, \dots, y_n\}\}$. For any elements $u_1, \dots, u_n \in k$, there is a unique k-linear homomorphism $\phi : k\{\{y_1, \dots, y_n\}\} \to k$ of differential rings such that $\phi(y_i) = u_i$ for all i. We will write $P(u_1, \dots, u_n)$ for $\phi(P)$. The *aim of this*

exercise is to show that there exist $u_1, \ldots, u_n \in k$ such that $\phi(P) \neq 0$.
(a) Show that it suffices to prove this result for $n = 1$.
(b) Let $v \in k,\ v' \neq 0$. Show that $wr(1, v, v^2, \ldots, v^m) \neq 0$ for $m \geq 1$.
(c) Let $v \in k,\ v' \neq 0$ and let $A = W(1, v, v^2, \ldots, v^m)$, where $W(\ldots)$ is the wronskian matrix. Let $z_0, \ldots z_m$ be indeterminates. Define the k-algebra homomorphism $\Phi : k[y, y^{(1)}, \ldots, y^{(m)}] \to k[z_0, \ldots, z_m]$ by formulas for $\Phi(y^{(i)})$, symbolically given by $\Phi((y, y', \ldots, y^{(m)})^T) = A(z_0, z_1, \ldots, z_m)^T$. Prove that Φ is an isomorphism. Conclude that if $P \in k\{\{y\}\}$ has order m, then there exist constants $c_0, \ldots c_m \in C$ such that $\Phi(P)(c_0, \ldots, c_m) \neq 0$.
(d) Take $u = c_0 + c_1 v + c_2 v^2 + \cdots + c_m v^m$ and show that $P(u) = \Phi(P)(c_0, \ldots, c_m)$.
(e) Show that the condition that k contains a nonconstant is necessary.
This result appears in [247], p. 35 and in [162], Theorem 2, p. 96.

7. *Equations over algebraic extensions*
Let k be a differential field, K an algebraic extension of k with $[K : k] = m$ and let $u_1, \ldots, u_m$ be a k-basis of K. Let $Y' = AY$ be a differential equation of order n over K. Show that there exists a differential equation $Z' = BZ$ of order nm over k such that if $Z = (z_{1,1}, \ldots, z_{1,m}, z_{2,1}, \ldots, z_{2,m}, \ldots, z_{n,m})^T$ is a solution of $Z' = BZ$, then for $y_i = \sum_j z_{i,j} u_j$, $Y = (y_1, \ldots, y_n)^T$ is a solution of $Y' = AY$.
Let (M, ∂) be the differential module of dimension n over K for which $Y' = AY$ is an associated matrix differential equation. One can view (M, ∂) as a differential module over k of dimension nm. Find the basis of M over k such that the associated matrix equation is $Z' = BZ$. □

1.3 Picard-Vessiot Extensions

Throughout the rest of this chapter, k will denote a differential field with $\mathbf{Q} \subset k$ and with an algebraically closed field of constants C. We shall freely use the notation and results concerning varieties and linear algebraic groups contained in Appendix A.

Let R be a differential ring with derivation $'$. A *differential ideal* I in R is an ideal satisfying $f' \in I$ for all $f \in I$. If R is a differential ring over a differential field k and I is a differential ideal of R, $I \neq R$, then the factor ring R/I is again a differential ring over k (see Exercise 1.2.1). A *simple differential ring* is a differential ring whose only differential ideals are (0) and R.

Definition 1.15 A *Picard-Vessiot ring* over k for the equation $y' = Ay$, with $A \in \mathrm{M}_n(k)$, is a differential ring R over k satisfying:

1. R is a simple differential ring.
2. There exists a fundamental matrix F for $y' = Ay$ with coefficients in R, i.e., the matrix $F \in \mathrm{GL}_n(R)$ satisfies $F' = AF$.
3. R is generated as a ring by k, the entries of a fundamental matrix F and the inverse of the determinant of F.

A *Picard-Vessiot ring* for a differential module M over k is defined as a Picard-Vessiot ring of a matrix differential equation $y' = Ay$ associated to M. □

Exercises 1.16 *Picard-Vessiot rings for differential modules.*

(1) Let $y' = Ay$ and $f' = \tilde{A}f$ be two matrix differential equations associated to the same differential module M. Prove that a differential ring R over k is a Picard-Vessiot ring for $y' = Ay$ if and only if R is a Picard-Vessiot ring for $f' = \tilde{A}f$.
Note that this justifies the last part of the definition.

(2) Let M be a differential module over k of dimension n. Show that the following alternative definition of Picard-Vessiot ring R is equivalent to that of Definition 1.15.
The alternative definition:
(i) R is a simple differential ring.
(ii) $V := \ker(\partial, R \otimes_k M)$ has dimension n over C.
(iii) Let $e_1, \ldots, e_n$ denote any basis of M over k, then R is generated over k by the coefficients of all $v \in V$ w.r.t. the free basis $e_1, \ldots, e_n$ of $R \otimes_k M$ over R.

(3) The C-vector space V in part (2) is referred to as the *solution space of the differential module*. For two Picard-Vessiot rings R_1, R_2 there are two solution spaces V_1, V_2. Show that any isomorphism $\phi : R_1 \to R_2$ of differential rings over k induces a C-linear isomorphism $\psi : V_1 \to V_2$. Is ψ independent of the choice of ϕ? □

Lemma 1.17 *Let R be a simple differential ring over k.*

1. *R has no zero divisors.*
2. *Suppose that R is finitely generated over k, then the field of fractions of R has C as set of constants.*

Proof. 1. We will first show that any non-nilpotent element $a \in R, a \neq 0$ is a nonzero divisor. Consider the ideal $I = \{b \in R \mid \text{there exists a } n \geq 1 \text{ with } a^n b = 0\}$. This is a differential ideal not containing 1. Thus $I = (0)$ and a is not a zero divisor.

Let $a \in R, a \neq 0$ be nilpotent. We will show that a' is also nilpotent. Let $n > 1$ be minimal with $a^n = 0$. Differentiation yields $a'na^{n-1} = 0$. Since $na^{n-1} \neq 0$ we have that a' is a zero divisor and thus a' is nilpotent.

Finally, the ideal J consisting of all nilpotent elements is a differential ideal and thus equal to (0).

2. Let L be the field of fractions of R. Suppose that $a \in L, a \neq 0$ has derivative $a' = 0$. We have to prove that $a \in C$. The nonzero ideal $\{b \in R | ba \in R\}$ is a differential ideal and thus equal to R. Hence $a \in R$. We suppose that $a \notin C$. We then have that for every $c \in C$, the nonzero ideal $(a - c)R$ is a differential

ideal. This implies that $a - c$ is an invertible element of R for every $c \in C$. Let X denote the affine variety $(\max(R), R)$ over k. Then $a \in R$ is a regular function $X(\bar{k}) \to \mathbf{A}_k^1(\bar{k}) = \bar{k}$. By Chevalley's theorem, the image of a is a constructible set, i.e., a finite union of intersections of open and closed subsets. (See also the discussion following Exercises A.9.) In this special case, this means that the image of a is either finite or cofinite. Since $a - c$ is invertible for $c \in C$, the image of a has an empty intersection with C. Therefore the image is finite and there is a polynomial $P = X^d + a_{d-1}X^{d-1} + \cdots + a_0 \in k[X]$ of minimal degree such that $P(a) = 0$. Differentiation of the equality $P(a) = 0$ yields $a'_{d-1}a^{d-1} + \cdots + a'_0 = 0$. By the minimality of P, one has $a_i \in C$ for all i. Since C is algebraically closed one finds a contradiction. (Compare with Exercise 1.5.)

An alternative proof uses the fact that R is an integral domain (part 1 of this lemma) and Lemma A.4, which implies that a is algebraic over k. □

Example 1.18 $y' = a$ *with* $a \in k$.
One can verify that a Picard-Vessiot ring for the matrix equation $\binom{y_1}{y_2}' = \begin{pmatrix} 0 & a \\ 0 & 0 \end{pmatrix}\binom{y_1}{y_2}$ is generated by a solution of $y' = a$. We shall refer to this Picard-Vessiot ring as the Picard-Vessiot ring of the equation $y' = a$. If k contains a solution b of the scalar equation then $\begin{pmatrix} 1 & b \\ 0 & 1 \end{pmatrix}$ is a fundamental matrix and $R = k$ is a Picard-Vessiot ring for the equation.

We suppose now that the scalar equation has no solution in k. Define the differential ring $R = k[Y]$ with the derivation $'$ extending $'$ on k and $Y' = a$ (see Exercise 1.5(1)). Then R contains an obvious solution of the scalar equation and $\begin{pmatrix} 1 & Y \\ 0 & 1 \end{pmatrix}$ is a fundamental matrix for the matrix equation.

The minimality of the ring $R = k[Y]$ is obvious. We want to show that R has only trivial differential ideals. Let I be a proper ideal of $k[Y]$. Then I is generated by some $F = Y^n + \cdots + f_1Y + f_0$ with $n > 0$. The derivative of F is $F' = (na + f'_{n-1})Y^{n-1} + \cdots$. If I is a differential ideal then $F' \in I$ and thus $F' = 0$. In particular, $na + f'_{n-1} = 0$ and $\frac{-f_{n-1}}{n}' = a$. This contradicts our assumption. We conclude that $R = k[Y]$ is a Picard-Vessiot ring for $y' = a$. □

Example 1.19 $y' = ay$ *with* $a \in k^*$.
Define the differential ring $R = k[T, T^{-1}]$ with the derivation $'$ extending $'$ on k and $T' = aT$. Then R contains a nonzero solution of $y' = ay$. The minimality of R is clear and the ring R would be the answer to our problem if R has only trivial differential ideals. For the investigation of this we have to consider two cases:

(a) Assume that k contains no solution ($\neq 0$) of $y' = nay$ for all $n \in \mathbf{Z}$, $n \neq 0$. Let $I \neq 0$ be a differential ideal. Then I is generated by some $F = T^m + a_{m-1}T^{m-1} + \cdots + a_0$, with $m \geq 0$ and $a_0 \neq 0$. The derivative $F' = maT^m + ((m-1)aa_{m-1} + a'_{m-1})T^{m-1} + \cdots + a'_0$ of F belongs to I. This implies $F' = maF$. For $m > 0$ one obtains the contradiction $a'_0 = maa_0$. Thus $m = 0$ and $I = R$. We conclude that $R = k[T, T^{-1}]$ is a Picard-Vessiot ring for the equation $y' = ay$.

(b) Assume that $n > 0$ is minimal so that $y' = nay$ has a solution $y_0 \in k^*$. Then $R = k[T, T^{-1}]$ has a nontrivial differential ideal (F) with $F = T^n - y_0$. Indeed, $F' = naT^n - nay_0 = naF$. The differential ring $k[T, T^{-1}]/(T^n - y_0)$ over k will be written as $k[t, t^{-1}]$, where t is the image of T. One has $t^n = y_0$ and $t' = at$. Every element of $k[t, t^{-1}]$ can uniquely be written as $\sum_{i=0}^{n-1} a_i t^i$. We claim that $k[t, t^{-1}]$ is a Picard-Vessiot ring for $y' = ay$. The minimality of $k[t, t^{-1}]$ is obvious. We have to prove that $k[t, t^{-1}]$ has only trivial differential ideals.

Let $I \subset k[t, t^{-1}]$, $I \neq 0$ be a differential ideal. Let $0 \leq d < n$ be minimal such that I contains a nonzero F of the form $\sum_{i=0}^{d} a_i t^i$. Suppose that $d > 0$. We may assume that $a_d = 1$. The minimality of d implies $a_0 \neq 0$. Consider $F' = dat^d + ((d-1)aa_{d-1} + a'_{d-1})t^{d-1} + \cdots + a'_0$. The element $F' - daF$ belongs to I and is 0, since d is minimal. Then $a'_0 = daa_0$ contradicts our assumption. Thus $d = 0$ and $I = k[t, t^{-1}]$. □

Proposition 1.20 *Let $y' = Ay$ be a matrix differential equation over k.*

1. *There exists a Picard-Vessiot ring for the equation.*
2. *Any two Picard-Vessiot rings for the equation are isomorphic.*
3. *The constants of the quotient field of a Picard-Vessiot ring is again C.*

Proof. 1. Let $(X_{i,j})$ denote an $n \times n$ matrix of indeterminates and let "det" denote the determinant of $(X_{i,j})$. For any ring or field F one writes $F[X_{i,j}, \frac{1}{\det}]$ for the polynomial ring in these n^2 indeterminates, localized w.r.t. the element "det". Consider the differential ring $R_0 = k[X_{i,j}, \frac{1}{\det}]$ with the derivation, extending the one of k, given by $(X'_{i,j}) = A(X_{i,j})$. Exercise 1.5.1 shows the existence and unicity of such a derivation. Let $I \subset R_0$ be a maximal differential ideal. Then $R = R_0/I$ is easily seen to be a Picard-Vessiot ring for the equation.

2. Let R_1, R_2 denote two Picard-Vessiot rings for the equation. Let B_1, B_2 denote the two fundamental matrices. Consider the differential ring $R_1 \otimes_k R_2$ with derivation given by $(r_1 \otimes r_2)' = r'_1 \otimes r_2 + r_1 \otimes r'_2$ (see Sect. A.1.2 for basic facts concerning tensor products). Choose a maximal differential ideal $I \subset R_1 \otimes_k R_2$ and define $R_3 := (R_1 \otimes_k R_2)/I$. There are obvious morphisms of differential rings $\phi_i : R_i \to R_3$, $i = 1, 2$. Since R_i is simple, the morphism $\phi_i : R_i \to \phi_i(R_i)$ is an isomorphism. The image of ϕ_i is generated over k by the coefficients of $\phi_i(B_i)$ and $\phi_i(\det B_i^{-1})$. The matrices $\phi_1(B_1)$ and $\phi_2(B_2)$ are fundamental matrices over the ring R_3. Since the set of constants of R_3 is C one has $\phi_1(B_1) = \phi_2(B_2)M$, where M is an invertible matrix with coefficients in C. This implies that $\phi_1(R_1) = \phi_2(R_2)$ and so R_1 and R_2 are isomorphic.

3. Follows from Lemma 1.17. □

We note that the maximal differential ideal I of R_0 in the above proof is, in general, not a maximal ideal of R_0 (see Examples 1.18 and 1.19).

Definition 1.21 A *Picard-Vessiot field* for the equation $y' = Ay$ over k (or for a differential module M over k) is the field of fractions of a Picard-Vessiot ring for this equation. □

In the literature there is a slightly different definition of the Picard-Vessiot field of a linear differential equation. The equivalence of the two definitions is stated in the next proposition.

Proposition 1.22 *Let $y' = Ay$ be a matrix differential equation over k and let $L \supset k$ be an extension of differential fields. The field L is a Picard-Vessiot field for this equation if and only if the following conditions are satisfied.*

1. *The field of constants of L is C,*
2. *There exists a fundamental matrix $F \in \mathrm{GL}_n(L)$ for the equation, and*
3. *L is generated over k by the entries of F.*

The proof requires a lemma in which one considers an $n \times n$ matrix of indeterminates $(Y_{i,j})$ and its determinant, denoted simply by "det". For any field F one denotes by $F[Y_{i,j}, \frac{1}{\det}]$ the polynomial ring over F in these indeterminates, localized w.r.t. the element "det".

Lemma 1.23 *Let M be any differential field with field of constants C. The derivation $'$ on M is extended to a derivation on $M[Y_{i,j}, \frac{1}{\det}]$ by setting $Y'_{i,j} = 0$ for all i, j. One considers $C[Y_{i,j}, \frac{1}{\det}]$ as a subring of $M[Y_{i,j}, \frac{1}{\det}]$. The map $I \mapsto (I)$ from the set of ideals of $C[Y_{i,j}, \frac{1}{\det}]$ to the set of the differential ideals of $M[Y_{i,j}, \frac{1}{\det}]$ is a bijection. The inverse map is given by $J \mapsto J \cap C[Y_{i,j}, \frac{1}{\det}]$.*

Proof. Choose a basis $\{m_s\}_{s \in S}$, with $m_{s_0} = 1$, of M over C. Then $\{m_s\}_{s \in S}$ is also a free basis of the $C[Y_{i,j}, \frac{1}{\det}]$-module $M[Y_{i,j}, \frac{1}{\det}]$. The differential ideal (I) consists of the finite sums $\sum_s a_s m_s$ with all $a_s \in I$. Hence $(I) \cap C[Y_{i,j}, \frac{1}{\det}] = I$.

We finish the proof by showing that any differential ideal $J \subset M[Y_{i,j}, \frac{1}{\det}]$ is generated by $I := J \cap C[Y_{i,j}, \frac{1}{\det}]$. Let $\{e_\beta\}_{\beta \in \mathcal{B}}$ be a basis of $C[Y_{i,j}, \frac{1}{\det}]$ over C. Any element $f \in J$ can be uniquely written as a finite sum $\sum_\beta m_\beta e_\beta$ with the $m_\beta \in M$. By the length $l(f)$ we will mean the number of βs with $m_\beta \neq 0$. By induction on the length, $l(f)$, of f we will show that $f \in (I)$. When $l(f) = 0, 1$, the result is clear. Assume $l(f) > 1$. We may suppose that $m_{\beta_1} = 1$ for some $\beta_1 \in \mathcal{B}$ and $m_{\beta_2} \in M \backslash C$ for some $\beta_2 \in \mathcal{B}$. One then has that $f' = \sum_\beta m'_\beta e_\beta$ has a length smaller than $l(f)$ and so belongs to (I). Similarly, $(m_{\beta_2}^{-1} f)' \in (I)$. Therefore $(m_{\beta_2}^{-1})' f = (m_{\beta_2}^{-1} f)' - m_{\beta_2}^{-1} f' \in (I)$. Since C is the field of constants of M, one has $(m_{\beta_2}^{-1})' \neq 0$ and so $f \in (I)$. □

Proof of 1.22. According to Proposition 1.20, the conditions (1)–(3) are necessary. Suppose L satisfies these three conditions. As in Proposition 1.20, we consider the differential ring $R_0 = k[X_{i,j}, \frac{1}{det}]$ with $(X'_{i,j}) = A(X_{i,j})$. Consider the differential rings $R_0 \subset L \otimes_k R_0 = L[X_{i,j}, \frac{1}{\det}]$. Define a set of n^2 new variables $Y_{i,j}$ by $(X_{i,j}) = F \cdot (Y_{i,j})$. Then $L \otimes_k R_0 = L[Y_{i,j}, \frac{1}{\det}]$ and $Y'_{i,j} = 0$ for all i, j. We can identify $L \otimes_k R_0$ with $L \otimes_C R_1$ where $R_1 := C[Y_{i,j}, \frac{1}{\det}]$. Let P be a maximal differential ideal of R_0. The ideal P generates an ideal in $L \otimes_k R_0$ that is denoted by (P). Since $L \otimes R_0/(P) \cong L \otimes (R_0/P) \neq 0$, the ideal (P) is a proper differential ideal. Define the ideal $\tilde{P} \subset R_1$ by $\tilde{P} = (P) \cap R_1$. Lemma 1.23 implies that the ideal (P) is generated by $\tilde{P}$. If M is a maximal ideal of R_1 containing $\tilde{P}$ then $R_1/M = C$. The corresponding homomorphism of C-algebras $R_1 \to C$ extends to a differential homomorphism of L-algebras $L \otimes_C R_1 \to L$. Its kernel contains $(P) \subset L \otimes_k R_0 = L \otimes_C R_1$. Thus we have found a k-linear differential homomorphism $\psi : R_0 \to L$ with $P \subset \ker(\psi)$. The kernel of ψ is a differential ideal and so $P = \ker(\psi)$. The subring $\psi(R_0) \subset L$ is isomorphic to R_0/P and is therefore a Picard-Vessiot ring. The matrix $(\psi(X_{i,j}))$ is a fundamental matrix in $\mathrm{GL}_n(L)$ and must have the form $F \cdot (c_{i,j})$ with $(c_{i,j}) \in \mathrm{GL}_n(C)$, because the field of constants of L is C. Since L is generated over k by the coefficients of F one has that L is the field of fractions of $\psi(R_0)$. Therefore, L is a Picard-Vessiot field for the equation. □

Exercises 1.24

1. *Finite Galois extensions are Picard-Vessiot extensions*

Let k be a differential field with derivation $'$ and with algebraically closed field of constants C. Let K be a finite Galois extension of k with Galois group G. Exercise 1.5(3) implies that there is a unique extension of $'$ to K. The aim of this exercise is to show that K is a Picard-Vessiot extension of k.

(a) Show that for any $\sigma \in G$ and $v \in K$, $\sigma(v') = \sigma(v)'$. Hint: Consider the map $v \mapsto \sigma^{-1}(\sigma(v)')$.

(b) We may write $K = k(w_1, \ldots w_m)$, where G permutes the w_i. This implies that the C-span V of the w_i is invariant under the action of G. Let $v_1, \ldots, v_n$ be a C-basis of V.

(i) Let $W = W(v_1, \ldots, v_n)$ (cf. Definition 1.11) be the wronskian matrix of $v_1, \ldots, v_n$. Show that there exists for each $\sigma \in G$, a matrix $A_\sigma \in \mathrm{GL}_n(C)$ such that $\sigma(W) = WA_\sigma$.

(ii) Show that $wr(v_1, \ldots, v_n) \neq 0$ and so W is invertible.

(iii) Show that the entries of the matrix $B = W'W^{-1}$ are left fixed by the elements of G and that W is a fundamental matrix for the matrix differential equation $y' = By,\ B \in \mathrm{M}_n(k)$. Conclude that K is the Picard-Vessiot ring for this equation.

It may seem that the above construction of the matrix differential equation over k having K as Picard-Vessiot ring is somewhat arbitrary. However, the terminology of differential modules clarifies the matter. Define the differential module (M, ∂) by $M = K$ and ∂ is the unique differentiation on K, extending the one of k. The statement now reads:

K is the Picard-Vessiot extension of the differential module (M, ∂).

Try to prove in this terminology, using Chap. 2, that K is the Picard-Vessiot ring of M. Hints:
(i) Use Exercises 1.16.
(ii) Show that $\ker(\partial, K \otimes_k M)$ has dimension n over C by observing that ∂ is a differentiation of the *ring* $K \otimes_k K$ and by (iii).
(iii) Use the fact that $K \otimes_k K$ is a direct product of fields $Ke_1 \oplus Ke_2 \oplus \cdots \oplus Ke_n$. Prove that $e_i^2 = e_i$ implies $\partial e_i = 0$.
(iv) Show that for a proper subfield $L \subset K$, containing k the space $\ker(\partial, L \otimes_k K)$ has C-dimension $< n$.

2. *Picard-Vessiot extensions for scalar differential equations*
Let $L(y) = 0$ be a homogeneous scalar differential equation over k. We define the Picard-Vessiot extension ring or field for this equation to be the Picard-Vessiot extension ring or field associated to the matrix equation $Y' = A_L Y$, where A_L is the companion matrix.

(a) Show that a Picard-Vessiot ring for this equation is a simple differential ring over k containing a fundamental set of solutions of $L(y) = 0$ such that no proper differential subring contains a fundamental set of solutions of $L(y) = 0$.

(b) Using the comment following Definition 1.21, show that a Picard-Vessiot field for this equation is a differential field over k containing a fundamental set of solutions of $L(y) = 0$, whose field of constants is the same as that of k such that no proper subfield contains a fundamental set of solutions of $L(y) = 0$. □

1.4 The Differential Galois Group

In this section we introduce the (differential) Galois group of a linear differential equation in matrix form, or in module form, and develop theory to prove some of its main features.

Definition 1.25 *The differential Galois group* of an equation $y' = Ay$ over k, or of a differential module over k, is defined as the group $\mathrm{Gal}(R/k)$ of differential k-algebra automorphisms of a Picard-Vessiot ring R for the equation. More precisely, $\mathrm{Gal}(R/k)$ consists of the k-algebra automorphisms σ of R satisfying $\sigma(f') = \sigma(f)'$ for all $f \in R$. □

As we have seen in Exercises 1.24, a finite Galois extension R/k is the Picard-Vessiot ring of a certain matrix differential equation over k. This exercise also states that the ordinary Galois group of R/k coincides with the differential Galois group. Therefore our notation for the differential Galois does not lead to confusion.

Observations 1.26 *The differential Galois group as group of matrices.*
Let M be a differential module over k and let $y' = Ay$ be an associated matrix

differential equation obtained by choosing a basis of M over k. Let R/k denote the Picard-Vessiot extension.

(1) The differential Galois group $G = \mathrm{Gal}(R/k)$ can be made more explicit as follows. As in Exercises 1.16 one considers the solution space $V := \ker(\partial, R \otimes_k M)$. The k-linear action of G on R extends to a k-linear action on $R \otimes_k M$. This action commutes with ∂ on $R \otimes_k M$. Thus, there is an induced C-linear action of G on the solution space V. This action is injective. Indeed, fix a basis of V over C and a basis of M over k and let F denote the matrix that expresses the first basis into the second basis. Then R is generated over k by the entries of F and the inverse of the determinant of F. In other words, there is a natural injective group homomorphism $G \to \mathrm{GL}(V)$.

(2) The above can be translated in terms of the matrix differential equation $y' = Ay$. Namely, let $F \in \mathrm{GL}_n(R)$ be a fundamental matrix. Then, for any $\sigma \in G$, also $\sigma(F)$ is a fundamental matrix and hence $\sigma(F) = FC(\sigma)$ with $C(\sigma) \in \mathrm{GL}_n(C)$. The map $G \to \mathrm{GL}_n(C)$, given by $\sigma \mapsto C(\sigma)$, is an injective group homomorphism (because R is generated over k by the entries of F and $\frac{1}{\det F}$). This is just a translation of (1) above since the columns of F form a basis of the solution space V.

(3) Let L denote the field of fractions of R. Then one can also consider the group $\mathrm{Gal}(L/k)$ consisting of the k-linear automorphisms of L, commuting with the differentiation on L. Any element in $\mathrm{Gal}(R/k)$ extends in a unique way to an automorphism of L of the required type. Thus there is an injective homomorphism $\mathrm{Gal}(R/k) \to \mathrm{Gal}(L/k)$. This homomorphism is bijective. Indeed, an element $\sigma \in \mathrm{Gal}(L/k)$ acts upon $L \otimes_k M$ and $\ker(\partial, L \otimes_k M)$. The latter is equal to V. With the notations of (1) or (2), R is generated by the entries of a matrix F and the inverse of its determinant. Further $\sigma(F) = FC(\sigma)$ for some constant matrix $C(\sigma)$. Therefore $\sigma(R) = R$. Hence σ is the image of the restriction of σ to R. □

What makes differential Galois groups a powerful tool is that they are linear algebraic groups and, moreover, establish a Galois correspondence, analogous to the classical Galois correspondence. Torsors will explain the connection between the Picard-Vessiot ring and the differential Galois group. The tannakian approach to linear differential equations provides new insight and useful methods. Some of this is rather technical in nature. We will try to explain theorems and proofs on various levels of abstraction.

Theorem 1.27 *Let $y' = Ay$ be a differential equation of degree n over k, having Picard-Vessiot field $L \supset k$ and differential Galois group $G = \mathrm{Gal}(L/k)$. Then*
(1) G, considered as a subgroup of $\mathrm{GL}_n(C)$, is an algebraic group.
(2) The Lie algebra of G coincides with the Lie algebra of the derivations of L/k that commute with the derivation on L.
(3) The field L^G of G-invariant elements of L is equal to k.

Proof. *An intuitive proof of (1) and (2).*
L is the field of fractions of $R := k[X_{i,j}, \frac{1}{\det}]/q$, where q is a maximal differential ideal. Using Observation 1.26 one can identify G with the group of matrices $M \in \mathrm{GL}_n(C)$ such that the automorphism σ_M of $R_0 := k[X_{i,j}, \frac{1}{\det}]$, given by $(\sigma X_{i,j}) = (X_{i,j})M$, has the property $\sigma_M(q) \subset q$. One has to verify that the property $\sigma_M(q) \subset q$ defines a Zariski-closed subset of $\mathrm{GL}_n(C)$. This can be seen as follows. Let $q_1, \ldots, q_r$ denote generators of the ideal q and let $\{e_i\}_{i\in I}$ be a C-basis of R. Then $\sigma_M(q_j)\mathrm{mod}\ q$ can be expressed as a finite sum $\sum_i C(M, j, i)e_i$ with coefficients $C(M, i, j) \in C$ depending on M. It is not difficult to verify that $C(M, i, j)$ is, in fact, a polynomial expression in the entries of M and $\frac{1}{\det M}$. Thus G is the Zariski-closed subset of $\mathrm{GL}_n(C)$ given by the set of equations $\{C(M, i, j) = 0\}_{i,j}$.

According to Sect. A.2.2, the Lie algebra of G can be described as the set of matrices $M \in \mathrm{M}_n(C)$ such that $1+\epsilon M$ lies in $G(C[\epsilon])$. This property of M translates into the k-linear derivation $D_M : R_0 \to R_0$, given by $(D_M X_{i,j}) = (X_{i,j})M$, has the property $D_M(q) \subset q$. Clearly D_M commutes with the differentiation of R_0. Thus the property $D_M(q) \subset q$ is equivalent to D_M induces a k-linear derivation on R commuting with $'$. The latter extends uniquely to a k-linear derivation of L commuting with $'$. One can also start with a k-linear derivation of L commuting with $'$ and deduce a matrix $M \in \mathrm{M}_n(C)$ as above.

Formalization of the proof of (1) and (2).
Instead of working with G as a group of matrices, one introduces a functor $\mathcal{G}$ from the category of C-algebras to the category of groups. Furthermore, $\mathcal{G}(C) = G$. It will be shown that this functor is representable by a certain finitely generated C-algebra U. It follows that Max(U) (or Spec(U)) is a linear algebraic group and G is identified with the set of C-valued points of this linear algebraic group. We refer to the appendices for the terminology used here.

For any C-algebra B (always commutative and with a unit element) one defines differential rings $k \otimes_C B,\ R \otimes_C B$ with derivations given by $(f \otimes b)' = f' \otimes b$ for $f \in k$ or R. The ring of constants of the two differential rings is B. The group $\mathcal{G}(B)$ is defined to be the group of the $k \otimes B$-linear automorpisms of $R \otimes_C B$ commuting with the derivation. It is evident that $\mathcal{G}$ is a functor. As above for the case $B = C$, one can describe the elements of $\mathcal{G}(B)$ as the group of matrices $M \in \mathrm{GL}_n(B)$ such that the differential automorphism σ_M of $k[X_{i,j}, \frac{1}{\det}] \otimes B$, given by the formula $(\sigma_M X_{i,j}) = (X_{i,j})M$, has the property $\sigma_M(q) \subset (q)$. Here, (q) is the ideal of $k[X_{i,j}, \frac{1}{\det}] \otimes B$ generated by q.

In order to show that $\mathcal{G}$ is representable we make for B the choice $C[Y_{s,t}, \frac{1}{\det}]$ (with the usual sloppy notation) and we consider the matrix $M_0 = (Y_{s,t})$ and write $\sigma_{M_0}(q_j)\mathrm{mod}\ (q) \in R \otimes_C C[Y_{s,t}, \frac{1}{\det}]$ as a finite sum

$$\sum_i C(M_0, i, j)e_i \text{ with all } C(M_0, i, j) \in C[Y_{s,t}, \frac{1}{\det}].$$

Let $I \subset C[Y_{s,t}, \frac{1}{\det}]$ denote the ideal generated by all $C(M_0, i, j)$. Now we claim that $U := C[Y_{s,t}, \frac{1}{\det}]/I$ represents $\mathcal{G}$.

Let B be any C-algebra and $\sigma \in \mathcal{G}(B)$ identified with σ_M for some $M \in \mathrm{GL}_n(B)$. One defines the C-algebra homomorphism $\phi : C[Y_{s,t}, \frac{1}{\det}] \to B$ by $(\phi Y_{s,t}) = M$. The condition on M implies that the kernel of ϕ contains I. Thus we find a unique C-algebra homomorphism $\psi : U \to B$ with $\psi(M_0 \bmod I) = M$. This proves the claim. According to Appendix B the fact that $\mathcal{G}$ is a functor with values in the category of groups implies that $\mathrm{Spec}(U)$ is a linear algebraic group. A result of Cartier ([302], Sect. 11.4) states that linear algebraic groups over a field of characteristic 0 are reduced. Hence I is a radical ideal.

Finally, the Lie algebra of the linear algebraic group is equal to the kernel of $\mathcal{G}(C[\epsilon]) \to \mathcal{G}(C)$ (where $\epsilon^2 = 0$ and $C[\epsilon] \to C$ is given by $\epsilon \mapsto 0$). The elements in this kernel are identified with the differential automorphisms of $R \otimes_C C[\epsilon]$ over $k \otimes_C C[\epsilon]$ having the form $1 + \epsilon D$. The set of Ds described here is easily identified with the k-linear derivations of R commuting with the differentiation on R.

(3) Let $a = \frac{b}{c} \in L \backslash k$ with $b, c \in R$ and let $d = b \otimes c - c \otimes b \in R \otimes_k R$. From Exercise A.15, one has that $d \neq 0$. Lemma A.16 implies that the ring $R \otimes_k R$ has no nilpotent elements since the characteristic of k is zero. Let J be a maximal differential ideal in the differential ring $(R \otimes_k R)[\frac{1}{d}]$, where the derivation is given by $(r_1 \otimes r_2)' = r_1' \otimes r_2 + r_1 \otimes r_2'$. Consider the two obvious morphisms $\phi_i : R \to N := (R \otimes_k R)[\frac{1}{d}]/J$. The images of the ϕ_i are generated (over k) by fundamental matrices of the same matrix differential equation. Therefore both images are equal to a certain subring $S \subset N$ and the maps $\phi_i : R \to S$ are isomorphisms. This induces an element $\sigma \in G$ with $\phi_1 = \phi_2 \sigma$. The image of d in N is equal to $\phi_1(b)\phi_2(c) - \phi_1(c)\phi_2(b)$. Since the image of d in N is nonzero, one finds $\phi_1(b)\phi_2(c) \neq \phi_1(c)\phi_2(b)$. Therefore $\phi_2((\sigma b)c) \neq \phi_2((\sigma c)b)$ and so $(\sigma b)c \neq (\sigma c)b$. This implies $\sigma(\frac{b}{c}) \neq \frac{b}{c}$. □

Now we give a *geometric formulation* of the Picard-Vessiot ring and the action of the differential Galois group. The notations of the proof of Theorem 1.27 will be used. The Picard-Vessiot ring R is written as $k[X_{i,j}, \frac{1}{\det}]/q$. Define $Z = \max(R)$. We have shown that Z is a reduced, irreducible subspace of $\mathrm{GL}_{n,k} := \max(k[X_{i,j}, \frac{1}{\det}])$. The differential Galois group $G \subset \mathrm{GL}_n(C)$ has been identified with the group consisting of the elements $g \in \mathrm{GL}_n(C)$ such that $Zg = Z$ (or equivalently g leaves the ideal q invariant). The multiplication on $\mathrm{GL}_{n,k}$ induces a morphism of k-affine varieties, $m : Z \times_C G \to Z$, given by $(z, g) \mapsto zg$. The morphism m is a group action in the sense that $(zg_1)g_2 = z(g_1g_2)$ for $z \in Z$ and $g_1, g_2 \in G$.

The *next technical step* is to prove that the morphism $Z \times_C G \to Z \times_k Z$, given by $(z, g) \mapsto (zg, z)$, is an isomorphism of affine varieties over k. This is precisely the definition of "Z is a G-torsor over k"(cf. Sect. A.2.6). Put $G_k = G \otimes_C k$. This abuse of notation means that G_k is the algebraic group over k, whose coordinate ring is $C[G] \otimes_C k$. Then one has $Z \times_C G = Z \times_k G_k$. Since both Z and G_k are contained in $\mathrm{GL}_{n,k}$ and the G_k-action on Z is multiplication on the right, the statement that Z is a G-torsor roughly means that $Z \subset \mathrm{GL}_{n,k}$ is a right coset for the subgroup G_k.

If Z happens to have a k-rational point p, i.e., $Z(k) \neq \emptyset$, then Z is a G-torsor, if and only if $Z = pG_k$. In this case Z is called a trivial torsor. In the general situation with $Z \subset \mathrm{GL}_{n,k}$ and $G \subset \mathrm{GL}_{n,C}$, the statement that Z is a G-torsor means that for some field extension $F \supset k$, one has that $Z_F := Z \otimes_k F$ is a right coset of $G_F := G \otimes_C F$ in $\mathrm{GL}_{n,F}$. See the appendices for more information.

Theorem 1.28 *Let R be a Picard-Vessiot ring with differential Galois group G. Then $Z = \max(R)$ is a G-torsor over k.*

Proof. We keep the above notation. We will show that Z_L is a right coset for G_L, where L is the Picard-Vessiot field, equal to the field of fractions of R. This will prove the theorem. Consider the following rings

$$k[X_{i,j}, \frac{1}{\det}] \subset L[X_{i,j}, \frac{1}{\det}] = L[Y_{s,t}, \frac{1}{\det}] \supset C[Y_{s,t}, \frac{1}{\det}],$$

where the relation between the variables $X_{i,j}$ and $Y_{s,t}$ is given by the formula $(X_{i,j}) = (r_{a,b})(Y_{s,t})$. The elements $r_{a,b} \in L$ are the images of $X_{a,b}$ in $k[X_{i,j}, \frac{1}{\det}]/q \subset L$. The three rings have a differentiation and a $\mathrm{Gal}(L/k)$-action. The differentiation is given by the known differentiation on L and by the formula $(X'_{i,j}) = A(X_{i,j})$. Since $(r_{a,b})$ is a fundamental matrix for the equation one has $Y'_{s,t} = 0$ for all s, t and the differentiation on $C[Y_{s,t}, \frac{1}{\det}]$ is trivial. The $\mathrm{Gal}(L/k)$-action is induced by the $\mathrm{Gal}(L/k)$-action on L. Thus $\mathrm{Gal}(L/k)$ acts trivially on $k[X_{i,j}, \frac{1}{\det}]$. For any $\sigma \in \mathrm{Gal}(L/k)$ one has $(\sigma r_{a,b}) = (r_{a,b})M$ for a certain $M \in G(C)$. Then $(\sigma Y_{s,t}) = M^{-1}(Y_{s,t})$. In other words, the $\mathrm{Gal}(L/k)$-action on $C[Y_{s,t}, \frac{1}{\det}]$ is translated into an action of the algebraic subgroup $G \subset \mathrm{GL}_{n,C}$ defined by the ideal I, constructed in the proof of Theorem 1.27. Let us admit for the moment the following lemma.

Lemma 1.29 *The map $I \mapsto (I)$ from the set of ideals of $k[X_{i,j}, \frac{1}{\det}]$ to the set of $\mathrm{Gal}(L/k)$-invariant ideals of $L[X_{i,j}, \frac{1}{\det}]$ is a bijection. The inverse map is given by $J \mapsto J \cap k[X_{i,j}, \frac{1}{\det}]$.*

Combining this with the similar Lemma 1.23, one finds a bijection between the differential ideals of $k[X_{i,j}, \frac{1}{\det}]$ and the $\mathrm{Gal}(L/k)$-invariant ideals of $C[Y_{s,t}, \frac{1}{\det}]$. A maximal differential ideal of the first ring corresponds to a maximal $\mathrm{Gal}(L/k)$-invariant ideal of the second ring. Thus $r := qL[X_{i,j}, \frac{1}{\det}] \cap C[Y_{s,t}, \frac{1}{\det}]$ is a maximal $\mathrm{Gal}(L/k)$-invariant ideal of the second ring. By this maximality r is a radical ideal and its zero set $W \subset \mathrm{GL}_n(C)$ is minimal w.r.t. $\mathrm{Gal}(L/k)$-invariance. Thus W is a left coset in $\mathrm{GL}_n(C)$ for the group $G(C)$, seen as a subgroup of $\mathrm{GL}_n(C)$. The matrix 1 belongs to W. Indeed, q is contained in the ideal of $L[X_{i,j}, \frac{1}{\det}]$ generated by $\{X_{i,j} - r_{i,j}\}_{i,j}$. This ideal is also generated by $\{Y_{s,t} - \delta_{s,t}\}_{s,t}$. The intersection of this ideal with $C[Y_{s,t}, \frac{1}{\det}]$ is the ideal defining $\{1\} \subset \mathrm{GL}_{n,C}$. Thus $W = G$.

One concludes that

$$L \otimes_k R = L \otimes_k (k[X_{i,j}, \frac{1}{\det}]/q) \cong L \otimes_C (C[Y_{s,t}, \frac{1}{\det}]/r) \cong L \otimes_C U.$$

This isomorphism translates into $Z_L = (r_{a,b})G_L$. A proof of Lemma 1.29 finishes the proof of the theorem. □

Proof of Lemma 1.29.
The proof is rather similar to that of Lemma 1.23. The only thing that we have to verify is that every $\mathrm{Gal}(L/k)$-invariant ideal J of $L[X_{i,j}, \frac{1}{\det}]$ is generated by $I := J \cap k[X_{i,j}, \frac{1}{\det}]$. Choose a basis $\{e_a\}_{a \in A}$ of $k[X_{i,j}, \frac{1}{\det}]$ over k. Any $f \in J$ can uniquely be written as a finite sum $\sum_a \ell_a e_a$ with all $\ell_a \in L$. The length $l(f)$ of f is defined as the number of $a \in A$ with $\ell_a \neq 0$. By induction on the length we will show that $f \in (I)$.

For $l(f) = 0$ or 1, this is trivial. Suppose $l(f) > 1$. We may, after multiplication by a nonzero element of L suppose that $\ell_{a_1} = 1$ for some a_1. If all $\ell_a \in k$, then $f \in (I)$. If not, then there exists an a_2 with $\ell_{a_2} \in L \setminus k$. For any $\sigma \in \mathrm{Gal}(L/k)$, the length of $\sigma(f) - f$ is less than $l(f)$. Thus $\sigma(f) - f \in (I)$.

According to Theorem 1.27, there exists a σ with $\sigma(\ell_{a_2}) \neq \ell_{a_2}$. As above, one finds that $\sigma(\ell_{a_2}^{-1} f) - \ell_{a_2}^{-1} f \in (I)$. Then

$$\sigma(\ell_{a_2}^{-1} f) - \ell_{a_2}^{-1} f = \sigma(\ell_{a_2}^{-1})(\sigma(f) - f) + (\sigma(\ell_{a_2}^{-1}) - \ell_{a_2}^{-1})f.$$

From $\sigma(\ell_{a_2}^{-1}) - \ell_{a_2}^{-1} \in L^*$, it follows that $f \in (I)$. □

Corollary 1.30 *Let R be a Picard-Vessiot ring for the equation $y' = Ay$ over k. Let L be the field of fractions of R. Put $Z = \mathrm{Spec}(R)$. Let G denote the differential Galois group and $C[G]$ the coordinate ring of G and let $\mathfrak{g}$ denote the Lie algebra of G. Then:*
(1) There is a finite extension $\tilde{k} \supset k$ such that $Z_{\tilde{k}} \cong G_{\tilde{k}}$.
(2) Z is smooth and connected.
(3) The transcendence degree of L/k is equal to the dimension of G.
(4) Let H be a subgroup of G with Zariski closure $\overline{H}$. Then $L^H = k$ if and only if $\overline{H} = G$.

Proof. (1) Take a $B \in Z(\overline{k})$. Then B is defined over some finite extension $\tilde{k}$ of k. Over this extension the torsor becomes trivial.
(2) By Proposition 1.20, Z is connected. The algebraic group G is smooth over C. Using the fact that smoothness is preserved in "both directions" by field extensions, one has the fact that Z is smooth over k.
(3) The transcendence degree of L/k is equal to the Krull dimension of R and that of $\tilde{k} \otimes_k R \cong \tilde{k} \otimes C[G]$. The latter is equal to the dimension of G.
(4) It is easily seen that $L^H = L^{\overline{H}}$. Therefore we may assume that H is Zariski closed. From Theorem 1.27, $L^G = k$.

Suppose now $L^H = k$. Fix a finite extension $\tilde{k} \supset k$ such that $\tilde{k} \otimes_k R \cong \tilde{k} \otimes_C C[G]$. Let $Qt(C[G])$ be the total ring of fractions of $C[G]$. Then the total rings of fractions of $\tilde{k} \otimes_k R$ and $\tilde{k} \otimes_C C[G]$ are $\tilde{k} \otimes_k L$ and $\tilde{k} \otimes_C Qt(C[G])$. Taking H-invariants leads to $\tilde{k} \otimes_k L^H \cong \tilde{k} \otimes_C Qt(C[G])^H$. The ring $Qt(C[G])^H$ consists of the H-invariant rational functions on G. The latter is the same as the ring of the rational functions on G/H (see [141], Chap. 12). Therefore $L^H = k$ implies $H = G$. □

The proof of Theorem 1.27 is not constructive; although it tells us that the Galois group is a linear algebraic group it does not give us a way to calculate this group. Nonetheless, the following proposition yields some restrictions on this group.

Proposition 1.31 *Consider the equation $y' = Ay$ over k with Galois group G and torsor Z. Let $\mathfrak{g}$ denote the Lie algebra of G.*
(1) Let $H \subset \mathrm{GL}_{n,C}$ be a connected algebraic subgroup with Lie algebra $\mathfrak{h}$. If $A \in \mathfrak{h}(k)$, then G is contained in (a conjugate of) H.
(2) Z is a trivial torsor if and only if there is an equivalent equation $v' = \tilde{A}v$ such that $\tilde{A} \in \mathfrak{g}(k)$.

Proof. (1) Let $H \subset \mathrm{GL}_{n,C}$ by given by the radical ideal $I \subset C[X_{i,j}, \frac{1}{\det}]$. Let (I) denote the ideal in $k[X_{i,j}, \frac{1}{\det}]$ generated by I. As before, one defines a derivation on $k[X_{i,j}, \frac{1}{\det}]$ by the formula $(X'_{i,j}) = A(X_{i,j})$. We claim that (I) is a differential ideal.

It suffices to show that for any $f \in I$ the element f' lies in (I). Since det is invertible, we may suppose that f is a polynomial in the n^2 variables $X_{i,j}$ with coefficients in C. The element f is seen as a map from $\mathrm{M}_n(\bar{k})$ to $\bar{k}$, where $\bar{k}$ denotes an algebraic closure of k. The ideal (I) is a radical ideal, since $(C[X_{i,j}, \frac{1}{\det}]/I) \otimes_C k$ has no nilpotent elements. Therefore $f' \in (I)$ if $f'(B) = 0$ for all $B \in H(\bar{k})$.
Now we use the terminology of Sect. A.2.2. One has $1 + \epsilon A \in H(\bar{k}[\epsilon])$ and $B + \epsilon AB \in H(\bar{k}[\epsilon])$. Hence, $0 = f(B + \epsilon AB) = \epsilon \sum_{i,j} (AB)_{i,j} \frac{\partial f}{\partial X_{i,j}}(B)$.
Furthermore, $f' = \sum_{i,j} X'_{i,j} \frac{\partial f}{\partial X_{i,j}} = \sum_{i,j} (A \cdot (X_{s,t}))_{i,j} \frac{\partial f}{\partial X_{i,j}}$. Hence, $f'(B) = 0$.

Let $q \supset (I)$ be a maximal differential ideal of $k[X_{i,j}, \frac{1}{\det}]$. Let $Z \subset H_k \subset \mathrm{GL}_{n,k}$ be the reduced, irreducible subspace defined by q. For any M in the differential Galois group and any $B \in Z(\bar{k})$ one has $BM \in Z(\bar{k})$ and thus $M \in H(\bar{k})$. Furthermore, $H(\bar{k}) \cap \mathrm{GL}_n(C) = H(C)$.

(2) If $\tilde{A} \in \mathfrak{g}(k)$, then the proof of part (1) yields the fact that G_k is its torsor and BG_k is the torsor of $y' = Ay$.

If Z is a trivial torsor, then $Z = BG_k$ for some $B \in Z(k)$. The equivalent differential equation $v' = \tilde{A}v$, obtained by the substitution $y = Bv$, has the property that the ideal $\tilde{q} \subset k[Z_{i,j}, \frac{1}{\det}]$ of G_k, where $(X_{i,j}) = B(Z_{i,j})$, is a maximal differential ideal. Let $z_{i,j}$ denote the image of $Z_{i,j}$ in the Picard-Vessiot ring $k[Z_{i,j}, \frac{1}{\det}]/\tilde{q}$ of $v' = \tilde{A}v$. Then $F := (z_{i,j})$ is a fundamental matrix and lies in $G(L)$, where L is the

Picard-Vessiot field. As in the proof of part (1) one verifies that $F + \epsilon F' \in G(L[\epsilon])$. It follows that $\tilde{A} = F^{-1}F'$ lies in $\mathfrak{g}(L) \cap \mathrm{M}_n(k) = \mathfrak{g}(k)$. □

For a differential field that is a C_1-field, there is a (partial) converse of Proposition 1.31. Examples of such fields are $C(z)$, $C((z))$ and $\mathbf{C}(\{z\})$ for any algebraically closed field C.

Corollary 1.32 *Let the differential field k be a C_1-field. Assume that the differential Galois group G of the equation $y' = Ay$ over k is connected. Let $\mathfrak{g}$ be the Lie algebra of G. Let a connected algebraic group $H \supset G$ with Lie algebra $\mathfrak{h}$ be given such that $A \in \mathfrak{h}(k)$. Then there exists $B \in H(k)$ such that the equivalent differential equation $f' = \tilde{A}f$, with $y = Bf$ and $\tilde{A} = B^{-1}AB - B^{-1}B'$, satisfies $\tilde{A} \in \mathfrak{g}(k)$.*

Proof. The assumptions that G is connected and k is a C_1-field imply that Z is a trivial torsor. Now apply Proposition 1.31. □

Remarks 1.33

(1) The condition that G is connected is necessary for Corollary 1.32. Indeed, consider the case $H = G$. If $\tilde{A} \in \mathfrak{h}(k) = \mathfrak{g}(k)$ can be found, then by Proposition 1.31 part (1), $G \subset H^o$ and thus $G = G^o$.

(2) We recall that an *algebraic Lie subalgebra* of the Lie algebra $\mathrm{M}_n(C)$ of $\mathrm{GL}_n(C)$ is the Lie algebra of an algebraic subgroup of $\mathrm{GL}_n(C)$. Assume that k is a C_1-field and that the differential Galois group of $y' = Ay$ is connected. Let $\mathfrak{h} \subset \mathrm{M}_n(C)$ be a minimal algebraic Lie subalgebra such that there exists an equivalent equation $v' = \tilde{A}v$ with $\tilde{A} \in \mathfrak{h}(k)$. Then, by Corollary 1.32, $\mathfrak{h}$ is the Lie algebra of the differential Galois group. This observation can be used to find the differential Galois group or to prove that a proposed group is the differential Galois group. □

Proposition 1.34 The Galois Correspondence

Let $y' = Ay$ be a differential equation over k with Picard-Vessiot field L and write $G := \mathrm{Gal}(L/k)$. Consider the two sets

$\mathcal{S}$:= the closed subgroups of G and

$\mathcal{L}$:= the differential subfields M of L, containing k.

Define $\alpha : \mathcal{S} \to \mathcal{L}$ by $\alpha(H) = L^H$, the subfield of L consisting of the H-invariant elements. Define $\beta : \mathcal{L} \to \mathcal{S}$ by $\beta(M) = \mathrm{Gal}(L/M)$, which is the subgroup of G consisting of the M-linear differential automorphisms. Then

1. *The maps α and β are inverses of each other.*
2. *The subgroup $H \in \mathcal{S}$ is a normal subgroup of G if and only if $M = L^H$ is, as a set, invariant under G. If $H \in \mathcal{S}$ is normal then the canonical map $G \to \mathrm{Gal}(M/k)$ is surjective and has kernel H. Moreover, M is a Picard-Vessiot field for some linear differential equation over k.*
3. *Let G^o denote the identity component of G. Then $L^{G^o} \supset k$ is a finite Galois extension with Galois group G/G^o and is the algebraic closure of k in L.*

Proof. Since the elements of G commute with the derivation, L^H is a differential subfield of L. One observes that the Picard-Vessiot field of the equation $y' = Ay$ over M is again L and thus $\beta(M) = \mathrm{Gal}(L/M)$ is its differential Galois group. In particular $\beta(M)$ is a closed subgroup of G and belongs to $\mathcal{S}$.

1. For $M \in \mathcal{L}$ one has $\alpha\beta(M) = L^{\mathrm{Gal}(L/M)}$. By applying Theorem 1.27 to the Picard-Vessiot extension L/M for $y' = Ay$ over M, one sees that the last field is equal to M.
Let $H \subset G$ be a closed subgroup. The inclusion $H \subset H_1 := \mathrm{Gal}(L/L^H) = \beta\alpha(H)$ is obvious. One applies Corollary 1.30 with G replaced by H_1 and k replaced by $L^H = L^{H_1}$. We conclude that $H = H_1$.

2. Assume that $M = L^H$ is left invariant by all elements of G. One can then define a map $G \to \mathrm{Gal}(M/k)$ by restricting any $\sigma \in G$ to M. The kernel of this map is H, so H is normal in G. Furthermore, this map defines an injective homomorphism of the group G/H into $\mathrm{Gal}(M/k)$. To show that this map is surjective, one needs to show that any differential automorphism of M over k extends to a differential automorphism of L over k. Consider, more generally, $M \in \mathcal{L}$ and a k-homomorphism of differential fields $\psi : M \to L$. The Picard-Vessiot field for $y' = Ay$ over M is L. The Picard-Vessiot field for $y' = \psi(A)y$ (note that $\psi(A) = A$) over $\psi(M)$ is also L. The unicity of the Picard-Vessiot field yields a k-isomorphism of differential fields $\tilde{\psi} : L \to L$, extending ψ.

Now assume that there is an element $\tau \in G$ such that $\tau(M) \neq M$. The Galois group of L over $\tau(M)$ is $\tau H \tau^{-1}$. Since $\tau(M) \neq M$, part (1) of the proposition implies that $\tau H \tau^{-1} \neq H$. Therefore H is not normal in G.

It is more difficult to see that M is a Picard-Vessiot field for some linear differential equation over K and we postpone the proof of this fact to the next section (see Corollary 1.40).

3. G/G^o is a finite group. The property that $(L^{G^o})^{G/G^o} = k$ together with the Galois theory of algebraic extensions (cf. [170], Chap. VII.1, Artin's Theorem), implies that $L^{G^o} \supset k$ is a Galois extension with Galois group G/G^o. If u is algebraic over k, then the orbit of u under the action of G is finite. Therefore, the group $\mathrm{Aut}(L/k(u))$ is an algebraic subgroup of G of finite index. This implies that $G^o \subset \mathrm{Aut}(L/k(u))$ and so $k(u) \subset L^{G^o}$. □

Exercises 1.35

1. *The Galois group of $y' = a,\ a \in k$*
Show that the Galois group of this equation is either the additive group over C, i.e., $\mathbf{G}_{a,C} = (C, +)$ or the trivial group. Hint: Compare with Example 1.18.

2. *The Galois group of $y' = ay,\ a \in k^*$*
Show that the Galois group of this equation is either $(C^*, \times)$ or a finite cyclic group. Is the torsor trivial? Hint: Compare with Example 1.19.

3. *The Galois group of* $y'' = c^2 y,\ c \in C^*$
Show that the differential ring $C(z)[Y, Y^{-1}]$ given by $Y' = cY$ is a Picard-Vessiot ring for this equation over $C(z),\ z' = 1$. Calculate the differential Galois group and the torsor of this equation.

4. *The generic Picard-Vessiot extension and its Galois group*
Let k be a differential field with algebraically closed field of constants C, let $R = k\{\{y_1, \ldots, y_n\}\}$ be the ring of differential polynomials with coefficients in k and let F be the quotient field of R.
(a) Show that the constant subfield of F is C.
(b) Let $L(Y)$ be the linear scalar differential equation given by

$$L(Y) := \frac{wr(Y, y_1, \ldots, y_n)}{wr(y_1, \ldots, y_n)} = Y^{(n)} + a_{n-1}Y^{(n-1)} + \cdots + a_0 Y.$$

Show that $a_{n-1} = \dfrac{(wr(y_1, \ldots, y_n))'}{wr(y_1, \ldots, y_n)}$.
(c) Let E be the smallest differential subfield of F containing k and the elements $a_i,\ i = 0, \ldots, n-1$. Show that for any $A = (c_{i,j}) \in \mathrm{GL}_n(C)$, the map $\phi_A : F \to F$ defined by $(\phi_A(y_1), \ldots, \phi_A(y_n)) = (y_1, \ldots, y_n)A$ is a k-differential automorphism of F leaving all elements of E fixed. Hint: $wr(\phi_A(y_1), \ldots, \phi_A(y_n)) = \det(A)wr(y_1, \ldots, y_n)$.
(d) Using Exercise 1.24.2(b), show that F is a Picard-Vessiot extension of E with Galois group $\mathrm{GL}_n(C)$. Is the torsor of this equation trivial?

5. *Unimodular Galois groups*
(a) Let $y' = Ay$ be an $n \times n$ matrix differential equation over k, let L be its Picard-Vessiot field over k and let G be its Galois group. Let F be a fundamental matrix for $y' = Ay$ with coefficients in L. Show that $G \subset \mathrm{SL}_n(C)$ if and only if $\det(F) \in k$. Conclude that $G \subset \mathrm{SL}_n$ if and only if $u' = (\mathrm{tr}A)u$ has a nonzero solution in k. Hint: Use Exercise 1.14.5.
(b) Let $L(y) = y^{(n)} + a_{n-1}y^{(n-1)} + \cdots + a_0 y = 0$ be a homogeneous scalar linear differential equation over K. Show that the Galois group of $L(y) = 0$ is a subgroup of $\mathrm{SL}_n(C)$ if and only if $z' = -a_{n-1}z$ has a nonzero solution in k.
(c) Let $L(y) = y^{(n)} + a_{n-1}y^{(n-1)} + \cdots + a_0 y = 0$ be a homogeneous scalar linear differential equation over K. Setting $z = e^{1/n \int a_{n-1}} y$, show that z satisfies a homogeneous scalar linear differential equation of the form $z^{(n)} + \tilde{a}_{n-2}z^{(n-2)} + \cdots + \tilde{a}_0 y = 0$ and that this latter equation has a unimodular Galois group. □

Consider the differential field $C(z)$ with C algebraically closed and of characteristic 0 and derivation $\frac{d}{dz}$. We consider a scalar differential equation of the form $y'' = ry$. The Picard-Vessiot field will be denoted by L and the differential Galois group will be denoted by G. The following exercise will show how one can determine in many cases the Galois group of such an equation. A fuller treatment is given in [167] and [272, 273, 274].

The rather short list of the algebraic subgroups (up to conjugation) of $\mathrm{SL}_2(C)$ is the following (see, for instance, [167]):

(i) Reducible subgroups G, i.e., there exists a G-invariant line. In other terms, the subgroups of $\{\left(\begin{smallmatrix} a & b \\ 0 & a^{-1} \end{smallmatrix}\right) | \ a \in C^*, b \in C\}$.

(ii) Irreducible and imprimitive groups G, i.e., there is no G-invariant line but there is a pair of lines permuted by G. In other terms G is an irreducible subgroup of the infinite dihedral group D_∞, consisting of all $A \in \mathrm{SL}_2(C)$ such that A permutes the two lines $C(1,0)$, $C(0,1)$ in C^2.

(iii) Three finite primitive (i.e., irreducible but not imprimitive) groups: the tetrahedral, the octahedral, and the icosahedral group.

(iv) $\mathrm{Sl}_2(C)$.

Exercises 1.36 ([232])
1. *The equation* $y'' = ry$
(a) Using Exercise 1.35.5, show that the Galois group of $y'' = ry$ is a subgroup of $\mathrm{SL}_2(C)$.
(b) Associated to the equation $y'' = ry$ there is the nonlinear Riccati equation $u' + u^2 = r$. Let L be the Picard-Vessiot extension of k corresponding to this equation and let $V \subset L$ denote the vector space of solutions of $y'' = ry$. Then V is a two-dimensional vector space over C. The group G acts on V. Show that $u \in L$ is a solution of the Riccati equation $u' + u^2 = r$ if and only if $u = \frac{y'}{y}$ for some $y \in V$, $y \neq 0$.
(c) Show that G is reducible if and only if the Riccati equation has a solution in $C(z)$.
(d) Show that if G is irreducible and imprimitive, then the Riccati equation has a solution u that is algebraic over $C(z)$ of degree 2. Hint: There are two lines $Cy_1, Cy_2 \subset V$ such that G permutes $\{Cy_1, Cy_2\}$. Put $u_1 = \frac{y_1'}{y_1}$, $u_2 = \frac{y_2'}{y_2}$. Show that $u_1 + u_2$ and $u_1 u_2$ belong to $C(z)$.

2. *The equation* $y'' = (\frac{5}{16}z^{-2} + z)y$
(a) The field extension $C(t) \supset C(z)$ is defined by $t^2 = z$. Verify that $u_1 = -\frac{1}{4}z^{-1} + t \in C(t)$ is a solution of the Riccati equation. Find a second solution $u_2 \in C(t)$ of the Riccati equation.
(b) Prove that the differential ring $R = C(t)[y_1, y_1^{-1}]$, defined by $y_1' = u_1 y_1$, is a Picard-Vessiot ring for the equation. Hint: Verify that R is a simple differential ring. Prove that R is generated over $C(z)$ by the entries of a fundamental matrix for the equation.
(c) Determine the differential Galois group G of the equation.
(d) Verify that the Lie algebra of G is equal to the Lie algebra of the K-linear derivations $D : R \to R$ that commute with $'$.
(e) What can one say about the solutions of the equation?

3. *Liouville's differential equation* $y'' = ry$ *with* $r \in \mathbf{C}[z] \setminus \mathbf{C}$.
(a) Show that the Galois group of this equation is connected. Hint: Standard existence theorems imply that there are two linearly independent *entire* solutions y_1, y_2 of $y'' = ry$. Show that the subfield $K = \mathbf{C}(z, y_1, y_2, y_1', y_2')$ of the field of meromorphic functions on $\mathbf{C}$, is a Picard-Vessiot field for the equation. Show that if $u \in K$ is algebraic over $C(z)$, then u is meromorphic on the Riemann sphere and so in $\mathbf{C}(z)$. Deduce that $G = G^o$.
(b) Suppose that $r \in \mathbf{C}[z]$ has odd degree. Prove that the Riccati equation has no solution $u \in \mathbf{C}(z)$. Hint: Expand u at $z = \infty$ and find a contradiction.
(c) Suppose again that $r \in \mathbf{C}[z]$ has odd degree. Prove that $G = \mathrm{SL}_2(C)$ and give an explicit description of the Picard-Vessiot ring.
(d) Consider the equation $y'' = (z^2 + 1)y$. Find a solution $u \in \mathbf{C}(z)$ of the Riccati equation. Construct the Picard-Vessiot ring and calculate the differential Galois group. Hint: Consider first the equation $y' = uy$. A solution $y_1 \neq 0$ is also a solution of $y_1'' = (z^2 + 1)y_1$. Find a second solution y_2 by "variation of constants".

4. *Liouville's theorem (1841) for* $y'' = ry$ *with* $r \in \mathbf{C}[z] \setminus \mathbf{C}$
Prove the following slightly deformed version of Liouville's theorem:

Consider the differential equation $y'' = ry$ *with* $r \in \mathbf{C}[z] \setminus \mathbf{C}$. *The differential Galois group of this equation (over the differential field* $\mathbf{C}(z)$*) is equal to* $\mathrm{SL}_2(\mathbf{C})$ *unless* r *has even degree* $2n$ *and there are polynomials* v, F *with* $\deg v = n$ *such that* $u := v + \frac{F'}{F}$ *is a solution of the Riccati equation* $u' + u^2 = r$.
In the last case, the differential Galois group is conjugate to the group $\{\left(\begin{smallmatrix} a & b \\ 0 & a^{-1} \end{smallmatrix}\right) | a \in \mathbf{C}^*,\ b \in \mathbf{C}\}$.

Hints:
(i) Use part 3 of the exercise and the classification of the Zariski-closed subgroups of $\mathrm{SL}_2(\mathbf{C})$, to prove that the differential Galois group can only be (up to conjugation) $\mathrm{SL}_2(\mathbf{C})$, $\{\left(\begin{smallmatrix} a & b \\ 0 & a^{-1} \end{smallmatrix}\right) | a \in \mathbf{C}^*,\ b \in \mathbf{C}\}$ or $\{\left(\begin{smallmatrix} a & 0 \\ 0 & a^{-1} \end{smallmatrix}\right) | a \in \mathbf{C}^*\}$.
(ii) Show that the three cases correspond to 0, 1 or 2 solutions $u \in \mathbf{C}(z)$ of the Riccati equation $u' + u^2 = r$.
(iii) Assume that $u \in \mathbf{C}(z)$ is a solution of the Riccati equation. Make the observation that for any point $c \in \mathbf{C}$, the Laurent expansion of u at c has the form $\frac{\epsilon}{z-c} + * + *(z-c) + \cdots$ with $\epsilon = 0, 1$. Show that u must have the form $v + \frac{F'}{F}$, where F is a polynomial of degree $d \geq 0$ with simple zeros and v is a polynomial of degree n.
(iv) Show that there is at most one rational solution of the Riccati equation $u' + u^2 = r$ by expanding $u = v + \frac{F'}{F}$ at ∞, i.e., as a Laurent series in z^{-1}. Note that the expansion of $\frac{F'}{F}$ is $dz^{-1} + *z^{-2} + \cdots$. □

Exercise 1.37 *Algebraically independent solutions of differential equations.*
Let $r \in \mathbf{C}[z]$ be a polynomial of odd degree. Let $y_1 = 1 + \sum_{n \geq 2} a_n z^n$, $y_2 = z + \sum_{n \geq 2} b_n z^n$ be entire solutions of the equation $y'' = ry$. Show that the "only" polynomial relation over $\mathbf{C}$ between z, y_1, y_2, y_1, y_2' is $y_1 y_2' - y_1' y_2 = 1$. Hint: See Exercise 1.36. □

Theorem 1.28 allows us to identify the Picard-Vessiot ring inside the Picard-Vessiot field. This is the result of the following Corollary (see [34, 183, 267]).

Corollary 1.38 *Let $y' = Ay$ be a differential equation over k with Picard-Vessiot field L, differential Galois group G and Picard-Vessiot ring $R \subset L$. The following properties of $f \in L$ are equivalent.*

(1) $f \in R$.

(2) *The C-vector space $< Gf >$, spanned by the orbit $Gf := \{g(f) |\ g \in G\}$ has finite dimension m over C.*

(3) *The k-vector space $< f, f', f'', \ldots >$ spanned by f and all its derivatives has finite dimension m over k.*

Proof. (1)⇒(2). By Theorem 1.28, there is a finite extension $\tilde{k} \supset k$ such that $\tilde{k} \otimes_k R \cong \tilde{k} \otimes_C C[G]$. Here $C[G]$ denotes the coordinate ring of G. It is well known, see [141], that the G-orbit of any element in $C[G]$ spans a finite dimensional vector space over C. This property is inherited by $\tilde{k} \otimes_C C[G]$ and also by R.
(2)⇒(3). Choose a basis $v_1, \ldots, v_m$ of $< Gf >$ over C. There is a unique scalar differential equation $P(y) = y^{(m)} + a_{m-1}y^{(m-1)} + \cdots + a_1 y^{(1)} + a_0 y$ with all $a_i \in L$ such that $P(v_i) = 0$ for all i (see, for instance, the proof of Lemma 1.12). Then $< Gf >$ is the solution space of P. The G-invariance of this space implies that all $a_i \in L^G = k$. From $P(f) = 0$ it follows that $< f, f', f'', \cdots >$ has dimension $\leq m$ over k. Let Q be the monic scalar equation of minimal degree $n \leq m$ over k such that $Q(f) = 0$. The solution space of Q in L contains f and the m-dimensional C-vector space $< Gf >$. Hence $m = n$.
(3)⇒(1). Suppose that $W =< f, f', f'', \cdots >$ has dimension m over k. Then f is a solution of a monic linear scalar differential equation P over k of order m. Consider the nonzero ideal $I \subset R$ consisting of the elements $a \in R$ such that $aW \subset R$. For $a \in I$ and $w \in W$, one has $a'w = (aw)' - aw'$. Since both R and W are invariant under differentiation, one finds $a'w \in R$. Thus I is a differential ideal. Now R is a simple differential ring and therefore $I = R$. This proves that $f \in R$. □

Exercise 1.39 *Solutions of differential equations and their reciprocals.*
k is a differential field with algebraically closed field of constants C. Let $R \supset k$ be a Picard-Vessiot ring with field of fractions L. The goal of this exercise is to show:

Let $f \in L^*$. Then both f and f^{-1} satisfy a scalar linear differential equation over k if and only if $\frac{f'}{f}$ is algebraic over k

For the proof one needs a result of Rosenlicht [249] (see also [181, 267]) that states:

Let G be a connected linear algebraic group over an algebraically closed field K and let $f \in K[G]$ (i.e., the coordinate ring of G) be an invertible element such that $f(1) = 1$. Then f is a character, i.e., $f(g_1 g_2) = f(g_1)f(g_2)$ for all $g_1, g_2 \in G$.

(1) Show that it suffices to consider the case where k is algebraically closed. Hint: Replace k by its algebraic closure $\overline{k}$ and L by $\overline{k}L$.

(2) Prove that $\frac{f'}{f} \in k$ implies that $f, f^{-1} \in R$.

(3) Show that $R \cong k \otimes_C C[G]$ and that G is connected.

(4) Suppose that f is an invertible element of R. Show that f considered as an element of $k \otimes_C C[G]$ has the form $b \cdot \chi$, where $\chi : G_k \to k^*$ is a character and $b \in k^*$. Conclude that $\sigma(f) = \chi(\sigma) f$ for any $\sigma \in G$.

(5) Prove that any character $\chi : G_k \to k^*$ has the property $\chi(\sigma) \in C^*$ for all $\sigma \in G$. Hint: Two proofs are possible. The first one shows that any character of G_k comes from a character of G. We suggest a second proof. Any character χ belongs to R and satisfies, according to Corollary 1.38, a linear differential equation over k. Let $y^{(m)} + a_{m-1} y^{(m-1)} + \cdots + a_1 y^{(y)} + a_0 y$ be the differential equation of minimal degree over k, satisfied by χ. Fix $\sigma \in G$ and define $a \in k^*$ by $\sigma(\chi) = a\chi$. Since σ commutes with the differentiation, the same equation is the scalar linear differential equation of minimal degree over k satisfied by $\sigma(\chi) = a\chi$. Prove that $a_{m-1} = m\frac{a'}{a} + a_{m-1}$ and conclude that $a \in C^*$.

(6) Prove that $\frac{f'}{f} \in k$.

(7) Show that $\sin z$ satisfies a linear differential equation over $\mathbf{C}(z)$ and that $\frac{1}{\sin z}$ does not. Hint: A periodic function cannot be algebraic over $\mathbf{C}(z)$ (why?).

The main result of this exercise was first proved in [123]. See also [267] and [279]. □

We now use Theorem 1.28 to give a proof that a normal subgroup corresponds to a subfield that is also a Picard-Vessiot extension, thereby finishing the proof of Proposition 1.34.

Corollary 1.40 *Let $L \supset k$ be the Picard-Vessiot field of the equation $y' = Ay$ over k. Let $G := \mathrm{Gal}(L/k)$ be the differential Galois group of the equation and let $H \subset G$ be a closed normal subgroup. Then $M = L^H$ is a Picard-Vessiot field for some linear differential equation over k.*

Proof. This proof depends on the following three facts from the theory of linear algebraic groups. Let G be a linear algebraic group and H a Zariski-closed normal subgroup.

1. The G-orbit of any element $f \in C[G]$ spans a finite dimensional C-vector space.
2. The group G/H has a structure of an affine group and its coordinate ring $C[G/H]$ is isomorphic to the ring of invariants $C[G]^H$.
3. The two rings $Qt(C[G])^H$ and $Qt(O[G]^H)$ are naturally isomorphic.

These facts can be found in [141], §11 and 12, and [36]. Let L be the quotient field of the Picard-Vessiot ring R. Let $\tilde{k}$ be a finite Galois extension of k with (ordinary) Galois group U such that the torsor corresponding to R becomes trivial over $\tilde{k}$. This means that $\tilde{k} \otimes_k R \simeq \tilde{k} \otimes_C C[G]$. Note that U acts on $\tilde{k} \otimes_k R$ by acting on the left factor as the Galois group and on the right factor as the identity. The group G acts on $\tilde{k} \otimes_k R \simeq \tilde{k} \otimes_C C[G]$ by acting trivially on the left factor and acting on R via the Galois action (or equivalently, on $C[G]$ via the natural action of G on its coordinate ring). Using the above facts, we have the fact that $\tilde{k} \otimes_k R^H \simeq \tilde{k} \otimes_C C[G/H]$ and that $\tilde{k} \otimes_k L^H$ is equal to $\tilde{k} \otimes_C Qt(C[G]^H)$. Since $C[G/H]$ is a finitely generated C-algebra, there exist $r_1, \ldots, r_m \in R^H$ that generate $\tilde{k} \otimes_k R^H$ as a $\tilde{k}$-algebra. Taking invariants under U, one finds that R^H is a finitely generated k-algebra whose field of fractions is L^H. We may, furthermore, assume that that R^H is generated by a basis $y_1, \ldots, y_n$ of a finite dimensional C-vector space that is G/H-invariant. Lemma 1.12 implies that the wronskian matrix $W = W(y_1, \ldots, y_n)$ is invertible. Furthermore, the matrix $A = W'W^{-1}$ is left invariant by G/H and so has entries in k. Since the constants of L^H are C and L^H is generated by a fundamental set of solutions of the linear differential equation $y' = Ay$, Proposition 1.22 implies that L^H is a Picard-Vessiot field. □

Exercises 1.41 Let G be a connected solvable linear algebraic group. In this exercise the fact that any G-torsor over k is trivial will be used. For this, see the comments following Lemma A.51.

1. *Picard-Vessiot extensions with Galois group* $(\mathbf{G}_a)^r$.
Suppose that K is a Picard-Vessiot extension of k with Galois group $(\mathbf{G}_a)^r$. Show that there exist $t_1, \ldots, t_r \in K$ with $t_i' \in k$ such that $K = k(t_1, \ldots, t_r)$. Hint: Consider the Picard-Vessiot subring of K and use $C[\mathbf{G}_a^r] = C[t_1, \ldots, t_r]$.

2. *Picard-Vessiot extensions with Galois group* $(\mathbf{G}_m)^r$.
Show that if K is a Picard-Vessiot extension of k with Galois group $(\mathbf{G}_m)^r$, then there exist nonzero $t_1, \ldots, t_r \in K$ with $t_i'/t_i \in k$ such that $K = k(t_1, \ldots, t_r)$.

3. *Picard-Vessiot extensions whose Galois groups have solvable identity component.*
Let K be a Picard-Vessiot extension of k whose Galois group has solvable identity component. Show that there exists a tower of fields $k \subset K_1 \subset \cdots \subset K_n = K$ such that K_1 is an algebraic extension of k and for each $i = 2, \ldots, n$, $K_i = K_{i-1}(t_i)$ with t_i transcendental over K_{i-1} and either $t_i' \in K_{i-1}$ or $t_i'/t_i \in K_{i-1}$.
Hint: Produce a tower of closed subgroups $\{1\} = G_0 \subset G_1 \subset \cdots \subset G^o \subset G$, where G^o is the identity component of the Galois group G and each G_i is a normal subgroup of G_{i+1} such that G_{i+1}/G_i is either $\mathbf{G}_a$ or $\mathbf{G}_m$. (Compare Chap. 17, Exercise 7 and Theorem 19.3 of [141]). Apply Corollary 1.40.

In the next section, an elementary proof of the above statement will be given that does not use Theorem 1.28. □

1.5 Liouvillian Extensions

In this section we show how one can formalize the notion of solving a linear differential equation in "finite terms", that is solving in terms of algebraic combinations and iterations of exponentials and integrals, and give a Galois theoretic characterization of this property.

In classical Galois theory, one formalizes the notion of solving a polynomial equation in terms of radicals by using towers of fields. A similar approach will be taken here.

Definition 1.42 The differential field k is assumed to have an algebraically closed field of constants C. An extension $K \supset k$ of differential fields is called a *liouvillian extension* of k if the field of constants of K is C and if there exists a tower of fields $k = K_0 \subset K_1 \subset \ldots \subset K_n = K$ such that $K_i = K_{i-1}(t_i)$ for $i = 1, \ldots, n$, where either

1. $t_i' \in K_{i-1}$, that is t_i is an *integral (of an element of K_{i-1})*,
2. $t_i \neq 0$ and $t_i'/t_i \in K_{i-1}$, that is t_i is an *exponential (of an integral of an element of K_{i-1})*, or
3. t_i is algebraic over K_{i-1}.

If K is a liouvillian extension of k and each of the t_i is an integral (respectively, exponential), we say that K is an *extension by integrals* (respectively, *extension by exponentials*) of k. □

The main result of this section is the following theorem.

Theorem 1.43 *Let K be a Picard-Vessiot extension of k with differential Galois group G. The following are equivalent:*
(1) G^o is a solvable group.
(2) K is a liouvillian extension of k.
(3) K is contained in a liouvillian extension of k.

Proof.
(1)⇒(2). (In fact a stronger statement follows from Exercise 1.41.3 but we present here a more elementary proof, not depending on the theory of torsors, of this weaker statement.)
Let K be the Picard-Vessiot extension of a scalar differential equation $L(y) = 0$ of order n over k. Let G be the differential Galois group of the equation and G^o be its identity component. Let $V \subset K$ be the solution space of L. Let k_0 be the fixed field of G^o. Then K is the Picard-Vessiot field for the equation $L(y) = 0$ over k_0 and its Galois group is G^o. The Lie-Kolchin Theorem (Theorem A.46) implies that V has a basis $y_1, \ldots, y_n$ over C such that $G^o \subset \mathrm{GL}(V)$ consists of upper triangular

matrices w.r.t. the basis $y_1, \ldots, y_n$. We will use induction on the order n of L and on the dimension of G^o.

Assume that $y_1 \notin k_0$. For any $\sigma \in G^o$, there is a constant $c(\sigma) \in C^*$ with $\sigma y_1 = c(\sigma) y_1$. Hence $\frac{y_1'}{y_1} \in k_0$. Now $K \supset k_0(y_1)$ is the Picard-Vessiot field for the equation $L(y) = 0$ over $k_0(y_1)$ and its differential Galois group is a proper subgroup of G^o. By induction, $K \supset k_0(y_1)$ is a liouvillian extension and so is $K \supset k$.

Assume that $y_1 \in k_0$. Let $L(y) = 0$ have the form $a_n y^{(n)} + \cdots + a_0 y = 0$. Then $L(yy_1) = b_n y^{(n)} + \cdots + b_1 y^{(1)} + b_0 y$. The term b_0 is zero since $L(y_1) = 0$. Consider the scalar differential equation $M(f) = b_n f^{(n-1)} + \cdots + b_1 f = 0$. Its solution space in K is $C(\frac{y_2}{y_1})' + \cdots + C(\frac{y_n}{y_1})'$. Hence the Picard-Vessiot field $\tilde{K}$ of M lies in K and its differential Galois group is a connected solvable group. By induction, $k \subset \tilde{K}$ is a liouvillian extension. Moreover, $K = \tilde{K}(t_2, \ldots, t_n)$ and $t_i' = (\frac{y_i}{y_1})'$ for $i = 2, \ldots, n$. Thus $K \supset \tilde{K}$ is liouvillian and so is $K \supset k$.

(3)⇒(1). Let $M = k(t_1, \ldots, t_m)$ be a liouvillian extension of k containing K. We shall show that G^o is solvable using induction on m.

The subfield $K(t_1)$ of M is the Picard-Vessiot field of the equation $L(y) = 0$ over $k(t_1)$. Indeed, $K(t_1)$ is generated over $k(t_1)$ by the solutions y of $L(y) = 0$ and their derivatives. The differential Galois group $H = \text{Gal}(K(t_1)/k(t_1))$ is a closed subgroup of G. The field of invariants $K^H = K(t_1)^H \cap K = k(t_1) \cap K$. Since K is also the Picard-Vessiot field of the equation $L(y) = 0$ over $k(t_1) \cap K$, one has the fact that $H = \text{Gal}(K/k(t_1) \cap K)$. By induction, H^o is solvable.

If $k(t_1) \cap K = k$, then $H = G$ and we are done. Suppose that $k(t_1) \cap K \neq k$. We now deal with the three possibilities for t_1. If t_1 is algebraic over k, then $k(t_1) \cap K$ is algebraic over k and lies in the fixed field K^{G^o}. Hence $H^o = G^o$ and we are done. Suppose that t_1 is transcendental over k and that $k(t_1) \cap K \neq k$. If $t_1' = a \in k^*$, then $k(t_1) \supset k$ has differential Galois group $\mathbf{G}_{a,C}$. This group has only trivial algebraic subgroups and so $k(t_1) \subset K$. The equation $t_1' = a \in k^*$ shows that $k(t_1)$ is set-wise invariant under $G = \text{Gal}(K/k)$. Thus there is an exact sequence of algebraic groups

$$1 \to \text{Gal}(K/k(t_1)) \to \text{Gal}(K/k) \to \text{Gal}(k(t_1)/k) \to 1.$$

Because H^o is solvable and $\text{Gal}(k(t_1)/k) = \mathbf{G}_{a,C}$, one easily deduces that G^o is solvable.

If $t_1' = at_1$ with $a \in k^*$, then $\text{Gal}(k(t_1)/k) = \mathbf{G}_{m,C}$. The only nontrivial closed subgroups of $\mathbf{G}_{m,C} = C^*$ are the finite groups of roots of unity. Hence $k(t_1) \cap K = k(t_1^d)$ for some integer $d \geq 1$. As above, this yields the result that G^o is solvable. □

Exercise 1.44 Using Exercise A.44, modify the above proof to show that if G is a torus, then K can be embedded in an extension by exponentials. (This can also be deduced from Exercise 1.41.) □

In general, one can detect from the Galois group if a linear differential equation can be solved in terms of only integrals or only exponentials or only algebraics or any combination of these. We refer to Kolchin's original paper [161] or [162] for a discussion of this. Finally, using the fact that a connected solvable group can be written as a semidirect product of a unipotent group U and a torus T one can show: *If the identity component of the Galois group of a Picard-Vessiot extension K of k is solvable, then there is a chain of subfields $k = K_0 \subset K_1 \subset \cdots \subset K_n = K$ such that $K_i = K_{i-1}(t_i)$ where*

1. *t_1 is algebraic over k,*
2. *for $i = 2, \ldots, n-m$, $m = \dim U$, t_i is transcendental over K_{i-1} and $t_i'/t_i \in K_{i-1}$,*
3. *for $i = n-m+1, \ldots, n$, t_i is transcendental over K_{i-1} and $t_i' \in K_{i-1}$.*

We refer to [183], Proposition 6.7, for a proof of this result.

Theorem 1.43 describes the Galois groups of linear differential equations, all of whose solutions are liouvillian. It will be useful to discuss the case when only some of the solutions are liouvillian.

Proposition 1.45 *Let $L(y) = 0$ be a scalar differential equation with coefficients in k and with Picard-Vessiot field K. Suppose that $L(y) = 0$ has a nonzero solution in some liouvillian extension of k. Then there is a solution $y \in K$, $y \neq 0$ of $L(y) = 0$ such that $\frac{y'}{y}$ is algebraic over k.*

Proof. Let $k(t_1, \ldots, t_n)$ be a liouvillian extension of k and let $y \in k(t_1, \ldots, t_n)$, $y \neq 0$ satisfy $L(y) = 0$. We will show the statement by induction on n.

Let $n = 1$ and t_1 be algebraic over k. Then y and $\frac{y'}{y}$ are algebraic over k. Assume that t_1 is transcendental over k and $t_1' = a \in k^*$. The element y satisfies a differential equation over k and lies therefore in the Picard-Vessiot ring $k[t_1]$ (see Corollary 1.38). The elements $\sigma \in \mathrm{Gal}(k(t_1)/k)$ have the form $\sigma(t_1) = t_1 + c$ (with arbitrary $c \in C$). Furthermore, $\sigma(y)$ and $\sigma(y) - y$ are also solutions of L. One concludes that L has a nonzero solution in k itself. Assume that t_1 is transcendental over k and that $t_1' = at_1$ for some $a \in k^*$. Then y lies in the Picard-Vessiot ring $k[t_1, t_1^{-1}]$. The elements $\sigma \in \mathrm{Gal}(k(t_1)/k)$ act by $\sigma(t_1) = ct_1$ with $c \in C^*$ arbitrary. Also $\sigma(y) - dy$, with $\sigma \in \mathrm{Gal}(k(t_1)/k)$ and $d \in C$ are solutions of L. It follows that $k(t_1)$ contains a solution of L of the form $y = bt_1^d$ with $b \in k^*$ and $d \in \mathbf{Z}$. For such a y, one has $\frac{y'}{y} \in k$.

Assume that $y \in k(t_1, \ldots, t_{n+1})$, $y \neq 0$ is a solution of L. The induction hypothesis implies that the algebraic closure $\overline{k(t)}$, with $t = t_1$, contains solutions of the Riccati equation of L. If t is algebraic over k, then we are done. If t is transcendental over k, then one considers, as in the last part of the proof of Theorem 1.43, the Picard-Vessiot field of L over $k(t)$, which is denoted by $Kk(t)$ or $K(t)$.

Furthermore, $K\overline{k(t)}$ denotes the Picard-Vessiot field of L over $\overline{k(t)}$. Let $V \subset K$ denote the solution space of L (in K and also in $K\overline{k(t)}$). Let a $y \in V,\ y \neq 0$ be given such that $\frac{y'}{y}$ is algebraic over $k(t)$. For any $\sigma \in \mathrm{Gal}(K/k)$ the element $\sigma(y)$ has the same property. Choose $\sigma_1, \ldots, \sigma_s \in \mathrm{Gal}(K/k)$, with s maximal, such that the elements $\sigma_1 y, \ldots, \sigma_s y \in V$ are linearly independent over C. The vector space $W \subset V$ spanned by $\sigma_1 y, \ldots, \sigma_s y$ is clearly invariant under the action of $\mathrm{Gal}(K/k)$. Let $f^{(s)} + a_{s-1} f^{(s-1)} + \cdots + a_0 f$ be the unique differential equation M over K with $M(\sigma_i y) = 0$ for $i = 1, \ldots, s$. For any $\sigma \in \mathrm{Gal}(K/k)$, the transformed equation σM has the same space W as its solution space. Hence $\sigma M = M$ and we conclude that M has coefficients in k. We now replace L by M. Consider the liouvillian field extension $k(t, u_1, \ldots, u_s, \sigma_1 y, \ldots, \sigma_s y) \subset K\overline{k(t)}$ of k, where the $u_i := \frac{\sigma_i y'}{\sigma_i y}$ are algebraic over $k(t)$. This field contains the Picard-Vessiot field of the equation of M over k. By Theorem 1.43, the differential Galois group H of M over k has the property that H^o is solvable. Let $f \in W,\ f \neq 0$ be an eigenvector for H^o. Then $\frac{f'}{f}$ is invariant under H^o and is therefore algebraic over k. Since $W \subset V$, also $L(f) = 0$. □

Exercise 1.46 Show that the equation $y''' + zy = 0$ has no nonzero solutions liouvillian over $C(z)$. Hint: As in Exercise 1.36(3), show that the Galois group of this equation is connected. If $\exp(\int u)$ is a solution of $y''' + zy = 0$ then u satisfies $u'' + 3uu' + u^3 + z = 0$. By expanding at ∞, show that this latter equation has no nonzero solution in $C(z)$. □

Exercises 1.47 *The "normality" of a Picard-Vessiot extension.*
(1) In the classical Galois theory a finite extension $K \supset k$ is called *normal* if every irreducible polynomial over k that has one root in K has all its roots in K. Prove the following analogous property for Picard-Vessiot fields:

Assume that $K \supset k$ is a Picard-Vessiot extension and let $f \in K$ be a solution of an irreducible scalar differential equation P over k of order m. Show that the solution space of P in K has dimension m (over the field of constants C of k).
We note that some results of Chap. 2 are needed for this exercise, namely the definition of "irreducible operator" and Exercise 2.4.3.

Liouvillian extensions are very different from Picard-Vessiot extensions.
(2) Consider the liouvillian extension $k(t, f)$ of k defined by: t is transcendental over k and $\frac{t'}{t} \in k^*$. Furthermore, f is algebraic over $k(t)$ with equation $f^2 = 1 - t^2$. Show that $k(t, f)$ is not a differential subfield of a Picard-Vessiot extension of k.
Hint: Let $k \subset k(t, f) \subset K$ with K/k a Picard-Vessiot extension. For every $c \in C^*$ there exists an element $\sigma_c \in \mathrm{Gal}(K/k)$ such that $\sigma_c t = ct$. Now $\sigma_c(f)^2 = 1 - c^2 t^2$. Show that the algebraic field extension of $k(t)$ generated by all $\sigma_c(f)$ is infinite. Now use the fact that $K/k(t)$ is also a Picard-Vessiot extension. □

2 Differential Operators and Differential Modules

2.1 The Ring $\mathcal{D} = k[\partial]$ of Differential Operators

In this chapter k is a differential field such that its subfield of constants C is different from k and has characteristic 0. The skew (i.e., noncommutative) ring $\mathcal{D} := k[\partial]$ consists of all expressions $L := a_n\partial^n + \cdots + a_1\partial + a_0$ with $n \in \mathbf{Z}$, $n \geq 0$ and all $a_i \in k$. These elements L are called *differential operators*. The degree of L $\deg L$ above is m if $a_m \neq 0$ and $a_i = 0$ for $i > m$. In the case $L = 0$ we define the degree to be $-\infty$. The addition in $\mathcal{D}$ is obvious. The multiplication in $\mathcal{D}$ is completely determined by the prescribed rule $\partial a = a\partial + a'$. Since there exists an element $a \in k$ with $a' \neq 0$, the ring $\mathcal{D}$ is not commutative. One calls $\mathcal{D}$ *the ring of linear differential operators with coefficients in k.*

A differential operator $L = a_n\partial^n + \cdots + a_1\partial + a_0$ acts on k and on differential extensions of k, with the interpretation $\partial(y) := y'$. Thus the equation $L(y) = 0$ has the same meaning as the scalar differential equation $a_n y^{(n)} + \cdots + a_1 y^{(1)} + a_0 y = 0$. In connection with this one sometimes uses the expression *order of* L, instead of the degree of L.

The ring of differential operators shares many properties with the ordinary polynomial ring in one variable over k.

Lemma 2.1 *For $L_1, L_2 \in \mathcal{D}$ with $L_1 \neq 0$, there are unique differential operators $Q, R \in \mathcal{D}$ such that $L_2 = QL_1 + R$ and* $\deg R < \deg L_1$.

The proof is not different from the usual division with remainder for the ordinary polynomial ring over k. The version where left and right are interchanged is equally valid. An interesting way to interchange left and right is provided by the "involution" $i : L \mapsto L^*$ of $\mathcal{D}$ defined by the formula $i(\sum a_i\partial^i) = \sum(-1)^i\partial^i a_i$. The operator L^* is often called the *formal adjoint* of L.

Exercise 2.2 The term "involution" means that i is an additive bijection, $i^2 = id$ and $i(L_1L_2) = i(L_2)i(L_1)$ for all $L_1, L_2 \in \mathcal{D}$. Prove that i, as defined above, has these properties. Hint: Let $k[\partial]^\star$ denote the additive group $k[\partial]$ made into a ring by the opposite multiplication given by the formula $L_1 \star L_2 = L_2L_1$. Show that $k[\partial]^\star$ is also a skew polynomial ring over the field k and with variable $-\partial$. Observe that $(-\partial) \star a = a \star (-\partial) + a'$. □

Corollary 2.3 *For any left ideal $I \subset k[\partial]$ there exists an $L_1 \in k[\partial]$ such that $I = k[\partial]L_1$. Similarly, for any right ideal $J \subset k[\partial]$ there exists an $L_2 \in k[\partial]$ such that $J = L_2 k[\partial]$.*

From these results one can define the *least common left multiple*, $\mathrm{LCLM}(L_1, L_2)$, of $L_1, L_2 \in k[\partial]$ as the unique monic generator of $k[\partial]L_1 \cap k[\partial]L_2$ and the *greatest common left divisor*, $\mathrm{GCLD}(L_1, L_2)$, of $L_1, L_2 \in k[\partial]$ as the unique monic generator of $L_1 k[\partial] + L_2 k[\partial]$. The *least common right multiple* of $L_1, L_2 \in k[\partial]$, $\mathrm{LCRM}(L_1, L_2)$ and the *greatest common right divisor* of $L_1, L_2 \in k[\partial]$, $\mathrm{GCRD}(L_1, L_2)$ can be defined similarly. We note that a modified version of the Euclidean algorithm can be used to find the $\mathrm{GCLD}(L_1, L_2)$ and the $\mathrm{GCRD}(L_1, L_2)$.

Exercises 2.4 *The ring $k[\partial]$*
1. Show that for any nonzero operators $L_1, L_2 \in k[\partial]$, with $\deg(L_1) = n_1$, $\deg(L_2) = n_2$ we have the fact that $\deg(L_1L_2 - L_2L_1) < n_1 + n_2$. Show that $k[\partial]$ has no two-sided ideals other than (0) and $k[\partial]$.

2. Let M be a $\mathcal{D} = k[\partial]$-submodule of the free left module $F := \mathcal{D}^n$. Show that F has a free basis $e_1, \ldots, e_n$ over $\mathcal{D}$ such that M is generated by elements $a_1e_1, \ldots, a_ne_n$ for suitable $a_1, \ldots, a_n \in \mathcal{D}$. Conclude that M is also a free $\mathcal{D}$-module. Hints:
(a) For any element $f = (f_1, \ldots, f_n) \in F$ there is a free basis $e_1, \ldots, e_n$ of F such that $f = ce_n$, with $c \in \mathcal{D}$ such that $\mathcal{D}c = \mathcal{D}f_1 + \cdots + \mathcal{D}f_n$.
(b) Choose $m = (b_1, \ldots, b_n) \in M$ such that the degree of the $c \in \mathcal{D}$ with $\mathcal{D}c = \mathcal{D}b_1 + \cdots + \mathcal{D}b_n$ is minimal. Choose a new basis, called $e_1, \ldots, e_n$ of F, such that $m = ce_n$. Prove that M is the direct sum of $M \cap (\mathcal{D}e_1 \oplus \cdots \oplus \mathcal{D}e_{n-1})$ and $\mathcal{D}ce_n$.
(c) Use induction to finish the proof.

3. Let $L_1, L_2 \in k[\partial]$ with $\deg(L_1) = n_1$, $\deg(L_2) = n_2$. Let K be a differential extension of k having the same constants C as k and let $\mathrm{Soln}_K(L_i)$ denote the C-space of solutions of $L_i(y) = 0$ in K. Assume that $\dim_C(\mathrm{Soln}_K(L_2)) = n_2$. Show that:
(a) Suppose that every solution in K of $L_2(y) = 0$ is a solution of $L_1(y) = 0$. Then there exists a $Q \in k[\partial]$ such that $L_1 = QL_2$.
(b) Suppose that L_1 divides L_2 on the right, then $\mathrm{Soln}_K(L_1) \subset \mathrm{Soln}_K(L_2)$ and $\dim_C(\mathrm{Soln}_K(L_1)) = n_1$. □

Lemma 2.5 Finitely generated left $k[\partial]$-modules.
Every finitely generated left $k[\partial]$-module is isomorphic to a finite direct sum $\oplus M_i$, where each M_i is isomorphic to either $k[\partial]$ or $k[\partial]/k[\partial]L$ for some $L \in k[\partial]$ with $\deg L > 0$.

Proof. Let M be a finitely generated left $k[\partial]$-module. Then there is a surjective homomorphism $\phi : k[\partial]^n \to M$ of $k[\partial]$-modules. The kernel of ϕ is a submodule of the free module $k[\partial]^n$. Exercises 2.4.2 applied to $\ker(\phi)$ yields the required direct sum decomposition of M. □

Observation 2.6 *A differential module M over k is the same object as a left $k[\partial]$-module such that* $\dim_k M < \infty$.

Exercise 2.7 Let $y' = Ay$ be a matrix differential equation over k of dimension n with corresponding differential module M. Show that the following properties are equivalent:
(1) There is a fundamental matrix F for $y' = Ay$ with coefficients in k.
(2) $\dim_C \ker(\partial, M) = n$.
(3) M is a direct sum of copies of $\mathbf{1}_k$, where $\mathbf{1}_k$ denotes the 1-dimensional differential module ke with $\partial e = 0$.

A differential module M over k is called *trivial* if the equivalent properties (2) and (3) hold for M. Assume now that C is algebraically closed. Prove that M is a trivial differential module if and only if the differential Galois group of M is $\{1\}$. □

Intermezzo on multilinear algebra.
Let F be any field. For vector spaces of finite dimension over F there are "constructions of linear algebra" that are used very often in connection with differential modules. Apart from the well-known "constructions" *direct sum* $V_1 \oplus V_2$ *of two vector spaces, subspace* $W \subset V$, *quotient space* V/W, *dual space* V^* *of* V, there are the less elementary constructions:

The *tensor product* $V \otimes_F W$ (or simply $V \otimes W$) of two vector spaces. Although we have already used this construction many times, we recall its categorical definition. A bilinear map $b : V \times W \to Z$ (with Z any vector space over F) is a map $(v, w) \to b(v, w) \in Z$ that is linear in v and w separately. The tensor product $(t, V \otimes W)$ is defined by $t : V \times W \to V \otimes W$ is a bilinear map such that there exists for each bilinear map $b : V \times W \to Z$ a unique linear map $\ell : V \otimes W \to Z$ with $\ell \circ t = b$. The elements $t(v, w)$ are denoted by $v \otimes w$. It is easily seen that bases $\{v_1, \ldots, v_n\}$ of V and $\{w_1, \ldots, w_m\}$ of W give rise to a basis $\{v_i \otimes w_j\}_{i=1,\ldots,n;\, j=1,\ldots,m}$ of $V \otimes W$. The tensor product of several vector spaces $V_1 \otimes \cdots \otimes V_s$ can be defined in a similar way by multilinear maps. A basis of this tensor product can be obtained in a similar way from bases for every V_i.

The *vector space of the homomorphisms* $\mathrm{Hom}(V, W)$ consist of the F-linear maps $\ell : V \to W$. Its structure as an F-vector space is given by $(\ell_1 + \ell_2)(v) := \ell_1(v) + \ell_2(v)$ and $(f\ell)(v) := f\ell(v)$. There is a natural isomorphism $\alpha : V^* \otimes W \to \mathrm{Hom}(V, W)$, given by the formula $\alpha(\ell \otimes w)(v) := \ell(v) \cdot w$.

The *symmetric powers* $\mathrm{sym}^d V$ of a vector space V. Consider the d-fold tensor product $V \otimes \cdots \otimes V$ and its subspace W generated by the vectors $(v_1 \otimes \cdots \otimes v_d) - (v_{\pi(1)} \otimes \cdots \otimes v_{\pi(d)})$, with $v_1, \ldots, v_d \in V$ and $\pi \in S_d$, the group of all permutations on $\{1, \ldots, d\}$. Then $\mathrm{sym}^d V$ is defined as the quotient space $(V \otimes \cdots \otimes V)/W$. The notation for the elements of $\mathrm{sym}^d V$ is often the same as for the elements of $V \otimes \cdots \otimes V$, namely finite sums of expressions $v_1 \otimes \cdots \otimes v_d$. For the symmetric powers, one has (by definition) $v_1 \otimes \cdots \otimes v_d = v_{\pi(1)} \otimes \cdots \otimes v_{\pi(d)}$ for any $\pi \in S_d$. Sometimes one omits the tensor product in the notation for the elements in the symmetric powers. Thus $v_1 v_2 \cdots v_d$ is an element of $\mathrm{sym}^d V$. Let $\{v_1, \ldots, v_n\}$ be a basis of V, then $\{v_1^{a_1} v_2^{a_2} \cdots v_n^{a_n} |\text{ all } a_i \geq 0 \text{ and } \sum a_i = d\}$ is a basis of $\mathrm{sym}^d V$. One extends this definition by $\mathrm{sym}^1 V = V$ and $\mathrm{sym}^0 V = F$.

The *exterior powers* $\Lambda^d V$. One considers again the tensor product $V \otimes \cdots \otimes V$ of d copies of V. Let W be the subspace of this tensor product generated by the expressions $v_1 \otimes \cdots \otimes v_d$, where there are (at least) two indices $i \neq j$ with $v_i = v_j$. Then $\Lambda^d V$ is defined as the quotient space $(V \otimes \cdots \otimes V)/W$. The image of the element $v_1 \otimes \cdots \otimes v_d$ in $\Lambda^d V$ is denoted by $v_1 \wedge \cdots \wedge v_d$. If $\{v_1, \ldots, v_n\}$ is a basis of V, then the collection $\{v_{i_1} \wedge \cdots \wedge v_{i_d} |\ 1 \leq i_1 < i_2 < \cdots < i_d \leq n\}$ is a basis of $\Lambda^d V$. In particular, $\Lambda^d V = 0$ if $d > n$ and $\Lambda^n V \cong F$. This isomorphism is made explicit by choosing a basis of V and mapping $w_1 \wedge \cdots \wedge w_n$ to the determinant in F of the matrix with columns the expressions of the w_i as linear combinations of the given basis. One extends the definition by $\Lambda^1 V = V$ and $\Lambda^0 V = F$. We note that for $1 \leq d \leq n$ one has $w_1 \wedge \cdots \wedge w_d \neq 0$ if and only if $w_1, \ldots, w_d$ are linearly independent over F.

Both the symmetric powers and the exterior powers can also be defined in a categorical way using symmetric multilinear maps and alternating multilinear maps.

Definition 2.8 *Cyclic vector.*
Let M be a differential module over k. An element $e \in M$ is called a *cyclic vector* if M is generated over k by the elements $e, \partial e, \partial^2 e, \ldots$. □

The following proposition extends Lemma 2.5.

Proposition 2.9 *Every finitely generated left $k[\partial]$-module has the form $k[\partial]^n$ or $k[\partial]^n \oplus k[\partial]/k[\partial]L$ with $n \geq 0$ and $L \in k[\partial]$.*

Proof. The only thing that we have to show is that a differential module M of dimension n over k is isomorphic to $k[\partial]/k[\partial]L$ for some L. This translates into the existence of an element $e \in M$ such that M is generated by $e, \partial e, \ldots, \partial^{n-1} e$. In other words, e is a cyclic vector for M.

Any $k[\partial]$-linear map $\phi : k[\partial] \to M$ is determined by $e := \phi(1) \in M$, where $1 \in k[\partial]$ is the obvious element. The map ϕ is surjective if and only if e is a cyclic element. If the map is surjective, then its kernel is a left ideal in $k[\partial]$ and has the form $k[\partial]L$. Thus $k[\partial]/k[\partial]L \cong M$. On the other hand, an isomorphism $k[\partial]/k[\partial]L \cong M$ induces a surjective $k[\partial]$-linear map $k[\partial] \to M$. The proof of the existence of a cyclic vector for M is reproduced from N. Katz's paper [155].

Choose an element $h \in k$ with $h' \neq 0$ and define $\delta = \frac{h}{h'}\partial$. Then $k[\partial] = \mathcal{D}$ is also equal to $k[\delta]$. Furthermore, $\delta h = h\delta + h$ and $\delta h^k = h^k \delta + k h^k$ for all $k \in \mathbf{Z}$. Take an $e \in M$. Then $\mathcal{D}e$ is the subspace of M generated over k by $e, \delta e, \delta^2 e, \ldots$.

Let $\mathcal{D}e$ have dimension m. If $m = n$ then we are finished. If $m < n$ then we will produce an element $\tilde{e} = e + \lambda h^k f$, where $\lambda \in \mathbf{Q}$ and $k \in \mathbf{Z}$ and $f \in M \setminus \mathcal{D}e$, such that $\dim \mathcal{D}\tilde{e} > m$. This will prove the existence of a cyclic vector. We will work in the exterior product $\Lambda^{m+1}M$ and consider the element

$$E := \tilde{e} \wedge \delta(\tilde{e}) \wedge \cdots \wedge \delta^m(\tilde{e}) \in \Lambda^{m+1}M.$$

The multilinearity of the $\wedge$ and the rule $\delta h^k = h^k\delta + kh^k$ lead to a decomposition of E of the form

$$E = \sum_{0\leq a\leq m} (\lambda h^k)^a (\sum_{0\leq b} k^b \omega_{a,b}), \text{ with } \omega_{a,b} \in \Lambda^{m+1}M \text{ independent of } \lambda, k.$$

Suppose that E is zero for every choice of λ and k. Fix k. For every $\lambda \in \mathbf{Q}$ one finds a linear dependence of the $m+1$ terms $\sum_{0\leq b} k^b\omega_{a,b}$. One concludes that for every a the term $\sum_{0\leq b} k^b\omega_{a,b}$ is zero for all choices of $k \in \mathbf{Z}$. The same argument shows that each $\omega_{a,b} = 0$. However, one easily calculates that $\omega_{1,m} = e\wedge\delta(e)\wedge\cdots\wedge\delta^{m-1}(e)\wedge f$. This term is not zero by our choice of f. □

There are other proofs of the existence of a cyclic vector, relevant for algorithms. These proofs produce a set $S \subset M$ of small cardinality such that S contains a cyclic vector. We will give two of those statements. The first one is due to Kovacic [168] (with some similarities to Cope [72, 73]).

Lemma 2.10 *Let M be a differential module with k-basis $\{e_1, \ldots, e_n\}$ and let $\eta_1, \ldots, \eta_n \in k$ be linearly independent over C, the constants of k. Then there exist integers $0 \leq c_{i,j} \leq n$, $1 \leq i, j \leq n$, such that $m = \sum_{i=1}^n a_i e_i$ is a cyclic vector for M, where $a_i = \sum_{j=1}^n c_{i,j}\eta_j$. In particular, if $z \in k$, $z' \neq 0$, then $a_i = \sum_{j=1}^n c_{i,j} z^{j-1}$ is, for suitable $c_{i,j}$ as above, a cyclic vector.*

The second one is due to Katz [155].

Lemma 2.11 *Assume that k contains an element z such that $z' = 1$. Let M be a differential module with k-basis $\{e_0, \ldots, e_{n-1}\}$. There exists a set $S \subset C$ with at most $n(n-1)$ elements such that if $a \notin S$ the element*

$$\sum_{j=0}^{n-1} \frac{(z-a)^j}{j!} \sum_{p=0}^{j} (-1)^p \binom{j}{p} \partial^p(e_{j-p}) \text{ is a cyclic vector.}$$

We refer to the literature for the proofs of these and to [80, 144, 237, 6, 30, 31]. For a generalization of the cyclic vector construction to systems of nonlinear differential equations, see [70].

2.2 Constructions with Differential Modules

The constructions with vector spaces (direct sums, tensor products, symmetric powers, etc.) extend to several other categories. The first interesting case concerns a finite group G and a field F. The category has as objects the representations G in finite dimensional vector spaces over F. A representation (ρ, V) is a homomorphism $\rho : G \to \mathrm{GL}(V)$, where V is a finite dimensional vector space over F. The tensor product $(\rho_1, V_1) \otimes (\rho_2, V_2)$ is the representation (ρ_3, V_3) with $V_3 = V_1 \otimes_F V_2$ and ρ_3 given by the formula $\rho_3(v_1 \otimes v_2) = (\rho_1 v_1) \otimes (\rho_2 v_2)$. In a similar way one defines direct sums, quotient representations, symmetric powers and exterior powers of a representation.

A second interesting case concerns a linear algebraic group G over F. A representation (ρ, V) consists of a finite dimensional vector space V over F and a homomorphism of algebraic groups over F, $\rho : G \to \mathrm{GL}(V)$. The formulas for tensor products and other constructions are the same as for finite groups. This example (and its extension to affine group schemes) is explained in the appendices.

A third example concerns a Lie algebra L over F. A representation (ρ, V) consists of a finite dimensional vector space V over F and an F-linear map $\rho : L \to \mathrm{End}(V)$ satisfying the property $\rho([A, B]) = [\rho(A), \rho(B)]$. The tensor product (ρ_1, V_1) $\otimes(\rho_2, V_2) = (\rho_3, V_3)$ with again $V_3 = V_1 \otimes_F V_2$ and with ρ_3 given by the formula $\rho_3(v_1 \otimes v_2) = (\rho_1 v_1) \otimes v_2 + v_1 \otimes (\rho_2 v_2)$.

As we will see, the above examples are related with constructions with differential modules. The last example is rather close to the constructions with differential modules.

The category of all differential modules over k will be denoted by Diff_k. *Now we start the list of constructions of linear algebra for differential modules.*

The *direct sum* $(M_1, \partial_1) \oplus (M_2, \partial_2)$ is (M_3, ∂_3), where $M_3 = M_1 \oplus M_2$ and $\partial_3(m_1 \oplus m_2) = \partial_1(m_1) \oplus \partial_2(m_2)$.

A *(differential) submodule* N of (M, ∂) is a k-vector space $N \subset M$ such that $\partial(N) \subset N$. Then $N = (N, \partial|_N)$ is a differential module.

Let N be a submodule of (M, ∂). Then M/N, provided with the induced map ∂, given by $\partial(m + N) = \partial(m) + N$, is the *quotient differential module*.

The *tensor product* $(M_1, \partial_1) \otimes (M_2, \partial_2)$ is (M_3, ∂_3) with $M_3 = M_1 \otimes_k M_2$ and ∂_3 is given by the formula $\partial_3(m_1 \otimes m_2) = (\partial_1 m_1) \otimes m_2 + m_1 \otimes (\partial_2 m_2)$. We note that this is not at all the tensor product of two $k[\partial]$-modules. In fact, the tensor product of two left $k[\partial]$-modules does not exist since $k[\partial]$ is not commutative.

A *morphism* $\phi : (M_1, \partial_1) \to (M_2, \partial_2)$ is a k-linear map such that $\phi \circ \partial_1 = \partial_2 \circ \phi$. If we regard differential modules as special left $k[\partial]$-modules, then the above translates into ϕ is a $k[\partial]$-linear map. We will sometimes write $\mathrm{Hom}_{k[\partial]}(M_1, M_2)$ (omitting ∂_1 and ∂_2 in the notation) for the C-vector space of all morphisms. This object is *not* a differential module over k, but it is $\mathrm{Mor}(M_1, M_2)$ the C-linear vector space of the morphisms in the category Diff_k.

The *internal Hom*, $\mathrm{Hom}_k((M_1, \partial_1), (M_2, \partial_2))$ of two differential modules is the k-vector space $\mathrm{Hom}_k(M_1, M_2)$ of the k-linear maps form M_1 to M_2 provided with a ∂ given by the formula $(\partial\ell)(m_1) = \ell(\partial_1 m_1) - \partial_2(\ell(m_1))$. This formula leads to the observation that

$$\mathrm{Hom}_{k[\partial]}(M_1, M_2) \text{ is equal to } \{\ell \in \mathrm{Hom}_k(M_1, M_2) | \ \partial\ell = 0\}.$$

In particular, the C-vector space $\mathrm{Mor}(M_1, M_2) = \mathrm{Hom}_{k[\partial]}(M_1, M_2)$ has dimension at most $\dim_k M_1 \cdot \dim_k M_2$.

The *trivial differential module of dimension 1* over k is again denoted by $\mathbf{1}_k$ or $\mathbf{1}$. A special case of internal Hom is the *dual M^* of a differential module M* defined by $M^* = \mathrm{Hom}_k(M, \mathbf{1}_k)$.

Symmetric powers and exterior powers are derived from tensor products and the formation of quotients. Their structure can be made explicit. The exterior power $\Lambda^d M$, for instance, is the k-vector space $\Lambda_k^d M$ provided with the operation ∂ given by the formula $\partial(m_1 \wedge \cdots \wedge m_d) = \sum_{i=1}^{d} m_1 \wedge \cdots \wedge (\partial m_i) \wedge \cdots \wedge m_d$.

The next collection of exercises presents some of the many properties of the above constructions and their translations into the language of differential operators and matrix differential equations.

Exercises 2.12 *Properties of the constructions*
1. Show that the tensor product of differential modules as defined above is indeed a differential module.

2. Show that, for a differential module M over k, the natural map $M \to M^{**}$ is an isomorphism of differential modules.

3. Show that the differential modules $\mathrm{Hom}_k(M_1, M_2)$ and $M_1^* \otimes M_2$ are "naturally" isomorphic.

4. Show that the k-linear map $M^* \otimes M \to 1_k$, defined by $\ell \otimes m = \ell(m)$, is a morphism of differential modules. Conclude that $M^* \otimes M$ has a nontrivial submodule if $\dim_k M > 1$.

5. Suppose that M is a trivial differential module. Show that all the constructions of linear algebra applied to M again produce trivial differential modules. Hint: Show that M^* is trivial; show that the tensor product of two trivial modules is trivial; show that any submodule of a trivial module is trivial too.

6. Suppose that $M \cong k[\partial]/k[\partial]L$. Show that $M^* \cong k[\partial]/k[\partial]L^*$. Here $L \mapsto L^*$ is the involution defined in Exercise 2.2. Hint: Let L have degree n. Show that the element $e \in \mathrm{Hom}_k(k[\partial]/k[\partial]L, \mathbf{1}_k)$ given by $e(\sum_{i=0}^{n-1} b_i\partial^i) = b_{n-1}$ is a cyclic vector and that $L^*e = 0$.

7. *The differential module M_L associated to the differential operator L.*
Consider an operator $L = \partial^n + a_{n-1}\partial^{n-1} + \cdots + a_0 \in k[\partial]$. As in Sect. 1.2, one associates to L a matrix differential equation $Y' = A_L Y$, where A_L is the companion matrix

$$A_L = \begin{pmatrix} 0 & 1 & 0 & 0 & \dots & 0 \\ 0 & 0 & 1 & 0 & \dots & 0 \\ \vdots & \vdots & \vdots & \vdots & \dots & \vdots \\ 0 & 0 & 0 & 0 & \dots & 1 \\ -a_0 & -a_1 & \dots & \dots & \dots & -a_{n-1} \end{pmatrix}.$$

This matrix differential equation induces a differential module M_L and we call this the *differential module associated with the operator L.*
(a) Prove that the differential modules M_L and $(k[\partial]/k[\partial]L)^*$ are isomorphic.
(b) *Operators of the same type.*
Let $L_1, L_2 \in k[\partial]$ be monic of degree n. Prove that M_{L_1} and M_{L_2} are isomorphic if and only if there are elements $R, S \in k[\partial]$ of degree $< n$ such that $L_1 R = SL_2$ and $GCRD(R, L_2) = 1$.
Hint: Describe an isomorphism $\phi : k[\partial]/k[\partial]L_1 \to k[\partial]/k[\partial]L_2$ by an operator of degree $< n$ representing the element $\phi(1) \in k[\partial]/k[\partial]L_2$.
In the classical literature, operators L_1, L_2 such that $M_{L_1} \cong M_{L_2}$ are called *of the same type*. This concept appears in the 19th century literature (for references to this literature as well as more recent references, see [271]).
(c) Prove that every matrix differential equation is equivalent to an equation of the form $Y' = A_L Y$.

8. *The matrix differential of the dual M^*.*
Let M be a differential equation and let $y' = Ay$ be an associated matrix differential equation by the choice of a basis $\{e_1, \dots, e_n\}$. Find the matrix differential equation for M^* associated to the dual basis $\{e_1^*, \dots, e_n^*\}$ of M^*.

9. *Extensions of differential fields.*
Let $K \supset k$ be an extension of differential fields. For any differential module (M, ∂) over k one considers the K-vector space $K \otimes_k M$. One defines ∂ on $K \otimes_k M$ by $\partial(a \otimes m) = a' \otimes m + a \otimes (\partial m)$. Show that this definition makes sense and that $(K \otimes_k M, \partial)$ is a differential module over K. Prove that the formation $M \mapsto K \otimes_k M$ commutes with all constructions of linear algebra.

10. *The characterization of the "internal hom".*
For the reader familiar with representable functors this exercise, which shows that the "internal hom" is derived from the tensor product, might be interesting. Consider two differential modules M_1, M_2. Associate to this the contravariant functor $\mathcal{F}$ from Diff_k to the category of sets given by the formula $\mathcal{F}(T) = \mathrm{Hom}_{k[\partial]}(T \otimes M_1, M_2)$. Show that $\mathcal{F}$ is a representable functor and that it is represented by $\mathrm{Hom}_k(M_1, M_2)$. Compare also the definition of tannakian category given in the appendices. □

Now we continue Exercise 2.12.7 and the set of morphisms between two differential modules in terms of differential operators. An operator $L \in k[\partial]$ is said to be *reducible over* k if L has a nontrivial right-hand factor. Otherwise L is called *irreducible*. Suppose that L is reducible, say $L = L_1L_2$. Then there is an obvious exact sequence of differential modules

$$0 \to \mathcal{D}/\mathcal{D}L_1 \stackrel{.L_2}{\to} \mathcal{D}/\mathcal{D}L_1L_2 \to \mathcal{D}/\mathcal{D}L_2 \to 0,$$

where the first nontrivial arrow is multiplication on the right by L_2 and the second nontrivial arrow is the quotient map. In particular, the monic right-hand factors of L correspond bijectively to the quotient modules of $\mathcal{D}/\mathcal{D}L$ (and at the same time to the submodules of $\mathcal{D}/\mathcal{D}L$).

Proposition 2.13 *For* $L_1, L_2 \in k[\partial]$, *one defines* $\mathcal{E}(L_1, L_2)$ *to consist of the* $R \in k[\partial]$ *with* $\deg R < \deg L_2$, *such that there exists an* $S \in k[\partial]$ *with* $L_1R = SL_2$.
(1) *There is a natural* C*-linear bijection between* $\mathcal{E}(L_1, L_2)$ *and* $\mathrm{Hom}_{k[\partial]}(k[\partial]/k[\partial]L_1, k[\partial]/k[\partial]L_2)$.
(2) $\mathcal{E}(L, L)$ *or* $\mathcal{E}(L)$ *is called the* (right) eigenring *of* L. *This eigenring* $\mathcal{E}(L)$ *is a finite dimensional* C*-subalgebra of* $\mathrm{End}_k(k[\partial]/k[\partial]L)$, *which contains* $C.id$. *If* L *is irreducible and* C *is algebraically closed, then* $\mathcal{E}(L) = C.id$.

Proof. (1) A $k[\partial]$-linear map $\phi : k[\partial]/k[\partial]L_1 \to k[\partial]/k[\partial]L_2$ lifts uniquely to a $k[\partial]$-linear map $\psi : k[\partial] \to k[\partial]$ such that $R := \psi(1)$ has degree $< \deg L_2$. Furthermore, $\psi(k[\partial]L_1) \subset k[\partial]L_2$. Hence, $\psi(L_1) = L_1R \in k[\partial]L_2$ and $L_1R = SL_2$ for some $S \in k[\partial]$. On the other hand, an R and S with the stated properties determine a unique ψ that induces a $k[\partial]$-linear map $\phi : k[\partial]/k[\partial]L_1 \to k[\partial]/k[\partial]L_2$.
(2) The first statement is obvious. The kernel of any element of $\mathcal{E}(L)$ is a submodule of $k[\partial]/k[\partial]L$. If L is irreducible, then any nonzero element of $\mathcal{E}(L)$ is injective and therefore also bijective. Hence $\mathcal{E}(L)$ is a division ring. Since C is algebraically closed, one has $\mathcal{E}(L) = C$. □

Exercise 2.14 *The Eigenring.*
The eigenring provides a method to obtain factors of a reducible operator, see [136, 271] and Sect. 4.2. However, even if C is algebraically closed, a reducible operator L may satisfy $\mathcal{E}(L) = C.id$. In this case no factorization is found. The aim of this exercise is to provide an example.

1. The field C of the constants of k is supposed to be algebraically closed. Let $M = k[\partial]/k[\partial]L$ be a differential module over k of dimension 2. Prove that $\mathcal{E}(L) \neq C.id$ if and only if M has submodules N_1, N_2 of dimension 1 such that N_2 and M/N_1 are isomorphic. Hint: $\mathcal{E}(L) \neq C.id$ implies that there is a morphism $\phi : M \to M$ such that $N_1 := \ker(\phi)$ and $N_2 := \mathrm{im}(\phi)$ have dimension 1.

2. Take $k = C(z)$, $z' = 1$ and $L = (\partial + 1 + z^{-1})(\partial - 1)$. Show that $M := k[\partial]/k[\partial]L$ has only one submodule N of dimension 1 and that N and M/N are not isomorphic. Hint: The submodules of dimension 1 correspond to right-hand factors $\partial - u$ of L,

with $u \in k$. Perform Kovacic's algorithm to obtain the possibilities for u. This works as follows (see also Chap. 4). Derive the equation $u^2 + z^{-1}u + u' - (1 + z^{-1}) = 0$. Expand a potential solution u at $z = 0$ and $z = \infty$ as a Laurent series and show that u has no poles at $z = 0$ and $z = \infty$. At any point $c \in C^*$, the Laurent series of u has the form $\frac{\epsilon}{z-c} + \cdots$ with $\epsilon = 0, 1$. Calculate that $u = 1$ is the only possibility. □

We end this section with a discussion of the "solution space" of a differential module. To do this we shall need a *universal differential extension field* of a field k. This is defined formally (and made explicit in certain cases) in Sect. 3.2 but for our purposes it is enough to require this to be a field $\mathcal{F} \supset k$ with the same field of constants of k such that any matrix differential equation $Y' = AY$ over k has a solution in $\mathrm{GL}_n(\mathcal{F})$. Such a field can be constructed as a direct limit of all Picard-Vessiot extensions of k and we shall fix one and denote it by $\mathcal{F}$. We note that Kolchin [162] uses the term universal extension to denote a field containing solutions of ALL differential equations but our restricted notion is sufficient for our purposes.

Definition 2.15 Let M be a differential module over k with algebraically closed constants C and $\mathcal{F}$ a universal differential extension of k. The *covariant solution space* of M is the C-vector space $\ker(\partial, \mathcal{F} \otimes_k M)$. The *contravariant solution space* is the C-vector space $\mathrm{Hom}_{k[\partial]}(M, \mathcal{F})$. □

The terms "covariant" and "contravariant" reflect the following properties. Let $\phi : M_1 \to M_2$ be a morphism of differential modules. Then there are induced homomorphisms of C-vector spaces $\phi_* : \ker(\partial, \mathcal{F} \otimes_k M_1) \to \ker(\partial, \mathcal{F} \otimes_k M_2)$ and $\phi^* : \mathrm{Hom}_{k[\partial]}(M_2, \mathcal{F}) \to \mathrm{Hom}_{k[\partial]}(M_1, \mathcal{F})$. Let $0 \to M_1 \to M_2 \to M_3 \to 0$ be an exact sequence of differential modules, then so is

$$0 \to \ker(\partial, \mathcal{F} \otimes_k M_1) \to \ker(\partial, \mathcal{F} \otimes_k M_2) \to \ker(\partial, \mathcal{F} \otimes_k M_3) \to 0.$$

This follows easily from the exactness of the sequence

$$0 \to \mathcal{F} \otimes_k M_1 \to \mathcal{F} \otimes_k M_2 \to \mathcal{F} \otimes_k M_3 \to 0$$

and the observation that $\dim_C \ker(\partial, \mathcal{F} \otimes_k M) = \dim_k M$ for any differential module M over k. The contravariant solution space also induces a contravariant exact functor from differential modules to finite dimensional C-vector spaces.

Lemma 2.16 *Let M be a differential modules with basis $e_1, \ldots, e_n$ and let $\partial e_i = -\sum_j a_{j,i} e_j$ and $A = (a_{i,j})$. Then*
1. $\ker(\partial, \mathcal{F} \otimes_k M) \simeq \{y \in \mathcal{F}^n \mid y' = Ay\}$.
2. *There are natural C-vector space isomorphisms*

$$\mathrm{Hom}_{k[\partial]}(M, \mathcal{F}) \simeq \mathrm{Hom}_{\mathcal{F}[\partial]}(\mathcal{F} \otimes_k M, \mathcal{F}) \simeq \mathrm{Hom}_C(\ker(\partial, \mathcal{F} \otimes_k M), C).$$

3. *Let $e \in M$ and let $L \in k[\partial]$ be its minimal monic annihilator. Let $W = \{y \in \mathcal{F} \mid L(y) = 0\}$. The map $\mathrm{Hom}_{k[\partial]}(M, \mathcal{F}) \to W \subset \mathcal{F}$, given by $\phi \mapsto \phi(e)$, is surjective.*

Proof. 1. The basis $e_1, \ldots, e_n$ yields an identification of $\mathcal{F} \otimes M$ with $\mathcal{F}^n$ and of ∂ with the operator $\frac{d}{dz} - A$ on $\mathcal{F}^n$.

2. Any $k[\partial]$-linear map $\phi : M \to \mathcal{F}$ extends to an $\mathcal{F}[\partial]$-linear map $\mathcal{F} \otimes_k M \to \mathcal{F}$. This gives the first isomorphism. Any ϕ in $\mathrm{Hom}_{\mathcal{F}[\partial]}(\mathcal{F} \otimes_k M, \mathcal{F})$ defines by restriction a C-linear map $\tilde{\phi} : \ker(\partial, \mathcal{F} \otimes_k M) \to C$. The map $\phi \mapsto \tilde{\phi}$ is a bijection since the natural map $\mathcal{F} \otimes_C \ker(\partial, \mathcal{F} \otimes_k M) \to \mathcal{F} \otimes_k M$ is an isomorphism.

3. The natural morphism $\mathrm{Hom}_{k[\partial]}(M, \mathcal{F}) \to \mathrm{Hom}_{k[\partial]}(k[\partial]e, \mathcal{F})$ is surjective, since these spaces are contravariant solution spaces and $k[\partial]e$ is a submodule of M. The map $\mathrm{Hom}_{k[\partial]}(k[\partial]e, \mathcal{F}) \to W$, given by $\phi \mapsto \phi(e)$, is bijective since the map $k[\partial]/k[\partial]L \to k[\partial]e$ (with $1 \mapsto e$) is bijective. □

2.3 Constructions with Differential Operators

Differential operators do not form a category where one can perform constructions of linear algebra. However, in the literature tensor products, symmetric powers, etc., of differential operators are often used. In this section we will explain this somewhat confusing terminology and relate it with the constructions of linear algebra on differential modules.

A pair (M, e) of a differential module M and a cyclic vector $e \in M$ determines a monic differential operator L, namely the operator of smallest degree with $Le = 0$. Two such pairs (M_i, e_i), $i = 1, 2$ define the same monic operator if and only if there exists an isomorphism $\psi : M_1 \to M_2$ of differential modules such that $\psi e_1 = e_2$. Moreover, this ψ is unique. For a monic differential operator L one chooses a corresponding pair (M, e). On M and e one performs the construction of linear algebra. This yields a pair $(\mathrm{constr}(M), \mathrm{constr}(e))$. Now $\mathrm{constr}(L)$ is defined as the monic differential operator of minimal degree with $\mathrm{constr}(L)\mathrm{constr}(e) = 0$. This procedure extends to constructions involving several monic differential operators. There is one complicating factor, namely $\mathrm{constr}(e)$ is, in general, not a cyclic vector for $\mathrm{constr}(M)$.

There is another interpretation of a monic differential operator L. Let, as before, $\mathcal{F} \supset k$ denote a fixed universal differential field. One can associate to L its solution space $\mathrm{Sol}(L) := \{y \in \mathcal{F} | L(y) = 0\}$. This space determines L. Indeed, assume that $L = \partial^n + a_{n-1}\partial^{n-1} + \cdots + a_1\partial + a_0$. Then $\mathrm{Sol}(L)$ has dimension n over C. Let $y_1, \ldots, y_n$ be a basis of $\mathrm{Sol}(L)$. Then $a_{n-1}, \ldots, a_0$ satisfy the linear equations

$$y_i^{(n)} + a_{n-1}y_i^{(n-1)} + \cdots + a_1 y_i^{(1)} + a_0 y_i = 0 \text{ for } i = 1, \ldots, n.$$

The wronskian matrix of $y_1, \ldots, y_n$ has nonzero determinant and thus the equations determine $a_{n-1}, \ldots, a_0$. Let $\mathrm{Gal}(\mathcal{F}/k)$ denote the group of the differential automorphisms of $\mathcal{F}/k$, i.e., the automorphisms of the field $\mathcal{F}$ that are k-linear and commute with the differentiation on $\mathcal{F}$. For a Picard-Vessiot extension $K \supset k$ the group $\mathrm{Gal}(K/k)$ of differential automorphisms of K/k has the property that $K^{\mathrm{Gal}(K/k)} = k$. The universal differential extension $\mathcal{F}$ is the direct limit of all Picard-Vessiot field extensions of k. It follows from this that $\mathcal{F}^{\mathrm{Gal}(\mathcal{F}/k)} = k$. This leads to the following result.

Lemma 2.17 *Let $V \subset \mathcal{F}$ be a vector space over C of dimension n. There exists a (unique) monic differential operator $L \in k[\partial]$ with* $\mathrm{Sol}(L) = V$ *if and only if V is (set-wise) invariant under* $\mathrm{Gal}(\mathcal{F}/k)$.

Proof. As above, one observes that any V determines a unique monic differential operator $L \in \mathcal{F}[\partial]$ such that $V = \{y \in \mathcal{F} \mid L(y) = 0\}$. Then V is invariant under $\mathrm{Gal}(\mathcal{F}/k)$ if and only if L is invariant under $\mathrm{Gal}(\mathcal{F}/k)$. The latter is equivalent to $L \in k[\partial]$. □

We note that the lemma remains valid if $\mathcal{F}$ is replaced by a Picard-Vessiot field extension $K \supset k$ and $\mathrm{Gal}(\mathcal{F}/k)$ by $\mathrm{Gal}(K/k)$.

This leads to another way, omnipresent in the literature, of defining a construction of linear algebra to a monic differential operator L. One applies this construction to $\mathrm{Sol}(L)$ and finds a new subspace V of $\mathcal{F}$. This subspace is finite dimensional over C and invariant under G. By the above lemma this determines a new monic differential operator. This procedure extends to constructions with several monic differential operators.

The link between these two ways of making new operators is given by the contravariant solution space. Consider a monic differential operator L and a corresponding pair (M, e). By Definition 2.15 and Lemma 2.16, $\mathrm{Sol}(L)$ is the image of the contravariant solution space $\mathrm{Hom}_{k[\partial]}(M, \mathcal{F})$ of M under the map $\phi \mapsto \phi(e)$. We will make the above explicit for various constructions of linear algebra. Needless to say, this section is only concerned with the language of differential equations and does not contain new results.

Tensor Products. Let $(M_i, e_i),\ i = 1, 2$ denote two differential modules with cyclic vectors. The tensor product construction is $(M_1 \otimes M_2, e_1 \otimes e_2)$. In general, $e_1 \otimes e_2$ need not be a cyclic vector of $M_1 \otimes M_2$ (see Exercise 2.21). Our goal is to describe the contravariant solution space of $M_1 \otimes M_2$, the minimal monic annihilator of $e_1 \otimes e_2$ and its solution space in $\mathcal{F}$.

Lemma 2.18 *The canonical isomorphism*

$$\mathrm{Hom}_{k[\partial]}(M_1, \mathcal{F}) \otimes_C \mathrm{Hom}_{k[\partial]}(M_2, \mathcal{F}) \simeq \mathrm{Hom}_{k[\partial]}(M_1 \otimes M_2, \mathcal{F})$$

is described by $\phi_1 \otimes \phi_2 \mapsto \overline{\phi_1 \otimes \phi_2}$ *where* $\overline{\phi_1 \otimes \phi_2}(m_1 \otimes m_2) := \phi(m_1)\phi_2(m_2)$.

Proof. The canonical isomorphism $c : (\mathcal{F} \otimes_k M_1) \otimes_{\mathcal{F}} (\mathcal{F} \otimes_k M_2) \to \mathcal{F} \otimes_k (M_1 \otimes_k M_2)$ of differential modules over $\mathcal{F}$ is given by $(f_1 \otimes m_1) \otimes (f_2 \otimes m_2)$

$\mapsto f_1 f_2 \otimes m_1 \otimes m_2$. This c induces an isomorphism of the covariant solution spaces

$$\ker(\partial, \mathcal{F} \otimes_k M_1) \otimes_C \ker(\partial, \mathcal{F} \otimes_k M_2) \to \ker(\partial, \mathcal{F} \otimes_k (M_1 \otimes_k M_2)).$$

We write again c for this map. By taking duals as C-vector spaces and after replacing c by c^{-1} one obtains the required map $(c^{-1})^*$ (cf. Lemma 2.16). The formula for this map is easily verified. □

Corollary 2.19 *Let the monic differential operators L_i correspond to the pairs (M_i, e_i) for $i = 1, 2$. Let L be the monic operator of minimal degree such that $L(e_1 \otimes e_2) = 0$. Then the solution space of L in $\mathcal{F}$, i.e., $\{y \in \mathcal{F} | L(y) = 0\}$, is equal to the image of the contravariant solution space $\mathrm{Hom}_{k[\partial]}(M_1 \otimes M_2, \mathcal{F})$ under the map $\phi \mapsto \phi(e_1 \otimes e_2)$. In particular, L is the monic differential operator of minimal degree such that $L(y_1 y_2) = 0$ for all pairs $y_1, y_2 \in \mathcal{F}$ such that $L_1(y_1) = L_2(y_2) = 0$.*

Proof. Apply Lemma 2.16.3 to $e_1 \otimes e_2$. The image of the contravariant solution space of $M_1 \otimes M_2$ in $\mathcal{F}$ under the map $\phi \mapsto \phi(e_1 \otimes e_2)$ is generated as a vector space over C by the products $\phi_1(e_1)\phi_2(e_2)$, according to Lemma 2.18. □

It is hardly possible to compute the monic operator L of minimal degree satisfying $L(e_1 \otimes e_2) = 0$ by the previous corollary. Indeed, $\mathcal{F}$ is, in general, not explicit enough. The obvious way to find L consists of computing the elements $\partial^n(e_1 \otimes e_2)$ in $M_1 \otimes M_2$ and to find a linear relation over k between these elements. In the literature one finds the following definition (or an equivalent one) of the tensor product of two monic differential operators.

Definition 2.20 Let L_1 and L_2 be two differential operators. The minimal monic annihilating operator of $1 \otimes 1$ in $k[\partial]/k[\partial]L_1 \otimes k[\partial]/k[\partial]L_2$ is the *tensor product* $L_1 \otimes L_2$ *of* L_1 *and* L_2.

Exercise 2.21 Prove that $\partial^3 \otimes \partial^2 = \partial^4$. □

Similar definitions and results hold for tensor products with more than two factors.

Symmetric Powers. The d-th *symmetric power* $\mathrm{sym}^d M$ of a differential module is a quotient of the ordinary d-fold tensor product $M \otimes \cdots \otimes M$. The image of $m_1 \otimes m_2 \otimes \cdots \otimes m_d$ in this quotient will be written as $m_1 m_2 \cdots m_d$. In particular, m^d denotes the image of $m \otimes \cdots \otimes m$. This construction applied to (M, e) produces $(\mathrm{sym}^d M, e^d)$.

Lemma 2.22 *Let M be a differential module over k. The canonical isomorphism of contravariant solution spaces*

$$\mathrm{sym}^d(\mathrm{Hom}_{k[\partial]}(M, \mathcal{F})) \to \mathrm{Hom}_{k[\partial]}(\mathrm{sym}^d M, \mathcal{F})$$

is given by the formula $\phi_1\phi_2 \cdots \phi_d \mapsto \overline{\phi_1\phi_2 \cdots \phi_d}$,
where $\overline{\phi_1\phi_2 \cdots \phi_d}(m_1 m_2 \cdots m_d) := \phi(m_1)\phi_2(m_2) \cdots \phi_d(m_d)$.

The proof is similar to that of Lemma 2.18. The same holds for the next corollary.

Corollary 2.23 *Let L correspond to the pair (M, e). The image of the map $\phi \mapsto \phi(e^d)$ from $\mathrm{Hom}_{k[\partial]}(\mathrm{sym}^d M, \mathcal{F})$ to $\mathcal{F}$ is the C-vector space generated by $\{f_1 f_2 \cdots f_d \mid L(f_i) = 0\}$.*

Definition 2.24 Let L be a monic differential operator and let $e = 1$ be the generator of $k[\partial]/k[\partial]L$. The minimal monic annihilating operator of e^d in $\mathrm{sym}^d(k[\partial]/k[\partial]L)$ is the d-th symmetric power $\mathrm{Sym}^d(L)$ of L. □

Exercise 2.25
(1) Show that $\mathrm{Sym}^2(\partial^3) = \partial^5$.
(2) Show that $\mathrm{Sym}^d(L)$ has degree $d+1$ if L has degree 2. Hint: Use Proposition 4.26. □

Exterior Powers. One associates to a pair (M, e) (with e a cyclic vector) the pair $(\Lambda^d M, e \wedge \partial e \wedge \cdots \wedge \partial^{d-1} e)$.

Definition 2.26 Let L be a differential operator and let $e = 1$ be the generator of $k[\partial]/k[\partial]L$. The minimal monic annihilating operator of $e \wedge \partial e \wedge \ldots \wedge \partial^{d-1} e$ in $\Lambda^d(k[\partial]/k[\partial]L)$ is the d-th exterior power $\Lambda^d(L)$ of L. □

We denote by $\mathcal{S}_d$ the permutation group of d elements. Similar to the previous constructions one has the following lemma.

Lemma 2.27 *Let M be a differential module over k. The natural isomorphism of contravariant solution spaces*

$$\Lambda_C^d \mathrm{Hom}_{k[\partial]}(M, \mathcal{F}) \to \mathrm{Hom}_{k[\partial]}(\Lambda_k^d M, \mathcal{F})$$

is given by $\phi_1 \wedge \cdots \wedge \phi_d \mapsto \overline{\phi_1 \wedge \cdots \wedge \phi_d}$, where

$$\overline{\phi_1 \wedge \cdots \wedge \phi_d}(m_1 \wedge \cdots \wedge m_d) := \sum_{\pi \in \mathcal{S}_d} \mathrm{sgn}(\pi) \phi_1(m_{\pi(1)}) \phi_2(m_{\pi(2)}) \cdots \phi_d(m_{\pi(d)}).$$

Note that for $e \in M, \phi_1 \ldots, \phi_d \in \mathrm{Hom}_{k[\partial]}(M, \mathcal{F})$ and $y_i := \phi_i(e)$, we have

$$\overline{\phi_1 \wedge \cdots \wedge \phi_d}(e \wedge \partial e \wedge \cdots \wedge \partial^{d-1} e) = \det \begin{pmatrix} y_1 & \cdots & y_d \\ y_1' & \cdots & y_d' \\ \vdots & \cdots & \vdots \\ y_1^{(d-1)} & \cdots & y_d^{(d-1)} \end{pmatrix}$$
$$= wr(y_1, \ldots, y_d).$$

One therefore has the following corollary.

Corollary 2.28 *Let e be a cyclic vector for M with minimal annihilating operator L. Let $W \subset \mathcal{F}$ be the C-span of $\{wr(y_1, \ldots, y_d) \mid L(y_i) = 0\}$. Then the map $\phi \mapsto \phi(e \wedge \partial e \wedge \ldots \wedge \partial^{d-1}e)$ defines a surjection of* $\mathrm{Hom}_{k[\partial]}(\Lambda^d M, \mathcal{F})$ *onto W and W is the solution space of the minimal annihilating operator of $e \wedge \partial e \wedge \ldots \wedge \partial^{d-1}e$.*

The calculation of the d-th exterior power of L is similar to the calculations in the previous two constructions. Let $v = e \wedge \partial e \wedge \cdots \wedge \partial^{d-1}e$. Differentiate v $\binom{n}{d}$ times and use L to replace occurrences of ∂^j, $j \geq n$ with linear combinations of ∂e^i, $i < n$. This yields a system of $\binom{n}{d} + 1$ equations

$$\partial^i v = \sum_{\substack{J = (j_1, \ldots, j_d) \\ 0 \leq j_1 < \cdots < j_d \leq n-1}} a_{i,J}\, \partial^{j_1} e \wedge \cdots \wedge \partial^{j_d} e \tag{2.1}$$

in the $\binom{n}{d}$ quantities $\partial^{j_1} e \wedge \cdots \wedge \partial^{j_d} e$ with $a_{i,J} \in k$. These equations are linearly dependent and a linear relation among the first t of these (with t as small as possible) yields the exterior power.

We illustrate this with one example. (A more detailed analysis and simplification of the process to calculate the associated equations is given in [58, 60].)

Example 2.29 Let $L = \partial^3 + a_2\partial^2 + a_1\partial + a_0$, $a_i \in k$ and $M = k[\partial]/k[\partial]L$. Letting $e = 1$, we have that $\Lambda^2 M$ has a basis $\{\partial^i \wedge \partial^j \mid 1 \leq i < j \leq 2\}$. We have

$$\begin{aligned} v &= e \wedge \partial e, \\ \partial v &= e \wedge d^2 e, \\ \partial^2 v &= e \wedge (-a_2\partial^2 e - a_1 \partial e -_0 e) + \partial e \wedge \partial^2 e. \end{aligned}$$

Therefore $(\partial^2 + a_2\partial + a_1)v = \partial e \wedge \partial^2 e$ and so $\partial(\partial^2 + a_2\partial + a_1)v = \partial e \wedge (-a_2\partial^2 e$ $-a_1\partial e - a_0 e)$. This implies that the minimal annihilating operator of v is $(\partial + a_2)$ $\cdot(\partial^2 + a_2\partial + a_1) - a_0$. □

It is no accident that the order of the $(n-1)$st exterior power of an operator of order n is also n. The following exercise outlines a justification.

Exercise 2.30 *Exterior powers and adjoint operators*
Let $L = \partial^n + a_{n-1}\partial^{(n-1)} + \cdots + a_0$ with $a_i \in k$. Let K be a Picard-Vessiot extension of k associated with L and let $\{y_1, \ldots, y_n\}$ be a fundamental set of solutions of $L(y) = 0$. The set $\{u_1, \ldots, u_n\}$, where $u_i = wr(y_1, \ldots, \hat{y}_i, \ldots, y_n\}$, spans the solution space of $\Lambda^{n-1}(L)$. The aim of this exercise is to show that the set $\{u_1, \ldots, u_n\}$ is linearly independent and so $\Lambda^{n-1}(L)$ always has order n. We, furthermore, show that the operators $\Lambda^{n-1}(L)$ and L^* (the adjoint of L, see Exercise 2.1) are related in a special way (cf. [255], pages 167–171).

1. Show that $v_i = u_i/wr(y_1, \ldots, y_n)$ satisfies $L^*(v_i) = 0$. Hint: Let A_L be the companion matrix of L and $W = Wr(y_1, \ldots, y_n)$. Since $W' = A_L W$, we have the

fact that $U = (W^{-1})^T$ satisfies $U' = -A_L^T U$. Let $(f_0, \dots, f_{n-1})^T$ be a column of U. Note that $f_{n-1} = v_i$ for some i. One has (cf. Exercise 2.1),

$$\begin{aligned} -f'_{n-1} + a_{n-1} f_{n-1} &= f_{n-2} \\ -f'_i + a_i f_{n-1} &= f_{i-1} \quad 1 \le i \le n-2 \\ -f'_0 + a_0 f_{n-1} &= 0, \end{aligned}$$

and so

$$\begin{aligned} -f'_{n-1} + a_{n-1} f_{n-1} &= f_{n-2} \\ (-1)^2 f''_{n-1} - a_{n-1} f'_{n-1} + a_{n-2} f_{n-1} &= f_{n-3} \\ \vdots \quad &\vdots \quad \vdots \\ (-1)^n f^{(n)}_{n-1} + (-1)^{n-1} (a_{n-1} f_{n-1})^{(n-1)} + \dots + a_0 f_{n-1} &= 0. \end{aligned}$$

This last equation implies that $0 = L^*(f_n) = L^*(v_i)$.

2. Show that $wr(v_1, \dots, v_n) \neq 0$. Therefore the map $z \mapsto z/wr(y_1, \dots, y_n)$ is an isomorphism of the solution space of $\Lambda^{n-1}(L)$ onto the solution space of L^* and, in particular, the order of $\Lambda^{n-1}(L)$ is always n. Hint: Standard facts about determinants imply that $\sum_{i=1}^n v_i y_i^j = 0$ for $j = 0, 1, \dots, n-2$ and $\sum_{i=1}^n v_i y_i^{(n-1)} = 1$. Use these equations and their derivatives to show that $Wr(v_1, \dots, v_n) Wr(y_1, \dots, y_n) = 1$. □

Exercise 2.31 Show that $\Lambda^2(\partial^4) = \partial^5$. Therefore the d-th exterior power of an operator of order n can have order less than $\binom{n}{d}$. Hint: Show that the solution space of $\Lambda^2(\partial^4)$ is the space of polynomials of degree at most 4. □

We note that in the classical literature (cf. [255], p. 167), the d-th exterior power of an operator is referred to as the $(n-d)$th *associated operator*.

In connection with Chap. 4, a generalization of $\Lambda^d(L)$ is of interest. This generalization is present in the algorithms developed by Tsarev, Grigoriev et al. that refine Beke's algorithm for finding factors of a differential operator. Let $\mathcal{I} = (i_1, \dots, i_d)$, $0 \le i_1 < \dots < i_d \le n-1$. Let $e = 1$ in $k[\partial]/k[\partial]L$. We define the d-th *exterior power of L with respect to* $\mathcal{I}$, denoted by $\Lambda^d_{\mathcal{I}}(L)$, to be the minimal annihilating operator of $\partial^{i_1} e \wedge \dots \wedge \partial^{i_d} e$ in $\Lambda^d(k[\partial]/k[\partial]L)$. One sees as above that the solution space of $\Lambda^d_{\mathcal{I}}(L)$ is generated by $\{w_{\mathcal{I}}(y_1, \dots, y_d) \mid L(y_i) = 0\}$, where $w_{\mathcal{I}}(y_1, \dots, y_d)$ is the determinant of the $d \times d$ matrix formed from the rows $i_1 + 1, \dots i_d + 1$ of the $n \times d$ matrix

$$\begin{pmatrix} y_1 & y_2 & \dots & y_d \\ y'_1 & y'_2 & \dots & y'_d \\ \vdots & \vdots & \dots & \vdots \\ y_1^{(n-1)} & y_2^{(n-1)} & \dots & y_d^{(n-1)} \end{pmatrix}.$$

This operator is calculated by differentiating the element $v = \partial^{i_1} e \wedge \cdots \wedge \partial^{i_d} e$ as above. The following lemma is useful.

Lemma 2.32 *Let k and L be as above and assume that $\Lambda^d(L)$ has order $\nu = \binom{n}{d}$. For any $\mathcal{I}$ as above, there exist $b_{\mathcal{I},0}, \ldots, b_{\mathcal{I},\nu-1} \in k$ such that*

$$w_{\mathcal{I}}(y_1, \ldots, y_d) = \sum_{j=0}^{\nu-1} b_{\mathcal{I},j} wr(y_1, \ldots, y_d)^{(i)}$$

for any solutions $y_1, \ldots, y_d$ of $L(y) = 0$.

Proof. If $\Lambda^d(L)$ has order ν, then this implies that the system of equations (2.1) has rank ν. Furthermore, $\partial^{i_1} e \wedge \cdots \wedge \partial^{i_d} e$ appears as one of the terms in this system. Therefore we can solve for $\partial^{i_1} e \wedge \cdots \wedge \partial^{i_d} e$ as a linear function $\sum_{i=0}^{\nu-1} b_{\mathcal{I},i} \partial^i v$ of $v = e \wedge \partial e \wedge \cdots \wedge \partial^{d-1} e$ and its derivatives up to order $\nu - 1$. This gives the desired equation. □

We close this section by noting that MAPLE V contains commands in its DEtools package to calculate tensor products, symmetric powers, and exterior powers of operators.

2.4 Differential Modules and Representations

Throughout this section k will denote a differential field with algebraically closed subfield of constants C.

We recall that Diff_k denotes the category of all differential modules over k. Fix a differential module M over k. For integers $m, n \geq 0$ one defines the differential module $M_n^m = M \otimes \cdots M \otimes M^* \otimes \cdots \otimes M^*$, i.e., the tensor product of n copies of M and m copies of the dual M^* of M. For $m = n = 0$ the expression M_0^0 is assumed to mean $\mathbf{1} = \mathbf{1}_k$, the trivial 1-dimensional module over k. A *subquotient* of a differential module N is a differential module of the form N_1/N_2 with $N_2 \subset N_1 \subset N$ submodules. The subcategory $\{\{M\}\}$ of Diff_k is defined by: The objects of this category are the subquotients of finite direct sums of the M_n^m. For objects A, B of $\{\{M\}\}$ one defines $\mathrm{Hom}(A, B)$ to be $\mathrm{Hom}_{k[\partial]}(A, B)$. Thus $\mathrm{Hom}(A, B)$ has the same meaning in $\{\{M\}\}$ and in Diff_k. This is usually expressed as "$\{\{M\}\}$ is a full subcategory of Diff_k." It is easily seen that $\{\{M\}\}$ is the smallest full subcategory of Diff_k that contains M and is closed under all operations of linear algebra (i.e., direct sums, tensor products, duals, subquotients).

For a linear algebraic group G over C one considers the category Repr_G that consists of the representations of G on finite dimensional vector spaces over C (see the beginning of Sect. 2.2 and the appendices). A finite dimensional vector space W over C together with a representation $\rho : G \to \mathrm{GL}(W)$ is also called a G-module.

In the category Repr_G one can also perform all operations of linear algebra (i.e., direct sums, tensor products, duals, subquotients). The strong connection between the differential module M and its differential Galois group G is given in the following theorem.

Theorem 2.33 *Let M be a differential module over k and let G denote its differential Galois group. There is a C-linear equivalence of categories*

$$S : \{\{M\}\} \to \text{Repr}_G,$$

which is compatible with all constructions of linear algebra.

Proof. We start by *explaining the terminology*. First of all, S is a functor. This means that S associates to every object A of the first category an object $S(A)$ of the second category. Likewise, S associates to every morphism $f \in \text{Hom}(A, B)$ of the first category a morphism $S(f) \in \text{Hom}(S(A), S(B))$ of the second category. The following rules should be satisfied:
$S(1_A) = 1_{S(A)}$ and $S(f \circ g) = S(f) \circ S(g)$.
The term C-linear means that the map from $\text{Hom}(A, B)$ to $\text{Hom}(S(A), S(B))$, given by $f \mapsto S(f)$, is C-linear. The term "equivalence" means that the map $\text{Hom}(A, B) \to \text{Hom}(S(A), S(B))$ is bijective and that there exists for every object B of the second category an object A of the first category such that $S(A)$ is isomorphic to B. The compatibility of S with, say, tensor products means that there are isomorphisms $i_{A,B} : S(A \otimes B) \to S(A) \otimes S(B)$. These isomorphisms should be "natural" in the sense that for any morphisms $f : A \to A',\ g : B \to B'$ the following diagram is commutative.

$$\begin{array}{ccc} S(A \otimes B) & \stackrel{S(f\otimes g)}{\longrightarrow} & S(A' \otimes B') \\ i_{A,B} \downarrow & & \downarrow i_{A',B'} \\ S(A) \otimes S(B) & \stackrel{S(f)\otimes S(g)}{\longrightarrow} & S(A') \otimes S(B'). \end{array}$$

Thus, the compatibility with the constructions of linear algebra means that S maps a construction in the first category to one object in the second category that is in a "natural" way isomorphic to the same construction in the second category. For the S that we will construct almost all these properties will be obvious.

For the definition of S we need the Picard-Vessiot field $K \supset k$ of M. The differential module $K \otimes_k M$ over K is trivial in the sense that there is a K-basis $e_1, \ldots, e_d$ of $K \otimes_k M$ such that $\partial e_i = 0$ for all i. In other words, the obvious map $K \otimes_C \ker(\partial, K \otimes_k M) \to K \otimes_k M$ is a bijection. Indeed, this is part of the definition of the Picard-Vessiot field. Also, every $K \otimes_k M_n^m$ is a trivial differential module over K. We conclude that for every object N of $\{\{M\}\}$ the differential module $K \otimes_k N$ is trivial. One defines S by $S(N) = \ker(\partial, K \otimes_k N)$. This object is a finite dimensional vector space over C. The action of G on $K \otimes_k N$ (induced by the action of G on K) commutes with ∂ and thus G acts on the kernel of ∂ on $K \otimes_k N$. From

Theorem 1.27 one easily deduces that the action of G on $\ker(\partial, K \otimes_k N)$ is algebraic. In other words, $S(N)$ is a representation of G on a finite dimensional vector space over C. Let $f : A \to B$ be a morphism in $\{\{M\}\}$. Then f extends to a K-linear map $1_K \otimes f : K \otimes_k A \to K \otimes_k B$, which commutes with ∂. Therefore, f induces a C-linear map $S(f) : S(A) \to S(B)$ which commutes with the G-actions.

We will omit the straightforward and tedious verification that S commutes with the constructions of linear algebra. It is not a banality to show that $\mathrm{Hom}(A, B) \to \mathrm{Hom}(S(A), S(B))$ is a bijection. Since $\mathrm{Hom}(A, B)$ is equal to $\ker(\partial, A^* \otimes B) = \mathrm{Hom}(\mathbf{1}_k, A^* \otimes B)$ we may assume that $A = \mathbf{1}_k$ and that B is arbitrary. Clearly $S(\mathbf{1}_k) = \mathbf{1}_G$, where the latter is the 1-dimensional trivial representation of G. Now $\mathrm{Hom}(\mathbf{1}_k, B)$ is equal to $\{b \in B | \ \partial(b) = 0\}$. Furthermore, $\mathrm{Hom}(\mathbf{1}_G, S(B))$ is equal to $\{v \in \ker(\partial, K \otimes_k B) | \ gv = v \text{ for all } g \in G\}$. Since $K^G = k$, one has $(K \otimes_k B)^G = B$. This implies that $\{b \in B | \ \partial b = 0\} \to \mathrm{Hom}(\mathbf{1}_G, S(B))$ is a bijection.

Finally, we have to show that any representation B of G is equivalent to the representation $S(A)$ for some $A \in \{\{M\}\}$. This follows from the following fact on representations of any linear algebraic group G (see [302] and the appendices):

Suppose that V is a faithful representation of G (i.e., $G \to \mathrm{GL}(V)$ is injective). Then every representation of G is a direct sum of subquotients of the representations $V \otimes \cdots \otimes V \otimes V^* \otimes \cdots \otimes V^*$.

In our situation we take for V the representation $S(M)$ that is by definition faithful. Since S commutes with the constructions of linear algebra, we have the fact that any representation of G is isomorphic to $S(N)$ for some N that is a direct sum of subquotients of the M_n^m. In other words, N is an object of $\{\{M\}\}$. □

Remarks 2.34

(1) In the terminology of tannakian categories, Theorem 2.33 states that the category $\{\{M\}\}$ is a neutral tannakian category and that G is the corresponding affine group scheme (see the appendices).

(2) The functor S has an "inverse". We will describe this inverse by constructing the differential module N corresponding to a given representation W. One considers the trivial differential module $K \otimes_C W$ over K with ∂ defined by $\partial(1 \otimes w) = 0$ for every $w \in W$. The group G acts on $K \otimes_C W$ by $g(f \otimes w) = g(f) \otimes g(w)$ for every $g \in G$. Now one takes the G-invariants $N := (K \otimes_C W)^G$. This is a vector space over k. The operator ∂ maps N to N, since ∂ commutes with the action of G. One has now to show that N has finite dimension over k, that N is an object of $\{\{M\}\}$, and that $S(N)$ is isomorphic to W.

We know already that $W \cong S(A)$ for some object A in $\{\{M\}\}$. Let us write $W = S(A)$ for convenience. Then by the definition of S one has $K \otimes_C W = K \otimes_k A$ and the two objects have the same G-action and the same ∂. Then $(K \otimes_C W)^G = A$ and this finishes the proof.

(3) Let H be a closed normal subgroup of G. Choose a representation W of G such that the kernel of $G \to \mathrm{GL}(W)$ is H. Let N be an object of $\{\{M\}\}$ with

$S(N) = W$. The field K contains a Picard-Vessiot field L for N, since $K \otimes_k N$ is a trivial differential module over K. The action of the subgroup H on L is the identity since by construction the differential Galois group of N is G/H. Hence $L \subset K^H$. Equality holds by Galois correspondence, see Proposition 1.34 part 1. Thus, we have obtained a more natural proof of the statement in loc. cit. part 2, namely that K^H is the Picard-Vessiot field of some differential equation over k. □

Corollary 2.35 *Let $L \in k[\partial]$ be a monic differential operator of degree ≥ 1. Let K be the Picard-Vessiot field of $M := k[\partial]/k[\partial]L$ and G its differential Galois group. Put $V = \ker(\partial, K \otimes_k M)$ (This is the covariant solution space of M.) There are natural bijections between:*
(a) *The G-invariant subspaces of V.*
(b) *The submodules of M.*
(c) *The monic right-hand factors of L.*

The only thing to explain is the correspondence between (b) and (c). Let $e = 1$ be the cyclic element of M and let N be a submodule of M. There is a unique monic operator R of minimal degree such that $Re \in N$. This is a right-hand factor of L. Moreover, $M/N = \mathcal{D}/\mathcal{D}R$ (compare the exact sequence before Proposition 2.13). Of course R also determines a unique left-hand factor of L. We note that the above corollary can also be formulated for the contravariant solution space $\mathrm{Hom}_{k[\partial]}(M, K)$.

We recall that an operator $L \in k[\partial]$ is *reducible over* k if L has a nontrivial right-hand factor. Otherwise L is called *irreducible*. The same terminology is used for differential equations in matrix form or for differential modules or for representations of a linear algebraic group over C.

Exercise 2.36 Show that a matrix differential equation is reducible if and only if it is equivalent to an equation $Y' = BY,\ B \in M_n(k)$ where B has the form

$$B = \begin{pmatrix} B_1 & 0 \\ B_2 & B_3 \end{pmatrix} .$$

□

Definition 2.37 A differential module M is called *completely reducible* or *semi-simple* if there exists for every submodule N of M a submodule N' such that $M = N \oplus N'$. □

The same terminology is used for differential operators and for representations of a linear algebraic group G over C. We note that the terminology is somewhat confusing because an irreducible module is, at the same time, completely reducible.

A G-module W and a G-submodule W_1 has a *complementary submodule* if there is a G-submodule W_2 of W such that $W = W_1 \oplus W_2$. Thus a (finite dimensional) G-module V is *completely reducible* if every G-submodule has a complementary submodule. This is equivalent to V being a direct sum of irreducible submodules (compare with Exercise 2.38 part 1).

The *unipotent radical* of a linear algebraic group G is the largest normal unipotent subgroup G_u of G (see [141] for definitions of these notions). The group G is called *reductive* if G_u is trivial. We note that for this terminology G is reductive if and only G^o is reductive.

When G is defined over an algebraically closed field of characteristic zero, it is known that G is reductive if and only if it has a *faithful* completely reducible G-module (cf. the Appendix of [32]). In this case, all G-modules will be completely reducible.

Exercise 2.38 *Completely reducible modules and reductive groups.*
(1) Show that M is completely reducible if and only if M is a direct sum of irreducible modules. Is this direct sum unique?

(2) Let M be a differential module. Show that M is completely reducible if and only if its differential Galois group is reductive. Hint: Use the above information on reductive groups.

(3) Let M be a completely reducible differential module. Prove that every object N of $\{\{M\}\}$ is completely reducible. Hint: Use the above information on reductive groups.

(4) Show that the tensor product $M_1 \otimes M_2$ of two completely reducible modules is again completely reducible. Hint: Apply (2) and (3) with $M := M_1 \oplus M_2$. We note that a direct proof (not using reductive groups) of this fact seems to be unknown. □

Exercise 2.39 *Completely reducible differential operators.*
(1) Let $R_1, \dots, R_s$ denote irreducible monic differential operators (of degree ≥ 1). Let L denote LCLM$(R_1, \dots, R_s)$, the least common left multiple of $R_1, \dots, R_s$. In other terms, L is the monic differential operator satisfying $k[\partial]L = \cap_{i=1}^{s} k[\partial]R_i$. This generalizes the LCLM of two differential operators, defined in Sect. 2.1. Show that the obvious map $k[\partial]/k[\partial]L \to k[\partial]/k[\partial]R_1 \oplus \cdots \oplus k[\partial]/k[\partial]R_s$ is injective. Conclude that L is completely reducible.

(2) Suppose that L is monic and completely reducible. Show that L is the LCLM of suitable (distinct) monic irreducible operators $R_1, \dots, R_s$. Hint: By definition $k[\partial]/k[\partial]L = M_1 \oplus \cdots \oplus M_s$, where each M_i is irreducible. The element $\bar{1} \in k[\partial]/k[\partial]L$ is written as $\bar{1} = m_1 + \cdots + m_s$ with each $m_i \in M_i$. Let L_i be the monic operator of smallest degree with $L_i m_i = 0$. Show that L_i is irreducible and that $L = \text{LCLM}(L_1, \dots, L_s)$.

(3) Let $k = C$ be a field of constants and let L be a linear operator in $C[\partial]$. We may write $L = p(\partial) = \prod p_i(\partial)^{n_i}$ where the p_i are distinct irreducible polynomials and $n_i \geq 0$. Show that L is completely reducible if and only if all the $n_i \leq 1$.

(4) Let $k = C(z)$. Show that the operator $L = \partial^2 + (1/z)\partial \in C(z)[\partial]$ is not completely reducible. Hint: The operator is reducible since $L = (\partial + (1/z))(\partial)$ and ∂ is the only first order right factor. □

Proposition 2.40 *Let M be a completely reducible differential module. Then M can be written as a direct sum $M = M_1 \oplus \cdots \oplus M_r$, where each M_i is a direct sum of n_i copies of an irreducible module N_i. Moreover, $N_i \not\cong N_j$ for $i \neq j$. This unique decomposition is called the* isotypical decomposition *of M. Then the eigenring $\mathcal{E}(M)$ (i.e., the ring of the endomorphisms of M) is equal to the product $\prod_{i=1}^{r} \mathrm{M}_{n_i}(C)$ of matrix algebras over C.*

Proof. The first part of the proposition is rather obvious. For $i \neq j$, every morphism $N_i \to N_j$ is zero. Consider an endomorphism $f : M \to M$. Then $f(M_i) \subset M_i$ for every i. This shows already that the isotypical decomposition is unique. Furthermore, M_i is isomorphic to $N_i \otimes L_i$, where L_i is a trivial differential module over k of dimension n_i. One observes that $\mathcal{E}(N_i) = C.1_{N_i}$ follows from the irreducibility of N_i. From this one easily deduces that $\mathcal{E}(M_i) \cong \mathrm{M}_{n_i}(C)$. □

We note that the above proposition is a special case of a result on semisimple modules over a suitable ring (compare [170], Chap. XVII.1, Proposition 1.2).

The Jordan-Hölder Theorem is also valid for differential modules. We recall its formulation. A tower of differential modules $M_1 \supset M_2 \supset \ldots \supset M_r = \{0\}$ is called a *composition series* if the set of quotients $(M_i/M_{i+1})_{i=1}^{r-1}$ consists of irreducible modules. Two composition series for M yield isomorphic sets of irreducible quotients, up to a permutation of the indices.

A (monic) differential operator L can be written as a product $L_1 \cdots L_r$ of irreducible monic differential operators. For any other factorization $L = R_1 \cdots R_s$ with irreducible operators R_i, one has $r = s$ and there exists a permutation π of $\{1, \ldots, r\}$ such that L_i is equivalent to $R_{\pi(i)}$. Indeed, the factorization $L = L_1 \cdots L_r$ induces for the module $k[\partial]/k[\partial]L$ the composition series $k[\partial]/k[\partial]L \supset k[\partial]L_r/k[\partial]L \supset \cdots$.

A monic differential operator has, in general, many factorizations into irreducible monic operators. Consider $k = C(z)$ and $L = \partial^2$. Then all factorizations are $\partial^2 = (\partial + \frac{f'}{f})(\partial - \frac{f'}{f})$ with f a monic polynomial in z of degree ≤ 1.

3 Formal Local Theory

In this chapter we will classify linear differential equations over the field of formal Laurent series $\widehat{K} = k((z))$ and describe their differential Galois groups. Here k is an algebraically closed field of characteristic 0. For most of what follows the choice of the field k is immaterial. In the first two sections one assumes that $k = \mathbf{C}$. This has the advantage that the roots of unity have the convenient description $e^{2\pi i\lambda}$ with $\lambda \in \mathbf{Q}$. Moreover, for $k = \mathbf{C}$ one can compare differential modules over $\widehat{K}$ with differential modules over the field of convergent Laurent series $\mathbf{C}(\{z\})$. In the third section k is an arbitrary algebraically closed field of characteristic 0. Unless otherwise stated the term differential module will refer in this chapter to a differential module over $\widehat{K}$.

3.1 Formal Classification of Differential Equations

This classification can be given in various ways:

1. A factorization of $L \in \widehat{K}[\partial]$ into linear factors over the algebraic closure of $\widehat{K}$.

2. Finding a canonical form in each equivalence class of matrix differential equations $v' = Av$.

3. Description of the isomorphism classes of left $\widehat{K}[\partial]$-modules of finite dimension over $\widehat{K}$.

4. Description of a fundamental matrix F for a matrix differential equation in canonical form.

5. Description of a structure on the solution space V of the differential equation.

The problem is somewhat analogous to the classification (or Jordan normal form) of linear maps A acting on a vector space V of finite dimension over the field of real numbers $\mathbf{R}$. Let us recall how this is done. The eigenvalues of A are, in general, complex and therefore we need to make of V the complex vector space $W = \mathbf{C} \otimes V$. Let $\alpha_1, \cdots, \alpha_s$ denote the distinct eigenvalues of A. The generalized eigenspace for the eigenvalue α_i is defined by;

$$W(\alpha_i) := \{w \in W \mid (A - \alpha_i)^m w = 0 \text{ for sufficiently large } m\}.$$

One finds a decomposition $W = \oplus W(\alpha_i)$ of W into A-invariant subspaces. For each subspace $W(\alpha_i)$ the operator $B_i := A - \alpha_i$ is nilpotent and one can decompose $W(\alpha_i)$ as a direct sum of subspaces $W(\alpha_i)_k$. Each such subspace has a basis $e_1, \cdots, e_r$ such that $B_i(e_1) = e_2, \cdots, B_i(e_{r-1}) = e_r$, $B_i(e_r) = 0$. Writing the matrix of A with respect to this decompositions and these bases one finds the familiar Jordan normal form for this matrix. The given fact that A is a linear map on a real vector space implies now that for every complex α_i its conjugate is some α_j and the "block-decompositions" of $W(\alpha_i)$ and $W(\alpha_j)$ are the same.

To classify differential equations over $\widehat{K}$ we will need to first work over the algebraic closure of $\widehat{K}$. In the next section we shall show that every finite algebraic extension of $\widehat{K}$ of degree m over $\widehat{K}$ is of the form $\widehat{K}_m := \widehat{K}(v)$ with $v^m = z$. In the sequel we will often write $v = z^{1/m}$. The main result of this chapter is the following theorem

Theorem 3.1
1. For every monic (skew) polynomial

$$L = \partial^d + a_1\partial^{d-1} + \cdot + a_{d-1}\partial + a_d \in \widehat{K}[\partial]$$

there is some integer $m \geq 1$ and an element $u \in \widehat{K}_m$ such that L has a factorization of the form $L = L_2(\partial - u)$.

2. After replacing $\widehat{K}$ by a finite field extension $\widehat{K}_m$ the differential equation in matrix form $v' = Av$ (where $' := z\frac{d}{dz}$) is equivalent to a differential equation $u' = Bu$, where the matrix B has a "decomposition" into square blocks $B_{i,a}$ with $i = 1, \ldots, s$ and $1 \leq a \leq m_i$ of the form

$$\begin{pmatrix} b_i & 0 & . & . & . & 0 \\ 1 & b_i & 0 & . & . & 0 \\ 0 & 1 & b_i & 0 & . & 0 \\ . & . & . & . & . & . \\ . & . & . & . & . & . \\ 0 & . & . & 0 & 1 & b_i \end{pmatrix}.$$

Furthermore, $b_i \in \mathbf{C}[z^{-1/m}]$ and for $i \neq j$ one has $b_i - b_j \notin \mathbf{Q}$.

3. Let M denote a left $\widehat{K}[\partial]$ module of finite dimension. There is a finite field extension $\widehat{K}_m$ of $\widehat{K}$ and there are distinct elements $q_1, \ldots, q_s \in z^{-1/m}\mathbf{C}[z^{-1/m}]$ such that $\widehat{K}_m \otimes_{\widehat{K}} M$ decomposes as a direct sum $\oplus_{i=1}^s M_i$. For each i there is a vector space W_i of finite dimension over $\mathbf{C}$ and a linear map $C_i : W_i \to W_i$ such that $M_i = \widehat{K}_m \otimes_{\mathbf{C}} W_i$, and the operator $\delta := z\partial$ on M_i is given by the formula

$$\delta(f \otimes w) = (q_i f \otimes w) + (f' \otimes w) + (f \otimes C_i(w)).$$

In the sequel we prefer to work with $\delta = z\partial$ instead of ∂. Of course, $\widehat{K}[\partial] = \widehat{K}[\delta]$ holds. Furthermore, we will go back and forth between the skew polynomial L and the left $\widehat{K}[\delta]$ module $M = \widehat{K}[\delta]/\widehat{K}[\delta]L$. By induction on the degree it suffices to find some factorization of L or equivalently some decomposition of M. Furthermore, we note that the formulations 2 and 3 in the theorem are equivalent by using the ordinary Jordan normal forms of the maps C_i of part 3. We shall treat questions of uniqueness and descent to $\widehat{K}$ later in the chapter.

Exercise 3.2 *Solutions of differential equations over* $\widehat{K}$
Let E be a differential extension of $\widehat{K}$ containing:

1. all fields $\widehat{K}_m$,
2. for any m and any $b \in \widehat{K}_m^*$, a nonzero solution of $y' = by$,
3. a solution of $y' = 1$.

Show, assuming Theorem 3.1, that E contains a fundamental matrix for any equation $Y' = AY$ with $A \in \mathrm{M}_n(\widehat{K})$. □

In this section, Theorem 3.1 will be proved by means of differential analogs of Hensel's Lemma. In the third section another proof will be given based upon Newton polygons. We will start by recalling how the classical form of Hensel's Lemma allows one to prove that fields of the form $\widehat{K}_n$ are the only finite algebraic extensions of $\widehat{K}$.

We begin by noting that the field $\widehat{K}_n = \widehat{K}(z^{1/n}) = \mathbf{C}((z^{1/n}))$ is itself the field of formal power series over $\mathbf{C}$ in the variable $z^{1/n}$. This field extension has degree n over $\widehat{K}$ and is a Galois extension of $\widehat{K}$. The Galois automorphisms σ are given by the formula $\sigma(z^{1/n}) = \zeta z^{1/n}$ with $\zeta \in \{e^{2\pi i k/n} |\ 0 \leq k < n\}$. The Galois group is isomorphic to $\mathbf{Z}/n\mathbf{Z}$. We note that $\widehat{K}_n \subset \widehat{K}_m$ if n divides m. Therefore it makes sense to speak of the union $\overline{\widehat{K}} = \cup_n \widehat{K}_n$ and our statement concerning algebraic extensions of $\widehat{K}$ implies that $\overline{\widehat{K}}$ is the algebraic closure of $\widehat{K}$.

We will also need the valuation v on $\widehat{K}$. This is defined as a map

$$v : \widehat{K} \to \mathbf{Z} \cup \{\infty\}$$

with $v(0) = \infty$ and $v(f) = m$ if $f = \sum_{i \geq m} a_i z^i$ and $a_m \neq 0$. We note that $v(fg) = v(f) + v(g)$ and $v(f+g) \geq \min(v(f), v(g))$. This valuation is extended to each field $\widehat{K}_n$ as a map $v : \widehat{K}_n \to (1/n)\mathbf{Z} \cup \{\infty\}$ in the obvious way: $v(f) = \lambda$ if $f = \sum_{\mu \geq \lambda; n\mu \in \mathbf{Z}} a_\mu z^\mu$ and $a_\lambda \neq 0$. Finally, v is extended to a valuation $v : \overline{\widehat{K}} \to \mathbf{Q} \cup \{\infty\}$. Furthermore, we will write $O_n = \mathbf{C}[[z^{1/n}]] = \{f \in \widehat{K}_n |\ v(f) \geq 0\}$ and $\overline{O} := \{f \in \overline{\widehat{K}} |\ v(f) \geq 0\}$. It is easily seen that O_n and $\overline{O}$ are rings with fields of quotients $\widehat{K}_n$ and $\overline{\widehat{K}}$. The element $\pi := z^{1/n} \in O_n$ has the property that πO_n is the unique maximal ideal of O_n and that $O_n/\pi O_n \cong \mathbf{C}$. On $\widehat{K}_n$ one can also introduce a metric as follows $d(f, g) = e^{-v(f-g)}$. With respect to this metric $\widehat{K}_n$ is complete. In the sequel we will talk about limits with respect to this metric. Most of the statements that we made about the algebraic and topological structure of $\widehat{K}$ are rather obvious. The only not so obvious statement is that every finite extension of $\widehat{K}$ is some field $\widehat{K}_n$. This will follow from the next proposition.

Proposition 3.3 *Every polynomial $T^d + a_1 T^{d-1} + \cdots + a_{d-1} T + a_d \in \widehat{K}[T]$ has a root in some $\widehat{K}_n$.*

Proof. Define $\lambda := \min\{\frac{v(a_i)}{i} \mid 1 \le i \le d\}$ and make the substitution $T = z^{-\lambda} E$, where E is a new indeterminate. The new monic polynomial that arises has the form

$$P = E^d + b_1 E^{d-1} + \cdots + b_{d-1} E + b_d,$$

with $b_1, \ldots, b_d \in \widehat{K}_m$, where m is the denominator of λ. Now $\min v(b_i) = 0$. We have $P \in O_m[E]$ and we write $\overline{P} \in \mathbf{C}[E]$ for the reduction of P modulo $\pi := z^{1/m}$ (i.e., reducing all the coefficients of P modulo π). Note that the fact that $\min v(b_i) = 0$ implies that $\overline{P}$ has at least two nonzero terms. Note that $v(b_i) = 0$ precisely for those i with $\frac{v(a_i)}{i} = \lambda$. Therefore if $v(b_1) = 0$, we have that λ is an integer and $m = 1$. The key for finding decompositions of P is now the following lemma.

Lemma 3.4 Classical Hensel's Lemma
If $\overline{P} = F_1 F_2$ with $F_1, F_2 \in \mathbf{C}[E]$ monic polynomials with g.c.d.$(F_1, F_2) = 1$ then there is a unique decomposition $P = P_1 P_2$ of P into monic polynomials such that $\overline{P_i} = F_i$ for $i = 1, 2$.

Proof. Suppose that we have already found monic polynomials $Q_1(k), Q_2(k)$ such that $\overline{Q_i}(k) = F_i$ (for $i = 1, 2$) and $P \equiv Q_1(k) Q_2(k)$ modulo π^k. Then define $Q_i(k+1) = Q_i(k) + \pi^k R_i$ where $R_i \in \mathbf{C}[E]$ are the unique polynomials with degree $R_i <$ degree F_i and

$$R_1 F_2 + R_2 F_1 = \frac{P - Q_1(k) Q_2(k)}{\pi^k} \text{ modulo } \pi.$$

One easily sees that $P \equiv Q_1(k+1) Q_2(k+1)$ modulo π^{k+1}. Define now $P_i = \lim_{k \to \infty} Q_i(k)$ (the limit is taken here for every coefficient separately). It is easily seen that P_1, P_2 have the required properties. □

Example 3.5 Let $P = y^2 - 2zy - 1 + z^2$. Then $\overline{P} = y^2 - 1 = (y-1)(y+1)$. We let $Q_1(0) = y - 1$ and $Q_2(0) = y + 1$ and define $Q_1(1) = Q_0(0) + zR_1$ and $Q_2(1) = Q_2(0) + zR_2$. We then have $P - Q_1(1) Q_2(1) = -2zy - z(y+1)R_1 - z(y-1)R_2 + z^2 R_1 R_2$. Solving $-2y = (y+1)R_1 - z(y-1)R_2 \bmod z$, we get $R_1 = R_2 = -1$. Therefore $Q_1(1) = y - 1 - z$ and $Q_2(1) = y + 1 - z$. At this point we have $Q_1(1) Q_2(1) = P$ so the procedure stops. □

Continuation of the proof of Proposition 3.3: We use induction on the degree d. If $\overline{P}$ has at least two different roots in $\mathbf{C}$ then induction finishes the proof. If not then

$\overline{P} = (E - c_0)^d$ for some $c_0 \in \mathbf{C}$. As we have noted, $\overline{P}$ has at least two nonzero terms so we have $c_0 \neq 0$. This, furthermore, implies that $\overline{P}$ has $d + 1$ nonzero terms and so $m = 1$ and λ is an integer. One then writes

$$P = (E - c_0)^d + e_1(E - c_0)^{d-1} + \cdots + e_{d-1}(E - c_0) + e_d,$$

with all $v(e_i) > 0$. Put λ_1 =min $\{\frac{v(e_i)}{i} | \ 1 \leq i \leq d\}$ and make the substitution $E = c_0 + z^{\lambda_1} E^*$. It is then possible that an application of Lemma 3.4 yields a factorization and we will be done by induction. If not, we can conclude as above that λ_1 is an integer. We then make a further substitution $E = c_0 + c_1 z^{\lambda_1} + z^{\lambda_2} E^{**}$ with $0 < \lambda_1 < \lambda_2$ and continue. If we get a factorization at any stage using Lemma 3.4, then induction finishes the proof. If not, we will have generated an infinite expression $f := \sum_{n=0}^{\infty} c_n z^{\lambda_n}$ with $0 < \lambda_1 < \lambda_2 <$ .. a sequence of integers such that $P = (E - f)^d$. This finishes the proof of Proposition 3.3. □

Example 3.6 Let $P = E^2 - 2zE + z^2 - z^3$. We have that $\overline{P} = E^2$ and (using the above notation) that $e_1 = -2z$ and $e_2 = z^2 - z^3$. Furthermore, $\lambda_1 = \min\{\frac{1}{1}, \frac{2}{2}\} = 1$. We then let $E = zE^*$, so $Q = z^2 E^{*2} - 2z^2 E^* + z^2 - z^3$. Let $Q_1 = E^{*2} - 2E^* + 1 - z$. We see that $\overline{Q_1} = E^{*2} - 2E^* + 1 = (E^* - 1)^2$. We write $Q_1 = (E^* - 1)^2 - z$ and so $\lambda_2 = \min\{\frac{\infty}{1}, \frac{1}{2}\} = 1/2$. We let $E^* = 1 + z^{1/2} E^{**}$ and so $Q_1 = (z^{1/2} E^{**})^2 - z$ $= zE^{**2} - z$. Letting $Q_2 = E^{**2} - 1$, we have that $E^{**} = \pm 1$. The process stops at this point and we have that the two roots of Q are $1 + z(1 \pm z^{1/2})$. □

3.1.1 Regular Singular Equations

We will now develop versions of Hensel's Lemma for differential modules and differential equations that will help us prove Theorem 3.1. We start by introducing some terminology. Let M be a finite dimensional vector space over $\widehat{K}$. Let, as before, $O := \{f \in \widehat{K} | \ v(f) \geq 0\}$.

Definition 3.7 A lattice *is a subset N of M of the form $N = Oe_1 + \cdots + Oe_d$ where $e_1, \ldots, e_d$ is a $\widehat{K}$-basis of M.* □

The lattice is itself an O-module. One can prove that any finitely generated O-module N (i.e., there are elements $f_1, \ldots, f_m$ with $N = Of_1 + \cdots + Of_m$) of M that contains a basis of M is a lattice. For a lattice N we introduce the space $\overline{N} = N/\pi N$ where $\pi = z$. This is a vector space over $\mathbf{C}$ with dimension d. The image of $n \in N$ in $\overline{N}$ will be denoted by $\overline{n}$. Properties that we will often use are the following.

Exercise 3.8 *Lattices.*
(1) $f_1, \ldots, f_m \in N$ are generators of N over O if and only if $\overline{f}_1, \ldots, \overline{f}_m$ are generators of the vector space $\overline{N}$ over $\mathbf{C}$. Hint: Nakayama's Lemma ([170], Chap. X, §4).

(2) $f_1, \ldots, f_d \in N$ is a free basis of N over O if and only if $\overline{f}_1, \ldots, \overline{f}_d$ is a basis of the vector space $\overline{N}$ over $\mathbf{C}$. □

Although lattices are ubiquitous, only special differential modules have δ-invariant lattices.

Definition 3.9 A differential module M is said to be a *regular singular* module if there exists a δ-invariant lattice N in M. A differential equation $Y' = AY$, A an $n \times n$ matrix with coefficients in $\widehat{K}$, is said to be regular singular if the associated module is regular singular. If M is not regular singular then we say it is *irregular singular*. □

The differential module associated with an equation of the form $\delta Y = AY$, where $A \in \mathrm{M}_n(\mathbf{C}[[z]])$ is a regular singular module. In particular, any equation of the form $\delta Y = AY$, where $A \in \mathrm{M}_n(\mathbf{C})$, is a regular singular equation. In Proposition 3.12 we will show that all regular singular modules are associated with such an equation.

Lemma 3.10 *If M_1 and M_2 are regular singular modules, then the same holds for $M_1 \oplus M_2$, $M_1 \otimes M_2$ and M_1^*. Furthermore, any $\widehat{K}[\delta]$ submodule and quotient module of a regular singular module is regular singular.*

Proof. Let N_1 and N_2 be δ-invariant lattices in M_1 and M_2. A calculation shows that $N_1 \oplus N_2$, $N_1 \otimes N_2$ and N_1^* are δ-invariant lattices in the corresponding $\widehat{K}[\partial]$ modules. If M is a regular singular module with δ-invariant lattice N and M' is a submodule of M, then $N \cap M'$ is a δ-invariant lattice of M'. Using duals and applying this result, we obtain a similar result for quotients. □

Let M be a regular singular module and let N be a δ-invariant lattice. We have that πN is invariant under δ and hence δ induces a $\mathbf{C}$ linear map $\overline{\delta}$ on $\overline{N}$. Let $c_1, \cdots, c_s$ denote the distinct eigenvalues of $\overline{\delta}$ and let $\overline{N} = \overline{N}(c_1) \oplus \cdots \oplus \overline{N}(c_s)$ denote the decomposition of $\overline{N}$ into generalized eigenspaces. One can choose elements $e_{i,j} \in N$ with $1 \le i \le s$ and $1 \le j \le m_i$ such that $\{\overline{e}_{i,j} | \ 1 \le j \le m_i\}$ forms a basis of $\overline{N}(c_i)$ for every i. Then we know that $\{e_{i,j}\}$ is a free basis of the O-module N. We now define another δ-invariant lattice N_1 generated over O by the set $\{ze_{1,1}, \cdots, ze_{1,m}, e_{2,1}, \cdots, e_{s,m_s}\}$. The linear map $\overline{\delta}$ on $\overline{N_1}$ has as eigenvalues $\{c_1 + 1, c_2, \cdots, c_s\}$. We come now to the following conclusion.

Lemma 3.11 *If M is a regular singular differential module, then there exists a δ-invariant lattice N in M such that the eigenvalues $c_1, \cdots, c_s$ of $\overline{\delta}$ on $\overline{N}$ have the property: If $c_i - c_j \in \mathbf{Z}$ then $c_i = c_j$.*

Proposition 3.12 *A regular singular equation $\delta Y = AY$ is equivalent to an equation of the form $\delta Y = A_0 Y$ with $A_0 \in \mathrm{M}_n(\mathbf{C})$ and such that the distinct eigenvalues of A_0 do not differ by integers.*

Proof. We begin with a well-known fact from linear algebra. Let $U, V \in \mathrm{M}_n(\mathbf{C})$ and assume that U and V have no eigenvalues in common. We claim that the map $X \mapsto UX - XV$ is an isomorphism on $\mathrm{M}_n(\mathbf{C})$. To prove this it is enough to show that the map is injective. If $UX - XV = 0$ then for any $P \in \mathbf{C}[T]$, $P(U)X - XP(V) = 0$. If P_U is the characteristic polynomial of U, then the assumptions imply that $P_U(V)$ is invertible. Therefore $X = 0$.

We now turn to the proof of the proposition. With respect to the basis of a δ-invariant lattice, we can assume the associated equation is of the form $\delta Y = AY$ with $A \in \mathbf{C}[[z]]$. Let $A = A_0 + A_1 z + \cdots$, $A_i \in \mathrm{M}_n(\mathbf{C})$. Furthermore, by Lemma 3.11, we may assume that the distinct eigenvalues of A_0 do not differ by integers. We will construct a matrix $P = I + P_1 z + \cdots$, $P_i \in \mathrm{M}_n(\mathbf{C})$ such that $PA_0 = AP - \delta P$. This will show that $\delta Y = A\,Y$ is equivalent to $\delta Y = A_0 Y$. Comparing powers of t, one sees that

$$A_0 P_i - P_i(A_0 + iI) = -(A_i + A_{i-1}P_1 + \cdots + A_1 P_{i-1}) .$$

Our assumption on the eigenvalues of A_0 implies that we can solve these equations recursively yielding the desired P. □

The above proposition combined with the Jordan form of the matrix A_0 proves part 2 and part 3 of Theorem 3.1 for the special case where the differential equation is regular singular. We will give another proof using a form of Hensel's Lemma for regular singular differential modules. This prepares the way for the irregular singular case.

Exercise 3.13 *Solutions of regular singular equations.*
The following result, in a somewhat different form, is attributed to Frobenius. Let E be a differential extension of $\widehat{K}$ containing a solution of $\delta y = 1$ and such that for any $c \in \mathbf{C}^*$, E contains a nonzero solution of $\delta y = cy$. This solution will be denoted by z^c. Show that any regular singular differential equation $\delta Y = AY,\ A \in \mathrm{M}_n(\widehat{K})$ has a nonzero solution of the form $z^a \phi$ where $\phi \in \widehat{K}^n$ and a fundamental matrix with entries in E. Hint: Use Proposition 3.12 and Jordan forms. *A converse of this Exercise 3.13 is given in Exercise 3.29.* □

For differential operators one can also define the property "regular singular".

Definition 3.14 A differential operator $L = \delta^d + a_1\delta^{d-1} + \cdots + a_{d-1}\delta + a_d \in \widehat{K}[\delta]$ is said to be a *regular singular operator* if all $v(a_i) \geq 0$.

Exercise 3.15 *Factors of regular singular operators.*
In this exercise we indicate the proof of a Gauss Lemma for operators in $\widehat{K}[\delta]$. This result is, in fact, a special case of Lemma 3.45. As before $a' := z\frac{da}{dz}$ and $a^{(i+1)} := z\frac{da^{(i)}}{dz}$ for $i \geq 0$.

(1) Prove for $a \in \widehat{K}$ the formula

$$\delta^s a = a\delta^s + \binom{s}{1} a^{(1)}\delta^{s-1} + \binom{s}{2} a^{(2)}\delta^{s-2} + \cdots + \binom{s}{s} a^{(s)}.$$

(2) Let L_1, L_2 be monic differential operators such that $L_1 L_2$ is regular singular (i.e., has its coefficients in O). Show that L_1 and L_2 are both regular singular. Hint: Choose non-negative powers a_m, b_n of z such that all coefficients of $a_m L_1 = \sum_{i=0}^{m} a_i \delta^i$ and of $L_2 b_n = \sum_{i=0}^{n} b_i \delta^i$ are in O and, moreover, $(a_m, a_{m-1}, \ldots, a_0) = O$ and $(b_n, b_{n-1}, \ldots, b_0) = O$. Write $a_m L_1 L_2 b_n = \sum_{k=0}^{m+n} c_k \delta^k$. Use (1) to show that all $c_i \in O$. Prove that $(c_{m+n}, \ldots, c_0) = O$ by reducing the coefficients modulo the maximal ideal (z) of O. Conclude that $a_m = b_n = 1$.

(3) Verify that (2) remains valid if the field $\widehat{K}$ is replaced by the field $\mathbf{C}(\{z\})$ of convergent Laurent series. □

Proposition 3.16 *Let M be a differential module of dimension d over $\widehat{K}$ with cyclic vector e. Let L be the monic polynomial of minimal degree with $Le = 0$. Then M is regular singular if and only if L is regular singular. The same statement holds with $\widehat{K}$ replaced by the field $\mathbf{C}(\{z\})$ of convergent Laurent series.*

Proof. Suppose that L is regular singular, then $e, \delta(e), \cdots, \delta^{d-1}(e)$ is a basis of M. The lattice $N := Oe + O\delta(e) + \cdots + O\delta^{d-1}(e)$ is invariant under δ. Indeed, $\delta^d e \in N$ since the coefficients of the monic L are in O. Thus M is regular singular.

Assume that M is regular singular and let N be a δ-invariant lattice. For any $f \in \widehat{K}^*$, the lattice fN is also δ-invariant. Therefore we may assume that $e \in N$. Consider the O-submodule N' of N generated by all $\delta^m e$. Since O is noetherian, N' is finitely generated and thus a δ-invariant lattice. From Exercise 3.8 there are indices $i_1 < i_2 < \cdots < i_d$ such that $\delta^{i_1} e, \ldots, \delta^{i_d} e$ is a free basis of N' over O. Then $\delta\delta^{i_d} e$ is an O-linear combination of $\delta^{i_1} e, \ldots, \delta^{i_d} e$. In other words, there is a monic differential operator $\tilde{L}$ with coefficients in O such that $\tilde{L}e = 0$. The operator L is a monic right-hand factor of $\tilde{L}$. From Exercise 3.15, L is a regular singular operator.

For the last part of the proposition, we have to define regular singular for an operator L and for a differential module M over $\mathbf{C}(\{z\})$. The obvious definitions are $L = \delta^n + a_{n-1}\delta^{n-1} + \cdots + a_0$ with all $a_i \in \mathbf{C}\{z\}$ and M has a $\mathbf{C}\{z\}$-lattice that is invariant under δ. □

We now return to regular singular modules and prove the following proposition.

Proposition 3.17 Hensel's Lemma for regular singular modules
Let N denote a δ-invariant lattice of the left $\widehat{K}[\delta]$ module M of finite dimension over $\widehat{K}$. Let a direct sum decomposition of $\overline{N}$ into $\overline{\delta}$-invariant subspaces F_1, F_2 be given such that for any eigenvalue c of $\overline{\delta}$ on F_1 and any eigenvalue d of $\overline{\delta}$ on F_2 one has $c - d \notin \mathbf{Z}$. Then there exists a unique decomposition $N = N_1 \oplus N_2$ of N into δ-invariant O-modules such that $\overline{N}_i = F_i$ for $i = 1, 2$. In particular, M admits a direct sum decomposition as a left $\widehat{K}[\delta]$-module.

Proof. For each n we shall construct **C**-subspaces $F_1(n)$, $F_2(n)$ of $N/\pi^{n+1}N$ such that

1. $N/\pi^{n+1}N = F_1(n) \oplus F_2(n)$,
2. The $F_i(n)$ are invariant under δ and multiplication by π,
3. The map $N/\pi^{n+1}N \to N/\pi^n N$ maps $F_i(n)$ onto $F_i(n-1)$.

Once we have shown this, the spaces N_i constructed by taking the limits of the $F_i(n)$ give the desired direct sum decomposition of N.

Let S_1 and S_2 be the set of eigenvalues of $\bar{\delta}$ acting on F_1 and F_2, respectively. Since $\pi^{n+1}N$ is invariant under δ, the map δ induces a **C**-linear map on $N/\pi^{n+1}N$. We will again denote this map by δ. We will first show that the eigenvalues of δ on $N/\pi^{n+1}N$ lie in $(S_1 + \mathbf{Z}) \cup (S_2 + \mathbf{Z})$. Since each $V(n) = \pi^n N/\pi^{n+1}N$ is invariant under the action of δ, it is enough to show this claim for each $V(n)$. If $\pi^n v$, $v \in V(0)$ is an eigenvalue of δ, then

$$\delta(\pi^n v) = n\pi^n v + \pi^n \delta(v) = c\pi^n v$$

for some $c \in \mathbf{C}$. Therefore, $c \in (S_1 + \mathbf{Z}) \cup (S_2 + \mathbf{Z})$. We therefore define $F_1(n)$ to be the sum of the generalized eigenspaces of δ corresponding to eigenvalues in $S_1 + \mathbf{Z}$ and $F_2(n)$ to be the sum of the generalized eigenspaces of δ corresponding to eigenvalues in $S_2 + \mathbf{Z}$. By the assumptions of the lemma and what we have just shown, $N/\pi^{n+1}N = F_1(n) \oplus F_2(n)$. Items 2 and 3 above are easily checked.

The uniqueness follows from the fact that the image of each N_i in $\pi^n N/\pi^{n+1}N$ is the image of F_i under the map $F_i \to \pi^n F_i$. □

We are now in a position to prove part 3 of Theorem 3.1 under the additional assumption that the module M is regular singular. Lemma 3.11 and Proposition 3.17 imply that M can be decomposed as a direct sum of modules M_i such that M_i admits a δ-invariant lattice N_i such that $\bar{\delta}$ has only one eigenvalue c_i on $\overline{N}_i$. The next step will be to decompose each M_i into indecomposable pieces.

From now on let M denote a regular singular module with a δ-invariant lattice such that $\bar{\delta}$ has only one eigenvalue c on $\overline{N}$. By changing δ into $\delta - c$ one may suppose that $c = 0$. Therefore, $\bar{\delta}$ is a nilpotent linear map on $\overline{N}$ and there is a "block decomposition" of $\overline{N}$ as a direct sum of $\bar{\delta}$- invariant subspaces $\overline{N}(i)$ with $i = 1, \ldots, a$ such that each $\overline{N}(i)$ has a basis $\{f_{i,1}, \ldots, f_{i,s_i}\}$ with

$$\bar{\delta} f_{i,1} = f_{i,2}, \cdots, \bar{\delta} f_{i,s_i-1} = f_{i,s_i}, \bar{\delta} f_{i,s_i} = 0.$$

One tries to lift this decomposition to N. Suppose that one has found elements $e_{i,j}$ such that $\bar{e}_{i,j} = f_{i,j}$ and such that $\delta(e_{i,j}) \equiv e_{i,j+1}$ modulo π^k for all i, j and where $e_{i,j} = 0$ for $j > s_i$. One then needs to determine elements $\tilde{e}_{i,j} = e_{i,j} + \pi^k a_{i,j}$ with

$a_{i,j} \in N$ such that the same congruences hold now modulo π^{k+1}. A calculation shows that the $a_{i,j}$ are determined by congruences of the form

$$(\delta + k)a_{i,j} = \frac{\delta(e_{i,j}) - e_{i,j+1}}{\pi^k} + a_{i,j+1} \text{ modulo } \pi.$$

Since $\delta + k$ is invertible modulo π when $k > 0$, these congruences can be recursively solved. Taking the limit of this sequence of liftings of $f_{i,j}$ one finds elements $E_{i,j}$ such that $\overline{E}_{i,j} = f_{i,j}$ with $\delta(E_{i,j}) = E_{i,j+1}$ for all i, j and where again $E_{i,j} = 0$ for $j > s_i$. We will leave the construction of the $a_{i,j}$ to the reader. This finishes the study of the regular singular case.

Remark 3.18 We will discuss the Galois group of a regular singular module in Sect. 3.2 and return to the study of regular singular equations in Chaps. 5 and 6. □

3.1.2 Irregular Singular Equations

We now turn to the general case. Let e denote a cyclic element of a left $\widehat{K}[\partial]$ module M of finite dimension and let the minimal equation of e be $Le = 0$, where

$$L = \delta^d + a_1\delta^{d-1} + \cdots + a_{d-1}\delta + a_d \in \widehat{K}[\partial].$$

We may assume that $\lambda := \min \{\frac{v(a_i)}{i} | \ 1 \leq i \leq d\}$ is negative since we have already dealt with the regular singular case. Now we imitate the method of Proposition 3.3 and write $\delta = z^{-\lambda}E$. The skew polynomial L is then transformed into a skew polynomial

$$P := E^d + b_1E^{d-1} + \cdots + b_{d-1}E + b_d,$$

with $\min v(b_i) = 0$ and so $P \in \mathbf{C}[[z^{1/m}]][E]$ where m is the denominator of λ. Consider the lattice $N = O_m e + O_m E(e) + \cdots + O_m E^{d-1}(e)$ in $\widehat{K}_m \otimes M$, where $O_m := \mathbf{C}[[z^{1/m}]]$. The lattice N is E-invariant. Let π denote $z^{1/m}$. Also πN is E-invariant and E induces a $\mathbf{C}$-linear map, called $\overline{E}$, on the d-dimensional vector space $\overline{N} = N/\pi N$. As in the regular singular case there is a lemma about lifting $\overline{E}$-invariant subspaces to E-invariant submodules of N. We will formulate this for the ground field $\widehat{K}$, although a similar statement holds over $\widehat{K}_n$.

Proposition 3.19 Hensel's Lemma for irregular singular modules
Let M denote a left $\widehat{K}[\partial]$-module of finite dimension; let $E = z^{\alpha}\delta$ with $\alpha \in \mathbf{Z}$ and $\alpha > 0$; let N denote an E-invariant lattice and let $\overline{N} := N/\pi N$ where $\pi = z$. Let $\overline{N} = F_1 \oplus F_2$ be a direct sum decomposition, where F_1, F_2 are $\overline{E}$-invariant subspaces such that $\overline{E}|_{F_1}$ and $\overline{E}|_{F_2}$ have no common eigenvalue. Then there are unique E-invariant O-submodules N_1, N_2 of N with $N = N_1 \oplus N_2$ and $\overline{N_i} = F_i$ for $i = 1, 2$.

Proof. The proof is similar to the proof of Proposition 3.17. Let S_1 and S_2 be the set of eigenvalues of $\overline{E}$ acting on F_1 and F_2, respectively. Since $\pi^n N$ is invariant under E, the map E induces a $\mathbf{C}$ linear map on $N/\pi^{n+1}N$. We will again denote this map by E. A calculation similar to that given in the proof of Proposition 3.17 shows that the eigenvalues of E on $N/\pi^{n+1}N$ are again $S_1 \cup S_2$. We therefore define $F_1(n)$ to be the sum of the generalized eigenspaces of E corresponding to eigenvalues in S_1 and $F_2(n)$ to be the sum of the generalized eigenspaces of E corresponding to eigenvalues in S_2. By the assumptions of the lemma and what we have just shown, $N/\pi^{n+1}N = F_1(n) \oplus F_2(n)$. Taking limits as before yields the N_i. □

We are now ready to prove Theorem 3.1 in its full generality. If we can apply Proposition 3.19 to get a decomposition of $\widehat{K}_m \otimes M$, then the proof can be finished using induction. If no decomposition occurs then the characteristic polynomial of $\overline{E}$ has the form $(T - c)^d$ for some $c \in \mathbf{C}$ and $m = 1$ follows as in the proof of Proposition 3.3. Now make the substitution $\delta = cz^\lambda + t^\mu E^{**}$ with a suitable choice for $\mu > \lambda$. If, for the operator, E^{**} still no decomposition occurs then μ is an integer and one continues. Either one will be able to apply Proposition 3.19 or one will generate a sequence of *integers* $\lambda_1 < \lambda_2 < \cdots$. These integers must eventually become positive, at which point the operator $D = \delta - \sum_{i=1}^r c_i z^{i/m}$ acts on $\widehat{K}_m \otimes M$ so that this module is regular singular. In this case we are in a situation that we have already studied. The process that we have described yields a decomposition of $\widehat{K}_m \otimes M$ as a direct sum $\oplus M_i$ such that for each i there is some $q_i \in z^{-1/m}\mathbf{C}[z^{-1/m}]$ with $\delta - q_i$ acting in a regular singular way on M_i. Our discussion of regular singular modules now proves part 3 of the theorem. After choosing a basis of each space W_i such that C_i has Jordan normal form one finds statement 2 of the theorem. Finally, for every M there exists an integer $m \geq 1$ such that $\widehat{K}_m \otimes M$ has a submodule of dimension 1. This proves part 1 of the theorem.

Remarks 3.20

1. Theorem 3.1 and its proof are valid for any differential field $k((z))$, where k is an algebraically closed field of chararteristic 0. Indeed, in the proof given above, we have used no more than the fact that $\mathbf{C}$ is algebraically closed and has characteristic 0.

2. Let k be any field of characteristic 0 and let $\overline{k}$ denotes its algebraic closure. The above proof of Theorem 3.1 can be applied to a differential module M over $k((z))$. In some steps of the proof a finite field extension of k is needed. It follows that Theorem 3.1 remains valid in this case with $\widehat{K}_m$ replaced by $k'((z^{1/m}))$ for a suitable finite field extension k' of k. Furthermore, the $q_, \ldots, q_s$ in part 3 are now elements in $z^{-1/m}k'[z^{-1/m}]$.

3. Concerning part 1 of the Theorem one can say that the module $\widehat{K}[\partial]/\widehat{K}[\partial]L$ has, after a finite field extension, at least one (and possibly many) 1-dimensional submodules. Hence there are elements u algebraic over $\widehat{K}$ such that L decomposes as $L = L_2(\partial - u)$. Any such u can be seen as $u = \frac{y'}{y}$, where y is a solution of $Ly = 0$. The element u itself satisfies a nonlinear equation of order $d - 1$. This

equation is called the *Riccati equation* of L and has the form

$$P_d + a_{d-1}P_{d-1} + \cdots + a_1P_1 + a_0P_0 = 0,$$

where the P_i are defined by induction as follows: $P_0 = 1$; $P_i = P'_{i-1} + uP_{i-1}$. One has $P_1 = u$, $P_2 = u' + u^2$, $P_3 = u'' + 3uu' + u^3$ etc.

4. The proof given above of Theorem 3.1 does not readily yield an efficient method for factoring an operator L over $\overline{\widehat{K}}$. In Sect. 3.3 we shall present a *second proof* that gives a more efficient method.

5. In parts 2 and 3 of Theorem 3.1 an extra condition is needed to assure that the given decomposition actually comes of something over $\widehat{K}$ and not of an equation or a module that can only be defined over some proper extension of $\widehat{K}$. Another point is to know some unicity of the decompositions. Let us already state that the $q_1, \ldots, q_s$ in 3 are unique. We see these elements as "eigenvalues" of the operator δ on M. We will return to those questions after the introduction, in the next section, of a universal Picard-Vessiot ring $\text{UnivR}_{\widehat{K}} \supset \widehat{K}$.

6. A left $\overline{\widehat{K}}[\delta]$ module M of finite dimension over $\overline{\widehat{K}}$ is called irreducible if M has no proper submodules. From the theorem one can deduce that any such irreducible M must have dimension 1 over $\overline{\widehat{K}}$ and so $M = \overline{\widehat{K}}e$ for some element e. Then $\delta(e) = Fe$ for some $F \in \overline{\widehat{K}}$. A change of e into ge with $g \in \overline{\widehat{K}}$ and $g \neq 0$ changes F into $f = F + \frac{g'}{g}$. Hence we can choose the basis of M such that $f \in \cup_{n\geq 1}\mathbf{C}[z^{-1/n}]$. Let us call $M(f)$ the module $\overline{\widehat{K}}e$ with $\delta(e) = fe$ and $f \in \cup_{n\geq 1}\mathbf{C}[z^{-1/n}]$. Then $M(f_1) \cong M(f_2)$ if and only $f_1 - f_2 \in \mathbf{Q}$.

7. Another statement that follows from the theorem is that every irreducible element of $\overline{\widehat{K}}[\delta]$ has degree 1.

8. Bouffet [46, 47] gives version of Hensel's Lemma for operators with coefficients in liouvillian extensions of $\mathbf{C}((z))$. □

Exercise 3.21 Let k be any field of characteristic 0 and let $\overline{k}$ denote its algebraic closure. Put $K = \overline{k}\otimes_k k((z))$ and $\widehat{K} = \overline{k}((z))$. Then K is in a natural way a differential subfield of $\widehat{K}$.
(a) Prove that $K = \widehat{K}$ if and only if $[\overline{k} : k] < \infty$.
(b) Suppose that $\overline{k}$ is an infinite extension of k. Prove that Theorem 3.1 remains valid for differential modules over K.
(c) Prove the following more precise formulation of part (b), namely:
The functor $M \mapsto \widehat{K} \otimes_K M$ from the category of the differential modules over K to the category of the differential modules over $\widehat{K}$ is an equivalence of (tannakian) categories. □

3.2 The Universal Picard-Vessiot Ring of $\widehat{K}$

The aim is to construct a differential extension $\mathrm{UnivR}_{\widehat{K}}$ of $\widehat{K}$, such that the differential *ring* $\mathrm{UnivR}_{\widehat{K}}$ has the following properties:

1. $\mathrm{UnivR}_{\widehat{K}}$ is a simple differential ring, i.e., the only differential ideals of $\mathrm{UnivR}_{\widehat{K}}$ are 0 and $\mathrm{UnivR}_{\widehat{K}}$.
2. Every matrix differential equation $y' = Ay$ over $\widehat{K}$ has a fundamental matrix $F \in \mathrm{GL}_n(\mathrm{UnivR}_{\widehat{K}})$.
3. $\mathrm{UnivR}_{\widehat{K}}$ is minimal in the sense that $\mathrm{UnivR}_{\widehat{K}}$ is generated over $\widehat{K}$ by all the entries of F and $\frac{1}{\det F}$ of the fundamental matrices F of all matrix differential equations $y' = Ay$ over $\widehat{K}$.

One can prove that for *any* differential field, with an algebraically closed field C of constants of characteristic 0, such a ring exists and is unique up to isomorphism (see Chap. 10). The ring $\mathrm{UnivR}_{\widehat{K}}$ can be constructed as the direct limit of all Picard-Vessiot rings of matrix differential equations. Moreover, $\mathrm{UnivR}_{\widehat{K}}$ is a domain and its field of fractions has again C as field of constants. The situation is rather similar to the existence and uniqueness of an algebraic closure of a field. Let us call UnivR the *universal Picard-Vessiot ring* of the differential field. The interesting feature is that $\mathrm{UnivR}_{\widehat{K}}$ can be constructed explicitly for the differential field $\widehat{K} = \mathbf{C}((z))$.

Intuitive idea for the construction of $\mathrm{UnivR}_{\widehat{K}}$

As before we will use the derivation $\delta = z\frac{\partial}{\partial z}$ and the notation y' shall refer to δy. Since $\mathrm{UnivR}_{\widehat{K}}$ must contain the entries of fundamental matrices for linear differential equations over $\widehat{K}$, $\mathrm{UnivR}_{\widehat{K}}$ must contain solutions to all equations of the form $y' = \frac{1}{m}y$ for $m \in \mathbf{Z}$. Any matrix differential equation (of size n) over the field $\widehat{K}(z^{1/m})$ can be rewritten as a matrix differential equation (of size nm) over $\widehat{K}$ (see Exercise 1.14.7). Thus every order one equation $y' = ay$ with a in the algebraic closure of $\widehat{K}$ must have a solution $y \in \mathrm{UnivR}^*_{\widehat{K}}$. Furthermore, $\mathrm{UnivR}_{\widehat{K}}$ must contain a solution of the equation $y' = 1$. From the formal classification (see Exercise 3.2), we conclude that no more is needed for the existence of a fundamental matrix for any matrix equation $y' = Ay$ over $\widehat{K}$ (and over the algebraic closure of $\widehat{K}$).

To insure that we construct $\mathrm{UnivR}_{\widehat{K}}$ correctly we will need to understand the relations among solutions of the various $y' = ay$. Therefore, we need to classify the order one equations $y' = ay$ over the algebraic closure $\overline{\widehat{K}}$ of $\widehat{K}$. Two equations $y' = ay$ and $y' = by$ are equivalent if and only if $b = a + \frac{f'}{f}$ for some $f \in \overline{\widehat{K}},\ f \neq 0$. The set $Log := \{\frac{f'}{f} |\ f \in \overline{\widehat{K}},\ f \neq 0\}$ is easily seen to consist of the elements of $\overline{\widehat{K}}$ of the form $c + \sum_{n>0} c_n z^{n/m}$, with $c \in \mathbf{Q}$, $c_n \in \mathbf{C}$ and $m \in \mathbf{Z}_{>0}$. The quotient group $\overline{\widehat{K}}/Log$ classifies the order one homogeneous equations over $\overline{\widehat{K}}$. One chooses a $\mathbf{Q}$-vector space $M \subset \mathbf{C}$ such that $M \oplus \mathbf{Q} = \mathbf{C}$. Put $\mathcal{Q} = \cup_{m\geq 1} z^{-1/m}\mathbf{C}[z^{-1/m}]$. Then $M \oplus \mathcal{Q} \subset \overline{\widehat{K}}$ maps bijectively to $\overline{\widehat{K}}/Log$, and classifies the order one homogeneous equations over $\overline{\widehat{K}}$. For each element in $\overline{\widehat{K}}/Log$, the ring $\mathrm{UnivR}_{\widehat{K}}$ must contain an invertible element that is the solution of the corresponding order one homogeneous equation. We separate the equations corresponding to M and to $\mathcal{Q}$. We note that this separation is immaterial for differential equations over $\widehat{K}$. In contrast, the separation is very important for the study of equations over the field of convergent Laurent series $\mathbf{C}(\{z\})$. The equations corresponding to M turn out to be regular singular. The elements in $\mathcal{Q}$ form the basis for the study of asymptotic properties of differential equations over $\mathbf{C}(\{z\})$.

The ring $\mathrm{UnivR}_{\widehat{K}}$ must then have the form $\overline{\widehat{K}}[\{z^a\}_{a\in M}, \{e(q)\}_{q\in\mathcal{Q}}, l]$, with the following rules:

1. The *only relations* between the symbols are $z^0 = 1,\ z^{a+b} = z^a z^b,\ e(0) = 1,$ $e(q_1 + q_2) = e(q_1)e(q_2)$.
2. The differentiation in $\mathrm{UnivR}_{\widehat{K}}$ is given by $(z^a)' = az^a$, $e(q)' = qe(q)$, $l' = 1$.

One may object to the $\mathbf{Q}$-vector space $M \subset \mathbf{C}$, since it is not constructive. Indeed, the following equivalent definition of $\mathrm{UnivR}_{\widehat{K}}$ is more natural. Let $\mathrm{UnivR}_{\widehat{K}} = \widehat{K}[\{z^a\}_{a\in\mathbf{C}}, \{e(q)\}_{q\in\mathcal{Q}}, l]$, with the following rules:

1. The *only relations* between the symbols are $z^{a+b} = z^a z^b$, $z^a = z^a \in \widehat{K}$ for $a \in \mathbf{Z}$, $e(q_1 + q_2) = e(q_1)e(q_2)$, $e(0) = 1$.
2. The differentiation in $\mathrm{UnivR}_{\widehat{K}}$ is given by $(z^a)' = az^a$, $e(q)' = qe(q)$, $l' = 1$.

We prefer the first description since it involves fewer relations. The *intuitive interpretation* of the symbols is:

1. z^a is the function $e^{a\log(z)}$,
2. l is the function $\log(z)$, and
3. $e(q)$ is the function $\exp(\int q\frac{dz}{z})$.

In a sector S at $z = 0$, $S \neq \mathbf{S}^1$, this interpretation makes sense.

Formal construction of the universal Picard-Vessiot ring

Define the ring $\mathcal{R} = \overline{\widehat{K}}[\{Z^a\}_{a\in M}, \{E(q)\}_{q\in\mathcal{Q}}, L]$ as the polynomial ring over $\widehat{K}$ in the infinite collection of variables $\{Z^a\}_{a\in M} \cup \{E(q)\}_{q\in Q} \cup \{L\}$. Define the differentiation $'$ on $\mathcal{R}$ by: $'$ is $z\frac{d}{dz}$ on $\overline{\widehat{K}}$, $(Z^a)' = aZ^a$, $E(q)' = qE(q)$ and $L' = 1$. Let $I \subset \mathcal{R}$ denote the ideal generated by the elements

$$Z^0 - 1,\ Z^{a+b} - Z^a Z^b,\ E(0) - 1,\ E(q_1 + q_2) - E(q_1)E(q_2).$$

It is easily seen that I is a differential ideal and $I \neq \mathrm{UnivR}_{\widehat{K}}$. Put $\mathrm{UnivR}_{\widehat{K}} := \mathcal{R}/I$. Then $\mathrm{UnivR}_{\widehat{K}}$ coincides with the intuitive description that we made above. By construction, $\mathrm{UnivR}_{\widehat{K}}$ has the properties 2 and 3 defining a universal Picard-Vessiot ring. We want to prove that $\mathrm{UnivR}_{\widehat{K}}$ also satisfies property 1 and has some more pleasant features:

Proposition 3.22 Properties of $\mathrm{UnivR}_{\widehat{K}}$.

1. $\mathrm{UnivR}_{\widehat{K}}$ *has no differential ideals, different from* 0 *and* $\mathrm{UnivR}_{\widehat{K}}$.
2. $\mathrm{UnivR}_{\widehat{K}}$ *is a domain.*
3. *The field of fractions* $\mathrm{UnivF}_{\widehat{K}}$ *of* $\mathrm{UnivR}_{\widehat{K}}$ *has* $\mathbf{C}$ *as field of constants.*

Proof. Consider elements $m_1, \ldots, m_s \in M$ and $q_1, \ldots, q_t \in \mathcal{Q}$, linearly independent over $\mathbf{Q}$. Consider the differential subring

$$\tilde{R} := \overline{\widehat{K}}[z^{m_1}, z^{-m_1}, \ldots, z^{m_s}, z^{-m_s}, e(q_1), e(-q_1), \ldots, e(q_t), e(-q_t), l]$$

of $\mathrm{UnivR}_{\widehat{K}}$. The ring $\mathrm{UnivR}_{\widehat{K}}$ is the union of differential subrings of the type $\tilde{R}$. It suffices to prove that $\tilde{R}$ has only trivial differential ideals, that $\tilde{R}$ is a domain and that the field of constants of the field of fractions of $\tilde{R}$ is $\mathbf{C}$. One observes that $\tilde{R}$ is the localization of the "free" polynomial ring $\overline{\widehat{K}}[z^{m_1}, \ldots, z^{m_s}, e(q_1), \ldots, e(q_t), l]$ with respect to the element $z^{m_1} \cdot z^{m_2} \cdots z^{m_s} \cdot e(q_1) \cdot e(q_2) \cdots e(q_t)$. Thus $\tilde{R}$ has no zero divisors. Let $J \neq (0)$ be a differential ideal in $\tilde{R}$. We have to show that $J = R$.

This is a combinatorial exercise. Let (only for this proof) a "monomial m" be a term $z^a e(q)$ with $a \in \mathbf{Z}m_1 + \cdots + \mathbf{Z}m_s$ and $q \in \mathbf{Z}q_1 + \cdots + \mathbf{Z}q_t$. Let $\mathcal{M}$ be the set of all monomials. We note that $m' = \alpha(m)m$ holds with $\alpha(m) \in \overline{\widehat{K}}^*$. Any $f \in \tilde{R}$ can be written as $\sum_{m \in \mathcal{M}, n \geq 0} f_{m,n} m l^n$. The derivative of f is then $\sum (f'_{m,n} + \alpha(m) f_{m,n}) m l^n + \sum n f_{m,n} m l^{n-1}$. Let us first prove that a differential ideal $J_0 \neq (0)$ of the smaller ring

$$\tilde{R}_0 := \overline{\widehat{K}}[z^{m_1}, z^{-m_1}, \ldots, z^{m_s}, z^{-m_s}, e(q_1), e(-q_1), \ldots, e(q_t), e(-q_t)]$$

is necessarily equal to $\tilde{R}_0$.

Choose $f \in J_0,\ f \neq 0$ with $f = \sum_{i=1}^N f_i m(i)$ and $N \geq 1$ minimal. After multiplying f by an invertible element of the ring $\tilde{R}_0$, we may assume that $f_1 = 1$ and $m(1) = 1$. If N happens to be 1, then the proof ends. For $N > 1$, the derivative f' lies in J_0 and must be zero according to the minimality of N. Then $f_N \in \overline{\widehat{K}}^*$ satisfies $f'_N + \alpha(m(N)) f_N = 0$. Since f'_N / f_N has a rational constant term and no terms of negative degree, this is in contradiction with the construction of $M \oplus \mathcal{Q}$. Thus $\tilde{R}_0$ has only trivial differential ideals.

We continue with a differential ideal $J \subset \tilde{R},\ J \neq (0)$. Choose $n_0 \geq 0$ minimal such that J contains an expression that has degree n_0 with respect to the variable l. If $n_0 = 0$, then $J \cap \tilde{R}_0$ is a non zero differential ideal of $\tilde{R}_0$ and the proof ends. Suppose that $n_0 > 0$. Let $J_0 \subset \tilde{R}_0$ denote the set of coefficients of l^{n_0} of all elements in J that have degree $\leq n_0$ with respect to the variable l. Then J_0 is seen to be a differential ideal of $\tilde{R}_0$ and thus $J_0 = \tilde{R}_0$. Therefore J contains an element of the form $f = l^{n_0} + hl^{n_0-1} + \cdots$, with $h \in \tilde{R}_0$. The derivative f' must be zero, according to the minimality of n_0. Thus $n_0 + h' = 0$. Write $h = \sum_{m \in \mathcal{M}} h_m m$, with coefficients $h_m \in \overline{\widehat{K}}$. Then $n_0 + h' = 0$ implies that $n_0 + h_0' = 0$ for some $h_0 \in \overline{\widehat{K}}$. This is again a contradiction.

Consider the collection of equations

$$y_1' = m_1 y_1, \ldots, y_s' = m_s y_s, \;\; f_1' = q_1 f_1, \ldots, f_t' = q_t f_t, \;\; g'' = 0.$$

This can be seen as a matrix differential equation of size $s + t + 2$. We have, in fact, proven above that the ring $\tilde{R}$ is the Picard-Vessiot ring for this matrix equation over $\overline{\widehat{K}}$. It follows from the Picard-Vessiot theory that $\tilde{R}$ is a domain and that its field of fractions has $\mathbf{C}$ as a set of constants. □

Exercise 3.23 Modify the intuitive reasoning for the construction of $\mathrm{UnivR}_{\widehat{K}}$ to give a proof of the uniqueness of $\mathrm{UnivR}_{\widehat{K}}$. □

Remarks 3.24 1. A matrix differential equation $y' = Ay$ over $\widehat{K} = \mathbf{C}((z))$, or over its algebraic closure $\overline{\widehat{K}}$ will be called *canonical* if the matrix A is a direct sum of square blocks A_i and each block A_i has the form $q_i I + C_i$, where the q_i are distinct elements of $\mathcal{Q}$ and C_i is a constant matrix. One can refine this block decomposition by replacing each block $q_i I + C_i$ by blocks $q_i I + C_{i,j}$, where the constant matrices $C_{i,j}$ are the blocks of the usual Jordan decomposition of the C_i.
The matrices C_i and $C_{i,j}$ are not completely unique since one may translate the eigenvalues of C_i and $C_{i,j}$ over rational numbers. If one insists on using only eigenvalues in the $\mathbf{Q}$-vector space $M \subset \mathbf{C}$, then the matrices C_i and $C_{i,j}$ are unique up to conjugation by constant matrices.

2. Let $y' = Ay$ be a differential equation over $\widehat{K} = \mathbf{C}((z))$ or over its algebraic closure $\overline{\widehat{K}}$. Then there exists a $H \in \mathrm{GL}_n(\overline{\widehat{K}})$ that transforms this equation to the canonical form $y' = A^c y$. This means that that $A^c = H^{-1}AH - H^{-1}H'$. For the canonical equation $y' = A^c y$ one has a "symbolic" fundamental matrix, $\mathrm{fund}(A^c)$ with coefficients in $\mathrm{UnivR}_{\widehat{K}}$, which uses only the symbols $z^a, e(q), l$. The fundamental matrix for the original equation is then $H \cdot \mathrm{fund}(A^c)$. A fundamental matrix of a similar form appears in the work of Turrittin [288, 289] where the symbols are replaced by the multivalued functions $z^a, \exp(\int q \frac{dz}{z})$, $\log(z)$, and the fundamental matrix has the form $Hz^L e^Q$, where H is an invertible matrix with entries in $\overline{\widehat{K}}$, where L is a constant matrix (i.e., with coefficients in $\mathbf{C}$), where z^L means $e^{\log(z)L}$, where Q is a diagonal matrix with entries in $\mathcal{Q}$ and such that the matrices L and Q commute.

We note that Turrittin's formulation is a priori somewhat vague. One problem is that a product $f\exp(\int q \frac{dz}{z})$, with $f \in \overline{\widehat{K}}$ and $q \in \mathcal{Q}$ is not given a meaning. The multivalued functions may also present problems. The form of the fundamental matrix is not unique. Finally, one does not distinguish between canonical forms over

$\widehat{K}$ and over $\overline{\widehat{K}}$. The above presentation formalizes Turrittin's work and also allows us to classify differential equations over $\widehat{K}$ by giving a structure on the solution space of the equations. We shall do this in the next section. □

A structure on the solution space V

The field $\overline{\widehat{K}}$ has many $\widehat{K}$-automorphisms. One of them is γ given by the formula $\gamma(z^\lambda) = e^{2\pi i\lambda} z^\lambda$ for all rational numbers λ (and extended to Laurent series in the obvious way). This γ and its further action on various spaces and rings is called *the formal monodromy*. One can show that the Galois group of $\overline{\widehat{K}}$ over $\widehat{K}$ is equal to $\widehat{\mathbf{Z}}$, the inverse limit of the family $\{\mathbf{Z}/m\mathbf{Z}\}$ ([170], Chap. VIII, §11, Example 20), and that γ is a topological generator of this compact group. The latter statement follows from the easily verified fact that the set of γ-invariant elements of $\overline{\widehat{K}}$ is precisely $\widehat{K}$.

The γ as defined above also acts on $\mathcal{Q}$, seen as a subset of $\overline{\widehat{K}}$. We define the formal monodromy γ of the universal Picard-Vessiot ring $\text{UnivR}_{\widehat{K}}$ by:

1. γ acts on $\overline{\widehat{K}}$ as explained above.
2. $\gamma z^a = e^{2\pi i a} z^a$ for $a \in \mathbf{C}$.
3. $\gamma e(q) = e(\gamma q)$ for $q \in \mathcal{Q}$.
4. $\gamma l = l + 2\pi i$.

It is not difficult to see that γ is a well-defined differential automorphism of $\text{UnivR}_{\widehat{K}}$ (and also of its field of fractions $\text{UnivF}_{\widehat{K}}$). We introduce still other differential automorphisms of $\text{UnivR}_{\widehat{K}}$ over $\widehat{K}$. Let $\text{Hom}(\mathcal{Q}, \mathbf{C}^*)$ denote the group of the homomorphisms of $\mathcal{Q}$ to the (multiplicative) group $\mathbf{C}^*$. In other words, $\text{Hom}(\mathcal{Q}, \mathbf{C}^*)$ is the group of the characters of $\mathcal{Q}$. Let an element h in this group be given. Then one defines a differential automorphism σ_h of $\text{UnivR}_{\widehat{K}}$ by

$$\sigma_h(l) = l,\ \sigma_h(z^a) = z^a,\ \sigma_h e(q) = h(q)e(q) \text{ for } a \in \mathbf{C},\ q \in \mathcal{Q}.$$

The group of all σ_h introduced by Ramis [202, 203], is called *the exponential torus* and we will denote this group by $\mathcal{T}$. It is a large commutative group. γ does not commute with the elements of $\mathcal{T}$. Indeed, one has the following relation: $\gamma\sigma_{h'} = \sigma_h\gamma$, where h' is defined by $h'(q) = h(\gamma q)$ for all $q \in \mathcal{Q}$.

Proposition 3.25 *Let* $\text{UnivF}_{\widehat{K}}$ *denote the field of fractions of* $\text{UnivR}_{\widehat{K}}$. *Suppose that* $f \in \text{UnivF}_{\widehat{K}}$ *is invariant under* γ *and* $\mathcal{T}$. *Then* $f \in \widehat{K}$.

Proof. The element f belongs to the field of fractions of a free polynomial subring $P := \overline{\widehat{K}}[z^{m_1}, \ldots, z^{m_s}, e(q_1), \ldots, e(q_t), l]$ of $\text{UnivR}_{\widehat{K}}$, where the $m_1, \ldots, m_s \in M$ and the $q_1, \ldots, q_t \in \mathcal{Q}$ are linearly independent over $\mathbf{Q}$. Write $f = \frac{f_1}{f_2}$ with $f_1, f_2 \in P$ and with g.c.d. 1. One can normalize f_2 such that it contains a term $(z^{m_1})^{n_1} \cdots (z^{m_s})^{n_s} \cdot e(q_1)^{b_1} \cdots e(q_t)^{b_t} l^n$ with coefficient 1. For $h \in \text{Hom}(\mathcal{Q}, \mathbf{C}^*)$ one has $\sigma_h(f_1) = c(h)f_1$ and $\sigma_h(f_2) = c(h)f_2$, with a priori $c(h) \in \overline{\widehat{K}}^*$. Due to the normalization of f_2, we have that $c(h) = h(b_1q_1 + \cdots + b_tq_t)$. One concludes that f_1 and f_2 cannot contain the variables $e(q_1), \ldots, e(q_t)$. Thus f lies in the field of fractions of $\overline{\widehat{K}}[z^{m_1}, \ldots, z^{m_s}, l]$. Applying γ to $f = \frac{f_1}{f_2}$ we find at once that l is not present in f_1 and f_2. A similar reasoning as above shows that in fact $f \in \widehat{K}$. □

We consider a differential equation over $\widehat{K}$ and want to associate with it a solution space with additional structure. For convenience, we suppose that this differential equation is given as a scalar equation $Ly = 0$ of order d over $\widehat{K}$. The set of all solutions $V(L)$ in the universal Picard-Vessiot ring $\mathrm{UnivR}_{\widehat{K}}$ is a vector space over $\mathbf{C}$ of dimension d. The ring $\mathrm{UnivR}_{\widehat{K}}$ has a decomposition as $\mathrm{UnivR}_{\widehat{K}} = \oplus_{q\in\mathcal{Q}} R_q$, where $R_q := \overline{\widehat{K}}[\{z^a\}, l]e(q)$. Put $V(L)_q := V(L) \cap R_q$. Since the action of L on $\mathrm{UnivR}_{\widehat{K}}$ leaves each R_q invariant, one has $V(L) = \oplus_{q\in\mathcal{Q}} V(L)_q$. This is a direct sum of vector spaces over $\mathbf{C}$, and of course $V(L)_q$ can only be nonzero for finitely many elements $q \in \mathcal{Q}$. The formal monodromy γ acts on $\mathrm{UnivR}_{\widehat{K}}$ and leaves $V(L)$ invariant. Thus we find an induced action γ_L on $V(L)$. From $\gamma(e(q)) = e(\gamma q)$ it follows that $\gamma_L V(L)_q = V(L)_{\gamma q}$.

Definition 3.26 An element $q \in \mathcal{Q}$ is called an *eigenvalue of L* if $V(L)_q \neq 0$.

Exercise 3.27 *Eigenvalues* I
Let L_1 and L_2 be equivalent operators with coefficients in $\overline{\widehat{K}}$. Show that the eigenvalues of L_1 and L_2 are the same. □

The previous exercise implies that we can make the following definition.

Definition 3.28 The *eigenvalues of a differential equation or module* are the eigenvalues of any linear operator associated with these objects. □

Exercise 3.29 *Eigenvalues* II
Let M be a differential module over $\widehat{K}$. Show that the eigenvalues of M are all 0 if and only if the module is regular singular. Therefore, if a singular differential equation has a fundamental matrix with entries in $\overline{\widehat{K}}[\{z^a\}, l]$, then it is regular singular. This gives a converse to Exercise 3.13. □

We now introduce a category Gr_1, whose objects are the triples $(V, \{V_q\}, \gamma_V)$ satisfying:

1. V is a finite dimensional vector space over $\mathbf{C}$.
2. $\{V_q\}_{q\in\mathcal{Q}}$ is a family of subspaces such that $V = \oplus V_q$.
3. γ_V is a $\mathbf{C}$-linear automorphism of V such that $\gamma_V(V_q) = V_{\gamma q}$ for all $q \in \mathcal{Q}$.

A morphism $f : (V, \{V_q\}, \gamma_V) \to (W, \{W_q\}, \gamma_W)$ is a $\mathbf{C}$-linear map $f : V \to W$ such that $f(V_q) \subset W_q$ (for all q) and $\gamma_W f = \gamma_V f$. One can define tensor products and duals (and, more generally, all constructions of linear algebra) for the objects in the category Gr_1.

The above construction associates to a scalar equation L over $\widehat{K}$ an object of this category Gr_1. We will now do this more generally. Let N be a differential module over $\widehat{K}$ of dimension n. Then one considers the tensor product $\mathrm{UnivR}_{\widehat{K}} \otimes_{\widehat{K}} N$ and defines $V(N) := \ker(\partial, \mathrm{UnivR}_{\widehat{K}} \otimes_{\widehat{K}} N)$. This is a vector space of dimension n over C, again seen as the covariant solution space for the differential module. Letting $V(N)_q := \ker(\partial, R_q \otimes_{\widehat{K}} N)$, we then again have $V(N) = \oplus V(N)_q$. The action of γ on $\mathrm{UnivR}_{\widehat{K}}$ induces an action γ_N on $V(N)$ and the formula $\gamma_N V(N)_q = V(N)_{\gamma q}$ holds. This construction leads to the following statement.

Proposition 3.30 *The category of the differential modules* $\mathrm{Diff}_{\widehat{K}}$ *over* $\widehat{K}$ *is equivalent with the category* Gr_1. *The equivalence acts* $\mathbf{C}$*-linearly on* Hom*'s and commutes with all constructions of linear algebra, in particular with tensor products.*

Proof. Let Trip denote the functor from the first category to the second. It is clear that Trip commutes with tensor products etc. The two things that one has to prove are:

1. Every object $(V, \{V_q\}, \gamma_V)$ of Gr_1 is isomorphic to $\mathrm{Trip}(N)$ for some differential module over $\widehat{K}$.
2. The $\mathbf{C}$-linear map $\mathrm{Hom}(N_1, N_2) \to \mathrm{Hom}(\mathrm{Trip}(N_1), \mathrm{Trip}(N_2))$ is an isomorphism.

Proof of 1. On $W := \mathrm{UnivR}_{\widehat{K}} \otimes_{\mathbf{C}} V$ one considers the natural additive maps ∂, γ and σ_h for $h \in \mathrm{Hom}(\mathcal{Q}, \mathbf{C}^*)$ defined by the following formulas (where $r \in \mathrm{UnivR}_{\widehat{K}}$ and $v \in V_q$):
$\partial(r \otimes v) = r' \otimes v$,
$\gamma(r \otimes v) = (\gamma(r)) \otimes (\gamma_V(v))$, and
$\sigma_h(r \otimes v) = (\sigma_h(r)) \otimes (h(q)v)$.

Let N be the set of elements of W that are invariant under γ and all σ_h. Then N is clearly a vector space over $\widehat{K}$. The map ∂ on $\mathrm{UnivR}_{\widehat{K}}$ commutes with γ and all σ_h, and induces therefore a map $\partial : N \to N$ having the usual properties. In order to prove that N is a differential module over $\widehat{K}$ it suffices to verify that its dimension is finite. Let $q_1, \ldots, q_r$ denote the elements such that $V_{q_i} \neq 0$. Then the invariant of W under the group of all σ_h is equal to $W_1 := \oplus_{i=1}^{r} R_0 e(-q_i) \otimes V_{q_i}$. Furthermore, N is the set on elements of W_1 invariant under γ. Let $m \geq 1$ be minimal such that all $q_i \in z^{-1/m}\mathbf{C}[z^{-1/m}]$. Consider $W_1^{\gamma^m}$, the set of invariants of W_1 under γ^m. It suffices to prove that this is a finite dimensional vector space over $\widehat{K}_m$. Each term $R_0 e(-q_i) \otimes V_{q_i}$ is set-wise invariant under γ^m. Thus we may restrict our attention to only one such term. Furthermore, we may assume that the action of γ^m on V_{q_i} has only one Jordan block, say with eigenvalue λ and with length s. One observes that the γ^m-invariant elements of $R_0 e(-q_i) \otimes V_{q_i}$ lie in $\widehat{K}_m[l]_s z^b e(-q_i) \otimes V_{q_i}$, where b

is chosen such that $e^{-2m\pi ib} = \lambda$ and where $\widehat{K}_m[l]_s$ denotes the set of polynomials of degree $< s$. This proves that the space of invariants has finite dimension over $\widehat{K}_m$. Thus N is a differential module over $\widehat{K}$.

The verification that the natural map $\mathrm{UnivR}_{\widehat{K}} \otimes N \to W = \mathrm{UnivR}_{\widehat{K}} \otimes V$ is a bijection is straightforward. It follows that Trip(N) is isomorphic to the given object $(V, \{V_q\}, \gamma_V)$.

Proof of 2. One uses $\mathrm{Hom}(N_1, N_2) = \mathrm{Hom}(\mathbf{1}, N_1^* \otimes N_2)$, where $\mathbf{1}$ denotes the 1-dimensional trivial module $\widehat{K}e$ with $\partial e = 0$ and where $*$ stands for the dual. Then 2 reduces to proving that the map $\ker(\partial, N) \to \{v \in V |\ v \in V_0,\ \gamma_V(v) = v\}$, where $(V, \{V_q\}, \gamma_V) = \mathrm{Trip}(N)$, is a bijection. This easily follows from $\{r \in \mathrm{UnivR}_{\widehat{K}} \mid r \in R_0,\ \gamma(r) = r\} = \widehat{K}$. □

Remark 3.31 Consider a differential module N over $\widehat{K}$ with $\mathrm{Trip}(N) = (V, \{V_q\}, \gamma_V)$. The space $V := \ker(\partial, \mathrm{UnivR}_{\widehat{K}} \otimes N)$ is invariant under any element σ_h of the exponential torus $\mathcal{T}$. The action of σ_h on V is explicitly given by requiring that σ_h is multiplication by $h(q)$ on the subspaces V_q of V. The image of $\mathcal{T}$ in $\mathrm{GL}(V)$ is called the *exponential torus of N or of* Trip(N). It is actually an algebraic torus in $\mathrm{GL}(V)$. □

Corollary 3.32 *Let the differential module N define the triple* $(V, \{V_q\}, \gamma_V)$ *in* Gr_1. *Then the differential Galois group of N is, seen as an algebraic subgroup of* $\mathrm{GL}(V)$, *generated by the exponential torus and the formal monodromy. Furthermore, N is regular singular if and only if the exponential torus is trivial.*

Proof. The Picard-Vessiot field $L \supset \widehat{K}$ of N is the subfield of $\mathrm{UnivF}_{\widehat{K}}$ generated over $\widehat{K}$ by all the coordinates of a basis of $V \subset \mathrm{UnivR}_{\widehat{K}} \otimes_{\widehat{K}} N$ with respect to a basis of N over $\widehat{K}$. The exponential torus and the formal monodromy are seen as elements in $\mathrm{GL}(V)$. At the same time, they act as differential automorphisms of L and therefore belong to the differential Galois group of N. We have already proven that an element of $\mathrm{UnivF}_{\widehat{K}}$ that is invariant under the exponential torus and the formal monodromy belongs to $\widehat{K}$. The same holds then for the subfield $L \subset \mathrm{UnivF}_{\widehat{K}}$. By Picard-Vessiot theory, the differential Galois group is the smallest algebraic subgroup of $\mathrm{GL}(V)$ containing the exponential torus and the formal monodromy.

If N is a regular singular module, then the exponential torus $\mathcal{T}$ acts trivially on the solution space so the exponential torus of N is trivial. Conversely, if the exponential torus of N is trivial then 0 is the only eigenvalue of M. Exercise 3.29 implies that N is regular singular. □

Example 3.33 *The Airy equation* $y'' = zy$.

This equation has a singular point at ∞. One could write everything in the local variable $t = \frac{1}{z}$ at ∞. However, we prefer to keep the variable z. The solution space V at ∞ has a direct sum decomposition $V = V_{z^{3/2}} \oplus V_{-z^{3/2}}$ in spaces of dimension 1 (we shall show this in Sect. 3.3, Example 3.52.2). The formal monodromy γ interchanges the two spaces $V_{z^{3/2}}$ and $V_{-z^{3/2}}$. If v_1 generates $V_{z^{3/2}}$, $v_2 = \gamma(v_1)$ generates $V_{-z^{3/2}}$.

Since the Galois group of the equation is a subgroup of $\mathrm{SL}_2(\mathbf{C})$, the matrix of γ with respect to $\{v_1, v_2\}$ is $\begin{pmatrix} 0 & -1 \\ 1 & 0 \end{pmatrix}$. The exponential torus has the form $\{\begin{pmatrix} t & 0 \\ 0 & t^{-1} \end{pmatrix} | t \in \mathbf{C}^*\}$. The differential Galois group of the Airy equation over the field $\mathbf{C}((z^{-1}))$ is then the infinite dihedral group $D_\infty \subset \mathrm{SL}_2(\mathbf{C})$. □

Remark 3.34 *Irreducible differential modules over* $\widehat{K}$.
Consider a differential module N over $\widehat{K}$ and let (V, V_q, γ_V) be the corresponding triple. Then N is irreducible if and only if this triple is irreducible. It is not difficult to verify that the triple is irreducible if and only if the nonzero V_qs have dimension 1 and form one orbit under the action of γ_V. To see this, note that a γ_V-orbit of V_qs defines a subobject. Hence, there is only one γ_V-orbit, say of length m and consisting of $q_1, \ldots, q_m$. Take a 1-dimensional subspace W of V_{q_1}, invariant under γ_V^m. Then $W \oplus \gamma_V W \oplus \cdots \oplus \gamma_V^{m-1} W$ is again a subobject. Hence, the dimension of V_{q_1} and the other V_{q_i} is 1.
This translates into:
N is irreducible if and only if there exists an integer $m \geq 1$ such that $\widehat{K}_m \otimes N$ has a basis $e_1, \ldots, e_m$ with the properties:
(i) $\partial e_i = Q_i e_i$ for $i = 1, \ldots, m$ and all $Q_i \in \mathbf{C}[z^{-1/m}]$.
(ii) $\{Q_1, \ldots, Q_m\}$ is one orbit under the action of γ on $\mathbf{C}[z^{-1/m}]$.
From this explicit description of $\widehat{K}_m \otimes N$ one can obtain an explicit description of $N = (\widehat{K}_m \otimes N)^\gamma$, by computing the vector space of the γ-invariant elements.

Another way to make the module N explicit is to consider the map $N \to \widehat{K}_m \otimes N \xrightarrow{pr} \widehat{K}_m e_1$. The first arrow is the map $n \mapsto 1 \otimes n$ and the second arrow is the projection on the direct summand $\widehat{K}_m e_1$. The composite map $N \to \widehat{K}_m e_1$ is a non-zero morphism of differential modules over $\widehat{K}$. Since N is irreducible, this morphism is an isomorphism. In other words, an irreducible differential module of dimension m over $\widehat{K}$ has the form $\widehat{K}_m e$ with $\partial e = Qe$, where $Q \in \mathbf{C}[z^{-1/m}]$ has the property $\widehat{K}_m = \widehat{K}[Q]$. Furthermore, two elements $Q_1, Q_2 \in \mathbf{C}[z^{-1/m}]$, algebraic of degree m over $\widehat{K}$ define isomorphic irreducible differential modules over $\widehat{K}$ if and only if there is an integer i such that $\gamma^i(Q_1) - Q_2 \in \frac{1}{m}\mathbf{Z}$.

We illustrate the above with an example. Let N be irreducible of dimension two over $\widehat{K}$. Then $\widehat{K}_2 \otimes N = \widehat{K}_2 e_1 + \widehat{K}_2 e_2$ with, say, $\partial e_1 = (z^{-1/2} + z^{-1})e_1$ and $\partial e_2 = (-z^{-1/2} + z^{-1})e_2$. A basis for $(\widehat{K}_2 \otimes N)^\gamma$ is $f_1 := e_1 + e_2,\ f_2 := z^{-1/2}(e_1 - e_2)$. On this basis one can calculate the action of ∂, namely: $\partial f_1 = z^{-1} f_1 + f_2$ and $\partial f_2 = z^{-1} f_1 + (z^{-1} - 1/2) f_2$. The other possibility is to identify N with $\widehat{K}_2 e_1$. Then $f_1 := e_1,\ f_2 := z^{-1/2} e_1$ is a basis of N over $\widehat{K}$ and one can calculate that the action of ∂ on this basis is given by the same formulas.

We note that the sufficiency of the above irreducibility criterion also appears in [154], where it is stated in terms of the slopes of N (see the next section for this concept): N is irreducible if it has just one slope and that this is a rational number with exact denominator equal to the dimension of N. □

Exercise 3.35 *An observation on automorphisms made by M. van Hoeij.*
Let N be a differential module over $\widehat{K}$ such that its group of automorphisms is equal to $\mathbf{C}^*$. Prove that N is irreducible. Hint: Consider the triple $(V, \{V_q\}, \gamma_V)$ associated to N. An automorphism of the triple is a bijective linear $A : V \to V$ such that $A(V_q) = V_q$ for all q and $A\gamma_V = \gamma_V A$. By assumption this implies that A is a multiple of the identity. Prove first that the set $\{q|\ V_q \neq 0\}$ is one orbit under the action of γ. Then show that $V_q \neq 0$ implies that V_q has dimension 1 and compare with Remark 3.34. □

Exercise 3.36 *Semisimple differential modules over $\widehat{K}$.*
We recall, see Definition 2.37, that a differential module M is semisimple (or completely reducible) if every submodule of M is a direct summand. As before $\widehat{K} = \mathbf{C}((z))$. Let $\mathcal{C}$ denote the full subcategory of the category $\mathrm{Diff}_{\widehat{K}}$ of all differential modules over $\widehat{K}$, whose objects are the semisimple differential modules. Prove that $\mathcal{C}$ has the properties stated in Sect. 10.1. Show that the universal differential ring for $\mathcal{C}$ is equal to $\mathbf{C}((z))[\{z^a\}_{a\in\mathbf{C}}, \{e(q)\}_{q\in\mathcal{Q}}]$. (Note that l is missing in this differential ring.) □

Remarks 3.37 *Triples for differential modules over the field $k((z))$.*
(1) We consider first the case of an algebraically closed field k (of characteristic 0). As remarked before, the classification of differential modules over $k((z))$ is completely similar to the case $\mathbf{C}((z))$. The universal differential ring for the field $k((z))$ also has the same description, namely $k((z))[\{z^a\}_{a\in k}, l, \{e(q)\}_{q\in\mathcal{Q}}]$ with $\mathcal{Q} = \cup_{m\geq 1} z^{-1/m}k[z^{-1/m}]$. For the definition of the differential automorphism γ of this universal differential ring, one needs an isomorphism of groups, say, $\exp : k/\mathbf{Z} \to k^*$. For $k = \mathbf{C}$, we have used the natural isomorphism $\exp(c) = e^{2\pi i c}$. In the general case, an isomorphism exp exists. Indeed, the group $k/\mathbf{Z}$ is isomorphic to $\mathbf{Q}/\mathbf{Z} \oplus A$ where A is a vector space over $\mathbf{Q}$ of infinite dimension. The group k^* is isomorphic to $\mathbf{Q}/\mathbf{Z} \oplus B$ with B a vector space over $\mathbf{Q}$ of infinite dimension. The vector spaces A and B are isomorphic since they have the same cardinality. However, there is no *natural* candidate for exp. Nevertheless, this suffices to define the differential automorphism γ as before by:
(i) $\gamma(z^a) = \exp(a)z^a$ for all $a \in k$,
(ii) $\gamma(e(q)) = e(\gamma q)$, and
(iii) $\gamma(l) = l + 1$ (here 1 replaces the $2\pi i$ of the complex case).
With these changes, Proposition 3.30 and its proof remain valid.

(2) Consider now any field k of characteristic 0 and let $\overline{k}$ denote its algebraic closure. The classification of differential modules M over $k((z))$ in terms of "tuples" is rather involved. Let K denote the differential field $\overline{k}\otimes_k k((z))$ (compare with Exercise 3.21). For the differential field K there is an obvious description of the universal differential ring, namely again $R := K[\{z^a\}_{a\in\overline{k}}, l, \{e(q)\}_{q\in\mathcal{Q}}]$ where $\mathcal{Q} = \cup_{m\geq 1} z^{-1/m}\overline{k}[z^{-1/m}]$. On this ring there is an obvious action of the Galois group $\mathrm{Gal}(\overline{k}/k)$. One associates to M the solution space $V = \ker(\partial, R \otimes_{k((z))} M)$. This solution space has a direct sum decomposition $\oplus_{q\in\mathcal{Q}} V_q$, an action of γ (defined in (1)), called γ_V and an action of the Galois group $\mathrm{Gal}(\overline{k}/k)$, called ρ_V. Thus we can associate to M the tuple

$(V, \{V_q\}, \gamma_V, \rho_V)$. This tuple satisfies the compatibilities of the triple $(V, \{V_q\}, \gamma_V)$ and, moreover, staisfies a compatibility of ρ_V with respect to the $\{V_q\}$s and γ_V. One can show, as in Proposition 3.30, that the functor $M \mapsto (V, \{V_q\}, \gamma_V, \rho_V)$ is an equivalence between the (tannakian) categories of the differential modules over $k((z))$ and the one of tuples described above. This description is probably too complicated to be useful. □

Observations 3.38 *Irreducible differential modules over* $k((z))$.
The field k has characteristic 0 and is not necessarily algebraically closed. We present here the description of the irreducible differential modules over $k((z))$, with differentiation $\delta = z\frac{d}{dz}$, given by Sommeling [278]. The ideas and methods are an extension of Remark 3.34.

We will first describe the finite extensions of $k((z))$. Let $K \supset k((z))$ be a finite field extension. The field K is again complete w.r.t. a discrete valuation. The differentiation of $k((z))$ extends uniquely to K. We will either write $\delta(a)$ or a' for the derivative of an element $a \in K$. The minimal monic polynomial of any constant c of K has coefficients in k. Thus the field of constants k' of K is the algebraic closure of k in K. Since one works in characteristic zero this is also the unique coefficient field of K containing k. Thus $K = k'((u))$ for a suitable element u. The element z is equal to some expression $c^{-1}u^m(1 + c_1u + c_2u^2 + \cdots)$. The number $m \geq 1$ is called the ramification index. After replacing u by $t(1 + c_1u + c_2u^2 + \cdots)^{-1/m}$ one finds $K = k'((t))$ and $cz = t^m$ with $c \in k'$. We note that $\delta(t) = \frac{1}{m}t$. Furthermore, t is unique up to multiplication by a nonzero element in k' and c is unique up to the n-th power of this element in k'.

Consider the 1-dimensional differential module Ke given by $\partial e = Qe$. One normalizes Q such that $Q \in k'[t^{-1}]$ (this normalization does not depend on the choice of t). The thesis of R. Sommeling contains the following results:

(1) *Suppose that* $K = k((z))[Q]$, *then* Ke *with* $\partial e = Qe$, *considered as a differential module over* $k((z))$, *is irreducible.*

(2) *Two irreducible differential modules over* $k((z))$, *of the form considered in (1) and given by* Q_1 *and* Q_2, *are isomorphic if and only if there exists an* $k((z))$-*isomorphism* $\sigma : k((z))[Q_1] \to k((z))[Q_2]$ *with* $\sigma(Q_1) - Q_2 \in \frac{1}{m}\mathbf{Z}$, *where* $m \geq 1$ *is the ramification index of* $k((z))[Q_1]$.

(3) *Every irreducible differential over* $k((z))$ *is isomorphic to a differential module of the form considered in (1).*

Proof. (1) K is seen as a subfield of a fixed algebraic closure $\overline{k((z))}$ of $k((z))$. Put $M = Ke$ and take any nonzero element $v = fe \in M$. Then $\partial(v) = (\frac{f'}{f} + Q)v$. Let $L \in k((z))[\delta]$ denote the minimal monic operator with $Lv = 0$. Then L, seen as operator in $\overline{k((z))}[\delta]$ has right-hand factor $\delta - (\frac{f'}{f} + Q)$. For every automorphism σ of $\overline{k((z))}$ over $k((z))$, the operator $\delta - \sigma(\frac{f'}{f} + Q)$ is also a right-hand divisor of L. Let $\sigma_1, \ldots, \sigma_n$ denote the set of the $k((z))$-linear homomorphisms of K into $\overline{k((z))}$.

Then $n = [K : k((z))]$ and since Q is normalized, the operators $\delta - \sigma_i(\frac{f'}{f} + Q)$, $i = 1, \ldots, n$ are pair-wise inequivalent. The least common left multiple L_1 of these operators in $\overline{k((z))}[\delta]$ is, in fact, a monic operator in $k((z))[\delta]$, since it is invariant under the action of the Galois group of $\overline{k((z))}$ over $k((z))$. Clearly L is a left multiple of L_1 and by minimality one has $L = L_1$ and L has degree n. This shows that the differential module M over $k((z))$ has no proper submodules.
Furthermore, we note that the $\delta - \sigma_i(\frac{f'}{f} + Q)$ are the only possible monic right-hand factors of degree one of L in $\overline{k((z))}[\delta]$, since they are pair-wise inequivalent.

(2) Suppose that $\sigma(Q_1) - Q_2 \in \frac{1}{m}\mathbf{Z}$. Then $K = k((z))[Q_1] = k((z))[Q_2]$. Let $M = Ke$ with $\partial e = Q_1 e$, then for a suitable power f of t (i.e., the element defined in the above description of K) one has $\partial fe = Q_2 fe$. Thus the two differential modules over $k((z))$ are isomorphic.

On the other hand, suppose that the two differential modules M_1 and M_2 over $k((z))$, given by Q_1 and Q_2, are isomorphic. Then M_1 and M_2 contain nonzero elements v_1, v_2 such that the minimal monic operators $L_i \in k((z))[\delta]$ with $L_i v_i = 0$ are equal. In (1) we have seen that these operators are least common left multiples of conjugates of $\delta - (\frac{f_i'}{f_i} + Q_i)$ for $i = 1, 2$. The unicity of these sets of monic right-hand factors of degree one in $\overline{k((z))}[\delta]$, implies that there exists a σ with $\sigma(\frac{f_1'}{f_1} + Q_1) = \frac{f_2'}{f_2} + Q_2$. It follows that $k((z))[Q_1] = k((z))[Q_2]$. Let $m \geq 1$ denote the ramification of the latter field. Then $\frac{f_i'}{f_i}$ is modulo the maximal ideal of the ring of integers of K equal to some element in $\frac{1}{m}\mathbf{Z}$. Thus $\sigma(Q_1) - Q_2 \in \frac{1}{m}\mathbf{Z}$.

(3) Let M be an irreducible differential module over $k((z))$. One considers a field extension $K \supset k((z))$, lying in $\overline{k((z))}$, of minimal degree, such that $K \otimes_{k((z))} M$ contains a submodule Ke of dimension one. As above, one writes $K = k'((t))$ with $t^m = cz$. Furthermore, one normalizes e such that $\partial e = Qe$ with $Q \in k'[t^{-1}]$. By minimality, $K = k((z))[Q]$. Let $\sigma_1, \ldots, \sigma_n$ denote the $k((z))$-linear embeddings of K into $\overline{k((z))}$. This leads to a differential submodule $N := \oplus_{i=1}^n \overline{k((z))}\sigma_i(e)$ of $\overline{k((z))} \otimes_{k((z))} M$, with an action of the Galois group G of $\overline{k((z))}$ indicated by the notation and ∂ given by $\partial\sigma_i(e) = \sigma_i(Q)\sigma_i(e)$. Since N is stable under the action of G, one has that the space of invariants N^G is a nonzero $k((z))$-differential submodule of M. Since M is irreducible, one has that $M = N^G$. The latter translates into M is isomorphic as $k((z))$-differential module with Ke with $\partial e = Qe$. □

In R. Sommeling's thesis the above results are extended to a description of all semisimple differential modules over $k((z))$ by means of certain equivalence classes of monic polynomials over the field $\overline{k((z))}$. □

Split and quasi-split equations over $K_{conv} = \mathbf{C}(\{z\})$

We now turn to equations with meromorphic coefficients. We let K_{conv} be the field of convergent Laurent series in z and $K_{conv,m}$ be the field of convergent Laurent series in $z^{1/m}$.

Definition 3.39 A differential equation $y' = Ay$ over $\mathbf{C}(\{z\})$ will be called *split* if it is the direct sum of equations $y' = (q_i + C_i)y$ with $q_i \in z^{-1}\mathbf{C}[z^{-1}]$ and C_i constant matrices. The equation is called *quasi-split* if it is split over $\mathbf{C}(\{z^{1/m}\})$ for some $m \geq 1$. □

We translate the notions in terms of differential modules. A differential module M over the field K_{conv} of convergent Laurent series is *split* if M is a direct sum $\oplus_{i=1}^{s} E(q_i) \otimes N_i$, where $q_1, \ldots, q_s \in z^{-1}\mathbf{C}[z^{-1}]$, where $E(q)$ denotes the one-dimensional module $K_{conv}e_q$ over K_{conv} with $\partial e_q = qe_q$ and where the N_i are regular singular differential modules over K_{conv}. The differential module M over K_{conv} is called *quasi-split* if for some $m \geq 1$ the differential module $K_{conv,\, m} \otimes M$ is split over $K_{conv,\, m}$.

One has that the Picard-Vessiot extension of $\mathbf{C}(\{z\})$ corresponding to a quasi-split equation can be taken to lie in the subfield of $\mathrm{UnivF}_{\widehat{K}}$ generated over $\mathbf{C}(\{z\})$ by the elements l, $\{z^a\}_{a \in \mathbf{C}}$, $\{e(q)\}_{q \in \mathcal{Q}}$. The argument of Corollary 3.32 implies the following proposition.

Proposition 3.40 *The differential Galois groups of a quasi-split differential equation $y' = Ay$ over $\mathbf{C}(\{z\})$ and $\mathbf{C}((z))$ are the same. This group is the smallest linear algebraic group containing the exponential torus and the formal monodromy.*

For equations that are not quasi-split, the Galois group over $\mathbf{C}(\{z\})$ will, in general, be larger. We will give a complete description of the Galois group in Chap. 8. The starting point in this description is the following proposition.

Proposition 3.41 *Every differential equation $y' = Ay$ with coefficients in $\widehat{K}$ is, over the field $\widehat{K}$, equivalent with a unique (up to isomorphism over K_{conv}) quasi-split equation over K_{conv}. The translation of this statement in terms of differential modules over $\widehat{K}$ is:*

For every differential module M over $\widehat{K}$, there is a unique $N \subset M$, such that:

1. *N is a quasi-split differential module over the field K_{conv}.*
2. *The natural $\widehat{K}$-linear map $\widehat{K} \otimes_{K_{conv}} N \to M$ is an isomorphism.*

To prove this proposition, we need the following result that will allow us to strengthen the results of Proposition 3.12.

Lemma 3.42 *Let $A \in \mathrm{M}_n(K_{conv})$ and assume that the equation $Y' = AY$ is equivalent over $\widehat{K}$ to an equation with constant coefficients. Then $Y' = AY$ is equivalent over K_{conv} to an equation with constant coefficients.*

Proof. By assumption, there is a matrix $B \in \mathrm{GL}_n(\widehat{K})$ such that $B^{-1}AB - B^{-1}B'$ is a constant matrix. By truncating B after a suitably high power, we may assume that

A is equivalent (over K_{conv}) to a matrix in $\mathrm{M}_n(\mathbf{C}\{z\})$, and so, from the start assume that $A \in \mathrm{M}_n(\mathbf{C}\{z\})$. Following the argument of Lemma 3.11, we may assume that $A = A_0 + A_1 z + \cdots$, where the distinct eigenvalues of A_0 do not differ by integers. As in Proposition 3.12, we wish to construct a matrix $P = I + P_1 z + \cdots$, $P_i \in \mathrm{M}_n(\mathbf{C})$ such that the power series defining P is convergent in a neighborhood of the origin and $PA_0 = AP - P'$. Comparing powers of z, one sees that

$$A_0 P_i - P_i(A_0 + iI) = -(A_i + A_{i-1}P_1 + \cdots + A_1 P_{i-1}) .$$

Proposition 3.12 implies that these equations have a unique solution. Let L_{n+1} denote the linear map $X \mapsto A_0 X - X A_0 - (n+1)X$. Using the norm $\| (a_{i,j}) \| = \max |a_{i,j}|$, one sees that $\| L_{n+1}^{-1} \| = O(\frac{1}{n})$. Using this bound, one can show that the series defining P converges. □

Proof of Proposition 3.41. We give a proof using differential modules and return later to matrices. The first case that we study is that of a differential module M over $\widehat{K}$, which has only 0 as eigenvalue. In other words, M is *regular singular over* $\widehat{K}$. As we have seen before, M has a basis $e_1, \ldots, e_m$ over $\widehat{K}$ such that the matrix C of ∂, with respect to this basis, has coefficients in $\mathbf{C}$. Using the argument before Lemma 3.11, we may even assume that the (distinct) eigenvalues λ_i, $i = 1, \ldots, r$ (with multiplicities $k_1, \ldots, k_r$) of this constant matrix satisfy $0 \le Re(\lambda_i) < 1$. It is clear that $N := K_{conv}e_1 + \cdots + K_{conv}e_m$ has the properties 1. and 2. We now want to prove that N is unique.

A small calculation shows that the set of solutions $m \in M$ of the equation $(\delta - \lambda_i)^{k_i} m = 0$ is a $\mathbf{C}$-linear subspace W_i of $\mathbf{C}e_1 + \cdots + \mathbf{C}e_m$. Moreover, $\mathbf{C}e_1 + \cdots + \mathbf{C}e_m$ is the direct sum of the W_i. For a complex number μ such that $\mu - \lambda_i \notin \mathbf{Z}$ for all i, one calculates that the set of the $m \in M$ with $(\delta - \mu)^k m = 0$ (any $k \ge 1$) is just 0. Consider now another $\tilde{N} \subset M$ having the properties 1 and 2. Then $\tilde{N}$ is regular singular over K_{conv} and we know, from Lemma 3.42, that there is a basis $f_1, \ldots, f_m$ of $\tilde{N}$ over K_{conv}, such that the matrix D of ∂, with respect to this basis, is constant and all its eigenvalues μ satisfy $0 \le Re(\mu) < 1$. From the calculation above it follows that the eigenvalues of D are also eigenvalues for C (and also the converse). We conclude now that $\mathbf{C}f_1 + \cdots + \mathbf{C}f_m = \mathbf{C}e_1 + \cdots + \mathbf{C}e_m$. In particular, $N = \tilde{N}$.

The next case that we consider is a differential module M over $\widehat{K}$, such that all its eigenvalues belong to $z^{-1}\mathbf{C}[z^{-1}]$. Again we want to show the existence and the uniqueness of a $N \subset M$ with properties 1 and 2, such that N is *split*. M decomposes (uniquely) over $\widehat{K}$ as a direct sum of modules having only one eigenvalue. It is easily seen that it suffices to prove the proposition for the case of only one eigenvalue q. One considers the one-dimensional module $F(q) := \widehat{K} \otimes_{K_{conv}} E(q)$. Thus $F(q) = \widehat{K}e_q$ and $\partial e_q = qe_q$. The module $F(-q) \otimes_{\widehat{K}} M$ again has only one eigenvalue and this eigenvalue is 0. This is the regular singular case that we have treated above.

Finally, we take a general differential module M over $\widehat{K}$. Take $m \ge 1$ such that all its eigenvalues belong to $\widehat{K}_m = \widehat{K}[z^{1/m}]$. Then the module $\widehat{K}_m \otimes M$ has a unique subset $\tilde{N}$, which is a split differential module over $K_{conv,\,m}$ and such that the natural map $\widehat{K}_m \otimes_{K_{conv,\,m}} \tilde{N} \to \widehat{K}_m \otimes_{\widehat{K}} M$ is an isomorphism. Let σ be a generator of the Galois group of $\widehat{K}_m$ over $\widehat{K}$. Then σ acts on $\widehat{K}_m \otimes M$ by the formula $\sigma(f \otimes m) = \sigma(f) \otimes m$. Clearly $\sigma(\tilde{N})$ has the same property as $\tilde{N}$. The uniqueness implies that $\sigma(\tilde{N}) = \tilde{N}$. Thus σ acts on $\tilde{N}$. This action is semilinear, i.e., $\sigma(f\tilde{n}) = \sigma(f)\sigma(\tilde{n})$. Let N denote the set of the σ-invariant elements of $\tilde{N}$. Then it is easily seen that the natural maps $K_{conv,\,m} \otimes_{K_{conv}} N \to \tilde{N}$ and $\widehat{K} \otimes_{K_{conv}} N \to M$ are isomorphisms. Thus we have found an N with properties 1 and 2. The uniqueness of N follows from its construction.

We now return to the matrix formulation of the proposition. For a matrix equation $y' = Ay$ over $\widehat{K}$ (with module M over $\widehat{K}$), such that the eigenvalues are in $z^{-1}\mathbf{C}[z^{-1}]$, it is clear that the module N over K_{conv} has a matrix representation $y' = By$ that is a direct sum of equations $y' = (q_i + C_i)y$ with $q_i \in z^{-1}\mathbf{C}[z^{-1}]$ and constant matrices C_i. In the case that $y' = Ay$ has eigenvalues that are not in $z^{-1}\mathbf{C}[z^{-1}]$, one can again take a basis of the module N and consider the matrix equation $y' = By$ obtained in this way. □

Remarks 3.43 1. It is more difficult to give this matrix B, defined in the final paragraph of the above proof, explicitly. This problem is somewhat analogous to the formulation of the real Jordan decomposition of real matrices. We will give an example. Consider a two-dimensional equation $y' = Ay$ with eigenvalues q_1, q_2 that are not in $z^{-1}\mathbf{C}[z^{-1}]$. Then the eigenvalues are in $z^{-1/2}\mathbf{C}[z^{-1/2}]$ and they are conjugate. The module $\tilde{N}$ over $K_{conv,\,2}$, of the proof of the proposition, has a basis e_1, e_2 such that $\partial e_i = q_i e_i$. Let σ be a generator of the Galois group of $\widehat{K}_2$ over $\widehat{K}$. Then one easily sees that $\sigma e_1 = e_2$ and $\sigma e_2 = e_1$. The elements $f_1 = e_1 + e_2$ and $f_2 = z^{-1/2}(e_1 - e_2)$ form a basis of N over K_{conv} and the matrix of ∂ with respect to this basis is equal to $\begin{pmatrix} \lambda & z^{-1}\mu \\ \mu & \lambda - 1/2 \end{pmatrix}$, where $q_1 = \lambda + z^{-1/2}\mu$, $q_2 = \lambda - z^{-1/2}\mu,\ \lambda, \mu \in z^{-1}\mathbf{C}[z^{-1}]$.

The issue of finding B explicitly is also addressed in [179] where a version of Proposition 3.41 is also proven. Proposition 3.41 also appears in [17].

2. For the study of the asymptotic theory of differential equations, we will use Proposition 3.41 as follows. Let a matrix differential equation $y' = Ay$ over K_{conv} be given. Then there exists a quasi-split equation $y' = By$ over K_{conv} and an $\widehat{F} \in \mathrm{GL}_n(\mathbf{C}((z)))$ such that $\widehat{F}^{-1}A\widehat{F} - \widehat{F}^{-1}\widehat{F}' = B$. The equation $y' = By$ is unique up to equivalence over K_{conv}. For a fixed choice of B the formal transformation $\widehat{F}$ is almost unique. Any other choice for the formal transformation has the form $\widehat{F}C$ with $C \in \mathrm{GL}_n(\mathbf{C})$ such that $C^{-1}BC = B$. The asymptotic theory is concerned with lifting $\widehat{F}$ to an invertible meromorphic matrix F on certain sectors at $z = 0$, such that $F^{-1}AF - F^{-1}F' = B$ holds. The above matrix C is irrelevant for the asymptotic liftings F. □

3.3 Newton Polygons

In this section we present another approach to the classification of differential modules over a field that is complete w.r.t. a discrete valuation. Let k denote a field of characteristic 0 and let $\mathcal{D} := k((z))[\delta]$ denote the skew ring of differential operators over $k((z))$, where $\delta := z\partial_z$. Note that $\delta z = z\delta + z$. For a finite field extension $K \supset k((z))$ we also have the skew ring $K[\delta]$. For every $f \in K$ one has $\delta f - f\delta = f'$, where $f \mapsto f'$ is the unique extension of $z\frac{d}{dz}$ to K.

The Newton polygon $N(L)$ of an operator

$$L = \sum_{i=0}^{n} a_i\delta^i = \sum_{i,j} a_{i,j}z^j\delta^i \in k((z))[\delta] \text{ with } a_n \neq 0$$

is a convex subset of $\mathbf{R}^2$ that contains useful combinatorial information of L. The slopes $k_1 < \cdots < k_r$ of the line segments forming the boundary of the Newton polygon are important in many discussions concerning L and will be crucial when we discuss the notion of multisummation. In this section we will use Newton polygons for the formal decomposition of L, following the work of Malgrange [189] and Ramis [236]. We begin by recalling some facts concerning polyhedral subsets of $\mathbf{R}^2$, [97].

A subset of $\mathbf{R}^2$ that is the intersection of a finite number of closed half-planes is said to be a *polyhedral set*. We will only consider connected polyhedral sets with nonempty interior. The boundary of such a set is the union of a finite number of (possibly infinite) closed line segments called *edges*. The endpoints of the edges are called *vertices* or *extremal points*. The vertices and edges of such a set are collectively referred to as the *faces* of the set. Given two subsets N and M of $\mathbf{R}^2$ we define the (Minkowski) sum of these sets to be $M + N = \{m + n \mid m \in M,\ n \in N\}$. Any face of the sum of two polyhedral sets M and N is the sum of faces of M and N, respectively. In particular, any vertex of $M + N$ is the sum of vertices of M and N.

On $\mathbf{R}^2$ one defines a partial order, namely $(x_1, y_1) \geq (x_2, y_2)$ is defined as $y_1 \geq y_2$ and $x_1 \leq x_2$. We can now make the following definition.

Definition 3.44 The elements of $\mathcal{D} = k((z))[\delta]$ of the form $z^m\delta^n$ will be called monomials. The *Newton polygon* $N(L)$ of $L \neq 0$ is the convex hull of the set

$$\{(x, y) \in \mathbf{R}^2 \mid \text{there is a monomial } z^m\delta^n \text{ in } L \text{ with } (x, y) \geq (n, m)\}.$$

□

$N(L)$ has finitely many extremal points $\{(n_1, m_1), \ldots, (n_{r+1}, m_{r+1})\}$ with $0 \leq n_1 < n_2 < \cdots < n_{r+1} = n$. The positive slopes of L are $k_1 < \cdots < k_r$ with $k_i = \frac{m_{i+1}-m_i}{n_{i+1}-n_i}$. It is also useful to introduce the notation $k_{r+1} = \infty$. If $n_1 > 0$ then one adds a slope $k_0 = 0$ and in this case we put $n_0 = 0$. The interesting part of the boundary of $N(L)$ is the graph of the function $f : [0, n] \to \mathbf{R}$, given by

1. $f(n_0) = f(n_1) = m_1$.
2. $f(n_i) = m_i$ for all i.
3. f is (affine) linear on each segment $[n_i, n_{i+1}]$.

The slopes are the slopes of this graph. The *length* of the slope k_i is $n_{i+1} - n_i$. We reserve the term *special polygon* for a convex set that is the Newton polygon of some differential operator.

Let $b(L)$ or $b(N(L))$ denote the graph of f. The *boundary part* $B(L)$ of L is defined as $B(L) = \sum_{(n,m)\in b(L)} a_{n,m} z^m \delta^n$. Write $L = B(L) + R(L)$. We say that $L_1 > L_2$ if the points of $b(L_1)$ either lie in the interior of $N(L_2)$ or on the vertical ray $\{(n_{r+1}, y) \mid y > m_{r+1}\}$. Clearly $R(L) > B(L)$ and $R(L) > L$. We note that the product of two monomials $M_1 := z^{m_1}\delta^{n_1}$, $M_2 := z^{m_2}\delta^{n_2}$ is not a monomial. In fact, the product is $z^{m_1+m_2}(\delta + m_2)^{n_1}\delta^{n_2}$. However, $B(M_1M_2) = z^{m_1+m_2}\delta^{n_1+n_2}$. This is essential for the following result.

Lemma 3.45 *1. $N(L_1L_2) = N(L_1) + N(L_2)$.*

2. The set of slopes of L_1L_2 is the union of the sets of slopes of L_1 and L_2.

3. The length of a slope of L_1L_2 is the sum of the lengths of the same slope for L_1 and L_2.

Proof. 1. Write $L_1 = \sum a_{i,j} z^j \delta^i$ and $L_2 = \sum b_{i,j} z^j \delta^i$. From the above it follows that $L_1L_2 = L_3 + R$ with

$$L_3 := \sum_{(i_1,j_1)\in b(L_1),(i_2,j_2)\in b(L_2)} a_{i_1,j_1} b_{i_2,j_2} z^{j_1+j_2}\delta^{i_1+i_2},$$

and one has $R > L_3$. This shows at once that $N(L_1L_2) \subset N(L_1) + N(L_2)$. The boundary part of L_3 can be written as

$$\sum_{(s_1,s_2)\in b(L_1L_2)} (\sum a_{n_1,m_1} b_{n_2,m_2}) z^{s_2}\delta^{s_1},$$

where the second sum is taken over all $(n_1, m_1) \in b(L_1)$, $(n_2, m_2) \in b(L_2)$ with $(n_1, m_1)+(n_2, m_2) = (s_1, s_2)$. By making a drawing one easily verifies the following statement:

Assume that v is a vertex of $N(L_1) + N(L_2)$ and $v = v_1 + v_2$ with $v_i \in N(L_i)$, $i = 1, 2$. Then v_i is a vertex of $N(L_i)$ for $i = 1, 2$. Moreover, v determines v_1 and v_2.

From this statement we see that for a vertex $v = (s_1, s_2)$ of $N(L_1) + N(L_2)$ the coefficient of $z^{s_1}\delta^{s_2}$ in L_3 does not vanish. Therefore $N(L_1) + N(L_2) \subset N(L_1L_2)$. This proves the first part of the lemma.
The two other parts follow easily from the above facts concerning the faces of $N(L_1) + N(L_2)$. □

Example 3.46 The operator $L = z\delta^2 + \delta - 1$ factors as $L = L_1 L_2$ where $L_1 = \delta - 1$ and $L_2 = z\delta + 1$. Figure 3.1 show the corresponding Newton polygons. □

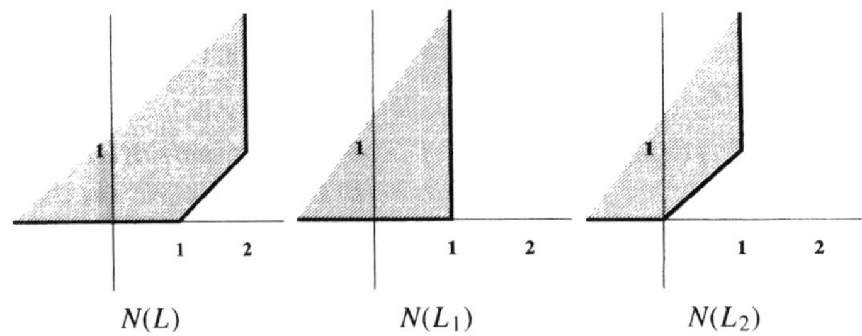

Fig. 3.1. Newton polygons for Example 3.46

Exercises 3.47 *Newton polygons and regular singular points*
1. Show that 0 is a regular singular point of an operator L if and only if the corresponding Newton polygon has only one slope and this slope is 0.
2. Show that if 0 is a regular singular point of an operator L, then it is a regular singular point of any factor of L. □

The next statement is a sort of converse of the lemma.

Theorem 3.48 *Suppose that the Newton polygon of a monic differential operator L can be written as a sum of two special polygons P_1, P_2 that have no slope in common. Then there are unique monic differential operators L_1, L_2 such that P_i is the Newton polygon of L_i and $L = L_1 L_2$. Moreover,*

$$\mathcal{D}/\mathcal{D}L \cong \mathcal{D}/\mathcal{D}L_1 \oplus \mathcal{D}/\mathcal{D}L_2.$$

Proof. For the Newton polygon $N(L)$ of L we use the notations above. We start by proving three special cases.

(1) Suppose that $n_1 > 0$ and that P_1 has only one slope and that this slope is 0. In particular, this implies that P_2 has no slope equal to zero. We would then like to find the factorization $L = L_1 L_2$. Every element $M \in \mathcal{D} = k((z))[\delta]$ is given an expansion $M = \sum_{i>>-\infty} z^i M(i)(\delta)$ where the $M(i)(\delta) \in k[\delta]$ are polynomials of bounded degree. Let $L = \sum_{k \geq m} z^k L(k)$. The $L_1 = \sum_{i \geq 0} z^i L_1(i)$ that we want to find satisfies: $L_1(0)$ is monic of degree n_1 and the $L_1(i)$ have degree $< n_1$ for $i \neq 0$. Furthermore, if we write $L_2 = \sum_{i \geq m} z^i L_2(i)$, we will have the fact that $L_2(m)$ is constant since P_2 has no slope equal to zero. The equality $L_1 L_2 = L$ and the formula $z^{-j} L_1(i)(\delta) z^j = L_1(i)(\delta + j)$ induces the following formula:

$$\sum_{k\geq m} z^k \sum_{i+j=k, i\geq 0, j\geq m} L_1(i)(\delta+j)L_2(j)(\delta) = \sum_{k\geq m} z^k L(k)(\delta).$$

From $L_1(0)(\delta+m)L_2(m)(\delta) = L(m)(\delta)$ and $L_1(0)$ monic and $L_2(m)$ constant, one finds $L_1(0)$ and $L_2(m)$. For $k = m+1$ one finds an equality

$$L_1(0)(\delta+m+1)L_2(m+1)(\delta) + L_1(1)(\delta+m)L_2(m)(\delta) = L(m+1)(\delta).$$

This equality is, in fact, the division of $L(m+1)(\delta)$ by $L_1(0)(\delta+m+1)$ with remainder $L_1(1)(\delta+m)L_2(m)(\delta)$ of degree less than n_1 = the degree of $L_1(0)$ $(\delta+m+1)$. Hence $L_1(1)$ and $L_2(m+1)$ are uniquely determined. Every new value of k determines two new terms $L_1(\dots)$ and $L_2(\dots)$. This proves the existence and uniqueness in this special case.

(2) Suppose now that $n_1 = 0$ and that P_1 has only one slope s, which is the minimal slope of L. Write $s = \frac{b}{a}$ with $a, b \in \mathbf{Z}$; $a, b > 0$ and g.c.d.$(a, b) = 1$. We allow ourselves the field extension $k((z)) \subset k((t))$ with $t^a = z$. Write $\Delta = t^b\delta$. After multiplying L with a power of t we may suppose that $L \in k((t))[\Delta]$ is monic. Note that the Newton polygon of L now has minimal slope 0 and that this slope has length n_2. Every $M \in k((t))[\Delta]$ can be written as $M = \sum_{i>>-\infty} t^i M(i)$, where the $M(i) \in k[\Delta]$ are polynomials of bounded degree. We want to find $L_1, L_2 \in k((t))[\Delta]$ with $L_1L_2 = L$; $L_1(0)$ is monic of degree $n_2 - n_1 = n_2$; $L_1(i)$ has degree less than n_2 for $i > 0$. Using the fact that $\Delta t = t\Delta + \frac{1}{a}t^{b+1}$, one finds for every index k an equation of the form

$$\sum_{i+j=k} L_1(i)L_2(j) + \text{“lower terms”} = L(k).$$

Here, “lower terms” means terms coming from a product $L_1(i)L_2(j)$ with $i+j<k$. The form of the exhibited formula strongly uses the fact that $b > 0$. It is clear now that there is a unique solution for the decomposition $L = L_1L_2$. We then normalize L, L_1, L_2 again to be monic elements of $k((t))[\delta]$. Consider the automorphism τ of $k((t))[\delta]$ that is the identity on $k((z))[\delta]$ and satisfies $\tau(t) = \zeta t$, where ζ is a primitive a^{th} root of unity. Since the decomposition is unique, one finds $\tau L_i = L_i$ for $i = 1, 2$. This implies that the L_i are in $k((z))[\delta]$. This finishes the proof of the theorem in this special case.

(3) The bijective map $\phi : k((z))[\delta] \to k((z))[\delta]$, given by $\phi(\sum a_i\delta^i) = \sum(-\delta)^i a_i$ is an anti-isomorphism, i.e., ϕ is $k((z))$-linear and $\phi(L_1L_2) = \phi(L_2)\phi(L_1)$. Using this ϕ and (1),(2) one finds another new case of the theorem, namely: Suppose that $N(L) = P_1 + P_2$ where P_2 has only one slope and this slope is the minimal slope (≥ 0) of L. Then there is a unique decomposition $L = L_1L_2$ with the properties stated in the theorem.

(4) Existence in the general case. The smallest slope $s \geq 0$ of L belongs either to P_1 or P_2. Suppose that it belongs to P_1 (the other case is similar). According to (1) and

(2) we can write $L = AB$ with A, B monic and such that A has only s as slope and B does not have s as slope. By induction on the degree we may suppose that B has a decomposition $B = B_1B_2$ with $N(B_2) = P_2$ and B_1, B_2 monic. Then $L_1 := AB_1$ and $L_2 := B_2$ is the required decomposition of L.

(5) The unicity. Suppose that we find two decompositions $L = L_1L_2 = \tilde{L}_1\tilde{L}_2$ satisfying the properties of the theorem. Suppose that the smallest slope $s \geq 0$ of L occurs in P_1. Write $L_1 = AB$ and $\tilde{L}_1 = \tilde{A}\tilde{B}$ where A and $\tilde{A}$ have as unique slope the minimal slope of L and where B, $\tilde{B}$ have no slope s. Then $L = ABL_2 = \tilde{A}\tilde{B}\tilde{L}_2$ and the unicity proved in (1) and (2) implies that $A = \tilde{A}$ and $BL_2 = \tilde{B}\tilde{L}_2$. Induction on the degree implies that $B = \tilde{B}$ and $L_2 = \tilde{L}_2$. This finishes the proof of the first part of the theorem.

(6) There is an exact sequence of $k((z))[\delta]$-modules

$$0 \to \mathcal{D}/\mathcal{D}L_1 \stackrel{.L_2}{\to} \mathcal{D}/\mathcal{D}L \stackrel{\pi_1}{\to} \mathcal{D}/\mathcal{D}L_2 \to 0$$

corresponding to the decomposition $L = L_1L_2$. It suffices to show that π_1 splits. There is also a decomposition $L = \tilde{L}_2\tilde{L}_1$ with $N(\tilde{L}_i) = P_i$. This gives another exact sequence

$$0 \to \mathcal{D}/\mathcal{D}\tilde{L}_2 \stackrel{.\tilde{L}_1}{\to} \mathcal{D}/\mathcal{D}L \stackrel{\pi_2}{\to} \mathcal{D}/\mathcal{D}\tilde{L}_1 \to 0.$$

It suffices to show that

$$\psi : \mathcal{D}/\mathcal{D}\tilde{L}_2 \stackrel{.\tilde{L}_1}{\to} \mathcal{D}/\mathcal{D}L \stackrel{\pi_1}{\to} \mathcal{D}/\mathcal{D}L_2$$

is an isomorphism. Since the two spaces have the same dimension, it suffices to show that ψ is injective. Let $A \in \mathcal{D}$ have degree less than d = the degree of L_2 and $\tilde{L}_2$. Suppose that $A\tilde{L}_1$ lies in $\mathcal{D}L_2$. Hence, $A\tilde{L}_1 = BL_2$. We note that $\tilde{L}_1$ and L_2 have no slopes in common. This means that $N(A)$ must contain $N(L_2)$. This implies that the degree of A is at least d. This contradicts our hypothesis. □

Examples 3.49

1. We consider the operator $L(y) = z\delta^2 + \delta + 1$ of Example 3.46. One sees from Fig. 3.1 that the Newton polygon of this operator is the sum of two special polygons P_1, having a unique slope 0, and P_2, having a unique slope 1. Using the notation of part (1) of the proof of Theorem 3.48, we have $n_1 = 1$ and $m = 0$. We let

$$L_1 = L_1(0) + zL_1(1) + \cdots,$$
$$L_2 = L_2(0) + zL_2(1) + \cdots,$$

where $L_1(0)$ is monic of degree 1, the $L_1(i)$ have degree 0 for $i > 0$ and $L_2(0) = 1$. Comparing the coefficients of z^0 in $L = L_1L_2$ we have

$$L_1(0)L_2(0) = L_1(0) = \delta - 1.$$

Comparing coefficients of z^1 we have

$$L_1(0)(\delta+1)L_2(1)(\delta) + L_1(1)(\delta)L_2(0)(\delta) = \delta L_2(1)(\delta) + L_1(1) = \delta^2.$$

This implies that $L_2(1) = \delta$ and $L_1(1) = 0$. One can show by induction that $L_1(i) = L_2(i) = 0$ for $i \geq 2$. This yields the factorization given in Example 3.46.

2. We consider the operator

$$L = \delta^2 + (\frac{1}{z^2} + \frac{1}{z})\delta + \frac{1}{z^3} - \frac{2}{z^2}.$$

The Newton polygon of this operator can be written as the sum of two special polygons P_1 and P_2 (see Fig. 3.2).

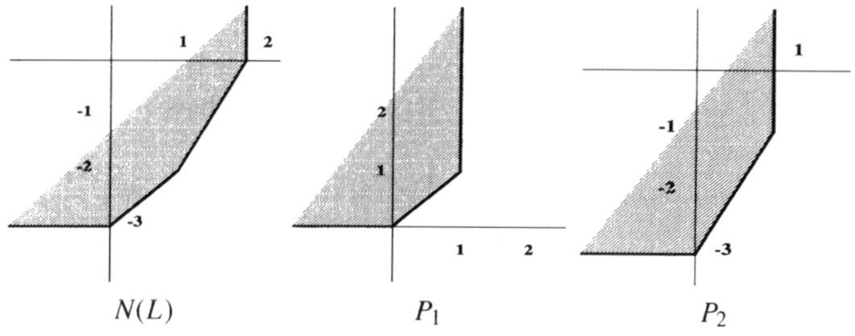

Fig. 3.2. Newton polygons for Example 3.49.2

The polygon P_1 has minimal slope 1 so, using the notation of part (2) of the proof Theorem 3.48, we have $a = b = 1$ and $t = z$. Letting $\Delta = z\delta$ we have

$$L = \frac{1}{z}\Delta^2 + (\frac{1}{z^3} + \frac{1}{z^2} - \frac{1}{z})\Delta + \frac{1}{z^3} - \frac{2}{z}.$$

Dividing by z to make this operator monic, we now consider the operator

$$\tilde{L} = \Delta^2 + (\frac{1}{z} + 1 - z)\Delta + \frac{1}{z} - 2$$

whose Newton polygon is given in Fig. 3.3.

We write $\tilde{L} = L_1L_2$, where

$$L_1 = L_1(0) + zL_1(1) + z^2L_1(2) + \cdots,$$
$$L_2 = z^{-1}L_2(-1) + L_2(0) + zL_2(1) + \cdots,$$

where $L_1(0)$ has degree 1 (i.e., $L_1(0) = r\Delta + s$), $L_1(i)$ is constant for $i > 0$ and $L_1(-1) = 1$. Composing and equating coefficients of powers of z we get

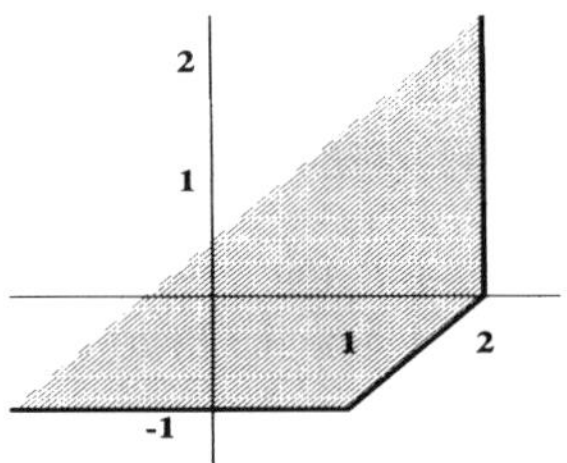

Fig. 3.3. Newton polygon for $\tilde{L}$

$$\begin{aligned}
r\Delta + s &= \Delta + 1 && \text{coefficients of } z^{-1}, \\
-r + (\Delta+1)L_2(0) + L_1(1) &= \Delta^2 + \Delta - 2 && \text{coefficients of } z^{0}, \\
(\Delta+1)L_2(1) + L_1(1)L_2(0) + L_1(2) &= -\Delta && \text{coefficients of } z^{1}.
\end{aligned}$$

These imply that $r = s = 1$, $L_2(0) = \Delta$, $L_1(1) = -1$ and $L_2(1) = L_1(2) = 0$. One can show by induction that $L_2(i) = L_2(i+1) = 0$ for $i > 1$. This gives a factorization $\tilde{L} = (\Delta + 1 - z)(\Delta + z^{-1})$. We therefore have

$$\begin{aligned}
L = \delta^2 + (\frac{1}{z^2} + \frac{1}{z})\delta + \frac{1}{z^3} - \frac{2}{z^2} &= \frac{1}{z}\Delta^2 + (\frac{1}{z^3} + \frac{1}{z^2} - \frac{1}{z})\Delta + \frac{1}{z^3} - \frac{2}{z} \\
&= z^{-2}(\Delta^2 + (\frac{1}{z} + 1 - z)\Delta + \frac{1}{z} - 2) \\
&= z^{-2}(\Delta + 1 - z)(\Delta + z^{-1}) \\
&= z^{-2}(z\delta + 1 - z)(z\delta + z^{-1}) \\
&= z^{-2}(z\delta + 1 - z)z(\delta + z^{-2}) \\
&= z^{-2}(z^2\delta + z)(\delta + z^{-2}) \\
&= (\delta + z^{-1})(\delta + z^{-2}).
\end{aligned}$$

This gives a factorization of L. □

Theorem 3.48 allows us to factor linear operators whose Newton polygons have at least two slopes. We now turn to operators with only one *positive* slope s. Write as before $s = \frac{b}{a}$ with g.c.d$(a, b) = 1$ and $a, b \in \mathbf{Z}$; $a, b > 0$. We make the field extension $k((t)) \supset k((z))$ with $t^a = z$ and we write $\Delta = t^b\delta$. After normalization we may assume that L is monic with respect to Δ. Write $L = \sum_{i\geq 0} t^i L(i)(\Delta)$, where the $L(i)$ are polynomials in Δ such that $L(0)$ is monic of degree n and the $L(i)$ have degree less than n for $i \neq 0$. The following result is a restatement of Hensel's Lemma for irregular differential operators.

Proposition 3.50 *Suppose (using the above notation) that $L \in k[[t]][\Delta]$ is monic of degree n. Suppose that $L(0) \in k[\Delta]$ factors into relative prime monic polynomials $L(0) = PQ$. Then there is a unique factorization $L = AB$ with A, B monic and $A(0) = P$, $B(0) = Q$. Moreover,*

$$k((t))[\delta]/k((t))[\delta]L \cong k((t))[\delta]/k((t))[\delta]A \oplus k((t))[\delta]/k((t))[\delta]B\ .$$

Proof. Write $A = \sum_{i \geq 0} t^i A(i)$; $B = \sum_{j \geq 0} t^j B(j)$. Then

$$AB = \sum_{m \geq 0} t^m (\sum_{i+j=m} A(i)B(j) + \text{"lower terms"}) = \sum_{m \geq 0} t^m L(m).$$

Again "lower terms" means some expression involving $A(i)$ and $B(j)$ with $i+j < m$. Clearly one can solve this set of equations, using the fact that $A(0)$ and $B(0)$ are relatively prime, step by step in a unique way. This proves the first part of the proposition. The second part is proved as in Theorem 3.48. □

Remark 3.51 The hypothesis that $s > 0$ is crucial in Proposition 3.50. If $s = 0$, then the point zero is a regular singular point and the exhibited equation in the proof of Proposition 3.50 becomes

$$AB = \sum_{m \geq 0} z^m (\sum_{i+j=m} A(i)(\delta + j)B(j)(\delta) + \quad \text{"lower terms"}) = \sum_{m \geq 0} z^m L(m).$$

In order to proceed, one needs to assume that $A(0)(\Delta + j)$ and $B(0)(\Delta)$ are relatively prime for $j = 0, 1, 2, \ldots$. With this assumption, one can state a result similar to the Hensel Lemma for regular singular points given in the previous section. □

Examples 3.52 1. Consider the operator $\tilde{L} = \delta^2 - \frac{3}{2}\delta + \frac{2z-1}{4z}$ whose Newton polygon is given in Fig. 3.4.

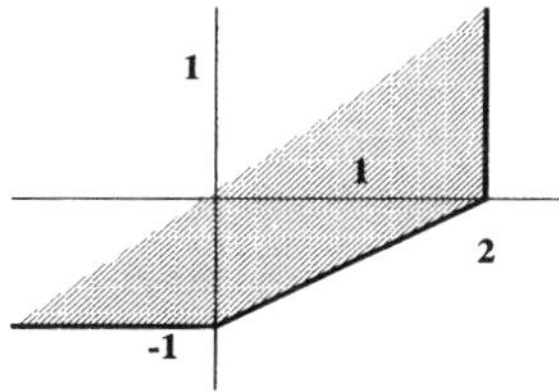

Fig. 3.4. Newton polygon for Example 3.52.1

Using the above notation, we have $t^2 = z$ and $\Delta = t\delta$. Rewriting $\tilde{L}$ in terms of t and Δ, we have $\tilde{L} = \frac{1}{t^2}L$ where

$$\begin{aligned} L &= \Delta^2 - 2t\Delta + \frac{1}{4}(2t^2 - 1) \\ &= L(0) + tL(1) + t^2L(2) \\ &= (\Delta^2 - \frac{1}{4}) + t(-2\Delta) + t^2(\frac{1}{2}). \end{aligned}$$

Since $\Delta^2 - \frac{1}{4} = (\Delta + \frac{1}{2})(\Delta - \frac{1}{2})$ we can apply Proposition 3.50. Let $L_1 = \Delta + \frac{1}{4} + tL_1(1) + t^2L_1(2) + \cdots$ and $L_2 = \Delta - \frac{1}{4} + tL_2(1) + t^2L_2(2) + \cdots$. Comparing the powers of t in $L = L_1L_2$, the coefficients of t^0 and t^2 are, respectively

$$\begin{aligned} L_1(1)(\Delta - \tfrac{1}{2}) + L_2(1)(\Delta + \tfrac{1}{2}) &= -2\Delta, \\ L_2(2)(\Delta - \tfrac{1}{2}) + L_1(2)(\Delta - \tfrac{1}{2}) + L_1(1)L_2(1) + \tfrac{1}{2}L_2(1) &= \tfrac{1}{2}. \end{aligned}$$

Therefore, $L_1(1) = L_2(1) = -1$ and $L_1(2) = L_2(2) = 0$. One sees that this implies that $L_1(i) = L_2(i) = 0$ for all $i \geq 2$. Therefore,

$$\begin{aligned} \tilde{L} &= \frac{1}{t^2}L \\ &= \frac{1}{t^2}(\Delta + \frac{1}{2} - t)(\Delta - \frac{1}{2} - t) \\ &= \frac{1}{t^2}(t\delta + \frac{1}{2} - t)t(\delta - 1 - \frac{1}{2t}) \\ &= (\delta - \frac{1}{2} + \frac{1}{2t})(\delta - 1 - \frac{1}{2t}). \end{aligned}$$

2. We consider the Airy equation $y'' - zy = 0$ mentioned in Example 3.33. We wish to consider the behavior at infinity so we make the change of variable $t = \frac{1}{z}$ and write the resulting equation in terms of $\delta = t\frac{d}{dt}$. This yields the equation

$$\tilde{L} = \delta^2 - \delta - \frac{1}{t^3},$$

which has the Newton polygon given in Fig. 3.5.

The unique slope is $\frac{3}{2}$ so we let $\tau = t^{1/2}$ and $\Delta = \tau^3\delta$. Rewriting $\tilde{L}$ in terms of τ and Δ we have $L = \tau^{-6}\Delta^2 - \frac{1}{2}\tau^{-3}\Delta - \tau^{-6}$. Dividing by τ^{-6} yields the equation

$$L = \Delta^2 - \frac{1}{2}\tau^3\Delta - 1.$$

Since $L(0) = \Delta^2 - 1$ we may write $L = L_1L_2$, where $L_1 = (\Delta - 1) + \tau L_1(1) + \cdots$ and $L_2 = (\Delta + 1) + \tau L_2(1) + \cdots$. Composing these operators and comparing

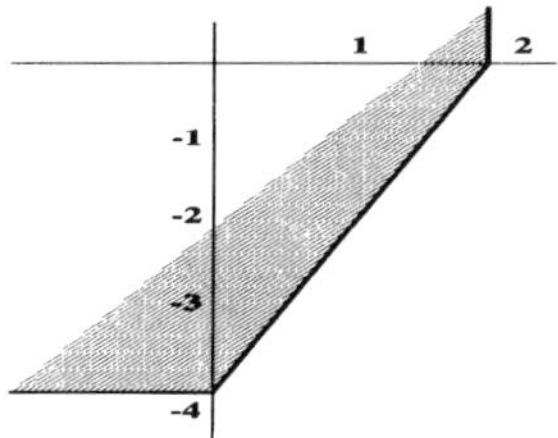

Fig. 3.5. Newton polygon for Example 3.52.2

coefficients of powers of τ shows that $L_1(1) = L_1(2) = L_2(1) = L_2(2) = 0$. Therefore,

$$\begin{aligned}\tilde{L} &= \tau^{-6}(\Delta - 1 + \tau^3(\ldots))(\Delta + 1 + \tau^3(\ldots)) \\ &= (\delta - \tau^{-3} + \cdots)(\delta + \tau^{-3} + \text{ non-negative powers of } \tau).\end{aligned}$$

The form of the last factor shows that the Airy equation has a solution in $R_{z^{3/2}}$. Reversing the roles of $\Delta + 1$ and $\Delta - 1$ shows that it also has a solution in $R_{-z^{3/2}}$. This verifies the claim made in Exercise 3.33. □

In order to factor a general L as far as possible, one uses the algebraic closure $\overline{k}$ of k and fractional powers of z. Suppose that L has only one slope and that this slope is positive. If Proposition 3.50 does not give a factorization then $L(0)$ must have the form $(\Delta + c)^n$ for some $c \in \overline{k}^*$ (note that $c \neq 0$ since $L(0)$ must have at least two terms). This implies that the original Newton polygon must have a point of the form $(1, m)$ on its boundary, that is on the line $bx - ay = 0$. Therefore, $a = 1$ and $\Delta = z^b\delta$ in this case. One makes a change of variables $\delta \mapsto \delta + cz^{-b}$. One then sees that the Newton polygon N' of the new equation is contained in the Newton polygon N of the old equation. The bottom edge of N' contains just one point of N and this is the point (n, bn) that must be a vertex of N'. Therefore, the slopes of N' are strictly less than b. If no factorization, due to Theorem 3.48 or Proposition 3.50 occurs then L has again only one slope and this slope is an integer b' with $0 \leq b' < b$. For $b' = 0$ one stops the process. For $b' > 0$ one repeats the method above. The factorization of L stops if each factor $\tilde{L}$ satisfies:

> There is an element $q \in t^{-1}k'[t^{-1}]$, where k' is a finite extension of k and $t^m = z$ for some $m \geq 1$, such that $\tilde{L}$ has only slope zero with respect to $\delta - q$. This can be restated as $\tilde{L} \in k'[[t]][(\delta - q)]$ and $\tilde{L}$ is monic in $(\delta - q)$.

Example 3.53 Consider the operator

$$L = \delta^2 + \frac{4 + 2z - z^2 - 3z^3}{z^2}\delta + \frac{4 + 4z - 5z^2 - 8z^3 - 3z^4 + 2z^6}{z^4},$$

whose Newton polygon is given in Fig. 3.6.

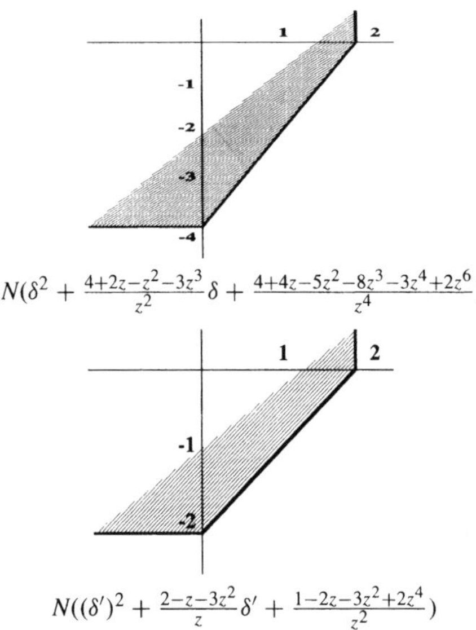

Fig. 3.6. Newton polygons for Example 3.53

Since this has only one slope and this is 2, we let $\Delta = z^2\delta$. Rewriting the equation in terms of Δ and dividing by a suitable power of z to make the resulting operator monic we have $L(0) = (\Delta + 2)^2$. We then let $\delta' = \delta + 2z^{-2}$ and have

$$L = (\delta')^2 + \frac{2 - z - 3z^2}{z}\delta' + \frac{1 - 2z - 3z^2 + 2z^4}{z_2},$$

whose Newton polygon is given in Fig. 3.6. Rewriting this operator in terms of $\Delta' = z\delta'$ and making the resulting operator monic, one has $L(0) = (\Delta' + 1)^2$, Therefore, we continue and let $\delta'' = \delta' + z^{-1}$. One then has

$$L = (\delta'')^2 - (3z + 1)\delta'' + 2z^2 .$$

This operator is regular and can be factored as $L = (\delta'' - (2z+1))(\delta'' - z)$. Therefore,

$$L = (\delta + \frac{2}{2z^2} + \frac{1}{z} - (2z + 1))(\delta + \frac{2}{2z^2} + \frac{1}{z} - z).$$

□

We continue the discussions in Remarks 3.37 and Observations 3.38, concerning the classification of differential modules over more general differential fields than $\mathbf{C}((z))$. Let, as before, k be any field of characteristic 0 and let $k((z))$ be the differential field with derivation $\delta = z\frac{d}{dz}$. A finite field extension K of $k((z))$ is again presented as $K = k'((t))$ with $k \subset k'$ and t with $t^m = cz$ for some nonzero $c \in k'$.

As in the case $k((z))$, a monic operator $L \in K[\delta]$, is called *regular singular* if we have $L \in k'[[t]][\delta]$. The Definition 3.9 of a regular singular differential module is, in an obvious way, extended to the case of the more general field K. One can show that this notion is equivalent to: $M \cong K[\delta]/K[\delta]L$ for a regular singular L. As in Proposition 3.12, one shows that for a regular singular differential module M over K there exists a basis $\{e_1, \ldots, e_n\}$ of M over K such that the matrix of δ with respect to $\{e_1, \ldots, e_n\}$ is constant. In other words, the corresponding matrix equation is $\delta y = Ay$ with A a matrix with coefficients in k'. It is not easy to decide when two equations $\delta y = A_i y,\ i = 1, 2$ with coefficients in k' are equivalent over K. In the case $K = k((t))$ with k algebraically closed and $t^m = z$, one chooses a set $S \subset k$ of representatives of $k/(\frac{1}{m}\mathbf{Z})$. Any matrix equation with constant coefficients, can be normalized into an equation $\delta y = Ay$ where the eigenvalues of the constant matrix A are in S. Two "normalized" equations $\delta y = A_i y,\ i = 1, 2$ are equivalent over $K = k((t))$ if and only if A_2 is a conjugate of A_1.

For the field $K = \mathbf{C}((t))$ with $t^m = z$, one associates to a matrix equation with constant coefficients $\delta y = Ay$ the matrix $e^{2m\pi i A}$. This matrix (or its conjugacy class) is called the topological monodromy of the equation (w.r.t. the field K). Using Proposition 3.30, one can show that two equations $\delta y = A_i y$ with constant matrices A_i are isomorphic if and only if $e^{2m\pi i A_1}$ is a conjugate of $e^{2m\pi i A_2}$ (see also Theorem 5.1).

For $q \in t^{-1}k'[t^{-1}]$ we write $E(q)$ for the $k'((t))[\delta]$-module generated over $k'((t))$ by one element v such that $\delta v = qv$. Let M be a regular singular module with cyclic vector e and minimal monic equation $Le = 0$ where $L = \sum a_i \delta^i$. Then $M \otimes E(q)$ has the cyclic vector $e \otimes v$.

The minimal monic equation for this cyclic vector is $\sum a_i(\delta - q)^i$. Furthermore, for any operator of the form $L = \sum a_i \delta^i$, the $k'((t))[\delta]$-module $k'((t))[\delta]/k'((t))[\delta]L$ is of the form $M \otimes E(q)$. In particular, this is true for each $\tilde{L}$ described in the exhibited paragraph preceding Exercise 3.53. We can now state the following theorem.

Theorem 3.54 *Let $L \in k((z))[\delta]$ be a monic differential operator. There exist a finite field extension k' of k, an integer $m \geq 1$, elements $q_1, \ldots, q_s \in t^{-1}k'[t^{-1}]$ with $t^m = cz$ (some nonzero $c \in k'$), and $L_1, \ldots, L_s \in k'((t))[\delta]$ such that:*

1. *If $i \neq j$ then $q_i \neq q_j$.*
2. *$L_i \in k'[[t]][\delta - q_i]$ and is monic in $\delta - q_i$.*
3. *$L = L_1 \cdots L_s$.*

Moreover, one has

$$k'((t))[\delta]/k'((t))[\delta]L \cong \oplus M_i \otimes E(q_i),$$

where the M_i are regular singular $k'((t))[\delta]$-modules.

Proof. The above methods allow one to factor L and give a factorization $L = R_1 \cdots R_a$ that yields a direct sum decomposition $k'((t))[\delta]/k'((t))[\delta]L = \oplus k'((t))[\delta]/k'((t))[\delta]R_i$. According to the above discussion, each factor has the form $N_q \otimes E(q)$ with N_q regular singular. The qs need not be distinct. Let $\{q_1, \ldots, q_s\}$ denote the distinct qs occurring. Put $M_i = \oplus_{q=q_i} N_q$. This proves the second part of the theorem.

To prove the first part of the theorem, we let e be a cyclic vector of $k'((t))[\delta]/k'((t))[\delta]L$ annihilated by L and let $e = e_1 + \cdots + e_s$ with each $e_i \in M_i \otimes E(q_i)$. One sees that each e_i is a cyclic vector of $M_i \otimes E(q_i)$ and that $L(e_i) = 0$. If L_s is the minimal monic annihilator of e_s, then L_s must divide L on the right. Furthermore, since $(M_i \otimes E(q_i)) \otimes E(-q_s)$ is regular, Proposition 3.16 implies that $L_s(\delta + q_i)$ is a regular operator and so is in $k'[[t]]$. Therefore, $L_s \in k[[t]][\delta - q_s]$. An induction on s finishes the proof of the first part of the theorem. □

Remarks 3.55 1. We have seen in Proposition 3.41 that the module $M = \mathcal{D}/\mathcal{D}L$ determines uniquely the direct sum decomposition Theorem 3.54 part (2). In particular the q_i and the dimensions d_i of the M_i (as vector spaces over $k'((t))$) are determined by M. From this information one can reconstruct the Newton polygon of L.

Indeed, L_i has one slope, namely $-v(q_i)$ with length d_i = the order of L_i. Since $N(L) = N(L_1) + \cdots + N(L_s)$ one finds the following:

> λ is a slope of $N(L)$ if and only if $\lambda = -v(q_i)$ for some i. Moreover, the length of the slope λ is equal to $\sum_{\lambda=-v(q_i)} d_i$.

In particular, the Newton polygon of M does not depend on the choice of a cyclic vector.

2. We also note that the methods described in this section yield an algorithm to calculate the q_i of Proposition 3.41. Moreover, these methods produce a set of at most n such q_i. More efficient algorithms are presented in the works of Barkatou [19, 20, 21], Barkatou and Jung [23], Barkatou and Pflügel [24], Chen [66], Della Dora, di Crescenzo and Tournier [83], Hilali [128], Hilali and Wazner [129, 130, 131] van Hoeij [138], Pflügel [220, 221] and Tournier [281]. □

We end the chapter by noting that the formal classification of general linear differential equations has a long history going back to the nineteenth century with the works of Fuchs [103, 104] (see also [112, 113]) and Fabry [99], who wrote down a fundamental set of local solutions of regular singular equations and general linear equations, respectively. In the early twentieth century, Cope [72, 73] also considered these issues. Besides the works of Deligne, Katz and Malgrange (see [187, 190]), Ramis and Turrittin (already mentioned), this problem has been considered by Babbitt and Varadarajan [12], Balser et al. [17], Levelt [172], Robba [248] and Wasow [301] (who attribute the result to Turrittin). The papers of Varadarajan [298] and Babitt and Varadarajan [13, 297] give a more detailed exposition of the recent history of the problem.

4 Algorithmic Considerations

Linear differential equations over the differential field $C((z))$ (with C an algebraically closed field of characteristic 0, in particular $C = \mathbf{C}$) were classified in Chap. 3. When the standard form of such a differential equation is known, then its Picard-Vessiot ring, its differential Galois group, the formal solutions etc., are known. The methods of Chap. 3 have been transformed into algorithms and are implemented. In this chapter we consider "global" linear differential equations, i.e., equations over the differential field $C(z)$. Here C is a field of characteristic 0 and the differentiation on $C(z)$ is the usual one, namely $f \mapsto f' = \frac{df}{dz}$. We, furthermore, assume that there are algorithms to perform the field operations in C as well as algorithms to factor polynomials over $C(z)$ (see [102, 234] for a formalization of this concept). Natural choices for C are $\mathbf{Q}$, any number field or the algebraic closure of $\mathbf{Q}$.

It is no longer possible to transform any linear differential equation over $C(z)$ into some standard equation from which one can read off its Picard-Vessiot ring, its differential Galois group, etc. Instead we will present algorithmic methods to find global solutions that are rational, exponential or liouvillian. Factoring linear differential operators over $C(z)$ is, in fact, the main theme of this chapter. One has to distinguish between "theoretical" algorithms and efficient ones. Especially the latter category is progressing quickly and we will only indicate some of its features. We observe that the language of differential operators and the one of differential modules (or matrix differential equations) have both their advantages and disadvantages. In this chapter we choose between the two for the purpose of simplifying the exposition.

The last part of this chapter is concerned with the inverse problem for finite groups. An effective algorithm is explained that produces for a representation of a finite group a corresponding differential equation.

4.1 Rational and Exponential Solutions

Rational Solutions

Let

$$L = \partial^n + a_{n-1}\partial^{n-1} + \cdots + a_0 \tag{4.1}$$

be a linear differential operator with coefficients in $C(z)$ and $\partial = \frac{d}{dz}$. The problem of finding the solutions $y \in C(z)$ of $L(y) = 0$ has a simpler analog namely, finding solutions $a \in \mathbf{Q}$ of $p(z) = a_n z^n + \cdots + a_0 = 0,\ p(z) \in \mathbf{Z}[z]$. If a is written as $\frac{u}{v}$ with $u, v \in \mathbf{Z}$ and $(u, v) = 1$, then u divides a_0 and v divides a_n. This obviously solves this problem. Consider a nonzero solution $y = \frac{u}{v}$, with $u, v \in C[z]$ and $(u, v) = 1$, of the differential equations $a_n y^{(n)} + \cdots + a_0 y = 0$ with $a_i \in C[z]$, $a_n \neq 0$. This equation is regular at any point $c \in \overline{C}$ (i.e., the algebraic closure of C), which is not a root of a_n. Hence y has no pole at such point c. It follows that any irreducible factor q of v is a divisor of a_n. The problem that we have to solve is to determine the exact power q^m that divides v. As an example, the equation $zy' + 5y = 0$ has solution z^{-5}.

More generally, consider the equation $y^{(n)} + a_{n-1}y^{(n-1)} + \cdots + a_1 y^{(1)} + a_0 y = 0$, where some of the $a_{n-1}, \ldots, a_0 \in C(z)$ have a pole at 0. Now we make a calculation in the differential field $C((z))$ and write $y = \frac{u}{v} = z^\alpha + \cdots$ (where $\alpha \in \mathbf{Z}$ has to be found) and $a_i = \sum_{m \geq \alpha_i} a_{i,m} z^m$ for $i = 0, \ldots, n$ (where $a_n = 1$) for their Laurent series. We consider among the Laurent series $y^{(n)}, a_{n-1}y^{(n-1)}, \ldots, a_1 y^{(1)}, a_0 y$ the ones with (potentially) the smallest order at 0 (this does not depend on α). The sum of the leading coefficients of these Laurent series must be zero. This yields an equation

$$\sum_{i \in S} a_{i,\alpha_i} \alpha(\alpha - 1) \cdots (\alpha - i + 1) = 0,$$

where the sum is taken over the subset S of $\{0, \ldots, n\}$ corresponding to the selected Laurent series. $I(T) := \sum_{i \in S} a_{i,\alpha_i} T(T-1) \cdots (T - i + 1)$ is called the *indicial polynomial* of the equation at 0. This polynomial is nonzero and its roots (in an algebraic closure of C) are called the *local exponents* of the equation at 0. We conclude that the possible values $m > 0$ for the exact power z^m dividing v are the negative integers $-m$ with $I(-m) = 0$.

Now we perform a similar calculation at ∞. This means that we work in the Laurent series field $C((z^{-1}))$ and develop the putative solution $y = \frac{u}{v} \in C(z)^*$ and the $a_0, \ldots, a_n$ as Laurent series in the variable z^{-1}. The Laurent series of y in z^{-1} has the form $y = z^\alpha + *z^{\alpha-1} + \cdots$ with $\alpha = \deg_z v - \deg_z u$. There results an indicial polynomial equation for ∞ of which α is a root. We conclude that the possible values for $\deg_z v - \deg_z u$ are found.

We suppose now that the largest possible denominator $\tilde{v}$ of the putative solution y has been found. Then for the degree of the numerator u there are finitely many possibilities. One chooses again the largest possibility d, and writes u as a polynomial $u_0+u_1z+\cdots+u_{d-1}z^{d-1}+u_dz^d$ with as yet unknown coefficients. The differential equation for y translates into a set of homogeneous linear equations for $u_0,\dots,u_d$. Let U denote the C-linear subspace of polynomials u of degree $\leq d$ satisfying these linear equations. Then $\{\frac{u}{v}|u\in U\}$ is the C-vector space of all solutions $y\in C(z)$ of our differential equation.

Therefore, the algorithm will be completed once we have generalized the above example of a power of z dividing the denominator to the case of a monic irreducible $q\in C[z]$. Furthermore, we are also interested in the solutions y in the field $\overline{C}(z)$. Propositions 4.1 and 4.3 give the formalities of this approach.

Let an irreducible monic polynomial $q\in C[z]$ be given. One associates to q a map $v_q : C(z)\to \mathbf{Z}\cup\{\infty\}$ by $v_q(0)=\infty$ and $v_q(f)=m$ if $f\neq 0$ can be written as $f=\frac{a}{b}q^m$, where $a,b\in C[z]$, $(a,q)=(b,q)=1$ and $m\in\mathbf{Z}$. This map is called a discrete valuation of $C(z)$ over C. The map $v_\infty : C(z)\to\mathbf{Z}\cup\{\infty\}$ defined by $v_\infty(0)=\infty$ and $v_\infty(\frac{a}{b})=\deg_z b-\deg_z a$ for $a,b\in C[z],\ a,b\neq 0$, is also a discrete valuation of $C(z)$ over C. The integers $v_q(f)$ and $v_\infty(f)$ for $f\in C(z)^*$ are called the *order of f at the place q and the order of f at infinity*. The above examples are, in fact, all discrete valuations of $C(z)$ over C. One can complete the field $C(z)$ with respect to any discrete valuation. The resulting fields will be denoted by k_q or k_∞ ([170], Chap. XII). For $q=z-a$ with $a\in C$, this completion is easily seen to be the field of formal Laurent series $C((z-a))$. Furthermore, $k_\infty=C((z^{-1}))$. For a q of degree >1 the field k_q is isomorphic to $k'((t))$ with $k'=C[z]/(q)$ and t an indeterminate. The derivation on $C(z)$ uniquely extends to a continuous derivation on k_q and on k_∞. The elements $f\in k_q$ can also be uniquely represented as an infinite sum

$$f_mq^m+f_{m+1}q^{m+1}+\cdots,$$

where each $f_i\in C[z]$ satisfies $\deg_z f_i<\deg_z q$. This is called the *q-adic expansion of f*. One sees by induction that

$$f^{(j)}=u_jq^{m-j}+\cdots,$$

where $u_j\equiv m(m-1)\dots(m-j+1)f_n\cdot(q')^j \bmod q$. Since f_n and q' are relatively prime to q, we see that $u_j\neq 0$ if $m<0$. The elements of the *completion at infinity* $k_\infty=C((z^{-1}))$, can uniquely be written as infinite sums

$$f=f_mz^m+f_{m-1}z^{m-1}+\cdots,$$

where the f_i are constants and this is called the *expansion at infinity of f*. For the j-th derivative of f one has the formula

$$f^{(j)}=m(m-1)\cdots(m-j+1)f_mz^{m-j}+\cdots.$$

We begin by describing the C-space of solutions of $Ly=0$ in $C(z)$.

Proposition 4.1 *Let $L = \partial^n + a_{n-1}\partial^{n-1} + \cdots + a_0$ be a linear differential operator with coefficients in $C(z)$. One can find, in a finite number of steps, a C-basis of V, the space of solutions in $C(z)$ of $Ly = 0$.*

Proof. For convenience of notation, we let $a_n = 1$. Let $y \neq 0$ be a putative solution of $Ly = 0$ and let q be a monic irreducible element of $C[z]$. We let

$$y = y_\alpha q^\alpha + \ldots$$
$$a_i = a_{i,\alpha_i} q^{\alpha_i} + \cdots$$

be the q-adic expansions of y and the a_i. We are only interested in the case $\alpha < 0$. As remarked before, this implies that q divides the denominator of some a_i. Thus the finite set of qs that we have to consider is known. For each q we have to find the possibilities for the exact power of q dividing the denominator of y. As before, we consider the q expansions of the elements $y^{(n)}, a_{n-1}y^{(n-1)}, \ldots, a_0 y$ with lowest order. The sum of their leading coefficients must be 0, since $L(y) = 0$. Thus for some subset S of $\{0, 1, \ldots, n\}$, independent of α one has

$$\sum_{i \in S} a_{i,\alpha_i} \alpha(\alpha - 1) \cdots (\alpha - i + 1) y_\alpha (q')^i \equiv 0 \bmod q\ .$$

Dividing by y_α and replacing α by T yields a nonzero polynomial

$$I(T) := \sum_{i \in S} a_{i,\alpha_i} T(T - 1) \cdots (T - i + 1)(q')^i \bmod q = 0 \in C[z]/(q)[T],$$

called (as before) the *indicial polynomial* of L at the place q. The roots of the indicial polynomial (in an algebraic extension of $C[z]/(q)$) are called (as before) the *local exponents* of L at the place q. We conclude that the negative integer α should be a root of the indicial polynomial. The assumption on the field C guarantees that one can calculate the possible αs. This completes the exposition of the algorithm. We note that when the indicial polynomial for some q has no negative integer as root, then there are no rational solutions $\neq 0$ of L. □

Exercises 4.2 *Polynomial and rational solutions*

1. Find a basis of the space of polynomial solutions of

$$y''' - \frac{z^2 + 4z}{z^2 + 2z - 2} y'' + \frac{2z + 4}{z^2 + 2z - 2} y' - \frac{2}{z^2 + 2z - 2} y = 0.$$

2. Find a basis of the space of rational solutions of

$$y'' + \frac{4}{(z + 1)} y' + \frac{2}{(z + 1)^2} y = 0.$$

3. Let L be as in Proposition 4.1 and $f \in C(z)$. Modify the method given in Proposition 4.1 to show how one can decide if $Ly = f$ has a solution in $C(z)$, and find one if it does. □

We shall now show that the $\overline{C}$-vector space $\overline{V}$ of solutions of $Ly = 0$ in $\overline{C}(z)$ has a $\overline{C}$-basis of elements in $C(z)$. This follows from the general result

Proposition 4.3 *Let K be a differential field of characteristic zero with subfield of constants C. Consider a linear differential operator $L = \partial^n + a_{n-1}\partial^{n-1} + \cdots + a_0$ over K and let $V \subset K$ denote the C-vector space of the solutions of L in K. Let $\overline{C}$ be an algebraic closure of C and let $\overline{V} \subset \overline{C}K$ be the solution space of L on $\overline{C}K$. The natural $\overline{C}$-linear map $\overline{C} \otimes_C V \to \overline{V}$ is an isomorphism.*

Proof. Let $v_1, \ldots, v_m$ be a $\overline{C}$-basis of $\overline{V}$. There exists a $c \in \overline{C}$ such that $K(v_1, \ldots, v_m) \subset K(c)$. Let $[K(c) : K] = t$. For each i, $1 \leq i \leq m$, there exist $v_{i,j} \in K$ such that $v_i = \sum_{j=0}^{t-1} v_{i,j}c^j$. Since $0 = L(v_i) = \sum_{j=0}^{t-1} L(v_{i,j})c^j$, we have that the $v_{i,j}$ span $\overline{V}$ and therefore, $\overline{V}$ has a basis in K. Corollary 1.13 implies that any C-basis of V remains linearly independent over $\overline{C}$. Therefore $\dim_{\overline{C}} \overline{V} = \dim_C V$. □

Exercise 4.4 *Inhomogeneous equations.* Let L be as in Proposition 4.3 and $f \in K$. Show that $Ly = f$ has a solution in $\overline{C}K$ if and only if it has a solution in K. Hint: $\overline{C}K$ is an algebraic extension of K. Consider for a solution $y \in \overline{C}K$ of $Ly = f$ all its conjugates. □

Remarks 4.5 1. A *C-structure* on a vector space W over $\overline{C}$ is a C-subspace W_0 of W such that $W = \overline{C} \otimes_C W_0$. The previous proposition gives a C-structure on $\overline{V}$. In [126], the authors show how one can put a C-structure on the entire solution space contained in a Picard-Vessiot extension of $\overline{C}(z)$ associated with a linear differential equation with coefficients in $C(z)$. This is used to understand the smallest subfield of $\overline{C}(z)$ needed when one is searching for a solution of the Riccati equation (cf. Definition 4.6) in $\overline{C}(z)$. We note that Proposition 4.3 also appears in [57] and [126].

2. The algorithm in the proof of Proposition 4.1 can be improved in several ways. For example, there are more efficient algorithms to find polynomial solutions of linear differential equations. These and related matters are discussed in [2, 4, 5, 57].

3. In many situations one is given a system $Y' = AY$ of differential equations where A is an $n \times n$ matrix with coefficients in $C(z)$ and asked to determine a basis for all solutions in $C(z)^n$. In theory, by finding a cyclic vector, one can reduce this problem to finding all solutions of an associated scalar equation $Ly = 0$ in $C(z)$ but finding this associated equation can be costly. An algorithm to find rational solutions of the system $Y' = AY$ directly has been given by Barkatou [22] and Abramov-Bronstein [3]. □

Exponential Solutions

We will keep the following notations. k is a differential field of characteristic 0 and let C be its field of constants. Fix $L = \partial^n + a_{n-1}\partial^{n-1} + \cdots + a_0 \in k[\partial]$ and a Picard-Vessiot extension K for L over the field $\overline{C}k$. Let $V := \{y \in K \mid L(y) = 0\}$ be the solution space of L in K and write $G \subset \mathrm{GL}(V)$ for the differential Galois group of L. A nonzero element $y \in V \subset K$ with $L(y) = 0$ is called an *exponential solution of* L if $u := \frac{y'}{y}$ lies in $\overline{C}k$. We will sometimes write, as a formal notation, $y = e^{\int u}$. Our aim is to compute the exponential solutions. We begin by reviewing some facts concerning the Riccati equation (cf. Remarks 3.20). Let $y, u \in K$ satisfy $y' = uy$. Formally differentiating this identity yields $y^{(i)} = P_i(u, u', \ldots u^{(i-1)})y$ where the P_i are polynomials with integer coefficients satisfying $P_0 = 1$ and $P_i = P'_{i-1} + uP_{i-1}$. Furthermore, $y \neq 0$ satisfies $Ly = 0$ if and only if $u := \frac{y'}{y}$ satisfies

$$R(u) = P_n(u, \ldots, u^{(n-1)}) + a_{n-1}P_{n-1}(u, \ldots, u^{(n-2)}) + \cdots + a_0 = 0. \quad (4.2)$$

Definition 4.6 *Equation (4.2) is called the* Riccati equation associated with $Ly = 0$. □

Exercise 4.7 *Riccati Equations.*
1. Show that $v \in \overline{C}k$ is a solution of the Riccati equation if and only if $\partial - v$ is a right-hand factor of L (i.e., $L = \tilde{L} \circ (\partial - v)$ for some $\tilde{L}$).

2. Show that $v \in K$ is a solution of the Riccati equation if and only if there is a $y \in V \subset K,\ y \neq 0$ with $y'/y = v$. □

The following gives the group theoretic interpretation of exponential solutions of a linear differential equation. Recall that a character of an algebraic group G over $\overline{C}$ is a homomorphism $\chi : G \to \overline{C}^*$ of algebraic groups.

Lemma 4.8 *With the above notations one has:*

1. *An element $y \in V \subset K,\ y \neq 0$ is an exponential solution if and only if there is a character χ of G such that $\sigma(y) = \chi(\sigma)y$ for all $\sigma \in G$.*
2. *If $u \in \overline{C}k$ is a solution of the Riccati equation then for some character χ there is a $y \in V_\chi := \{v \in V \mid \sigma(y) = \chi(\sigma)y$ for all $\sigma \in G\}$ such that $y \neq 0$ and $y'/y = u$.*
3. *The G-invariant lines of V are in a one-to-one correspondence with the solutions $u \in \overline{C}k$ of the Riccati equations.*
4. *The Riccati equation has an infinite number of solutions in $\overline{C}k$ if and only if, for some χ, $\dim_{\overline{C}} V_\chi \geq 2$. Furthermore, if the Riccati equation has finitely many solutions in $\overline{C}k$ then the number of solutions is at most n.*
5. *Let $y_1, y_2 \in V$ be two nonzero exponential solutions. Put $u_i = \frac{y_i'}{y_i}$. Then y_1, y_2 belong to the same V_χ if and only if $\frac{y_1}{y_2} \in \overline{C}k$. The latter is also equivalent to $u_1 - u_2$ having the form $\frac{f'}{f}$ for some $f \in \overline{C}k,\ f \neq 0$.*

Proof. 1. Consider any $\sigma \in G$. The element $y'/y \in \overline{C}k$ is invariant under G and thus $(\sigma(y)/y)' = 0$. Therefore, there is a $c_\sigma \in \overline{C}$ such that $\sigma(y) = c_\sigma y$. Clearly, $\sigma \mapsto c_\sigma$ is a character. Conversely, if $\sigma(y) = \chi(\sigma)y$ for all $\sigma \in G$, then y'/y is left fixed by G and so must be in $\overline{C}k$.

2. According to Exercise 4.7 $u = \frac{y'}{y}$ for some nonzero element y of V. Now apply part 1.
3. The condition $y \in V_\chi$ for some character χ of G is clearly equivalent to $\overline{C}y$ being a G-invariant line. Now use parts 1 and 2.
4. Let $\chi_1, \ldots, \chi_s$ denote the distinct characters of G such that the vector space V_{χ_j} is $\neq 0$. It is easily seen that the sum $\sum_{i=1}^{s} V_{\chi_i}$ is a direct sum. Using part 3, the statements easily follow.
5. y_1, y_2 belong to the same V_χ if and only if $\frac{y_1}{y_2}$ is invariant under G. The latter is equivalent to $\frac{y_1}{y_2} \in \overline{C}k$ and again (by logarithmic differentiation) with $u_1 - u_2 = \frac{f'}{f}$ for some nonzero element $f \in \overline{C}k$. □

Now we specialize to the case $k = C(z)$ and present an algorithm to find all exponential solutions for $L \in C(z)[\partial]$.

Proposition 4.9 *In addition to the above notations we assume that $k = C(z)$.*

1. *One can decide, in a finite number of steps, whether the Riccati equation $R(u) = 0$ has a solution in $\overline{C}(z)$.*
2. *Suppose that the Riccati equation has solution(s) in $\overline{C}(z)$. Let $\chi_1, \ldots \chi_s$ denote the distinct characters of G such that $V_{\chi_i} \neq 0$. Then one can calculate solutions $\{u_i\}_{i=1,\ldots,s} \in \overline{C}(z)$ of the Riccati equation and for each i a finite dimensional $\overline{C}$-vector space $W_i \subset \overline{C}[z]$ containing $\overline{C}$ such that for each i one has $V_{\chi_i} = y_i W_i$, where $y_i \in K$ is the exponential solution given by $u_i = \frac{y_i'}{y_i}$. Moreover, $\cup_{i=1}^{s}\{u_i + \frac{w'}{w} | w \in W_i,\ w \neq 0\}$ is the set of all solutions in $\overline{C}(z)$ is the Riccati equation.*

Proof. The idea of the proof is to solve the Riccati equation locally at every singular point and then glue the local solutions to a global solution. We consider first a local formal situation. Let 0 be a singular point of L. The solutions $u \in \overline{C}((z))$ of the Riccati equation of L can be derived from the classification of formal differential equations of Chap. 3. More precisely, one writes $u = \sum_{j \geq 2} \frac{c_j}{z^j} + r$ with $r \in z^{-1}\overline{C}[[z]]$. Then the "truncation" $[u]_0 := \sum_{j \geq 2} \frac{c_j}{z^j}$ of u has the property that $z[u]_0$ is an eigenvalue $q \in \mathcal{Q}$, as defined in Definition 3.27, which happens to lie in $z^{-1}\overline{C}[z^{-1}]$. The Newton polygon method presented in Chap. 3.3 actually computes the possibilities for these eigenvalues q (see Remarks 3.55). In Exercise 4.10 we outline how the Newton polygon techniques can be specialized and simplified to give this result directly.

Next, we consider a putative solution $u \in \overline{C}(z)$ of the Riccati equation. Let S be the set of the singular points of L (possibly including ∞). For each $\alpha \in S$, one calculates the finitely many possibilities for the truncated Laurent expansion $[u]_\alpha$ at α. After choosing for each α one of these possibilities one has $u = \tilde{u} + r$ where $\tilde{u} = \sum_{\alpha \in S}[u]_\alpha$ and the remainder r has the form $\sum_{\alpha \in \overline{C}} \frac{c_\alpha}{z-\alpha}$. One shifts ∂ to $\partial - \tilde{u}$ and computes the new operator $\tilde{L} := L(\partial - \tilde{u})$.

We now have to investigate whether the Riccati equation of $\tilde{L}$ has a solution $r \in \overline{C}(z)$ of the above form. For a singular point $\alpha \in S$ the coefficient c_α is seen to be a zero of the indicial polynomial of $\tilde{L}$ at α (compare with the case of rational solutions). At a regular point of L the putative solution u has locally the form $\frac{y'}{y}$, where $y \neq 0$ is a formal local solution of L. The order of y at the regular point lies in $\{0, 1, \dots, n-1\}$ and thus $c_\alpha \in \{0, 1, \dots, n-1\}$. We note that it is, a priori, not possible to find the regular points α for L where $c_\alpha \neq 0$. After choosing for each singular point α a possibility for c_α, the putative r has the form $r = \sum_{\alpha \in S} \frac{c_\alpha}{z-\alpha} + \frac{F'}{F}$, where F is a polynomial in $\overline{C}[z]$. The possible degree of F can be found by calculating a truncated local solution of the Riccati equation of $\tilde{L}$ at ∞. Let d be a possible degree for F. Then one puts $F = f_0 + f_1 z + \cdots + f_d z^d$, with as yet unknown coefficients $f_0, \dots, f_d$. The Riccati equation for r translates into a linear differential equation for F, which is equivalent to a system of homogenous linear equations for $f_0, \dots, f_d$. This ends the algorithm for the first part of the proposition. In trying all possibilities for the truncations $[u]_\alpha$ and the coefficients c_α for the singular points one obtains in an obvious way, the second part of the proposition. □

Exercise 4.10 *Rational solutions of the Riccati equation.*
In Proposition 4.9 we made use of the Newton polygon to find the possibilities for the truncation $[u]_\alpha$ of a rational solution u of the Riccati equation at the singular point α. In this exercise, the Newton polygon method is adapted to the present situation (cf. [269]). For convenience we suppose that $C = \overline{C}$.

1. Let $u \in C(z)$.
(i) Let $u = cz^{-\gamma} + \cdots \in C((z))$ with $c \in C^*, \gamma \geq 1$. Use the relation $P_{i+1} = P_i' + uP_i$ to show that:
(a) If $\gamma > 1$, then $P_i(u, u', \dots, u^{(i-1)}) = c^i z^{-i\gamma} + \cdots$.
(b) if $\gamma = 1$, then $P_i(u, u', \dots, u^{(i-1)}) = \prod_{j=0}^{i-1}(c-j)z^{-i} + \cdots$.
(ii) Find the translation of (i) at the point ∞. In other words, u is now considered as an element of $C((z^{-1}))$.

2. Let L be as in (4.1) and let $R(u) := \sum_{i=1}^{n} a_i P_i = 0$ be the associated Riccati equation. Let $u \in C(z)$ be a putative solution of $R(u) = 0$ and let $u = cz^{-\gamma} + \cdots \in C((z))$, with $c \in C^*$ and $\gamma > 1$, be its Laurent expansion. Derive from $R(u) = 0$ and part 1 an equation for γ and c. Show that there are only finitely many possibilities for $cz^{-\gamma}$.

3. Choose a possible term $cz^{-\gamma}$ from part 2. Indicate how one can find possible truncations $[u]_0$ of solutions $u = cz^{-\gamma} + \cdots \in C((z))$ by repeating part 2. Hint: Replace the operator $L = \sum a_i \partial^i$ by $\tilde{L}(\partial) = L(\partial + cz^{-\gamma})$.

4. Indicate how one can change the operator L into one or more operators $\tilde{L}$ such that the problem of finding rational solutions of the Riccati equation of L is translated into finding rational solutions of the Riccati equation of $\tilde{L}$ having the form $u = p + \sum u_\alpha/(z-\alpha)$, where $u_\alpha, \alpha \in C$ and $p \in C[z]$.

Now we concentrate on finding solutions u of $R(u) = 0$ having this form. Suppose that $\alpha \in C$ is a pole of some a_i, i.e., a singular point of L. Find an equation for u_α (this is again an indicial equation) and show that there are only finitely many possibilities for u_α. Show that one can modify L such that the putative u has the form $u = P'/P + p$, where $P, p \in C[z]$ and P has no roots in common with a denominator of any a_i.

5. Use part 1(ii) and calculations similar to those in part 2, to produce finitely many possibilities for the polynomial p. Modify the operator L such that $u = P'/P$. Now use Proposition 4.1 to find the polynomial solutions of the modified linear differential equation. □

Note that the proof of Proposition 4.9 (or the above exercise) implies that a solution u of the Riccati equation must be of the form

$$u = \frac{P'}{P} + Q + \frac{R}{S}, \tag{4.3}$$

where $P, Q, R, S \in \overline{C}[z]$, the zeroes of S are singular points and the zeroes of P are nonsingular points. We can therefore select S to be a product of the irreducible factors of the denominators of the a_i and so have it lie in $C[z]$. The next examples show that, in general, one cannot assume that $P, Q, R \in C[z]$.

Examples 4.11 1. The functions $\sqrt{z-i}, \sqrt{z+i}$ (with $i^2 = -1$) form a basis of the solution space of $y'' - \frac{1}{z^2+1}y' + \frac{1}{4(z^2+1)}y = 0$. One then sees that the only solutions in $\overline{\mathbf{Q}}(z)$ of the associated Riccati equation are $\frac{z \pm i}{2z^2+2}$. Thus the above R does not lie in $\mathbf{Q}[z]$.

2. The functions $(z+i)e^{iz}, (z-i)e^{-iz}$ form a basis of the solution space of $y'' - \frac{2}{z}y' + y = 0$. The only solutions in $\overline{\mathbf{Q}}(z)$ of the associated Riccati equation are $\{\frac{1}{z+i} + i, \frac{1}{z-i} - i\}$. Thus the above P and Q do not belong to $\mathbf{Q}[z]$. □

The algorithm in Proposition 4.9 goes back to Beke [28] (see also [255], §177). There are two aspects that contribute to the computational complexity of the above algorithm. The first is combinatorial. At each singular point one selects a candidate for terms of degree less than or equal to -1. If one uses the Newton polygon method described in Chap. 3, one generates at most n distinct candidates, where n is the order of the differential operator (see Remarks 3.55). If there are m singular points then one may need to try n^m possibilities and test n^m transformed differential equations to see if they have polynomial solutions. The second is the apparent need to work in algebraic extensions of C of large degree over C.

In [137], van Hoeij gives methods to deal with the combinatorial explosion in this algorithm and the problem of large field extensions of C (as well as a similar problem encountered when one tries to factor linear operators). The method also avoids the use of the Gröbner basis algorithm. Roughly speaking it works as follows. One makes a good choice of a singular point of the operator L and a formal local right-hand factor of degree 1 at this point. After a translation of the variable ($z \mapsto z + c$ or $z \mapsto z^{-1}$) and a shift $\partial \mapsto \partial + f$ with $f \in \overline{C}(z)$, the operator L has a right-hand factor of the form $\partial - \frac{y'}{y}$ with an explicit $y \in \overline{C}[[z]]$. Now one tries to find out whether $\frac{y'}{y}$ belongs to $\overline{C}(z)$. Equivalently, one tries to find a linear relation between y and y' over $\overline{C}[z]$. This is carried out by a Padé approximation. The method extends to finding right-hand factors of higher degree and applies in that case a generalization of the Padé approximation. This local-to-global approach works very well in practice and has been implemented in MAPLE V.5.

One can also proceed as follows (cf. [56, 220]). Let α be a fixed singular point. We may write a rational solution of the Riccati equation as

$$u = e_\alpha + f_\alpha,$$

where $e_\alpha = \frac{a_{n_\gamma,\gamma}}{(z-\alpha)^{n_\gamma}} + \cdots + \frac{a_{1,\gamma}}{z-\alpha}$ and $f_\alpha = b_{0,\gamma} + b_{1,\gamma}(z-\alpha) + \cdots$. One can calculate (at most) n possibilities for e_α. We shall refer to e_α as a *principal part at* α. One then considers the new differential equation $\tilde{L}(\partial) = L(\partial - e_\alpha)$. The term f_α will be of the form y'/y for some power series solution y of $\tilde{L}y = 0$. One can use the classical Frobenius algorithm to calculate (to arbitrary precision) a basis $y_1, \ldots, y_t$ of these power series solutions. Since f_α is a *rational* function, one must decide if there are any constants $c_1, \ldots, c_t$ such that $\frac{(c_1y_1+\cdots+c_ty_t)'}{(c_1y_1+\ldots+c_ty_t)}$ is rational, and such that $e_\alpha + \frac{(c_1y_1+\cdots+c_ty_t)'}{(c_1y_1+\cdots+c_ty_t)}$ is a solution of the Riccati equation. This can be done as follows.

One first calculates a bound N (see the next paragraph) on the degrees of the numerators and denominators of possible rational solutions of the Riccati equation. One then uses the first $2N + 1$ terms of the power series expansions of $\frac{(c_1y_1+\cdots+c_ty_t)'}{(c_1y_1+\cdots+c_ty_t)}$ to find a *Padé approximant* $\tilde{f}_\alpha$ [27] of $\frac{(c_1y_1+\cdots+c_ty_t)'}{(c_1y_1+\cdots+c_ty_t)}$ and then one substitutes $e_\alpha + \tilde{f}_\alpha$ into the Riccati equation and determines if there are any c_i that make this equation vanish. More concretely, given N, we may assume that the value of $c_1y_1 + \cdots + c_ty_t$ at $z = \alpha$ is 1 and write

$$\frac{(c_1y_1 + \cdots + c_ty_t)'}{(c_1y_1 + \cdots + c_ty_t)} = d_0(c_1, \cdots, c_t) + d_1(c_1, \cdots, d_t)(z - \alpha)$$

$$+ \cdots + d_{2N}(c_1, \cdots, c_t)(z - \alpha)^{2N} \bmod (z - \alpha)^{2N+1},$$

where the $d_1, \ldots, d_{2N}$ are polynomials in the c_i that can be calculated using the power series expansions of the y_i. One now must decide if there exist h_i, g_i such that

$$\tilde{f}_\alpha = \frac{h_N(z - \alpha)^N + \cdots + h_0}{g_N(z - \alpha)^N + \cdots + g_0} = d_0(c_1, \cdots, c_t) + d_1(c_1, \cdots, d_t)(z - \alpha)$$

$$+ \cdots + d_{2N}(c_1, \cdots, c_t)(z - \alpha)^{2N} \bmod (z - \alpha)^{2N+1}.$$

Multiplying both sides of the above equation by $g_N(z-\alpha)^N + \cdots + g_0$ and comparing the first $2N + 1$ powers of $z - \alpha$ yields a system $\mathcal{S}$ of polynomial equations in the

c_i, g_i, h_i that are linear in the g_i and h_i but nonlinear in the c_i. Substituting $u = e_\alpha + \tilde{f}_\alpha$ into the Riccati equation $R(u) = 0$, clearing denominators and equating powers of $z - \alpha$ yields another system of nonlinear polynomial equations $\tilde{\mathcal{S}}$. One can then use Gröbner basis methods to decide if there are c_i such that the system $\mathcal{S} \cup \tilde{\mathcal{S}}$ is solvable.

We now show how one can calculate a bound N on the degrees of the numerator and denominator of a rational solution of the Riccati equation. At each singular point $\alpha \in \overline{C}$ one can calculate the possible principal parts. In particular, this allows one to find the possible integers n_α and so bound the degrees of R and S in (4.3). At ∞, one can also calculate possible principal parts $e_\infty = \frac{a_{n_\infty,\infty}}{t^{n_\infty}} + \ldots + \frac{a_{1,\infty}}{t}$, where $t = \frac{1}{z}$. This allows one to bound the degree of Q in (4.3). Note that the constant $a_{1,\infty} = \deg P - \sum_\alpha a_{1,\alpha}$. Therefore, once we have bounded (or determined) all the residues $a_{1,\alpha}$ and $a_{1,\infty}$, we can bound (or determine) the possible degrees of P in (4.3). Therefore we can find the desired bound N. Note that although we have had to calculate mn principal parts, we have avoided the necessity of testing exponentially many combinations.

Both the algorithm in Proposition 4.9 and the above algorithm are presented in a way that has one work in (possibly large) extensions of C. Several ways to minimize this are given in [56, 57], and [137]. The examples above show that extensions of C cannot be avoided. For an even simpler example, let $p(z)$ be an irreducible polynomial over $\mathbf{Q}(z)$. The solutions of $p(\partial)y = 0$ are of the form $e^{\alpha z}$, where α is a root of $p(z) = 0$. Therefore, each solution of the Riccati equation is defined over an extension of $\mathbf{Q}$ of degree equal to the order of $p(\partial)$. Proposition 4.12 says that this is the worst that can happen.

Proposition 4.12 *Let L be a linear differential operator of order n with coefficients in $C(z)$ and let $R(u) = 0$ be the associated Riccati equation.*

1. *If there are only a finite number of solutions of $R(u) = 0$ in $\overline{C}(z)$ then each of them lies in a field of the form $C_0(z)$ where $[C_0 : C] \leq n$.*
2. *If $R(u) = 0$ has an infinite number of solutions in $\overline{C}(z)$ then there is a solution in a field of the form $C_0(z)$ where $[C_0 : C] \leq \frac{n}{2}$.*

Proof. We will let $k = \overline{C}(z)$ and use the notation of Lemma 4.8.
1. Let us assume that the Riccati equation has only a finite number of solutions. In this case, Lemma 4.8 implies that there are at most n of these. The group $\mathrm{Aut}(\overline{C}/C)$ acts on $\overline{C}(z)$ and permutes these solutions. Therefore, the orbit of any solution of the Riccati equation has size at most n and so is defined over a field of degree at most n over C.
2. One can prove this statement easily after introducing a $\mathrm{Gal}(\overline{C}/C)$ action on the solution space $V \subset K$ of the differential operator L. This operator has a regular point in C and for notational convenience we assume that 0 is a regular point for L. Then $W := \{y \in C((z)) |\ Ly = 0\}$ is a C-vector space of dimension n. The field $\overline{C}((z))$ contains a Picard-Vessiot field for L over $\overline{C}(z)$, namely the differential subfield generated over $\overline{C}(z)$ by all the elements of W. So we may identify K with this

subfield of $\overline{C}((z))$. The natural map $\overline{C} \otimes_C W \to V$, where $V \subset K$ is the solution space V of L in K, is clearly bijective. The group $\mathrm{Gal}(\overline{C}/C)$ acts on $\overline{C}((z))$ by $\sigma(\sum_{n>>-\infty} a_n z^n) = \sum_{n>>-\infty} \sigma(a_n) z^n$. This action induces on the subfield $\overline{C}(z)$ the natural action and the elements of W are fixed. Hence the subfield K is invariant under this action. Moreover, the action of $\mathrm{Gal}(\overline{C}/C)$ on V is the one given by the isomorphism $\overline{C} \otimes_C W \to V$.

Let $\chi_1, \ldots, \chi_s$ denote the distinct characters of the differential Galois group G such that the spaces V_{χ_i} are $\neq 0$. By assumption and by Lemma 4.8 one of these spaces, say V_{χ_1}, has dimension ≥ 2. The group $\mathrm{Gal}(\overline{C}/C)$ permutes the spaces V_{χ_i}. Therefore, the stabilizer $H \subset \mathrm{Gal}(\overline{C}/C)$ of V_{χ_1} is a closed subgroup of index $\leq n/2$. Let $C_0 \supset C$ denote the fixed field of H. Then $[C_0 : C] \leq n/2$ and the subspace V_{χ_1} is invariant under the action of $H = \mathrm{Gal}(\overline{C}/C_0)$. The action of H on V_{χ_1} yields a 1-cocycle class in $H^1(\mathrm{Gal}(\overline{C}/C_0), \mathrm{GL}_d(\overline{C}))$, where d is the dimension of V_{χ_1}. This cohomology set is well known to be trivial ([261]) and it follows that V_{χ_1} has a basis of elements in $C_0 \otimes_C W \subset C_0((z))$. For such a basis element y one has $\frac{y'}{y} \in C_0((z)) \cap \overline{C}(z) = C_0(z)$ as required. □

The above proposition appears in [126] and its proof applies to equations with coefficients in $C((z))$ as well. In this case the Riccati equation will always have a solution in a field whose degree over $C((z))$ is at most the order of L. In the latter case, the result also follows from a careful analysis of the Newton polygon or similar process (cf. [83, 137, 172, 281]). Despite Proposition 4.12, we know of no algorithm that, except in the case $n = 2$ (due to Berkenbosch [29] and, independently, to van Hoeij, who has included it in his modification and implementation of the Kovacic algorithm), will compute a rational solution of the Riccati equation that guarantees that all calculations are done in a field $C_0(z)$ with $[C_0 : C] \leq n$.

We end this section by noting that an algorithm for computing exponential solutions of linear differential systems is given in [220].

4.2 Factoring Linear Operators

Let a differential module M over the field $C(z)$, or equivalently a matrix differential operator $\partial - A$ over $C(z)$, be given. One final goal for algorithmic computations on M is to completely determine its Picard-Vessiot ring and its differential Galois group. For the case $C = \mathbf{C}$, many new questions arise, e.g., concerning monodromy groups, asymptotic behavior, Stokes matrices, etc. Here, we will restrict to the possibility of computing the Picard-Vessiot ring and the differential Galois group.

Let M_n^m denote the tensor product $M \otimes \cdots \otimes M \otimes M^* \otimes \cdots \otimes M^*$ (with n factors M and m factors M^*, and M^* denotes the dual of M). From the tannakian point of view, complete information on the Picard-Vessiot ring and the differential Galois group is equivalent to having a complete knowledge of all the differential submodules of finite direct sums of the M_n^m. Thus the basic problem is to find for a given differential module M all its submodules. We recall that M has a cyclic vector and is therefore isomorphic to $C(z)[\partial]/C(z)[\partial]L$ for some monic differential operator L. The submodules of M are in one-to-one correspondence with the monic right-hand factors of L. Therefore, the central problem is to factor differential operators. We will sketch a solution for this problem. This solution does not produce a theoretical algorithm for the computation of the Picard-Vessiot ring and the differential Galois group. Indeed, following this approach, one has to compute the submodules of infinitely many direct sums of the modules M_n^m. Nonetheless, algorithms modifying this approach have been given in [71] for the case when the differential Galois group is known to be reductive. An algorithm for the general case was recently presented in [140].

In order to simplify this exposition we will assume that C is algebraically closed. Computing the rational solutions for a differential operator $L \in C(z)[\partial]$ translates into finding the C-linear vector space $\{m \in M | \partial m = 0\}$, where M is the dual of the differential module $C(z)[\partial]/C(z)[\partial]L$. M.A. Barkatou and E. Pflügel, [22, 24], have developed (and implemented in their ISOLDE package) efficient methods to do this computation directly on the differential module (i.e., its associated matrix differential equation) without going to a differential operator by choosing a cyclic vector. Computing the exponential solutions of a differential operator translates into finding the 1-dimensional submodules of M. Again there is an efficient algorithm by Barkatou and Pflügel directly for the differential module (instead of an associated differential operator). Let $I_1, \ldots, I_s$ denote a maximal set of nonisomorphic 1-dimensional submodules of M. The sum of all 1-dimensional submodules of M is given as a direct sum $N_1 \oplus \cdots \oplus N_s \subset M$, where each N_i is a direct sum 1-dimensional submodules, isomorphic to I_i. This decomposition translates into the direct sum $\oplus V_\chi \subset V$, taken over all characters χ of the differential Galois group considered in Lemma 4.8.

4.2.1 Beke's Algorithm

Now we consider the problem of finding the submodules of dimension d of a given module M. We will explain the method, which goes back to Beke [28], in terms of differential modules. Let $N \subset M$ be a d-dimensional submodule. Then $\Lambda^d N$ is a 1-dimensional submodule of the exterior power $\Lambda^d M$. Of the latter we suppose that the 1-dimensional submodules are known. A 1-dimensional submodule P of $\Lambda^d M$ has the form $\Lambda^d N$ if and only if P is generated by a decomposable vector, i.e., a vector of the form $m_1 \wedge \cdots \wedge m_d$. Some multilinear algebra is needed to characterize the decomposable vectors in $\Lambda^d M$. We outline this, more information can be found in [114, 122, 133].

Let A be a vector space of dimension n over some field F. One denotes by A^* the dual vector space. There are contraction operators $i : \Lambda^k A^* \to \mathrm{Hom}_F(\Lambda^l A, \Lambda^{l-k} A)$ for $k \leq l$ and $i : \Lambda^k A^* \to \mathrm{Hom}_F(\Lambda^l A, \Lambda^{k-l} A^*)$ for $l \leq k$. For $k = 1$ and $l > 1$, the formula for the contraction operator i reads

$$i(L)(v_1 \wedge \cdots \wedge v_l) = \sum_{j=1}^{l} (-1)^{j-1} L(v_j) v_1 \wedge \cdots \widehat{v_j} \cdots \wedge v_l,$$

where L is an element of V^*, $v_1, \ldots, v_l \in V$ and where $\widehat{v_j}$ means that this term is removed. The formulas for the general case are similar. One shows that an element $a \in \Lambda^d A$ (with $1 < d < n$) is decomposable if and only if for every $b \in \Lambda^{d+1} A^*$ the expression $i(i(b)a)a$ is zero. These relation are called the *Plücker relations*. Choose a basis $e_1, \ldots, e_n$ for A and write $a = \sum_{i_1 < \cdots < i_d} a_{i_1,\ldots,i_d} e_{i_1} \wedge \cdots \wedge e_{i_d}$. Then for every $b \in \Lambda^{d+1} A^*$ the equation $i(i(b)a)a = 0$ is equivalent with a set of quadratic equations for the coefficients $a_{i_1,\ldots,i_d}$ of a. For the case $d = 2$ this simplifies to the element $a \wedge a \in \Lambda^4 A$ is zero. The latter is equivalent with $\binom{n}{4}$ quadratic equations for the coefficients of a. We note that for a decomposable $a = a_1 \wedge \cdots \wedge a_d \in \Lambda^d A$ the vector space generated by $a_1, \ldots, a_d \in A$ can also be found by applying $i(b)$ to a for all $b \in \Lambda^{d-1} A^*$.

We apply this to $\Lambda^d M$. As above, the 1-dimensional submodules of this space form a direct sum $N_1(d) \oplus \cdots \oplus N_s(d)$ (where the d indicates that we are working in $\Lambda^d M$). We pick one of these spaces, say $N_i(d)$, and give it a basis $w_1, \ldots, w_t$. The Plücker relations are applied to a general element $f_1 w_1 + \cdots + f_t w_t$ with all $f_i \in C(z)$. Solving these quadratic equations leads to all d-dimensional submodules of M. The quadratic equations are over the field $C(z)$. One can replace them by quadratic equations over C in the following way. Consider a regular point in C for M. For notational convenience we suppose that 0 is this regular point. Replace the module M by a matrix differential equation $\frac{d}{dz} + A$, with a matrix A that has no poles at 0. This matrix equation has a (unique) fundamental matrix F with coefficients in $C[[z]]$ such that $F(0) = 1$. Likewise, $\Lambda^d M$ has a fundamental matrix G with coefficients in $C[[z]]$ and $G(0) = 1$, obtained by taking the d-th exterior power of F. Evaluating the above elements $w_1, \ldots, w_t$ and $f_1, \ldots, f_t$ at $z = 0$ translates the Plücker relations over the field $C(z)$ to equivalent Plücker relations over the field C.

In order to make the above into an actual (and efficient) algorithm, one has to give the translation in terms of matrix differential operators. Let $e_1, \ldots, e_n$ be a basis of the differential module M. Let A be the matrix of ∂ w.r.t. this basis. Thus ∂ can be identified with the matrix operator $\frac{d}{dz} + A$ on the space $C(z)^n$. Then $\Lambda^d M$ has basis $\{e_{i_1} \wedge \cdots \wedge e_{i_d} | \ i_1 < \cdots < i_d\}$. The operator ∂ on $\Lambda^d M$ is defined by $\partial(w_1 \wedge \cdots \wedge w_d) = \sum_i w_1 \wedge \cdots \wedge (\partial w_i) \wedge \cdots \wedge w_d$. From this, one easily obtains the matrix differential operator for $\Lambda^d M$. The algorithms of Barkatou and Pflügel can now be put into action.

Remarks 4.13

(1) The original formulation of Beke's algorithm uses differential equations (or differential operators). This has several disadvantages. One has to use certain complicated minors of the wronskian matrix of a basis $y_1, \ldots, y_n$ of the solutions of the degree n operator $L \in C(z)[\partial]$. Let $M = C(z)[\partial]/C(z)[\partial]L$ be the differential module associated to L. Write e for the image of 1 in M. This is the cyclic vector cor-

responding to L. A natural element of the exterior product $\Lambda^d M$ is $e \wedge \partial e \wedge \cdots \wedge \partial^{d-1} e$. However, this element is not always a cyclic vector. Thus some work has to be done to produce a cyclic vector and a suitable differential operator for $\Lambda^d M$. This differential operator can be of high complexity, etc. We note also that Beke's original algorithm did not take the Plücker relation into account. Tsarev [284] gives essential improvements to Beke's original algorithm and puts the Plücker relation into action.

(2) One may insist on working with differential operators, and on producing rational and exponential solutions as explained above. There is a way out of the problem of the cyclic vector and its high complexity for the exterior power by applying the method of [134]. There, the matrix differential operator for a construction of linear algebra, applied to a differential operator, is used to produce the relevant information. Actually, one has to make a small variation on their method described for symmetric powers and eigenrings.

(3) Other improvements to the Beke algorithm have been given by several authors [56, 58, 60, 257]. In [116], Grigoriev also gives simplifications of the Beke algorithm as well as a detailed complexity analysis. An algorithm for determining the reducibility of a differential system is given in [115]. A method to enumerate all factors of a differential operator is given in [285].

(4) As remarked earlier, van Hoeij [137] gives methods to factor differential operators that are not based on Beke's algorithm. In this paper, he uses algorithms that find local factorizations (i.e., factors with coefficients in $\overline{C}((z))$) and applies an adapted version of Padé approximation to produce a global factorization. □

Example 4.14 We illustrate Beke's algorithm to find all the right-hand factors of order 2 of $L = \partial^4 - \partial^3$ over the field $\overline{\mathbf{Q}}(z)$.

The differential module $M := \overline{\mathbf{Q}}(z)[\partial]/\overline{\mathbf{Q}}(z)[\partial]L$ has basis $\{\partial^i e \mid i = 0, \ldots, 3\}$. It is easily seen to have a basis $e_1, \ldots, e_4$ such that $\partial e_i = 0$ for $i = 1, 2, 3$ and $\partial e_4 = e_4$. The differential module $\Lambda^2 M$ has basis $\{e_i \wedge e_j | 1 \le i < j \le 4\}$. We are looking for the 1-dimensional submodules. Such a module is generated by an element $a = \sum_{1 \le i < j \le 4} a_{i,j} e_i \wedge e_j$ with coefficients in $\overline{\mathbf{Q}}[z]$ such that ∂a is a multiple of a. Moreover, the Plücker relation $a \wedge a = 0$ translates into $a_{1,2}a_{3,4} - a_{1,3}a_{2,4} + a_{1,4}a_{2,3} = 0$. One writes

$$a = \sum_{1 \le i < j \le 3} a_{i,j} e_i \wedge e_j + \Big(\sum_{i=1,2,3} b_i e_i \Big) \wedge e_4,$$

and finds that $\partial a = \sum_{1 \le i < j \le 3} a'_{i,j} e_i \wedge e_j + (\sum_{i=1,2,3} ((b_i + b'_i) e_i) \wedge e_4$. Using degrees in z one finds that $\partial a = \lambda a$ implies that λ is a constant and, in fact, λ can only be 0 or 1. This yields two vector spaces of solutions, namely $\sum_{1 \le i < j \le 3} a_{i,j} e_i \wedge e_j$ with $a_{i,j}$ constants and $(\sum_{i=1}^3 b_i e_i) \wedge e_4$ with b_1, b_2, b_3 constants. Both families satisfy the Plücker relation. This yields two families of 2-dimensional submodules N of M, namely:

(i) N is generated over $\overline{\mathbf{Q}}(z)$ by a two-dimensional subspace of $\oplus_{i=1}^{3}\overline{\mathbf{Q}}e_i$.
(ii) N is generated over $\overline{\mathbf{Q}}(z)$ by e_4 and a 1-dimensional subspace of $\oplus_{i=1}^{3}\overline{\mathbf{Q}}e_i$.
Translating this back to monic right-hand factors of L, one finds two families, parametrized by $\mathbf{P}^2(\overline{\mathbf{Q}})$, namely:

$$\partial^2 - \frac{d_2 + 2d_3z}{d_1 + d_2z + d_3z^2}\partial + \frac{2d_2}{d_1 + d_2z + d_3z^2} \quad \text{and}$$

$$\partial^2 - \left(\frac{d_2 + 2d_3z + d_1 + d_2z + d_3z^2}{d_1 + d_2z + d_3z^2}\right)\partial + \frac{d_2 + 2d_3z}{d_1 + d_2z + d_3z^2}.$$

□

4.2.2 Eigenring and Factorizations

Another method, not based on Beke's algorithm, is given in [271]. This method uses the eigenring (see Proposition 2.13). It does not always factor reducible operators (see Exercise 2.14) but does often yield factors quickly. We will show that the method does factor all reducible completely reducible operators (see Definition 2.37).

We recall and continue the discussion of the eigenring in Sects. 2.2 and 2.4. Consider instead of a differential operator L of degree n, the associated differential module M over $C(z)$. We assume for convenience that C is algebraically closed. The eigenring $\mathcal{E}(M)$ of M consists of all $C(z)$-linear maps $B : M \to M$, which commute with ∂. One of the constructions of linear algebra applied to M is the differential module $\mathrm{Hom}(M, M)$ (isomorphic to $M^* \otimes M$). Then the eigenring of M is the C-algebra of the rational solutions of $\mathrm{Hom}(M, M)$. Clearly $1_M \in \mathcal{E}(M)$ and the dimension of $\mathcal{E}(M)$ is $\leq n^2$. This dimension is equal to n^2 if and only if M is a direct sum of n copies of a 1-dimensional module.

Suppose that $\mathcal{E}(M)$ contains an element B that is not a multiple of the identity. The elements $1, B, B^2, \ldots, B^{n^2} \in \mathcal{E}(M)$ are linearly dependent over C. One can easily calculate the monic polynomial $I(T) \in C[T]$ of minimal degree satisfying $I(B) = 0$. Let $c \in C$ be a root of $I(T)$. Then $B - c1_M$ is not invertible and is not 0. Hence the kernel of $B - c1_M$ is a nontrivial submodule of M.

An interpretation of the eigenring $\mathcal{E}(M)$ is the following. Let V denote the solution space of M provided with the action of the differential Galois group G. Then every $B \in \mathcal{E}(M)$ induces a C-linear map $\tilde{B} : V \to V$ commuting with the action of G. The above polynomial $I(T)$ is the minimum polynomial of $\tilde{B}$. Conversely, any C-linear map $V \to V$, commuting with the action of G, is a $\tilde{B}$ for a unique $B \in \mathcal{E}(M)$.

For the actual calculation of $\mathcal{E}(M)$ one replaces M by a matrix differential operator $\frac{d}{dz} + A$. The B that we are looking for are now the matrices commuting with $\frac{d}{dz} + A$. In other terms, they are the rational solutions of the matrix differential equation $B' = BA - AB$. Suppose for convenience that A has no poles at 0. Then one can give the solution space V of M the interpretation of the kernel of $\frac{d}{dz} + A$

operating on $C[[z]]^n$. Every solution is determined by its constant term. In this way, one finds the action of a $B \in \mathcal{E}(M)$ on V explicitly. This method is useful for determining the algebra structure of $\mathcal{E}(M)$.

Direct decompositions of a given differential module M correspond to idempotent elements of $\mathcal{E}(M)$. They can be computed as follows. Let $B_1, \ldots, B_r$ be a C-basis of $\mathcal{E}(M)$. Consider any C-linear combination $e := \lambda_1 B_1 + \cdots + \lambda_r B_r$. Then $e^2 = e$ yields a set of quadratic equations for $\lambda_1, \ldots, \lambda_r$ with, a priori coefficients in $C(z)$. The above method of evaluation at $z = 0$ turns these equations into quadratic equations over C.

Assume that the differential module M is completely reducible. According to Proposition 2.40 the eigenring $\mathcal{E}(M)$ is a direct product of the matrix algebras $\mathrm{M}_{n_i}(C)$ for $i = 1, \ldots, s$. Thus the above method will produce a complete decomposition of M as a direct sum of irreducible submodules.

We return now to the differential operator $L \in \mathcal{D} := C(z)[\partial]$ of degree n. We recall (see Proposition 2.13) that the eigenring $\mathcal{E}(L)$ is also the eigenring of the differential module $M := \mathcal{D}/\mathcal{D}L$. Moreover, $\mathcal{E}(L)$ is described by

$$\mathcal{E}(L) = \{R \in \mathcal{D} \mid \deg R < n \text{ and there exists } S \in \mathcal{D} \text{ with } LR = SL\}.$$

One can make the above condition on R explicit by writing $R = R_0 + R_1\partial + \cdots + R_{n-1}\partial^{n-1}$ and dividing LR on the right by L with a remainder. Thus $LR = FL + \tilde{R}_0 + \tilde{R}_1\partial + \cdots + \tilde{R}_{n-1}\partial^{n-1}$ Each $\tilde{R}_i$ is formally a linear homogeneous expression in $R_0, \ldots, R_{n-1}$ and their derivatives. Then $R \in \mathcal{E}(L)$ if and only if $\tilde{R}_i = 0$ for $i = 1, \ldots, n-1$. These equations are sometimes called the *eigenequations* of L. We note that these equations, written as a differential equation for the matrix $(R_i^{(j)})_{i,j=1}^n$, is equivalent to the matrix differential equation $B' = BA - AB$, which we have encountered above.

In [271] general methods are given for determining $\dim_C \mathcal{E}(L)$. For example, using Exercise 2.4.2, one can find operators $L_1, \ldots, L_n$ such that there is an effective correspondence between the solutions of $L_1(Z_1) = 0, \ldots, L_n(Z_n) = 0$ and the solutions of the eigenequations. One can then use the methods of Sect. 4.1 to find solutions of this former system in $C(z)$. Other techniques for finding $\mathcal{E}(L)$ are discussed in [18] and [136].

Exercise 4.15 Let L be the differential operator

$$\partial^4 + (2 + 2z^2)\partial^2 + 4z\partial + (4 + 2z^2 + z^4),$$

say over the differential field $\overline{\mathbf{Q}}(z)$. Try to prove that $\mathcal{E}(L)$ is isomorphic to the matrix algebra $\mathrm{M}_2(\overline{\mathbf{Q}})$. Hint: Use a computer. □

We will see in the next section that completely reducible operators arise naturally. A test for complete reducibility of operators over $C(z)$ (with C algebraically closed) is given in [271] and this is extended to algebraic extensions of $C(z)$ in [71].

We end this section with an exercise giving a version of the Eisenstein irreducibility criterion that can be applied to differential operators.

Exercise 4.16 *Factorization over $\overline{C}(z)$ versus factorization over $\overline{C}[z]$.*

(1) Show that $z\partial^2 + z^2\partial - z = (\partial + z)(z\partial - 1)$. Note that each of the two first-degree factors has coefficients with g.c.d. 1, while z divides the coefficients of the product. Therefore, a naive version of Gauss's Lemma is false for linear operators over the ring $\overline{C}[z]$.

(2) Let $L = \partial^2 + z\partial - 1$. Show that L factors over $\overline{C}(z)$ but that L cannot be written as the product of first-degree operators with coefficients in $\overline{C}[z]$. Hint: Show that z^{-1} is the only exponential solution of $Ly = 0$.

Despite these examples, Kovacic [166] gives the following Eisenstein-like criterion for the irreducibility of a differential operator:
Let R be a differential integral domain with quotient field F and let P be a prime differential ideal in R. Assume that the local ring R_P is principal. Let $L = \sum_{i=0}^{l} c_i\partial^i$ be a differential operator with coefficients in R such that $c_i \in P$ for $i = 1, \dots, l$, $c_0 \notin P$ and $c_l \notin P^2$. Then L is irreducible over F.

(3) Use the above criterion to show that if $L = \partial^2 + p$, where $p \in \overline{C}[z]$ is of odd degree, then L is irreducible over $\overline{C}(z)$. Hint: Let $\deg_z p = 2k + 1$, define $\delta = z^{-k}\partial$ and rewrite $z^{-2k-1}L$ as operator in δ with coefficients in the ring $R = \overline{C}[z^{-1}]$. The operator δ makes R into a differential domain with differentiation given by $r \mapsto z^{-k}\frac{dr}{dz}$. Show that the ideal $P = z^{-1}R$ is a prime ideal and a differential ideal (with respect to this differentiation on R). Now apply Kovacic's criterion. □

4.3 Liouvillian Solutions

In this section k is a differential field with an algebraically closed field of constants C (of characteristic 0). Proposition 1.45 in Sect. 1.5 states that a linear differential equation $Ly = 0$ of order n over k that has a nonzero liouvillian solution also has a liouvillian solution $y \neq 0$ such that $u := y'/y$ is algebraic over k. In other words, the Riccati equation associated to $Ly = 0$ has a solution u that is algebraic over k. Using some group theory we will prove that there is a constant $I(n)$, depending only on n, such that there is even an algebraic solution of this Riccati equation with degree $\leq I(n)$ over k. This leads to the following method of testing whether the equation $Ly = 0$ has liouvillian solutions. For each d with $1 \leq d \leq I(n)$ the existence of an algebraic solution u of degree d over k of the Riccati equation is tested by calculating (special) exponential solutions of the "d-th symmetric power of L". If no solutions are found then $Ly = 0$ has no liouvillian solutions $\neq 0$. In the opposite case we indicate how one determines the minimum polynomial of u over k. Special algorithms to find liouvillian solutions for second and third order operators will be discussed.

4.3.1 Group Theory

The group theory that we need is based on the following theorem of Jordan ([145, 146]; see also the exposition of Jordan's ideas given by Dieudonné [84]). It is interesting to note that Jordan proved this result in order to study algebraic solutions of linear differential equations.

Theorem 4.17 *Let C be an algebraically closed field of characteristic zero. There exists an integer-valued function $n \mapsto J(n)$ such that every finite subgroup of $\mathrm{GL}_n(C)$ contains an abelian normal subgroup of index at most $J(n)$.*

Various authors have given bounds for $J(n)$. Blichtfeldt [39] showed that $J(n) < n!(6^{n-1})^{\pi(n+1)+1}$ where $\pi(x)$ denoted the number of primes less than or equal to x (see [86] for a modern presentation). One also finds the following values of $J(n)$ in [39]: $J(2) = 12$, $J(3) = 360$, and $J(4) = 25920$. Schur [256] showed that $J(n) \le (\sqrt{8n}+1)^{2n^2} - (\sqrt{8n}-1)^{2n^2}$ (see [76] for a modern exposition). Other proofs can be found in [85] and [305].

Proposition 4.18 *C is an algebraically closed field of characteristic zero. A subgroup $G \subset \mathrm{GL}_n(C)$ acts on $\mathbf{P}(C^n) = \mathbf{P}^{n-1}(C)$, i.e., the set of lines in C^n through 0. Assume that G has some finite orbit on $\mathbf{P}(C^n)$, then it also has an orbit of length at most $I(n) := \max_{r \le n}\{[\frac{n}{r}]J(r)\}$.*

Proof. We may replace G by its Zariski closure in $\mathrm{GL}_n(C)$ and assume that G is a linear algebraic group. Assume that the line $Cw \subset V := C^n$ has a finite orbit $\{Cw_1, \dots, Cw_s\}$ under G. Then $H = \{h \in G |\ h(Cw_i) = Cw_i \text{ for all } i\}$ is a normal subgroup of G of index $\le s!$. Let $\chi_1, \dots, \chi_t$ denote the distinct characters $\chi_i : H \to C^*$ such that the vector space $V_{\chi_i} := \{v \in V | h(v) = \chi_i(h)v \text{ for all } h \in H\}$ is not 0. Then V has $\oplus_{i=1}^t V_{\chi_i}$ as subspace. Since H is normal in G, one has the fact that G permutes the spaces V_{χ_i} and $\oplus_{i=1}^t V_{\chi_i}$ is a G-invariant subspace. Consider the stabilizer $H_1 \subset G$ of V_{χ_1}. Then the index of H_1 in G is $\le t$. Thus $[G : H_1] \le [\frac{n}{r}]$, where r is the dimension of V_{χ_1}. If r happens to be 1, then the line V_{χ_1} has a G-orbit of length $\le n$.

Now suppose $r > 1$. The action of H_1 on V_{χ_1} induces an action of the finite group $H_1/H \subset \mathrm{PGL}_r(C) = \mathrm{PSL}_r(C)$ on the projective space $\mathbf{P}(V_{\chi_1}) = \mathbf{P}^{r-1}(C)$. Indeed, H acts on V_{χ_1} via its character χ_1. Let $H_2 \subset \mathrm{SL}_r(C)$ be the preimage of H/H_1. One applies Theorem 4.17 to the finite group H_2 and obtains a normal abelian subgroup $H_3 \subset H_2$ of index $\le J(r)$. The abelian subgroup H_3 stabilizes some line $L \subset V_{\chi_1}$. The H_2-orbit of L has length $\le [H_2 : H_3] \le J(r)$. The H_1-orbit of L coincides with the H_2-orbit of L. Finally, the G-orbit of L has length $\le [G : H_1]$ times the length of the H_1-orbit of L. Thus the length of the G-orbit of L is $\le [\frac{n}{r}]J(r)$. □

Proposition 4.19 *Suppose that the linear differential equation $Ly = 0$ of degree n over k has a nonzero liouvillian solution. Then there is a solution $y \ne 0$ such that $u := \frac{y'}{y}$ is algebraic over k of degree $\le I(n)$.*

Proof. Let $K \supset k$ denote a Picard-Vessiot extension for the equation $Ly = 0$ over the field k. One considers the action of the differential Galois group G of $Ly = 0$ over k, on the solution space $V := \{y \in K | \ L(y) = 0\}$. Suppose that $y \in V$, $y \neq 0$ is such that $u := \frac{y'}{y} \in K$ is algebraic over k and has minimum polynomial $P(T) \in k[T]$ of degree d. Then for any $\sigma \in G$, the element $\sigma(u)$ is again a solution of $P(T) = 0$. This implies that the connected component of the identity G^o of G acts trivially on u. In other words, $u \in K^{G^o}$. By Galois correspondence, $K^{G^o} \supset k$ is a Galois extension with a Galois group G/G^o. From ordinary Galois theory it follows that the G-orbit of u consists of all the zeros of $P(T)$ and has length d. This implies that the G-orbit of the line $Cy \subset V$ also has length d.

On the other hand, a line $Cy \subset V$ that has a finite G-orbit yields an element $u := \frac{y'}{y} \in K$, which has a finite G-orbit. Hence u is algebraic over k. From the above it follows that its degree over k is equal to the length of the G-orbit of Cy. Using this translation, an application of Proposition 4.18 finishes the proof. □

Weaker versions of Propositions 4.18 and 4.19 originally appeared in [265]. Proposition 4.18 can also be deduced from results of Platonov and Malcev (see Remark 11.12 and [303], Theorem 3.6, p.45 and Corollary 10.11, p.142). The present versions of Propositions 4.18 and 4.19 appear in [61]. In this paper, [290], and [291] other results concerning sharper bounds on the degrees of algebraic solutions of the Riccati equation for certain classes of differential Galois groups can be found.

4.3.2 Liouvillian Solutions for a Differential Module

Let k be again a differential field of characteristic 0 having a field of constants $C \neq k$, which is algebraically closed. For a differential operator $L \in k[\partial]$ one wants to determine the solutions $u = \frac{y'}{y}$ of the associated Riccati equation that are algebraic over k of a given degree d. This amounts to producing the monic minimal polynomial of u over k. As far as we know, the existing algorithms, which will be described in some detail, are written in the framework of differential operators.

It is helpful to reformulate the problem for differential modules (or matrix differential equations). The differential module M will be the *dual* of the differential modules $k[\partial]/k[\partial]L$. The *contravariant* solution space of L is $V := \{y \in K| \ Ly = 0\}$, where K is a Picard-Vessiot field for L over k. This space can now be identified with the *covariant* solution space $V := \ker(\partial, K \otimes M)$ of M. In the sequel we will use the canonical isomorphism $K \otimes_C V \to K \otimes_k M$. As remarked before, rational solutions of L correspond to $\{m \in M | \partial m = 0\}$ and exponential solutions of L (or equivalently rational solutions of the Riccati equation of L) correspond to 1-dimensional submodules of M.

Lemma 4.20 *Let $\bar{k}$ denote the algebraic closure of k and let $\ell \subset \bar{k}$ be any subfield containing k.*

The solutions in ℓ of the Riccati equation of L are in one-to-one correspondence with the one-dimensional ℓ-submodules of the differential module $\ell \otimes_k M$ over ℓ.

Proof. One replaces k by ℓ and M by $\ell \otimes_k M$ and regards L as an element of $\ell[\partial]$. Now the statement translates into the observation made above, namely: "The rational solutions of the Riccati equation of L are in one-to-one correspondence with the 1-dimensional submodules." □

Now we make a detailed investigation of this correspondence. Let $u = \frac{y'}{y}$, with $y \in V,\ y \neq 0$, be an algebraic solution of the Riccati equation of L, having degree d over k. According to the proof of Proposition 4.19, the line $Cy \subset V$ has a G-orbit of length d. As before, $G \subset \mathrm{GL}(V)$ denotes the differential Galois group.

Using the canonical isomorphism $K \otimes_C V \to K \otimes_k M$ as an identification, we have $Ky \subset K \otimes_k M$. We fix a basis $e_1, \ldots, e_n$ of M and write $y = \sum_{i=1}^n f_i e_i$ with all $f_i \in K$. For convenience, we suppose that $f_n \neq 0$. One normalizes y to $\tilde{y} := \frac{1}{f_n} y = \sum_{i=1}^n g_i e_i$ with $g_n = 1$. The line $Cy \subset V$ has a G-orbit $\{Cy_1 = Cy, \ldots, Cy_d\}$ of length d. The same holds for the 1-dimensional subspace $Ky = K\tilde{y}$. For any $\sigma \in G$ one has $\sigma(K\tilde{y}) = K\tilde{y}$ if and only if $\sigma\tilde{y} = \tilde{y}$. Indeed, $\sigma\tilde{y} = \sum_{i=1}^n \sigma(g_i)e_i$ and $\sigma(g_n) = 1 = g_n$. Write $\{\tilde{y} = \tilde{y}_1, \ldots, \tilde{y}_d\}$ for the G-orbit of $\tilde{y}$. Clearly the coefficients of $\tilde{y}_1$ with respect to the basis $e_1, \ldots, e_n$ generate a field extension of k of degree d. The lemma implies that this field extension is equal to $k(u)$.

Intermezzo on symmetric powers

For the next step we will need some information on *symmetric powers* of vector spaces (see also [170], Chap. XVI, §8 for information concerning the symmetric powers). Let F be any field and A a vector space over F of dimension n with basis $e_1, \ldots, e_n$. The d-th symmetric power of A, denoted by $\mathrm{sym}_F^d A = \mathrm{sym}^d A$, is the quotient of the ordinary tensor product $A \otimes \cdots \otimes A$ (of d copies of A) by the linear subspace generated by all elements $a_1 \otimes \cdots \otimes a_n - a_{\pi(1)} \otimes \cdots \otimes a_{\pi(n)}$ with $a_1, \ldots, a_n \in A$ and $\pi \in S_n$. We will not distinguish in notation between the elements of $\mathrm{sym}^d A$ and their preimages in $A \otimes \cdots \otimes A$. The space $\mathrm{sym}^d A$ has basis $\{e_{i_1} \otimes \cdots \otimes e_{i_d} |$ with $1 \leq i_1 \leq i_2 \leq \cdots \leq i_d \leq n\}$. A vector in $\mathrm{sym}^d A$ will be called *decomposable* if it has the form $a_1 \otimes \cdots \otimes a_d$ for certain elements $a_1, \ldots, a_d \in \overline{F} \otimes A$, where $\overline{F}$ denotes the algebraic closure of F. Here is an *important subtle point* that can be explained as follows. After choosing a basis for A over F one may identify A with the homogeneous polynomials in n variables $X_1, \ldots, X_n$ over F, having degree 1. This leads to an identification of $\mathrm{sym}^d A$ with the homogeneous polynomials of degree d in $X_1, \ldots, X_n$. An irreducible homogeneous polynomial of degree d in $F[X_1, \ldots, X_n]$ may factor as a product of d linear homogeneous polynomials in $\overline{F}[X_1, \ldots, X_n]$. Consider a homogeneous $H \in F[X_1, \ldots, X_n]$ of degree 3, which factors over $\overline{F}$ as a product of three linear terms. We may suppose that the coefficient of X_n^3 in H is 1. The factorization can be put in the form $H = (X_n - a)(X_n - b)(X_n - c)$ where a, b, c are homogeneous terms of degree 1 in $\overline{F}[X_1, \ldots, X_{n-1}]$. The Galois group $\mathrm{Gal}(\overline{F}/F)$ acts on this decomposition and permutes the a, b, c. Thus we have a (continuous) homomorphism $\mathrm{Gal}(\overline{F}/F) \to S_3$.

Consider the extreme case, where this homomorphism is surjective. The images $\tilde{a}, \tilde{b}, \tilde{c} \in \overline{F}$ under a suitable substitution $X_i \mapsto c_i \in F$, $i = 1, \ldots, n-1$, are the roots of the irreducible polynomial $(X_n - \tilde{a})(X_n - \tilde{b})(X_n - \tilde{c}) \in F[X_n]$. Then the linear factor $(X_n - a)$ lies in $F(\tilde{a})[X_1, \ldots, X_n]$.

The subset of the decomposable vectors in $\mathrm{sym}^d A$ are the F-rational points of an algebraic variety given by homogeneous equations. In terms of the chosen basis $e_1, \ldots, e_n$, one writes a vector of $\mathrm{sym}^d A$ as a linear expression

$$\sum_{i_1 \leq \cdots \leq i_d} x(i_1, \ldots, i_d) e_{i_1} \otimes \cdots \otimes e_{i_d}.$$

Let the indeterminates $X_{i_1,\ldots,i_d}$, with $i_1 \leq \cdots \leq i_d$, stand for the coordinate functions on the vector space $\mathrm{sym}^d A$ w.r.t. the given basis. There is a homogeneous prime ideal P in the polynomial ring $F[\{X_{i_1,\ldots,i_d}\}_{i_1 \leq \cdots \leq i_d}]$ such that the set of decomposable vectors is the zero set of P. A certain collection of generators of P is known under the name *Brill's equations*([108], pp. 120, 140; [53], p. 181). The existence of these equations follows from the observation that the morphism $\mathbf{P}(A) \times \cdots \times \mathbf{P}(A) \to \mathbf{P}(\mathrm{sym}^d A)$, given by $(Fa_1, Fa_2, \ldots, Fa_d) \mapsto Fa_1 \otimes \cdots \otimes a_d$, between projective spaces has a Zariski-closed image. We will not need the precise form of Brill's equations.

Now we return to G-orbit $\{\tilde{y}_1 = \tilde{y}, \ldots, \tilde{y}_d\}$ of $\tilde{y}$. The element $m(d, u) := \tilde{y}_1 \otimes \cdots \otimes \tilde{y}_d$ of $\mathrm{sym}^d_K(K \otimes_k M) = K \otimes_k \mathrm{sym}^d_k M$ is invariant under G and belongs therefore to $\mathrm{sym}^d M$. Moreover, it is a decomposable vector, since it is a decomposable vector over the field extension K of k. The covariant solution space $\ker(\partial, K \otimes_k \mathrm{sym}^d M)$ can clearly be identified with $\mathrm{sym}^d_C V$. The one-dimensional subspace $Cy_1 \otimes \cdots \otimes y_d$ is G-invariant. One observes that $Ky_1 \otimes \cdots \otimes y_d = K\tilde{y}_1 \otimes \cdots \otimes \tilde{y}_d = Km(d, u)$. Thus $Km(d, u)$ is invariant under ∂ and then the same holds for the 1-dimensional subspace $km(d, u)$ of $\mathrm{sym}^d M$. We conclude that the algebraic solution u of the Riccati equation yields a 1-dimensional submodule of $\mathrm{sym}^d M$, generated by a decomposable vector.

A converse holds as well. Suppose that $km(d)$ is a submodule of $\mathrm{sym}^d M$ generated by a decomposable vector. Then $m(d)$ gives rise to one or more algebraic solutions of the Riccati equation of L, having degree $\leq d$ over k. We want to indicate an algorithm with input a decomposable 1-dimensional submodule $km(d)$ of $\mathrm{sym}^d M$ and output one or more algebraic solutions of the Riccati equation. For notational convenience we take $d = 3$. One has to compute elements $m_1, m_2, m_3 \in \overline{k} \otimes_k M$ such that $m_1 \otimes m_2 \otimes m_3 = m$. In the extreme case, considered above, one computes an extension $k(r) \supset k$ of degree 3 and an $m_1 \in k(r) \otimes_k M$. The 1-dimensional space $k(r)m_1 \subset k(r) \otimes_k M$ is invariant under ∂. Thus $\partial m_1 = um_1$ for some $u \in k(r)$. Then u is an algebraic solution of the Riccati equation and $k(u) = k(r)$ and the minimal polynomial of degree 3 of u over k is found. We note that an algorithm based on symmetric powers, decomposable vectors and Brill's equations for determining liouvillian solutions of an operator is presented in [275] with improvements presented in [134].

4.3.3 Liouvillian Solutions for a Differential Operator

In this subsection, we will present a simple (but not very efficient) algorithm to decide if a linear differential operator L over $C(z)$ has a nonzero liouvillian solution and produce such a solution if it exists. This algorithm can be modified by applying Tsarev's refinements of the straightforward Beke algorithm for differential operators. At the end of the section we will discuss other refinements.

We begin by reviewing some facts about symmetric powers $\mathrm{Sym}^d(L)$ of an operator L. In Sect. 2.3, we showed that the solution space of this operator is spanned by $\{y_1 \cdots y_d \mid Ly_i = 0 \text{ for all } i\}$. Furthermore, we showed that $\mathrm{Sym}^d(L)$ can be calculated in the following manner: Let L have order n and let $e = 1$ be a cyclic vector of $k[\partial]/k[\partial]L$ with minimal annihilating operator L. One differentiates e^d, $\mu = \binom{n+d-1}{n-1}$ times. This yields a system of $\mu + 1$ equations:

$$\partial^j e^d = \sum a_{j,I}\mathcal{E}^I \quad j = 0, \ldots, \mu, \tag{4.4}$$

where the sum is over all $I = (i_0, i_1, \ldots, i_{n-1})$ with $i_0 + i_1 + \cdots + i_{n-1} = d$ and $\mathcal{E}^I = e^{i_0}(\partial e)^{i_1} \cdots (\partial^{n-1}e)^{i_{n-1}}$. The smallest t such that the first t of the forms on the right-hand side of these equations are linearly dependent over $C(z)$ yields a relation $\partial^t e^d + b_{t-1}\partial^{t-1}e^d + \cdots + b_0 e^d = 0$ and so $\mathrm{Sym}^d(L) = \partial^t + b_{t-1}\partial^{t-1} + \cdots + b_0$. The following example will be used several times in this chapter.

Example 4.21 Let $L = \partial^2 - \frac{1}{2z}\partial - z$ and $m = 2$. We shall calculate the equations (4.4) and $\mathrm{Sym}^2(L)$. Following the above procedure, we have

$$\begin{pmatrix} e^2 \\ \partial e^2 \\ \partial^2 e^2 \\ \partial^3 e^2 \end{pmatrix} = \begin{pmatrix} 1 & 0 & 0 \\ 0 & 2 & 0 \\ 2z & \frac{1}{z} & 2 \\ 3 & \frac{8z^3 - 1/2}{z^2} & \frac{3}{z} \end{pmatrix} \begin{pmatrix} e^2 \\ e\partial e \\ (\partial e)^2 \end{pmatrix}. \tag{4.5}$$

The above matrix B of coefficients has rank 3. A calculation shows that $(0, -\frac{4z^3-1}{z^2}, -\frac{3}{2z}, 1)B = 0$. Therefore, $\mathrm{Sym}^2(L) = \partial^3 - \frac{3}{2z}\partial^2 - \frac{4z^3-1}{z^2}\partial$. □

We will also need other auxiliary operators. These will be formed using the following definition.

Definition 4.22 *Let k be a differential field and $L \in k[\partial]$. The* derivative of L *denoted by* $\mathrm{Der}(L)$, *is defined to be the minimal monic annihilating operator of $\partial \in k[\partial]/k[\partial]L$.* □

As in Sect. 2.3, one can show that the solution space of $\mathrm{Der}(L)$ is $\{y' \mid Ly = 0\}$.

Example 4.23 Let $L = \partial^2 - \frac{1}{2z}\partial - z$ and let $e = \partial \in k[\partial]/k[\partial]L$. To calculate $\mathrm{Der}(L)$ we form the following system:

$$e = \partial$$
$$\partial e = \frac{1}{2z}\partial + z$$
$$\partial^2 e = (z - \frac{1}{4z^2})\partial + \frac{3}{2}.$$

Therefore, $\mathrm{Der}(L) = \partial^2 - \frac{3}{2z}\partial + (\frac{1}{z^2} - z)$. We shall also need in Example 4.25 that $\mathrm{Sym}^2(\mathrm{Der}(L)) = \partial^3 - \frac{9}{2z}\partial^2 - \frac{4z^3-10}{z^2}\partial + \frac{4s-10}{z^3}$. □

Proposition 4.24 *Let L be a linear differential operator of order n with coefficients in $k = C(z)$. One can decide, in a finite number of steps, if $Ly = 0$ has a nonzero solution liouvillian over k and, if so, find the minimal polynomial of an element u algebraic over k so that any y with $y'/y = u$, we have $Ly = 0$.*

Proof. We shall present an algorithm having its roots in [200] and given explicitly in [265].

The algorithm Proposition 4.19 implies that if $Ly = 0$ has a nonzero liouvillian solution then it has a solution $y \neq 0$ such that $u = y'/y$ is algebraic of order at most $I(n)$. The algorithm proceeds by searching for the minimal polynomial of such a u. Let m be a positive integer less than or equal to $I(n)$ and let

$$P(u) = u^m + b_{m-1}u^{m-1} + \cdots + b_0$$

be a putative minimal polynomial of the logarithmic derivative u of a nonzero solution of $Ly = 0$. Note that u satisfies the Riccati equation $R(u) = 0$ associated with L. Since the (ordinary) Galois group of $P(u)$ acts transitively on the roots of $P(u)$, all solutions of $P(u) = 0$ also satisfy the Riccati equation and therefore each of these roots is the logarithmic derivative of a solution of $Ly = 0$. Let $u_i = z_i'/z_i$, $i = 1, \ldots, m$ be the roots of $P(u) = 0$ where the z_i are solutions of $Ly = 0$. Since the coefficients of $P(u)$ are the elementary symmetric function of the u_i, we have that, for each $i = 1, \ldots, m-1$,

$$\binom{m}{i} b_{m-i} = \sum_{\sigma \in S_m} \frac{z'_{\sigma(1)} \cdots z'_{\sigma(i)}}{z_{\sigma(1)} \cdots z_{\sigma(i)}} \tag{4.6}$$

$$= \frac{\sum_{\sigma \in S_m} z'_{\sigma(1)} \cdots z'_{\sigma(i)} z_{\sigma(i+1)} \cdots z_{\sigma(m)}}{z_1 \cdots z_m}, \tag{4.7}$$

where S_m is the group of permutations on m elements. Note that $b_{m-1} = (z_1 \cdots z_m)'/(z_1 \cdots z_m)$ and so is the logarithmic derivative of a solution of the m-th symmetric power $\mathrm{Sym}^m(L)$ of L. Furthermore, for each $i = 2, \ldots, m$, the element $(z_1 \cdots z_m)b_{m-i}$ is a solution of $L_i := \mathrm{Sym}^{m-i}(L) \otimes \mathrm{Sym}^i(\mathrm{Der}(L))$. In particular, for each $i = 2, \ldots, m$, b_{m-i} is a rational solution of $L_i(\partial + b_{m-1})$. Note that this latter statement holds trivially for $i = 1$ as well.

Proposition 4.9 applied to the operator $\mathrm{Sym}^m(L)$ implies that one can find $v_1, \ldots, v_s$ such that for any exponential solution y of $\mathrm{Sym}^m(L)y = 0$ there exists a j such that some $y/y_j \in \overline{C}(z)$ for any solution of $y_j' = v_j y_j$. Therefore, for some j, we have that

b_{m-i} is a rational solution of $L_i(\partial + v_j)$

for $i = 1, \ldots, m$. Fix a value of j. Let $z_{i,1}, \ldots, z_{i,n_i}$ be a basis of the rational solutions of $L_i(\partial + v_j)$. Let

$$b_{m-i} = c_{i,1}z_{i,1} + \cdots + c_{i,n_i}z_{i,n_i}, \tag{4.8}$$

where the $c_{r,s}$ are indeterminate constants. To see if there exist constants $c_{r,s}$ such that the resulting polynomial is the minimal polynomial of a solution of the Riccati equation one proceeds as follows. The set of these constants for which the resulting $P(u)$ is irreducible over $C(z)$ forms a constructible set $\mathcal{I}$. Let us assume that $\mathcal{I}$ is nonempty. Assuming the $c_{r,s}$ take values in $C = \overline{C}$, one has $u' = P_1(u)$, where P_1 is a polynomial of degree at most $m - 1$ in u that can be calculated using the equality $P(u) = 0$. Similar expressions $u^{(i)} = P_i(u)$ can be calculated for all derivatives of u. Replacing $u^{(i)}$ in $R(u)$ with the $P_i(u)$ and then reducing modulo $P(u)$ yields an expression that must vanish if $P(u) = 0$ is the minimal polynomial of a solution of the Riccati equation. This yields algebraic conditions on the constants $\{c_{j,l}\}$ and defines a constructible set and standard techniques (e.g., Gröbner bases) can be used to decide if any of these are consistent. Repeating this for all j yields all possible minimal polynomials of algebraic solutions of degree m of the Riccati equation. □

Example 4.25 Consider the operator $L = \partial^2 - \frac{1}{2z}\partial - z$. We shall show that this operator has a solution y with y'/y algebraic of degree two over $C(z)$. Let $P(u) = u^2 + b_1u + b_0$ be the putative minimal polynomial of an algebraic solution of the Riccati equation. In Example 4.21, we showed that $\mathrm{Sym}^2(L) = \partial^3 - \frac{3}{2z}\partial^2 - \frac{4z^3-1}{z^2}\partial$. The only exponential solution of this equation is $y = 1$ so we must have $b_1 = 0$. To find b_0 we consider $\mathrm{Sym}^2(\mathrm{Der}(L)) = \partial^3 - \frac{9}{2z}\partial^2 - \frac{4z^3-10}{z^2}\partial + \frac{4s-10}{z^3}$ (see Example 4.23). This has a one-dimensional space of rational solutions and this is spanned by z. Therefore $P(u) = u^2 - cz$ for some constant c. The associated Riccati equation is $R(u) = u' + u^2 - \frac{1}{2}u - z$. From $P(u) = 0$, we have $u' = \frac{c}{2z}u$, so c is determined by $\frac{c}{2z}u + cz - \frac{1}{2z}u - z = 0$. Therefore $c = 1$ and $P(u) = u^2 - z$. This implies that L has a solution space with basis $\{e^{\int\sqrt{z}}, e^{\int -\sqrt{z}}\}$. □

We will not present the much more involved modifications needed to make the above algorithm efficient. One problem that occurs is that the order of the m-th symmetric power of the differential operator L is less than the maximal possible order. This can be avoided using the techniques in [134] and [135]. In these papers, the authors show how to construct matrix differential equations whose solution spaces are isomorphic to the symmetric powers of the solution space of $Ly = 0$. Using this, they are then able to construct, independent of the order of $\mathrm{Sym}^m(L)$, polynomials all of whose roots are logarithmic derivatives of solutions of $Ly = 0$ when such polynomials exist.

The algorithm presented in Proposition 4.24 is based on [265], where an algorithm to find *all* liouvillian solutions of a linear differential equation is presented. Many of the ideas in [265] already appear in [200]. In [290] and [291], Ulmer refines the group theoretic techniques to significantly improve the bounds in all cases and also develops conditions to further narrow down the set of possible degrees of algebraic logarithmic derivatives of solutions that can occur. The modifications needed for the algorithm that we presented appear in [275] and [276]. We also note that the case of inhomogeneous equations is discussed in [78] and the situation where the coefficients of the equation lie in more general fields (e.g., liouvillian extensions of $C(z)$) is discussed in [55] and [269].

The question of deciding if a linear differential equation has only algebraic solutions (or even one nonzero algebraic solution) has a long history. In 1872, Schwarz [258] gave a list of those parameters for which the hypergeometric equation has only algebraic solutions (for higher hypergeometric functions this was done by Beukers and Heckman [33]). An algorithm (with some mistakes) to find the minimal polynomial of an element u algebraic over $C(z)$ with $\exp(\int u)$ satisfying a given second order linear differential equation was found by Pepin [219] in 1881. Using invariant theory, Fuchs [105, 106] was able to find the minimal polynomial of an algebraic solution of a second order linear differential equation assuming that the Galois group is the finite imprimitive group of order 24, 48 or 120 (this method is generalized in [273]). In [158, 159], Klein shows that any second order linear differential equation with only algebraic solutions comes from some hypergeometric equation via a rational change in the independent variable $z := r(z)$. This approach was turned into an algorithm by Baldassarri and Dwork [14]. Jordan [145] considered the problem of deciding if a linear differential equation of order n has only algebraic solutions. As already mentioned, he showed that a finite subgroup of GL_n has an abelian normal subgroup of index bounded by a computable function $J(n)$ of n. This implies that such an equation has a solution whose logarithmic derivative is algebraic of degree at most $J(n)$. Jordan's approach was made algorithmic in [264] (see also [48] and [218] for similar but incomplete algorithms due to Boulanger and Painlevé). It should be noted that the algorithms of Boulanger, Klein, Painlevé, and Pépin, are all incomplete in at least one regard. Each of these algorithms, at one point or another, is confronted with the following problem: Given an element u, algebraic over $C(z)$, decide if $\exp(\int u)$ is also algebraic over $C(z)$. Boulanger refers to this as *Abel's Problem* ([48], p. 93) and none of these authors gives an algorithm to answer this question. In 1970, Risch [246] showed that this problem could be solved if one could decide if a given divisor on a given algebraic curve is of finite order. Risch showed how one could solve this latter problem by reducing the jacobian variety of the curve modulo two distinct primes and bounding the torsion of the resulting finite groups. For other work concerning Abel's Problem, see [14, 54, 77, 282, 299, 300]. The introduction to [200], the articles [112, 277] and the book [113] give historical accounts of work concerning algebraic solutions of linear differential equations.

One can also ask if one can solve linear differential equations in terms of other functions. The general problem of solving a linear differential equation in terms of lower order linear differential equations is given in [266] and [268].

4.3.4 Second Order Equations

Kovacic' influential paper [167] presented for the first time an efficient algorithm to find all liouvillian solutions of a second order linear differential equation. In this section we shall describe this algorithm in the context of the methods of the last two sections, originating in [265].

The general method for finding liouvillian solutions simplifies considerably for second order operators, due to the following observations.

Proposition 4.26 *([167, 59, 292]) Suppose that the field of constants C of the differential field k is algebraically closed. Let $L = \partial^2 + a\partial + b \in k[\partial]$ and let $K \supset k$ be its Picard-Vessiot extension.*

1. *The n-th symmetric power* $\mathrm{Sym}^n(L)$ *of L has order* $\binom{n+2-1}{2-1} = n+1$.
2. *Fix $n \geq 2$. Define operators L_i, by the recursion $L_0 = 1$, $L_1 = \partial$ and $L_{i+1} = \partial L_i + iaL_i + i(n-i+1)bL_{i-1}$. Then $L_{n+1} = \mathrm{Sym}^n(L)$.*
3. *Any solution of* $\mathrm{Sym}^n(L)y = 0$ *in K is the product of n solutions of $Ly = 0$.*
4. *Suppose that $L = \partial^2 - r$ with $r \in k$. Let $P(T) := -T^n + \sum_{i=0}^{n-1} \frac{a_i}{(n-i)!} T^i$ be an irreducible polynomial over k, u a solution of the Riccati equation $u' + u^2 = r$ of L and $P(u) = 0$. Then the a_i satisfy the recurrence relation*

$$a_{i-1} = -a_i' - a_{n-1}a_i - (n-i)(i+1)ra_{i+1} \text{ for } i = n, \ldots, 0,$$

and $a_n = -1$, $a_{-1} = 0$. In particular, the coefficient a_{n-1} determines the polynomial $P(T)$. Moreover, a_{n-1} is an exponential solution of $\mathrm{Sym}^n(L)$ *(or in other terms a solution in k of the Riccati equation of* $\mathrm{Sym}^n(L)$.

Proof. 1. Let differential module M with cyclic vector e associated to L be $k[\partial]/k[\partial]L$ and e be the image of 1. The operator $\mathrm{Sym}^n(L)$ is the minimal monic operator, annihilating the element $e \otimes \cdots \otimes e \in \mathrm{sym}^n M$. For notational convenience, we will write a tensor product as an ordinary product. In particular, the element $e \otimes \cdots \otimes e$ is written as e^n. The space M has basis e, ∂e and $\partial(\partial e) = -be - a\partial e$. The collection $\{e^{n-i}(\partial e)^i |\ i = 0, \ldots, n\}$ is a basis of $\mathrm{sym}^n M$. A straightforward calculation shows that the elements $\partial^i(e^n)$ for $i = 0, \ldots, n$ form a basis of $\mathrm{sym}^n M$. Thus $\mathrm{Sym}^n(L)$ has order $n+1$.
2. By induction, one shows that $L_i(e^n) = n(n-1)\cdots(n-i+1)e^{n-i}(\partial e)^i$ for $i = 0, \ldots, n+1$.
3. The solution space of L is $V = \{f \in K |\ Lf = 0\}$. Let y_1, y_2 be a basis. Any

homogeneous polynomial H in two variables X_1, X_2 over C is a product of linear homogeneous terms. Hence $H \neq 0$ implies that $H(y_1, y_2)$ is not zero. By counting dimensions one finds that $\{H(y_1, y_2)|\ H$ homogeneous of degree $n\}$ is the solution space of $\mathrm{Sym}^n(L)$.
4. The idea of the proof is to differentiate the polynomial relation $P(u) = 0$ in the subfield $k(u)$ of K and to use that $u' = -u^2 + r$. This yields the equality

$$\sum_{i=0}^{n-1} \frac{a_i'}{(n-i)!} u^i + (-nu^{n-1} + \sum_{i=0}^{n-1} \frac{a_i}{(n-i)!} iu^{i-1})(-u^2 + r) = 0.$$

The term u^n occurring in this expression is replaced by $\sum_{i=0}^{n-1} \frac{a_i}{(n-i)!} u^i$. There results a polynomial expression u of degree less than n. All its coefficients have to be zero. These coefficients yield the recurrence relations of part 4.
The equation $P(T) = 0$ has all its solutions $u = u_1, \ldots, u_n$ in K. Each u_i has the form $\frac{y_i'}{y_i}$ for suitable solutions in K of $Ly_i = 0$. Then $a_{n-1} = u_1 + \cdots + u_n = \frac{f'}{f}$ with $f = y_1 \cdots y_n$ a solutions of $\mathrm{Sym}^n(L)$ in K. □

Remarks 4.27 1. In [292], the authors show that the above recursion holds without the assumption that $P(T)$ is irreducible. They use this fact to give further improvements on Kovacic's algorithm. We will only use the above proposition in our presentation.
2. Kovacic' method for solving $(\partial^2 - r)y = 0$ is now almost obvious. For a suitable $n \geq 1$ one tries to find exponential solutions of $\mathrm{Sym}^n(L)$ with the methods explained earlier. If a solution is found then the polynomial $P(T)$ can be calculated and, by construction, any root of $P(T) = 0$ is a liouvillian solution of the required type. We still have to explain what the suitable ns are. □

Exercise 4.28 Let $k = \mathbf{C}(z)$ and let

$$Ly = y'' + \frac{3-4z}{16z^2} y .$$

1. Show that $Ly = 0$ has no exponential solution over $\mathbf{C}(z)$.
2. Use Proposition 4.26 to show that

$$\mathrm{Sym}^2(L)y = y''' - \frac{(-3+4z)}{4z^2} y' + \frac{2z-3}{4z^3} y,$$

and that this equation has $y = z^{\frac{1}{2}}$ as an exponential solution. Therefore, $Ly = 0$ has a solution whose logarithmic derivative is algebraic of degree 2.
3. Use Proposition 4.26 to show that

$$P(u) = u^2 - \frac{1}{2z} u + \frac{1}{16z^2} - \frac{1}{4z}$$

is the minimal polynomial of an algebraic solution of the associated Riccati equation. □

The final information for Kovacic' algorithm comes from the classification of the algebraic subgroups of $\mathrm{SL}_2(C)$. To be able to use this information one transforms a linear differential operator $\partial^2 + a\partial + b$ into the form $\partial^2 - r$ by means of the shift $\partial \mapsto \partial - a/2$. The liouvillian solutions $e^{\int u}$, with u algebraic over k, of the first operator are shifted by $u \mapsto u + a/2$ to the liouvillian solutions of the same type of the second operator. Thus we may and will restrict ourselves to operators of the form $\partial^2 - r$. The advantage is that the differential Galois group of $\partial^2 - r$ lies in $\mathrm{SL}_2(C)$ (see Exercise 1.35(5)).

The well-known classification of algebraic subgroups of $\mathrm{SL}_2(C)$ ([151], p.31; [167], p.7, 27) is the following theorem.

Theorem 4.29 *Let G be an algebraic subgroup of $\mathrm{SL}_2(C)$. Then, up to conjugation, one of the following cases occurs:*

1. *G is a subgroup of the Borel group*

$$B := \left\{ \begin{pmatrix} a & b \\ 0 & a^{-1} \end{pmatrix} \mid a \in C^*, b \in C \right\}.$$

2. *G is not contained in a Borel group and is a subgroup of the infinite dihedral group*

$$D_\infty = \left\{ \begin{pmatrix} c & 0 \\ 0 & c^{-1} \end{pmatrix} \mid c \in C^* \right\} \cup \left\{ \begin{pmatrix} 0 & c \\ -c^{-1} & 0 \end{pmatrix} \mid c \in C^* \right\}.$$

3. *G is one of the groups $A_4^{\mathrm{SL}_2}$ (the tetrahedral group), $S_4^{\mathrm{SL}_2}$ (the octahedral group) or $A_5^{\mathrm{SL}_2}$ (the icosahedral group). These groups are the preimages in $\mathrm{SL}_2(C)$ of the subgroups $A_4, S_4, A_5 \subset \mathrm{PSL}_2(C) = \mathrm{PGL}_2(C)$.*
4. *$G = \mathrm{SL}_2(C)$.*

We now present a rough version of the Kovacic algorithm.
Theorem 4.29 gives the following information:
Let $L = \partial^2 - r$ have solution space V and differential Galois group $G \subset \mathrm{GL}(V) = \mathrm{SL}_2(C)$. The smallest integer $n \geq 1$ such that the Riccati equation of L has an algebraic solution of degree n over k is equal to the smallest length of a G-orbit of a line $W \subset V$ in $\mathbf{P}(V) = \mathbf{P}^1(C)$. From the above classification one can read off this n. Namely, $n = 1$ for case 1; $n = 2$ for case 2; $n = 4, 6, 12$ for case 3. Indeed, one knows that the actions of A_4, S_4, A_5 on $\mathbf{P}^1(C)$ have orbits of minimal length 4, 6, 12, respectively (see [167]). Finally, in case 4 there is no finite G-orbit and there are no liouvillian solutions.

The algorithm computes whether there are algebraic solutions u of the Riccati equation $u' + u^2 = r$ for $n = 1, 2, 4, 6, 12$ (and in this order). If a solution u is found then essentially one finds a complete description of the solution space, the Picard-Vessiot field and the differential Galois group. Note that Example 4.28 illustrates this procedure.

Remarks 4.30 1. The algorithm that Kovacic presents in [167] (see also [232]) is more detailed (and effective). He does not calculate the symmetric powers but shows how one can determine directly an exponential solution of the prescribed symmetric powers. This is done by calculating local solutions of the second order equation at each singular point, that is, solutions in the fields $C((z-c))$ or $C((z^{-1}))$. This allows one to determine directly the possible principal parts at singular points of solutions of symmetric powers. Kovacic then develops techniques to determine if these principal parts can be glued together to give exponential solutions. The question of determining the local formal Galois group (i.e., over $C((z-c))$ or $C((z^{-1}))$) is considered in [232] where explicit simpler algorithms are also given to determine the global Galois groups of second order equations with one and two singular points.

2. Various improvements and modifications have been given for the Kovacic Algorithm since its original publication. Duval and Loday-Richaud [87] have given a more uniform treatment of the considerations concerning singular points and have also applied the algorithm to decide which parameters in the hypergeometric equations (as well as several other classes of second order equations) lead to liouvillian solutions. In [292], Ulmer and Weil show that except in the reducible case, one can decide if there is a liouvillian solution (and find one) by looking for solutions of appropriate symmetric powers *that lie in* $C(z)$. This eliminates some of the nonlinear considerations of Kovacic's algorithm. If the equation has coefficients in $C_0(x)$, where C_0 is not algebraically closed, it is important to know in advance how large an algebraic extension of C_0 is required. In [126] and [306] sharp results are given for Kovacic's algorithm as well as a general framework for higher order equations. In [291], sharp results are given concerning what constant fields are needed for equations of all prime orders. An algorithm to determine the Galois groups of second and third order equations and decide if they have liouvillian solutions (but not necessarily find these solutions) is given in [272] and [273]. This will be discussed in the next section.

3. We note that Kovacic's algorithm finds a solution of the form $\exp(\int u)$ where u is algebraic over $C(z)$ when the equation has liouvillian solutions. When the equation has only algebraic solutions, the algorithm does not find the minimal polynomials of such solutions, even when the Galois group is tetrahedral, octahedral or icosahedral. For these groups this task was begun in [105] and [106] and completed and generalized to third order equations in [272] and [273].

4. Applications of Kovacic's algorithm to questions concerning the integrability of Hamiltonian systems are given in [213] (see also the references given in this book). □

4.3.5 Third Order Equations

It is possible to extend Kovacic's algorithm to third order operators. We will suppose that the field of constants C, of the differential field k, is algebraically closed. By normalizing a third order operator to the form $L = \partial^3 + a\partial + b \in k[\partial]$, one essentially loses no information. The solution space of L will be called V and the differential Galois group G of L is contained in $\mathrm{SL}(V) = \mathrm{SL}_3(C)$. The nice properties of second order operators, given by Proposition 4.26, need not hold for L. It is possible that a symmetric power $\mathrm{Sym}^n(L)$ of L has an order strictly less than the dimension of the corresponding differential module $\mathrm{sym}^n(k[\partial]/k[\partial]L)$. Another possibility is that $\mathrm{Sym}^n(L)$ has the maximal possible dimension, but an exponential solution of $\mathrm{Sym}^n(L)$ is not decomposable. A third problem is that a decomposable exponential solution of a $\mathrm{Sym}^n(L)$ of maximal dimension docs not automatically yield the minimal polynomial of the corresponding algebraic solution u of degree n over k of the Riccati equation of L. The list of (conjugacy classes of) algebraic subgroups of $\mathrm{SL}_3(C)$ is quite long. A summary is the following.

L is reducible. This means that V contains a G-invariant subspace of dimension 1 or 2. In the first case, L has an exponential solution. In the second case, one replaces L by the dual operator $L^* := -\partial^3 - \partial \cdot a + b$ with solution space V^*, the dual of V. Now L^* has an exponential solution.

L is irreducible and imprimitive. In general, one calls a differential operator *imprimitive* if the action of the differential Galois group G on the solution space is irreducible and there is a direct sum decomposition $V = \oplus_{i=1}^{s} V_i$ that is respected by the action of G. Thus for every $g \in G$ there is a permutation $\pi \in S_s$ such that $g(V_i) = V_{\pi(i)}$ for all i. Moreover, since L is irreducible, G acts transitively on the set of subspaces $\{V_i\}$. L is called *primitive* if L is irreducible and not imprimitive.

Here $\dim V = 3$ and imprimitivity of L means that $V = Ce_1 + Ce_2 + Ce_3$ and G permutes the set of lines $\{Ce_i\}$. The line $Ce_1 \otimes e_2 \otimes e_3 \in \mathrm{sym}^3 V$ is G-invariant. Thus $\mathrm{Sym}^3(L)$ has an exponential solution that is decomposable.

L is primitive. The possibilities for G, up to conjugation, are:
(1) $\mathrm{SL}_3(C)$.
(2) $\mathrm{PSL}_2(C)$. This group is the image of $\mathrm{SL}_2(C)$ in $\mathrm{SL}_3(C)$ for its natural action on $\mathrm{sym}^2 C^2 \cong C^3$.
(3) $\mathrm{PSL}_2(C) \times C_3$ where $C_3 = \{\lambda I \mid \lambda \in C,\ \lambda^3 = 1\}$ is the center of $\mathrm{SL}_3(C)$.
(4) A list of eight finite primitive subgroups of $\mathrm{SL}_3(C)$:

- A_5 with its familiar interpretation as the group of the symmetries (of determinant 1) of the icosahedron. This group and its companion $A_5 \times C_3$ have minimal length 6 for an orbit on $\mathbf{P}(C^3)$.
- The Valentiner group $A_6^{\mathrm{SL}_3}$ of order 108. This is the preimage of A_6 for the map $\mathrm{SL}_3(C) \to \mathrm{PSL}_3(C)$. The minimal length of an A_6-orbit in $\mathbf{P}(C^3)$ is 36.
- The group $H_{216}^{\mathrm{SL}_3}$ of order 648, which is the preimage of the Hessian group $H_{216} \subset \mathrm{PSL}_3(C)$ of order 216. The minimal length of an H_{216}-orbit is 9.
- The subgroup $H_{72}^{\mathrm{SL}_3}$ of index 3 in $H_{216}^{\mathrm{SL}_3}$, has again minimal length 9 for an orbit in $\mathbf{P}(C^3)$.

- The subgroup $F_{36}^{\mathrm{SL}_3}$ of index 2 of $H_{72}^{\mathrm{SL}_3}$ has a minimal length 6 of an orbit in $\mathbf{P}(C^3)$.
- The group $G_{168} \cong \mathrm{PSL}_2(\mathbf{F}_7)$ of order 168 and its companion $G_{168} \times C_3$ have minimal length 21 for an orbit in $\mathbf{P}(C^3)$.

Thus it is interesting to calculate *decomposable* exponential solutions of $\mathrm{Sym}^n(L)$ for $n = 1, 3, 6, 9, 21, 36$. Exponential solutions that are not decomposable, in particular for $n = 2, 3, 4$, also give useful information on the group G. Algorithms for third order equations using the above ideas can be be found in [272, 273, 274, 292] and especially [134]. In the latter paper a complete algorithm for order three equations is presented (as of the writing of this book, there seems to be no complete implementation). It uses the above classification of algebraic subgroups and especially for the finite primitive groups invariants and semi-invariants. For order four differential equations similar results were obtained in [127]. Samples of these results are:

(Translated from [134]).
Suppose that $\mathrm{Sym}^n(L)$ *has no decomposable exponential solutions for* $n = 1, 2$ *but that there is an rational solution for* $n = 4$. *Then* $G = G_{168}$ *and an irreducible polynomial* $P(T) \in k[T]$ *of degree 21 is produced such that any solution* u *of* $P(u) = 0$ *is a solution of the Riccati equation of* L.

(Taken from [272]).
Suppose that $L = \partial^3 + a\partial + b$ *is irreducible. Then* $Ly = 0$ *has a liouvillian solution if and only if*
1. $\mathrm{Sym}^4(L)$ *has order strictly less than* 15 *or factors, and*
2. *one of the following holds:*
   (a) $\mathrm{Sym}^2(L)$ *has order* 6 *and is irreducible, or*
   (b) $\mathrm{Sym}^3(L)$ *has a factor of order* 4.

4.4 Finite Differential Galois groups

Until now, we have been concerned with algorithmic aspects of the direct problem for differential equations. Inverse problems are discussed in Chap. 11 and in Sect. 7 a construction is given for a linear differential equation having a prescribed *connected* differential Galois group. Here we will discuss the same algorithmic problem but now for finite groups. The problem can be stated as follows: Let G be a finite group and $\rho : G \to \mathrm{GL}(V)$ be a faithful representation (i.e., ρ is injective) with V an n-dimensional vector space over an algebraically closed field C of characteristic 0. The general *algorithmic problem* is to produce a differential operator $L \in C(z)[\partial]$ of degree n such that the representation ρ is isomorphic to the action of the differential Galois group of L on its solution space. This is the main theme of [233]. Instead of asking for a differential operator L, one may ask for a differential module (or a matrix differential operator) with the required properties. This, however, has not been implemented.

The construction of an algorithm (and its implementation) for the required operator L is quite different from what we have seen until now. In particular, Chaps. 5 and 6 will be used and some more theory is presented. Furthermore, we will narrow down on producing second and third order scalar fuchsian equations with three singular points $0, 1, \infty$. Since we will use some analytic theory, the field C is supposed to be a subfield of $\mathbf{C}$. Consider $\mathbf{P}^* = \mathbf{P}^1_{\mathbf{C}} \setminus \{0, 1, \infty\}$ with base point $1/2$ for the fundamental group π_1. This group has the presentation $\pi_1 =< a_0, a_1, a_\infty | \ a_0 a_1 a_\infty = 1 >$. A homomorphism $h : \pi_1 \to \mathrm{GL}_n(\mathbf{C})$ is given by the images $A_0, A_1, A_\infty \in \mathrm{GL}_n(\mathbf{C})$ of the a_0, a_1, a_∞, having the relation $A_0 A_1 A_\infty = 1$. The image of the homomorphism is a subgroup G of $\mathrm{GL}_n(\mathbf{C})$. We are interested in the situation where G is a given finite group.

The solution of the (strong) Riemann-Hilbert for finite groups (see Chap. 6) guarantees a fuchsian matrix differential equation with singularities at $0, 1, \infty$, with h as monodromy representation and G as differential Galois group (compare with Sects. 5.2 and 5.3). In many cases, one can also prove the existence of the scalar fuchsian differential equation with the same data. This is what we are looking for. First we provide information about the form of such a scalar differential equation.

4.4.1 Generalities on Scalar Fuchsian Equations

Consider the scalar fuchsian equation

$$L(y) := y^{(n)} + a_1 y^{(n-1)} + \cdots + a_{n-1} y^{(1)} + a_n y = 0,$$

with singular points $p_1, \ldots, p_s \in \mathbf{C}$ and possibly ∞. For each singular point, there is a set of *local exponents*. Let us consider for convenience the situation that $z = 0$ is a regular singular point. In general, the local solutions at $z = 0$ of the equation lie in the field $\mathbf{C}((z))(\{z^a\}_{a \in \mathbf{C}}, l)$, where the symbols z^a and l have the interpretation: the functions $e^{a \log(z)}$ and the function $\log(z)$ on a suitable sector at $z = 0$. We consider the situation that l is not present in the set of local solutions. (This is certainly the case when the monodromy group is finite.) Then there is a basis of local solutions

$$y_1 = z^{\lambda_1}(1 + *z + *z^2 + \cdots), \ldots, y_n = z^{\lambda_n}(1 + *z + *z^2 + \cdots),$$

with distinct $\lambda_1, \cdots, \lambda_n \in \mathbf{C}$. The λ_i are called the *local exponents of the equation at* $z = 0$. Let $\lambda \in \mathbf{C}$, then $L(z^\lambda) = I(\lambda) \cdot z^\lambda(1 + *z + *z^2 + \cdots)$, where $I(\lambda)$ is a polynomial in λ, seen as variable. The polynomial I is called the *indicial polynomial at* $z = 0$. The exponents at $z = 0$ are the zeros of the indicial polynomial. We recall that the equation L can be put in matrix form and locally at $z = 0$, this matrix equation is equivalent to the equation $\frac{dv}{dz} = \frac{D}{z} v$, where D is a diagonal matrix with diagonal entries $\lambda_1, \ldots, \lambda_n$ (Again under the assumption that l is not involved in the local solutions.) Furthermore, the local monodromy matrix of the equation at $z = 0$ is conjugated to $e^{2\pi i D}$. For other regular singular points one has similar definitions involving the local parameter $t = z - c$ or $t = \frac{1}{z}$.

We suppose that for each singular point p the set E_p of exponents consists of n elements. (This is again equivalent to assuming that l does not enter into the local solutions at p.)

Let $y_1, \dots, y_n$ be a basis of solutions (somewhere defined or in a Picard-Vessiot field) of the equation L. Recall that the wronskian wr is the determinant of the matrix

$$\begin{pmatrix} y_1 & y_2 & .. & y_n \\ y_1^{(1)} & y_2^{(1)} & .. & y_n^{(1)} \\ y_1^{(2)} & y_2^{(2)} & .. & y_n^{(2)} \\ . & . & .. & . \\ . & . & .. & . \\ y_1^{(n-1)} & y_2^{(n-1)} & .. & y_n^{(n-1)} \end{pmatrix}.$$

The wronskian is nonzero and is determined up to a multiplicative constant. Furthermore, wr is a nonzero solution of $f' = -a_1 f$. It is easily seen that the order μ_i of wr at p_i is the $-(n-1)n/2 + \sum E_{p_i}$ (i.e.,$-(n-1)n/2$ plus the sum of the exponents of L at p_i). The order μ_∞ of wr at ∞ is $(n-1)n/2 + \sum E_\infty$. Furthermore, $-a_1 dz$ has simple poles and the residues are μ_i at p_i and μ_∞ at ∞. The sum of the residues $-a_1 dz$ is zero. This implies the well-known *Fuchs relation*:

$$\sum_i \sum E_{p_i} + \sum E_\infty = \frac{(s-1)(n-1)n}{2}.$$

We note, in passing, that the Fuchs relation remains valid if at one or more regular singular points p the indicial polynomial has multiple roots. In this situation E_p is interpreted as the set of the roots of the indicial polynomial, counted with multiplicity. The proof is easily adapted to this situation (see also [225]).

Assume, furthermore, that the differential Galois group of the equation lies in $\mathrm{SL}_n(\mathbf{C})$. Then $w' = -a_1 w$ has a nontrivial solution in $\mathbf{C}(z)$. The residues of $a_i dz$ are integers and thus for all $p \in \{p_1, \dots, p_k, \infty\}$ one has $\sum E_p$ is an integer.

We specialize now to the situation of a fuchsian scalar equation L of order two with singular points $\{0, 1, \infty\}$. The form of L is

$$y^{(2)} + (\frac{a_0}{z} + \frac{a_1}{z-1})y^{(1)} + (\frac{b_0}{z^2} + \frac{b_1}{(z-1)^2} + \frac{b_2}{z(z-1)})y = 0, \text{ and one has}$$

$$t(t-1) + a_0 t + b_0 = \prod_{\alpha \in E_0} (t - \alpha),$$

$$t(t-1) + a_1 t + b_1 = \prod_{\alpha \in E_1} (t - \alpha),$$

$$t(t+1) - (a_0 + a_1)t + (b_0 + b_1 + b_2) = \prod_{\alpha \in E_\infty} (t - \alpha).$$

The three polynomials in t are the indicial polynomials at $0, 1, \infty$. One observes that L is determined by E_0, E_1, E_∞. Furthermore, $\sum_{j=0,1,\infty}\sum E_j = 1$. The differential Galois group is a subgroup of $\mathrm{SL}_2(\mathbf{C})$ if and only if $\sum E_j$ is an integer for $j = 0, 1, \infty$.

For a third order fuchsian differential equations (with singular points $0, 1, \infty$) we will use the normalized form

$$L = \partial^3 + (\frac{a_0}{z} + \frac{a_1}{z-1})\partial^2 + (\frac{b_0}{z^2} + \frac{b_1}{(z-1)^2} + \frac{b_2}{z(z-1)})\partial$$
$$+\frac{c_0}{z^3} + \frac{c_1}{(z-1)^3} + \frac{c_2(z-1/2)}{z^2(z-1)^2} + \frac{\mu}{z^2(z-1)^2}.$$

The indicial polynomials at $0, 1, \infty$ are:

$$t(t-1)(t-2) + a_0t(t-1) + b_0t + c_0 = \prod_{\alpha\in E_0}(t-\alpha),$$
$$t(t-1)(t-2) + a_1t(t-1) + b_1t + c_1 = \prod_{\alpha\in E_1}(t-\alpha),$$
$$t(t+1)(t+2) - (a_0+a_1)t(t+1) + (b_0+b_1+b_2)t - (c_0+c_1+c_2)$$
$$= \prod_{\alpha\in E_\infty}(t-\alpha).$$

Thus $a_0, a_1, b_0, b_1, b_2, c_0, c_1, c_2$ are determined from the exponent sets E_0, E_1, E_∞. We will call μ the *accessory parameter*. Thus L is determined by the exponents and the value of the accessory parameter. The dual (or adjoint) L^* of L has the data $-a_0, -a_1, b_0+2a_0, b_1+2a_1, b_2, -c_0-2a_0-2b_0, -c_1-2a_1-2b_1, -c_2-2b_2, -\mu$. The indicial equations for L^* at $0, 1, \infty$ are obtained from the indicial equations for L by the substitutions $t \mapsto -t+2, -t+2, -t-2$.
The substitution $z \mapsto 1-z$ applied to L, with exponents sets E_0, E_1, E_∞, and accessory parameter μ, produces a fuchsian equation M with exponents sets E_1, E_0, E_∞, and accessory parameter $-\mu$.

4.4.2 Restrictions on the Exponents

Let matrices A_0, A_1, A_∞, generating the finite group G, with $A_0A_1A_\infty = 1$, be given. We suppose that G is an irreducible subgroup of $\mathrm{GL}_n(\mathbf{C})$. From the definition of the exponents we conclude that the set $e^{2\pi i E_j}$ is the set of the eigenvalues of A_j, for $j = 0, 1, \infty$. In other words, the exponents are known modulo integers. This leaves too many possibilities for the exponents. Some algebraic geometry is needed to obtain lower bounds for the exponents. We will simply indicate what the statement is. Define the number $m = m(A_0, A_1, A_\infty)$ by the rather complicated formula

$$-n+\frac{1}{e_0}\sum_{s=1}^{e_0-1} s\ (\ \frac{1}{e_0}\sum_{j=0}^{e_0-1} tr(A_0^j)\zeta_0^{sj})$$
$$+\frac{1}{e_1}\sum_{s=1}^{e_1-1} s\ (\ \frac{1}{e_1}\sum_{j=0}^{e_1-1} tr(A_1^j)\zeta_1^{sj})$$
$$+\frac{1}{e_\infty}\sum_{s=1}^{e_\infty-1} s\ (\ \frac{1}{e_\infty}\sum_{j=0}^{e_\infty-1} tr(A_\infty^j)\zeta_\infty^{sj}),$$

where e_0, e_1, e_∞ are the orders of the matrices A_0, A_1, A_∞, $\zeta_0 = e^{2\pi i/e_0}$, $\zeta_1 = e^{2\pi i/e_1}$, $\zeta_\infty = e^{2\pi i/e_\infty}$ and $tr(B)$ means the trace of a matrix B.

The number m turns out to be an integer ≥ 0. If $m > 0$, then the following lower bounds for the exponents are valid: the exponents in E_0 and E_1 are > -1 and the exponents in E_∞ are > 1. We will use all this as a black box (the next subsection gives a hint for it and full details are given in [233]). The importance is that the lower bound (in the case $m > 0$) together with the Fuchs relation gives a finite set of possibilities for the exponents E_0, E_1, E_∞.

4.4.3 Representations of Finite Groups

For a better understanding of the examples, we will separate the finite group G and its embedding in some $\mathrm{GL}_n(\mathbf{C})$. We recall some facts from the representation theory of finite groups. A representation of the finite group G is a homomorphism $\rho : G \to \mathrm{GL}_n(\mathbf{C})$. The character χ of ρ is the function on G given by $\chi(g) = tr(\rho(g))$. There is a bijection between the representations of a group and the set of characters of representations.

A representation ρ and its character are called irreducible if the only invariant subspaces for $\rho(G)$ are $\{0\}$ and $\mathbf{C}^n$. Every representation is a direct sum of irreducible ones and every character is the sum of irreducible characters. Every character is constant on a conjugacy class of G. Moreover, the irreducible characters form a basis of the vector space of the functions on G that are constant on each conjugacy class. In particular, there are as many irreducible characters of a group as there are conjugacy classes. The character table of a group is a table giving the values of the irreducible characters as functions on the conjugacy classes of the group. For "small" finite groups, the character tables are known.

The data that we are given can also be described by:
(a) The finite group G.
(b) Three generators g_0, g_1, g_∞ of the group with $g_0 g_1 g_\infty = 1$.
(c) A faithful (i.e., "injective") representation ρ with character χ. We will assume that ρ and χ are irreducible.

The formula in Sect. 4.4.2 can be partially explained as follows. The data determine a finite Galois extension K of $\mathbf{C}(z)$ with Galois group G. In geometric terms, this corresponds to a Galois covering of curves $X \to \mathbf{P}^1$ with group G. The vector space of the holomorphic differentials on X has dimension g, which is the genus of the curve. On this vector space the group G acts. In other words, the holomorphic differentials on C form a representation of G. The number $m(A_0, A_1, A_\infty)$ of Sect. 4.4.2 is the number of times that the irreducible character ρ is present in this representation.

In the construction of examples of (irreducible) fuchsian differential equations of order n for a given group G, we will thus use the following data:
(a) A choice of generators g_0, g_1, g_∞ for G with $g_0 g_1 g_\infty = 1$. One calls (g_0, g_1, g_∞) an admissible triple. The orders (e_0, e_1, e_∞) will be called *the branch type*. A more precise definition of branch type $[e_0, e_1, e_\infty]$ could be given as follows. Consider the set S of all admissible triples (g_0, g_1, g_∞) with e_i being the order of g_i for $i = 0, 1, \infty$. Two admissible triples (g_0, g_1, g_∞) and (h_0, h_1, h_∞) will be called equivalent if there is an automorphism A of G such that $h_i = A(g_i)$ for $i = 0, 1, \infty$. The branch type $[e_0, e_1, e_\infty]$ can be defined as the set of the equivalence classes of S. In some cases a branch type $[e_0, e_1, e_\infty]$ contains only one equivalence class of admissible triples. In general, it contains finitely many equivalence classes.
(b) An irreducible faithful representation of G.

In the examples we will restrict ourselves to a few groups and to irreducible representations in $\mathrm{SL}_n(\mathbf{C})$ with $n = 2, 3$.

4.4.4 A Calculation of the Accessory Parameter

Suppose that we are trying to find a fuchsian order three equation L with known exponents. Then one has still to calculate the accessory parameter μ. We will not explain the procedure in the general case to obtain μ. There is a "lucky situation" where two exponents belonging to the same singular point differ by an integer. Let us make the *assumption* that for some $j \in \{0, 1, \infty\}$ the set E_j contains two elements with difference $m \in \mathbf{Z},\ m > 0$. We note that this situation occurs if and only if A_j has multiple eigenvalues.

Lemma 4.31 *Assume that the differential Galois group of L is finite then μ satisfies an explicitly known polynomial equation over $\mathbf{Q}$ of degree m.*

Proof. For notational convenience we suppose $j = 0$ and by assumption λ, $\lambda + m \in E_0$ with m a positive integer. The assumption that the differential Galois group of L is finite implies that there are three Puiseux series at $z = 0$, solutions of $L = 0$. One of these has the form $z^\lambda g$ with $g = 1 + c_1 z + c_2 z^2 + \cdots \in \mathbf{C}[[z]]$. There is a formula

$$L(z^t) = P_0(t) z^{t-3} + P_1(t) z^{t-2} + P_2(t) z^{t-1} + P_3(t) z^t + \ldots,$$

where the P_i are polynomials in t and μ. In fact, P_0 does not contain μ and the other P_i have degree 1 in μ. An evaluation of the equation $L(z^\lambda(1+c_1z+c_2z^2+\cdots))=0$ produces a set of linear equations for the coefficients c_i. In order to have a solution, a determinant must be zero. This determinant is easily seen to be a polynomial in μ of degree m. □

Explicit formulas:
$P_0(t)=t(t-1)(t-2)+a_0t(t-1)+b_0t+c_0$,
$P_1(t)=-a_1t(t-1)-b_2t-c_2/2+\mu$,
$P_2(t)=-a_1t(t-1)+(b_1-b_2)t+2\mu$,
$P_3(t)=-a_1t(t-1)+(2b_1-b_2)t+c_2/2+3\mu$.
The polynomials of the lemma for $m=1,2$ and 3 are the following.
If λ, $\lambda+1$ are exponents at 0, then $P_1(\lambda)=0$.
If λ, $\lambda+2$ are exponents at 0, then $P_1(\lambda)P_1(\lambda+1)-P_0(\lambda+1)P_2(\lambda)=0$.
If λ and $\lambda+3$ are exponents at 0 then the determinant of the matrix

$$\begin{pmatrix} P_1(\lambda) & P_0(\lambda+1) & 0 \\ P_2(\lambda) & P_1(\lambda+1) & P_0(\lambda+2) \\ P_3(\lambda) & P_2(\lambda+1) & P_1(\lambda+2) \end{pmatrix}$$ is zero.

4.4.5 Examples

The Tetrahedral Group $A_4^{\mathrm{SL}_2}$

This group has 24 elements. Its center has order two and the group modulo its center is equal to A_4. The group has 7 conjugacy classes $conj_1,\ldots,conj_7$. They correspond to elements of order 1, 2, 3, 3, 4, 6, 6. There is only one faithful (unimodular) character of degree 2, denoted by χ_4. The two values $0\leq\lambda<1$ such that the $e^{2\pi i\lambda}$ are the eigenvalues for the representation corresponding to χ_4 are given for each conjugacy class: $(\frac{0,0}{1},\frac{1,1}{2},\frac{1,2}{3},\frac{1,2}{3},\frac{1,3}{4},\frac{1,5}{6},\frac{1,5}{6})$. One can make a list of (conjugacy classes of) admissible triples. For each triple one can calculate that the integer m of Sect. 4.4.2 is equal to 1. This information leads to unique data for the exponents and the equations.

Branch type	Genera	Char	xp 0	exp 1	exp ∞	Schwarz triple
3,3,4	2,0	χ_4	$-2/3,-1/3$	$-2/3,-1/3$	5/4,7/4	1/2,1/3,1/3
3,3,6	3,1	χ_4	$-2/3,-1/3$	$-2/3,-1/3$	7/6,11/6	2/3,1/3,1/3
3,4,6	4,0	χ_4	$-2/3,-1/3$	$-3/4,-1/4$	7/6,11/6	1/2,1/3,1/3
4,6,6	6,0	χ_4	$-3/4,-1/4$	$-5/6,-1/6$	7/6,11/6	1/2,1/3,1/3
6,6,6	7,1	χ_4	$-5/6,-1/6$	$-5/6,-1/6$	7/6,11/6	1/2,1/3,1/3
6,6,6	7,1	χ_4	$-5/6,-1/6$	$-5/6,-1/6$	7/6,11/6	2/3,1/3,1/3

"Schwarz triple" compares the data with "the list of Schwarz", which is a classification of the second order differential equations with singular points $0,1,\infty$, and finite irreducible differential Galois group. This list has 15 items. Our lists are somewhat longer, due to Schwarz' choice of equivalence among equations!

The first item under "genera" is the genus of the curve $X \to \mathbf{P}^1$ corresponding to the finite Galois extension $K \supset \mathbf{C}(z)$, where K is the Picard-Vessiot field of the equation. The second item is the genus of the curve X/Z, where Z is the center of the differential Galois group.

The Octahedral Group $S_4^{\mathrm{SL}_2}$

This group has 48 elements. Its center has two elements and the group modulo the center is isomorphic to S_4. This group has 8 conjugacy classes $conj_1, .., conj_8$. They correspond to elements of order $1, 2, 3, 4, 4, 6, 8, 8$. There are two faithful (unimodular) irreducible representations of degree 2. Their characters are denoted by χ_4 and χ_5 and the eigenvalues of these characters are given for each conjugacy class:

$$\chi_4 : (\frac{0,0}{1}, \frac{1,1}{2}, \frac{1,2}{3}, \frac{1,3}{4}, \frac{1,3}{4}, \frac{1,5}{6}, \frac{1,7}{8}, \frac{3,5}{8}),$$
$$\chi_5 : (\frac{0,0}{1}, \frac{1,1}{2}, \frac{1,2}{3}, \frac{1,3}{4}, \frac{1,3}{4}, \frac{1,5}{6}, \frac{3,5}{8}, \frac{1,7}{8}).$$

Thus there is an automorphism of $S_4^{\mathrm{SL}_2}$ that permutes the classes $conj_7, conj_8$. For the admissible triple, representing the unique element of the branch type we make the choice that the number of times that $conj_7$ occurs is greater than or equal to the number of times that $conj_8$ is present. In all cases the number m of Sect. 4.4.2 is equal to 1. This information suffices to calculate all exponents and equations.

Branch t.	Genera	Char	exp 0	exp 1	exp ∞	Schwarz triple
3,4,8	8,0	χ_4	$-2/3,-1/3$	$-3/4,-1/4$	9/8,15/8	1/2,1/3,1/4
		χ_5	$-2/3,-1/3$	$-3/4,-1/4$	11/8,13/8	1/2,1/3,1/4
3,8,8	11,3	χ_4	$-2/3,-1/3$	$-5/8,-3/8$	9/8,15/8	2/3,1/4,1/4
		χ_5	$-2/3,-1/3$	$-7/8,-1/8$	11/8,13/8	2/3,1/4,1/4
4,6,8	13,0	χ_4	$-3/4,-1/4$	$-5/6,-1/6$	11/8,13/8	1/2,1/3,1/4
		χ_5	$-3/4,-1/4$	$-5/6,-1/6$	9/8,15/8	1/2,1/3,1/4
6,8,8	15,3	χ_4	$-5/6,-1/6$	$-5/8,-3/8$	11/8,13/8	2/3,1/4,1/4
		χ_5	$-5/6,-1/6$	$-7/8,-1/8$	9/8,15/8	2/3,1/4,1/4

The Icosahedral Group $A_5^{\mathrm{SL}_2}$

This group has 120 elements and is modulo its center isomorphic to A_5. There are 9 conjugacy classes $conj_1, \ldots, conj_9$ corresponding to elements of orders $1, 2, 3, 4, 5, 5, 6, 10, 10$. There are two faithful irreducible (unimodular) representations of degree 2. As before, their characters χ_2 and χ_3 are given on the conjugacy classes by the two eigenvalues:

$$\chi_2 : (\frac{0,0}{1}, \frac{1,1}{2}, \frac{1,2}{3}, \frac{1,3}{4}, \frac{2,3}{5}, \frac{1,4}{5}, \frac{1,5}{6}, \frac{3,7}{10}, \frac{1,9}{10}),$$

$$\chi_3 : (\frac{0,0}{1}, \frac{1,1}{2}, \frac{1,2}{3}, \frac{1,3}{4}, \frac{1,4}{5}, \frac{2,3}{5}, \frac{1,5}{6}, \frac{1,9}{10}, \frac{3,7}{10}).$$

Thus there is an automorphism of $A_5^{\mathrm{SL}_2}$ that permutes the two pairs of conjugacy classes $conj_5, conj_6$ and $conj_8, conj_9$. For the admissible triple representing the unique element of the branch type there are several choices. One can deduce from the table which choice has been made.

Branch type	Genera	Char	exp 0	exp 1	exp ∞	Schwarz triple
3,3,10	15,5	χ_2	$-2/3,-1/3$	$-2/3,-1/3$	11/10,19/10	2/3,1/3,1/5
		χ_3	$-2/3,-1/3$	$-2/3,-1/3$	13/10,17/10	2/5,1/3,1/3
3,4,5	14,0	χ_2	$-2/3,-1/3$	$-3/4,-1/4$	7/5,8/5	1/2,1/3,1/5
		χ_3	$-2/3,-1/3$	$-3/4,-1/4$	6/5,9/5	1/2,2/5,1/3
3,4,10	20,0	χ_2	$-2/3,-1/3$	$-3/4,-1/4$	11/10,19/10	1/2,1/3,1/5
		χ_3	$-2/3,-1/3$	$-3/4,-1/4$	13/10,17/10	1/2,2/5,1/3
3,5,5	17,9	χ_2	$-2/3,-1/3$	$-3/5,-2/5$	6/5,9/5	3/5,1/3,1/5
		χ_3	$-2/3,-1/3$	$-4/5,-1/5$	7/5,8/5	3/5,1/3,1/5
3,5,6	19,5	χ_2	$-2/3,-1/3$	$-3/5,-2/5$	7/6,11/6	2/3,1/3,1/5
		χ_3	$-2/3,-1/3$	$-4/5,-1/5$	7/6,11/6	2/5,1/3,1/3
3,5,10	23,9	χ_2	$-2/3,-1/3$	$-3/5,-2/5$	11/10,19/10	2/3,1/5,1/5
		χ_3	$-2/3,-1/3$	$-4/5,-1/5$	13/10,17/10	3/5,2/5,1/3
3,10,10	29,9	χ_2	$-2/3,-1/3$	$-9/10,-1/10$	13/10,17/10	3/5,1/3,1/5
		χ_3	$-2/3,-1/3$	$-7/10,-3/10$	11/10,19/10	3/5,1/3,1/5
4,5,5	22,4	χ_2	$-3/4,-1/4$	$-3/5,-2/5$	6/5,9/5	1/2,2/5,1/5
		χ_3	$-3/4,-1/4$	$-4/5,-4/5$	7/5,8/5	1/2,2/5,1/5
4,5,6	24,0	χ_2	$-3/4,-1/4$	$-3/5,-2/5$	7/6,11/6	1/2,1/3,1/5
		χ_3	$-3/4,-1/4$	$-4/5,-1/5$	7/6,11/6	1/2,2/5,1/3
4,5,10	28,4	χ_2	$-3/4,-1/4$	$-3/5,-2/5$	13/10,17/10	1/2,2/5,1/5
		χ_3	$-3/4,-1/4$	$-4/5,-1/5$	11/10,19/10	1/2,2/5,1/5
4,6,10	30,0	χ_2	$-3/4,-1/4$	$-5/6,-1/6$	11/10,19/10	1/2,1/3,1/5
		χ_3	$-3/4,-1/4$	$-5/6,-1/6$	13/10,17/10	1/2,2/5,1/3
		χ_3	$-3/4,-1/4$	$-7/10,-3/10$	11/10,19/10	1/2,2/5,1/5
5,5,6	27,9	χ_2	$-3/5,-2/5$	$-3/5,-2/5$	7/6,11/6	2/3,1/5,1/5
		χ_3	$-4/5,-1/5$	$-4/5,-1/5$	7/6,11/6	3/5,2/5,1/3
5,5,10	31,13	χ_2	$-3/5,-2/5$	$-3/5,-2/5$	11/10,19/10	4/5,1/5,1/5
		χ_3	$-4/5,-1/5$	$-4/5,-1/5$	13/10,17/10	2/5,2/5,2/5
5,6,10	33,9	χ_2	$-3/5,-2/5$	$-5/6,-1/6$	13/10,17/10	3/5,1/3,1/5
		χ_3	$-4/5,-1/5$	$-5/6,-1/6$	11/10,19/10	3/5,1/3,1/5
6,6,10	35,5	χ_2	$-5/6,-1/6$	$-5/6,-1/6$	11/10,19/10	2/3,1/3,1/5
		χ_3	$-5/6,-1/6$	$-5/6,-1/6$	13/10,17/10	2/5,1/3,1/3
6,10,10	39,9	χ_2	$-5/6,-1/6$	$-9/10,-1/10$	11/10,19/10	2/3,1/5,1/5
		χ_3	$-5/6,-1/6$	$-7/10,-3/10$	13/10,17/10	3/5,2/5,1/3
10,10,10	43,13	χ_2	$-9/10,1/10$	$-9/10,-1/10$	11/10,19/10	4/5,1/5,1/5
		χ_3	$-7/10,3/10$	$-7/10,-3/10$	13/10,17/10	2/5,2/5,2/5

The Data for G_{168}

The group G_{168} is the simple group $\mathrm{PSL}_2(\mathbf{F}_7)$ and has 168 elements. There are six conjugacy classes $conj_1, \ldots, conj_6$; they correspond to elements of order $\{1, 2, 3, 4, 7, 7\}$. There are two irreducible characters of degree three, called χ_2, χ_3. Both are faithful and unimodular. The three values $0 \le \lambda < 1$ such that $e^{2\pi i\lambda}$ are the eigenvalues for the representation are given for each conjugacy class as follows:

$$\chi_2 : (\frac{0,0,0}{1}, \frac{0,1,1}{2}, \frac{0,1,2}{3}, \frac{0,1,3}{4}, \frac{3,5,6}{7}, \frac{1,2,4}{7}),$$

$$\chi_3 : (\frac{0,0,0}{1}, \frac{0,1,1}{2}, \frac{0,1,2}{3}, \frac{0,1,3}{4}, \frac{1,2,4}{7}, \frac{3,5,6}{7}).$$

The character χ_3 is the dual of χ_2. We introduce some terminology. The *conjugacy triple i, j, k of an admissible triple* (g_0, g_1, g_∞) is defined by: the conjugacy classes of g_0, g_1, g_∞ are $conj_i, conj_j, conj_k$.

(1) *Branch type* [2, 3, 7] consists of one element, represented by the conjugacy triple 2,3,5. The genus of the curve X is 3. For χ_3 one calculates that $m = 1$. For this character, the lower bounds for the exponents add up to 3, so they are the actual exponents. From the exponent difference 1 at $z = 0$ one obtains all the data for L: $-1/2, 0, 1/2|| - 2/3, -1/3, 0||8/7, 9/7, 11/7||\mu = 12\,293/24\,696$. This equation was, in fact, found by Hurwitz [142]. Our theoretical considerations provide an "overkill" since the corresponding covering $X \to \mathbf{P}^1$ is well known. It is the Klein curve in $\mathbf{P}^2$ given by the homogeneous equation $x_0x_1^3 + x_1x_2^3 + x_3x_0^3 = 0$, having automorphism group G_{168}, or in another terminology, it is the modular curve $X(7)$ with automorphism group $\mathrm{PSL}_2(\mathbf{F}_7)$.

Exercise 4.32 We continue now with order three equations for the group G_{168}. The reader is asked to verify the following calculations.
(2). *Branch type* [2, 4, 7] with conjugacy triple 2,4,5. Prove that $m = 0$ for χ_2 and $m = 1$ for χ_3. For χ_3 the lower bounds for the exponents add up to 3 and are the actual values; at $z = 0$ there is an exponent difference 1. This leads to the data $-1/2, 0, 1/2|| - 3/4, -1/4, 0||8/7, 9/7, 11/7||\mu = 5273/10\,976$.

(3). *Branch type* [2, 7, 7] and conjugacy triple 2,5,5. Prove that $m = 0$ for χ_2 and $m = 2$ for χ_3. For χ_3 the lower bounds for the exponents add up to 2; there is an integer exponent difference at $z = 0$. From $m = 2$ one can conclude that one may add $+1$ to any of the nine exponents (whenever this does not come in conflict with the definition of exponents). We will not prove this statement. Verify now the following list of differential equations for G_{168}. The data for the exponents and μ are:

$$-1/2, 1, 1/2|| - 6/7, -5/7, -3/7||8/7, 9/7, 11/7||\mu = 1045/686$$
$$-1/2, 0, 3/2|| - 6/7, -5/7, -3/7||8/7, 9/7, 11/7||\mu = 2433/1372 \pm 3\sqrt{21}/392$$
$$-1/2, 0, 1/2|| \quad 1/7, -5/7, -3/7||8/7, 9/7, 11/7||\mu = 1317/2744$$
$$-1/2, 0, 1/2|| - 6/7, \quad 2/7, -3/7||8/7, 9/7, 11/7||\mu = 1205/2744$$
$$-1/2, 0, 1/2|| - 6/7, -5/7, \quad 4/7||8/7, 9/7, 11/7||\mu = 1149/2744$$
$$-1/2, 0, 1/2|| - 6/7, -5/7, -3/7||15/7, 9/7, 11/7||\mu = 3375/2744$$
$$-1/2, 0, 1/2|| - 6/7, -5/7, -3/7||8/7, 16/7, 11/7||\mu = 3263/2744$$
$$-1/2, 0, 1/2|| - 6/7, -5/7, -3/7||8/7, 9/7, 18/7||\mu = 3207/2744.$$

Analytic Theory

5 Monodromy, the Riemann-Hilbert Problem, and the Differential Galois Group

5.1 Monodromy of a Differential Equation

Let U be an open connected subset of the complex sphere $\mathbf{P}^1 = \mathbf{C} \cup \{\infty\}$ and let $Y' = AY$ be a differential equation on U, with A an $n \times n$ matrix with coefficients that are meromorphic functions on U. We assume that the equation is regular at every point $p \in U$. Thus, for any point $p \in U$, the equation has n independent solutions $y_1, \ldots, y_n$ consisting of vectors with coordinates in $\mathbf{C}(\{z - p\})$. It is known ([132], Chap. 9; [225], p. 5) that these solutions converge in a disk of radius ρ, where ρ is the distance from p to the complement of U. These solutions span an n-dimensional vector space denoted by V_p. If we let F_p be a matrix whose columns are the n independent solutions $y_1, \ldots, y_n$ then F_p is a fundamental matrix with entries in $\mathbf{C}(\{z-p\})$. One can normalize F_p such that $F_p(p)$ is the identity matrix. The question we are interested in is:

Does there exist on all of U, a solution space for the equation having dimension n?

The main tool for answering this question is *analytical continuation* which in turn relies on the notion of the *fundamental group* ([7], Chap. 8; [132], Chap. 9). These can be described as follows. Let $q \in U$ and let λ be a path from p to q lying in U (one defines a path from p to q in U as a continuous map $\lambda : [0, 1] \to U$ with $\lambda(0) = p$ and $\lambda(1) = q$). For each each point $\lambda(t)$ on this path, there is an open set $\mathcal{O}_{\lambda(t)} \subset U$ and fundamental solution matrix $F_{\lambda(t)}$ whose entries converge in $\mathcal{O}_{\lambda(t)}$. By compactness of $[0, 1]$, we can cover the path with a finite number of these open sets, $\{\mathcal{O}_{\lambda(t_i)}\}$, $t_0 = 0 < t_1 < \cdots < t_m = 1$. The maps induced by sending the columns of $F_{\lambda(i)}$ to the columns of $F_{\lambda(i+1)}$ induce $\mathbf{C}$-linear bijections $V_{\lambda(t_i)} \to V_{\lambda(t_{i+1})}$. The resulting $\mathbf{C}$-linear bijection $V_p \to V_q$ can be seen to depend only on the homotopy class of λ (we note that two paths λ_0 and λ_1 in U from p to q are homotopic if there exists a continuous $H : [0, 1] \times [0, 1] \to U$ such that $H(t, 0) = \lambda_0(t)$, $H(t, 1) = \lambda_1(t)$ and $H(0, s) = p$, $H(1, s) = q$). The $\mathbf{C}$-linear bijection $V_p \to V_q$ is called the *analytic continuation* along λ.

For the special case that $\lambda(0) = \lambda(1) = p$ we find an isomorphism that is denoted by $\mathbf{M}(\lambda) : V_p \to V_p$. The collection of all closed paths, starting and ending in p, divided out by homotopy, is called the *fundamental group* and denoted by $\pi_1(U, p)$. The group structure on $\pi_1(U, p)$ is given by "composing" paths. The resulting group homomorphism $\mathbf{M} : \pi_1(U, p) \to \mathrm{GL}(V_p)$ is called the *monodromy map*. The image of $\mathbf{M}$ in $\mathrm{GL}(V_p)$ is called the *monodromy group*. The open connected set U is called *simply connected* if $\pi_1(U, p) = \{1\}$. If U is simply connected then one sees that analytical continuation yields n independent solutions of the differential equation on U. Any open disk, $\mathbf{C}$ and also $\mathbf{P}^1$ are simply connected.

The fundamental group of $U := \{z \in \mathbf{C} | \ 0 < |z| < a\}$ (for $a \in (0, \infty]$) is generated by the circle around 0, say through $b \in \mathbf{R}$ with $0 < b < a$ and in the positive direction. Let us write λ for this generator. There are no relations and thus the fundamental group is isomorphic with the group $\mathbf{Z}$. The element $\mathbf{M}(\lambda) \in \mathrm{GL}(V_b)$ is called the *local monodromy*. As a first example, consider the differential equation $y' = \frac{c}{z}y$. The solution space V_b has basis z^c (for the usual determination of this function). Furthermore, $\mathbf{M}(\lambda)z^c = e^{2\pi i c}z^c$ and $e^{2\pi i c} \in \mathrm{GL}_1$ is the local monodromy.

5.1.1 Local Theory of Regular Singular Equations

In this subsection we continue the study of regular singular equations, now over the field $K = K_{conv} = \mathbf{C}(\{z\})$ of the convergent Laurent series. We give the following definition: a matrix differential operator, here also referred to as a "matrix differential equation", $\frac{d}{dz} - A$ over K_{conv} is called *regular singular* if the equation is equivalent over K_{conv} to $\frac{d}{dz} - B$ such that the entries of B have poles at $z = 0$ of order at most 1. Otherwise stated, the entries of zB are analytic functions in a neighborhood of $z = 0$. Recall (Sect. 1.2) that two equations $\frac{d}{dz} - A$ and $\frac{d}{dz} - B$ are equivalent if there is a $F \in \mathrm{GL}_n(\mathbf{C}(\{z\}))$ with $F^{-1}(\frac{d}{dz} - A)F = (\frac{d}{dz} - B)$.

One can express this notion of regular singular for matrix equations also in terms of $\delta := z\frac{d}{dz}$. A matrix differential equation over K_{conv} is regular singular if it is equivalent (over K_{conv}) to an equation $\delta - A$ such that the entries of A are holomorphic functions in a neighborhood of $z = 0$ (i.e., lie in $\mathbf{C}\{z\}$).

A differential module M over K_{conv} is called regular singular if M contains a lattice over $\mathbf{C}\{z\}$ that is invariant under δ (compare Definition 3.9 for the formal case). As in the formal case, M is regular singular if and only if an (or every) associated matrix differential equation is regular singular.

The following theorem gives a complete overview of the regular singular equations at $z = 0$. We will return to this theme in Sect. 10.2.

Theorem 5.1 *Let $\delta - A$ be regular singular at $z = 0$.*

1. *$\delta - A$ is equivalent over the field $K_{conv} = \mathbf{C}(\{z\})$ of convergent Laurent series to $\delta - C$, where C is a constant matrix. More precisely, there is a unique constant matrix C such that all its eigenvalues λ satisfy $0 \leq Re(\lambda) < 1$ and $\delta - A$ equivalent to $\delta - C$.*

2. *The local monodromies of the equations $\delta - A$ and $\delta - C$ with C as in 1. are conjugate (even without the assumption on the real parts of the eigenvalues). The local monodromy of $\delta - C$ has matrix $e^{2\pi i C}$.*

3. *$\delta - A$ is equivalent to a regular singular $\delta - \tilde{A}$, if and only if the local monodromies are conjugate.*

Proof. In Proposition 3.12, it is shown that $\delta - A$ is equivalent over $\hat{K} = \mathbf{C}((z))$ to $\delta - C$ with C as in statement 1. Lemma 3.42 states that this equivalence can be taken over K_{conv}. This implies that, with respect to any bases of the solution spaces, the local monodromies of the two equations are conjugate. At the point $1 \in \mathbf{C}$, the matrix $e^{C\log(z)}$ is a fundamental solution matrix for $\delta - C$. Since analytical continuation around the generator of the fundamental group maps $\log(z)$ to $\log(z) + 2\pi i$, the conclusion of part 2 follows.

If $\delta - A$ is equivalent to a regular singular $\delta - \tilde{A}$, then clearly their local monodromies are conjugate. To prove the reverse implication, assume that, with respect to suitable bases of the solution spaces, the local monodromy of $\delta - C_1$ is the same as the local monodromy of $\delta - C_2$, where C_1, C_2 are constant matrices. This implies that $e^{2\pi i C_1} = e^{2\pi i C_2}$. At the point 1 the matrix $e^{C_j \log(z)}$ is the fundamental matrix for $\delta - C_j$ for $j = 1, 2$. Let $B = e^{-C_1 \log(z)} e^{C_2 \log(z)}$. Analytic continuation around the generator of the fundamental group leaves B fixed, so the entries of this matrix are holomorphic functions in a punctured neighborhood of the origin. Furthermore, one sees that the absolute value of any such entry is bounded by $|z|^N$ for a suitable $N \in \mathbf{Z}$ in such a neighborhood. Therefore, the entries of B have at worst poles at $z = 0$ and so lie in K_{conv}. Therefore $\delta - C_1$ is equivalent to $\delta - C_2$. over K_{conv}. Part 3 follows from this observation. □

Corollary 5.2 *Let $\delta - A$ be regular singular at $z = 0$. The differential Galois group G of this equation over the differential field $\mathbf{C}(\{z\})$ is isomorphic to the Zariski closure in $\mathrm{GL}_n(\mathbf{C})$ of the group generated by the monodromy matrix. Moreover, the differential Galois group of $\delta - A$ over $\mathbf{C}((z))$ coincides with G.*

Proof. Theorem 5.1 implies that the equation $\delta - A$ is equivalent, over K_{conv}, to an equation $\delta - C$, where C is a constant matrix. We may assume that C is in Jordan normal form and so the associated Picard-Vessiot extension is of the form $F = K_{conv}(z^{a_1}, \ldots, z^{a_r}, \epsilon \log z)$, where $a_1, \ldots, a_r$ are the eigenvalues of C and with $\epsilon = 0$ if C is diagonizable and $\epsilon = 1$ otherwise. Any element f of F is meromorphic on any sector at $z = 0$ of opening less than 2π. If analytic continuation around $z = 0$ leaves such an element fixed, it must be analytic in a punctured neighborhood of $z = 0$. Furthermore, $|f|$ is bounded by $|z|^N$ for a suitable N in such a neighborhood and therefore must be meromorphic at the origin as well. Therefore, $f \in K_{conv}$. The Galois correspondence implies that the Zariski closure of the monodromy matrix must be the Galois group.

Let UnivR be the universal differential ring constructed in Sect. 3.2 and let UnivF be its field of fractions. One can embed F into UnivF. The action of the formal monodromy on F coincides with the action of analytic continuation. Therefore, we may assume that the monodromy matrix is in the Galois group of $\delta - A$ over $\mathbf{C}((z))$. Since this latter Galois group may be identified with a subgroup of the Galois group of $\delta - A$ over K, we have that the two groups coincide. □

Exercise 5.3 *Local Galois groups at a regular singular point*
The aim of this exercise is to show that the Galois group over K of a regular singular equation at $z = 0$ is of the form $\mathbf{G}_m^n \times \mathbf{G}_a^\epsilon \times C_d$, where n is a non-negative integer, $\epsilon = 0, 1$ and C_d is a cyclic group of order d. To do this it will be enough to show that a linear algebraic group $H \subset \mathrm{GL}_m(k)$, k algebraically closed of characteristic zero is of this type if and only if it is the Zariski closure of a cyclic group.

1. Let $H \subset \mathrm{GL}_m$ be the Zariski closure of a cyclic group generated by g. Using the Jordan decomposition of g, we may write $g = g_s g_u$ where g_s is diagonalizable, g_u is unipotent (i.e., $g_u - id$ is nilpotent) and $g_s g_u = g_u g_s$. It is, furthermore, known that $g_u, g_s \in H$ ([141], Chap. 15).
(a) Show that H is abelian and that $H \simeq H_s \times H_u$ where H_s is the Zariski closure of the group generated by g_s and H_u is the Zariski closure of the group generated by g_u.
(b) The smallest algebraic group containing a unipotent matrix (not equal to the identity) is isomorphic to $\mathbf{G}_a$ ([141], Chap. 15) so $H_u = \mathbf{G}_a$ or $\{1\}$.
(c) Show that H_s is diagonalizable and use Lemma A.45 to deduce that H_s is isomorphic to a group of the form $\mathbf{G}_m^n \times C_d$.

2. Let H be isomorphic to $\mathbf{G}_m^n \times \mathbf{G}_a^\epsilon \times C_d$. Show that H has a Zariski dense cyclic subgroup. Hint: If $p_1, \dots, p_n$ are distinct primes, the group generated by $(p_1, \dots, p_n)$ lies in no proper algebraic subgroup of $\mathbf{G}_m^n$.

3. Construct examples showing that any group of the above type is the Galois group over K of a regular singular equation. □

The ideas in the proof of Theorem 5.1 can be used to characterize regular singular points in terms of growth of analytic solutions near a singular point. An *open sector* $S(a, b, \rho)$ is the set of the complex numbers $z \neq 0$ satisfying $\arg(z) \in (a, b)$ and $|z| < \rho(\arg(z))$, where $\rho : (a, b) \to \mathbf{R}_{>0}$ is some continuous function. We say that a function $g(z)$ analytic in an open sector $S = S(a, b, \rho)$ is of *moderate growth on* S if there exists an integer N and real number $c > 0$ such that $|g(z)| < c|z|^N$ on S.

We say that a differential equation $\delta - A$, $A \in \mathrm{GL}_n(K)$ has *solutions of moderate growth at* $z = 0$ if on any open sector $S = S(a, b, \rho)$ with $|a - b| < 2\pi$ and sufficiently small ρ there is a fundamental solution matrix Y_S whose entries are of moderate growth on S. Note that if A is constant then it has solutions of moderate growth.

Theorem 5.4 *Let $\delta - A$ be a differential equation with $A \in \mathrm{GL}_n(K)$. A necessary and sufficient condition that $\delta - A$ have all of its solutions of moderate growth at $z = 0$ is that $\delta - A$ be regular singular at $z = 0$.*

Proof. If $\delta - A$ is regular singular at $z = 0$, then it is equivalent over K to an equation with constant matrix and so has solutions of regular growth at $z = 0$. Conversely, assume that $\delta - A$ has solutions of moderate growth at $z = 0$. Let $e^{2\pi i C}$ be the monodromy matrix. We will show that $\delta - A$ is equivalent to $\delta - C$. Let Y be a fundamental solution matrix of $\delta - A$ in some open sector containing 1 and let $B = Ye^{-C\log(z)}$. Analytic continuation around $z = 0$ will leave B invariant and so its entries will be analytic in a punctured neighborhood of $z = 0$. The moderate growth condition implies that the entries of B will, furthermore, be meromorphic at $z = 0$ and so $B \in \mathrm{GL}_n(K)$. Finally, $A = B'B^{-1} + BCB^{-1}$ implies that $\delta - A$ is equivalent to $\delta - C$ over K. □

As a corollary of this result, we can deduce what is classically known as *Fuchs' Criterion*.

Corollary 5.5 *Let* $L = \delta^n + a_{n-1}\delta^{n-1} + \cdots + a_0$ *with* $a_i \in K$. *The coefficients* a_i *are holomorphic at* 0 *if and only if for any sector* $S = S(a, b, \rho)$ *with* $|a - b| < 2\pi$ *and* ρ *sufficiently small,* $L(y) = 0$ *has a fundamental set of solutions holomorphic and of moderate growth on S. In particular, if* A_L *denotes the companion matrix of* L, *the* a_i *are holomorphic at* $z = 0$ *if and only if* $\delta - A_L$ *is regular singular at* $z = 0$.

Proof. By Proposition 3.16, the operator L is regular singular if and only if $M := K[\delta]/K[\delta]L$ is regular singular. Furthermore, $\delta - A_L$ is the matrix equation associated to M. Thus the corollary follows from Theorem 5.4. □

Exercise 5.6 Show that $L = \delta^n + a_{n-1}\delta^{n-1} + \cdots + a_0$ with a_i holomorphic at $z = 0$ if and only if $L = z^n(d/dz)^n + z^{n-1}b_{n-1}(d/dz)^{n-1} + \cdots + z^i b_i(d/dz)^i + \cdots + b_0$ where the b_i are holomorphic at 0. □

5.1.2 Regular Singular Equations on $\mathbf{P}^1$

A differential equation $\frac{d}{dz} - A$, where the matrix A has entries in the field $\mathbf{C}(z)$ has an obvious interpretation as an equation on the complex sphere $\mathbf{P}^1 = \mathbf{C} \cup \{\infty\}$. A point $p \in \mathbf{P}^1$ is singular for $\frac{d}{dz} - A$ if the equation cannot be made regular at p with a *local* meromorphic transformation. A singular point is called regular singular if some local transformation at p produces an equivalent equation with a matrix having poles of at most order 1. The equation $\frac{d}{dz} - A$ is called *regular singular* if every singular point is, in fact, regular singular. In the sequel we will work with regular singular equations and S will denote its (finite) set of singular points.

An example of a regular singular equation is $\frac{d}{dz} - \sum_{i=1}^{k} \frac{A_i}{z-a_i}$, where the A_i are constant matrices and $a_1, \ldots, a_k$ are distinct complex numbers.

Exercise 5.7 Calculate that ∞ is a regular singular point for the equation $\frac{d}{dz} - \sum_{i=1}^{n} \frac{A_i}{z-a_i}$. Prove that $\sum A_i = 0$ implies that ∞ is a regular (i.e., not a singular) point for this equation. Calculate in the "generic" case the local monodromy matrices of the equation. Why is this condition "generic" necessary? □

Let $S = \{s_1, \dots, s_k, \infty\}$, then the equation $\frac{d}{dz} - \sum_{i=1}^{k} \frac{A_i}{z-s_i}$ is called a *fuchsian differential equation for* S if each of the points in S is singular. In general, a regular singular differential equation $\frac{d}{dz} - A$ with the above S as its set of singular points cannot be transformed into the form $\frac{d}{dz} - \sum_{i=1}^{k} \frac{A_i}{z-s_i}$. One can find transformations of $\frac{d}{dz} - A$ that work well for each of the singular points, but, in general, there is no global transformation that works for all singular points at the same time and does not introduce poles outside the set S.

We consider the open set $U = \mathbf{P}^1 \setminus S$ and choose a point $p \in U$. Let $S = \{s_1, \dots, s_k\}$ and consider closed paths $\lambda_1, \dots, \lambda_k$, beginning and ending at p, and each λ_i forms a small "loop" around s_i. If the choice of the loops is correct (i.e., each loop contains a unique and distinct s_i and all are oriented in the same direction) then the fundamental group $\pi_1(U, p)$ is generated by the $\lambda_1, \dots, \lambda_k$ and the only relation between the generators is $\lambda_1 \circ \cdots \circ \lambda_k = 1$. In particular, the fundamental group is isomorphic to the free noncommutative group on $k-1$ generators. The monodromy map of the equation is the homomorphism $\mathbf{M} : \pi_1(U, p) \to \mathrm{GL}(V_p)$ and the monodromy group is the image in $\mathrm{GL}(V_p)$ of this map.

Theorem 5.8 *The differential Galois group of the regular singular equation* $\frac{d}{dz} - A$ *over* $\mathbf{C}(z)$, *is the Zariski closure of the monodromy group* $\subset \mathrm{GL}(V_p)$.

Proof. For any point $q \in U$ one considers, as before, the space V_q of the local solutions of $\frac{d}{dz} - A$ at q. The coordinates of the vectors in V_q generate over the field $\mathbf{C}(z)$ a subring $R_q \subset \mathbf{C}(\{z-q\})$, which is (by Picard-Vessiot theory) a Picard-Vessiot ring for $\frac{d}{dz} - A$. For a path λ from p to q, the analytical continuation induces a $\mathbf{C}$-bijection from V_p to V_q and also a $\mathbf{C}(z)$-algebra isomorphism $R_p \to R_q$. This isomorphism commutes with differentiation. For any closed path λ through p, one finds a differential automorphism of R_p that corresponds with $\mathbf{M}(\lambda) \in \mathrm{GL}(V_p)$. In particular, $\mathbf{M}(\lambda)$ is an element of the differential Galois group of $\frac{d}{dz} - A$ over $\mathbf{C}(z)$. The monodromy group is then a subgroup of the differential Galois group.

The field of fractions of R_p is a Picard-Vessiot field, on which the monodromy group acts. From the Galois correspondence in the differential case, the statement of the theorem follows from the assertion:

Let f belong to the field of fractions of R_p. If f is invariant under the monodromy group, then $f \in \mathbf{C}(z)$.

The meromorphic function f is, a priori, defined in a neighborhood of p. But it has an analytical continuation to every point q of $\mathbf{P}^1 \setminus S$. Moreover, by assumption this analytical continuation does not depend on the choice of the path from p to q. We conclude that f is a meromorphic function on $\mathbf{P}^1 \setminus S$. Since the differential equation is, at worst, regular singular at each s_i and infinity, it has solutions of moderate growth at each singular point. The function f is a rational expression in the coordinates of the solutions at each singular point and so has also moderate growth at each point in S. Thus f is a meromorphic function on all of $\mathbf{P}^1$ and therefore belongs to $\mathbf{C}(z)$. □

Exercise 5.9 Prove that the differential Galois group G of $\delta - C$, with C a constant matrix, over the field $\mathbf{C}(z)$ is equal to the Zariski closure of the subgroup of $\mathrm{GL}_n(\mathbf{C})$ generated by $e^{2\pi i C}$. Therefore the only possible Galois groups over $\mathbf{C}(z)$ are those given in Exercise 5.3. Give examples where G is isomorphic to $\mathbf{G}_m^n$, $\mathbf{G}_m^n \times \mathbf{G}_a$ and $\mathbf{G}_m^n \times \mathbf{G}_a \times C_d$, where C_d is the cyclic group of order d. □

Example 5.10 *The hypergeometric differential equation.*
In Chap. 6 (cf. Remarks 6.23.4, Example 6.31 and Lemma 6.11) we will show that any order two regular singular differential equation on $\mathbf{P}^1$ with singular locus in $\{0, 1, \infty\}$ is equivalent to a scalar differential equation of the form:

$$y'' + \frac{Az + B}{z(z-1)}y' + \frac{Cz^2 + Dz + E}{z^2(z-1)^2}y = 0.$$

Classical transformations ([225], Chap. 21) can be used to further transform this equation to the scalar hypergeometric differential equation:

$$y'' + \frac{(a+b+1)z - c}{z(z-1)}y' + \frac{ab}{z(z-1)}y = 0.$$

One can write this in matrix form and calculate at the points $0, 1, \infty$ the locally equivalent equations of Theorem 5.1:

$zv' = \begin{pmatrix} 0 & 0 \\ -ab & c \end{pmatrix} v$ at 0 (eigenvalues $0, c$).

$(z-1)v' = \begin{pmatrix} 0 & 0 \\ ab & a+b-c+1 \end{pmatrix} v$ at 1 (eigenvalues $0, a+b-c+1$).

$tv' = \begin{pmatrix} 0 & 1 \\ -ab & -a-b \end{pmatrix} v$ at ∞, with $t = z^{-1}$ and $' = \frac{d}{dt}$ (eigenvalues $-a, -b$).

This calculation is only valid if the eigenvalues for the three matrices do not differ by a nonzero integer. This is equivalent to assuming that none of the numbers $c, b, a, a+b-c$ is an integer. In the contrary case, one has to do some more calculations. The hypergeometric series

$$F(a, b, c; z) = \sum_{n \geq 0} \frac{(a)_n (b)_n}{n!(c)_n} z^n,$$

where the symbol $(x)_n$ means $x(x+1)\cdots(x+n-1)$ for $n > 0$ and $(x)_0 = 1$, is well defined for $c \neq 0, -1, -2, \ldots$. We will exclude those values for c. One easily computes that $F(a, b, c; z)$ converges for $|z| < 1$ and that it is a solution of the hypergeometric differential equation. Using the hypergeometric series one can "in principle" compute the monodromy group and the differential Galois group of the equation (the calculation of the monodromy group was originally carried out by Riemann ([245]; see also [297] and [225]). One takes $p = 1/2$. The fundamental group is generated by the two circles (in the positive direction) through the point $1/2$ and around 0 and 1. At the point $1/2$ we take a basis of the solution space: $u_1 = F(a,b,c;z)$ and $u_2 = z^{1-c}F(a-c+1, b-c+1, 2-c; z)$. The circle around 0 gives a monodromy matrix $\begin{pmatrix} 1 & 0 \\ 0 & e^{-2\pi i c} \end{pmatrix}$. The circle around 1 produces a rather complicated monodromy matrix $\begin{pmatrix} B_{1,1} & B_{1,2} \\ B_{2,1} & B_{2,2} \end{pmatrix}$ with:

$$B_{1,1} = 1 - 2ie^{\pi i(c-a-b)} \frac{\sin(\pi a)\sin(\pi b)}{\sin(\pi c)},$$

$$B_{1,2} = -2\pi i e^{\pi i(c-a-b)} \frac{\Gamma(2-c)\Gamma(1-c)}{\Gamma(1-a)\Gamma(1-b)\Gamma(1+a-c)\Gamma(1+b-c)},$$

$$B_{2,1} = -2\pi i e^{\pi i(c-a-b)} \frac{\Gamma(c)\Gamma(c-1)}{\Gamma(c-a)\Gamma(c-b)\Gamma(a)\Gamma(b)},$$

$$B_{2,2} = 1 + 2ie^{\pi i(c-a-b)} \frac{\sin(\pi(c-a))\sin(\pi(c-b))}{\sin(\pi c)}.$$

We refer for the calculation of the $B_{i,j}$ to ([96, 225, 297]). □

Exercise 5.11 Consider the case $a = b = 1/2$ and $c = 1$. Calculate that the two monodromy matrices are $\begin{pmatrix} 1 & 2 \\ 0 & 1 \end{pmatrix}$ and $\begin{pmatrix} 1 & 0 \\ -2 & 1 \end{pmatrix}$. (We note that, since $c = 1$ and $a+b-c+1 = 1$, one cannot quite use the preceeding formulas. A new calculation in this special case is needed.) Determine the monodromy group and the differential Galois group of the hypergeometric differential equation for the parameter values $a = b = 1/2$ and $c = 1$. □

Other formulas for generators of the monodromy group can be found in [160]. A systematic study of the monodromy groups for the generalized hypergeometric equations ${}_nF_{n-1}$ can be found in [33]. The basic observation, which makes computation possible and explains the explicit formulas in [160, 33], is that the monodromy of an irreducible generalized hypergeometric equation is *rigid*. The latter means that the monodromy group is, up to conjugation, determined by the local monodromies at the three singular points. Rigid equations and rigid monodromy groups are rather special and rare. In [157] a theory of rigid equations is developed. This theory leads to an algorithm that produces, in principle, all rigid equations.

5.2 A Solution of the Inverse Problem

The inverse problem for ordinary Galois theory asks what the possible Galois groups are for a given field. The most important problem is to find all possible finite groups that are Galois groups of a Galois extension of $\mathbf{Q}$. The inverse problem for a differential field K, with algebraically closed field of constants C, is the analogous question:

> *Which linear algebraic groups over C are the differential Galois groups of linear differential equations over K ?*

As we will show, the answer for $\mathbf{C}(z)$ is:

Theorem 5.12 *For any linear algebraic group G over $\mathbf{C}$, there is a differential equation $\frac{d}{dz} - A$ over $\mathbf{C}(z)$ with differential Galois group G.*

This answer was first given by C. Tretkoff and M. Tretkoff [283]. The simple proof is based upon two ingredients:

1. Every linear algebraic group $G \subset \mathrm{GL}_n(\mathbf{C})$ has a Zariski-dense, finitely generated subgroup H.
2. Let a finite set $S \subset \mathbf{P}^1$ be given and a homomorphism $M : \pi_1(U, p) \to \mathrm{GL}_n(\mathbf{C})$, where $U = \mathbf{P}^1 \setminus S$ and $p \in U$. Then there is a regular singular differential equation $\frac{d}{dz} - A$ over $\mathbf{C}(z)$ with singular locus S, such that the monodromy map $\mathbf{M} : \pi_1(U, p) \to \mathrm{GL}(V_p)$ is, with respect to a suitable basis of V_p, equal to the homomorphism M.

Proof. Assuming the two ingredients above, the proof goes as follows. Take elements $g_1, \dots, g_k \in G$ such that the subgroup generated by the $g_1, \dots, g_k$ is Zariski dense in G. Consider the singular set $S = \{1, 2, 3, \dots, k, \infty\}$ and let $U = \mathbf{P}^1 \setminus S$. Then the fundamental group $\pi_1(U, 0)$ is the free group generated by $\lambda_1, \dots, \lambda_k$, where λ_i is a loop starting and ending in 0, around the point i. The homomorphism $M \to G \subset \mathrm{GL}_n(\mathbf{C})$ is defined by $M(\lambda_i) = g_i$ for $i = 1, \dots k$. The regular singular differential equation $\frac{d}{dz} - A$ with monodromy map equal to M, has differential Galois group G, according to Theorem 5.8. □

We now turn to the two ingredients of the proof. We will prove the first in this section and give an outline of the proof of the second in the next section. A fuller treatment of this second ingredient is give in the next chapter.

Lemma 5.13 *Every linear algebraic group G has a Zariski-dense, finitely generated subgroup.*

Proof. Let G^o denote the connected component of the identity. Since G^o is a normal subgroup of finite index, it suffices to prove the lemma for G^o. In other words, we may suppose that $G \subset \mathrm{GL}_n(\mathbf{C})$ is connected and $G \neq \{id\}$. We will now use induction with respect to the dimension of G.

First, we want to show that G has an element g of infinite order and therefore contains a connected subgroup $\overline{< g >}^o$ of positive dimension.

Consider the morphism $f : G \to \mathbf{C}^n$ of algebraic varieties over $\mathbf{C}$, defined by $f(g) = (f_{n-1}(g), \dots, f_0(g))$ where $X^n + f_{n-1}(g)X^{n-1} + \cdots + f_0(g)$ is the characteristic polynomial of g.
Assume first that f is constant. Then every element of G has characteristic polynomial $(X - 1)^n$, the characteristic polynomial of the identity. The only matrix of finite order having this characteristic polynomial is the identity, so G must contain elements of infinite order.
Now assume that f is not constant. By Chevalley's theorem, the image I of f is a constructible subset of $\mathbf{C}^n$. Moreover, this image I is irreducible since G is connected. If all elements of G were of finite order, then the roots of the associated characteristic polynomials would be roots of unity. This would imply that the image I is countable, a contradiction.

In the above proof we have used the fact that $\mathbf{C}$ is not countable. The following proof is valid for any algebraically closed field C of characteristic 0. One observes that an element of G that has finite order is semisimple (i.e., diagonalizable). If every element of G has finite order, then every element of G is semisimple. A connected linear algebraic group of positive dimension all of whose elements are diagonalizable is isomorphic to a torus, i.e., a product of copies of $\mathbf{G}_m$ ([141], Exercise 21.4.2). Such groups obviously contain elements of infinite order.

We now finish the proof of the theorem. If the dimension of G is 1, then there exists an element $g \in G$ of infinite order. The subgroup generated by g is clearly Zariski dense in G.

Suppose now that the dimension of G is greater than 1. Let $H \subset G$ be a maximal proper connected subgroup. If H happens to be a normal subgroup then G/H is known to be a linear algebraic group. By induction we can take elements $a_1 \dots, a_n \in G$ such that their images in G/H generate a Zariski-dense subgroup of G/H. Take elements $b_1, \dots, b_m \in H$ that generate a Zariski-dense subgroup of H. Then the collection $\{a_1, \dots, a_n, b_1, \dots, b_m\}$ generates a Zariski dense subgroup of G.

If H is not a normal subgroup then there is a $g \in G$ with $gHg^{-1} \neq H$. Consider a finite set of elements $a_1, \dots, a_n \in H$ that generate a Zariski-dense subgroup of H. Let L denote the subgroup of G generated by $a_1, \dots, a_n, g$. The Zariski closure $\overline{L}$ of L contains both H and gHg^{-1}, so does $\overline{L}^o$ and $\overline{L}^o \neq H$. The maximality of H implies that $\overline{L}^o = G$ and therefore also $\overline{L} = G$. □

Remark 5.14 There has been much work on the inverse problem in differential Galois theory. Ramis [241, 242] has described how his characterization of the local Galois group can be used to solve the inverse problem over $\mathbf{C}(\{z\})$ and $\mathbf{C}(z)$. This is presented in the Chaps. 8, 10, and 11. In [210], it is shown that any connected linear algebraic group is a differential Galois group over a differential field k of characteristic zero with algebraically closed field of constants C and whose transcendence degree over C is finite and nonzero (see also [211]). This completed a program begun by Kovacic who proved a similar result for solvable connected groups ([164, 165]). A more complete history of the problem can be found in [210]. A description and recasting of the results of [210] and [241] can be found in [230]. We shall describe the above results more fully in Chap. 11. A method for effectively constructing linear differential equations with given finite group is presented in [233] (see Chap. 4). □

5.3 The Riemann-Hilbert Problem

Let $S \subset \mathbf{P}^1$ be finite. Assume for convenience that $S = \{s_1, \ldots, s_k, \infty\}$. Put $U = \mathbf{P}^1 \setminus S$, choose a point $p \in U$ and let $M : \pi_1(U, p) \to \mathrm{GL}_n(\mathbf{C})$ be a homomorphism. The Riemann-Hilbert problem (= Hilbert's 21st problem) asks whether there is a fuchsian differential equation $\frac{d}{dz} - \sum_{i=1}^{k} \frac{A_i}{z-s_i}$, with constant matrices A_i, such that the monodromy map $\mathbf{M} : \pi_1(U, p) \to \mathrm{GL}(V_p)$ coincides with the given M for a suitable basis of V_p. For many special cases, one knows that this problem has a positive answer (see [9, 26]):

1. Let $\lambda_1, \ldots, \lambda_k$ be generators of $\pi_1(U, p)$, each enclosing just one of the s_i (cf. Sect. 5.1.2). If one of the $M(\lambda_i)$ is diagonalizable, then the answer is positive (Plemelj [224]).
2. If all the $M(\lambda_i)$ are sufficiently close to the identity matrix, then the solution is positive (Lappo-Danilevskii [171]).
3. When $n = 2$, the answer is positive (Dekkers [79]).
4. If the representation M is irreducible, the answer is positive (Kostov [163] and Bolibruch [9, 42]).

The first counterexample to the Riemann-Hilbert problem was given by A.A. Bolibruch ([9, 41]) This counterexample is for $n = 3$ and S consisting of 4 points. In addition, Bolibruch [41] has characterized when the solution is positive for $n = 3$.

We will present proofs of the statements 2, 3, and 4 in Chap. 6 but in this section we shall consider a weaker version of this problem. The weaker version only asks for a regular singular differential equation with singular locus S and M equal to the monodromy map $\mathbf{M}$. Here the answer is always positive. The modern version of the proof uses machinery that we will develop in Chap. 6, but for now we will indicate the main ideas of the proof.

Theorem 5.15 *For any homomorphism $M : \pi_1(U, p) \to \mathrm{GL}_n(\mathbf{C})$, there is a regular singular differential equation with singular locus S and with monodromy map equal to M.*

Proof. We start with the simplest case: $S = \{0, \infty\}$. Then $U = \mathbf{C}^*$ and we choose $p = 1$. The fundamental group is isomorphic to $\mathbf{Z}$. A generator for this group is the circle in positive orientation through 1 and around 0. The homomorphism M is then given by a single matrix $B \in \mathrm{GL}_n(\mathbf{C})$, the image of the generator. Choose a constant matrix A with $e^{2\pi i A} = B$. Then the differential equation $\delta - A$ is a solution to the problem.

Assume now $\#S > 2$. We now introduce the concept of *a local system* $\mathbf{L}$ *on* U. This is a sheaf of $\mathbf{C}$-vector spaces on U such that $\mathbf{L}$ is locally isomorphic to the constant sheaf $\mathbf{C}^n$. Take any point $q \in U$ and a path λ from p to q. Using the fact that $\mathbf{L}$ is locally isomorphic to the constant sheaf $\mathbf{C}^n$, one finds by following the path λ a $\mathbf{C}$-linear bijection $\mathbf{L}_p \to \mathbf{L}_q$. This is completely similar to analytical continuation and can be seen to depend only on the homotopy class of the path. If $p = q$, this results in a group homomorphism $\Phi_L : \pi_1(U, p) \to \mathrm{GL}(\mathbf{L}_p)$. Using some algebraic topology (for instance, the universal covering of U) one shows that for any homomorphism $\Phi : \pi_1(U, p) \to \mathrm{GL}(\mathbf{C}^n)$ there is a local system $\mathbf{L}$ such that Φ_L is equivalent to Φ. In particular, there is a local system $\mathbf{L}$ such that $\Phi_L = M$.

The next step is to consider the sheaf $H := \mathbf{L} \otimes_{\mathbf{C}} O_U$, where O_U denotes the sheaf of analytic functions on U. On this sheaf one introduces a differentiation $'$ by $(l \otimes f)' = l \otimes f'$. Now we are already getting close to the solution of the weak Riemann-Hilbert problem. Namely, it is known that the sheaf H is isomorphic with the sheaf O_U^n. In particular, $H(U)$ is a free $O(U)$-module and has some basis $e_1, \ldots, e_n$ over $O(U)$. The differentiation with respect to this basis has a matrix A with entries in $O(U)$. Then we obtain the differential equation $\frac{d}{dz} + A$ on U, which has M as monodromy map. We note that $\mathbf{L}$ is, by construction, the sheaf of the solutions of $\frac{d}{dz} + A$ on U.

We want a little more, namely that the entries of A are in $\mathbf{C}(z)$. To do this we will extend the sheaf H to a sheaf on all of S. This is accomplished by gluing to H with its differentiation, for each point $s \in S$, another sheaf with differentiation that lives above a small neighborhood of s. To make this explicit, we suppose that $s = 0$. The restriction of H with its differentiation on the pointed disk $D^* := \{z \in \mathbf{C} | \ 0 < |z| < \epsilon\} \subset U$ can be seen to have a basis $f_1, \ldots, f_n$ over $O(V)$, such that the matrix of the differentiation with respect to this basis is $z^{-1}C$, where C is a constant matrix. On the complete disk $D := \{z \in \mathbf{C} | \ |z| < \epsilon\}$ we consider the sheaf O_D^n with differentiation given by the matrix $z^{-1}C$. The restriction of the latter differential equation to D^* is isomorphic to the restriction of H to D^*. Thus one can glue the two sheaves, respecting the differentiations. After doing all the gluing at the points of S we obtain a differential equation $\frac{d}{dz} - B$, where the entries of B are meromorphic functions on all of $\mathbf{P}^1$ and thus belong to $\mathbf{C}(z)$. By construction, S is the singular set of the equation and the monodromy map of $\frac{d}{dz} - B$ is the prescribed one. Furthermore, at any singular point s the equation is equivalent to an equation having at most a pole of order 1. □

Remarks 5.16 In Chap. 6 we will describe a more sophisticated formulation of a regular, or a regular singular differential equation on any open subset U of $\mathbf{P}^1$ (including the case $U = \mathbf{P}^1$). We give a preview of this formulation here.

As above, an analytic vector bundle M of rank n on U is a sheaf of O_U-modules that is locally isomorphic to the sheaf O_U^n. One considers also Ω_U^{an}, the sheaf of the holomorphic differential forms on U. This is an analytic vector bundle on U of rank 1. A *regular connection* on M is a morphism of sheaves $\nabla : M \to \Omega_U^{an} \otimes M$, which is $\mathbf{C}$-linear and satisfies the rule: $\nabla(fm) = df \otimes m + f\nabla(m)$ for any sections f of O_U and m of M above any open subset of U.

Let $S \subset U$ be a finite (or discrete) subset of U. Then $\Omega_U^{an}(S)$ denotes the sheaf of the meromorphic differential forms on U, which have poles of order at most 1 at the set S. A *regular singular connection* on M, with singular locus in S, is a morphism of sheaves $\nabla : M \to \Omega_U^{an}(S) \otimes M$, having the same properties as above.

In the case of a finite subset S of $U = \mathbf{P}^1$, one calls a regular singular connection on M *fuchsian* if, moreover, the vector bundle M is trivial, i.e., isomorphic to the direct sum of n copies of the structure sheaf O_U. For the case $U = \mathbf{P}^1$, there is a 1–1 correspondence between analytic and algebraic vector bundles (by the so-called GAGA theorem). This means that the analytic point of view for connections coincides with the algebraic point of view.

In the sketch of the proof of Theorem 5.15, we have, in fact, made the following steps. First, a construction of a regular connection ∇ on an analytic vector bundle M above $U := \mathbf{P}^1 \setminus S$, which has the prescribed monodromy. Then for each point $s \in S$, we have glued to the connection (M, ∇) a regular singular connection (M_s, ∇_s) living on a neighborhood of s. By this gluing one obtains a regular singular analytic connection (N, ∇) on $\mathbf{P}^1$ having the prescribed monodromy. Finally, this analytic connection is identified with an algebraic one. Taking the rational sections of the latter (or the meromorphic sections of N) one obtains the regular singular differential equation $\frac{d}{dz} - A$ with $A \in \mathrm{M}(n \times n, \mathbf{C}(z))$, which has the prescribed singular locus and monodromy. Suppose for notational convenience that $S = \{s_1, \ldots, s_k, \infty\}$. Then (N, ∇) is fuchsian (i.e., N is a trivial vector bundle) if and only if $\frac{d}{dz} - A$ has the form $\frac{d}{dz} - \sum_{i=1}^{k} \frac{A_i}{z - s_i}$ with constant matrices $A_1, \ldots, A_k$. □

6 Differential Equations on the Complex Sphere and the Riemann-Hilbert Problem

Let a differential field K with a derivation $f \mapsto f'$ be given. A differential module over K has been defined as a K-vector space M of finite dimension together with a map $\partial : M \to M$ satisfying the rules: $\partial(m_1 + m_2) = \partial(m_1) + \partial(m_2)$ and $\partial(fm) = f'm + f\partial(m)$. In this definition, one refers to the chosen derivation of K. We want to introduce the more general concept of *connection*, which avoids this choice. The advantage is that one can perform constructions, especially for the Riemann-Hilbert problem, without reference to local parameters. To be more explicit, consider the field $K = \mathbf{C}(z)$ of the rational functions on the complex sphere $\mathbf{P} = \mathbf{C} \cup \{\infty\}$. The derivations that we have used are $\frac{d}{dt}$ and $t^N \frac{d}{dt}$ where t is a local parameter on the complex sphere (say t is $z - a$ or $1/z$ or an even more complicated expression). The definition of connection (in its various forms) requires other concepts such as (universal) differentials, analytic and algebraic vector bundles, and local systems. We will introduce those concepts and discuss the properties that interest us here.

6.1 Differentials and Connections

All the rings that we will consider are supposed to be commutative, to have a unit element and to contain the field $\mathbf{Q}$. Let $k \subset A$ be two rings.

Definition 6.1 A *differential* (or derivation, or differential module) for A/k is a k-linear map $D : A \to M$, where M is an A-module, such that $D(ab) = aD(b) + bD(a)$. □

We note that $D : A \to M$, as above, is often called a differential module. This is, however, in conflict with the terminology introduced in Chap. 1. The same observation holds for the following terminology. There exists a *universal differential* (or universal differential module, or universal derivation), denoted by $d = d_{A/k} : A \to \Omega_{A/k}$. This object is supposed to have the property: for every derivation $D : A \to M$, there exists a unique A-linear map $l : \Omega_{A/k} \to M$ such that $D = l \circ d_{A/k}$. This property is easily seen to determine $d_{A/k} : A \to \Omega_{A/k}$ up to canonical isomorphism. The construction of the universal differential is similar to other general constructions such as the tensor product and we refer to ([170], Chap. XIX, §3) for the details.

Examples 6.2

1. Let k be a field and $A = k(z)$ a transcendental field extension. Then the universal differential $d : A \to \Omega_{A/k}$ can easily be seen to be: $\Omega_{A/k}$ the one-dimensional vector space over A with basis dz and d is given by $d(f) = \frac{df}{dz}dz$.

2. More generally let $k \subset A$ be a field extension such that A is an algebraic extension of a purely transcendental extension $k(z_1, \dots, z_n) \supset k$. Then $\Omega_{A/k}$ is a vector space over A with basis $dz_1, \dots, dz_n$. The universal differential d is given by $d(f) = \sum_{j=1}^{n} \frac{\partial f}{\partial z_j} dz_j$. The derivations $\frac{\partial}{\partial z_j}$ are defined as follows. On the field $k(z_1, \dots, z_n)$ the derivations $\frac{\partial}{\partial z_j}$ are defined as usual. Since the extension $k(z_1, \dots, z_n) \subset A$ is algebraic and separable, each derivation $\frac{\partial}{\partial z_j}$ uniquely extends to a derivation $A \to A$.

It is clear that what we have defined above is a differential. Now we will show that $d : A \to Adz_1 \oplus \cdots \oplus Adz_n$ is the universal differential. Let a derivation $D : A \to M$ be given. We have to show that there exists a unique A-linear map $l : \Omega_{A/k} \to M$ such that $D = l \circ d$. Clearly l must satisfy $l(dz_j) = D(z_j)$ for all $j = 1, \dots, n$ and thus l is unique. Consider now the derivation $E := D - l \circ d$. We have to show that $E = 0$. By construction $E(z_j) = 0$ for all j. Thus E is also 0 on $k(z_1, \dots, z_n)$. Since any derivation of $k(z_1, \dots, z_n)$ extends *uniquely* to A, we find that $E = 0$.

3. We consider now the case, k is a field and $A = k((z))$. One would like to define the universal differential as $d : A \to Adz$ with $d(f) = \frac{df}{dz}dz$. This is a perfectly natural differential module. Unfortunately, it does not have the universality property. The reason for this is that A/k is a transcendental extension of infinite transcendence degree. In particular there exists a nonzero derivation $D : A \to A$, which is 0 on the subfield $k(z)$. Still we prefer the differential module above which we will denote by $d : A \to \Omega^f_{A/k}$. It can be characterized among all differential modules by the more subtle property:

> For every differential $D : A \to M$, such that $D(k[[z]]) \subset M$ lies in a finitely generated $k[[z]]$-submodule of M, there exists a unique A-linear map $l : \Omega^f_{A/k} \to M$ with $D = l \circ d$.

For completeness, we will give a proof of this. The l that we need to produce must satisfy $l(dz) = D(z)$. Let l be the A-linear map defined by this condition and consider the derivation $E := D - l \circ d$. Then $E(z) = 0$ and also $E(k[[z]])$ lies in a finitely generated $k[[z]]$-submodule N of M. Consider an element $h \in k[[z]]$ and write it as $h = h_0 + h_1 z + \cdots + h_{n-1} z^{n-1} + z^n g$ with $g \in k[[z]]$. Then $E(h) = z^n E(g)$. As a consequence $E(h) \in \cap_{n \geq 1} z^n N$. From local algebra ([170], Chap. X, §5) one knows that this intersection is 0. Thus E is 0 on $k[[z]]$ and as a consequence also zero on A. One observes from the above that the differential does not depend on the choice of the local parameter z.

4. The next example is $k = \mathbf{C}$ and $A = \mathbf{C}(\{z\})$. The differential $d : A \to Adz$, with $d(f) = \frac{df}{dz}dz$, is again natural. It will be denoted by $d : A \to \Omega^f_{A/k}$. This differential

is not universal, but can be characterized by the more subtle property stated above. One concludes again that the differential does not depend on the choice of the local parameter z in the field A.

5. Let $k = \mathbf{C}$ and A be the ring of the holomorphic functions on the open unit disk (or any open subset of $\mathbf{C}$). The obvious differential $d : A \to Adz$, given by $d(f) = \frac{df}{dz}dz$, will be denoted by $\Omega^f_{A/k}$. Again it does not have the universal property, but satisfies a more subtle property analogous to 3. In particular, this differential does not depend on the choice of the variable z. □

In the following we will simply write $d : A \to \Omega$ for the differential that is suitable for our choice of the rings $k \subset A$. We note that $\mathrm{Hom}_A(\Omega, A)$, the set of the A-linear maps from Ω to A, can be identified with derivations $A \to A$ that are trivial on k. This identification is given by $l \mapsto l \circ d$. In the case that $\Omega = \Omega_{A/k}$ (the universal derivation) one finds an identification with all derivations $A \to A$ that are trivial on k. In the Examples 6.2.3–6.2.5, one finds all derivations of the type $h\frac{d}{dz}$ (with $h \in A$).

Definition 6.3 A *connection for A/k* is a map $\nabla : M \to \Omega \otimes_A M$, where:

1. M is a (finitely generated) module over A.
2. ∇ is k-linear and satisfies $\nabla(fm) = df \otimes m + f\nabla(m)$ for $f \in A$ and $m \in M$.

□

Let $l \in \mathrm{Hom}(\Omega, A)$ and $D = l \circ d$. One then defines $\nabla_D : M \to M$ as

$$\nabla : M \to \Omega \otimes M \stackrel{l\otimes 1_M}{\to} A \otimes M = M.$$

Thus $\nabla_D : M \to M$ is a differential module with respect to the differential ring A with derivation $f \mapsto D(f)$.

Examples 6.4

1. k is a field and $A = k(z)$. A connection $\nabla : M \to \Omega \otimes M$ gives rise to the differential module $\partial : M \to M$ with $\partial = \nabla_{\frac{d}{dz}}$ of $k(z)/k$ with respect to the derivation $\frac{d}{dz}$. On the other hand, a given differential module $\partial : M \to M$ (w.r.t. $\frac{d}{dz}$) can be made into a connection ∇ by the formula $\nabla(m) := dz \otimes \partial(m)$. We conclude that there is only a notational difference between connections for $k(z)/k$ and differential modules over $k(z)/k$.

2. Let k be a field and $A = k((z))$. As before Ω will be Adz and $d : A \to \Omega$ is the map $d(f) = \frac{df}{dz}dz$. Let M be a vector space over A of dimension n. A $k[[z]]$-lattice $\Lambda \subset M$ is a $k[[z]]$-submodule of M of the form $k[[z]]e_1 + \cdots + k[[z]]e_n$, where $e_1, \ldots e_n$ is a basis of M. Let (M, ∇) be a connection for A/k. The connection is

called *regular* if there is a lattice Λ such that $d(\Lambda) \subset dz \otimes \Lambda$. The connection is called *regular singular* if there is a lattice Λ such that $d(\Lambda) \subset dz \otimes z^{-1}\Lambda$.

Suppose now (for convenience) that k is algebraically closed. Let (M, ∇) be a connection for $k(z)/k$. For each point p of $k \cup \{\infty\}$ we consider the completion $\widehat{k(z)}_p$ of $k(z)$ with respect to this point. This completion is either $k((z-a))$ or $k((z^{-1}))$. The connection (M, ∇) induces a connection for $\widehat{k(z)}_p/k$ on $\widehat{M}_p := \widehat{k(z)}_p \otimes M$. One calls (M, ∇) *regular singular* if each of the $\widehat{M}_p$ is regular singular.

3. k is a field and $A = k(z_1, \ldots, z_n)$. A connection $\nabla : M \to \Omega \otimes M$ gives, for every $j = 1, \ldots, n$, to a differential module $\nabla_{\frac{\partial}{\partial z_j}} : M \to M$ with respect to the derivation $\frac{\partial}{\partial z_j}$. In other words a connection is a linear system of partial differential equations (one equation for each variable). See also Appendix D.

4. In parts 3–5 of Examples 6.2 a connection together with a choice of the derivation is again the same thing as a differential module with respect to this derivation. □

6.2 Vector Bundles and Connections

We consider a connected Riemann surface X. The sheaf of holomorphic functions on X will be called O_X. A *vector bundle* M of rank m on X can be defined as a sheaf of O_X-modules on X, such that M is locally isomorphic with the sheaf of O_X-modules O_X^m. The vector bundle M is called *free* (or trivial) if M is globally (i.e., on all of X) isomorphic to O_X^m. With vector bundles one can perform the operations of linear algebra: direct sums, tensor products, Hom's, kernels, etc. Vector bundles of rank one are also called *line bundles*. We will write $H^0(X, M)$, or sometimes $H^0(M)$, for the vector space of the global sections of M on X. It is known that any vector bundle on a noncompact Riemann surface is free, see [100]. For compact Riemann surfaces the situation is quite different. Below, we will describe the vector bundles on the Riemann sphere.

The *line bundle* Ω_X *of the holomorphic differentials* will be important for us. This sheaf can be defined as follows. For open $U \subset X$ and an isomorphism $t : U \to \{c \in \mathbf{C} |\ |c| < 1\}$, the restriction of Ω_X to U is $O_X dt$. Furthermore, there is a canonical morphism of sheaves $d : O_X \to \Omega_X$, which is defined on the above U by $d(f) = \frac{df}{dt}dt$ (see also Examples 6.2.5 and Examples 6.4).

In the literature, the term "vector bundle of rank m" refers sometimes to a closely related but somewhat different object. For the sake of completeness we will explain this. For the other object we will use the term *geometric vector bundle* of rank m on a Riemann surface X. This is a complex analytic variety V together with a morphism of analytic varieties $\pi : V \to X$. The additional data are: for each $x \in X$, the fibre $\pi^{-1}(x)$ has the structure of an m-dimensional complex vector space. Furthermore,

X has an open covering $\{U_i\}$ and for each i an isomorphism $f_i : \pi^{-1}(U_i) \to \mathbf{C}^m \times U_i$ of analytic varieties such that: $pr_2 \circ f_i$ is the restriction of π to $\pi^{-1}(U_i)$ and for each point $x \in U_i$ the map $\pi^{-1}(x) \to \mathbf{C}^m \times \{x\} \to \mathbf{C}^m$, induced by f_i, is an isomorphism of complex linear vector spaces.

The link between the two concepts can be given as follows. Let $\pi : V \to X$ be a geometric vector bundle. Define the sheaf M on X by letting $M(U)$ consist of the maps $s : U \to \pi^{-1}U$ satisfying $\pi \circ s$ is the identity on U. The additional structure on $V \to X$ induces a structure of $O_X(U)$-module on $M(U)$. The "local triviality" of $V \to X$ has as a consequence that M is locally isomorphic to the sheaf O_X^m. On the other hand, one can start with a vector bundle M on X and construct the corresponding geometric vector bundle $V \to X$.

Definition 6.5 A *regular connection* on a Riemann surface X is a vector bundle M on X together with a morphism of sheaves of groups $\nabla : M \to \Omega_X \otimes M$, which satisfies for every open U and for any $f \in O_X(U)$, $m \in M(U)$ the "Leibniz rule" $\nabla(fm) = df \otimes m + f\nabla(m)$. □

For an open U, which admits an isomorphism $t : U \to \{c \in \mathbf{C} |\ |c| < 1\}$ one can identify $\Omega_X(U)$ with $O_X(U)dt$ and $M(U)$ with $O_X^m(U)$. Then $\nabla(U) : M(U) \to O_X(U)dt \otimes M(U)$ is a connection in the sense of the definition given in Sect. 6.1. One can rephrase this by saying that a regular connection on X is the "sheafification" of the earlier notion of connection for rings and modules.

Examples 6.6 *Examples, related objects, and results.*
1. *Regular connections on a noncompact Riemann surface.*
According to ([100], Theorem 30.4)) every vector bundle M on a connected, non-compact Riemann surface is free. Let X be an open connected subset of $\mathbf{P}$ and suppose for notational convenience that $\infty \notin X$. We can now translate the notion of regular connection (M, ∇) on X in more elementary terms. The vector bundle M will be identified with O_X^m; the sheaf of holomorphic differentials is identified with $O_X dz$; furthermore, ∇ is determined by ∇ on $M(X)$ and by $\nabla_{\frac{d}{dz}}$ on $M(X)$. In this way, we find a matrix differential operator $\frac{d}{dz} + A$, where the coordinates of A are holomorphic functions on X. This matrix differential operator is "equivalent" with (M, ∇).

2. *Local systems on X.*
X will be a topological space that is connected and locally pathwise connected. A (complex) *local system* (of dimension n) on X is a sheaf L of complex vector spaces that is locally isomorphic to the constant sheaf $\mathbf{C}^n$. This means that X has a covering by open sets U such that the restriction of L to U is isomorphic to the constant sheaf $\mathbf{C}^n$ on U. For the space $[0, 1]$ any local system is trivial, which means that it is the constant sheaf $\mathbf{C}^n$. This can be seen by showing that n linearly independent sections above a neighborhood of 0 can be extended to the whole space. Let $\lambda : [0, 1] \to X$ be a path in X, i.e., a continuous function. Let L be a local system

on X. Then $\lambda^* L$ is a local system on $[0, 1]$. The triviality of this local system yields an isomorphism $(\lambda^* L)_0 \to (\lambda^* L)_1$. The two stalks $(\lambda^* L)_0$ and $(\lambda^* L)_1$ are canonically identified with $L_{\lambda(0)}$ and $L_{\lambda(1)}$. Thus we find an isomorphism $L_{\lambda(0)} \to L_{\lambda(1)}$ induced by λ. Let b be a base point for X and let π_1 denote the fundamental group of X with respect to this base point. Fix again a local system L on X and let V denote the stalk L_b. Then for any closed path λ through b we find an isomorphism of V. In this way, we have associated to L a representation $\rho_L : \pi_1 \to \mathrm{GL}(V)$ of the fundamental group.

We make this somewhat more systematic. Let $\mathrm{LocalSystems}(X)$ denote the category of the local systems on X and let Repr_{π_1} denote the category of the finite dimensional complex representations of π_1. Then we have defined a functor $\mathrm{LocalSystems}(X) \to \mathrm{Repr}_{\pi_1}$ that has many nice properties. We claim that:

The functor $\mathrm{LocalSystems}(X) \to \mathrm{Repr}_{\pi_1}$ *is an equivalence of categories.*

We will only sketch the (straightforward) proof. Let $u : U \to X$ denote the universal covering. On U every local system is trivial, i.e., isomorphic to a constant sheaf $\mathbf{C}^n$. This follows from U being simply connected (one defines n independent sections above any path connecting a base point to an arbitrary point, shows that this is independent of the path and so defines n independent global sections). Take a local system L on X and let $V = L_b$. Then the local system $u^* L$ is isomorphic to the constant sheaf V on U. The fundamental group π_1 is identified with the group of automorphisms of the universal covering $u : U \to X$. In particular, for any $\lambda \in \pi_1$ one has $\lambda \circ u = u$ and $\lambda^* \circ u^* L = u^* L$. This gives again the representation $\pi_1 \to \mathrm{GL}(V)$.

One can also define a functor in the other direction. Let $\rho : \pi_1 \to \mathrm{GL}(V)$ be a representation. This can be seen as an action on V considered as constant local system on U. In particular for any π_1-invariant open set $B \subset U$ we have an action of π_1 on $V(B)$. Define the local system L on X by specifying $L(A)$, for any open $A \subset X$, in the following way: $L(A) = V(u^{-1}A)^{\pi_1}$ (i.e., the elements of $V(u^{-1}A)$ invariant under the action of π_1). It can be verified that the two functors produce an equivalence between the two categories.

3. *Regular connections, local systems, and monodromy.*
We suppose that X is a connected noncompact Riemann surface. Let $\mathrm{Reg}(X)$ denote the category of the regular connections on X. For an object (M, ∇) of $\mathrm{Reg}(X)$ one considers the sheaf L given by $L(A) = \{m \in M(A) |\ \nabla(m) = 0\}$ for any open subset A. The set $L(A)$ is certainly a vector space. Since the connection is "locally trivial" it follows that L is locally isomorphic to the constant sheaf $\mathbf{C}^n$. Thus we found a functor from the category $\mathrm{Reg}(X)$ to the category $\mathrm{LocalSystems}(X)$. We claim that

The functor $\mathrm{Reg}(X) \to \mathrm{LocalSystems}(X)$ *is an equivalence.*

The essential step is to produce a suitable functor in the other direction. Let a local system L be given. Then the sheaf $N := L \otimes_{\mathbf{C}} O_X$ is a sheaf of O_X-modules. Locally, i.e., above some open $A \subset X$, the sheaf L is isomorphic to the constant sheaf $\mathbf{C}e_1 \oplus \cdots \oplus \mathbf{C}e_n$. Thus the restriction of N to A is isomorphic to $O_X e_1 \oplus \cdots \oplus O_X e_n$. This proves that N is a vector bundle. One defines ∇ on the restriction of N to A by the formula $\nabla(\sum f_j e_j) = \sum df_j \otimes e_j \in \Omega_X \otimes N$. These local definitions glue obviously to a global ∇ on N. This defines a functor in the other direction. From this construction it is clear that the two functors are each other's "inverses".

We note that the composition $\mathrm{Reg}(X) \to \mathrm{LocalSystems}(X) \to \mathrm{Repr}_{\pi_1}$ is, in fact, the functor that associates to each regular connection its monodromy representation. From the above it follows that this composition is also an equivalence of categories.

4. *The vector bundles on the complex sphere* **P**

These vector bundles have been classified (by G. Birkhoff [37], A. Grothendieck [117] et al.; see [216]). For a vector bundle M (or any sheaf) on $\mathbf{P}$ we will write $H^0(M)$ or $H^0(\mathbf{P}, M)$ for its set of global sections. For any integer n one defines the line bundle $O_{\mathbf{P}}(n)$ in the following way: Put $U_0 = \mathbf{P} \setminus \{\infty\}$ and $U_\infty = \mathbf{P} \setminus \{0\}$. Then the restrictions of $O_{\mathbf{P}}(n)$ to U_0 and U_∞ are free and generated by e_0 and e_∞. The two generators satisfy (by definition) the relation $z^n e_0 = e_\infty$ on $U_0 \cap U_\infty$.

The main result is that every vector bundle M on the complex sphere is isomorphic to a direct sum $O_{\mathbf{P}}(a_1) \oplus \cdots \oplus O_{\mathbf{P}}(a_m)$. One may assume that $a_1 \geq a_2 \geq \cdots \geq a_m$. Although this direct sum decomposition is not unique, one can show that the integers a_j are unique. One calls the sequence $a_1 \geq \cdots \geq a_m$ the *type of the vector bundle*. We formulate some elementary properties, which are easily verified:

(a) $O_{\mathbf{P}}(0) = O_{\mathbf{P}}$ and $O_{\mathbf{P}}(n) \otimes O_{\mathbf{P}}(m) = O_{\mathbf{P}}(n+m)$.

(b) $O_{\mathbf{P}}(n)$ has only 0 as global section if $n < 0$.

(c) For $n \geq 0$ the global sections of $O_{\mathbf{P}}(n)$ can be written as fe_0, where f runs in the space of polynomials of degree $\leq n$.

The unicity of the a_j above now follows from the calculation of the dimensions of the complex vector spaces $H^0(O_{\mathbf{P}}(n) \otimes M)$. We note that the above M is free if and only if all a_j are zero. Other elementary properties are:

(d) $\Omega_{\mathbf{P}}$ is isomorphic to $O_{\mathbf{P}}(-2)$.

(e) Let $D = \sum n_i[s_i]$ be a divisor on $\mathbf{P}$, i.e., a formal finite sum of points of $\mathbf{P}$ with integers as coefficients. The degree of the divisor D is, by definition, $\sum n_i$. One defines the sheaf $\mathcal{L}(D)$ on $\mathbf{P}$ by: For any open U in $\mathbf{P}$, the group $\mathcal{L}(D)(U)$ consists of the meromorphic functions f on U such that the divisor of f on U is $\geq$ the restriction of $-D$ to U. The sheaf $\mathcal{L}(D)$ is easily seen to be a line bundle and is, in fact, isomorphic to $O_{\mathbf{P}}(n)$, where $n = \sum n_i$ (i.e., the degree of the divisor D).

(f) Let M be any vector bundle on $\mathbf{P}$ and D a divisor. Then $M(D)$ is defined as $\mathcal{L}(D)\otimes M$. In particular, $\Omega_{\mathbf{P}}(D)$ is a sheaf of differential forms on $\mathbf{P}$ with prescribed zeros and poles by D. This sheaf is isomorphic to $O_{\mathbf{P}}(-2+\deg D)$. In the special case that the divisor is $S=[s_1]+\cdots+[s_m]$ (i.e., a number of distinct points with "multiplicity 1"), the sheaf $\Omega_{\mathbf{P}}(S)$ consists of the differential forms that have poles of order at most one at the points $s_1,\ldots,s_m$. The sheaf is isomorphic to $O_{\mathbf{P}}(-2+m)$ and for $m\geq 3$ the dimension of its vector space of global sections is $m-1$. Suppose that the points $s_1,\ldots,s_m$ are all different from ∞. Then $H^0(\Omega(S))$ consists of the elements $\sum_{j=1}^m \frac{a_j}{z-s_j}dz$ with $a_1,\ldots,a_j\in\mathbf{C}$ and $\sum a_j=0$.

5. *The GAGA principle for vector bundles on* **P**.
One can see $\mathbf{P}$ as the Riemann surface associated to the projective line $P^1:=\mathbf{P}^1_{\mathbf{C}}$ over $\mathbf{C}$. Also in the algebraic context one can define line bundles, vector bundles, connections, etc. The "GAGA" principle gives an equivalence between ("algebraic") vector bundles (or more generally coherent sheaves) on P^1 and ("analytic") vector bundles (or analytic coherent sheaves) on $\mathbf{P}$. We will describe some of the details and refer to [259] for proofs (see also [124] for more information concerning the notions of line bundles, vector bundles, etc., in the algebraic context).

We begin by describing the *algebraic* structure on projective line P^1, see [124]. The open sets of P^1, for the Zariski topology, are the empty set and the cofinite sets. The sheaf of regular functions on P^1 will be denoted by O. Thus for a finite set S we have that $O(P^1\setminus S)$ consists of the rational functions that have their poles in S. Let M be a vector bundle on P^1 of rank m. Then for any finite nonempty set S the restriction of M to $P^1\setminus S$ is a free bundle (because $O(P^1\setminus S)$ is a principal ideal domain and since $H^0(M|_{P^1\setminus S})$ is projective it must be free). In particular, $M(P^1\setminus S)$ is a free module of rank m over $O(P^1\setminus S)$. We want to associate to M a vector bundle M^{an} on $\mathbf{P}$.

One defines M^{an} by $M^{an}(\mathbf{P})=M(P^1)$ and for an open set $U\subset\mathbf{P}$, which has an empty intersection with a finite set $S\neq\emptyset$, one defines $M^{an}(U)$ as $O_{\mathbf{P}}(U)\otimes_{O(P^1\setminus S)} M(P^1\setminus S)$. It is not difficult to show that the latter definition is independent of the choice of $S\neq\emptyset$. Furthermore, it can be shown that M^{an} is a vector bundle on $\mathbf{P}$. The construction $M\mapsto M^{an}$ extends to coherent sheaves on P^1 and is "functorial".

In the other direction, we want to associate to a vector bundle N on $\mathbf{P}$ a vector bundle N^{alg} on P^1. One defines N^{alg} as follows. $N^{alg}(P^1)=N(\mathbf{P})$ and for any nonempty finite set S one defines $N^{alg}(P^1\setminus S)=\cup_{k\geq 1}H^0(N(k\cdot S))$. (We note that $k\cdot S$ is considered as a divisor on $\mathbf{P}$.) If one accepts the description of the vector bundles on $\mathbf{P}$, then it is easily seen that N^{alg} is indeed a vector bundle on P^1. The construction $N\mapsto N^{alg}$ extends to (analytic) coherent sheaves and is "functorial".

The two functors ^{an} and ^{alg} provide an equivalence between the vector bundles (or, more generally, analytic coherent sheaves) on $\mathbf{P}$ and the vector bundles (or coherent sheaves) on P^1.

The GAGA principle holds for projective complex varieties and, in particular, for the correspondence between nonsingular, irreducible, projective curves over $\mathbf{C}$ and compact Riemann surfaces. □

Exercise 6.7 *The sheaves $O_{\mathbf{P}}(n)^{alg}$ and $O(n)$.*
In order to describe the analytic line bundle $O_{\mathbf{P}}(n)$ in terms of meromorphic functions we identify $O_{\mathbf{P}}(n)$ with the line bundle $\mathcal{L}(n.[\infty])$ corresponding to the divisor $n.[\infty]$ on $\mathbf{P}$. Let $S = \{p_1, \dots, p_m\}$ be a finite set not containing ∞ and let $f_S = \prod_{i=1}^m (z-p_i)$. Show that for $U = P^1 \setminus S$, $O_{\mathbf{P}}(n)^{alg}(U)$ consists of all rational functions of the form g/f_S^k, where $k \geq 0$ and $\deg g \leq n + km$. Describe $O_{\mathbf{P}}(n)^{alg}(U)$, where $U = P^1 \setminus S$ and S contains the point at infinity. We denote the sheaf $O_{\mathbf{P}}(n)^{alg}$ by $O(n)$.

We note that the algebraic line bundle $O(n)$ on $\mathbf{P}^1$ is usually defined as follows. Put $U_0 = \mathbf{P}^1 \setminus \{\infty\}$ and $U_\infty = \mathbf{P}^1 \setminus \{0\}$. The restrictions of $O(n)$ to U_0 and U_∞ are the free sheaves $O_{U_0}e_0$ and $O_{U_\infty}e_\infty$ since both rings $O(U_0) = \mathbf{C}[z]$ and $O(U_\infty) = \mathbf{C}[z^{-1}]$ are unique factorization domains. The relation between the two generators in the restriction of $O(n)$ to $U_0 \cap U_\infty$ is given by $z^n e_0 = e_\infty$. It is obvious from this description that $O(n)^{an}$ is equal to $O_{\mathbf{P}}(n)$. □

We come now to the definition of a regular singular connection. Let X be a connected Riemann surface, S a finite subset of X.

Definition 6.8 A *regular singular connection on X with singular locus in S* is a pair (M, ∇) with M a vector bundle on X and $\nabla : M \to \Omega(S) \otimes M$ a morphism of sheaves of groups that satisfies for every open U and for any $f \in O_X(U)$, $m \in M(U)$ the "Leibniz rule" $\nabla(fm) = df \otimes m + f\nabla(m)$.

Here, S is seen as a divisor on X and $\Omega(S)$ is the sheaf of differential forms on X having poles of at most order 1 at the points of S. The difference from the earlier defined regular connections is clearly that we allow poles of order 1 at the points of S. We can make this explicit in the local situation: $X = \{c \in \mathbf{C} |\ |c| < 1\}$, $S = \{0\}$ and $M = O_X^m$. Then on X the map $\nabla_{\frac{d}{dz}} : O_X(X)^m \to z^{-1}O_X(X)^m$ identifies with a matrix differential operator $\frac{d}{dz} + A$, where the coefficients of A are meromorphic functions on X having a pole of order at most 1 at $z = 0$. One observes that the notion of regular singular connection is rather close to the definition of regular singular point of a matrix differential equation. One could also introduce *irregular connections* by replacing S by a divisor $\sum n_j[s_j]$ with integers $n_j \geq 1$.

Examples 6.9 *Some properties of regular singular connections.*
1. *The GAGA principle for regular singular connections on* $\mathbf{P}$.
For the sheaf of holomorphic differentials on P^1 we will use the notation Ω and for the analogous (analytic) sheaf on $\mathbf{P}$ we will write Ω^{an}. Let an "algebraic" regular singular connection on P^1 with singular locus in S be given, this is a $\nabla : M \to \Omega(S) \otimes M$, with M a vector bundle and ∇ with the obvious properties. We want to associate

a regular singular connection (M^{an}, ∇) on $\mathbf{P}$ with singular locus in S (see Examples 6.6.3). The only thing to verify is that the new ∇ is unique and well defined. Let U be an open set of $\mathbf{P}$ that has empty intersection with the finite set $T \neq \emptyset$. We have to verify that $\nabla : M^{an}(U) \to \Omega^{an}(S)(U) \otimes M^{an}(U)$ is unique and well defined. One has $M^{an}(U) = O_{\mathbf{P}}(U) \otimes_{O(P^1 \setminus T)} M(P^1 \setminus T)$ and $\Omega(S)^{an}(U) \otimes_{O_{\mathbf{P}}(U)} M^{an}(U)$ is canonically isomorphic to $\Omega(S)^{an}(U) \otimes_{O(P^1 \setminus T)} M(P^1 \setminus T)$. Consider an element $f \otimes m$ with $f \in O_{\mathbf{P}}(U)$ and $m \in M(P^1 \setminus T)$. Then the only possible definition for $\nabla(f \otimes m)$ is $df \otimes m + f\nabla(m)$. This expression lies in $\Omega(S)^{an}(U) \otimes_{O_{\mathbf{P}}(U)} M^{an}(U)$ since $df \in \Omega^{an}(U)$ and $\nabla(m) \in \Omega(S)(P^1 \setminus T) \otimes M(P^1 \setminus T)$.

On the other hand, let (N, ∇) be a regular singular connection with singular locus in S on $\mathbf{P}$. We have to show that N^{alg} inherits a regular singular connection with singular locus in S. Let T be a finite nonempty subset of $\mathbf{P}$. One considers $N(k \cdot T)$, where $k \cdot T$ is seen as a divisor. It is not difficult to see that ∇ on N induces a $\nabla : N(k \cdot T) \to \Omega(S)^{an} \otimes N((k+1) \cdot T)$. By construction $N^{alg}(P^1 \setminus T) = \cup_{k \geq 0} H^0(N(k \cdot T))$. Thus we find an induced map $\nabla : N^{alg}(P^1 \setminus T) \to \Omega(S)(P^1 \setminus T) \otimes N^{alg}(P^1 \setminus T)$. This ends the verification of the GAGA principle.

We now introduce three categories: RegSing$(\mathbf{P}, S)$, RegSing(P^1, S), and RegSing$(\mathbf{C}(z), S)$. The first two categories have as objects the regular singular connections with singular locus in S for $\mathbf{P}$ (i.e., analytic) and for P^1 (i.e., algebraic). The third category has as objects the connections for $\mathbf{C}(z)/\mathbf{C}$ (i.e., $\nabla : M \to C(z)dz \otimes M$, see Examples 6.4), which have at most regular singularities in the points of S (see Examples 6.4.2). We omit the obvious definition of morphism in the three categories. We have just shown that the first two categories are equivalent. There is a functor from the second category to the third one. This functor is given as follows. Let $\nabla : M \to \Omega(S) \otimes M$ be a connection on P^1 (regular singular with singular locus in S). The fibre M_η of M at the "generic point" η of $\mathbf{P}^1$ is defined as the direct limit of all $M(U)$, where U runs over the collection of the cofinite subsets of P^1. One finds a map $\nabla_\eta : M_\eta \to \Omega(S)_\eta \otimes M_\eta$. The expression M_η is a finite dimensional vector space over $\mathbf{C}(z)$ and $\Omega(S)_\eta$ identifies with $\Omega_{\mathbf{C}(z)/\mathbf{C}}$. Thus ∇_η is a connection for $\mathbf{C}(z)/\mathbf{C}$. Moreover, ∇_η has at most regular singularities at the points of S. We shall refer to (M_η, ∇_η) as the *generic fibre* of (M, ∇). We will show (Lemma 6.18) that the functor $\nabla \mapsto \nabla_\eta$ from RegSing(P^1, S) to RegSing$(\mathbf{C}(z), S)$ is surjective on objects. However, this functor is not an equivalence. In particular, nonisomorphic ∇_1, ∇_2 can have isomorphic generic fibres. We will be more explicit about this in Lemma 6.18.

2. *Regular singular connections on free vector bundles on* $\mathbf{P}$.
We consider $X = \mathbf{P}$, $S = \{s_1, \ldots, s_m\}$ with $m \geq 2$ and all s_i distinct from ∞. We want to describe the regular singular connections (M, ∇) with M a free vector bundle and with singular locus in S. From $M \cong O_{\mathbf{P}}^n$ it follows that the vector space of the global sections of M has dimension n. Let $e_1, \ldots, e_n$ be a basis. The global sections of $\Omega(S) \otimes M$ are then the expressions $\sum_{j=1}^n (\sum_k \frac{a_{k,j}}{z - s_k} dz) \otimes e_j$, where for each j we have $\sum_k a_{k,j} = 0$. The morphism ∇ is determined by the images $\nabla(e_j)$

of the global sections of M because M is also generated, locally at every point, by the $\{e_j\}$. Furthermore, we may replace $\nabla(e_j)$ by $\nabla_{\frac{d}{dz}}(e_j)$. This leads to the differential operator in matrix form $\frac{d}{dz}+\sum_{k=1}^{m}\frac{A_k}{z-s_k}$, where the A_j are constant square matrices of size n and satisfy $\sum_{k=1}^{m}A_k=0$. A matrix differential operator of this form will be called a *fuchsian differential equation with singular locus in S.*

For $S=\{s_1,\dots,s_{m-1},\infty\}$ one finds in a similar way an associated matrix differential equation $\frac{d}{dz}+\sum_{k=1}^{m-1}\frac{A_k}{z-s_k}$ (in this case there is no condition on the sum of the matrices A_k). We note that the notion of a fuchsian system with singular locus in S is, since it is defined by means of a connection, invariant under automorphisms of the complex sphere.

3. *A construction with regular singular connections.*
Let (M,∇) be a regular singular connection with singular locus in S. For a point $s\in S$ we will define a new vector bundle $M(-s)\subset M$. Let t be a local parameter at the point s. Then for U not containing s one defines $M(-s)(U)=M(U)$. If U is a small enough neighborhood of s then $M(-s)(U)=tM(U)\subset M(U)$. One can also define a vector bundle $M(s)$. This bundle can be made explicit by $M(s)(U)=M(U)$ if $s\notin U$ and $M(s)(U)=t^{-1}M(U)$ for a small enough neighborhood U of s. We claim that the vector bundles $M(-s)$ and $M(s)$ inherit from M a regular singular connection. For an open U that does not contain s, one has $M(s)(U)=M(-s)(U)=M(U)$ and we define the ∇s for $M(s)$ and $M(-s)$ to coincide with the one for M. For a small enough neighborhood U of s one defines the new ∇s by $\nabla(t^{-1}m)=-\frac{dt}{t}\otimes t^{-1}m+t^{-1}\nabla(m)$ (for $M(s)$ and m a section of M) and $\nabla(tm)=\frac{dt}{t}\otimes tm+t\nabla(m)$ (for $M(-s)$). This is well defined since $\frac{dt}{t}$ is a section of $\Omega(S)$. The ∇s on $M(-s)\subset M\subset M(s)$ are restrictions of each other.

More generally, one can consider any divisor D with support in S, i.e., $D=\sum m_j[s_j]$ for some integers m_j. A regular singular connection on M induces a "canonical" regular singular connection on $M(D)$.

Exercise 6.10 Let (M,∇) be a regular singular connection and let D be a divisor with support in S. Show that the induced regular singular connection on $M(D)$ has the same generic fibre as (M,∇) (see Example 6.9.1). □

4. The historically earlier notion of *fuchsian linear operator L of degree n and with singular locus in S* is defined in a rather different way. For the case $S=\{s_1,\dots,s_{m-1},\infty\}$ this reads as follows. Let $L=\partial^n+a_1\partial^{n-1}+\cdots+a_{n-1}\partial+a_n$, where $\partial=\frac{d}{dz}$ and the $a_j\in\mathbf{C}(z)$. One requires further that the only poles of the rational functions a_j are in S and that each singularity in S is "regular singular". The latter condition is that the associated matrix differential equation can locally at the points of S be transformed into a matrix differential equation with a pole of at most order 1. We will prove the following lemma.

Lemma 6.11 *L is a fuchsian scalar differential equation with singular locus in S if and only if the a_j have the form $\frac{b_j}{(z-s_1)^j\cdots(z-s_{m-1})^j}$ with b_j polynomials of degrees $\leq j(m-1)-j$.*

Proof. We first examine the order of each a_j, say at $z = s_i$. For notational convenience we suppose that $s_i = 0$. We consider $M = z^n L = z^n \partial^n + z a_1 z^{n-1} \partial^{n-1} + \cdots + z^{n-1} a_{n-1} z\partial + z^n a_n$, which can be written as $\delta^n + c_1 \delta^{n-1} + \cdots + c_n$ for certain $c_j \in \mathbf{C}(z)$. From the last expression one easily finds the Newton polygon at the point $z = 0$. The operator (or the corresponding matrix differential equation) is regular singular at $z = 0$ if and only if the Newton polygon has only slope 0. The last condition is equivalent to $ord_0(c_j) \geq 0$ for all j. From the obvious formula $z^m \partial^m = (\delta - m)(\delta - m + 1)\cdots(\delta - 1)\delta$ it follows that the condition on the c_j is equivalent to $ord_0(a_j) \geq -j$ for all j. A similar calculation at $z = \infty$ finishes the proof. □

We note that a scalar operator L, as in the statement, need not be singular at all the points of S. At some of the points of S the equation may have n independent local solutions. In that case, the point is sometimes called an *apparent singularity*. For example, the operator $\partial^2 - \frac{2}{z^2-2}$ is fuchsian with singular locus in $\{\sqrt{2}, -\sqrt{2}, \infty\}$. The point at infinity turns out to be regular.

The automorphisms ϕ of the complex sphere have the form $\phi(z) = \frac{az+b}{cz+d}$ with $\begin{pmatrix} a & b \\ c & d \end{pmatrix} \in \mathrm{PSL}_2(\mathbf{C})$. We extend this automorphism ϕ of $\mathbf{C}(z)$ to the automorphism, again denoted by ϕ, of $\mathbf{C}(z)[\partial]$ by $\phi(\partial) = \frac{1}{(cz+d)^2}\partial$. Suppose that (the monic) $L \in \mathbf{C}(z)[\partial]$ is a fuchsian operator with singular locus in S. Then one can show that $\phi(L) = fM$ with $f \in \mathbf{C}(z)^*$ and M a monic fuchsian operator with singular locus in $\phi(S)$. Thus the notion of a fuchsian scalar operator is also "invariant" under automorphisms of $\mathbf{P}$. □

6.3 Fuchsian Equations

The comparison between scalar fuchsian equations and fuchsian equations in matrix form is far from trivial. The next two sections deal with two results that are also present in [9]. In a later section we will return to this theme.

6.3.1 From Scalar Fuchsian to Matrix Fuchsian

C will denote an algebraically closed field of characteristic 0. Let an n-th order monic fuchsian operator $L \in C(z)[\partial]$ (where $\partial = \frac{d}{dz}$) with singular locus in S be given. We want to show that there is a fuchsian matrix equation of order n with singular locus in S, having a cyclic vector e, such that the minimal monic operator

$M \in C(z)[\partial]$ with $Me = 0$ coincides with L. This statement seems to be "classical". However, the only proof that we know of is that of ([9], Theorem 7.2.1). We present here a proof that is algebraic and even algorithmic.

If S consists of one point then we may, after an automorphism of $\mathbf{P}^1$, suppose that $S = \{\infty\}$. The fuchsian operator L can only be ∂^n and the statement is trivial. If S consists of two elements then we may suppose that $S = \{0, \infty\}$. Let us use the operator $\delta = z\partial$. Then $z^n L$ can be rewritten as operator in δ and it has the form $\delta^n + a_1\delta^{n-1} + \cdots + a_n$ with all $a_i \in C$. Let V be an n-dimensional vector space over C with basis $e_1, \ldots, e_n$. Define the linear map B on V by $B(e_i) = e_{i+1}$ for $i = 1, \ldots, n-1$ and $Be_n = -a_n e_n - a_{n-1}e_{n-1} - \cdots - a_1 e_1$. Then the matrix equation $\delta + B$ (or the matrix equation $\partial + \frac{B}{z}$) is fuchsian and the minimal monic operator M with $Me_1 = 0$ is equal to L. For a singular locus S with cardinality > 2 we may suppose that S is equal to $0, s_1, \ldots, s_k, \infty$.

Theorem 6.12 *Let $L \in C(z)[\partial]$ be a monic fuchsian operator with singular locus in $S = \{0, s_1, \ldots, s_k, \infty\}$. There are constant matrices $B_0, \ldots, B_k$*

$$\text{with } B_0 = \begin{pmatrix} * & & & & \\ 1 & * & & & \\ & \cdot & * & & \\ & & \cdot & * & \\ & & & 1 & * \end{pmatrix} \text{ and } B_1, \ldots, B_k \text{ upper triangular,}$$

$$\text{i.e., having the form } \begin{pmatrix} * & * & * & * & * \\ & * & * & * & * \\ & & \cdot & \cdot & \cdot \\ & & & \cdot & \cdot \\ & & & & * \end{pmatrix},$$

such that the first basis vector e_1 is cyclic for the fuchsian matrix equation $\partial + \frac{B_0}{z} + \sum_{i=1}^{k} \frac{B_i}{z-s_i}$ and L is the monic operator of smallest degree with $Le_1 = 0$.

Proof. Write $D = (z - s_1)\cdots(z - s_k)$ and $F = zD$. Consider the differential operator $\Delta = F\frac{d}{dz}$. One can rewrite $F^n L$ as a differential operator in Δ. It will have the form $\tilde{L} := \Delta^n + A_1\Delta^{n-1} + \cdots + A_{n-1}\Delta + A_n$, where the A_i are polynomials with degrees $\leq k.i$. Conversely, an operator of the form $\tilde{L}$ in Δ can be transformed into a fuchsian operator in ∂ with singular locus in S. Likewise, we multiply the matrix operator of the statement on the left-hand side by F and find a matrix operator of the form

$$\Delta = F\frac{d}{dz} + \begin{pmatrix} B_{11} & zB_{2,1} & \cdot\cdot & zB_{n,1} \\ D & B_{2,2} & \cdot\cdot & \cdot \\ & D & \cdot\cdot & \cdot \\ & & \cdot\cdot & zB_{n,n-1} \\ & & D & B_{n,n} \end{pmatrix}.$$

We note that the polynomials $B_{i,i}$ have degree $\leq k$ and the polynomials $B_{i,j}$ with $i > j$ have degree $\leq k-1$. Let $e_1, e_2, \ldots, e_n$ denote the standard basis, used in this presentation of the matrix differential operator Δ. For notational convenience, we write $e_{n+1} = 0$. For the computation of the minimal monic element $L_n \in C(z)[\Delta]$ with $L_n e_1 = 0$ we will use the notation:
$M_i = (\Delta - B_{i,i} - (i-1)zD')$. One defines a sequence of monic operators $L_i \in C[z][\Delta]$ as follows: $L_0 = 1$, $L_1 = M_1 = (\Delta - B_{1,1})$, $L_2 = M_2L_1 - FB_{2,1}L_0$ and recursively by

$$\begin{aligned} L_i = M_iL_{i-1} - FB_{i,i-1}L_{i-2} - FDB_{i,i-2}L_{i-3} \\ - \cdots - FD^{i-3}B_{i,2}L_1 - FD^{i-2}B_{i,1}L_0. \end{aligned}$$

One sees that the L_i are constructed such that $L_ie_1 = D^ie_{i+1}$. In particular, e_1 is a cyclic element for the matrix differential operator and L_n is the minimal monic operator in $C(z)[\Delta]$ with $L_ne_1 = 0$. Since L_n actually lies in $C[z][\Delta]$ and the coefficients of L_n w.r.t. Δ satisfy the correct bound on the degrees, it follows that L_n gives rise to a fuchsian scalar operator with the singular locus in S.

In order to prove that we can produce, by varying the coefficients of the matrices $B_0, B_1, \ldots, B_k$, any given element $T := \Delta^n + A_1\Delta^{n-1} + \cdots + A_{n-1}\Delta + A_n \in C[z][\Delta]$ with the degree of each A_i less than or equal to $k.i$, we have to analyze the formula for L_n a little further. We start by giving some explicit formulas:
$L_1 = M_1$ and $L_2 = M_2M_1 - FB_{2,1}$ and

$$\begin{aligned} L_3 &= M_3M_2M_1 - (M_3FB_{2,1} + FB_{3,2}M_1) - FDB_{3,1}, \\ L_4 &= M_4M_3M_2M_1 - (M_4M_3FB_{2,1} + M_4FB_{3,2}M_1 + FB_{4,3}M_2M_1) \\ &\quad -(M_4FDB_{3,1} + FDB_{4,2}M_1) - FD^2B_{4,1} + FB_{4,3}FB_{2,1}. \end{aligned}$$

By induction one derives the following formula for L_n:

$$\begin{aligned} M_n \cdots M_2M_1 - \sum_{i=1}^{n-1} M_n \cdots M_{i+2}FB_{i+1,i}M_{i-1} \cdots M_1 \\ - \sum_{i=1}^{n-2} M_n \cdots M_{i+3}FDB_{i+2,i}M_{i-1} \cdots M_1 \\ - \sum_{i=1}^{n-3} M_n \cdots M_{i+4}FD^2B_{i+3,i}M_{i-1} \cdots M_1 \\ - \cdots\cdots - M_nFD^{n-3}B_{n-1,1} - FD^{n-2}B_{n,1} + \text{ overflow terms.} \end{aligned}$$

The terms in this formula are polynomials of degrees $n, n-2, n-3, \ldots, 1, 0$ in Δ. By an "overflow term " we mean a product of, say $n-l$ of the M_is and involving two or more terms $B_{x,y}$ with $x - y \leq l - 2$.

We will solve the equation $L_n = T$ step-wise by solving modulo F, modulo FD, ..., modulo FD^{n-1}. At the j-th step we will determine the polynomials $B_{j+i-1,i}$, $1 \le i \le n-j+1$, i.e., the polynomials on the j-th diagonal. After the last step, one actually has the equality $L_n = T$ since the coefficients of $L_n - T$ are polynomials of degree $\le k.n$ and the degree of FD^{n-1} is $1+kn$. We note further that the left ideal I in $C[z][\Delta]$ generated by the element $a := z^{n_0}(z-s_1)^{n_1}\cdots(z-s_k)^{n_k}$ (for any $n_0,\ldots,n_k$) is, in fact, a two-sided ideal and thus we can work modulo I in the usual manner. We note further that M_i almost commutes with a in the sense that $M_i a = a(M_i + F\frac{a'}{a})$ and $F\frac{a'}{a} \in C[z]$.

The *first equation* that we want to solve is $L_n \equiv T$ modulo F. This is the same as $M_n \cdots M_1 \equiv T$ modulo F and again the same as $M_n \cdots M_1 \equiv T$ modulo each of the two-sided ideals $(z), (z-s_1), \ldots, (z-s_k)$ in $C[z][\Delta]$. This is again equivalent to the polynomials $\prod_{i=1}^n(\Delta - B_{i,i}(0))$ and, for each $s \in \{s_1,\ldots,s_k\}$, the $\prod_{i=1}^n(\Delta - B_{i,i}(s) - sD'(s))$ are prescribed as elements of $C[\Delta]$. For each i, this means that there are only finitely many possibilities for $B_{i,1,}(0), B_{i,i}(s_1), \ldots, B_{i,i}(s_k)$ and for each choice of these elements $B_{i,i}$ can be (uniquely) determined by interpolation. Therefore, there are finitely many possibilities for the polynomials $B_{1,1},\ldots,B_{n,n}$. In particular, for any $s \in \{s_1,\ldots,s_k\}$ one is allowed to permute the numbers $B_{n,n}(s)$ $+(n-1)sD'(s), \ldots, B_{2,2}(s)+sD'(s), B_{1,1}(s)$. After a suitable permutation for each $s \in \{s_1,\ldots,s_k\}$, the following "technical assumption" is satisfied: *For $i > j$, the difference*

$$\frac{B_{i,i}(s)+(i-1)sD'(s)}{sD'(s)} - \frac{B_{j,j}(s)+(j-1)sD'(s)}{sD'(s)}$$

is not a strictly positive integer. For example, we could permute the $B_{i,i}$ so that $Re(B_{i,i}(s)) \le Re(B_{j,j}(s))$ for $i > j$.

In *the second step*, we have to consider the equation $L_n \equiv T$ modulo FD. This can also be written as: produce polynomials $B_{i+1,i}$ of degrees $\le k-1$ such that the linear combination

$$(F)^{-1}(\sum_{i=1}^{n-1} M_n \cdots M_{i+2}FB_{i+1,i}M_{i-1}\cdots M_1)$$

is modulo D a prescribed element $C_{n-2}\Delta^{n-2}+C_{n-3}\Delta^{n-3}+\cdots+C_1\Delta+C_0 \in \mathbf{C}[z][\Delta]$ with the degrees of the C_i bounded by $k.i$ for all i. Again we can split this problem into an equivalence modulo $(z-s)$ for $s \in \{s_1,\ldots,s_k\}$. A sufficient condition for solving this problem (again using interpolation) is that for any such s the polynomials $F^{-1}M_n\cdots M_{i+2}FM_{i-1}\cdots M_1$ modulo $(z-s)$ in $\mathbf{C}[\Delta]$ (for $i = 1,\ldots,n-1$) are linearly independent. This will follow from our "technical assumption", as we will verify.

Write M_i^* for $F^{-1}M_iF$ and write $M_i^*(s), M_i(s) \in C[\Delta]$ for M_i^* and M_i modulo $(z-s)$. The zero of $M_i^*(s)$ is $B_{i,i}(s)+(i-1)sD'(s)-sD'(s)$ and the zero of $M_i(s)$

is $B_{i,i}(s) + (i-1)sD'(s)$. We calculate step-by-step the linear space V generated by the $n-1$ polynomials of degree $n-2$. The collection of polynomials contains $M_n^*(s)\cdots M_4^*(s)M_3^*(s)$ and $M_n^*(s)\cdots M_4^*(s)M_1(s)$. Since $M_3^*(s)$ and $M_1(s)$ have no common zero, we conclude that V contains $M_n^*(s)\cdots M_4^*(s)P_1$, where P_1 is any polynomial of degree ≤ 1. Furthermore, $M_n^*(s)\cdots M_5^*(s)M_2(s)M_1(s)$ belongs to the collection. Since $M_2(s)M_1(s)$ and $M_4^*(s)$ have no common zero we conclude that V contains all polynomials of the form $M_n^*(s)\cdots M_5^*(s)P_2$, where P_2 is any polynomial of degree ≤ 2. By induction, one finds that V consists of all polynomials of degree $\leq n-2$. Thus we can solve $L_n \equiv T$ modulo FD in a unique way (after the choice made in the first step). This ends the second step. The further steps, i.e., solving $L_n \equiv T$ modulo FD^j for $j = 2, \ldots, n$ are carried out in a similar way. In each step we find a unique solution. □

6.3.2 A Criterion for a Scalar Fuchsian Equation

In this section and Sect. 6.5, we shall consider regular singular connections $(\mathcal{M}, \nabla)$ with singular locus S whose generic fibres $(\mathcal{M}_\eta, \nabla_\eta)$ are irreducible connections for $\mathbf{C}(z)/\mathbf{C}$. We shall refer to such connections as *irreducible regular singular connections*. The connection $(\mathcal{M}_\eta, \nabla_\eta)$, furthermore, gives rise to a differential module. In the next proposition, we give a criterion for this module to have a cyclic vector with minimal monic annihilating operator that is fuchsian with singular locus S.

Proposition 6.13 *Let $\nabla : \mathcal{M} \to \Omega(S) \otimes \mathcal{M}$ be an irreducible regular singular connection of rank n on $\mathbf{P}^1$ with singular locus in S. Put $k = \#S - 2$. Suppose that the type of $\mathcal{M}$ is $b, b-k, b-2k, \ldots, b-(n-1)k$. Then there is an equivalent scalar fuchsian equation of order n having singular locus S.*

Proof. For any $s \in S$, $\mathcal{M}$ and $\mathcal{M}(-b[s])$ have the same generic fibre. Therefore, after replacing $\mathcal{M}$ by $\mathcal{M}(-b[s])$ for some $s \in S$, we may assume $b = 0$. If $k = 0$, then M is a free vector bundle. We may assume that $S = \{0, \infty\}$. As in Example 6.9.2, we see that this leads to a differential equation of the form $\frac{d}{dz} - \frac{A}{z}$, where $A \in \mathrm{M}_n(\mathbf{C})$. Since the connection is irreducible, the associated differential module M is also irreducible. This implies that A can have no invariant subspaces and so $n = 1$. The operator $\frac{d}{dz} - \frac{a}{z}$, $a \in \mathbf{C}$ is clearly fuchsian.

We now suppose that $k > 0$ and $S = \{0, \infty, s_1, \ldots, s_k\}$. As before, we write $\mathcal{L}(D)$ for the line bundle of the functions f with divisor $\geq -D$. We may identify $\mathcal{M}$ with the subbundle of $Oe_1 \oplus \cdots \oplus Oe_n$ given as

$$Oe_1 \oplus \mathcal{L}(-k[\infty])e_2 \oplus \mathcal{L}(-2k[\infty])e_3 \oplus \cdots \oplus \mathcal{L}(-(n-1)k[\infty])e_n.$$

Clearly e_1 is a basis of $H^0(\mathcal{M})$. We will show that the minimal monic differential operator $L \in C(z)[\partial]$ satisfying $Le_1 = 0$ has order n and is fuchsian. Actually, we will consider the differential operator $\Delta = z(z-s_1)\cdots(z-s_k)\frac{d}{dz}$ and show that the minimal monic operator $N \in C(z)[\Delta]$ such that $Ne_1 = 0$ has degree n and its coefficients are polynomials with degrees bounded by $k \cdot i$. (See the proof of Theorem 6.12.)

There is an obvious isomorphism $\Omega(S) \to \mathcal{L}(k \cdot [\infty])$, which sends $\frac{dz}{z}$ to $(z - s_1) \cdots (z - s_k)$. Define $\Delta : \mathcal{M} \to \mathcal{L}(k \cdot [\infty]) \otimes \mathcal{M}$ as the composition of $\nabla : \mathcal{M} \to \Omega(S) \otimes \mathcal{M}$ and the isomorphism $\Omega(S) \otimes \mathcal{M} \to \mathcal{L}(k \cdot [\infty]) \otimes \mathcal{M}$. One can extend Δ to a map $\Delta : \mathcal{L}(ik \cdot [\infty]) \otimes \mathcal{M} \to \mathcal{L}((i+1)k \cdot [\infty]) \otimes \mathcal{M}$. One has $\Delta(fm) = z(z - s_1) \cdots (z - s_k)\frac{df}{dz}m + f\Delta(m)$ for a function f and a section m of $\mathcal{M}$.

We observe that $\Delta(e_1)$ is a global section of $\mathcal{L}(k \cdot [\infty]) \otimes \mathcal{M}$ and has therefore the form $ae_1 + be_2$ with a a polynomial of degree $\leq k$ and b a constant. The constant b is nonzero, since the connection is irreducible. One changes the original $e_1, e_2, \ldots$ by replacing e_2 by $ae_1 + be_2$ and keeping the other e_js. After this change $\Delta(e_1) = e_2$. Similarly, Δe_2 is a global section of $\mathcal{L}(2k \cdot [\infty]) \otimes \mathcal{M}$ and has, therefore, the form $ce_1 + de_2 + ee_3$ with c, d, e polynomials of degrees $\leq 2k, k, 0$. The constant e is not zero since the connection is irreducible. One changes the element e_3 into $ce_1 + de_2 + ee_3$ and keeps the other e_js. After this change, one has $\Delta e_2 = e_3$. Continuing in this way one finds new elements $e_1, e_2, \ldots, e_n$ such that $\mathcal{M}$ is the subbundle of $Oe_1 \oplus \cdots \oplus Oe_n$, given as before, and such that $\Delta(e_i) = e_{i+1}$ for $i = 1, \ldots, n-1$. The final $\Delta(e_n)$ is a global section of $\mathcal{L}(nk \cdot [\infty]) \otimes \mathcal{M}$ and can therefore be written as $a_n e_1 + a_{n-1}e_2 + \cdots + a_1 e_n$ with a_i a polynomial of degree $\leq ki$. Then $N := \Delta^n - a_1\Delta^{n-1} - \cdots - a_{n-1}\Delta - a_n$ is the monic polynomial of minimal degree with $Ne_1 = 0$. □

We note that Proposition 6.13 and its converse, Proposition 6.14, are present or deducible from Bolibruch's work (Theorem 4.4.1 and Corollary 4.4.1 of [43], see also Theorems 7.2.1 and 7.2.2 of [9]).

Proposition 6.14 *Let L be a scalar fuchsian equation with singular locus S. Then there is an equivalent connection $(\mathcal{M}, \nabla)$ with singular locus S and of type $0, -k, -2k, \ldots, -(n-1)k$.*

Proof. We may suppose $S = \{0, s_1, \ldots, s_k, \infty\}$ and we may replace L by a monic operator $M \in C[z][\Delta]$, $M = \Delta^n - a_1\Delta^{n-1} - \cdots - a_{n-1}\Delta - a_n$ with a_i polynomials of degrees $\leq ki$. For the vector bundle $\mathcal{M}$ one takes the subbundle of $Oe_1 \oplus \cdots \oplus Oe_n$ given as

$$Oe_1 \oplus \mathcal{L}(-k \cdot [\infty])e_2 \oplus \mathcal{L}(-2k \cdot [\infty])e_3 \oplus \cdots \oplus \mathcal{L}(-(n-1) \cdot [\infty])e_n.$$

One defines $\Delta : \mathcal{M} \to \mathcal{L}(k \cdot [\infty]) \otimes \mathcal{M}$ by $\Delta(e_i) = e_{i+1}$ for $i = 1, \ldots, n-1$ and $\Delta(e_n) = a_n e_1 + a_{n-1}e_2 + \cdots + a_1 e_n$. The definition of ∇ on $\mathcal{M}$ follows from this and the type of $\mathcal{M}$ is $0, -k, \ldots, -(n-1)k$ as required. □

6.4 The Riemann-Hilbert Problem, Weak Form

We fix a finite subset S on the complex sphere $\mathbf{P}$ and a base point $b \notin S$ for the fundamental group π_1 of $\mathbf{P} \setminus S$. An object M of RegSing$(\mathbf{C}(z), S)$ (see part 1 of Examples 6.9) is a connection $\nabla : M \to \Omega \otimes M$, where M is a finite dimensional vector space over $\mathbf{C}(z)$, such that the singularities of the connection are regular singular and lie in S. Let V denote the local solution space of (M, ∇) at the point b. The monodromy of the connection is a homomorphism $\pi_1 \to \mathrm{GL}(V)$. Let Repr_{π_1} denote the category of the finite dimensional complex representations of π_1. Then we have attached to (M, ∇) an object of Repr_{π_1}. This extends, in fact, to a functor $\mathcal{M} : \mathrm{RegSing}(\mathbf{C}(z), S) \to \mathrm{Repr}_{\pi_1}$. A solution of the "weak form" of the Riemann-Hilbert problem is given in the following (see Appendix B for facts concerning tannakian categories).

Theorem 6.15 *The functor $\mathcal{M} : \mathrm{RegSing}(\mathbf{C}(z), S) \to \mathrm{Repr}_{\pi_1}$ is an equivalence of categories. This functor respects all "constructions of linear algebra" and is, in particular, an equivalence of tannakian categories.*

Proof. It is easy to see that $\mathcal{M}$ respects all constructions of linear algebra. We will first show that for two objects M_1, M_2 the $\mathbf{C}$-linear map $\mathrm{Hom}(M_1, M_2) \to \mathrm{Hom}(\mathcal{M}(M_1), \mathcal{M}(M_2))$ is an isomorphism. In proving this, it suffices to take $M_1 = \mathbf{1}$, i.e., the trivial connection of dimension 1. Then $\mathrm{Hom}(\mathbf{1}, M_2)$ consists of the elements $m_2 \in M_2$ with $\nabla(m_2) = 0$. The elements of $\mathrm{Hom}(\mathbf{1}, \mathcal{M}(M_2))$ are the vectors v in the solution space of M_2 at b, which are invariant under the monodromy of M_2. Such an element v extends to all of $\mathbf{P} \setminus S$. Since the connection has regular singularities v is bounded at each point s in S by a power of the absolute value of a local parameter at s. Thus v extends in a meromorphic way to all of $\mathbf{P}$ and is therefore an element of M_2 satisfying $\nabla(v) = 0$. This proves that the map under consideration is bijective.

The final and more difficult part of the proof consists of producing for a given representation $\rho : \pi_1 \to \mathrm{GL}_n(\mathbf{C})$ an object (M, ∇) of RegSing$(\mathbf{C}(z), S)$ such that its monodromy representation is isomorphic to ρ. From Example 6.6.3 the existence of a regular connection (N, ∇) on $\mathbf{P} \setminus S$ with monodromy representation ρ follows. The next step that one has to make is to extend N and ∇ to a regular singular connection on $\mathbf{P}$. This is done by a local calculation.

Consider a point $s \in S$. For notational convenience we suppose that $s = 0$. Put $Y^* := \{z \in \mathbf{C} |\ 0 < |z| < \epsilon\}$. Let V be the solution space of (N, ∇) at the point $\epsilon/2$. The circle through $\epsilon/2$ around 0 induces a monodromy map $B \in \mathrm{GL}(V)$. We choose now a linear map $A : V \to V$ such that $e^{2\pi i A} = B$ and define the regular singular connection (N_s, ∇_s) on $Y := \{z \in \mathbf{C} |\ |z| < \epsilon\}$ by the formulas: $N_s = O_Y \otimes V$ and $\nabla_s(f \otimes v) = df \otimes v + z^{-1} \otimes A(v)$. The restriction of (N_s, ∇_s) to $Y^* = Y \setminus \{0\}$ has local monodromy $e^{2\pi i A}$. From part 3 of Example 6.6 it follows that the restriction of the connections (N_s, ∇_s) and (N, ∇) to Y^* are isomorphic. We choose an isomorphism and use this to glue the connections (N, ∇) and (N_s, ∇_s) to a regular singular connection on $(\mathbf{P} \setminus S) \cup \{s\}$. This can be done for every point $s \in S$ and we arrive at a regular singular connection (M, ∇) on $\mathbf{P}$ with singular locus in S and with the prescribed monodromy representation ρ. From part 1 of Example 6.9 we know that (M, ∇) comes from an algebraic regular singular connection on P^1 with singular locus in S. The generic fibre of this algebraic connection is the object of RegSing$(\mathbf{C}(z), S)$, which has the required monodromy representation ρ. □

We note that the contents of the theorem are "analytic". Moreover, the proof of the existence of a regular connection for $(\mathbf{C}(z), S)$ with prescribed monodromy depends on the GAGA principle and is not constructive. Furthermore, one observes that the regular singular connection for $(\mathbf{P}, S)$ is not unique, since we have chosen matrices A with $e^{2\pi i A} = B$ and we have chosen local isomorphisms for the gluing. The Riemann-Hilbert problem in "strong form" requires a regular singular connection for $(\mathbf{P}, S)$ (or for (P^1, S)) such that the vector bundle in question is free. Given a weak solution for the Riemann-Hilbert problem, the investigation concerning the existence of a strong solution is then a purely algebraic problem.

In [9, 41, 44], Bolibruch has constructed counterexamples to the strong Riemann-Hilbert problem. He also gave a positive solution for the strong problem in the case that the representation is irreducible [9, 42] (see also the work of Kostov [163]). We will give an algebraic version of this proof in the next section.

6.5 Irreducible Connections

Let C denote an algebraically closed field of characteristic 0 and let (M, ∇) denote a regular singular connection for $C(z)/C$ with singular locus in $S \subset P^1$, where P^1 is the projective line over C. In this section we will show that, under the assumption that (M, ∇) is irreducible, there exists a regular singular connection $(\mathcal{M}, \nabla)$ on P^1, such that:

(a) The generic fibre of $(\mathcal{M}, \nabla)$ is (M, ∇).
(b) The singular locus of $(\mathcal{M}, \nabla)$ is contained in S.
(c) The vector bundle $\mathcal{M}$ is free.

Combining this result with Theorem 6.15 one obtains a solution of the Riemann-Hilbert problem in the strong sense for irreducible representations of the fundamental group of $\mathbf{P} \setminus S$. The proof that we give here relies on unpublished notes of O. Gabber and is referred to in the Bourbaki talk of A. Beauville [26]. We thank O. Gabber for making these notes available to us.

We have to do some preparations and to introduce some notations. The sheaf of regular functions on P^1 is denoted by O. By $O(n)$ we denote the line bundle of degree n on P^1 (see Exercise 6.7). For any point $p \in P^1$, one considers the stalk O_p of O at p. This is a discrete valuation ring lying in $C(z)$. Its completion is denoted by $\widehat{O}_p$ and the field of fractions of $\widehat{O}_p$ will be denoted by $\widehat{C(z)}_p$. This field is the completion of $C(z)$ with respect to the valuation ring O_p. A *lattice* in a finite dimensional vector space V over $\widehat{C(z)}_p$ is a free $\widehat{O}_p$-submodule of V with rank equal to the dimension of V. The following lemma describes a vector bundle on P^1 in terms of a basis of its generic fibre and lattices at finitely many points. We will use elementary properties of coherent sheaves and refer to [124] for the relevant facts.

Lemma 6.16 *Let M denote a vector space over $C(z)$ with a basis $e_1, \dots e_n$. Let U be a nontrivial open subset of P^1 and for each $p \notin U$ let Λ_p be a lattice of $\widehat{C(z)}_p \otimes M$. Then there exists a unique vector bundle $\mathcal{M}$ on P^1 such that:*
(a) *For every open $V \subset P^1$ one has $\mathcal{M}(V) \subset M$.*
(b) *$\mathcal{M}(U)$ is equal to $O(U)e_1 + \cdots + O(U)e_n \subset M$.*
(c) *For every $p \notin U$, the completion $\widehat{\mathcal{M}}_p := \widehat{O}_p \otimes \mathcal{M}_p$ coincides with Λ_p.*

Proof. For $p \in P^1 \setminus U$ we put $S_p := \widehat{O}_p e_1 + \cdots + \widehat{O}_p e_n$. Let for every $p \in P^1 \setminus U$ an integer A_p be given. Consider first the special case where each $\Lambda_p = t_p^{A_p} S_p$, where t_p denotes a local coordinate at p. Put $N = Oe_1 + \cdots + Oe_n$ and let A be the divisor $\sum A_p[p]$ (the sum extended over the $p \in P^1 \setminus U$). Then clearly the vector bundle $N(-A) = \mathcal{L}(-A) \otimes N$ solves the problem.

In the general case, there are integers A_p, B_p such that $t_p^{A_p} S_p \subset \Lambda_p \subset t_p^{B_p} S_p$ holds. Let B be the divisor $\sum B_p[p]$. Then $N(-A) \subset N(-B)$ are both vector bundles on P^1. Consider the surjective morphism of coherent sheaves $N(-B) \xrightarrow{q} N(-B)/N(-A)$. The second sheaf has support in $P^1 \setminus U$ and can be written as a skyscraper sheaf $\oplus_p t_p^{B_p} S_p / t_p^{A_p} S_p$ (see Example C.2.7 and Exercise C.11 and [124]). This skyscraper sheaf has the coherent subsheaf $T := \sum_p \Lambda_p / t_p^{A_p} S_p$. Define now $\mathcal{M}$ as the preimage under q of T. From the exact sequence $0 \to N(-A) \to \mathcal{M} \to T \to 0$ one easily deduces that $\mathcal{M}$ has the required properties (see [124], Chap. II.5 for the relevant facts about coherent sheaves). An alternative way of describing $\mathcal{M}$ is that the set $\mathcal{M}(V)$, for any open $V \neq \emptyset$, consists of the elements $m \in M$ such that for $p \in U \cap V$ one has $m \in O_p e_1 + \cdots + O_p e_n$ and for $p \in V,\ p \notin U$ one has $m \in \Lambda_p \subset \widehat{C(z)}_p \otimes M$. This shows the unicity of $\mathcal{M}$. □

Let $\mathcal{M}$ be a vector bundle on P^1. According to Grothendieck's classification (and the GAGA principle), $\mathcal{M}$ is equal to a direct sum $O(a_1) \oplus \cdots \oplus O(a_n)$ with integers $a_1 \geq \cdots \geq a_n$. This decomposition is not unique. However, there is a canonical filtration by subbundles $F^1 \subset F^2 \subset \dots$. One defines $F^1 := O(a_1) \oplus \cdots \oplus O(a_{s_1})$, where s_1 is the last integer with $a_{s_1} = a_1$. The subbundle is unique, since $O(-a_1) \otimes F^1$ is the subbundle of $O(-a_1) \otimes \mathcal{M}$ generated by the global sections $H^0(P^1, O(-a_1) \otimes \mathcal{M})$. In the case where not all a_j are equal to a_1 one defines s_2 to be the last integer with $a_{s_2} = a_{s_1+1}$. The term F^2, defined as the direct sum $O(a_1) \oplus \cdots \oplus O(a_{s_2})$, is again uniquely defined since it is the subbundle generated by the global sections of $O(-a_{s_2}) \otimes \mathcal{M}$. The other possible $F^i \subset \mathcal{M}$ are defined in a similar way. We will also need the notion of the *defect of the vector bundle* $\mathcal{M}$, which we define as $\sum(a_1 - a_i)$. In later parts of the proof we want to change a given vector bundle by changing the data of Lemma 6.16. The goal is to obtain a vector bundle with defect zero, i.e., $a_1 = a_2 = \cdots = a_n$. In the next lemma the effect of a small local change on the type of the vector bundle is given.

Lemma 6.17 *Let M, U, Λ_p, $\mathcal{M}$ be as in Lemma 6.16. Let the type of $\mathcal{M}$ be given by the integers $a_1 \geq \cdots \geq a_n$ and let $F^1 \subset F^2 \subset \ldots$ denote the canonical filtration of $\mathcal{M}$. We consider a $p_0 \in P^1 \setminus U$ with local parameter t and a nonzero vector $v \in V := \Lambda_{p_0}/t\Lambda_{p_0}$. Define a new lattice $\tilde{\Lambda}_{p_0} := \widehat{O}_p t^{-1}\tilde{v} + \Lambda_{p_0}$, where $\tilde{v} \in \Lambda_{p_0}$ has image $v \in V$. Let $\tilde{\mathcal{M}}$ denote the vector bundle on P^1 given by Lemma 6.16 using the same data as $\mathcal{M}$ with the exception that Λ_{p_0} is replaced by $\tilde{\Lambda}_{p_0}$.*

The vector space V has an induced filtration $F^1(V) \subset F^2(V) \subset \ldots$. Let i be the first integer such that $v \in F^i(V)$ and let j be the smallest integer such that $O(a_j)$ is present in $F^i \setminus F^{i-1}$. Then the type of $\tilde{\mathcal{M}}$ is obtained from the type of $\mathcal{M}$ by replacing a_j by $a_j + 1$.

Proof. Choose a direct sum decomposition $\mathcal{M} = O(a_1) \oplus \cdots \oplus O(a_n)$. Then $F^{i-1} = O(a_1) \oplus \cdots \oplus O(a_{j-1})$ and $F^i = O(a_1) \oplus \cdots \oplus O(a_k)$, where $a_1 \geq \cdots \geq a_{j-1} > a_j = \cdots = a_k$ (and $a_k > a_{k+1}$ if $k < n$). For $\tilde{v}$ we may choose an element in $F^i_{p_0}$ that does not lie in $F^{i-1}_{p_0}$. After changing the direct sum decomposition of F^i we can arrange that $\tilde{v} \in O(a_j)_{p_0}$. Then $\tilde{\mathcal{M}}$ is obtained from $\mathcal{M}$ by performing only a change to the direct summand $O(a_j)$ of $\mathcal{M}$. In this change the line bundle $O(a_j)$ is replaced by $\mathcal{L}(p_0) \otimes O(a_j)$. The latter bundle is isomorphic to $O(a_j + 1)$. □

We focus now on a regular singular connection (M, ∇) for $C(z)/C$ with singular locus in S. For every point $p \in P^1$ we choose a local parameter t_p. The induced connection on $\widehat{M}_p := \widehat{C(z)}_p \otimes M$ has the form $\nabla : \widehat{M}_p \to \widehat{C(z)}_p dt_p \otimes \widehat{M}_p$. For $p \notin S$, there exists a basis $e_1, \ldots, e_n$ of $\widehat{M}_p$ over $\widehat{C(z)}_p$ with $\nabla(e_j) = 0$ for all j. From this it follows that $\Lambda_p := \widehat{O}_p e_1 + \cdots + \widehat{O}_p e_n$ is the unique lattice such that $\nabla : \Lambda_p \to \widehat{O}_p dt_p \otimes \Lambda_p$. For $p \in S$ there is a basis $e_1, \ldots, e_n$ of $\widehat{M}_p$ over $\widehat{C(z)}_p$ such that the vector space $V = Ce_1 \oplus \cdots \oplus Ce_n$ satisfies $\nabla(V) \subset \frac{dt_p}{t_p} \otimes V$. Then $\Lambda_p := \widehat{O}_p \otimes V \subset \widehat{M}_p$ is a lattice satisfying $\nabla(\Lambda_p) \subset \frac{dt_p}{t_p} \otimes \Lambda_p$. We observe that there are many lattices in $\widehat{M}_p$ having the same property. We now want to extend Lemma 6.16 and Lemma 6.17 to the case of connections.

Lemma 6.18 *1. Let (M, ∇) be a regular singular connection for $C(z)/C$ with singular locus in S. For every $s \in S$ we choose a local parameter t_s. For every $s \in S$ let $\Lambda_s \subset \widehat{M}_s$ be a lattice that satisfies $\nabla(\Lambda_s) \subset \frac{dt_s}{t_s} \otimes \Lambda_s$. Then there is a unique regular singular connection $(\mathcal{M}, \nabla)$ on P^1 with singular locus in S such that:*

(a) *For every open $V \subset P^1$, one has $\mathcal{M}(V) \subset M$.*
(b) *The generic fibre of $(\mathcal{M}, \nabla)$ is (M, ∇).*
(c) *$\widehat{\mathcal{M}}_s := \widehat{O}_s \otimes \mathcal{M}_s$ coincides with Λ_s for all $s \in S$.*

2. *Let* $(\mathcal{M}, \nabla)$ *be any connection with singular locus in S and generic fibre isomorphic to* (M, ∇). *After identification of the generic fibre of* $\mathcal{M}$ *with* M, *the* $\widehat{\mathcal{M}}_s$ *are lattices* Λ_s *for* $\widehat{M}_s$ *satisfying* $\nabla(\Lambda_s) \subset \frac{dt_s}{t_s} \otimes \Lambda_s$. *Thus* $(\mathcal{M}, \nabla)$ *is the unique connection of part* 1.

Proof. We start with a basis $e_1, \ldots, e_n$ for the $C(z)$-vector space M and choose a nonempty open $U \subset P^1 \setminus \{\infty\}$ such that $\nabla(e_j) \in dz \otimes O(U)e_1 + \cdots + O(U)e_n$. For a point $p \notin U$ and $p \notin S$ we define the lattice Λ_p to be the unique lattice with $\nabla(\Lambda_p) \subset dt_p \otimes \Lambda_p$ (where t_p is again a local parameter). Lemma 6.16 produces a unique $\mathcal{M}$ with these data. The verification that the obvious ∇ on $\mathcal{M}$ has the property $\nabla : \mathcal{M} \to \Omega(S) \otimes \mathcal{M}$ can be done locally for every point p. In fact, it suffices to prove that ∇ maps $\widehat{M}_p$ into $dt_p \otimes \widehat{M}_p$ for $p \notin S$ and into $\frac{dt_p}{t_p} \otimes \widehat{M}_p$ for $p \in S$. The data that define $\mathcal{M}$ satisfy these properties. Part 2 of the lemma is an obvious consequence of part 1. □

Lemma 6.19 *We will use the notations of Lemma 6.18 and Lemma 6.17. Choose an* $s \in S$. *The map* $\nabla : \Lambda_s \to \frac{dt_s}{t_s} \otimes \Lambda_s$ *induces a C-linear map* $\delta_s : \Lambda_s/t_s\Lambda_s \to \frac{dt_s}{t_s} \otimes \Lambda_s/t_s\Lambda_s \to \Lambda_s/t_s\Lambda_s$, *which does not depend on the choice of* t_s. *Let* $v \in \Lambda_s/t_s\Lambda_s$ *be an eigenvector for* δ_s. *Define* $\tilde{\Lambda}_s$ *and* $\tilde{\mathcal{M}}$ *as in Lemma 6.17. Then:*

(a) ∇ *maps* $\tilde{\Lambda}_s$ *into* $\frac{dt_s}{t_s} \otimes \tilde{\Lambda}_s$.
(b) *The connection on* $\mathcal{M}$ *extends uniquely to* $\tilde{\mathcal{M}}$.
(c) *Let* $\tilde{\Lambda}_s$ *have an* $\widehat{O}_s$*-basis* $e_1, \ldots, e_n$ *such that* $\nabla(e_i) = \frac{dt_s}{t_s} \otimes \sum a_{i,j}e_j$ *with* $a_{i,j} \in t_s^N\widehat{O}_s$ *for* $i \neq j$ *and some* $N \geq 1$. *Suppose that the above* v *is equal to the image of* e_k *in* $\Lambda_s/t_s\Lambda_s$. *Then* $\tilde{\Lambda}_s$ *has the* $\widehat{O}_s$*-basis* $f_1, f_2, \ldots, f_n$ *with* $f_k = t^{-1}e_k$ *and* $f_l = e_l$ *for* $l \neq k$. *Define the matrix* $(b_{i,j})$ *by* $\nabla(f_i) = \frac{dt_s}{t_s} \otimes \sum b_{i,j}f_j$. *Then* $b_{k,k} = a_{k,k} - 1$ *and* $b_{l,l} = a_{l,l}$ *for* $l \neq k$. *Furthermore,* $b_{i,j} \in t^{N-1}\widehat{O}_s$ *for* $i \neq j$.

Proof. (a) Choose a representative $\tilde{v} \in \Lambda_s$ of v. Then $\nabla(\tilde{v}) \in \frac{dt_s}{t_s} \otimes (a\tilde{v} + t_s\Lambda_s)$ for some $a \in C$. Thus $\nabla(t_s^{-1}) \in \frac{dt_s}{t_s} \otimes (-t_s^{-1}\tilde{v} + at_s^{-1}\tilde{v} + \Lambda_s)$. This shows that $\tilde{\Lambda}_s = \widehat{O}_st_s^{-1}\tilde{v} + \Lambda_s$ has the property $\nabla(\tilde{\Lambda}_s) \subset \frac{dt_s}{t_s} \otimes \tilde{\Lambda}_s$. (b) follows from (a) and Lemma 6.18. A straightforward calculation shows (c). □

Lemma 6.20 *Let* (Z, ∇) *be a regular singular connection for* $C((z))/C$ *and let* $N > 0$ *be an integer. There exists a* $C[[z]]$*-lattice* Λ *with basis* $e_1, \ldots, e_n$ *such that* $\nabla(e_i) = \frac{dz}{z} \otimes \sum a_{i,j}e_j$ *with all* $a_{i,j} \in C[[z]]$ *and* $a_{i,j} \in z^NC[[z]]$ *for* $i \neq j$.

Proof. Write δ for the map $\nabla_{z\frac{d}{dz}} : Z \to Z$. According to the formal classification of regular singular differential equations it follows that Z has a basis $f_1, \ldots, f_n$ such that $\delta(f_i) = \sum c_{i,j}f_j$ for a matrix $(c_{i,j})$ with coefficients in C. If this matrix happens to be diagonizable, then one can choose a basis $e_1, \ldots, e_n$ such that $\nabla(e_i) = \frac{dz}{z} \otimes c_ie_i$ with all $c_i \in C$. In the general case, the Jordan normal form has one or several blocks of dimension > 1. It suffices to consider the case of one

Jordan block, i.e., $\delta(f_1) = cf_1,\ \delta(f_2) = cf_2 + f_1, \ldots, \delta(f_n) = cf_n + f_{n-1}$. One defines $e_1 = f_1,\ e_2 = t^N f_2,\ e_3 = t^{2N} f_3, \ldots$. One calculates that $\delta(e_1) = ce_1$, $\delta(e_2) = (c+N)e_2 + t^N e_1,\ \delta(e_3) = (c+2N)e_3 + t^N e_2, \ldots$. Thus the basis $e_1, \ldots, e_n$ has the required properties. □

Proposition 6.21 *Let $(\mathcal{M}, \nabla)$ be an irreducible regular singular connection on P^1 with singular locus in S. Let $a_1 \geq a_2 \geq \cdots \geq a_n$ denote the type of $\mathcal{M}$. Then $a_{j-1} - a_j \leq (-2 + \#S)$ for all $j \geq 1$. In particular, the defect of $\mathcal{M}$ is $\leq \frac{n(n-1)}{2} \cdot (-2 + \#S)$.*

Proof. $\mathcal{M}$ is written as a direct sum of the line bundles $O(a_1) \oplus \cdots \oplus O(a_n)$. Assume that $a_{j-1} > a_j$ and put $F = O(a_1) \oplus \cdots \oplus O(a_{j-1})$. Then F is one of the canonical subbundles of $\mathcal{M}$. One considers the morphism

$$L : F \subset \mathcal{M} \xrightarrow{\nabla} \Omega(S) \otimes \mathcal{M} \to \Omega(S) \otimes \mathcal{M}/F.$$

The morphism L is nonzero since $(\mathcal{M}, \nabla)$ is irreducible. Furthermore, L is an O-linear map and can therefore be considered as a nonzero global section of the vector bundle $F^* \otimes \Omega(S) \otimes \mathcal{M}/F$. This vector bundle has a direct sum decomposition isomorphic to $\sum_{k<j,\, l \geq j} O(-a_k) \otimes O(-2 + \#S) \otimes O(a_l)$. Since $L \neq 0$, we must have that some $-a_k - 2 + \#S + a_l \geq 0$. This is equivalent to $a_{j-1} - a_j \leq -2 + \#S$. □

Theorem 6.22 *Let (M, ∇) be an irreducible regular singular connection over $C(z)$ with singular locus contained in S. There exists a regular singular connection $(\mathcal{M}, \nabla)$ on P^1, such that:*

(a) *The generic fibre of $(\mathcal{M}, \nabla)$ is (M, ∇).*
(b) *The singular locus of $(\mathcal{M}, \nabla)$ is contained in S.*
(c) *The vector bundle $\mathcal{M}$ is free.*

Proof. Suppose that we have found an $(\mathcal{M}, \nabla)$ that has defect 0 and satisfies (a) and (b). The type of $\mathcal{M}$ is then $a_1 = \cdots = a_n$. Then $\mathcal{M}(-a_1[s])$ (for any $s \in S$) is free and still satisfies (a) and (b).

Let N be an integer $> \frac{n(n-1)}{2}(-2 + \#S)$. We start with a regular singular connection $(\mathcal{M}, \nabla)$ with singular locus in S such that:

(i) Its generic fibre is (M, ∇).
(ii) For some $s \in S$ the $\widehat{O}_s$-module $\widehat{\mathcal{M}}_s$ has a basis $e_1, \ldots, e_n$ such that $\nabla(e_i) = \frac{dt_s}{t_s} \otimes \sum a_{i,j} e_j$ with all $a_{i,j} \in \widehat{O}_s$ and $a_{i,j} \in t_s^N \widehat{O}_s$ for $i \neq j$.

The existence follows from Lemmas 6.20 and 6.18. We note that Lemma 6.21 implies that N will be greater than the defect of $(\mathcal{M}, \nabla)$. In the next steps we modify $\mathcal{M}$. Suppose that $\mathcal{M}$ has a defect > 0, then the canonical filtration $F^1 \subset F^2 \subset \ldots$ of $\mathcal{M}$ has at least two terms. Let i be defined by $F^{i-1} \neq \mathcal{M}$ and $F^i = \mathcal{M}$. The images of $e_1, \ldots, e_n$ in $V := \widehat{\mathcal{M}}_s / t_s \widehat{\mathcal{M}}_s$ form a basis of eigenvectors for the map δ_s (see Lemma 6.19 for the notation). Suppose that the image of e_k does not lie in $F^{i-1}(V)$. We apply Lemma 6.19 and find a new regular singular connection $\mathcal{M}(1)$, which has, according to Lemma 6.17, a strictly smaller defect. For $\widehat{\mathcal{M}(1)}_s$ the matrix of δ_s with respect to the $f_1, \ldots, f_n$ has again property (ii), but now with N replaced by $N-1$. Thus we can repeat this step to produce connections $\mathcal{M}(2)$, etc., until the defect of some $\mathcal{M}(i)$ is 0. □

Remarks 6.23

1. The proof of Theorem 6.22 fails for reducible regular singular connections (M, ∇) over $C(z)/C$, since there is no bound for the defect of the corresponding vector bundles $\mathcal{M}$. This prevents us from making an a priori choice of the number N used in the proof.

2. The proof of Theorem 6.22 also works under the assumption that for some singular point the differential module $\widehat{C(z)}_s \otimes M$ is "semisimple". By this we mean that there is a basis $e_1, \ldots, e_n$ of $\widehat{C(z)}_s \otimes M$ over $\widehat{C(z)}_s$ such that $\nabla(e_i) = \frac{dt_s}{t_s} \otimes a_i e_i$ for certain elements $a_i \in \widehat{O}_s$. In this case, condition (ii) in the proof holds for any $N > 1$ and, in particular, for any N greater than the defect D of the vector bundle. The proof then proceeds to produce connections of decreasing defect and halts after D steps. For the case $C = \mathbf{C}$, the connection $\widehat{\mathbf{C}(z)}_s \otimes M$ is semisimple if and only if the local monodromy map at the point s is semisimple. This gives a modern proof of the result of Plemelj [224].

3. Let the regular singular connection (M, ∇) with singularities in S be given. Take any point $p \notin S$ and consider $S' = S \cup \{p\}$. Since the local monodromy at p is trivial, one can follow the above remark 2 and conclude that there is a regular singular connection $(\mathcal{M}, \nabla)$ with singular locus in S' such that $\mathcal{M}$ is free.

4. The Riemann-Hilbert problem has a strong solution for a connection of dimension two, as noted by Dekkers [79]. Indeed, we have only to consider a reducible regular singular connection (M, ∇). After replacing M by the tensor product $N \otimes M$, where N is a 1-dimensional regular singular connection with singular locus in S, we may suppose that M contains a vector $e_1 \neq 0$ with $\nabla(e_1) = 0$. A second vector e_2 can be chosen such that $\nabla(e_2) = \omega_2 \otimes e_2 + \omega_3 \otimes e_1$, where $\omega_2 \in H^0(P^1, \Omega(S))$ and with ω_3 some meromorphic differential form. It suffices to find an $h \in C(z)$ such that $f_2 = e_2 + he_1$ satisfies $\nabla(f_2) = \omega_2 \otimes f_2 + \tilde{\omega}_3 \otimes e_1$ with $\tilde{\omega}_3 \in H^0(P^1, \Omega(S))$.

One calculates $\tilde{\omega}_3 = -h\omega_2 + dh + \omega_3$. For each point $p \in P^1$ we are given that the connection is regular singular (or regular) and this implies the existence of an $h_p \in \widehat{C(z)}_p$ such that the corresponding $\tilde{\omega}_3$ lies in $\widehat{\Omega(S)}_p$. One may replace this h_p by its "principal part $[h_p]_p$" at the point p. Take now $h \in C(z)$, which has for each point p the principal part $[h_p]_p$. Then for this h the expression $\tilde{\omega}_3$ lies in $H^0(P^1, \Omega(S))$. □

6.6 Counting Fuchsian Equations

One might hope that an even stronger result holds, namely that an irreducible regular singular connection M over $C(z)$ with singular locus in S can be represented by a scalar fuchsian equation with singular locus in S. By counting dimensions of moduli spaces we will show that, in general, any monic scalar "equation" $L \in C(z)[\partial]$ representing M, has singularities outside S. These new singular points for L are called *apparent*.

Definition 6.24 An *apparent singularity* p for any $L = \partial^n + a_1\partial^{n-1} + \cdots + a_n \in C(z)[\partial]$, is a pole of some a_i and such that L has n independent solutions in $C((z-p))$. □

Exercise 6.25 1. Show that, at an apparent singularity of L, there must be n distinct local exponents. Hint: To any basis $f_1, \ldots, f_n$ of the solution space of L at p, with $\text{ord}_p f_i \leq \text{ord}_p f_{i+1}$ associate the n-tuple $(\text{ord}_p f_1, \ldots, \text{ord}_p f_n)$. Show that there are only finitely many n-tuples that can arise in this way and that a maximal one (in the lexicographic order) has distinct entries.

2. Let $f_1, \ldots, f_n \in C((z-p))$ denote n independent solutions of L. Show that the wronskian of $f_1, \ldots, f_n$, which is an element of $C((z-p))^*$, has order $m_1 + \cdots + m_n - \frac{n(n-1)}{2}$. Hint: We may assume that each $f_i = x^{m_i} +$ higher order terms, where the m_i are the distinct exponents. Show that the term of lowest order in $wr(f_1, \ldots f_n)$ is $wr(x^{m_1}, \ldots, x^{m_n})$. □

Definition 6.26 Let p be an apparent singularity of $L \in C(z)[\partial]$ and let $\alpha_1 < \cdots < \alpha_n$ be the local exponents of L at the point p. One defines the *weight* of the apparent singularity to be

$$\text{weight}(L, p) = \alpha_1 + \cdots + \alpha_n - \frac{n(n-1)}{2}.$$

In the following, we will only consider apparent singularities such that $0 \leq \alpha_1 < \cdots < \alpha_n$. Under this assumption, $\text{weight}(L, p) = 0$ holds if and only if no a_i has a pole at p (in other words p is not a singularity at all).

Lemma 6.27 *Let V be a vector space of dimension n over C and let $C((t)) \otimes V$ be equipped with the trivial connection $\nabla(f \otimes v) = df \otimes v$ for all $f \in C((t))$ and $v \in V$. Consider a cyclic vector $e \in C[[t]] \otimes V$ and the minimal monic $L \in C((t))[\partial]$ with $Le = 0$. The weight of L is equal to the dimension over C of $(C[[t]] \otimes V)/(C[[t]]e + C[[t]]\partial e + \cdots + C[[t]]\partial^{n-1}e)$. This number is also equal to the order of the element $e \wedge \partial e \wedge \cdots \wedge \partial^{n-1}e \in C[[t]] \otimes \Lambda^n V \cong C[[t]]$.*

Proof. The element e can be written as $\sum_{m \geq 0} v_m t^m$ with all $v_m \in V$. One then has $\partial e = \sum_{m \geq 0} v_m m t^{m-1}$. Since e is a cyclic vector, its coefficients v_m generate the vector space V. Let us call m a "jump" if v_m does not belong to the subspace of V generated by the v_k with $k < m$. Let $\alpha_1 < \cdots < \alpha_n$ denote the jumps.

A straightforward calculation (as in Exercise 6.25.1) shows that the order of $e \wedge \partial e \wedge \cdots \wedge \partial^{n-1}e \in C[[t]] \otimes \Lambda^n V \cong C[[t]]$ is $\alpha_1 + \cdots + \alpha_n - \frac{n(n-1)}{2}$. A similar calculation shows that this number is also the dimension of the vector space $(C[[t]] \otimes V)/(C[[t]]e + C[[t]]\partial e + \cdots + C[[t]]\partial^{n-1}e)$. It suffices to show that $\alpha_1 < \cdots < \alpha_n$ are the local exponents of L. We note that $Le = \sum_{m\geq 0} v_m L(t^m) = 0$. Take any linear map $\phi : V \to C$. Then $L(y) = 0$ where $y = \sum_{m\geq 0} \phi(v_m)t^m \in C[[t]]$. By varying ϕ one obtains solutions $y \in C[[t]]$ of $L(y) = 0$ with orders $\alpha_1 < \cdots < \alpha_n$. □

We consider now an irreducible regular singular connection M over $C(z)$ whose dimension is n and singular locus in $S = \{s_0, s_1, \ldots, s_k, \infty\}$. There is a Fuchs system $\partial = \frac{d}{dz} + \sum_{j=0}^{k} \frac{A_j}{z-s_j}$ representing the connection. We denote the standard basis by $e_1, \ldots, e_n$. Let $R := C[z, \frac{1}{F}]$ with $F = (z - s_0)\cdots(z - s_k)$. The free R-module $Re_1 + \cdots + Re_n \subset M$ is invariant under the action of ∂.

Lemma 6.28 *Let $v \in M$, $v \neq 0$ and let L be the minimal monic operator with $Lv = 0$. Then L is fuchsian if and only if $v \in Re_1 + \cdots + Re_n$ and the elements $v, \partial v, \ldots, \partial^{n-1}v$ form a basis of the R-module $Re_1 + \cdots + Re_n$.*

Proof. Suppose that v satisfies the properties of the lemma. Then $\partial^n v$ is an R-linear combination of $v, \partial v, \ldots, \partial^{n-1}v$. Thus L has only singularities in S. Since M is regular singular it follows (as in the proof of Lemma 6.11) that L is a fuchsian operator.

On the other hand, suppose that L is fuchsian. Then $N := Rv + R\partial v + \cdots + R\partial^{n-1}v$ is an R-submodule of M, containing a basis of M over $C(z)$ and invariant under ∂. There is only one such object (as one concludes from Lemma 6.18) and thus $N = Re_1 + \cdots + Re_n$. □

Proposition 6.29 *Let $0 \neq v \in Re_1 + \cdots + Re_n \subset M$ and L with $Lv = 0$ be as in Lemma 6.28. Consider the operator $\Delta = F \cdot \partial$. Define the polynomial $P \in C[z]$, which has no zeros in $\{s_0, \ldots, s_k\}$, by the formula $v \wedge \Delta v \wedge \cdots \wedge \Delta^{n-1}v = (z - s_0)^{n_0} \cdots (z - s_k)^{n_k} P \cdot e_1 \wedge \cdots \wedge e_n$. Then the degree of P is equal to the sum of the weights of the apparent singularities of L (outside S).*

Proof. The dimension of the space $(Re_1 + \cdots + Re_n)/(Rv + R\partial v + \cdots + R\partial^{n-1}v)$ is equal to the degree of P. This dimension is the sum of the dimensions, taken over the apparent singular points p, of

$$(C[[z-p]]e_1 + \cdots + C[[z-p]]e_n)/(C[[z-p]]v + \cdots + C[[z-p]]\partial^{n-1}v).$$

Now the statement follows from Lemma 6.27. □

Proposition 6.30 *We use the notations above. There is a choice for the vector v such that for the monic operator L with $Lv = 0$ the sum of the weights of the apparent singular points is* $\leq \frac{n(n-1)}{2}k + 1 - n$.

Proof. Choose numbers $d_0, \ldots, d_k \in \{0, 1, \ldots, n-1\}$ such that $d_0 + \cdots + d_k = n-1$ and choose for each $j = 0, \ldots, k$ a subspace $V_j \subset Ce_1 + \cdots Ce_n$ of codimension d_j and invariant under A_j. For example, one may select $d_0 = n-1, d_1 = \ldots = d_k = 0$, V_0 to be spanned by an eigenvector of A_0 and $V_1 = \ldots = V_k = Ce_1 + \ldots + Ce_n$. For v we take a nonzero vector in the intersection $V_0 \cap V_1 \cap \cdots \cap V_k$ and consider the polynomial $Q(z)$ defined by $v \wedge \Delta v \wedge \cdots \wedge \Delta^{n-1} v = Q(z) e_1 \wedge \cdots \wedge e_n$. The degree of this polynomial is easily seen to be $\leq \frac{n(n-1)}{2}k$. We now give a local calculation at the point $z = s_j$ that shows that the polynomial Q has a zero of order $\geq d_j$ at s_j. Let t denote a local parameter at s_j. We may replace the operator Δ by $\delta := t\frac{d}{dt} + A_j + O(t)$, where $O(t)$ denotes terms divisible by t. Then $\delta^m v = A_j^m v + O(t)$. For $m \geq n - d_j$ one has that $A_j^m v$ is a linear combination of $v, A_j v, \ldots, A_j^{n-d_j-1} v$. Thus $v \wedge \delta v \wedge \cdots \wedge \delta^{n-1} v$ is divisible by t^{d_j}.

We conclude that Q is divisible by $(z - s_0)^{d_0} \cdots (z - s_k)^{d_k}$. We can now apply Proposition 6.29 with a polynomial P of degree $\leq \frac{n(n-1)}{2}k + 1 - n$. □

Example 6.31 *The irreducible fuchsian system* $\partial = \frac{d}{dz} + \frac{A_0}{z} + \frac{A_1}{z-1}$, *where* A_0, A_1 *are constant* 2×2*-matrices and* $S = \{0, 1, \infty\}$.
We will make the proof of Proposition 6.30 explicit and show that there exists a scalar fuchsian equation for this system without apparent singularities. Let e_1, e_2 denote the standard basis. Let R denote the ring $C[z, \frac{1}{z(z-1)}]$. The free R-module $Re_1 + Re_2$ is invariant under the action of ∂.

We take for $v \neq 0$ a constant vector, i.e., in $Ce_1 + Ce_2$, which is an eigenvector for the matrix A_0. Consider the determinant $v \wedge \partial v = v \wedge (\frac{A_0 v}{z} + \frac{A_1 v}{z-1}) = \frac{1}{z-1} v \wedge A_1 v$. From the irreducibility of the equation it follows that v is not an eigenvector for A_1. Thus the determinant has the form $\frac{c}{z-1} e_1 \wedge e_2$ with $c \in C^*$ and $v, \partial v$ form a basis for $Re_1 + Re_2$. This proves the claim. □

We will count "moduli", i.e., the number of parameters in certain families of differential equations. In the classical literature one uses the term *number of accessory parameters* for what is called "moduli" here. We start by considering the family of fuchsian operators L of degree n with regular singularities in the set $S = \{s_0, \ldots, s_k, \infty\}$. Let Δ denote the operator $(z - s_0) \cdots (z - s_k)\frac{d}{dz}$. Then L can be rewritten as a monic operator in Δ, namely $L = \Delta^n + C_1 \Delta^{n-1} + \cdots + C_{n-1}\Delta + C_n$. The coefficients are polynomials with $\deg C_j \leq j \cdot k$ (see Lemma 6.11). This family clearly has $\frac{n(n+1)}{2}k + n$ parameters.

Our next goal is to count the number of parameters of the family $\mathcal{F}$ (of the isomorphism classes) of the "generic" regular singular connections M over $C(z)$ of dimension n with singular locus in $S = \{s_0, \ldots, s_k, \infty\}$. Of course, the terms "family, generic, parameters" are somewhat vague. The term "generic" should at least imply that M is irreducible and thus can be represented by a Fuchs system $\partial + \sum \frac{A_j}{z-s_j}$. The matrices $A_0, \ldots, A_k$ with coefficients in C are chosen generically. In particular, for every point $s \in S$ there is a basis $e_1, \ldots, e_n$ of $\widehat{M}_s := \widehat{C(z)}_s \otimes M$ such that the action of $\delta_s = \nabla_{t_s \frac{d}{dt_s}}$ takes the form $\delta_s e_j = \lambda_j(s) e_j$ and $\lambda_i(s) - \lambda_j(s) \notin \mathbf{Z}$ for $i \neq j$. This property implies that for each point $s \in S$ there are only countably many lattices possible that give rise to a vector bundle with a connection (see Lemma 6.18). Furthermore, the lattices can be chosen such that the corresponding vector bundle with connection is free (see Remarks 6.23). Thus we may as well count the number of parameters of generic Fuchs systems of dimension n and with singular locus in S. Let V be a vector space over C of dimension n. Then we have to choose $k+1$ linear maps $A_j : V \to V$, up to simultaneous conjugation with elements of GL(V). This leads to the formula $kn^2 + 1$ for the number of parameters for $\mathcal{F}$.

We can now draw the following conclusion.

Corollary 6.32 *A general fuchsian system of rank n with $k+2$ singular points cannot be represented by a scalar Fuchs equation if $n^2k + 1 > \frac{n(n+1)}{2}k + n$. In other words, the only cases for which scalar fuchsian equations (without apparent singularities) exist are given by $kn \leq 2$.*

Remarks 6.33 *Counting moduli and the number of apparent singularities.*
1. Now we want to count the number of moduli for monic scalar operators L of degree n with $k+2$ regular singularities, i.e., S, and l apparent singular points $a_1, \ldots, a_l$ of weight 1 for which we do not fix the position. Let Δ denote the operator $(z - s_0) \cdots (z - s_k)(z - a_1) \cdots (z - a_l)\frac{d}{dz}$ and represent L as $L = \Delta^n + C_1 \Delta^{n-1} + \cdots + C_{n-1}\Delta + C_n$ with the C_js polynomials of degrees $\leq j(k+l)$. At each of the apparent singular points we fix the exponents to be $0, 1, \ldots, n-2, n$. This produces l equations. The condition that there are no logarithmic terms at any of the apparent singular point is given by $\frac{n(n-1)}{2}l$ equations (see [225], Chap. 8, §18). Assuming that the equations are independent and that they define a nonempty algebraic variety, one finds that this algebraic variety has dimension $\frac{n(n+1)}{2}k + n + l$. We note that it seems difficult to verify these assumptions and we have not done this in general.

2. Assuming that the algebraic variety in 1. has dimension $\frac{n(n+1)}{2}k + n + l$, we will show that the bound $\frac{n(n-1)}{2}k + 1 - n$ of Proposition 6.30 is sharp for a general regular singular connection M of dimension n over $C(z)$ with singular locus $S = \{s_0, \ldots, s_k, \infty\}$. Indeed, let A be the sharp bound. Take $l = A$ in (a) above and one finds the number of moduli $\frac{n(n+1)}{2}k + n + A$. This must be equal to $n^2k + 1$, the number of moduli for the family $\mathcal{F}$ above. Thus $A = \frac{n(n-1)}{2}k + 1 - n$.

3. Now assume that the bound $\frac{n(n-1)}{2}k + 1 - n$ of Proposition 6.30 is sharp. Then, as in part 2, a comparison of dimensions of moduli spaces yields that the formula in part 3 for the number of moduli is correct.

4. The counting of parameters that we have done, if correct, clarifies an observation made by Katz on accessory parameters in the introduction of his book ([157], p. 5–7). □

7 Exact Asymptotics

7.1 Introduction and Notation

Singularities of linear complex differential equations is a subject with a long history. New methods, often of an algebraic nature, have kept the subject youthful and growing. In this chapter we treat the asymptotic theory of divergent solutions and the more refined theory of multisummation of those solutions. The theory of multisummation has been developed by many authors, such as W. Balser, B.L.J. Braaksma, J. Écalle, W.B. Jurkat, D. Lutz, M. Loday-Richaud, B. Malgrange, J. Martinet, J.-P. Ramis, and Y. Sibuya. Excellent bibliographies can be found in [177] and [180]. Our aim is to give a complete proof of the multisummation theorem, based on what is called "the Main Asymptotic Existence Theorem" and some sheaf cohomology. In particular, the involved analytic theory of Laplace and Borel transforms has been avoided. However, the link between the cohomology groups and the Laplace and Borel method is made transparent in examples. This way of presenting the theory is close to that of Malgrange [195].

The problem can be presented as follows. Let $\mathbf{C}(\{z\})$ denote the field of the convergent Laurent series (in the variable z) and $\mathbf{C}((z))$ the field of all formal Laurent series. The elements of $\mathbf{C}(\{z\})$ have an interpretation as meromorphic functions on a disk $\{z \in \mathbf{C} |\ |z| < r\}$, for small enough $r > 0$, and having at most a pole at 0. Put $\delta := z\frac{d}{dz}$. Let A be an $n \times n$ matrix with entries in $\mathbf{C}(\{z\})$. The differential equation that concerns us is $(\delta - A)v = w$, where v, w are vectors with coordinates in either $\mathbf{C}(\{z\})$ or $\mathbf{C}((z))$, and where δ acts coordinate-wise on vectors. The differential equation is *(irregular) singular* at $z = 0$ if some entry of A has a pole at 0 and such that this remains the case after any $\mathbf{C}(\{z\})$-linear change of coordinates. For such a differential equation one encounters the following situation:

There is a formal (or divergent) solution $\hat{v}$ of $(\delta - A)\hat{v} = w$ with w convergent, i.e., $\hat{v}$ has coordinates in $\mathbf{C}((z))$ and w has coordinates in $\mathbf{C}(\{z\})$.

We have written here $\hat{v}$ to indicate that the solution is, in general, formal and not convergent. The standard example of this situation is the expression $\hat{v} = \sum_{n\geq 0} n!\, z^n$, which is a solution of *Euler's equation* $(\delta - (z^{-1} - 1))\hat{v} = -z^{-1}$. *The problem is to give $\hat{v}$ a meaning.* A naive way to deal with this situation is to replace $\hat{v}$ by a well-chosen truncation of the Laurent series involved. Our goal is to associate with $\hat{v}$ a meromorphic function defined in a suitable domain and having $\hat{v}$ as its "asymptotic expansion". We begin by giving a formal definition of this notion and some refinements.

Let ρ be a continuous function on the open interval (a, b) with values in the positive real numbers $\mathbf{R}_{>0}$, or in $\mathbf{R}_{>0} \cup \{+\infty\}$. An *open sector* $S(a, b, \rho)$ is the set of the complex numbers $z \neq 0$ satisfying $\arg(z) \in (a, b)$ and $|z| < \rho(\arg(z))$. The a, b are, in fact, elements of the circle $\mathbf{S}^1 := \mathbf{R}/2\pi\mathbf{Z}$. The positive (counter-clockwise) orientation of the circle determines the sector. In some situations it is better to introduce a function t with $e^{it} = z$ and to view a sector as a subset of the t-plane given by the relations $Re(t) \in (a, b)$ and $e^{-Im(t)} < \rho(Re(t))$. We will also have occasion to use *closed sectors* given by relations $\arg(z) \in [a, b]$ and $0 < |z| \leq c$, with $c \in \mathbf{R}_{>0}$.

Definition 7.1 A holomorphic function f on $S(a, b, \rho)$ is said to have the formal Laurent series $\sum_{n \geq n_0} c_n z^n$ as *asymptotic expansion* if for every $N \geq 0$ and every closed sector W in $S(a, b, \rho)$ there exists a constant $C(N, W)$ such that

$$|f(z) - \sum_{n_0 \leq n \leq N-1} c_n z^n| \leq C(N, W)|z|^N \text{ for all } z \in W.$$

One writes $J(f)$ for the formal Laurent series $\sum_{n \geq n_0} c_n z^n$. Let $\mathcal{A}(S(a, b, \rho))$ denote the set of holomorphic functions on this sector that have an asymptotic expansion. For an open interval (a, b) on the circle $\mathbf{S}^1$, one defines $\mathcal{A}(a, b)$ as the direct limit of the $\mathcal{A}(S(a, b, \rho))$ for all ρ. □

In more detail, this means that the elements of $\mathcal{A}(a, b)$ are equivalence classes of pairs $(f, S(a, b, \rho))$ with $f \in \mathcal{A}(S(a, b, \rho))$. The equivalence relation is given by $(f_1, S(a, b, \rho_1)) \sim (f_2, S(a, b, \rho_2))$ if there is a pair $(f_3, S(a, b, \rho_3))$ such that $S(a, b, \rho_3) \subset S(a, b, \rho_1) \cap S(a, b, \rho_2)$ and $f_3 = f_1 = f_2$ holds on $S(a, b, \rho_3)$. For any open $U \subset \mathbf{S}^1$, an element f of $\mathcal{A}(U)$ is defined by a covering by open intervals $U = \cup_i (a_i, b_i)$ and a set of elements $f_i \in \mathcal{A}(a_i, b_i)$ with the property that the restrictions of any f_i and f_j to $(a_i, b_i) \cap (a_j, b_j)$ coincide. One easily verifies that this definition makes $\mathcal{A}$ into a sheaf on $\mathbf{S}^1$. Let $\mathcal{A}^0$ denote the subsheaf of $\mathcal{A}$ consisting of the elements with asymptotic expansion 0. We let $\mathcal{A}_d, \ \mathcal{A}_d^0, \ \ldots$ denote the stalks of the sheaves $\mathcal{A}, \ \mathcal{A}^0, \ \ldots$ at a point $d \in \mathbf{S}^1$.

Exercises 7.2
1. Prove that $\mathcal{A}(\mathbf{S}^1) = \mathbf{C}(\{z\})$.

2. Show that $\mathcal{A}(S(a, b, \rho))$ is a differential $\mathbf{C}$-algebra, that is a $\mathbf{C}$-algebra closed under the operation of taking derivatives. Hint: (see [195]). The proofs that $\mathcal{A}(S(a, b, \rho))$ is closed under multiplication and sum are straightforward. To verify that this algebra is closed under differentiation, it suffices to show the following: *Let g be a function*

analytic in a sector W. If for any closed subsectors $W' \subset W$ one has that there exists a constant C such that for all $z \in W'$, $|g(z)| \leq C|z|^{n+1}$, then for any closed subsectors $W' \subset W$ one has that there exists a constant C' such that for all $z \in W'$, $|g'(z)| \leq C'|z|^n$. To prove this, let $W' \subsetneq W''$ be closed sectors and let δ be a positive integer so that for all $z \in W'$ the closed ball $\{w \mid |w - z| \leq |z|\delta\}$ lies entirely in W''. The Cauchy Integral Formula states that, for all $z \in W'$

$$g'(z) = \frac{1}{2\pi i}\int_\gamma \frac{g(\zeta)}{(\zeta - z)^2}d\zeta,$$

where γ is the circle of radius $|z|\delta$ centered at z. One then has, for all $z \in W'$,

$$|g'(z)| \leq \frac{\max_\gamma |g|}{|z|\delta} \leq C''|z|^{n+1}\frac{(1+\delta)^{n+1}}{|z|\delta} \leq C'|z|^n.$$

Apply this to $g = f - \sum_{k=0}^n a_k z^k$. Note that the asymptotic expansion of f' is the term-by-term derivative of the asymptotic expansion of f. □

The following result shows that *every* formal Laurent series is the asymptotic expansion of some function.

Theorem 7.3 Borel-Ritt
For every open interval $(a, b) \neq \mathbf{S}^1$, the map $J : \mathcal{A}(a, b) \to \mathbf{C}((z))$ is surjective.

Proof. We will prove this for the sector S given by $|\arg(z)| < \pi$ and $0 < |z| < +\infty$. Let $\sqrt{z}$ be the branch of the square-root function that satisfies $|\arg\sqrt{z}| < \pi/2$ for $z \in S$. We first note that for any real number b, the function $\beta(z) = 1 - e^{-b/\sqrt{z}}$ satisfies $|\beta(z)| \leq \frac{b}{\sqrt{|z|}}$ since $\mathrm{Re}(-\frac{b}{\sqrt{z}}) < 0$ for all $z \in S$. Furthermore, $\beta(z)$ has asymptotic expansion 0 on S.

Let $\sum a_n z^n$ be a formal Laurent series. By subtracting a finite sum of terms we may assume that this series has no negative terms. Let b_n be a sequence such that the series $\sum a_n b_n R^n$ converges for all real $R > 0$. For example, we may let $b_0 = 0$ and $b_n = 0$ if $a_n = 0$ and $b_n = 1/n!|a_n|$ if $a_n \neq 0$. Let W be a closed sector defined by $\arg(z) \in [a, b]$ and $0 < |z| \leq R$ in S. Let $\beta_n(z) = 1 - e^{-b_n/\sqrt{z}}$ and $f(z) = \sum a_n\beta_n(z)z^n$. Since $|a_n\beta_n(z)z^n| \leq |a_n|b_n|z|^{n-1/2}$, the function $f(z)$ is analytic on W. To see that $f(z) \in \mathcal{A}(S)$, note that, for $z \in S$

$$\begin{aligned}
|f(z) - \sum_{i=0}^n a_i z^i| &\leq |\sum_{i=0}^n a_i\beta_i(z)z^i - \sum_{i=0}^{n-1} a_i z^i| + |\sum_{i=n+1}^\infty a_i\beta_i(z)z^i| \\
&\leq C_1|z|^n + |z|^n\sum_{i=1}^\infty |a_i|b_i R^{i-n-\frac{1}{2}} \\
&\leq C|z|^n.
\end{aligned}$$

□

The *Main Asymptotic Existence Theorem* states the following:

Given is a formal solution $\hat{v}$ of an equation $(\delta - A)\hat{v} = w$ (with A and w convergent) and a direction $d \in \mathbf{S}^1$. Then there exists an interval (a, b) containing d and a $v \in (\mathcal{A}(a, b))^n$ such that $J(v) = \hat{v}$ and $(\delta - A)v = w$.

In the next section we will present an elementary proof of the main asymptotic existence theorem. We will call a v, having the properties of this theorem, *an asymptotic lift of* $\hat{v}$. The difference of two asymptotic lifts is a solution $g \in \mathcal{A}^0(a, b)$ of $(\delta - A)g = 0$. In general, nontrivial solutions g exist. In order to obtain a *unique* asymptotic lift v on certain sectors one has to refine the asymptotic theory by introducing *Gevrey functions* and *Gevrey series*.

Definition 7.4 Let k be a positive real number and let S be an open sector. A function $f \in \mathcal{A}(S)$, with asymptotic expansion $J(f) = \sum_{n \geq n_0} c_n z^n$, is said to be *a Gevrey function of order* k if the following holds: For every closed subsector W of S there are constants $A > 0$ and $c > 0$ such that for all $N \geq 1$ and all $z \in W$ and $|z| \leq c$ one has

$$|f(z) - \sum_{n_0 \leq n \leq N-1} c_n z^n| \leq A^N \Gamma(1 + \frac{N}{k})|z|^N. \qquad \square$$

We note that this is stronger than saying that f has asymptotic expansion $J(f)$ on S, since on any closed subsector one prescribes the form of the constants $C(N, W)$. Furthermore, we note that one may replace in this definition the (maybe mysterious) term $\Gamma(1 + \frac{N}{k})$ by $(N!)^{1/k}$. The set of all Gevrey functions on S of order k is denoted by $\mathcal{A}_{\frac{1}{k}}(S)$. One sees, as in Exercise 7.2, that this set is, in fact, an algebra over $\mathbf{C}$ and is invariant under differentiation. Moreover, $\mathcal{A}_{1/k}$ can be seen as a subsheaf of $\mathcal{A}$ on $\mathbf{S}^1$. We denote by $\mathcal{A}^0_{1/k}(S)$ the subset of $\mathcal{A}_{1/k}(S)$, consisting of the functions with asymptotic expansion 0. Again $\mathcal{A}^0_{1/k}$ can be seen as a subsheaf of $\mathcal{A}_{1/k}$ on $\mathbf{S}^1$. The following useful lemma gives an alternative description of the sections of the sheaf $\mathcal{A}^0_{1/k}$.

Lemma 7.5 *Let f be holomorphic on an open sector S. Then f belongs to $\mathcal{A}^0_{\frac{1}{k}}(S)$ if and only if for every closed subsector W there are positive constants A, B such that $|f(z)| \leq A \exp(-B|z|^{-k})$ holds for $z \in W$.*

Proof. We will use Stirling's formula:

$$\Gamma(1 + s) = \sqrt{2\pi}\, s^{s+1/2} e^{-s}(1 + o(s^{-1})) \text{ for } s \in \mathbf{R} \text{ and } s \to \infty.$$

If f belongs to $\mathcal{A}^0_{\frac{1}{k}}(S)$ then there is a constant C depending on W such that, for all $n \geq 1$ and $z \in W$, one has $|f(z)| \leq C^n \Gamma(1 + \frac{n}{k})|z|^n$. In other words

$$\log|f(z)| \leq \frac{n}{k}(-1 + \log|Cz|^k) + (\frac{n}{k} + 1/2)\log\frac{n}{k} + \text{ a constant}.$$

For a fixed $|z|$ the right-hand side has, as a function of the integer n, an almost minimal value if n is equal to the integer part of $\frac{k}{|Cz|^k}$. Substituting this value for n one finds that $\log|f(z)| \leq -B|z|^{-k} +$ a constant. This implies the required inequality.

For the other implication of the lemma, it suffices to show that for given k and B there is a positive D such that

$$\frac{r^{-n}\exp(-Br^{-k})}{\Gamma(1+\frac{n}{k})} \leq D^n \text{ holds for all } r \text{ and } n \geq 1.$$

Using Stirling's formula, the logarithm of the left-hand side can be estimated by

$$\frac{n}{k}(1+\log\, r^{-k} - \log\frac{n}{k}) - 1/2\log\frac{n}{k} - Br^{-k} + \text{ a constant.}$$

For a fixed n and variable r the maximal value of this expression is obtained for $r^{-k} = B^{-1}\frac{n}{k}$. Substitution of this value gives

$$\frac{n}{k}\log\, B^{-1} - 1/2\log\frac{n}{k} + \text{ a constant.}$$

This expression is bounded by a constant multiple of n. □

The notion of Gevrey function of order k does not have the properties that we will require for $k \leq 1/2$. In the following, we assume that $k > 1/2$. In the event of a smaller k one may replace z by a suitable root $z^{\frac{1}{m}}$ in order to obtain a new $k' = mk > 1/2$. We note further that the ks that interest us are slopes of the Newton polygon of the differential equation $\delta - A$. These ks are, in fact, rational and, after taking a suitable root of z, one may restrict to positive integers k.

Exercise 7.6 Let $f \in \mathcal{A}_{1/k}(S)$ with $J(f) = \sum_{n\geq n_0} c_n z^n$. Prove that for $N \geq 1$ the c_N satisfy the inequalities

$$|c_N| \leq A^N\Gamma(1+\frac{N}{k}), \text{ for a suitable constant } A \text{ and all } N \geq 1.$$

Hint: Subtract the two inequalities $|f(z) - \sum_{n=n_0}^{N-1} c_n z^n| \leq A^N\Gamma(1+\frac{N}{k})|z|^N$ and $|f(z) - \sum_{n=n_0}^{N} c_n z^n| \leq A^{N+1}\Gamma(1+\frac{N+1}{k})|z|^{N+1}$. □

Exercise 7.6 leads to the notion of *Gevrey series of order k*.

Definition 7.7 $f = \sum_{n\geq n_0} c_n z^n \in \mathbf{C}((z))$ is called a *Gevrey series of order k* if there is a constant $A > 0$ such that for all $n > 0$ one has $|c_n| \leq A^n\Gamma(1+\frac{n}{k})$. The set of all such series is denoted by $\mathbf{C}((z))_{\frac{1}{k}}$. The subset of the power series satisfying the above condition on the coefficients is denoted by $\mathbf{C}[[z]]_{\frac{1}{k}}$. □

As in the definition of Gevrey functions of order k, one can replace the condition $|c_n| \leq A^n\Gamma(1+\frac{n}{k})$ with $|c_n| \leq A^n\Gamma(n!)^{\frac{1}{k}}$.

Lemma 7.8 *1.* $\mathbf{C}[[z]]_{\frac{1}{k}}$ *is a differential ring with a unique maximal ideal, namely the ideal* (z).

2. $\mathbf{C}((z))_{\frac{1}{k}}$ *is the field of fractions of* $\mathbf{C}[[z]]_{\frac{1}{k}}$.
3. *If* $k < l$ *then* $\mathbf{C}[[z]]_{\frac{1}{k}} \supset \mathbf{C}[[z]]_{\frac{1}{l}}$.

Proof. 1. The set $A = \mathbf{C}[[z]]_{\frac{1}{k}}$ is clearly closed under addition. To see that it is closed under multiplication, let $f = \sum a_i z^i$ and $g = \sum b_i z^i$ be elements of this set and assume $|a_N| \leq A^N (N!)^{1/k}$ and $|b_N| \leq B^N (N!)^{1/k}$ for all $N \geq 1$. We then have $fg = \sum c_i z^i$ where $|c_N| = |\sum_{i=0}^N a_i b_{N-i}| \leq \sum_{i=0}^N A^i B^{N-i} (i!)^{1/k} (N-i)^{1/k}$ $\leq (AB)^N (N+1)(N!)^{1/k} \leq C^N (N!)^{1/k}$ for an appropriate C. The ring A is closed under taking derivatives because if $|a_N| \leq A^N (N!)^{1/k}$, then $|Na_N| \leq NA^N (N!)^{1/k}$ $\leq C^N ((N-1)!)^{1/k}$ for an appropriate C.

To prove the statement concerning the ideal (z), it suffices to show that any element $f = \sum a_i z^i$ not in the ideal (z) is invertible in $\mathbf{C}[[z]]_{\frac{1}{k}}$. Since $a_0 \neq 0$ such an element is clearly invertible in $\mathbf{C}[[z]]$. Let $g = \sum b_i z^i$ be the inverse of f. We have that $b_0 = 1/a_0$ and for $N \geq 1$, $b_N = -(1/a_0)(a_1 b_{N-1} + \ldots + a_N b_0)$. One then shows by induction that $|b_N| \leq C^N (N!)^{1/k}$ for an appropriate C. Parts 2 and 3 are clear. □

In a later section we will prove the following important properties of Gevrey functions.

1. If $|b-a| \leq \frac{\pi}{k}$ the map $J : \mathcal{A}_{\frac{1}{k}}(a,b) \to \mathbf{C}((z))_{\frac{1}{k}}$ is surjective but not injective. (Consequently, $\mathcal{A}^0_{1/k}(a,b) \neq 0$).
2. If $|b-a| > \frac{\pi}{k}$ the map $J : \mathcal{A}_{\frac{1}{k}}(a,b) \to \mathbf{C}((z))_{\frac{1}{k}}$ is injective but not surjective. (Consequently, $\mathcal{A}^0_{1/k}(a,b) = 0$).

We note that the above statements are false for $k \leq 1/2$, since $\mathcal{A}(\mathbf{S}^1) = \mathbf{C}(\{z\})$. This is the reason to assume $k > 1/2$. However, the case $k \leq 1/2$ can be treated by allowing ramification, i.e., replacing z by a suitable $z^{1/n}$.

Definition 7.9 Let $\hat{y} \in \mathbf{C}((z))$. Then $\hat{y}$ is called *k-summable in the direction d* if there is an $f \in \mathcal{A}_{\frac{1}{k}}(d - \frac{\alpha}{2}, d + \frac{\alpha}{2})$ with $J(f) = \hat{y}$ and $\alpha > \frac{\pi}{k}$. We note that f is unique. One says that $\hat{y} \in \mathbf{C}((z))_{1/k}$ is *k-summable* if there are only finitely many directions d such that $\hat{y}$ is *not* k-summable in the direction d. □

We can now formulate the results of the multisummation theory. A special case is the *k-summation Theorem* (cf. [237], Theorem 3.28, p. 80):

Suppose that the differential equation $(\delta - A)$ *has only one positive slope* k *(and* $k > 1/2$*) and consider a formal solution* $\hat{v}$ *of* $(\delta - A)\hat{v} = w$ *(with* A *and* w *convergent). Then (each coordinate of)* $\hat{v}$ *is k-summable.*

We draw some conclusions from this statement. The first is that the (in general) divergent solution $\hat{v}$ is not very divergent. Indeed, its coordinates lie in $\mathbf{C}((z))_{1/k}$. Let d be a direction for which $\hat{v}$ is k-summable. Then the element $v \in (\mathcal{A}_{1/k}(d - \frac{\alpha}{2}, d + \frac{\alpha}{2}))^n$ with image $J(v) = \hat{v} \in \mathbf{C}((z))^n$ is unique. Moreover, $g := (\delta - A)v$ is a vector with coordinates again in $\mathcal{A}_{1/k}(d - \frac{\alpha}{2}, d + \frac{\alpha}{2})$, with $\alpha > \frac{\pi}{k}$ and with $J(g) = w$. From the injectivity of $J : \mathcal{A}_{1/k}(d - \frac{\alpha}{2}, d + \frac{\alpha}{2}) \to \mathbf{C}((z))_{1/k}$, one concludes that $g = w$ and that v satisfies the differential equation $(\delta - A)v = w$. Thus v is the *unique* asymptotic lift, produced by the k-summation theorem. One calls v the *k-sum of $\hat{v}$ in the direction d.*

One possible formulation of the *Multisummation Theorem* is:

Suppose that $k_1 < k_2 < \cdots < k_r$ (with $k_1 > 1/2$) are the positive slopes of the equation $(\delta - A)$ and let $\hat{v}$ be a formal solution of the equation $(\delta - A)\hat{v} = w$ (with w convergent). There are finitely many "bad" directions, called the singular directions of $\delta - A$. *If d is not a singular direction, then $\hat{v}$ can be written as a sum $\hat{v}_1 + \hat{v}_2 + \cdots + \hat{v}_r$ where each $\hat{v}_i$ is k_i-summable in the direction d and, moreover, $(\delta - A)\hat{v}_i$ is convergent.*

We draw again some conclusions. First, $\hat{v} \in (\mathbf{C}((z))_{1/k_1})^n$. Let d be a direction that is not singular. Then each $\hat{v}_i$ is k_i-summable in the direction d and $w_i := (\delta - A)\hat{v}_i$ is convergent. There are unique elements v_i with coordinates in $\mathcal{A}_{1/k_i}(d - \frac{\alpha_i}{2}, d + \frac{\alpha_i}{2})$, with $\alpha_i > \frac{\pi}{k_i}$ and image $\hat{v}_i$ under J. Then $(\delta - A)v_i$ has coordinates in $\mathcal{A}_{1/k_i}(d - \frac{\alpha_i}{2}, d + \frac{\alpha_i}{2})$ and its asymptotic expansion is w_i, which is convergent. Since $\mathcal{A}^0_{1/k_i}(d - \frac{\alpha_i}{2}, d + \frac{\alpha_i}{2}) = 0$, it follows that $(\delta - A)v_i = w_i$. Then the sum $v = \sum_i v_i$ has coordinates in $\mathcal{A}(d - \frac{\alpha_r}{2}, d + \frac{\alpha_r}{2})$ and satisfies $J(v) = \hat{v}$. Moreover, $(\delta - A)v = w$. One calls v *the multisum of $\hat{v}$ in the direction d*. Note though that v depends on the decomposition of $\hat{v}$ as a sum $\hat{v}_1 + \hat{v}_2 + \cdots + \hat{v}_r$.

The multisummation theory also carries the name *exact asymptotics* because it refines the Main Asymptotic Existence Theorem by producing a uniquely defined asymptotic lift for all but finitely many directions. Since the multisum is uniquely defined, one expects an "explicit formula" for it. Indeed, the usual way to prove the multisummation theorem is based on a sequence of Borel and Laplace transforms and analytic continuations, which gives, in a certain sense, an "explicit formula" for the multisum . We will explain, in later sections, some details of this and of the related *Stokes phenomenon*.

7.2 The Main Asymptotic Existence Theorem

We recall the statement of this theorem.

Theorem 7.10 Main Asymptotic Existence Theorem
Let $\hat{v}$ be a formal solution of $(\delta - A)\hat{v} = w$, where A is an $n \times n$ matrix and w is

a vector of length n, both with coordinates in $\mathbf{C}(\{z\})$. *Let* $d \in \mathbf{S}^1$ *be a direction. Then there is an open interval* (a, b) *containing* d *and a* $v \in (\mathcal{A}(a, b))^n$ *with* $J(v) = \hat{v}$ *and* $(\delta - A)v = w$.

Remarks 7.11 1. Complete proofs of this theorem, originally due to Hukuhara and Turrittin, are given in [301] and [193]. Extensions of this theorem have been developed by J.-P. Ramis and Y. Sibuya [243].

2. Theorem 7.10 is an almost immediate consequence of the first part of Theorem 7.12 below. Indeed, by the Borel-Ritt theorem, we can choose a $\tilde{v} \in (\mathcal{A}_d)^n$ with $J(\tilde{v}) = \hat{v}$. Then $g = w - (\delta - A)\tilde{v} \in (\mathcal{A}_d^0)^n$ and, by the first part of Theorem 7.12, one can solve the equation $(\delta - A)f = g$ with some $f \in (\mathcal{A}_d^0)^n$. Recall that $\mathcal{A}_d, \mathcal{A}_d^0, \ldots$ denote the stalks of the sheaves $\mathcal{A}, \mathcal{A}^0, \ldots$ at a point $d \in \mathbf{S}^1$.

3. In this section we will give a complete and elementary proof of Theorem 7.10, inspired by [193], Appendix 1. First we study in detail the special case $n = 1$, i.e., inhomogeneous equations of order 1. The step from inhomogeneous equations of order 1 to "quasisplit" equations is rather straightforward. Finally, with a small calculation concerning norms on a linear space of analytic functions, the general case is proved. □

Theorem 7.12 *Let* A *be an* $n \times n$ *matrix with entries in* $\mathbf{C}(\{z\})$ *and let* $d \in \mathbf{S}^1$ *be a direction. The operator* $(\delta - A)$ *acts surjectively on* $(\mathcal{A}_d^0)^n$ *and on* $((\mathcal{A}_{1/k}^0)_d)^n$ *for any* $k > 0$.

It suffices to consider in the sequel the direction 0. We will first be concerned with the equation $(\delta - q)f = g$, with $q \in z^{-1}\mathbf{C}[z^{-1}]$ and $g \in \mathcal{A}_0^0$. The goal is to find a solution $f \in \mathcal{A}_0^0$. The general solution of the equation can be written, symbolically, as $e(q)(z)\int e(-q)(t)g(t)\frac{dt}{t} + ae(q)(z)$ where $e(q) = e^{\int q(t)\frac{dt}{t}}$. The problem is to find the correct value of the constant $a \in \mathbf{C}$. Moreover, we will need more precise information on this solution f. For this purpose we consider closed sectors $\Sigma = \Sigma(c, d) = \{z \in \mathbf{C} \mid 0 < |z| \leq c \text{ and } |\arg(z)| \leq d\}$ for $c, d > 0$. Let $\mathcal{F} = \mathcal{F}(\Sigma)$ denote the set of complex valued functions f on Σ, such that:

1. f is continuous on Σ.
2. f is holomorphic on the interior of Σ.
3. For every integer $N \geq 1$, there exists a constant C_N such that $|f(z)| \leq C_N|z|^N$ holds for all $z \in \Sigma$.

On $\mathcal{F}$ one considers a sequence of norms $\|\ \|_N$ defined by $\|f\|_N = \sup_{z\in\Sigma} |\frac{f(z)}{z^N}|$. We note that every element of $\mathcal{A}_0^0$ can be represented by an element in $\mathcal{F}$ for a suitable choice of c, d. On the other hand, any element of $\mathcal{F}$ determines an element of $\mathcal{A}_0^0$. In other words, $\mathcal{A}_0^0$ is the direct limit of the spaces $\mathcal{F}(\Sigma)$.

Lemma 7.13 *Let* $q = q_l z^{-l} + q_{l-1}z^{-l+1} + \cdots + q_1 z^{-1} \in z^{-1}\mathbf{C}[z^{-1}]$, *with* $q_l \neq 0$, *be given.*

1. *Suppose* $Re(q_l) < 0$. *For small enough* $c, d > 0$ *there is a linear operator* $K : \mathcal{F} \to \mathcal{F}$ *with* $\mathcal{F} = \mathcal{F}(\Sigma(c, d))$, *such that* $(\delta - q)K$ *is the identity on* $\mathcal{F}$ *and* K *is a contraction for every* $\| \ \|_N$ *with* $N \geq 2$, *i.e.,* $\|K(g)\|_N \leq c_N \|g\|_N$ *with* $c_N < 1$ *and all* $g \in \mathcal{F}$.
2. *Suppose* $Re(q_l) = 0$. *Then statement 1 remains valid.*
3. *Suppose* $Re(q_l) > 0$ *and let* $N > 0$ *be an integer. For small enough* $c, d > 0$ *there is a linear operator* $K : \mathcal{F} \to \mathcal{F}$ *such that* $(\delta - q)K$ *is the identity on* $\mathcal{F}$ *and* K *is a contraction for* $\| \ \|_N$.

Corollary 7.14 *Let* q *be as in Lemma 7.13.*

1. $(\delta - q)$ *acts surjectively on* $\mathcal{A}^0_0$.
2. $(\delta - q)$ *acts surjectively on* $(\mathcal{A}^0_{1/k})_0$.

Proof. 1. The existence of K in Lemma 7.13 proves that $(\delta - q)$ is surjective on $\mathcal{A}^0_0$. We note that this result remains valid if q is a finite sum of terms $q_s z^{-s}$ with $s \in \mathbf{R}_{>0}$.

2. Lemma 7.5 easily yields that $(\mathcal{A}^0_{1/k})_0$ is the union of $\mathcal{A}^0_0 e(Bz^{-k})$, taken over all $B \in \mathbf{R}_{>0}$. It suffices to show that $(\delta - q)$ is surjective on each of the spaces $\mathcal{A}^0_0 e(Bz^{-k})$. The observation $e(Bz^{-k})^{-1}(\delta - q)e(Bz^{-k}) = (\delta - q - kBz^{-k})$, reduces the latter to the first part of this corollary. □

The Proof of Lemma 7.13

(1) The function $e(q)$, defined by $e(q)(z) = e^{\int q(t)\frac{dt}{t}}$, is a solution of the homogeneous equation $(\delta - q)e(q) = 0$. The expression $\int q(t)\frac{dt}{t}$ is chosen to be $\frac{q_l}{-l}z^{-l} + \frac{q_{l-1}}{-l+1}z^{-l+1} + \cdots + \frac{q_1}{-1}z^{-1}$. For $z = re^{i\phi} \in \Sigma$, the logarithm of the absolute value of $e(q)(z)$ is equal to

$$r^{-l}\left(\frac{Re(q_l)}{-l}\cos(l\phi) + \frac{Im(q_l)}{-l}\sin(l\phi)\right)$$
$$+r^{-l+1}\left(\frac{Re(q_{l-1})}{-l+1}\cos((l-1)\phi) + \frac{Im(q_{l-1})}{-l+1}\sin((l-1)\phi)\right) + \cdots .$$

The coefficient of r^{-l} is positive for $\phi = 0$. One can take $d > 0$ small enough such that the coefficient of r^{-l} is positive for all $|\phi| \leq d$ and $0 < c < 1$ small enough such that the function $|e(q)(se^{i\phi})|$ is, for any fixed $|\phi| \leq d$, a decreasing function of $s \in (0, c]$. With these preparations we define the operator K by $K(g)(z) = e(q)(z)\int_0^z e(-q)(t)g(t)\frac{dt}{t}$. The integral makes sense, since $e(-q)(t)$ tends to zero for $t \in \Sigma$ and $t \to 0$. Clearly $(\delta - q)Kg = g$ and we are left with a computation of $\|K(g)\|_N$. One can write $K(g)(z) = e(q)(z)\int_0^1 e(-q)(sz)g(sz)\frac{ds}{s}$ and by the above choices one has $|e(-q)(sz)| \leq |e(-q)(z)|$ for all $s \in [0, 1]$. This produces the estimate $\int_0^1 \|g\|_N s^N |z|^N \frac{ds}{s} = \frac{\|g\|_N}{N}|z|^N$. Thus $K : \mathcal{F} \to \mathcal{F}$ and K is a contraction for $\| \ \|_N$ with $N \geq 2$.

2. Let $q_l = ip$ with $p \in \mathbf{R},\ p \neq 0$. We consider the case $p < 0$. The situation $p > 0$ is treated in a similar way. For $\log |e(-q)(se^{i\phi})|$ one has the formula

$$s^{-l}(\ \frac{p}{l}\sin(l\phi)\ )$$
$$+s^{-l+1}(\ \frac{Re(q_{l-1})}{l-1}\cos((l-1)\phi) + \frac{Im(q_{l-1})}{l-1}\sin((l-1)\phi)\ ) + \cdots .$$

We can now choose small enough $c, d > 0$ such that

(a) The function $s \mapsto |e(-q)(se^{id})|$ is increasing for $s \in [0, c]$.
(b) The function $\phi \mapsto |e(-q)(se^{i\phi})|$ is for any any fixed s, with $0 < s \leq c$, a decreasing function of $\phi \in [-d, d]$. For every point $z \in \Sigma$ we take a path from 0 to $z = re^{i\phi_0}$, consisting of two pieces. The first is the line segment $\{sre^{id}|0 \leq s \leq 1\}$ and the second one is the circle segment $\{re^{i\phi}|\phi_0 \leq \phi \leq d\}$. The operator K is defined by letting $K(g)(z)$ be the integral $e(q)(z)\int_0^z e(-q)(t)g(t)\frac{dt}{t}$ along this path. It is clear that the integral is well defined and that $(\delta - q)K(g) = g$. We now have to make an estimate for $\|K(g)\|_N$. The first part of the path can be estimated by

$$|e(q)(z)|\ \ |\int_0^1 e(-q)(sre^{id})g(sre^{id})\frac{ds}{s}|$$
$$\leq |e(q)(z)|\ \ |e(-q)(re^{id})|\ \ \|g\|_N r^N \int_0^1 s^N\frac{ds}{s} \leq \frac{1}{N}|z|^N\|g\|_N.$$

The second part can be estimated by

$$|e(q)(z)|\ \ |\int_{\phi_0}^d e(-q)(re^{i\phi})g(re^{i\phi})id\phi| \leq \int_{\phi_0}^d \|g\|_N r^N d\phi \leq 2d|z|^N\|g\|_N.$$

Thus $\|K(g)\|_N \leq (\frac{1}{N} + 2d)\|g\|_N$ and for $N \geq 2$ and d small enough we find that K is a contraction with respect to $\|\ \|_N$.

3. First, we take d small enough such that the coefficient of r^{-l} in the expression for $\log |e(q)(re^{i\phi})|$ is strictly negative for $|\phi| \leq d$. Furthermore, one can take $c > 0$ small enough such that for any fixed ϕ with $|\phi| \leq d$, the function $r \mapsto |e(q)(re^{i\phi})|$ is increasing on $[0, c]$.

The operator K is defined by letting $K(g)(z)$ be the integral $e(q)(z)\int_c^z e(-q)(t)$ $g(t)\frac{dt}{t}$ along any path in Σ from c to z. It is clear that $(\delta - q)K(g) = g$. For $z \in \Sigma$ with $|z| \leq c/2$ and any integer $M \geq 1$, one can estimate $|K(g)(z)|$ by

$$|e(q)(z)\int_c^{2z} e(-q)(t)g(t)\frac{dt}{t}| + |e(q)(z)\int_z^{2z} e(-q)(t)\frac{dt}{t}|,$$

and this is bounded by $|e(q)(z)e(q)(2z)^{-1}|\ \ \|g\|_M c^M + |z|^M\|g\|_M\frac{2^M-1}{M}$. Since the limit of $\frac{|e(q)(z)e(q)(2z)^{-1}|}{|z^M|}$ for $|z| \to 0$ is 0, one finds that there is some constant C_M

with $\|K(g)\|_M \leq C_M \|g\|_M$. In particular, $K(g) \in \mathcal{F}$. For the fixed integer $N \geq 1$ we have to be more precise and show that for small enough $c, d > 0$ there is an estimate $\|K(g)\|_N \leq C_N \|g\|_N$ with $C_N < 1$ (and for all $g \in \mathcal{F}$).

Set $f(z) = \frac{e(q)(z)}{z^N} \int_c^z e(-q)(t)g(t)\frac{dt}{t}$. We then want to show that $|f(z)| \leq C(c,d)\|g\|_N$ for $z \in \Sigma$, where $C(c,d)$ is a constant that is < 1 for small enough $c, d > 0$.

Let $z = re^{i\phi}$. We split $|f(z)|$ into two pieces. The first one is $|\frac{e(q)(re^{i\phi})}{r^N} \int_z^{ce^{i\phi}} e(-q)(t)g(t)\frac{dt}{t}|$ and the second is $|\frac{e(q)(re^{i\phi})}{r^N} \int_c^{ce^{i\phi}} e(-q)(t)g(t)\frac{dt}{t}|$. For the estimate of the first integral we introduce the function $E(t) := |e(q)(te^{i\phi})|$ and the first integral is bounded by $h(r)\|g\|_N$, where $h(r) := \frac{E(r)}{r^N} \int_r^c E(t)^{-1} t^N \frac{dt}{t}$. We want to show that for small enough $c > 0$, one has $h(r) \leq 1/2$ for all r with $0 < r \leq c$.

For the boundary point $r = c$ one has $h(c) = 0$. For the other boundary point $r = 0$ we will show that the limit of $h(r)$ for $r \to 0$ is zero. Take any $\alpha > 1$ and consider $0 < r$ with $\alpha r \leq c$. Then $h(r) = \frac{E(r)}{r^N} \int_r^{\alpha r} E(t)^{-1} t^N \frac{dt}{t} + \frac{E(r)}{r^N} \int_{\alpha r}^c E(t)^{-1} t^N \frac{dt}{t}$. Since $E(t)$ is an increasing function of t we can estimate $h(r)$ by $\frac{1}{r^N} \int_r^{\alpha r} t^N \frac{dt}{t}$ $+\frac{E(r)E(\alpha r)^{-1}}{r^N} \int_{\alpha r}^c t^N \frac{dt}{t}$ and thus by $\frac{\alpha^N - 1}{N} + \frac{E(r)E(\alpha r)^{-1}}{r^N} \frac{c^N}{N}$. The limit of $\frac{E(r)E(\alpha r)^{-1}}{r^N}$ for $r \to 0$ is 0. Since $\alpha > 1$ was arbitrary, this implies that the limit of $h(r)$ for $r \to 0$ is 0. The maximum value of $h(r)$ is therefore obtained for $r_0 \in (0, c)$. The function $h(r)$ satisfies the differential equation $rh'(r) = (\frac{rE'(r)}{E(r)} - N)h(r) - 1$. The expression $\log E(t)$ is equal to $c_l t^{-l} + c_{l-1} t^{-l+1} + \cdots$ with $c_l < 0$ and c_l depending on ϕ. Thus $h(r_0) = \frac{1}{-lc_l r_0^{-l} + \cdots - N}$ and this is, for small enough c, bounded by $\frac{1}{-c_l c^{-l} + \cdots - N} \leq 1/3$. The second part is bounded by $\|g\|_N F(\phi_0)$, where

$$F(\phi_0) := |e(q)(ce^{i\phi_0})| \; |\int_0^{\phi_0} |e(-q)(ce^{i\phi})| \, d\phi|.$$

The function F is continuous and $F(0) = 0$. Therefore, we can take $d > 0$ small enough such that $F(\phi) \leq 1/3$ for all ϕ with $|\phi| \leq d$. Thus the second part is bounded by $1/3\|g\|_N$ and $\|K(g)\|_N \leq 2/3\|g\|_N$. □

We now recall the following definition (see Definition 3.39).

Definition 7.15 A differential operator $(\delta - A)$, with A an $n \times n$ matrix with coefficients in $\mathbf{C}(\{z\})$ is called *split* if it is equivalent, by a transformation in $\mathrm{GL}_n(\mathbf{C}(\{z\}))$, to a direct sum of operators of the form $\delta - q + C$, where $q \in z^{-1}\mathbf{C}[z^{-1}]$ and C is a constant matrix. The operator $(\delta - A)$ is called *quasisplit* if it becomes split after replacing z by a suitable m-th root of z. □

Corollary 7.16 *Let $(\delta - A)$ be a quasisplit linear differential operator of order n and let $d \in \mathbf{S}^1$ be a direction. Then $(\delta - A)$ acts surjectively on $(\mathcal{A}_d^0)^n$ and on $((\mathcal{A}_{1/k}^0)_d)^n$ for all $k > 0$.*

Proof. For the proof we may suppose that the operator is split and even that it has the form $\delta - q + C$, where C is a constant matrix. Let T be a fundamental matrix for the equation $\delta y = Cy$. The equation $(\delta - q + C)f = g$ can be rewritten as $(\delta - q)Tf = Tg$. The transformation T induces a bijection on the spaces $(\mathcal{A}_d^0)^n$ and $((\mathcal{A}_{1/k}^0)_d)^n$. Thus we are reduced to proving that the operator $(\delta - q)$ acts surjectively on $\mathcal{A}_d^0$ and $(\mathcal{A}_{1/k}^0)_d$. For $d = 0$ this follows at once from Corollary 7.14. □

The proof of Theorem 7.12 for the general case (and the direction 0) follows from the next lemma.

Lemma 7.17 *Let B be an $n \times n$ matrix with entries in $\mathcal{A}_0$. Suppose that $S = J(B)$ has entries in $\mathbf{C}[z^{-1}]$ and that $\delta - S$ is a quasisplit equation. Then there exists an $n \times n$ matrix T with coefficients in $\mathcal{A}_0^0$ such that $(1 + T)^{-1}(\delta - B)(1 + T) = \delta - S$.*

Indeed, consider $(\delta - A)$ and a formal transformation $F \in \mathrm{GL}_n(\mathbf{C}((z)))$ such that $F^{-1}(\delta - A)F = (\delta - S)$, where S has entries in $\mathbf{C}[z^{-1}]$ and $(\delta - S)$ is quasisplit. The existence of F and S is guaranteed by the classification of differential equations over $\mathbf{C}((z))$, cf. Proposition 3.41. Let $\tilde{F} \in \mathrm{GL}_n(\mathcal{A}_0)$ satisfy $J(\tilde{F}) = F$. Define the $n \times n$ matrix B, with entries in $\mathcal{A}_0$, by $(\delta - B) = \tilde{F}^{-1}(\delta - A)\tilde{F}$. Since $\tilde{F}$ acts as a bijection on the spaces $(\mathcal{A}_0^0)^n$ and $((\mathcal{A}_{1/k}^0)_0)^n$, it suffices to consider the operator $(\delta - B)$ instead of $(\delta - A)$. By construction $J(B) = S$ and we can apply the above lemma. Also, $(1 + T)$ acts as a bijection on the spaces $(\mathcal{A}_0^0)^n$ and $((\mathcal{A}_{1/k}^0)_0)^n$. Thus Lemma 7.17 and Corollary 7.16 complete the proof of Theorem 7.12.

The Proof of Lemma 7.17

Using the arguments of the proof of Corollary 7.16, we may already assume that S is a diagonal matrix $\mathrm{diag}(q_1, \ldots, q_n)$ with the diagonal entries $q_i \in z^{-1}\mathbf{C}[z^{-1}]$. We note that T itself is supposed to be a solution of the equation $\delta(T) - ST + TS = B - S + (B - S)T$, having entries in $\mathcal{A}_0^0$. The differential operator $L : T \mapsto \delta(T) - ST + TS$ acting on the space of the $n \times n$-matrices is, on the usual standard basis for matrices, also in diagonal form with diagonal entries $q_i - q_j \in z^{-1}\mathbf{C}[z^{-1}]$.

Take a suitable closed sector $\Sigma = \Sigma(c, d)$ and consider the space $\mathcal{M}$ consisting of the matrix functions $z \mapsto M(z)$ satisfying:

(a) $M(z)$ is continuous on Σ and holomorphic on the interior of Σ.
(b) For every integer $N \geq 1$ there is a constant C_N such that $|M(z)| \leq C_N |z|^N$ holds on Σ. Here $|M(z)|$ denotes the l_2-norm on matrices, given by $|M(z)| := (\sum |M_{i,j}(z)|^2)^{1/2}$.

We note that for two matrices $M_1(z)$ and $M_2(z)$ one has $|M_1(z)M_2(z)| \leq |M_1(z)||M_2(z)|$. The space $\mathcal{M}$ has a sequence of norms $\| \ \|_N$, defined by $\|M\|_N := \sup_{z \in \Sigma} \frac{|M(z)|}{|z|^N}$. Using Lemma 7.13 and the diagonal form of L, one finds that the operator L acts surjectively on $\mathcal{M}$. Let us now fix an integer $N_0 \geq 1$. For small enough $c, d > 0$, Lemma 7.13, furthermore, states there is a linear operator K acting on $\mathcal{M}$, which has the properties:

(1) LK is the identity and
(2) K is a contraction for $\| \; \|_{N_0}$, i.e., $\|K(M)\|_{N_0} \leq c_{N_0}\|M\|_{N_0}$ with $c_{N_0} < 1$ and all $M \in \mathcal{M}$.

Define now a sequence of elements $T_k \in \mathcal{M}$ by $T_0 = K(B - S)$ and $T_k = K((B - S)\ T_{k-1})$ for $k \geq 1$. Since $\|B - S\|_N < 1$ for all integers $N \geq 1$, one can deduce from (2) that $\sum_{k=0}^{\infty} T_k$ converges uniformly on Σ to a matrix function T that is continuous on Σ, holomorphic on the interior of Σ and satisfies $|T(z)| \leq D|z|^{N_0}$ for a certain constant $D > 0$ and all $z \in \Sigma$. Then $L(T) = L(K(B-S) + K((B-S)T_0) + \cdots) = (B-S) + (B-S)T$. Thus we have found a certain solution T for the equation $\delta(T) - ST + TS = (B-S) + (B-S)T$. We want to show that the element T belongs to $\mathcal{M}$.

The element $(B - S)(1 + T)$ belongs to $\mathcal{M}$ and thus $L(K((B - S)(1 + T)))$ $= (B - S)(1 + T)$. Therefore $\tilde{T} := T - K((B - S)(1 + T))$ satisfies $L(\tilde{T}) = 0$ and, moreover, $\tilde{T}$ is continuous on Σ, holomorphic at the interior of Σ and $|\tilde{T}(z)| \leq D_N|z|^{N_0}$ holds for $z \in \Sigma$ and some constant D_{N_0}. From the diagonal form of L one deduces that the kernel of L consists of the matrices $\mathrm{diag}(e(-q_1), \ldots, e(-q_n)) \cdot C \cdot \mathrm{diag}(e(q_1), \ldots, e(q_n))$ with C a constant matrix. The entries of $\tilde{T}$ are, therefore, of the form $ce(q_i - q_j)$ with $c \in \mathbf{C}$ and satisfy inequalities $\leq \tilde{D}|z|^{N_0}$ for some constant $\tilde{D}$ and our choice of $N_0 \geq 1$. Thus the nonzero entries of $\tilde{T}$ are in $\mathcal{A}_0^0$. It follows that $\tilde{T} \in \mathcal{M}$ (again for $c, d > 0$ small enough) and thus $T \in \mathcal{M}$. □

7.3 The Inhomogeneous Equation of Order One

Let $q \in \mathbf{C}[z^{-1}]$ have degree k in the variable z^{-1}. In this section we consider the inhomogeneous equation

$$(\delta - q)\hat{f} = g \text{ with } g \in \mathbf{C}(\{z\}) \text{ and } \hat{f} \in \mathbf{C}((z)).$$

According to Theorem 7.10, there is for every direction $d \in \mathbf{S}^1$ an asymptotic lift of $\hat{f}$ in $\mathcal{A}(a, b)$, with $d \in (a, b)$ and $|b - a|$ "small enough". The aim of this section is to study the obstruction for the existence of an asymptotic lift on large intervals (or sectors). As quite often happens, the obstruction from local existence to global existence is measured by some cohomology group. In the present situation, we will show that the obstruction is the first cohomology group of the sheaf $\ker(\delta - q, \mathcal{A}^0)$. We refer to Appendix C for the definitions and concepts from sheaf theory that we shall need.

Let U be a nonempty open subset of $\mathbf{S}^1$ (including the case $U = \mathbf{S}^1$). There is a covering of U by "small" intervals S_i, such that there exists for i an $f_i \in \mathcal{A}(S_i)$ with asymptotic expansion $\hat{f}$ and $(\delta - q)f_i = g$. The difference $f_i - f_j$ belongs to $\ker(\delta - q, \mathcal{A}^0)(S_i \cap S_j)$. Hence the collection $\{g_{i,j}\} := \{f_i - f_j\}$ is a 1-cocycle for the sheaf $\ker(\delta - q, \mathcal{A}^0)$, since $g_{i,j} + g_{j,k} + g_{k,i} = 0$ holds on the intersection $S_i \cap S_j \cap S_k$. The image of this 1-cocycle in $H^1(U, \ker(\delta - q, \mathcal{A}^0))$ is easily seen to depend only on $\hat{f}$. Moreover, this image is zero if and only if $\hat{f}$ has an asymptotic lift on U. The practical point of this formalism is that we can actually calculate the cohomology group $H^1(U, \ker(\delta - q, \mathcal{A}^0))$, say for $U = \mathbf{S}^1$ or U an open interval.

Write $q = q_0 + q_1 z^{-1} + \cdots + q_k z^{-k}$ with $q_k \neq 0$ and let $e(q) := \exp(q_0 \log z + \frac{q_1}{-1} z^{-1} + \cdots + \frac{q_k}{-k} z^{-k})$ be a "symbolic solution" of $(\delta - q)e(q) = 0$. On a sector $S \neq \mathbf{S}^1$ one can give $e(q)$ a meaning by choosing the function $\log\ z$. For $k = 0$ one observes that $\ker(\delta - q, \mathcal{A}^0)$ is zero. This implies that any formal solution $\hat{f}$ of $(\delta - q)\hat{f} = g \in \mathbf{C}(\{z\})$ has an asymptotic lift in $\mathcal{A}(\mathbf{S}^1) = \mathbf{C}(\{z\})$. In other words $\hat{f}$ is, in fact, a convergent Laurent series.

From now on we will suppose that $k > 0$. We will introduce some terminology.

Definition 7.18 Let $q = q_0 + q_1 z^{-1} + \cdots + q_k z^{-k}$ with $q_k \neq 0$ and $k > 0$ and let $e(q) := exp(\ q_0 \log\ z + \frac{q_1}{-1} z^{-1} + \cdots + \frac{q_k}{-k} z^{-k})$. A *Stokes direction* $d \in \mathbf{S}^1$ *for* q is a direction such that $Re(\frac{q_k}{-k} z^{-k}) = 0$ for $z = |z|e^{id}$. A *Stokes pair* is a pair $\{d_1, d_2\}$ of Stokes directions such that $|d_2 - d_1| = \frac{\pi}{k}$, i.e., d_1, d_2 are consecutive Stokes directions. The Stokes pair $\{d_1, d_2\}$ is called *positive* if $Re(\frac{q_k}{-k} z^{-k}) > 0$ for z with $\arg(z) \in (d_1, d_2)$. The Stokes pair is called *negative* if $Re(\frac{q_k}{-k} z^{-k}) < 0$ for z with $\arg(z) \in (d_1, d_2)$. □

This terminology reflects the behavior of $|e(q)(z)|$ for small $|z|$. For $d \in (d_1, d_2)$, where $\{d_1, d_2\}$ is a positive Stokes pair, the function $r \mapsto |e(q)(re^{id})|$ explodes for $r \in \mathbf{R}_{>0},\ r \to 0$. If $\{d_1, d_2\}$ is a negative Stokes pair, then the function $r \mapsto |e(q)(re^{id})|$ tends rapidly to zero for $r \in \mathbf{R}_{>0},\ r \to 0$. The asymptotic behavior of $|e(q)(re^{id})|$ changes at the Stokes directions. The above notions can be extended to a q, which is a finite sum of terms $c_s z^{-s}$, with $s \in \mathbf{R}_{\geq 0}$ and $c_s \in \mathbf{C}$. However, in that case it is better to consider the directions d as elements of $\mathbf{R}$.

The sheaf $\ker(\delta - q, \mathcal{A}^0)$ is a sheaf of vector spaces over $\mathbf{C}$. For any interval (a, b), where $\{a, b\}$ is a negative Stokes pair, the restriction of $\ker(\delta - q, \mathcal{A}^0)$ to (a, b) is the constant sheaf with stalk $\mathbf{C}$. For a direction d that does not lie in such an interval the stalk of $\ker(\delta - q, \mathcal{A}^0)$ is zero. One can see $\ker(\delta - q, \mathcal{A}^0)$ as a subsheaf of $\ker(\delta - q, \mathcal{O})$, where $\mathcal{O}$ denotes the sheaf on $\mathbf{S}^1$ (of germs) of holomorphic functions. If $q_0 \in \mathbf{Z}$ then $\ker(\delta - q, \mathcal{O})$ is isomorphic to the constant sheaf $\mathbf{C}$ on $\mathbf{S}^1$. If $q_0 \notin \mathbf{Z}$, then the restriction of $\ker(\delta - q, \mathcal{O})$ to any proper open subset of $\mathbf{S}^1$ is isomorphic to the constant sheaf. Thus $\ker(\delta - q, \mathcal{A}^0)$ can always be identified with the subsheaf $\mathcal{F}$ of the constant sheaf $\mathbf{C}$ determined by its stalks $\mathcal{F}_d$: equal to $\mathbf{C}$ if d lies in an open interval (a, b) with $\{a, b\}$ a negative Stokes pair, and 0 otherwise.

More generally, consider a proper open subset $O \subset \mathbf{S}^1$ with complement F and let $i : F \to \mathbf{S}^1$ denote the inclusion. Let V be an abelian group (in our case this will always be a finite dimensional vector space over $\mathbf{C}$). Let V also denote the constant sheaf on $\mathbf{S}^1$ with stalk V. Then there is a natural surjective morphism of abelian sheaves $V \to i_*i^*V$. The stalk $(i_*i^*V)_d$ is zero for $d \in O$ and for $d \notin O$, the natural map $(V)_d \to (i_*i^*V)_d$ is a bijection. Write $V_F := i_*i^*V$ and define the sheaf V_O to be the kernel of $V \to V_F = i_*i^*V$. Then one can identify $\ker(\delta - q, \mathcal{A}^0)$ with $\mathbf{C}_O$, where O is the union of the k open intervals (a_i, b_i) such that $\{a_i, b_i\}$ are all the negative Stokes pairs. Clearly $\mathbf{C}_O$ is the direct sum of the sheaves $\mathbf{C}_{(a_i,b_i)}$. We are therefore interested in calculating $H^1(U, \mathbf{C}_I)$, with I an open interval and U either an open interval or $\mathbf{S}^1$. Consider the exact sequence of sheaves

$$0 \to V_I \to V \to V_F \to 0 \text{ on } \mathbf{S}^1.$$

For the sheaf V_F one knows that $H^i(U, V_F) = H^i(U \cap F, V)$ for all $i \geq 0$. Thus $H^0(U, V_F) \cong V^e$, where e is the number of connected components of $U \cap F$, and $H^i(U, V_F) = 0$ for all $i \geq 1$ (cf. the comments following Theorem C.27). Consider any open subset $U \subset \mathbf{S}^1$. The long exact sequence of cohomology reads

$$0 \to H^0(U, V_I) \to H^0(U, V) \to H^0(U, V_F) \to H^1(U, V_I) \to H^1(U, V) \to 0.$$

Lemma 7.19 *Let the notation be as above with $V = \mathbf{C}$. If $U = \mathbf{S}^1$ and for $U = (a, b)$ and the closure of I contained in U, then $H^1(U, \mathbf{C}_I) \cong \mathbf{C}$. In all other cases $H^1(U, \mathbf{C}_I) = 0$.*

Proof. Let $U = \mathbf{S}^1$. We have $H^0(\mathbf{S}^1, \mathbf{C}_I) = 0$, $H^0(\mathbf{S}^1, \mathbf{C}) \cong H^0(\mathbf{S}^1, \mathbf{C}_F) \cong \mathbf{C}$ (by the remarks preceeding the lemma) and $H^1(\mathbf{S}^1, \mathbf{C}) \cong \mathbf{C}$ (by Example C.22). Therefore the long exact sequence implies that $H^1(\mathbf{S}^1, \mathbf{C}_I) \cong \mathbf{C}$.

Let $U = (a, b)$ and assume that the closure of I is contained in U. We then have that $U \cap F$ has two components so $H^0(U, \mathbf{C}_F) = H^0(U \cap F, \mathbf{C}) \cong \mathbf{C} \oplus \mathbf{C}$. Furthermore, $H^0(U, \mathbf{C}_I) \cong 0$ and $H^0(U, \mathbf{C}) \cong \mathbf{C}$. Therefore $H^1(U, C_I) \cong \mathbf{C}$.

The remaining cases are treated similarly. □

The following lemma easily follows from the preceeding lemma.

Lemma 7.20 *Let $U \subset \mathbf{S}^1$ be either an open interval (a, b) or $\mathbf{S}^1$. Then $H^1(U, \ker(\delta - q, \mathcal{A}^0)) = 0$ if and only if U does not contain a negative Stokes pair. More generally, the dimension of $H^1(U, \ker(\delta - q, \mathcal{A}^0))$ is equal to the number of negative Stokes pairs contained in U. In particular, the dimension of $H^1(\mathbf{S}^1, \ker(\delta - q, \mathcal{A}^0))$ is k.*

This lemma can be easily generalized to characterize $H^1(U, \ker(\delta - B, \mathcal{A}^0))$ where $\delta - B$ is a quasisplit equation. We shall only need a weak form of this, which we state below. We refer to Definition 3.28 for the definition of the eigenvalue of a differential equation.

Corollary 7.21 *Let $U \subset \mathbf{S}^1$ be an open interval (a, b) and $\delta - B$ a quasisplit differential operator. Then $H^1(U, \ker(\delta - B, \mathcal{A}^0)) = 0$ if and only if U does not contain a negative Stokes pair of some eigenvalue of $\delta - B$.*

Proof. We may assume that the operator is split and it is the sum of operators of the form $\delta - q + C$, where C is a constant matrix. Therefore it is sufficient to prove this result when the operator is of this form. Let T be a fundamental matrix for the equation $\delta y = Cy$. The map $y \mapsto Ty$ gives an isomorphism of sheaves $\ker(\delta - q, \mathcal{A}^0)$ and $\ker(\delta - q + C, \mathcal{A}^0)$. The result now follows from Lemma 7.20. □

The map $\delta - q$ is bijective on $\mathbf{C}((z))$. This follows easily from $(\delta - q)z^n = -q_k z^{n-k} + \cdots$ for every integer n. Thus the obstruction for lifting the unique formal solution $\hat{f}$ of $(\delta - q)\hat{f} = g$ depends only on $g \in \mathbf{C}(\{z\})$. This produces the $\mathbf{C}$-linear map $\beta : \mathbf{C}(\{z\}) \to H^1(\mathbf{S}^1, \ker(\delta - q, \mathcal{A}^0))$, which associates to every $g \in \mathbf{C}(\{z\})$ the obstruction $\beta(g)$, for lifting $\hat{f}$ to an element of $\mathcal{A}(\mathbf{S}^1)$. From $\mathcal{A}(\mathbf{S}^1) = \mathbf{C}(\{z\})$ it follows that the kernel of β is the image of $\delta - q$ on $\mathbf{C}(\{z\})$.

Corollary 7.22 *After a transformation $(\delta - \tilde{q}) = z^{-n}(\delta - q)z^n$, we may assume that $0 \leq Re(q_0) < 1$. The elements $\{\beta(z^i) | \ i = 0, \ldots, k-1\}$ form a basis of $H^1(\mathbf{S}^1, \ker(\delta - q, \mathcal{A}^0))$. In particular, β is surjective and one has an exact sequence*

$$0 \to \mathbf{C}(\{z\}) \stackrel{\delta - q}{\to} \mathbf{C}(\{z\}) \stackrel{\beta}{\to} H^1(\mathbf{S}^1, \ker(\delta - q, \mathcal{A}^0)) \to 0.$$

Proof. According to Lemma 7.20 it suffices to show that the elements are independent. In other words, we have to show that the existence of a $y \in \mathbf{C}(\{z\})$ with $(\delta - q)y = a_0 + a_1 z + \cdots + a_{k-1}z^{k-1}$ implies that all $a_i = 0$. The equation has only two singular points, namely 0 and ∞. Thus y has an analytic continuation to all of $\mathbf{C}$ with at most a pole at 0. The singularity at ∞ is regular singular. Thus y has bounded growth at ∞, i.e., $|y(z)| \leq C|z|^N$ for $|z| >> 0$ and with certain constants C, N and so y is, in fact, a rational function with at most poles at 0 and ∞. Then $y \in \mathbf{C}[z, z^{-1}]$. Suppose that $y \neq 0$, then one can write $y = \sum_{i=n_0}^{n_1} y_i z^i$ with $n_0 \leq n_1$ and $y_{n_0} \neq 0 \neq y_{n_1}$. The expression $(\delta - q)y \in \mathbf{C}[z, z^{-1}]$ is seen to be $-q_k y_{n_0} z^{n_0 - k} + (n_1 - q_0)y_{n_1} z^{n_1} + \sum_{n_0 - k < i < n_1} * z^i$. This cannot be a polynomial in z of degree $\leq k - 1$. This proves the first part of the corollary. The rest is an easy consequence. □

We would like to show that the solution $\hat{f}$ of $(\delta - q)\hat{f} = g$ is k-summable. The next lemma gives an elementary proof of $\hat{f} \in \mathbf{C}((z))_{1/k}$.

Lemma 7.23 *The formal solution $\hat{f}$ of $(\delta - q)\hat{f} = g$ lies in $\mathbf{C}((z))_{1/k}$. More generally, $\delta - q$ is bijective on $\mathbf{C}((z))_{1/k}$.*

Proof. We give here an elementary proof depending on simple estimates. Write $\hat{f} = \sum c_n z^n$ and $g = \sum_n g_n z^n$. For the coefficients of $\hat{f}$ one finds a recurrence relation

$$c_{n+k} = -\frac{q_{k-1}}{q_k} c_{n+k-1} - \cdots - \frac{q_1}{q_k} c_{n+1} - \frac{q_0 - n}{q_k} c_n - \frac{1}{q_k} g_n.$$

There exists a constant $B > 0$ with $|g_n| \le B^n$ for $n > 0$. We must find an estimate of the form $|c_n| \le A^n \Gamma(1 + \frac{n}{k})$ for all $n > 0$ and some $A > 0$. We try to prove by induction that $\frac{|c_n|}{A^n \Gamma(1+\frac{n}{k})} \le 1$, for a suitable $A > 0$ and all $n > 0$. The induction step should follow from the bound for $\frac{|c_{n+k}|}{A^{n+k}\Gamma(1+\frac{n+k}{k})}$, given by the recurrence relation. This bound is the expression

$$\frac{*\Gamma(1+\frac{n+k-1}{k})}{A\Gamma(1+\frac{n+k}{k})} + \cdots + \frac{*\Gamma(1+\frac{1+n}{k})}{A^{k-1}\Gamma(1+\frac{n+k}{k})}$$
$$+\frac{(*+n)\Gamma(1+\frac{n}{k})}{A^k\Gamma(1+\frac{n+k}{k})} + \frac{*B^n}{A^{n+k}\Gamma(1+\frac{n+k}{k})},$$

where the $*$s denote fixed constants. From $\Gamma(1+\frac{n+k}{k}) = \frac{n+k}{k}\Gamma(1+\frac{n}{k})$ one easily deduces that a positive A can be found such that this expression is ≤ 1 for all $n > 0$. The surjectivity of $\delta - q$ follows by replacing the estimate B^n for $|g_n|$ by $B^n\Gamma(1+\frac{n}{k})$. The injectivity follows from the fact that $\delta - q$ is bijective on $\mathbf{C}((z))_{1/k}$ (see the discussion following Corollary 7.21). □

For a direction d such that $\{d - \frac{\pi}{2k}, d + \frac{\pi}{2k}\}$ is *not* a negative Stokes pair, Lemma 7.20 produces an asymptotic lift in $\mathcal{A}(d - \frac{\alpha}{2}, d + \frac{\alpha}{2})$, for some $\alpha > \frac{\pi}{k}$, of the formal solution $\hat{f}$ of $(\delta - q)\hat{f} = g$. This lift is easily seen to be unique. If we can show that this lift is, in fact, a section of the subsheaf $\mathcal{A}_{1/k}$, then the proof that $\hat{f}$ is k-summable would be complete. In the next section we will develop the necessary theory for the sheaf $\mathcal{A}_{1/k}$.

7.4 The Sheaves $\mathcal{A}, \mathcal{A}^0, \mathcal{A}_{1/k}, \mathcal{A}^0_{1/k}$

We start by examining the sheaves $\mathcal{A}$ and $\mathcal{A}^0$.

Proposition 7.24 *Consider the exact sequence of sheaves on* $\mathbf{S}^1$*:*

$$0 \to \mathcal{A}^0 \to \mathcal{A} \to \mathbf{C}((z)) \to 0,$$

where $\mathbf{C}((z))$ *denotes the constant sheaf on* $\mathbf{S}^1$ *with stalk* $\mathbf{C}((z))$*.*

1. *For every open* $U \neq \mathbf{S}^1$ *the cohomology group* $H^1(U, \,.\,)$ *is zero for the sheaves* $\mathcal{A}^0$, $\mathcal{A}$ *and* $\mathbf{C}((z))$*.*
2. *The natural map* $H^1(\mathbf{S}^1, \mathcal{A}^0) \to H^1(\mathbf{S}^1, \mathcal{A})$ *is the zero map. As a consequence, one has*

$$H^1(\mathbf{S}^1, \mathcal{A}) \stackrel{\sim}{\to} H^1(\mathbf{S}^1, \mathbf{C}((z))) \stackrel{\sim}{\to} \mathbf{C}((z)),$$

and there is an exact sequence

$$0 \to \mathbf{C}(\{z\}) \to \mathbf{C}((z)) \to H^1(\mathbf{S}^1, \mathcal{A}^0) \to 0.$$

Proof. We note that the circle has topological dimension one and for any abelian sheaf F and any open U one has $H^i(U, F) = 0$ for $i \geq 2$ (see Theorem C.28). We want to show that for any open $U \subset \mathbf{S}^1$ (including the case $U = \mathbf{S}^1$), the map $H^1(U, \mathcal{A}^0) \to H^1(U, \mathcal{A})$ is the zero map. Assume that this is true and consider the long exact sequence of cohomology:

$$0 \to H^0(U, \mathcal{A}^0) \to H^0(U, \mathcal{A}) \to H^0(U, \mathbf{C}((z))) \to H^1(U, \mathcal{A}^0)$$
$$\to H^1(U, \mathcal{A}) \to H^1(U, \mathbf{C}((z))) \to 0.$$

If $U \neq \mathbf{S}^1$, then the Borel-Ritt Theorem implies that the map $H^0(U, \mathcal{A}) \to H^0(U, \mathbf{C}((z)))$ is surjective so the map $H^0(U, \mathbf{C}((z))) \to H^1(U, \mathcal{A}^0)$ is the zero map. Combining this with the fact that $H^1(U, \mathcal{A}^0) \to H^1(U, \mathcal{A})$ is the zero map, we have $H^1(U, \mathcal{A}^0) \cong 0$ and $H^1(U, \mathcal{A}) \cong H^1(U, \mathbf{C}((z))) \cong 0$. Since each component of U is contractible (and so simply connected), Theorem C.27 implies that $H^1(U, \mathbf{C}((z))) \cong 0$ and part 1 follows. If $U = \mathbf{S}^1$ then $H^0(U, \mathcal{A}) \cong \mathbf{C}((z))$ and $H^0(U, \mathbf{C}((z))) \cong H^1(U, \mathbf{C}((z))) \cong \mathbf{C}((z))$ (cf. Example C.22). Since $H^1(U, \mathcal{A}^0) \to H^1(U, \mathcal{A})$ is the zero map, part 2 follows from the long exact sequence as well.

We start by considering the simplest covering: $U = (a_1, b_1) \cup (a_2, b_2)$ with $(a_1, b_1) \cap (a_2, b_2) = (a_2, b_1)$, i.e., inequalities $a_1 < a_2 < b_1 < b_2$ for the directions on $\mathbf{S}^1$ and $U \neq \mathbf{S}^1$. A 1-cocycle for $\mathcal{A}^0$ and this covering is given by a single element $f \in \mathcal{A}^0(a_2, b_1)$. Take a small positive ϵ such that $(a_1, b_1 - \epsilon) \cup (a_2 + \epsilon, b_2) = U$ and consider the integral $\frac{1}{2\pi i} \int_\gamma \frac{f(\zeta)}{\zeta - z} d\zeta$, where the path γ consists of three pieces γ_i for $i = 1, 2, 3$. The path γ_1 is the line segment from 0 to $re^{i(a_2+\epsilon/2)}$, γ_2 is the circle segment from $re^{i(a_2+\epsilon/2)}$ to $re^{i(b_1-\epsilon/2)}$ and γ_3 is the line segment from $re^{i(b_1-\epsilon/2)}$ to 0. The $r > 0$ is adapted to the size of the sector where f lives. We conclude that for z with $|z| < r$ and $\arg(z) \in (a_2 + \epsilon, b_1 - \epsilon)$ this integral is equal to $f(z)$. The path is divided into two pieces γ_+, which are γ_1 and the first half of γ_2 and the remaining part γ_-. The integral over the two pieces will be called $f_+(z)$ and $-f_-(z)$. We will show that $f_+ \in \mathcal{A}(a_2 + \epsilon, b_2)$ and $f_- \in \mathcal{A}(a_1, b_1 - \epsilon)$. From this it follows that our 1-cocycle for $\mathcal{A}^0$ has image 0 in $H^1(U, \mathcal{A})$.

By symmetry, it suffices to prove the statement for f_+. This function lives, in fact, on the open sector $V := \mathbf{S}^1 \setminus \{a_2 + \epsilon/2\}$ (and say $|z| < r$). The function $\frac{f(\zeta)}{\zeta - z}$ can be developed as a power series in z, namely $\sum_{n \geq 0} f(\zeta)\zeta^{-1-n} z^n$. We consider the formal power series $\hat{F} = \sum_{n \geq 0} (\frac{1}{2\pi i} \int_{\gamma_+} f(\zeta)\zeta^{-1-n} d\zeta)\, z^n$ and want to prove that f_+ has asymptotic expansion $\hat{F}$ on the open sector V. From $\frac{1}{1-z/\zeta} = \frac{1-(z/\zeta)^N}{1-z/\zeta} + \frac{(z/\zeta)^N}{1-z/\zeta}$, one concludes that the difference of f_+ and $\sum_{0 \leq n < N} \frac{1}{2\pi i} \int_{\gamma_+} f(\zeta)\zeta^{-1-n} d\zeta \;\, z^n$ is the integral $\frac{1}{2\pi i} \int_{\gamma_+} \frac{(z/\zeta)^N f(\zeta)}{\zeta(1-z/\zeta)} d\zeta$. For any closed subsector W of V one has $\inf_{z \in W} |1 - z/\zeta|$ is strictly positive. By assumption, there are constants C_{N+1} such that $|f(\zeta)| \leq C_{N+1}|\zeta|^{N+1}$ for all $N > 0$. One concludes that the last integral is bounded by $D_N |z|^N$ for some constant D_N.

The next case that we consider is a covering $(a_1, b_1), (a_2, b_2)$ of $\mathbf{S}^1$. The intersection $(a_1, b_1) \cap (a_2, b_2)$ is supposed to have two components (a_2, b_1) and (a_1, b_2). Let the 1-cocycle be given by $f \in \mathcal{A}^0(a_2, b_1)$ and $0 \in \mathcal{A}^0(a_1, b_2)$. Define $f_+ \in \mathcal{A}(a_2 + \epsilon, b_2)$ and $f_- \in \mathcal{A}(a_1, b_1 - \epsilon)$ as in the first case. Then $f_+ - f_-$ coincides with f on $(a_2 + \epsilon, b_1 - \epsilon)$ and is zero on (a_1, b_2).

The following case is a "finite special covering" of U, which is either an open interval or $\mathbf{S}^1$. We will define this by giving a sequence of directions $d_1 < d_2 < \cdots < d_n$ in U and intervals $(d_i - \epsilon, d_{i+1} + \epsilon)$ with small $\epsilon > 0$. In the case $U = \mathbf{S}^1$, the interval $(d_n - \epsilon, d_1 + \epsilon)$ is also present. A 1-cocycle ξ is given by a sequence of functions $f_i \in \mathcal{A}^0(d_i - \epsilon, d_i + \epsilon)$. One writes ξ as a sum of 1-cocycles ζ, which have only one nonzero f_i. It suffices to show that such a ζ is a trivial 1-cocycle for the sheaf $\mathcal{A}$. This follows from the first two cases, since one can see ζ also as a 1-cocycle for a covering of U by two open intervals.

Every covering of $\mathbf{S}^1$ and every finite covering of an open interval U can be refined to a finite special covering. We are left with studying infinite coverings of an open interval $U =: (a, b)$. Any infinite covering can be refined to what we will call a "special infinite covering" of U. The latter is defined by a sequence $d_n,\ n \in \mathbf{Z}$ of points in U, such that $d_i < d_{i+1}$ for all i. Moreover, $\cup[d_i, d_{i+1}] = U$. The covering of U by the closed intervals is replaced by a covering with open intervals (d_i^-, d_{i+1}^+), where $d_i^- < d_i < d_i^+$ and $|d_i^+ - d_i^-|$ is very small. A cocycle ξ is again given by elements $f_i \in \mathcal{A}^0(d_i^-, d_i^+)$. Using the argument above, one can write $f_i = g_i - h_i$ with g_i and h_i sections of the sheaf $\mathcal{A}$ above, say, the intervals $(a, (d_i + d_i^+)/2)$ and $((d_i^- + d_i)/2, b)$. Define, first formally, $F_i := \sum_{j \geq i} g_j - \sum_{j \leq i} h_j$ as a function on the interval $((d_i^- + d_i)/2, (d_{i+1}^+ + d_{i+1})/2)$. Then clearly $F_i - F_{i-1} = g_i - h_i = f_i$ on $((d_i^- + d_i)/2, (d_i^+ + d_i)/2)$. There is still one thing to prove, namely that the infinite sums appearing in F_i converge to a section of $\mathcal{A}$ on the given interval. This can be done using estimates on the integrals defining the g_i and h_i given above. We will skip the proof of this statement. □

Remark 7.25 1. The calculation of the cohomology of $\ker(\delta - q, \mathcal{A}^0)$ and $\ker(\delta - A, \mathcal{A}^0)$ was initiated by Malgrange [188] and Deligne and further developed by Loday-Richaud, Malgrange, Ramis and Sibuya (cf. [13, 180, 195]).

2. The first statement of Proposition 7.24.2 is sometimes referred to as the Cauchy-Heine Theorem (cf. [195], Theorem 1.3.2.1.i and ii). However, the name "Heine's Theorem" seems more appropriate. □

Lemma 7.26 The Borel-Ritt Theorem for $\mathbf{C}((z))_{1/k}$ *Suppose that* $k > 1/2$. *Then the map* $J : \mathcal{A}_{1/k}(a, b) \to \mathbf{C}((z))_{1/k}$ *is surjective if* $|b - a| \leq \frac{\pi}{k}$.

Proof. After replacing z by $e^{id} z^{1/k}$ for a suitable d we have to prove that the map $J : \mathcal{A}_1(-\pi, \pi) \to \mathbf{C}((z))_1$ is surjective. It suffices to show that an element $\hat{f} = \sum_{n \geq 1} c_n n! z^n$ with $|c_n| \leq (2r)^{-n}$ for some positive r is in the image of J. One could refine Proposition 7.3 to prove this. A more systematic procedure is the following. For any half-line γ, of the form $\{se^{id}|s \geq 0\}$ and $|d| < \pi$ one has $n! = \int_\gamma \zeta^n exp(-\zeta)d\zeta$. Thus for $z \neq 0$ and $\arg(z) \in (-\pi, \pi)$ one has $n!z^n = \int_0^\infty \zeta^n exp(-\frac{\zeta}{z})d(\frac{\zeta}{z})$, where the path of integration is the positive real line. This integral is written as a sum of two parts $F(n,r)(z) = \int_0^r \zeta^n exp(-\frac{\zeta}{z})d(\frac{\zeta}{z})$ and $R(n,r)(z) = \int_r^\infty \zeta^n exp(-\frac{\zeta}{z})d(\frac{\zeta}{z})$. The claim is that $F(z) := \sum_{n\geq 1} c_n F(n,r)(z)$ converges locally uniformly on $\{z \in \mathbf{C}|\ z \neq 0\}$, belongs to $\mathcal{A}_1(-\pi,\pi)$ and satisfies $J(F) = \hat{f}$.

The integral $\int_0^r (\sum_{n\geq 1} c_n\zeta^n)exp(-\frac{\zeta}{z})d(\frac{\zeta}{z})$, taken over the closed interval $[0, r] \subset \mathbf{R}$, exists for all $z \neq 0$ since $\sum_{n\geq 1} c_n\zeta^n$ has radius of convergence $2r$. Interchanging $\sum$ and $\int$ proves the first statement on F. To prove the other two statements we have to give for every closed subsector of $\{z \in \mathbf{C}|\ 0 < |z| \text{ and } \arg(z) \in (-\pi,\pi)\}$ an estimate of the form $E := |F(z) - \sum_{n=1}^{N-1} c_n n! z^n| \leq A^N N! |z|^N$ for some positive A, all $N \geq 1$ and all z in the closed sector.

Now $E \leq \sum_{n=1}^{N-1} |c_n||R(n,r)(z)| + |\int_0^r (\sum_{n\geq N} c_n\zeta^n)exp(-\frac{\zeta}{z})d(\frac{\zeta}{z}|$. The last term of this expression can be estimated by $r^{-N}\int_0^r \zeta^N |exp(-\frac{\zeta}{z})|\frac{d\zeta}{|z|}$, because one has the inequality $|\sum_{n\geq N} c_n\zeta^n| \leq r^{-N}\zeta^N$ for $\zeta \leq r$. Thus the last term can be estimated by $r^{-N}\int_0^\infty \zeta^N|exp(-\frac{\zeta}{z})|\frac{d\zeta}{|z|}$. The next estimate is $|R(n,r)(z)| \leq \int_r^\infty \zeta^n |exp(-\frac{\zeta}{z})|d\frac{\zeta}{|z|}$. Furthermore, $\zeta^n \leq r^{n-N}\zeta^N$ for $r \leq \zeta$. Thus $|R(n,r)(z)| \leq r^{n-N}\int_0^\infty \zeta^N |exp(-\frac{\zeta}{z})|\frac{d\zeta}{|z|}$. Now $r^{-N} + \sum_{n=1}^{N-1}|c_n|r^{n-N} \leq 2r^{-N}$ and we can estimate E by $2r^{-N}\int_0^\infty \zeta^N|exp(-\frac{\zeta}{z})|\frac{d\zeta}{|z|}$. For $z = |z|e^{i\theta}$ one has $|exp(-\frac{\zeta}{z})| = exp(-\frac{\zeta}{|z|}\cos\theta)$. The integral is easily computed to be $\frac{|z|^N}{(\cos\theta)^{N+1}}N!$. This gives the required estimate for E. □

For $k > 1/2$, the function $exp(-z^{-k})$ belongs to $\mathcal{A}^0_{1/k}(-\frac{\pi}{2k},\frac{\pi}{2k})$. The next lemma states that this is an extremal situation. For sectors with larger "opening" the sheaf $\mathcal{A}^0_{1/k}$ has only the zero section. This important fact, Watson's Lemma, provides the uniqueness for k-summation in a given direction.

Lemma 7.27 Watson's Lemma.
$\mathcal{A}^0_{1/k}(a,b) = 0$ *if* $|b-a| > \frac{\pi}{k}$.

Proof. After replacing z by $z^{1/k}e^{id}$ for a suitable d the statement reduces to $\mathcal{A}^0_1(-\alpha,\alpha) = 0$ for $\alpha > \frac{\pi}{2}$. We will prove the following slightly stronger statement (cf. Lemma 7.5):

Let S denote the open sector given by the inequalities $|\arg(z)| < \frac{\pi}{2}$ and $0 < |z| < r$. Suppose that f is holomorphic on S and that there are positive constants A, B such that $|f(z)| \leq A\ exp(-B|z|^{-1})$ holds for all $z \in S$. Then $f = 0$.

We start by choosing $M > B$ and $\epsilon > 0$ and defining β by $0 < \beta < \frac{\pi}{2}$ such that $\cos\beta = \frac{B}{M}$ and $\delta > 0$ by $(1+\delta)\beta < \frac{\pi}{2}$ and $\cos((1+\delta)\beta) = \frac{B}{2M}$. Define the function $F(z)$, depending on M and ϵ, by $F(z) := f(z)\ exp(-\epsilon z^{-1-\delta} + Mz^{-1})$. Let $\tilde{S}$ denote the closed sector given by the inequalities $|\arg(z)| \leq \beta$ and $0 < |z| \leq r/2$.

The limit of $F(z)$ for $z \to 0$ and $z \in \tilde{S}$ is 0 and thus $F(z)$ is bounded on $\tilde{S}$. According to the maximum principle, the maximum of $|F(z)|$ is assumed at the boundary of $\tilde{S}$. For $0 < |z| \leq r/2$ and $\arg(z) = \beta$ one can bound $|F(z)|$ by

$$\leq A\, exp(-B|z|^{-1})\, exp(-\epsilon|z|^{-1-\delta}\cos((1+\delta)\beta) + M|z|^{-1}\cos(\beta)) \leq A.$$

For the boundary $0 < |z| \leq r/2$ and $\arg(z) = -\beta$ one finds the same estimate. For z with $|\arg(z)| \leq \beta$ and $|z| = r/2$, one finds the estimate $|F(z)| \leq A\, exp((M - B)(r/2)^{-1})$. We conclude that for any $z \in \tilde{S}$ the inequality $|F(z)| \leq A\, exp((M - B)(r/2)^{-1})$ holds. Thus we find for $z \in \tilde{S}$ the inequality

$$|f(z)| \leq A\, exp((M - B)(r/2)^{-1})\ |exp(-Mz^{-1})|\ |exp(+\epsilon z^{-1-\delta})|.$$

Since $\epsilon > 0$ is arbitrary, we conclude also that

$$|f(z)| \leq A\, exp(-B(r/2)^{-1})\ |exp(M((r/2)^{-1} - z^{-1}))|$$

holds for all $z \in \tilde{S}$. For a fixed z with $|\arg(z)| < \frac{\pi}{2}$ and small enough $|z| > 0$ such that $Re((r/2)^{-1} - z^{-1}) < 0$, this inequality holds for all sufficiently large M. Since $|exp(M((r/2)^{-1} - z^{-1})|$ tends to 0 for $M \to \infty$, we conclude that $f(z) = 0$. □

Proposition 7.28

1. *The following sequence of sheaves on* $\mathbf{S}^1$ *is exact.*

$$0 \to \mathcal{A}^0_{1/k} \to \mathcal{A}_{1/k} \to \mathbf{C}((z))_{1/k} \to 0.$$

2. *For every open* $U \subset \mathbf{S}^1$, *including* $U = \mathbf{S}^1$, *the canonical map* $H^1(U, \mathcal{A}^0_{1/k}) \to H^1(U, \mathcal{A}_{1/k})$ *is the zero map.*
3. $H^1(U, \mathcal{A}_{1/k})$ *is zero for* $U \neq \mathbf{S}^1$ *and equal to* $\mathbf{C}((z))_{1/k}$ *for* $U = \mathbf{S}^1$.
4. $H^1((a, b), \mathcal{A}^0_{1/k}) = 0$ *for* $|b - a| \leq \frac{\pi}{k}$.
5. *For* (a, b) *with* $|b - a| > \frac{\pi}{k}$, *the following sequence is exact.*

$$0 \to \mathcal{A}_{1/k}(a, b) \to \mathbf{C}((z))_{1/k} \to H^1((a, b), \mathcal{A}^0_{1/k}) \to 0.$$

6. *The following sequence is exact.*

$$0 \to \mathbf{C}(\{z\}) \to \mathbf{C}((z))_{1/k} \to H^1(\mathbf{S}^1, \mathcal{A}^0_{1/k}) \to 0.$$

7. *There is a canonical isomorphism* $\mathbf{C}((z))_{1/k} \to H^0(\mathbf{S}^1, \mathcal{A}/\mathcal{A}^0_{1/k})$.

Proof. Part 1 follows from Lemma 7.26. The proof of part 2 of Proposition 7.24 extends to a proof of part 2 of the present proposition. One only has to verify that the functions f_+ and f_- are now sections of the sheaf $\mathcal{A}_{1/k}$. Furthermore, parts 3 to 6 are immediate consequences of parts 1 and 2, the known cohomology of the constant sheaf $\mathbf{C}((z))_{1/k}$, Lemma 7.27 and the long exact sequence of cohomology. We identify the constant sheaf $\mathbf{C}((z))_{1/k}$ with $\mathcal{A}_{1/k}/\mathcal{A}^0_{1/k}$. Thus there is an exact sequence of sheaves

$$0 \to \mathbf{C}((z))_{1/k} \to \mathcal{A}/\mathcal{A}^0_{1/k} \to \mathcal{A}/\mathcal{A}_{1/k} \to 0.$$

Taking sections above $\mathbf{S}^1$ we find an exact sequence

$$0 \to \mathbf{C}((z))_{1/k} \to H^0(\mathbf{S}^1, \mathcal{A}/\mathcal{A}^0_{1/k}) \to H^0(\mathbf{S}^1, \mathcal{A}/\mathcal{A}_{1/k}). \tag{7.1}$$

The exact sequence

$$0 \to \mathcal{A}_{1/k} \to \mathcal{A} \to \mathcal{A}/\mathcal{A}_{1/k} \to 0,$$

induces the long exact sequence of cohomology above $\mathbf{S}^1$:

$$0 \to \mathbf{C}(\{z\}) \to \mathbf{C}(\{z\}) \to H^0(\mathbf{S}^1, \mathcal{A}/\mathcal{A}_{1/k}) \to \mathbf{C}((z))_{1/k} \to \mathbf{C}((z)) \cdots .$$

This implies $H^0(\mathbf{S}^1, \mathcal{A}/\mathcal{A}_{1/k}) = 0$ and so, from the sequence (7.1), we conclude part 7. □

Remark 7.29 Proposition 7.28.2 is the Ramis-Sibuya Theorem (see [195], Theorem 2.1.4.2 and Corollaries 2.1.4.3 and 2.1.4.4).

7.5 The Equation $(\delta - q)\hat{f} = g$ Revisited

Some of the results of Sect. 7.3 can be established using the methods of Sect. 7.4.

Exercise 7.30 Give an alternative proof of the surjectivity of $\beta : \mathbf{C}(\{z\}) \to H^1(\mathbf{S}^1, \ker(\delta - q, \mathcal{A}^0))$ (see Corollary 7.22) by using Proposition 7.24. Hint: An element $\xi \in H^1(\mathbf{S}^1, \ker(\delta - q, \mathcal{A}^0))$ induces an element of $H^1(\mathbf{S}^1, \mathcal{A}^0)$. By Proposition 7.24.2, this element is zero in $H^1(\mathbf{S}^1, \mathcal{A})$ so, for some covering $\{S_i\}$ of $\mathbf{S}^1$, there exist $f_i \in H^0(S_i, \mathcal{A})$ such that $f_i - f_j = \xi_{i,j}$, where $\xi_{i,j}$ is a representative of ξ on $S_i \cap S_j$. Show that the $(\delta - q)f_i$ glue together to give an element $g \in H^0(\mathbf{S}^1, \mathcal{A}) = \mathbf{C}(\{x\})$ and that the f_i are lifts of some $\hat{f} \in \mathbf{C}((x))$ such that $(\delta - q)\hat{f} = g$. □

Exercise 7.31 Give an alternative proof of the fact that $(\delta - q)\hat{f} = g \in \mathbf{C}(\{z\})$ implies $\hat{f} \in \mathbf{C}((z))_{1/k}$ (see Lemma 7.23) by using the last statement of Proposition 7.28. Hint: g maps to an element $\beta(g) \in H^1(\mathbf{S}^1, \ker(\delta - q, \mathcal{A}^0))$. Observe that $\ker((\delta - q), \mathcal{A}^0) = \ker(\delta - q, \mathcal{A}^0_{1/k})$. Thus $\hat{f}$ can be seen as an element of $H^0(\mathbf{S}^1, \mathcal{A}/\mathcal{A}^0_{1/k})$. □

Proposition 7.32 *The element $\hat{f} \in \mathbf{C}((z))$ satisfying $(\delta - q)\hat{f} = g \in \mathbf{C}(\{z|\})$ is k-summable. More precisely, $\hat{f}$ is k-summable in the direction d if $\{d - \frac{\pi}{2k}, d + \frac{\pi}{2k}\}$ is not a negative Stokes pair.*

Proof. We know by Lemma 7.23, or by Exercise 7.31, that $\hat{f} \in \mathbf{C}((z))_{1/k}$. Take a direction d. By Proposition 7.28 there is an $h \in (\mathcal{A}_{1/k})_d$ with $J(h) = \hat{f}$. Clearly $(\delta - q)h - g = g_0 \in (\mathcal{A}^0_{1/k})_d$. By Theorem 7.12 there is an $h_0 \in (\mathcal{A}^0_{1/k})_d$ with $(\delta - q)h_0 = g_0$ and thus $(\delta - q)(h - h_0) = g$. In other words, the formal solution $\hat{f}$ lifts for small enough sectors S to a solution in $\mathcal{A}_{1/k}(S)$ of the same equation. This yields a 1-cocycle in the sheaf $\ker(\delta - q, \mathcal{A}^0_{1/k}) = \ker(\delta - q, \mathcal{A}^0)$. This 1-cocycle is trivial for an open interval $(d - \frac{\pi}{2k} - \epsilon, d + \frac{\pi}{2k} + \epsilon)$ (for some positive ϵ) when $\{d - \frac{\pi}{2k}, d + \frac{\pi}{2k}\}$ is not a negative Stokes pair (see Lemma 7.20). □

Definition 7.33 Consider $q = q_k z^{-k} + q_{k-1} z^{-k+1} + \cdots + q_1 z^{-1} \in z^{-1}\mathbf{C}[z^{-1}]$ with $q_k \neq 0$. A direction d will be called *singular* for q (or for the operator $\delta - q$) if $q_k e^{-idk}$ is a positive real number. □

One immediately sees that d is a singular direction for $\delta - q$ if and only if $\{d - \frac{\pi}{2k}, d + \frac{\pi}{2k}\}$ is a negative Stokes pair. Thus one can reformulate Proposition 7.32 by saying that $\hat{f}$ is k-summable in the direction d if d is not a singular direction.

7.6 The Laplace and Borel Transforms

The *formal Borel transformation* $\hat{\mathcal{B}}_k$ *of order* k is the operator $\mathbf{C}[[z]] \to \mathbf{C}[[\zeta]]$ defined by the formula

$$\hat{\mathcal{B}}_k\Big(\sum_{n\geq 0} c_n z^n\Big) = \sum_{n\geq 0} \frac{c_n}{\Gamma(1+\frac{n}{k})}\zeta^n.$$

The *Laplace transform* $\mathcal{L}_{k,d}$ *of order* k *in the direction* d is defined by the formula

$$(\mathcal{L}_{k,d} f)(z) = \int_d f(\zeta) exp(-(\frac{\zeta}{z})^k)\, d(\frac{\zeta}{z})^k.$$

The path of integration is the half-line through 0 with direction d. The function f is supposed to be defined and continuous on this half-line and have a suitable behavior at 0 and ∞ in order to make this integral convergent for z in some sector at 0, that is, $|f(\zeta)| \leq Ae^{B|z|^k}$ for positive constants A, B. We note that we have slightly deviated from the usual formulas for the formal Borel transformation and the Laplace transformation (although these agree with the definitions in [15]).

A straightforward calculation shows that the operator $\mathcal{L}_{k,d} \circ \hat{\mathcal{B}}_k$ has the property $\mathcal{L}_{k,d} \circ \hat{\mathcal{B}}_k(z^n) = z^n$ for any $n \geq 0$ and more generally $\mathcal{L}_{k,d} \circ \hat{\mathcal{B}}_k f = f$ for any $f \in \mathbf{C}\{z\}$. Suppose now that $\hat{f} \in \mathbf{C}[[z]]_{1/k}$. Then $(\hat{\mathcal{B}}_k \hat{f})(\zeta)$ is by definition a convergent power series at $\zeta = 0$. One can try to apply $\mathcal{L}_{k,d}$ to this function in order to obtain an asymptotic lift of $\hat{f}$ to some sector. The following theorem makes this precise. We define a function, analytic in a sector $\{\zeta \in \mathbf{C} |\ 0 < |\zeta| < \infty$ and $|\arg(\zeta) - d| < \epsilon\}$, to have *exponential growth of order* $\leq k$ *at* ∞ if there are constants A, B such that $|h(\zeta)| \leq A\ exp(\ B|\zeta|^k)$ holds for large $|\zeta|$ and $|\arg(\zeta) - d| < \epsilon$.

Theorem 7.34 *Let $\hat{f} \in \mathbf{C}[[z]]_{1/k}$ and let d be a direction. Then the following are equivalent:*

1. *$\hat{f}$ is k-summable in the direction d.*
2. *The convergent power series $\hat{\mathcal{B}}_k \hat{f}$ has an analytic continuation h in a full sector $\{\zeta \in \mathbf{C} |\ 0 < |\zeta| < \infty$ and $|\arg(\zeta) - d| < \epsilon\}$. In addition, this analytic continuation has exponential growth of order $\leq k$ at ∞ on this sector.*

Proof. We give here a sketch of the proof and refer to ([15], Chap. 3.1) for the missing details concerning the estimates that we will need. We may assume $k = 1$ and $d = 0$. Write $\hat{f} = \sum_{n \geq 0} c_n z^n$. We will start by proving that part 2 implies part 1. Let d be a direction with $|d| < \epsilon$. The integral

$$f(z) := (\mathcal{L}_{1,d} h)(z) = \int_d h(\zeta) exp(-\frac{\zeta}{z}) d(\frac{\zeta}{z})$$

converges for $z \neq 0$ with $|z|$ small enough and $|\arg(z) - d| < \frac{\pi}{2}$. Moreover, this integral is analytic and does not depend on the choice of d. Thus f is an analytic function on a sector $(-\frac{\pi}{2} - \epsilon, \frac{\pi}{2} + \epsilon)$. Write $h(\zeta) = \sum_{i=0}^{N-1} \frac{c_i}{i!} \zeta^i + h_N(\zeta)$. Then $f(z) = \sum_{i=0}^{N-1} c_i z^i + (\mathcal{L}_{1,d} h_N)(z)$. One can show (but we will not give details) that there exists a constant $A > 0$, independent of N, such that the estimate $|(\mathcal{L}_{1,d} h_N)(z)| \leq A^N N! |z|^N$ holds. In other words, f lies in $\mathcal{A}_1(-\frac{\pi}{2} - \epsilon, \frac{\pi}{2} + \epsilon)$ and has asymptotic expansion $\hat{f}$.

Suppose now that part 1 holds and let $f \in \mathcal{A}_1(-\frac{\pi}{2} - \epsilon, \frac{\pi}{2} + \epsilon)$ have asymptotic expansion $\hat{f}$. Then we will consider the integral

$$h(\zeta) := (\mathcal{B}_1 f)(\zeta) = \int_\lambda f(z) z\ exp(\frac{\zeta}{z})\ dz^{-1}$$

over the contour λ, which consists of the three parts $\{se^{i(-\frac{\pi+\epsilon}{2})} |\ 0 \leq s \leq r\}$, $\{re^{id} |\ -\frac{\pi+\epsilon}{2} \leq d \leq \frac{\pi+\epsilon}{2}\}$, and $\{se^{i(+\frac{\pi+\epsilon}{2})} |\ r \geq s \geq 0\}$.

For ζ with $0 < |\zeta| < \infty$ and $|\arg(\zeta)| < \epsilon/4$ this integral converges and is an analytic function of ζ. It is easily verified that h has exponential growth of order ≤ 1. The integral transform $\mathcal{B}_1$ is called the Borel transform of order 1. It is easily seen that for $f = z^n$ the Borel transform $\mathcal{B}_1(f)$ is equal to $\frac{\zeta^n}{n!}$. We now write $f = \sum_{i=0}^{N-1} c_i z^i + f_N$. Then $|f_N(z)| \leq A^N N! |z|^N$ holds for some constant $A > 0$, independent of N. Then $h(\zeta) = \sum_{i=0}^{N-1} \frac{c_i}{i!} \zeta^i + \mathcal{B}_1(f_N)(\zeta)$. One can prove (but we will not give details) an estimate of the form $|\mathcal{B}_1(f_N)(\zeta)| \leq A^N |\zeta^N|$ for small enough $|\zeta|$. Using this, one can identify the above h for ζ with $|\zeta|$ small and $|\arg(\zeta)| < \epsilon/4$ with the function $\hat{\mathcal{B}}_1 \hat{f}$. In other words, $\hat{\mathcal{B}}_1 \hat{f}$ has an analytic continuation, in a full sector $\{\zeta \in \mathbf{C} |\ 0 < |\zeta| < \infty$ and $|\arg(\zeta)| < \epsilon/4\}$, which has exponential growth of order ≤ 1. □

Remarks 7.35 1. In general, one can define the *Borel transform of order k in the direction d* in the following way. Let d be a direction and let S be a sector of them for

$\{z \mid |z| < R, \ |\arg(z) - d| < \rho\}$ where $\rho > \frac{\pi}{2k}$. Let f be analytic in S and bounded at 0. We then define the Borel transform of f of order k in the direction d to be

$$(\mathcal{B}_k f)(\zeta) := \int_\lambda f(z) z^k exp(\frac{\zeta^k}{z^k}) d(z^{-k}),$$

where λ is a suitable wedge-shaped path in S and ζ lies in the interior of this path (see [15], Chap. 2.3 for the details). The function $\mathcal{B}_k f$ can be shown to be analytic in the sector $\{\zeta \mid |\zeta| < \infty, \ |\arg(\zeta) - d| < \rho - \frac{\pi}{2k}\}$. Furthermore, applying $\mathcal{B}$ to each term of a formal power series $\hat{f} = \sum c_n z^n$ yields $\hat{\mathcal{B}} \hat{f}$.

2. The analytic way to prove the k-summation theorem for a solution $\hat{v}$ of an equation $(\delta - A)\hat{v} = w$, which has only $k > 0$ as positive slope, consists of a rather involved proof that $\hat{\mathcal{B}}_k \hat{v}$ satisfies part 2 of Theorem 7.34. The equivalence with part 1 yields then the k-summability of $\hat{v}$. In our treatment of the k-summation theorem (and the multisummation theorem later) the basic ingredient is the cohomology of the sheaf $\ker(\delta - A, (\mathcal{A}^0)^n)$ and the main asymptotic existence theorem. □

We illustrate this theorem with an example of the type $(\delta - q)\hat{v} = w$, which is chosen such that $\hat{\mathcal{B}}_k \hat{v}$ can actually be calculated. This example also produces for the image of $\hat{v}$ in the cohomology group $H^1(U, \ker(\delta - q, \mathcal{A}^0))$ of Lemma 7.20, an explicit 1-cocycle by the Laplace and Borel method.

Example 7.36 *The equation* $(\delta - kz^{-k} + k)\hat{v} = w$ *with* $w \in \mathbf{C}[z, z^{-1}]$.
Write $\hat{v} \in \mathbf{C}((z))$ as $\sum v_n z^n$. Then for $n >> 0$ one finds the relation $v_{n+k} = \frac{n+k}{k} v_n$. Thus for $n \gg 0$ one has $v_n = a_i \Gamma(1 + \frac{n}{k})$, where the constant a_i only depends on n modulo k. In other words the possibilities for $\hat{v}$ are $p + \sum_{i=0}^{k-1} a_i \sum_{n \geq 0} \Gamma(1 + \frac{nk+i}{k}) z^{nk+i}$ with $p \in \mathbf{C}[z, z^{-1}]$ and $a_0, \ldots, a_{k-1} \in \mathbf{C}$. It suffices to consider $\hat{v}$ with $p = 0$, and thus

$$(\delta - kz^{-k} + k)\hat{v} = \sum_{i=0}^{k-1} -a_i k \Gamma(1 + \frac{i}{k}) z^{-k+i}.$$

The formal Borel transform $\mathcal{B}_k \hat{v}$ is equal to $f := \frac{a_0 + a_1\zeta + \cdots + a_{k-1}\zeta^{k-1}}{1-\zeta^k}$. The radius of convergence of f is 1 (if $\hat{v} \neq 0$). For any direction d, not in the set $\{\frac{2\pi j}{k} | j = 0, \ldots, k-1\}$, the function f has a suitable analytic continuation on the half-line d. Consider a direction d with $0 < d < \frac{2\pi}{k}$. The integral $v(z) := (\mathcal{L}_{k,d} f)(z)$ $= \int_d f(\zeta) exp(-(\frac{\zeta}{z})^k) \, d(\frac{\zeta}{z})^k$ is easily seen to be an analytic function of z for $z \neq 0$ and $\arg(\frac{\zeta}{z})^k \in (-\frac{\pi}{2}, \frac{\pi}{2})$. Thus v is analytic for $z \neq 0$ and $\arg(z) \in (d - \frac{\pi}{2k}, d + \frac{\pi}{2k})$. Moreover, v does not depend on d, as long as $d \in (0, \frac{2\pi}{k})$. Thus we conclude that v is a holomorphic function on the sector, defined by the relation $\arg(z) \in (-\frac{\pi}{2k}, \frac{2\pi}{k} + \frac{\pi}{2k})$.

Exercise 7.37 Prove that the above v lies in $\mathcal{A}_{1/k}(-\frac{\pi}{2k}, \frac{2\pi}{k} + \frac{\pi}{2k})$ and has asymptotic expansion $\hat{v}$. Hint: Subtract from $f(\zeta)$ a truncation of its Taylor series at $\zeta = 0$. □

Let w be the Laplace transform $\mathcal{L}_{k,d}f$ for $d \in (-\frac{2\pi}{k}, 0)$. Then by the Cauchy residue formula one has

$$\begin{aligned}(v-w)(z) &= -2\pi i \; Res_{\zeta=1}(f(\zeta)exp(-(\frac{\zeta}{z})^k)\; d(\frac{\zeta}{z})^k) \\ &= 2\pi i \; (a_0 + a_1 + \cdots + a_{k-1})h,\end{aligned}$$

in which the function $h := z^{-k}exp(-z^{-k})$ is a solution of $(\delta - kz^{-k} + k)h = 0$. More generally, consider a direction $d_j := \frac{2\pi j}{k}$ and let d_j^+ and d_j^- denote directions of the form $d_j \pm \epsilon$ for small $\epsilon > 0$. Let v_{j+} and v_{j-} denote the Laplace integrals $\mathcal{L}_{k,d_{j+}}f$ and $\mathcal{L}_{k,d_{j-}}f$. Then one has the formula

$$(v_{j+} - v_{j-})(z) = 2\pi i \; (a_0 + a_1\zeta + \cdots + a_{k-1}\zeta^{k-1})h \text{ with } \zeta = e^{2\pi i j/k}.$$

We compare this with Sect. 7.3. The directions $\frac{2\pi j}{k}$ are the singular directions for $\delta - kz^{-k} + k$. The negative Stokes pairs are the pairs $\{\frac{2\pi j}{k} - \frac{\pi}{2k}, \frac{2\pi j}{k} + \frac{\pi}{2k}\}$. The Laplace-Borel method produces the asymptotic lifts of $\hat{v}$ on the maximal intervals, i.e., the maximal intervals not containing a negative Stokes pair. Consider, as in Sect. 7.3, the map $\beta : \mathbf{C}(\{z\}) \to H^1(\mathbf{S}^1, \ker(\delta - kz^{-k} + k, \mathcal{A}^0))$, that associates to each $w \in \mathbf{C}(\{z\})$ the image in $H^1(\mathbf{S}^1, \ker(\delta - kz^{-k} + k, \mathcal{A}^0))$ of the unique formal solution $\hat{v}$ of $(\delta - kz^{-k} + k)\hat{v} = w$. For w of the form $\sum_{i=0}^{k-1} b_i z^{-k+i}$ the above residues give the explicit 1-cocycle for $\beta(w)$.

Exercise 7.38 Extend the above example and the formulas to the case of a formal solution $\hat{v}$ of $(\delta - kz^{-k} + k)\hat{v} = w$ with $w = \sum w_n z^n \in \mathbf{C}(\{z\})$. In particular, give an explicit formula for the 1-cocycle $\beta(w)$ and find the conditions on the coefficients w_n of w that are necessary and sufficient for $\hat{v}$ to lie in $\mathbf{C}(\{z\})$. □

7.7 The *k*-Summation Theorem

This theorem can be formulated as follows. The notion of an eigenvalue of a differential equation is defined in Definition 3.28.

Theorem 7.39 *Consider a formal solution $\hat{v}$ of the inhomogeneous matrix equation $(\delta - A)\hat{v} = w$, where w and A have coordinates in $\mathbf{C}(\{z\})$ and such that the only positive slope of $\delta - A$ is k. Then $\hat{v}$ is k-summable (i.e., every coordinate of $\hat{v}$ is k-summable). Let $q_1, \ldots, q_s$ denote the distinct eigenvalues of $\delta - A$. Then $\hat{v}$ is k-summable in the direction d if d is not singular for any of the $q_1, \ldots, q_s$.*

We note that the q_i are, in fact, polynomials in $z^{-1/m}$ for some integer $m \geq 1$. The set of singular directions of a single q_i may not be well defined. The set $\{q_1, \ldots, q_s\}$ is invariant under the action on $\mathbf{C}[z^{-1/m}]$, given by $z^{-1/m} \mapsto e^{-2\pi i/m} z^{-1/m}$. Thus the set of the singular directions of all q_i is well defined. We start the proof of Theorem 7.39 with a lemma.

Lemma 7.40 *Let $\hat{v}$ be a formal solution of $(\delta - A)\hat{v} = w$, where A and w have coordinates in $\mathbf{C}(\{z\})$ and let $k > 0$ be the smallest positive slope of $\delta - A$. For every direction d there is an asymptotic lift v_d of $\hat{v}$ with coordinates in $(\mathcal{A}_{1/k})_d$.*

Proof. We will follow, to a great extent, the proof of Proposition 7.32. There exists a quasisplit equation $(\delta - B)$ that is formally equivalent to $(\delta - A)$, i.e., $\hat{F}^{-1}(\delta - A)\hat{F} = (\delta - B)$ and $\hat{F} \in \mathrm{GL}_n(\mathbf{C}((z)))$. The equation $(\delta - B)$ is a direct sum of $(\delta - q_i - C_i)$, where $q_1, \ldots, q_s$ are the distinct eigenvalues and the C_i are constant matrices. After replacing z by a root $z^{1/m}$, we are in the situation that $k > 0$ is an integer. Furthermore, we can use the method of Corollary 7.16 to reduce to the case where all the C_i are 0. The assumption that k is the smallest positive slope is equivalent to: if q_i is $\neq 0$ then the degree of q_i in z^{-1} is $\geq k$. Let d be a direction. By Theorem 7.10, there is an $F_d \in \mathrm{GL}_n(\mathcal{A}_d)$ with $J(F_d) = \hat{F}$ and $F_d^{-1}(\delta - A)F_d = (\delta - B)$. Since $\ker(\delta - q_i, \mathcal{A}_d^0) = \ker(\delta - q_i, (\mathcal{A}_{1/k}^0)_d)$, the kernel $\ker(\delta - B, ((\mathcal{A}^0)_d)^n)$ lies in $((\mathcal{A}_{1/k}^0)_d)^n$. Since F_d acts bijectively on $((\mathcal{A}_{1/k}^0)_d)^n$, one also has that the kernel of $\delta - A$ on $((\mathcal{A}^0)_d)^n$ lies in $((\mathcal{A}_{1/k}^0)_d)^n$. The element $\hat{v}$ has an asymptotic lift in $((\mathcal{A})_d)^n$, which is determined modulo the kernel of $(\delta - A)$ and thus defines a unique element of $((\mathcal{A}/\mathcal{A}_{1/k}^0)_d)^n$. By gluing, one finds a global section, i.e., over $\mathbf{S}^1$, of the corresponding sheaf. The last statement of Proposition 7.28 implies that the coordinates of $\hat{v}$ are in $\mathbf{C}((z))_{1/k}$. For a direction d one can first lift $\hat{v}$ to an element of $((\mathcal{A}_{1/k})_d)^n$ and then, using Theorem 7.12, we conclude that there is a lift $v_d \in ((\mathcal{A}_{1/k})_d)^n$ satisfying the equation $(\delta - A)v_d = w$. □

The obstruction to lifting $\hat{v}$ to a solution of the equation with coordinates in $((\mathcal{A}_{1/k})(a, b))^n$ is given by a 1-cocycle with image in the group $H^1((a, b), \ker(\delta - A, (\mathcal{A}_{1/k}^0)^n))$. The theorem will now follow from the known cohomology of the sheaf $\mathcal{K}_B := \ker(\delta - B, (\mathcal{A}_{1/k}^0)^n))$ (see Lemma 7.20), and the construction in the next lemma of an isomorphism between restrictions of the two sheaves $\mathcal{K}_A := \ker(\delta - A, (\mathcal{A}_{1/k}^0)^n))$ and $\mathcal{K}_B$ to suitable open intervals (a, b).

Lemma 7.41 *Suppose that d is not a singular direction for any of the q_i, then for some positive ϵ the restrictions of the sheaves $\mathcal{K}_A$ and $\mathcal{K}_B$ to the open interval $(d - \frac{\pi}{2k} - \epsilon, d + \frac{\pi}{2k} + \epsilon)$ are isomorphic.*

Proof. We may suppose that the q_i are polynomials in z^{-1}. As before $\delta - A$ is formally equivalent to $\delta - B$, which is a direct sum of $\delta - q_i + C_i$ and we may suppose that the C_i are 0. Let f be any direction. The formal $\hat{F}$ with $\hat{F}^{-1}(\delta - A)\hat{F} = (\delta - B)$ satisfies the differential equation $\delta(\hat{F}) = A\hat{F} - \hat{F}B$. By Theorem 7.10, $\hat{F}$ lifts to an $F_f \in \mathrm{GL}_n(\mathcal{A}_f)$ with $F_f^{-1}(\delta - A)F_f = (\delta - B)$. This produces locally at the direction f an isomorphism $(\mathcal{K}_A)_f \to (\mathcal{K}_B)_f$. The asymptotic lift F_f is not unique. Two asymptotic lifts differ by a $G \in \mathrm{GL}_n(\mathcal{A}_f)$ with $J(G) = 1$ and $G^{-1}(\delta - B)G = (\delta - B)$. We have to investigate $\mathcal{K}_B$ and the action of G on $\mathcal{K}_B$ in detail.

We note that $\mathcal{K}_B$ is the direct sum of $\mathcal{K}_B(i) := \ker(\delta - q_i, (\mathcal{A}^0_{1/k})^{n_i})$ over all nonzero q_i. The action of G on $(\mathcal{K}_B)_f$ has the form $1 + \sum_{i \neq j} l_{i,j}$, where 1 denotes the identity and $l_{i,j} \in \mathrm{Hom}_{\mathbf{C}}(\mathcal{K}_B(i), \mathcal{K}_B(j))_f$. For any $p = p_l z^{-l} + \cdots \in z^{-1}\mathbf{C}[z^{-1}]$ with $p_l \neq 0$, we will call the direction f *flat* if $Re(p_l e^{-ifl}) > 0$. With this terminology one has: $l_{i,j}$ can only be nonzero if the direction f is flat for $q_i - q_j$ (and f is, of course, also a flat direction for q_i and q_j).

Let us call $\mathcal{S}$ the sheaf of all the automorphisms of $\mathcal{K}_B$, defined by the above conditions. The obstruction for constructing an isomorphism between the restrictions of $\mathcal{K}_A$ and $\mathcal{K}_B$ to (a, b) is an element of the cohomology set $H^1((a, b), \mathcal{S})$. We will show that this cohomology set is trivial, i.e., it is just one element, for $(a, b) = (d - \frac{\pi}{2k} - \epsilon, d + \frac{\pi}{2k} + \epsilon)$ with small $\epsilon > 0$ and d not a singular direction. Although $\mathcal{S}$ is a sheaf of non-abelian groups, it is very close to sheaves of abelian groups. For any direction f, define $q_i <_f q_j$ if f is a flat direction for $q_i - q_j$.

Lemma 7.42 *Let $\mathcal{S}$ be as above.*

1. *For any $f \in \mathbf{S}^1$, every element of the stalk $\mathcal{S}_f$ is unipotent.*
2. *There exists a finite sequence of subsheaves $\mathcal{S}(r)$ of $\mathcal{S}$, given by $1 + \sum l_{j_1,j_2}$ belongs to $\mathcal{S}(r)_f$ if $l_{j_1,j_2} \neq 0$ implies that there are $s_1, \ldots, s_r$ with $q_{j_1} <_f q_{s_1} <_f \cdots <_f q_{s_r} <_f q_{j_2}$.*

Proof. 1. Let $G = I + N \in \mathcal{S}_f$ where $N = (l_{i,j})$. As noted above, if $l_{i,j} \neq 0$ then $q_i < q_j$. For any $r \geq 0$ let $N^r = (l_{i,j,r})$. One shows by induction that if $l_{i,j,r} \neq 0$, then there exist $s_1, \ldots, s_{r-1}$ such that $q_i <_f q_{s_1} <_f \cdots <_f q_{s_{r-1}} <_f q_j$. Therefore, $N^r = 0$ for sufficiently large r.

2. We define a sequence of subsheaves $\mathcal{S}(r)$ of $\mathcal{S}$, given by $1 + \sum l_{j_1,j_2}$ belonging to $\mathcal{S}(r)_f$ if $l_{j_1,j_2} \neq 0$ implies that there are $s_1, \ldots, s_r$ with $q_{j_1} <_f q_{s_1} <_f \cdots <_f q_{s_r} <_f q_{j_2}$. The quotients sheaves $\mathcal{S}/\mathcal{S}(1), \ldots, \mathcal{S}(i)/\mathcal{S}(i+1), \ldots$ are easily seen to be abelian sheaves. We now use the notation introduced in Sect. 7.3 before Lemma 7.19. Each quotient is a direct sum of sheaves $\mathcal{H}_H$, where $\mathcal{H} := \mathrm{Hom}_{\mathbf{C}}(\mathcal{K}_B(j_1), \mathcal{K}_B(j_2))$ and H is the open interval consisting of the directions g that are flat for $q_{j_1} - q_{j_2}$ (and for certain pairs $j_1 \neq j_2$). □

Thus the proof of Lemma 7.41 is reduced to proving that each sheaf $\mathcal{H}_H$ has a trivial H^1 on the proposed open intervals. The sheaves $\mathcal{K}_B(j)$ are direct sums of sheaves $\mathbf{C}_I$, with I an open interval of length $\frac{\pi}{k}$. If I, J are both open intervals of length $\frac{\pi}{k}$ and H is another open interval (I, J, H are determined by q_i, q_j, and $q_i - q_j$), then it suffices to show that the sheaf $\mathcal{T} := \mathrm{Hom}_{\mathbf{C}}(\mathbf{C}_I, \mathbf{C}_J)_H$ has a trivial H^1 on the proposed intervals $(d - \frac{\pi}{2k} - \epsilon, d + \frac{\pi}{2k} + \epsilon)$.

First, we will determine the sheaf $\mathrm{Hom}_{\mathbf{C}}(\mathbf{C}_I, \mathbf{C}_J)$. Let us recall the definition of the sheaf $\mathrm{Hom}_{\mathbf{C}}(F, G)$ for two sheaves of complex vector spaces F and G on, say, the circle $\mathbf{S}^1$. The sheaf $\mathrm{Hom}_{\mathbf{C}}(F, G)$ is defined as the sheaf associated to the presheaf P given requiring $P(U)$ to consist of the $\mathbf{C}$-linear homomorphisms h

between the restrictions $F|_U$ and $G|_U$. The element h consists of a family of $\mathbf{C}$-linear maps $h_V : F(V) \to G(V)$, for all open $V \subset U$, satisfying for all pairs of open sets $W \subset V \subset U$ the relation $res_{G,V,W} \circ h_V = h_W \circ res_{F,V,W}$. Here $res_{*,*,*}$ denote the restrictions of the sheaves F and G with respect to the sets $W \subset V$. A straightforward use of this definition leads to a $\mathbf{C}$-linear homomorphism of the sheaves $\phi : \mathbf{C} \to \mathrm{Hom}_{\mathbf{C}}(\mathbf{C}_I, \mathbf{C})$. Let $\overline{I}$ denote the closure of I. A small calculation shows that the stalk of the second sheaf at a point outside $\overline{I}$ is 0 and the stalk at any point in $\overline{I}$ is isomorphic to $\mathbf{C}$. Moreover, for any d, ϕ_d is surjective. One concludes that $\mathrm{Hom}_{\mathbf{C}}(\mathbf{C}_I, \mathbf{C})$ is isomorphic to $\mathbf{C}_{\overline{I}}$. We recall the exact sequence

$$0 \to \mathbf{C}_J \to \mathbf{C} \to \mathbf{C}_{\mathbf{S}^1 \setminus J} \to 0.$$

We then have that $\mathrm{Hom}_{\mathbf{C}}(\mathbf{C}_I, \mathbf{C}_J)$ is the subsheaf of $\mathrm{Hom}_{\mathbf{C}}(\mathbf{C}_I, \mathbf{C})$, consisting of the h such that the composition $\mathbf{C}_I \stackrel{h}{\to} \mathbf{C} \to \mathbf{C}_{\mathbf{S}^1 \setminus J}$ is the zero map. Thus $\mathrm{Hom}_{\mathbf{C}}(\mathbf{C}_I, \mathbf{C}_J)$ can be identified with $(\mathbf{C}_{\overline{I}})_{J \cap \overline{I}}$. The sheaf $\mathcal{T}$ can therefore be identified with $(\mathbf{C}_{\overline{I}})_{J \cap H \cap \overline{I}}$.

Let q_i, q_j and $q_i - q_j$ have leading terms a, b, and c with respect to the variable z^{-1} and let the degree of $q_i - q_j$ in z^{-1} be l. The intervals I, J, H are connected components of the set of directions f such that $Re(ae^{-ifk})$, $Re(be^{-ifk})$ and $Re(ce^{-ifl})$ are positive. We must consider two cases.

Suppose first that $I \neq J$. Then one sees that $J \cap H \cap \overline{I} = H \cap \overline{I}$ and, moreover, the complement of this set in $\overline{I}$ has only one component. In this case, the sheaf $\mathcal{T}$ has trivial H^1 for any open subset of $\mathbf{S}^1$.

Now suppose that $I = J$. The complement of $J \cap H \cap \overline{I}$ in $\overline{I}$ can have two components, namely the two endpoints of the closed interval $\overline{I}$. In this case, the H^1 of the sheaf $\mathcal{T}$ on $\overline{I}$ is not trivial. However, the midpoint of $\overline{I}$ is a singular direction. Thus only one of the two endpoints can belong to the open interval $(d - \frac{\pi}{2k} - \epsilon, d + \frac{\pi}{2k} + \epsilon)$ and the H^1 of $\mathcal{T}$ on this interval is trivial. □

We now deduce the following corollary. Note that we are continuing to assume that there is only one positive slope.

Corollary 7.43 *The sheaves $\mathcal{K}_A$ and $\mathcal{K}_B$ are isomorphic on $\mathbf{S}^1$.*

Proof. Let (a, b) be a (maximal) interval, not containing a negative Stokes pair for any of the q_i. The proof of Lemma 7.41 shows, in fact, that the restrictions of $\mathcal{K}_A$ and $\mathcal{K}_B$ to (a, b) are isomorphic. The sheaf $\mathcal{K}_B$ has a direct sum decomposition $\oplus_{i=1}^{s} K_{B,i}$ with $K_{B,i} := \mathbf{C}_{I_i}^{a_i}$, where the $a_i \geq 1$ are integers and the intervals I_i are distinct and have length $\frac{\pi}{k}$. We may assume that $I_i = (d_i - \frac{\pi}{2k}, d_i + \frac{\pi}{2k})$ and that $d_1 < d_2 < \cdots < d_s < d_1(+2\pi)$ holds on the circle $\mathbf{S}^1$. The intervals $J_1 := (d_s - \frac{\pi}{2k}, d_1 + \frac{\pi}{2k})$, $J_2 := (d_1 - \frac{\pi}{2k}, d_2 + \frac{\pi}{2k}), \ldots$ are maximal with respect to the condition that they do not contain a negative Stokes pair. Choose isomorphisms $\sigma_i : \mathcal{K}_B|_{J_i} \to \mathcal{K}_A|_{J_i}$ for $i = 1, 2$. Then $\sigma_{1,2} := \sigma_2^{-1}\sigma_1$ is an isomorphism of $\mathcal{K}_B|I_1$. We note that $H^0(I_1, \mathcal{K}_B) = H^0(I_1, K_{B,1}) = \mathbf{C}^{a_1}$ and $\sigma_{1,2}$ induces an automorphism of $\mathbf{C}^{a_1}$ and of $K_{B,1}$. The latter can be extended to an automorphism of $\mathcal{K}_B$ on $\mathbf{S}^1$. After changing σ_2 with this automorphism one may assume that $\sigma_{1,2}$ acts as the identity on $\mathbf{C}^{a_1}$. This implies that the restrictions of σ_1 and σ_2 to the sheaf $K_{B,1}$ coincide on $J_1 \cap J_2$. Thus we find a morphism of sheaves $K_{B,1}|_{J_1 \cup J_2} \to \mathcal{K}_A|_{J_1 \cup J_2}$. Since the support of $K_{B,1}$ lies in $J_1 \cup J_2$ we have a morphism $\tau_1 : K_{B,1} \to \mathcal{K}_A$. In a similar way one constructs morphisms $\tau_i : K_{B,i} \to \mathcal{K}_A$. The sum $\oplus \tau_i$ is a morphism $\tau : \mathcal{K}_B \to \mathcal{K}_A$. This is an isomorphism since it is an isomorphism for every stalk. □

***k*-summability for a scalar differential equation** In this subsection we will reformulate Theorem 7.39 for a scalar differential equation, i.e., an equation $L\hat{f} = g$ with a differential operator $L \in \mathbf{C}(\{z\})[\frac{d}{dz}]$, $g \in \mathbf{C}(\{z\})$, and $\hat{f} \in \mathbf{C}((z))$.

Instead of $\frac{d}{dz}$, we will use the operator $\Delta = \frac{1}{k} z \frac{d}{dz} z^k$. An operator L of order n can be written as $\sum_{i=0}^n a_i \Delta^i$ with $a_n = 1$ and all $a_i \in \mathbf{C}(\{z\})$. In the following we will assume that the only slope present in L is $k > 0$ and that k is an integer. In other words, all the eigenvalues q_i of L (or of the associated matrix equation $\delta - A$) are in $z^{-1}\mathbf{C}[z^{-1}]$ and have degree k in z^{-1}. A small calculation shows that these conditions are equivalent to L having the form

$$L = \sum_{i=0}^{n} a_i \Delta^i \text{ with } a_n = 1,\ a_i \in \mathbf{C}\{z\} \text{ and } a_0(0) \neq 0.$$

Define the *initial polynomial of L with respect to Δ* to be $P(T) = \sum_{i=0}^n a_i(0) T^i$. One easily calculates that the eigenvalues of L are of the form $cz^{-k} + \cdots$, where c is a zero of the initial polynomial. Then, Theorem 7.39 has the following corollary.

Corollary 7.44 The k-summation Theorem for Scalar Differential Equations.
Consider the equation $L\hat{f} = g$ with L as above, $g \in \mathbf{C}(\{z\})$, and $\hat{f} \in \mathbf{C}((z))$. Then $\hat{f}$ is k-summable. More precisely:

1. *A direction d is singular if and only if d is the argument of some ζ satisfying $P(\zeta^k) = 0$. The negative Stokes pairs are the pairs $\{d - \frac{\pi}{2k}, d + \frac{\pi}{2k}\}$ with d a singular direction.*
2. *$\hat{f}$ is k-summable in the direction d if d is not singular.*
3. *Suppose that the open interval (a, b) does not contain a negative Stokes pair and that $|b - a| > \frac{\pi}{k}$, then there is a unique $f \in \mathcal{A}_{\frac{1}{k}}(a, b)$ with $J(f) = \hat{f}$. Moreover, $Lf = g$.*

Example 7.45 *The method of Borel and Laplace applied to $L\hat{f} = g$.*
For the special case $L = P(\Delta)$ (i.e., all $a_i \in \mathbf{C}$), we will give here an independent proof of Corollary 7.44, using the formal Borel transformation and the Laplace transformation. This works rather well because one obtains an explicit and easy formula

for $\hat{\mathcal{B}}_k \hat{f}$. The general case can be seen as a "perturbation" of this special case. However, the proof for the general case, using the method of Borel and Laplace, is rather involved. The main problem is to show that $\hat{\mathcal{B}}_k \hat{f}$ satisfies part 2 of Theorem 7.34.

The formal Borel transform $\hat{\mathcal{B}}_k$ is only defined for formal power series. After subtracting from $\hat{f}$ a suitable first part of its Laurent series, we may suppose that $\hat{f} \in \mathbf{C}[[z]]$ and $g \in \mathbf{C}\{z\}$. Put $\phi = \hat{\mathcal{B}}_k(\hat{f})$. A small calculation yields $\hat{\mathcal{B}}_k(\Delta \hat{f})(\zeta) = \zeta^k \phi(\zeta)$. The equation $L\hat{f} = g$ is equivalent to $P(\zeta^k)\phi(\zeta) = (\hat{\mathcal{B}}_k g)(\zeta)$ and has the unique solution $\phi(\zeta) = \frac{\hat{\mathcal{B}}_k g}{P(\zeta^k)}$. The function $g = \sum_{n\geq 0} g_n z^n$ is convergent at 0, and thus $|g_n| \leq CR^n$ for suitable positive C, R. The absolute value of $\hat{\mathcal{B}}_k g(\zeta) = \sum \frac{g_n}{\Gamma(1+\frac{n}{k})}\zeta^n$ can be bounded by

$$\leq C\sum_{n\geq 0} \frac{R^n|\zeta|^n}{\Gamma(1+\frac{n}{k})} \leq C\sum_{i=0}^{k-1}\sum_{m\geq 0} \frac{(R|\zeta|)^{mk+i}}{\Gamma(1+m+\frac{i}{k})} \leq C\sum_{i=0}^{k-1} R^i|\zeta|^i exp(R^k|\zeta|^k).$$

Thus $\hat{\mathcal{B}}_k g$ is an entire function on $\mathbf{C}$ and has an exponential growth of order $\leq k$, i.e., is bounded by $\leq A\ exp(B|\zeta|^k)$ for suitable positive A, B.

The power series ϕ is clearly convergent and so $\hat{f} \in \mathbf{C}[[z]]_{\frac{1}{k}}$. Consider a direction d with $d \notin S := \{\arg(\zeta_1), ..., \arg(\zeta_{dk})\}$, where $\{\zeta_1, ..., \zeta_{dk}\}$ are the roots of $P(\zeta^k) = 0$. Let a, b be consecutive elements of S with $d \in (a, b)$. The function ϕ has, in the direction d, an analytic continuation with exponential growth of order $\leq k$. It follows that the integral $f(z) := (\mathcal{L}_{k,d}\phi)(z) = \int_d \phi(\zeta) exp(-(\frac{\zeta}{z})^k)\ d(\frac{\zeta}{z})^k$ converges for $\arg(z) \in (d - \frac{\pi}{2k}, d + \frac{\pi}{2k})$ and small enough $|z|$. One can vary d in the interval (a, b), without changing the function f. Thus f is defined on the open sector $I := (a - \frac{\pi}{2k}, b + \frac{\pi}{2k})$. It is not difficult to show that $f \in \mathcal{A}_{1/k}(I)$ with $J(f) = \hat{f}$. Indeed, let $\phi(\zeta) = \sum_{i\geq 0} c_i\zeta^i$ and write $\phi = \sum_{i=0}^{N-1} c_i\zeta^i + R_N(\zeta)\zeta^N$. Put $\hat{f} = \sum_{i\geq 0} f_i z^i$. Then $\mathcal{L}_{k,d}(\sum_{i=0}^{N-1} c_i\zeta^i) = \sum_{i=0}^{N-1} f_i z^i$ and one has to verify the required estimate for $|\mathcal{L}_{k,d}(R_N(\zeta)\zeta^N)(z)|$. Interchanging Δ and $\int_d$ easily leads to $Lf = g$. This proves the k-summability of $\hat{f}$ and the properties 1, 2, and 3.

More detailed information can be obtained by using the factorization $P(T) = \prod_{i=1}^{s}(T - c_i)^{n_i}$, with c_i the distinct roots of $P(T)$. Then L has a similar factorization and one finds that the eigenvalues of L are $q_i = kc_i z^{-k} - k$, with multiplicity n_i, for $i = 1, \ldots, s$. Write $P(T)^{-1} = \sum_i \frac{A_i(T)}{(T-c_i)^{n_i}}$. Then $\phi(\zeta) = \frac{\hat{\mathcal{B}}_k}{P(\zeta^k)}$ decomposes as $\sum \phi_i$, where $\phi_i(\zeta) = \frac{A_i(\zeta^k)}{(\zeta^k - c_i)^{n_i}}(\hat{\mathcal{B}}_k g)(\zeta)$. Consider a singular direction d, which is the argument of a ζ_i with $\zeta_i^k = c_i$. Let d^+, d^- denote directions with $d^- < d < d^+$ and $d^+ - d^-$ small. Then $\mathcal{L}_{k,d^+}\phi$ and $\mathcal{L}_{k,d^-}\phi$ exist and the difference $\mathcal{L}_{k,d^+}\phi - \mathcal{L}_{k,d^-}\phi$ is equal to

$$-(2\pi i)\ Res_{\zeta=\zeta_i}(\frac{A_i(\zeta^k)\hat{\mathcal{B}}_k g(\zeta)}{(\zeta^k - c_i)} d\zeta^k)\ z^{-k} exp(-c_i z^{-k}).$$

As in Example 7.36, this formula gives an explicit 1-cocycle for the image of $\hat{f}$ in $H^1(\mathbf{S}^1, \ker(L, \mathcal{A}^0))$. □

7.8 The Multisummation Theorem

Definition 7.46 $\underline{k}$ will denote a sequence of positive numbers $k_1 < k_2 < \cdots < k_r$ with $k_1 > 1/2$. Let $\hat{v} \in \mathbf{C}((z))$ and let d be a direction. Then $\hat{v}$ is called *$\underline{k}$-summable, or multisummable w.r.t. $\underline{k}$ in the direction d* if there is a sequence of elements $v_0, v_1, \ldots, v_r$ and a positive ϵ such that:

1. $v_0 \in H^0(\mathbf{S}^1, \mathcal{A}/\mathcal{A}^0_{1/k_1})$ and has image $\hat{v}$ under the isomorphism of Proposition 7.28.7.
2. $v_i \in H^0((d - \frac{\pi}{2k_i} - \epsilon, d + \frac{\pi}{2k_i} + \epsilon), \mathcal{A}/\mathcal{A}^0_{1/k_{i+1}})$ for $i = 1, \ldots, r-1$ and $v_r \in H^0((d - \frac{\pi}{2k_r} - \epsilon, d + \frac{\pi}{2k_r} + \epsilon), \mathcal{A})$.
3. For $i = 0, \ldots, r-1$, the images of v_i and v_{i+1} in $H^0((d - \frac{\pi}{2k_{i+1}} - \epsilon, d + \frac{\pi}{2k_{i+1}} + \epsilon), \mathcal{A}/\mathcal{A}^0_{1/k_{i+1}})$ coincide. The $\underline{k}$-sum of $\hat{v}$ in the direction d is the sequence $(v_1, \ldots, v_r)$.

One calls $\hat{v}$ *multisummable* or *$\underline{k}$-summable* if $\hat{v}$ is $\underline{k}$-summable in all but finitely many directions.

This definition is extended in an obvious way to elements of $\mathbf{C}((z))^n$. □

Remarks concerning Definition 7.46
The notion of multisummability is due to Écalle. His innovating manuscript "L'accélération des fonctions résurgentes", remains unpublished. References to this unpublished material can be found in [198, 195] and [93]. Independently, and for a different purpose, this notion was also introduced by Tougeron.

1. Condition 1 is, of course, the same as stating that $\hat{v} \in \mathbf{C}((z))_{1/k_1}$.

2. For any positive k, one sees the sheaf $\mathcal{A}/\mathcal{A}^0_{1/k}$ as a sheaf of "k-precise quasi-functions". Indeed, a section f of this sheaf above an open interval (a, b) can be represented by a covering of (a, b) by intervals (a_i, b_i) and elements $f_i \in \mathcal{A}(a_i, b_i)$, such that $f_i - f_j$ is, in general, not zero but lies in $\mathcal{A}^0_{1/k}((a_i, b_i) \cap (a_j, b_j))$.

3. The idea of the definition is that $\hat{v}$, seen as an element of $H^0(\mathbf{S}^1, \mathcal{A}/\mathcal{A}^0_{1/k_1})$, is lifted successively to the elements $v_1, v_2, \ldots,$ living each time on a smaller interval and being more precise. Finally, the last one v_r is really a function on the corresponding interval.

4. The size of the intervals with bisector d is chosen in a critical way. Indeed, for $1/2 < k < l$, one can consider the natural map

$$R : H^0((a, b), \mathcal{A}/\mathcal{A}^0_{1/l}) \to H^0((a, b), \mathcal{A}/\mathcal{A}^0_{1/k}).$$

The kernel of R is $H^0((a, b), \mathcal{A}^0_{1/k}/\mathcal{A}^0_{1/l})$. According to Theorem 7.47, the kernel is 0 if $|b-a| > \frac{\pi}{k}$. For $|b-a| \leq \frac{\pi}{k}$ the map is surjective according to Lemma 7.48. In particular, the elements $v_1, \ldots, v_r$ are uniquely determined by $\hat{v}$ and the direction d.

□

In general, one can show, using Theorem 7.47 below, that the multisum is unique, if it exists. We have unfortunately not found a direct proof in the literature. The proofs given in [198] use integral transformations of the Laplace and Borel type. However, a slight modification of the definition of the multisum for a formal solution of a linear differential equation yields uniqueness without any reference to Theorem 7.47 (see Theorem 7.50 and Remark 7.57).

Theorem 7.47 A relative form of Watson's Lemma.
Let $0 < k < l$ *and* $|b-a| > \frac{\pi}{k}$. *Then* $H^0((a,b), \mathcal{A}^0_{1/k}/\mathcal{A}^0_{1/l}) = 0$.

Lemma 7.48 *Suppose* $1/2 < k < l$ *and* $|b-a| \le \frac{\pi}{k}$. *Then the canonical map*

$$R : H^0((a,b), \mathcal{A}/\mathcal{A}^0_{1/l}) \to H^0((a,b), \mathcal{A}/\mathcal{A}^0_{1/k})$$

is surjective.

Proof. The map $H^0((a,b), \mathcal{A}) \to H^0((a,b), \mathcal{A}/\mathcal{A}^0_{1/k})$ is surjective, since by Proposition 7.28 the group $H^1((a,b), \mathcal{A}^0_{1/k})$ is zero. This map factors as

$$H^0((a,b), \mathcal{A}) \to H^0((a,b), \mathcal{A}/\mathcal{A}^0_{1/l}) \xrightarrow{R} H^0((a,b), \mathcal{A}/\mathcal{A}^0_{1/k}).$$

Thus R is surjective. □

Exercise 7.49 Let $\underline{k} = k_1 < \cdots < k_r$ with $1/2 < k_1$. Suppose that $\hat{v}$ is the sum of elements $F_1 + \cdots + F_r$, where each $F_i \in \mathbf{C}((z))$ is k_i-summable. Prove that $\hat{v}$ is $\underline{k}$-summable. Hint: Prove the following statements
(a) If $r = 1$, then k_1-summable is the same thing as $\underline{k}$-summable.
(b) If F and G are $\underline{k}$-summable then so is $F + G$.
(c) Let $\underline{k}'$ be obtained from $\underline{k}$ by leaving out k_i. If F is $\underline{k}'$-summable then F is also $\underline{k}$-summable. □

Theorem 7.50 The Multisummation Theorem.
Let w *be a vector with coordinates in* $\mathbf{C}(\{z\})$ *and let* $\hat{v}$ *be a formal solution of the equation* $(\delta - A)\hat{v} = w$. *Let* $\underline{k} = k_1 < k_2 < \cdots < k_r$ *with* $1/2 < k_1$ *denote the positive slopes of the differential operator* $\delta - A$. *Then* $\hat{v}$ *is* $\underline{k}$-*summable in any direction* d *that is not a singular direction for any of the eigenvalues of* $\delta - A$. *In particular* $\hat{v}$ *is* $\underline{k}$-*summable.*

Proof. The formal equivalence $\hat{F}^{-1}(\delta - A)\hat{F} = (\delta - B)$, where $(\delta - B)$ is an equation that is equivalent to a quasisplit differential equation for $\hat{F}$ (Proposition 3.41). One proves, as in Lemma 7.40, that $\ker(\delta - A, (\mathcal{A}^0)^n)$ is equal to $\ker(\delta - A, (\mathcal{A}^0_{1/k_1})^n)$. It follows that $\hat{v}$ has coordinates in $\mathbf{C}((z))_{1/k_1}$. Define for $i = 1, \ldots, r$ the sheaves $\mathcal{V}_i = \ker(\delta - A, (\mathcal{A}^0_{1/k_i})^n)$ and the sheaves $\mathcal{W}_i = \ker(\delta - B, (\mathcal{A}^0_{1/k_i})^n)$. For notational convenience we define $\mathcal{V}_{r+1}$ and $\mathcal{W}_{r+1}$ to be zero. Take a direction d, which is not a singular direction for any of the eigenvalues of $\delta - A$. The method of the proof of Lemma 7.41 yields that the restrictions of the sheaves $\mathcal{V}_1/\mathcal{V}_2$ and $\mathcal{W}_1/\mathcal{W}_2$ to $(d - \frac{\pi}{2k_1} - \epsilon, d + \frac{\pi}{2k_1} + \epsilon)$ are isomorphic. More generally, the proof of Lemma 7.41 can be modified to show that the restrictions of the sheaves $\mathcal{V}_i/\mathcal{V}_{i+1}$ and $\mathcal{W}_i/\mathcal{W}_{i+1}$ to $(d - \frac{\pi}{2k_i} - \epsilon, d + \frac{\pi}{2k_i} + \epsilon)$, are isomorphic. From Lemma 7.20 and Corollary 7.21 one concludes that the sheaves $\mathcal{W}_i/\mathcal{W}_{i+1}$ have a trivial H^1 and also H^0 on the interval $(d - \frac{\pi}{2k_i} - \epsilon, d + \frac{\pi}{2k_i} + \epsilon)$ (note that the sheaf $\ker(\delta - B, (\mathcal{A}^0)n)$ decomposes as a direct sum of similar sheaves where only one level (or one q_i) is present). The same holds then for the sheaves $\mathcal{V}_i/\mathcal{V}_{i+1}$.

Now $v_0 \in H^0(\mathbf{S}^1, (\mathcal{A}/\mathcal{A}^0_{1/k_1})^n)$ is simply the statement that $\hat{v} \in \mathbf{C}((z))^n_{1/k_1}$. The element $v_1 \in H^0((d - \frac{\pi}{2k_1} - \epsilon, d + \frac{\pi}{2k_1} + \epsilon), (\mathcal{A}/\mathcal{A}^0_{1/k_2})^n)$ is assumed to satisfy: $(\delta - A)v_1 \equiv w$ modulo $\mathcal{A}^0_{1/k_2}$ and v_0 and v_1 have the same image in $H^0((d - \frac{\pi}{2k_1} - \epsilon, d + \frac{\pi}{2k_1} + \epsilon), (\mathcal{A}/\mathcal{A}^0_{1/k_1})^n)$. The obstruction for the existence of v_1 is an element of the group $H^1((d - \frac{\pi}{2k_1} - \epsilon, d + \frac{\pi}{2k_1} + \epsilon), \mathcal{V}_1/\mathcal{V}_2)$. Since this group is 0, the element v_1 exists. Suppose that $\tilde{v}_1$ has the same properties. Then $\tilde{v}_1 - v_1$ is a section of the sheaf $\mathcal{V}_1/\mathcal{V}_2$ on the interval $(d - \frac{\pi}{2k_1} - \epsilon, d + \frac{\pi}{2k_1} + \epsilon)$. Since we also have that the H^0 of $\mathcal{V}_1/\mathcal{V}_2$ on this interval is 0, we find $\tilde{v}_1 = v_1$.

The existence and uniqueness of v_i with $(\delta - A)v_i \equiv w$ modulo $\mathcal{A}^0_{1/k_{i+1}}$ and v_i and v_{i-1} have the same image in $H^0((d - \frac{\pi}{2k_i} - \epsilon, d + \frac{\pi}{2k_i} + \epsilon), (\mathcal{A}/\mathcal{A}^0_{1/k_i})^n)$, follows from H^1 and H^0 of $\mathcal{V}_i/\mathcal{V}_{i+1}$ being for the open interval under consideration. Thus $\hat{v}$ is $\underline{k}$-summable in the direction d. □

Corollary 7.51 *We use the notation of Theorem 7.50 and its proof.*
For every i the sheaves $\mathcal{V}_i/\mathcal{V}_{i+1}$ and $\mathcal{W}_i/\mathcal{W}_{i+1}$ are isomorphic on $\mathbf{S}^1$. In particular, the spaces $H^1(\mathbf{S}^1, \ker(\delta - A, (\mathcal{A}^0)^n))$ and $H^1(\mathbf{S}^1, \ker(\delta - B, (\mathcal{A}^0)^n))$ have the same dimension. Let $(\delta - B)$ be the direct sum of $(\delta - q_i - C_i)$, where C_i is an $n_i \times n_i$ matrix and the degree of q_i in z^{-1} is k_i. Then the dimension of $H^1(\mathbf{S}^1, \ker(\delta - A, (\mathcal{A}^0)^n))$ is equal to $\sum_i k_i n_i$.

Proof. The first statement has the same proof as Corollary 7.43. The dimension of the cohomology group H^1 of the sheaf $\ker(\delta - A, (\mathcal{A}^0)^n)$ is easily seen to be the sum of the dimensions of the H^1 for the sheaves $\mathcal{V}_i/\mathcal{V}_{i+1}$. A similar statement holds for $\delta - B$ and thus the equality of the dimensions follows. From the direct sum decomposition of $\delta - B$ one easily derives the formula for the dimension. Indeed, if the k_i are integers then Lemma 7.20 implies the formula. In the general case, the k_i are rational numbers. One takes an integer $m \geq 1$ such that all mk_i are integers and considers the map $\pi_m : \mathbf{S}^1 \to \mathbf{S}^1$, given by $z \mapsto z^m$. The H^1 on $\mathbf{S}^1$ of $F := \ker(\delta - B, (\mathcal{A}^0)^n)$ is equal to $H^1(\mathbf{S}^1, \pi_m^* F)^G$, where G is the cyclic group with generator $z \to e^{2\pi i/m} z$ acting on $\mathbf{S}^1$. From this, the general case follows. □

We now define a number that measures the difference between formal and convergent solutions of $\delta - A$. Although we define this in terms of cohomology we will show in Corollary 7.54 that this number is just dim $\ker(\delta - A, \mathbf{C}((z))^n/\mathbf{C}(\{z\})^n)$. This number and its properties are also described in [187].

Definition 7.52 The dimension of $H^1(\mathbf{S}^1, \ker(\delta - A, (\mathcal{A}^0)^n))$ is called the *irregularity of* $\delta - A$. □

We note that the irregularity of $\delta - A$ depends only on the formal normal form $\delta - B$ of $\delta - A$. Furthermore, Corollary 7.51 implies the following corollary.

Corollary 7.53 *The irregularity of* $\delta - A$ *is zero if and only if* $\delta - A$ *is regular singular.*

Corollary 7.54 *Let the matrix A have coordinates in* $\mathbf{C}(\{z\})$. *Then* $\delta - A$ *has a finite dimensional kernel and cokernel for its action on both* $\mathbf{C}((z))^n$ *and* $(\mathbf{C}(\{z\}))^n$. *Define the Euler characteristics (or indices)*

$$\chi(\delta - A, \mathbf{C}((z))) = \dim \ker(\delta - A, \mathbf{C}((z))^n) - \dim \operatorname{coker}(\delta - A, \mathbf{C}((z))^n)$$
$$\chi(\delta - A, \mathbf{C}(\{z\})) = \dim \ker(\delta - A, \mathbf{C}(\{z\})^n) - \dim \operatorname{coker}(\delta - A, \mathbf{C}(\{z\})^n).$$

Then the irregularity of $\delta - A$ *is equal to* $\chi(\delta - A, \mathbf{C}((z))) - \chi(\delta - A, \mathbf{C}(\{z\})) = \dim \ker(\delta - A, \mathbf{C}((z))^n/\mathbf{C}(\{z\})^n)$.

Proof. Using Proposition 7.24.2, one sees that the exact sequence of sheaves

$$0 \to \mathcal{A}^0 \to \mathcal{A} \to \mathbf{C}((z)) \to 0$$

induces an exact sequence

$$0 \to \mathbf{C}(\{z\}) \to \mathbf{C}((z)) \to H^1(\mathbf{S}^1, \mathcal{A}^0) \to 0,$$

and we can identify the group $H^1(\mathbf{S}^1, \mathcal{A}^0)$ with $\mathcal{Q} := \mathbf{C}((z))/\mathbf{C}(\{z\})$.
According to Theorem 7.12 the map $(\delta - A) : (\mathcal{A}^0)^n \to (\mathcal{A}^0)^n$ is surjective and one finds an exact sequence of sheaves

$$0 \to \ker(\delta - A, (\mathcal{A}^0)^n) \to (\mathcal{A}^0)^n \to (\mathcal{A}^0)^n \to 0.$$

Taking cohomology on $\mathbf{S}^1$ one finds the exact sequence

$$0 \to H^1(\mathbf{S}^1, \ker(\delta - A, (\mathcal{A}^0)^n)) \to \mathcal{Q}^n \stackrel{\delta - A}{\to} \mathcal{Q}^n \to 0. \tag{7.2}$$

Let $\delta - A$ act on the exact sequence

$$0 \to (\mathbf{C}(\{z\})^n \to (\mathbf{C}((z)))^n \to \mathcal{Q}^n \to 0.$$

Let $\delta - A$ map each term in the exact sequence to itself. The sequence (7.2) implies that $\operatorname{coker}(\delta - A, \mathcal{Q}^n) = 0$. The Snake Lemma ([170], Lemma 9.1, Chap. III.10) applied to the last equivalence yields

$$\begin{aligned}0 &\to \ker(\delta - A, (\mathbf{C}(\{z\})^n) \to \ker(\delta - A, (\mathbf{C}((z)))^n) \to \ker(\delta - A, \mathcal{Q}^n) \quad (7.3)\\ &\to \operatorname{coker}(\delta - A, \mathbf{C}(\{z\})^n) \to \operatorname{coker}(\delta - A, \mathbf{C}((z)))^n) \to 0.\end{aligned}$$

The two kernels in this exact sequence have a finite dimension. We shall show below that the cokernel of $\delta - A$ on $\mathbf{C}((z))^n$ has finite dimension. Thus the other cokernel also has finite dimension and the formula for the irregularity of $\delta - A$ follows.

To see that the cokernel of $\delta - A$ on $\mathbf{C}((z))^n$ has finite dimension, note that $\delta - A$ is formally equivalent to a quasisplit $\delta - B$. We claim that it is enough to prove this claim for equations of the form $\delta - q + C$, where $q = q_N z^{-N} + \ldots + q_1 z^{-1}$, $q_N \neq 0$ and C is a matrix of constants. Since $\delta - B$ is quasisplit, if we establish the claim, then $\delta - A$ will have a finite dimensional cokernel of $\mathbf{C}((z^{1/m}))$ for some $m \geq 1$. If $v \in \mathbf{C}((z))^N$ is in the image of $\mathbf{C}((z^{1/m}))$ under $\delta - Z$ then it must be in the image of $\mathbf{C}((z))$ under this map. Therefore, the claim would prove that $\delta - A$ would have finite cokernel.

To prove the claim, first assume that $N > 0$. Then for any $v \in \mathbf{C}^n$ and any m, $(\delta - q + C)z^m v = q_N z^{m-N} v +$ higher order terms, so $\delta - A$ has 0 cokernel. If $N = 0$ (i.e., $q = 0$) then $(\delta - q + C)z^m v = (mI + C)z^m v$. Since, for sufficiently large m, $mI + C$ is invertible, we have that $\delta - A$ has 0 cokernel on $x^m \mathbf{C}[[z]]^n$ and, therefore, finite cokernel of $\mathbf{C}((x))^n$. □

Remark 7.55 Corollaries 7.52 and 7.54 imply that if $\delta - A$ is regular singular and $w \in \mathbf{C}(\{z\})^n)$ then any solution $v \in \mathbf{C}((z)))^n)$ of $(\delta - A)v = w$ is convergent.

Exercise 7.56 Consider a differential operator $L = \sum_{i=0}^{n} a_i \delta^i \in \mathbf{C}(\{z\})[\delta]$ with $a_n = 1$. Let $\delta - A$ be the associated matrix differential operator. Prove that L as an operator on $\mathbf{C}((z))$ and $\mathbf{C}(\{z\})$ has the same Euler characteristic as the operator $\delta - A$ on $\mathbf{C}((z))^n$ and $\mathbf{C}(\{z\})^n$. Prove that the irregularity of L, defined as the irregularity of $\delta - A$, is equal to $-\min_{0 \leq j \leq n} v(a_j)$. Here, v is the additive valuation on $\mathbf{C}(\{z\})$ (or on $\mathbf{C}((z))$) defined by $v(0) = +\infty$ and $v(b) = m$ if $b = \sum_{n \geq m} b_n z^n$ with $b_m \neq 0$. Hint: Note that $-\min_{0 \leq j \leq n} v(a_j)$ is the difference in the y-coordinates of the first and last corner of the Newton polygon of L. Now use Corollary 7.51 and Remark 3.55.1.

The result of this exercise appears in [187] where a different proof is presented. The result is also present in [109]. A more general version (and other references) appears in [180]. □

Remark 7.57 *The uniqueness of the multisum $v_1, \ldots, v_r$.*
The uniqueness of the multisum for any formal $\hat{v}$, is an immediate consequence of Theorem 7.47. In our situation $\hat{v}$ is a formal solution of $(\delta - A)\hat{v} = w$ and d is a nonsingular direction. Theorem 7.47 implies, moreover, that $(\delta - A)v_i \equiv w$ modulo $\mathcal{A}^0_{1/k_{i+1}}$.

We now *modify* the definition of the multisum of a formal solution $\hat{v}$ of $(\delta - A)\hat{v} = w$ by requiring that the sequence $v_1, \ldots, v_r$ satisfies $(\delta - A)v_i \equiv w \bmod \mathcal{A}^0_{1/k_{i+1}}$

for all i. The proof of Theorem 7.50 yields a *unique* sequence $v_1, \ldots, v_r$ satisfying this modified definition. In particular, for this modified definition we avoid the use of Theorem 7.47. □

Proposition 7.58 *Consider a formal solution $\hat{v}$ of the equation $(\delta - A)\hat{v} = w$. Let $\underline{k} = k_1 < \cdots < k_r$ with $1/2 < k_1$ denote the slopes of $\delta - A$ and let the direction d be not singular for $\delta - A$. Then, there are $F_1, \ldots, F_r \in \mathbf{C}((z))$ such that $\hat{v} = F_1 + \cdots + F_r$, each F_i is k_i-summable in the direction d and $(\delta - A)F_i$ convergent for each i.*

Proof. For convenience we consider only the case $r = 2$. It will be clear how to extend the proof to the case $r > 2$.

Let $\mathcal{V}_i$ for $i = 1, 2$ denote, as in the proof of Theorem 7.50, the sheaf $\ker(\delta - A, (\mathcal{A}^0_{1/k_i})^n)$. Let I denote the interval $(d - \frac{\pi}{2k_1} - \epsilon, d + \frac{\pi}{2k_1} + \epsilon)$ for suitable positive ϵ. Since d is not a singular direction, one has $H^1(I, \mathcal{V}_1/\mathcal{V}_2) = 0$. The obstruction for having an asymptotic lift of $\hat{v}$ on the sector I is an element $\xi_1 \in H^1(I, \mathcal{V}_1)$. From $H^0(I, \mathcal{V}_1/\mathcal{V}_2) = 0$ and $H^1(I, \mathcal{V}_1/\mathcal{V}_2) = 0$ one concludes that the map $H^1(I, \mathcal{V}_2) \to H^1(I, \mathcal{V}_1)$ is an isomorphism. Let $\xi_2 \in H^1(I, \mathcal{V}_2)$ map to ξ_1. The element ξ_2 can be given by a 1-cocycle with respect to a finite covering of I, since $H^1(J, \mathcal{V}_2) = 0$ if the length of the interval J is $\leq \frac{\pi}{k_2}$. Clearly, the covering and the 1-cocycle can be completed to a 1-cocycle for $\mathcal{V}_2$ on $\mathbf{S}^1$. In this way, one finds a $\xi_3 \in H^1(\mathbf{S}^1, \mathcal{V}_2)$ that maps to ξ_2.

One considers $\mathcal{V}_2$ as a subsheaf of $(\mathcal{A}^0_{1/k_2})^n$. According to Proposition 7.28, there is an element $F_2 \in \mathbf{C}((z))^n_{1/k_2}$ that maps to ξ_3. Furthermore, $(\delta - A)F_2$ maps to $(\delta - A)\xi_3 = 0$. Thus $w_2 := (\delta - A)F_2$ is convergent. The obstruction for having an asymptotic lift of F_2 to any interval J is an element of $H^1(J, \mathcal{V}_2)$ (in fact, the image of ξ_3). Since d is not a singular direction, this obstruction is 0 for an interval $(d - \frac{\pi}{2k_2} - \epsilon, d + \frac{\pi}{2k_2} + \epsilon)$ for small enough positive ϵ. This means that F_2 is k_2-summable in the direction d.

Define $F_1 := \hat{v} - F_2$ and $w_1 := w - w_2$. Then $(\delta - A)F_1 = w_1$. One can lift F_1, locally, to a solution in $(\mathcal{A}_{1/k_1})^n$ of the equation. The obstruction for a "global" asymptotic lift on the sector I is an element of $H^1(I, \mathcal{V}_1)$, namely the difference between ξ_1 and the image of ξ_3. By construction, this difference is 0 and it follows that F_1 is k_1-summable in the direction d. □

The next lemma is rather useful. We will give a proof using Laplace and Borel transforms (see [15], p. 30).

Lemma 7.59 *Let $1/2 < k_1 < k_2$ and assume that the formal power series $\hat{f}$ is k_1-summable and lies in $\mathbf{C}[[z]]_{1/k_2}$. Then $\hat{f} \in \mathbf{C}\{z\}$.*

Proof. It suffices to show that $\hat{f}$ is k_1-summable for every direction d, since the unique k_1-sums in the various directions glue to an element of $H^0(\mathbf{S}^1, \mathcal{A}_{1/k_1})$, which is equal to $\mathbf{C}(\{z\})$. In what follows we assume for convenience that $k_1 = 1$ and we consider the direction 0 and an interval (a, b) with $a < 0 < b$ and such that $\hat{f}$ is 1-summable in every direction $d \in (a, b),\ d \neq 0$. We now consider the formal Borel transform $g := \hat{\mathcal{B}}_1 \hat{f}$. If we can show that this defines an analytic function in a full sector containing $d = 0$ and having exponential growth of order ≤ 1, then Theorem 7.34 implies that $\hat{f}$ is 1-summable in the direction $d = 0$.

One sees that $g := \hat{\mathcal{B}}_1 \hat{f}$ is an entire function of exponential growth $\leq k$ with $\frac{1}{k} = 1 - \frac{1}{k_2}$. Indeed, let $\hat{f} = \sum_{n \geq 0} c_n z^n$. There are positive constants A_1, A_2 such that $|c_n| \leq A_1 A_2^n (n!)^{1/k_2}$ holds for all $n \geq 0$. The coefficients $\frac{c_n}{n!}$ of g satisfy the inequalities $|\frac{c_n}{n!}| \leq A_1 A_2^n (n!)^{-1+\frac{1}{k_2}}$ and this implies the exponential growth at ∞ of g of order $\leq k$. Moreover, according to Theorem 7.34, the function g has exponential growth of order ≤ 1 for any direction $d \in (a, b),\ d \neq 0$. The Phragmén-Lindelöf Theorem ([40], Chap. 33) implies that g also has exponential growth at ∞ of order ≤ 1 in the direction 0. In fact, one can prove this claim directly and in order to be complete, we include the proof.

Consider the closed sector S at ∞, given by the inequalities $R \leq |\zeta| < \infty$ and $|\arg(\zeta)| \leq \alpha$ with a fixed small $\alpha > 0$. On the boundaries $\arg(\zeta) = \pm\alpha$ the inequality $|g(\zeta)| \leq A\ exp(\ B|\zeta|)$ is given. Consider now the function $h(\zeta) = g(\zeta) exp(\ M\zeta - \epsilon\zeta^{k+\delta})$, with δ and ϵ positive and small and we take $M < 0$ such that $M \leq -\frac{B}{\cos\alpha}$. The presence of the term $exp(-\epsilon\zeta^{k+\delta})$ guarantees that $h(\zeta)$ tends to zero for $\zeta \in S,\ |\zeta| \to \infty$. Thus h is bounded on S and its maximum is obtained on the boundary of S. For $\zeta \in S$ with $\arg(\zeta) = \pm\alpha$ one estimates $|h(\zeta)|$ by

$$\leq A\ exp(B|\zeta|) exp(M\cos(\alpha)|\zeta| - \epsilon\cos((k+\delta)\alpha)|\zeta|)| \leq A.$$

For $\zeta \in S$ with $|\zeta| = R$ one can estimate $|h(\zeta)|$ by $\max\{|g(\zeta)|\ |\ |\zeta| = R \text{ and } \zeta \in S\}$. Thus there is a constant $C > 0$, not depending on our choices for M, ϵ, δ, with $|h(\zeta)| \leq C$ for all $\zeta \in S$. The inequality $|g(\zeta)| \leq C|exp(-M\zeta)|\ |exp(\epsilon\zeta^{k+\delta})|$ holds for fixed $\zeta \in S$ and all $\epsilon > 0$. Thus $|g(\zeta)| \leq C|exp(-M\zeta)|$ holds on S and g has exponential growth in the direction 0 of order ≤ 1. □

Example 7.60 *The equation* $(\delta - A)\hat{v} = w$ *with* $A = \begin{pmatrix} q_1 & 0 \\ 1 & q_2 \end{pmatrix}$ *with* $q_1, q_2 \in z^{-1}\mathbf{C}[z^{-1}]$ *of degrees* $k_1 < k_2$ *in the variable* z^{-1}.
We start with some observations.

- The equation $\delta - A$ is formally, but not analytically, equivalent to $\begin{pmatrix} q_1 & 0 \\ 0 & q_2 \end{pmatrix}$. Indeed, the formal equivalence is given by the matrix $\begin{pmatrix} 1 & 0 \\ f & 1 \end{pmatrix}$, where f is a solution of $(\delta + q_1 - q_2) f = 1$. According to Corollary 7.22, the unique solution f is divergent.
- The irregularity of $\delta - A$ is $k_1 + k_2$ and $\delta - A$ acts bijectively on $\mathbf{C}((z))^2$. According to Corollary 7.54, the cokernel of $\delta - A$ acting upon $\mathbf{C}(\{z\})^2$ has dimension $k_1 + k_2$. Using Corollary 7.22, one concludes that the cokernel of $\delta - A$ on $\mathbf{C}(\{z\})^2$ is represented by the elements $\begin{pmatrix} f_1 \\ f_2 \end{pmatrix}$ with f_1, f_2 polynomials in z of degrees $< k_1$ and $< k_2$.

- As in the proof of Theorem 7.50, we consider the sheaves $\mathcal{V}_1 := \ker(\delta - A, (\mathcal{A}^0)^2) = \ker(\delta - A, (\mathcal{A}^0_{1/k_1})^2)$ and the subsheaf $\mathcal{V}_2 := \ker(\delta - A, (\mathcal{A}^0_{1/k_2})^2)$ of $\mathcal{V}_1$. The sheaf $\mathcal{V}_2$ is isomorphic to $\ker(\delta - q_2, \mathcal{A}^0_{1/k_2})$, by the map $f \mapsto \binom{0}{f}$. The sheaf $\mathcal{V}_1/\mathcal{V}_2$ is isomorphic to $\ker(\delta - q_1, \mathcal{A}^0_{1/k_1})$.

We want to show two results:

In general, the exact sequence $0 \to \mathcal{V}_2 \to \mathcal{V}_1 \to \mathcal{V}_1/\mathcal{V}_2 \to 0$ does not split.

In general, the decomposition of Proposition 7.58 depends on the chosen direction d.

Indeed, we will consider the above family of examples with $q_1 = z^{-1}$ and $q_2 = z^{-2}$ and show that the exact sequence does not split and prove that the formal solution $\hat{v}$ of $(\delta - A)\hat{v} = \binom{1}{0}$ cannot globally, i.e., on all of $\mathbf{S}^1$, be written as a sum $F_1 + F_2$ with k_i-summable F_i for $i = 1, 2$.

It is, furthermore, easily seen that the computations in this special case extend to the general case of the above family of examples.

From now on we assume $q_1 = z^{-1}$ and $q_2 = z^{-2}$. Let $e(q)$ denote the standard solution of $(\delta - q)e(q) = 0$, i.e. $q \in z^{-1}\mathbf{C}[z^{-1}]$ and $e(q) = exp(\int q\frac{dz}{z})$ with again $\int q\frac{dz}{z} \in z^{-1}\mathbf{C}[z^{-1}]$. The interval where $e(q_1)$ is flat is $I_1 := (-\frac{\pi}{2}, \frac{\pi}{2})$ and the two intervals where q_2 is flat are $I_2 := (-\frac{\pi}{4}, \frac{\pi}{4})$ and $I_3 := (\frac{3\pi}{4}, \frac{5\pi}{4})$. The sheaf $\mathcal{V}_1/\mathcal{V}_2$ is isomorphic to $\mathbf{C}_{I_1}$ and the sheaf $\mathcal{V}_2$ is isomorphic to $\mathbf{C}_{I_2} \oplus \mathbf{C}_{I_3}$. The exact sequence

$$0 \to \mathcal{V}_2 \to \mathcal{V}_1 \to \mathcal{V}_1/\mathcal{V}_2 \to 0$$

of course splits locally. Two local splittings in a direction d differ by a morphism of $(\mathcal{V}_1/\mathcal{V}_2)_d \to (\mathcal{V}_2)_d$. The obstruction to global splitting is therefore an element of $H^1(\mathbf{S}^1, \mathrm{Hom}_{\mathbf{C}}(\mathcal{V}_1/\mathcal{V}_2, \mathcal{V}_2))$. The sheaf appearing in this cohomology group is, according to the proof of Lemma 7.41, isomorphic to $(\mathbf{C}_{\overline{I}_1})_{I_2}$. Since I_2 is contained in $\overline{I}_1$, the above cohomology group is isomorphic to $\mathbf{C}$. This is why we do not expect the sequence to be split. Of course we have to make a computation in order to show that the obstruction is actually nontrivial. It suffices to show that $H^0(I_1, \mathcal{V}_1) = 0$. Indeed, suppose that the exact sequence of sheaves splits above I_1. Then

$$0 \to H^0(I_1, \mathcal{V}_2) \to H^0(I_1, \mathcal{V}_1) \to H^0(I_1, \mathcal{V}_1/\mathcal{V}_2) \to 0$$

would be exact and thus $H^0(I_1, \mathcal{V}_1) \cong \mathbf{C}$.

A nonzero element of $H^0(I_1, \mathcal{V}_1)$ is a nonzero multiple of $\binom{e(q_1)}{f}$ where f would be flat on I_1 and satisfies $(\delta - q_2)f = e(q_1)$. This equation has a unique flat solution F_1 on the sector $(-\frac{\pi}{4}, \frac{\pi}{2})$ and a unique flat solution F_2 on the sector $(-\frac{\pi}{2}, \frac{\pi}{4})$. According to the proof of Lemma 7.13, those two solutions are given by integrals $F_i(z) = e(q_2)(z)\int_{\lambda_i} e(-q_2 + q_1)(t)\frac{dt}{t}$. The first path λ_1 from 0 to z consists of two pieces $\{re^{i\phi_1} | \ 0 \leq r \leq |z|\}$ (for any ϕ_1 such that $\frac{\pi}{4} \leq \phi_1 < \frac{\pi}{2}$) and $\{|z|e^{i\phi} | \ \phi$ from ϕ_1 to $\arg(z)\}$. The second path λ_2 consists of the two pieces $\{re^{i\phi_2} |\ 0 \leq r \leq |z|\}$ (for any ϕ_2 such that $-\frac{\pi}{2} < \phi_2 \leq -\frac{\pi}{4}$) and $\{|z|e^{i\phi}|\ \phi$ from ϕ_2 to $\arg(z)\}$. We want to prove that $F_1 \neq F_2$, because that implies that the equation $(\delta - q_2)f = e(q_1)$ does not have a flat solution on I_1 and so $H^0(I_1, \mathcal{V}_1) = 0$.

The difference $e(-q_2)(F_2 - F_1)$ is a constant, i.e., independent of z, and therefore equal to the integral $\int_{\lambda_R} e(-q_2+q_1)(t)\frac{dt}{t}$ for $R > 0$, where λ_R is a path consisting of three pieces $\{re^{-i\frac{\pi}{4}} |\ 0 \leq r \leq R\}$, $\{Re^{i\phi} |\ -\frac{\pi}{4} \leq \phi \leq \frac{\pi}{4}\}$ and $\{re^{i\frac{\pi}{4}} |\ R \geq r \geq 0\}$. After parametrization of λ_R one computes that the integral is equal to

$$2i \int_0^R e^{\frac{-\sqrt{2}}{2r}} \sin(\frac{1}{2r^2} - \frac{\sqrt{2}}{2r})\frac{dr}{r} + i\int_{-\frac{\pi}{4}}^{\frac{\pi}{4}} exp(\frac{e^{-2i\phi}}{2R^2} - \frac{e^{-i\phi}}{R})\, d\phi.$$

The second integral has limit $i\frac{\pi}{2}$ for $R \to \infty$. The first integral also has a limit for $R \to \infty$, namely $2ia$ with

$$a := \int_0^\infty e^{\frac{-\sqrt{2}}{2r}} \sin(\frac{1}{2r^2} - \frac{\sqrt{2}}{2r})\frac{dr}{r}.$$

Numerical integration gives $a = -0.2869...$ and thus the total integral is not 0.

We consider now the equation $(\delta - A)\hat{v} = \binom{1}{0}$ and suppose that $\hat{v} = \hat{v}_1 + \hat{v}_2$ with a k_i-summable $\hat{v}_i$ for $i = 1, 2$. Then $(\delta - A)\hat{v}_1$ is k_1-summable and belongs, moreover, to $\mathbf{C}((z))_{1/k_2}$. According to Lemma 7.59, $w_1 := (\delta - A)\hat{v}_1$ is convergent. Then, also $w_2 := (\delta - A)\hat{v}_2$ is convergent. Since $\hat{v}_2$ is k_2-summable it follows that w_2 is modulo the image of $(\delta - A)$ on $\mathbf{C}(\{z\})^2$ an element of the form $\binom{0}{h}$ with h a polynomial of degree ≤ 1. After changing $\hat{v}_2$ by a convergent vector, we may suppose that $(\delta - A)\hat{v}_2 = \binom{0}{h}$. Thus $(\delta - A)\hat{v}_1 = \binom{1}{-h}$. Thus we have found a k_1-summable $\hat{F}$ with $(\delta - A)\hat{F} = \binom{1}{k}$ with k a polynomial of degree ≤ 1.

By definition, $\hat{F}$ is k_1-summable in all but finitely many directions. There is some $\epsilon > 0$ such that $\hat{F}$ is k_1-summable in all directions in $(-\epsilon, 0) \cup (0, \epsilon)$. Using the first interval one finds an $f_1 \in (\mathcal{A}_{1/k_1})^2(-\epsilon - \frac{\pi}{2}, \frac{\pi}{2})$ with asymptotic expansion $\hat{F}$. Then $(\delta - A)f_1$ is k_1-summable with convergent asymptotic expansion $\binom{1}{k}$ on the same interval. Thus, by Lemma 7.27, one has $(\delta - A)f_1 = \binom{1}{k}$. Similarly there is an $f_2 \in (\mathcal{A}_{1/k_1})^2(-\frac{\pi}{2}, \frac{\pi}{2} + \epsilon)$ with asymptotic expansion $\hat{F}$ and $(\delta - A)f_2 = \binom{1}{k}$. The difference $f_1 - f_2$ lies in $H^0(I_1, \mathcal{V}_1)$ and is therefore 0. Thus there is an element $f_3 \in (\mathcal{A}_{1/k_1})^2(-\epsilon - \frac{\pi}{2}, \frac{\pi}{2} + \epsilon)$ with asymptotic expansion $\hat{F}$ and with $(\delta - A)f_3 = \binom{1}{k}$. The first coordinate g of f_3 lies in $\mathcal{A}_{1/k_1}(-\epsilon - \frac{\pi}{2}, \frac{\pi}{2} + \epsilon)$ and satisfies the equation $(\delta - q_1)g = 1$. The formal solution $\hat{u}$ of $(\delta - q_1)\hat{u} = 1$ also has a unique asymptotic lift $\tilde{g}$ on the sector $\mathbf{S}^1 \setminus \{0\}$. The difference $g - \tilde{g}$ is zero on the two sectors $(0, \frac{\pi}{2} + \epsilon)$ and $(-\epsilon - \frac{\pi}{2}, 0)$, since the sheaf $\ker(\delta - q_1, \mathcal{A}^0)$ has only the zero section on the two sectors. Thus g and $\tilde{g}$ glue to a convergent solution of $(\delta - q_1)h = 1$ and $h = \hat{u}$ is convergent. However, by Corollary 7.22, one knows that $\hat{u}$ is divergent. This ends the proof of our two claims.

We make some further comments on this example. From $H^0(I_1, \mathcal{V}_1) = 0$ it follows that also $H^1(I_1, \mathcal{V}_1) = 0$. This has the rather curious consequence that any formal solution of $(\delta - A)\hat{v} = w$ has a unique asymptotic lift above the sector I_1. This asymptotic lift is, in general, not a $\underline{k}$-sum in a direction.

We note that a small change of q_1 and q_2 does not affect the above calculation in the example. Similarly, one sees that q_1 and q_2 of other degrees $k_1 < k_2$ (in the variable z^{-1}) will produce, in general, the same phenomenon as above. Only rather special relations between the coefficients of q_1 and q_2 will produce a sheaf $\mathcal{V}_1$ that is isomorphic to the direct sum of $\mathcal{V}_2$ and $\mathcal{V}_1/\mathcal{V}_2$.

The first result gives a negative answer to the question posed by B. Malgrange in [193] (2.1) p. 138. The second result gives a negative answer to another open question. □

Remarks 7.61 *Multisummation and the Borel and Laplace transforms.*
The translation of k-summability in terms of Borel and Laplace transforms, given in Theorem 7.34, has an analog for multisummation. We will not use this formulation of multisummability, but present the highly complicated result for the information of the reader. More information can be found in [15, 16]. Given are $\hat{f} \in \mathbf{C}[[z]]$, a direction d and $\underline{k} = k_1 < \cdots < k_r$ with $k_1 > 1/2$. Then $\hat{f}$ is $\underline{k}$-summable in the direction d if the formula

$$\mathcal{L}_{k_r} a(\kappa_r)\mathcal{B}_{k_r} \cdots \mathcal{L}_{k_j} a(\kappa_j)\mathcal{B}_{k_j} \cdots a(\kappa_2)\mathcal{B}_{k_2}\mathcal{L}_{k_1} a(\kappa_1)\mathcal{B}_{k_1}\hat{f} \text{ is meaningful.}$$

We will explain what is meant by this.

- The κ_i are defined by $\frac{1}{k_i} = \frac{1}{k_{i+1}} + \frac{1}{\kappa_i}$. For notational convenience we write $k_{r+1} = \infty$ and hence $\kappa_r = k_r$. Moreover, $\mathcal{A}^0_{1/k_{r+1}}$ is by definition 0.
- The first $\mathcal{B}_{k_1}$ is by definition the formal Borel transform $\hat{\mathcal{B}}_{k_1}$ of order k_1. The first condition is that $\hat{\mathcal{B}}_{k_1}\hat{f}$ is convergent, in other words $\hat{f} \in \mathbf{C}[[z]]_{1/k_1}$.
- The $\mathcal{B}_{k_j}$ are "extended" Borel transforms of order k_j in the direction d for $j = 2, \ldots, r$. They can be seen as maps from $\mathcal{A}/\mathcal{A}^0_{1/k_j}(d - \frac{\pi}{2k_j} - \epsilon, d + \frac{\pi}{2k_j} + \epsilon)$ to $\mathcal{A}(d - \epsilon, d + \epsilon)$.
- The $\mathcal{L}_{k_j}$ are "extended" Laplace transforms of order k_j in the direction d. They map the elements in $\mathcal{A}(d - \epsilon, d + \epsilon)$, having an analytic continuation with exponential growth of order $\leq \kappa_{k_j}$, to elements of $\mathcal{A}/\mathcal{A}^0_{1/k_{j+1}}(d - \frac{\pi}{2k_j} - \epsilon, d + \frac{\pi}{2k_j} + \epsilon)$.
- The symbol $a(\kappa)\phi$ is not a map. It means that one supposes the holomorphic function ϕ to have an analytic continuation in a suitable full sector containing the direction d. Moreover, this analytic continuation is supposed to have exponentional growth of order $\leq \kappa$.

The Borel transform of order k in direction d, applied to a function h, is defined by the formula $(\mathcal{B}_k h)(\zeta) = \frac{1}{2\pi i}\int_\lambda h(z)z^k exp((\frac{\zeta}{z})^k)\, dz^{-k}$. The path of integration λ consists of the three parts $\{ae^{id_1} \mid 0 \le a \le r\}$, $\{e^{is} \mid d_1 \ge s \ge d_2\}$ and $\{ae^{id_2} \mid r \ge a \ge 0\}$, where $d + \frac{\pi}{2k} < d_1 < d + \frac{\pi}{2k} + \epsilon$ and $d - \frac{\pi}{2k} - \epsilon < d_2 < d - \frac{\pi}{2k}$ and with ϵ, r positive and small.

The expression "extended" means that the integral transforms $\mathcal{L}_*$ and $\mathcal{B}_*$, originally defined for functions, are extended to the case of "k-precise quasi-functions", i.e., sections of the sheaf $\mathcal{A}/\mathcal{A}^0_{1/k}$.

The multisum $(f_1, \ldots, f_r)$ itself is defined by $f_j = \mathcal{L}_{k_j} \cdots \mathcal{B}_{k_1} \hat{f}$ for $j = 1, \ldots, r$.

We note that [176] and [178] also contain a discussion of the relationship between k-summability and Borel and Laplace transforms as well as illustrative examples. □

Exercise 7.62 Consider the matrix differential operator $\delta - A$ of size n. Let $k_1 < \cdots < k_r$, with $1/2 < k_1$, denote the slopes of $\delta - A$. As in Theorem 7.50 one defines for $i = 1, \ldots, r$ the sheaves $\mathcal{V}_i = \ker(\delta - A, (\mathcal{A}^0_{1/k_i})^n)$. For notational convenience we put $\mathcal{V}_{r+1} = 0$. Prove that there is a canonical isomorphism

$$\phi : \{\hat{v} \in \mathbf{C}((z))^n \mid (\delta - A)\hat{v} \text{ is convergent }\}/\{\hat{v} \in \mathbf{C}(\{z\})^n\} \to H^1(\mathbf{S}^1, \mathcal{V}_1).$$

Furthermore, show that ϕ induces isomorphisms

$$\{\hat{v} \in \mathbf{C}((z))^n_{1/k_i} \mid (\delta - A)\hat{v} \text{ is convergent }\}/\{\hat{v} \in \mathbf{C}(\{z\})^n\} \to H^1(\mathbf{S}^1, \mathcal{V}_i),$$

and also isomorphisms between

$$\{\hat{v} \in \mathbf{C}((z))^n_{1/k_i} \mid (\delta - A)\hat{v} \text{ is convergent }\}/\{\hat{v} \in \mathbf{C}((z))^n_{1/k_{i+1}}\}$$

and $H^1(\mathbf{S}^1, \mathcal{V}_i/\mathcal{V}_{i+1})$. □

8 Stokes Phenomenon and Differential Galois Groups

8.1 Introduction

We will first sketch the contents of this chapter. Let $\delta - A$ be a matrix differential equation over $\mathbf{C}(\{z\})$. Then there is a unique (up to isomorphism over $\mathbf{C}(\{z\})$) quasi-split equation $\delta - B$ that is isomorphic, over $\mathbf{C}((z))$, to $\delta - A$ (cf. Proposition 3.41). This means that there is a $\hat{F} \in \mathrm{GL}_n(\mathbf{C}((z)))$ such that $\hat{F}^{-1}(\delta - A)\hat{F} = \delta - B$. In the following, $\delta - A$, $\delta - B$, and $\hat{F}$ are fixed and the eigenvalues of $\delta - A$ and $\delta - B$ are denoted by $q_1, \ldots, q_s$.

The aim is to find the differential Galois group of $\delta - A$ over the field $\mathbf{C}(\{z\})$ in terms of $\delta - B$ and $\hat{F}$. Since $\delta - B$ is a quasi-split equation, we have seen in Proposition 3.40 that the differential Galois groups over $\mathbf{C}(\{z\})$ and $\mathbf{C}((z))$ coincide. The latter group is known. From the formal matrix $\hat{F}$ one deduces, by means of multisummation, a collection of Stokes matrices (also called Stokes multipliers) for the singular directions for the set of elements $\{q_i - q_j\}$. These Stokes matrices are shown to be elements in the differential Galois group of $\delta - A$. Finally, it will be shown that the differential Galois group is generated, as a linear algebraic group, by the Stokes matrices and the differential Galois group of $\delta - B$. This result is originally due to J.-P. Ramis.

There are only a few examples where one can actually calculate the Stokes matrices. However, the above theorem of Ramis gains in importance from the following three additions:

1. The Stokes matrix associated to a singular directions (for the collection $\{q_i - q_j\}$) has a special form. More precisely, let V denote the space of solutions of $\delta - A$ in the universal differential extension of $\mathbf{C}((z))$ (see Sect. 3.2), let $V = \oplus_{i=1}^{s} V_{q_i}$ be its canonical decomposition with respect to eigenvalues of $\delta - A$ and let $\gamma \in \mathrm{GL}(V)$ denote the formal monodromy. Then the Stokes matrix St_d for the singular direction $d \in \mathbf{R}$, considered as an element of $\mathrm{GL}(V)$ has the form $id + \sum A_{i,j}$, where $A_{i,j}$ denotes a linear map of the form

$$V \overset{\text{projection}}{\rightarrow} V_{q_i} \overset{\text{linear}}{\rightarrow} V_{q_j} \overset{\text{inclusion}}{\rightarrow} V,$$

and where the sum is taken over all pairs i, j, such that d is a singular direction for $q_i - q_j$. Furthermore, $\gamma^{-1} St_d \gamma = St_{d+2\pi}$ holds.

2. Let $d_1 < \cdots < d_t$ denote the singular directions (for the collection $\{q_i - q_j\}$), then the product $\gamma \circ St_{d_t} \cdots St_{d_1}$ is conjugate to the topological monodromy, that is the change of basis resulting from analytic continuation around the singular points, of $\delta - A$, considered as an element of $\mathrm{GL}(V)$.
3. Suppose that $\delta - B$ is fixed, i.e., V, the decomposition $\oplus_{i=1}^{s} V_{q_i}$ and γ are fixed. Given any collection of automorphisms $\{C_d\}$ satisfying the conditions in part 1, there is a differential equation $\delta - A$ and a formal equivalence $\hat{F}^{-1}(\delta - A)\hat{F} = \delta - B$ (unique up to isomorphism over $\mathbf{C}(\{z\})$), which has the collection $\{C_d\}$ as Stokes matrices.

In this chapter, we will give the rather subtle proof of part 1 and the easy proof of part 2. In Chap. 9 (Corollary 9.8), we will also provide a proof of part 3 with the help of tannakian categories. We note that part 3 has rather important consequences, namely Ramis's solution for the inverse problem of differential Galois groups over the field $\mathbf{C}(\{z\})$.

The expression "*the Stokes phenomenon*" needs some explication. In Chap. 7 we have seen that any formal solution $\hat{v}$ of an analytic differential equation $(\delta - A)\hat{v} = w$ can be lifted to a solution $v \in \mathcal{A}(a, b)^n$ for suitably small sectors (a, b). The fact that the various lifts do not glue to a lift on $\mathbf{S}^1$, is called the Stokes phenomenon. One can formulate this differently. Let, again, $v \in \mathcal{A}(a, b)^n$ be an asymptotic lift of $\hat{v}$. Then the analytical continuation of v in another sector is still a solution of the differential equation but will, in general, not have $\hat{v}$ as asymptotic expansion. G.G. Stokes made this observation in his study of the Airy equation $y'' = zy$, which has the point ∞ as an irregular singularity.

8.2 The Additive Stokes Phenomenon

We recall the result from *the Multisummation Theorem*, Theorem 7.50. Let $\delta - A$ be given, with positive slopes $\underline{k} = k_1 < \cdots < k_r$ (and $1/2 < k_1$) and with eigenvalues $q_1, \ldots, q_s$. The collection of singular directions $d_1 < \cdots < d_m < d_1(+2\pi)$ of $\delta - A$ is the union of the singular directions for each q_i. Consider a formal solution $\hat{v}$ of $(\delta - A)\hat{v} = w$ (with w convergent). For a direction d that is not singular for $\delta - A$, the multisummation theorem provides a unique asymptotic lift, denoted by $S_d(\hat{v})$, which lives in $\mathcal{A}(d - \frac{\pi}{2k_r} - \epsilon, d + \frac{\pi}{2k_r} + \epsilon)^n$ for small enough positive ϵ. Suppose that $d_i < d < d_{i+1}$, with for notational convenience $d_{m+1} = d_1(+2\pi)$. The uniqueness of the multisum $S_d(\hat{v})$, implies that there is a unique asymptotic lift above the sector $(d_i - \frac{\pi}{2k_r}, d_{i+1} + \frac{\pi}{2k_r})$, which coincides with $S_d(\hat{v})$ for any $d \in (d_i, d_{i+1})$.

For a singular direction d, say $d = d_i$, the multisum $S_d(\hat{v})$ does not exist. However, for directions d^-, d^+, with $d^- < d < d^+$ and $|d^+ - d^-|$ small enough, the multisums $S_{d^+}(\hat{v})$ and $S_{d^-}(\hat{v})$ do exist. They are independent of the choices for d^+, d^- and can be analytically continued to the sectors $(d_i - \frac{\pi}{2k_r}, d_{i+1} + \frac{\pi}{2k_r})$ and $(d_{i-1} - \frac{\pi}{2k_r}, d_i + \frac{\pi}{2k_r})$.

The difference $S_{d^-}(\hat{v}) - S_{d^+}(\hat{v})$ is certainly a section of the sheaf $\ker(\delta - A, (\mathcal{A}^0)^n)$ above the sector $(d_i - \frac{\pi}{2k_r}, d_i + \frac{\pi}{2k_r})$, and, in fact, a rather special one. The fact that this difference is, in general, not 0, is again the "Stokes phenomenon", but now in a more precise form.

Definition 8.1 For a singular direction d and multisums $S_{d^-}(\hat{v})$, $S_{d^+}(\hat{v})$ as defined above, we will write $st_d(\hat{v})$ for $S_{d^-}(\hat{v}) - S_{d^+}(\hat{v})$. □

We will make this definition more precise. We fix a formal equivalence between $\delta - A$ and $\delta - B$, where $\delta - B$ is quasi-split. This formal equivalence is given by an $\hat{F} \in \mathrm{GL}_n(\mathbf{C}((z)))$ satisfying $\hat{F}^{-1}(\delta - A)\hat{F} = \delta - B$. Let us write $\mathcal{K}_A$ and $\mathcal{K}_B$ for the sheaves $\ker(\delta - A, (\mathcal{A}^0)^n)$ and $\ker(\delta - B, (\mathcal{A}^0)^n)$. Let W denote the solution space of $\delta - B$ (with coordinates in the universal ring UnivR) and with its canonical decomposition $W = \oplus W_{q_i}$. The operator $\delta - B$ is a direct sum of operators $\delta - q_i + C_i$ (after taking a root of z) and W_{q_i} is the solution space of $\delta - q_i + C_i$. For each singular direction d of q_i, we consider the interval $J = (d - \frac{\pi}{2k(q_i)}, d + \frac{\pi}{2k(q_i)})$, where $k(q_i)$ is the degree of q_i in the variable z^{-1}. From Chap. 7 it is clear that $\mathcal{K}_B$ is (more or less canonically) isomorphic to the sheaf $\oplus_{i,J}(W_{q_i})_J$ on $\mathbf{S}^1$. Let V denote the solution space of $\delta - A$ (with coordinates in the universal ring) with its decomposition $\oplus V_{q_i}$. The formal equivalence, given by $\hat{F}$, produces an isomorphism between W and V respecting the two decompositions and the formal monodromy. Locally on $\mathbf{S}^1$, the two sheaves $\mathcal{K}_B$ and $\mathcal{K}_A$ are isomorphic. Thus $\mathcal{K}_A$ is locally isomorphic to the sheaf $\oplus_{i,J}(V_{q_i})_J$.

Let us first consider the special case where $\delta - A$ has only one positive slope k. In that case it is proven in Chap. 7 that the sheaves $\mathcal{K}_B$ and $\mathcal{K}_A$ are isomorphic, however, not in a canonical way. Thus $\mathcal{K}_A$ is isomorphic to $\oplus_{i,J}(V_{q_i})_J$, but not in a canonical way. We will rewrite the latter expression. Write $J_1, \ldots, J_m$ for the distinct open intervals involved. They have the form $(d - \frac{\pi}{2k}, d + \frac{\pi}{2k})$, where d is a singular direction for one of the q_i. We note that d can be a singular direction for several q_is. Now the sheaf $\mathcal{K}_A$ is isomorphic to $\oplus_{j=1}^m (D_j)_{J_j}$, with D_j some vector space. This decomposition is canonical, as one easily verifies. But the identification of the vector space D_j with $\oplus_i V_{q_i}$, the direct sum taken over the i such that the middle of J_j is a singular direction for q_i, is not canonical.

Now we consider the general case. The sheaf $\mathcal{K}_A$ is given a filtration by subsheaves $\mathcal{K}_A = \mathcal{K}_{A,1} \supset \mathcal{K}_{A,2} \supset \cdots \supset \mathcal{K}_{A,r}$, where $\mathcal{K}_{A,i} := \ker(\delta - A, (\mathcal{A}^0_{1/\mathcal{K}_i})^n)$. For notational convenience we write $\mathcal{K}_{A,r+1} = 0$. The quotient sheaf $\mathcal{K}_{A,i}/\mathcal{K}_{A,i+1}$ can be identified with $\ker(\delta - A, (\mathcal{A}^0_{1/k_i}/\mathcal{A}^0_{1/k_{i+1}})^n)$ for $i = 1, \ldots r - 1$. Again, for notational convenience we write $k_{r+1} = \infty$ and $\mathcal{A}^0_{1/k_{r+1}} = 0$. For the sheaf $T := \oplus_{i,J}(V_{q_i})_J$ we also introduce a filtration $T = T_1 \supset T_2 \supset \cdots \supset T_r$ with $T_j = \oplus_{i,J}(V_{q_i})_J$, where the direct sum is taken over all i such that the degree of q_i in the variable z^{-1} is $\geq k_i$. For convenience we put $T_{r+1} = 0$. Then it is shown in Chap. 7 that there are (noncanonical) isomorphisms $\mathcal{K}_{A,i}/\mathcal{K}_{A,i+1} \cong T_i/T_{i+1}$ for $i = 1, \ldots, r$. Using those isomorphisms, one can translate sections and cohomology classes of $\mathcal{K}_A$ in terms of the sheaf T. In particular, for any open interval $I \subset \mathbf{S}^1$ of length $\leq \frac{\pi}{k_r}$, the sheaves $\mathcal{K}_A$ and T are isomorphic and $H^0(I, \mathcal{K}_A)$ can be identified with $H^0(I, T) = \oplus_{i,J} H^0(I, (V_{q_i})_J)$. As we know $H^0(I, (V_{q_i})_J)$ is zero, unless $I \subset J$. In the latter case, $H^0(I, (V_{q_i})_J) = V_{q_i}$.

We return now to the "additive Stokes phenomenon" for the equation $(\delta - A)\hat{v} = w$. For a singular direction d we have considered $st_d(\hat{v}) := S_{d^-}(\hat{v}) - S_{d^+}(\hat{v})$ as an element of $H^0((d - \frac{\pi}{2k_r}, d + \frac{\pi}{2k_r}), \mathcal{K}_A) \cong H^0((d - \frac{\pi}{2k_r}, d + \frac{\pi}{2k_r}), T)$. The following proposition gives a precise meaning to the earlier assertion that $st_d(\hat{v})$ is a rather special section of the sheaf T.

Proposition 8.2 *The element $st_d(\hat{v})$, considered as a section of T above $(d - \frac{\pi}{2k_r}, d + \frac{\pi}{2k_r})$, belongs to $\oplus_{i \in I_d} V_{q_i}$, where I_d is the set of indices i such that d is a singular direction for q_i.*

Proof. We consider first the case that $\delta - A$ has only one positive slope k (and $k > 1/2$). Then $st_d(\hat{v}) \in H^0((d - \frac{\pi}{2k}, d + \frac{\pi}{2k}), T)$. The only direct summands of $T = \oplus_{i,J}(V_{q_i})_J$ that give a nonzero contribution to this group H^0 are the pairs (i, J) with $J = (d - \frac{\pi}{2k}, d + \frac{\pi}{2k})$. For such a direct summand the contribution to the group H^0 is canonical isomorphic to V_{q_i}. This ends the proof in this special case. The proof for the general case, i.e., $r > 1$, is for $r > 2$ quite similar to the case $r = 2$. For $r = 2$ we will provide the details.

Let the direction d be nonsingular. The multisum in the direction d is, in fact, a pair (v_1, v_2) with v_1 a section of $(\mathcal{A}/\mathcal{A}^0_{1/k_2})^n$ satisfying $(\delta - A)v_1 = w$ (as sections of the sheaf $(\mathcal{A}/\mathcal{A}^0_{1/k_2})^n$). This section is defined on an interval $(d - \frac{\pi}{2k_1} - \epsilon, d + \frac{\pi}{2k_1} + \epsilon)$. The unicity of v_1 proves that v_1 is, in fact, defined on an open $(e - \frac{\pi}{2k_1}, f + \frac{\pi}{2k_1})$, where $e < f$ are the consecutive singular directions *for the slope* k_1 with $e < d < f$. The element v_2 is a section of the sheaf $(\mathcal{A})^n$ satisfying $(\delta - A)v_2 = w$. This section is defined above the interval $(d - \frac{\pi}{2k_2} - \epsilon, d + \frac{\pi}{2k_2} + \epsilon)$. As above, v_2 is defined on the interval $(e^* - \frac{\pi}{2k_2}, f^* + \frac{\pi}{2k_2})$ where $e^* < f^*$ are the consecutive singular directions *for the slope* k_2 such that $e^* < d < f^*$. Moreover, v_1 and v_2 have the same image as the section of the sheaf $(\mathcal{A}/\mathcal{A}^0_{1/k_2})^n$ above $(e - \frac{\pi}{2k_1}, f + \frac{\pi}{2k_1}) \cap (e^* - \frac{\pi}{2k_2}, f^* + \frac{\pi}{2k_2})$.

Let d now be a singular direction. We apply the above for the two directions d^+ and d^- and write (v_1^+, v_2^+) and (v_1^-, v_2^-) for the two pairs. Then $st_d(\hat{v}) = v_2^- - v_2^+$ is a section of $\mathcal{K}_{A,1}$ above the interval $I := (d - \frac{\pi}{2k_2}, d + \frac{\pi}{2k_2})$. Using the isomorphism of $\mathcal{K}_A = \mathcal{K}_{A,1}$ with $T = T_1$ above this interval we can identify $st_d(\hat{v})$ with an element of $H^0(I, T_1)$. One considers the exact sequence

$$0 \to H^0(I, T_2) \to H^0(I, T_1) \to H^0(I, T_1/T_2) \to 0.$$

The element $v_1^- - v_1^+$ lives in the sheaf $\mathcal{K}_{A,1}/\mathcal{K}_{A,2} \cong T_1/T_2$ above the interval $J = (d - \frac{\pi}{2k_1}, d + \frac{\pi}{2k_1})$. Furthermore, the images of $st_d(\hat{v})$ and $v_1^- - v_1^+$ in $H^0(I, T_1/T_2)$ are the same. The group $H^0(J, T_1/T_2)$ can be identified with the direct sum $\oplus V_{q_i}$, taken over all q_i with slope k_1 and d singular for q_i. In the same way, $H^0(I, T_2)$ can be identified with the direct sum $\oplus V_{q_i}$, taken over all q_i with slope k_2 and d as singular direction. Thus we conclude that $st_d(\hat{v})$ lies in the direct sum $\oplus V_{q_i}$, taken over all q_i such that d is a singular direction for q_i. □

Corollary 8.3 *The additive Stokes phenomenon yields isomorphisms between the following* **C**-*vector spaces:*
(a) $\{\hat{v} \in \mathbf{C}((z))^n \mid (\delta - A)\hat{v}$ *is convergent*$\}/\{\hat{v} \in \mathbf{C}(\{z\})^n\}$.
(b) $H^1(\mathbf{S}^1, \ker(\delta - A, (\mathcal{A}^0)^n))$.
(c) $\oplus_{d \text{ singular}} \oplus_{i \in I_d} V_{q_i}$.

Proof. Consider the (infinite dimensional) vector space M consisting of the $\hat{v} \in \mathbf{C}((z))^n$ such that $w := (\delta - A)\hat{v}$ is convergent. According to Chap. 7 every $\hat{v}$ has asymptotics lift v_S, on small enough sectors S, satisfying $(\delta - A)v_S = w$. The differences $v_S - v_{S'}$ determine a 1-cocycle for the sheaf $\ker(\delta - A, (\mathcal{A}^0)^n)$. The kernel of the resulting linear surjective map $M \to H^1(\mathbf{S}^1, \ker(\delta - A, (\mathcal{A}^0)^n))$ is $\mathbf{C}(\{z\})^n$.

One also considers the linear map $M \to \oplus_{d \text{ singular}} \oplus_{i \in I_d} V_{q_i}$, which maps any $\hat{v} \in M$ to the element

$$\{st_d(\hat{v})\}_{d \text{ singular}} \in \oplus_{d \text{ singular}} \oplus_{i \in I_d} V_{q_i}.$$

From the definition of st_d it easily follows that the kernel of this map is again $\mathbf{C}(\{z\})^n$. Finally, one sees that the spaces $\oplus_{d \text{ singular}} \oplus_{i \in I_d} V_{q_i}$ and $H^1(\mathbf{S}^1, \ker(\delta - A, (\mathcal{A}^0)^n))$ have the same dimension. □

Remark 8.4 1. Corollary 8.3 produces an isomorphism

$$\psi : \oplus_{d \text{ singular}} \oplus_{i \in I_d} V_{q_i} \to H^1(\mathbf{S}^1, \ker(\delta - A, (\mathcal{A}^0)^n)).$$

In the case where there is only one positive slope k (and $k > 1/2$), we will make this isomorphism explicit. One considers the singular directions $d_1 < \cdots < d_m < d_{m+1} := d_1(+2\pi)$ and the covering of $\mathbf{S}^1$ by the intervals $S_j := (d_{j-1} - \epsilon, d_j + \epsilon)$, for $j = 2, \ldots, m+1$ (and $\epsilon > 0$ small enough such that the intersection of any three distinct intervals is empty). For each j, the group $\oplus_{i \in I_{d_j}} V_{q_i}$ is equal to $H^0((d_j - \frac{\pi}{2k}, d_j + \frac{\pi}{2k}), \ker(\delta - A, (\mathcal{A}^0)^n))$ and maps to $H^0(S_j \cap S_{j+1}, \ker(\delta - A, (\mathcal{A}^0)^n))$. This results in a linear map of $\oplus_{d \text{ singular}} \oplus_{i \in I_d} V_{q_i}$ to the first Cech cohomology group of the sheaf $\ker(\delta - A, (\mathcal{A}^0)^n))$ for the covering $\{S_j\}$ of the circle. It is not difficult to verify that the corresponding linear map

$$\oplus_{d \text{ singular}} \oplus_{i \in I_d} V_{q_i} \to H^1(\mathbf{S}^1, \ker(\delta - A, (\mathcal{A}^0)^n))$$

coincides with ψ.

For the general case, i.e., $r > 1$, one can construct a special covering of the circle and a linear map from $\oplus_{d \text{ singular}} \oplus_{i \in I_d} V_{q_i}$ to the first Cech cohomology of the sheaf $\ker(\delta - A, (\mathcal{A}^0)^n)$ with respect to this covering, which represents ψ.

2. The equivalence of (a) and (b) is due to Malgrange and (c) is due to Deligne (cf. [175], Théorème 9.10 and [180], Proposition 7.1). □

Lemma 8.5 *Consider, as before, a formal solution $\hat{v}$ of the equation $(\delta - A)\hat{v} = w$. Let the direction d be nonsingular and let v be the multisum of $\hat{v}$ in this direction. The coordinates of $\hat{v}$ and v are denoted by $\hat{v}_1, \ldots, \hat{v}_n$ and $v_1, \ldots, v_n$. The two differential rings $\mathbf{C}(\{z\})[v_1, \ldots, v_n]$ and $\mathbf{C}(\{z\})[\hat{v}_1, \ldots, \hat{v}_n]$ are defined as subrings of $\mathcal{A}(S)$ and $\mathbf{C}((z))$, where S is a suitable sector around d. The canonical map $J : \mathcal{A}(S) \to \mathbf{C}((z))$ induces an isomorphism of the differential ring*

$$\phi : \mathbf{C}(\{z\})[v_1, \ldots, v_n] \to \mathbf{C}(\{z\})[\hat{v}_1, \ldots, \hat{v}_n].$$

Proof. It is clear that the morphism of differential rings is surjective, since each v_i is mapped to $\hat{v}_i$. In showing the injectivity of the morphism, we consider first the easy case where $\delta - A$ has only one positive slope k (and $k > 1/2$). The sector S then has the form $(d - \frac{\pi}{2k} - \epsilon, d + \frac{\pi}{2k} + \epsilon)$ and, in particular, its length is $> \frac{\pi}{k}$. The injectivity of $J : \mathcal{A}_{1/k}(S) \to \mathbf{C}((z))$ proves the injectivity of ϕ.

Now we consider the case of two positive slopes $k_1 < k_2$ (and $k_1 > 1/2$). The situation of more than two slopes is similar. Each v_i is a multisum and corresponds with a pair $(v_i(1), v_i(2))$, where $v_i(1)$ is a section of the sheaf $\mathcal{A}/\mathcal{A}^0_{1/k_2}$ above a sector $S_1 := (d - \frac{\pi}{2k_1} - \epsilon, d + \frac{\pi}{2k_1} + \epsilon)$. Furthermore, $v_i(2)$ is a section of the sheaf $\mathcal{A}$ above an interval of the form $S_2 := (d - \frac{\pi}{2k_2} - \epsilon, d + \frac{\pi}{2k_2} + \epsilon)$. Moreover, $v_i(1)$ and $v_i(2)$ have the same image in $\mathcal{A}/\mathcal{A}^0_{1/k_2}(S_2)$. The v_i of the lemma is, in fact, the element $v_i(2)$. Any $f \in \mathbf{C}(\{z\})[v_1, \ldots, v_n]$ is also multisummable, since it is a linear combination of monomials in the $v_1, \ldots, v_n$ with coefficients in $\mathbf{C}(\{z\})$. This f is represented by a pair $(f(1), f(2))$ as above with $f = f(2)$. Suppose that the image of f under J is 0, then $f(1) = 0$ because $J : \mathcal{A}/\mathcal{A}^0_{1/k_2}(S_1) \to \mathbf{C}((z))$ is injective. Thus $f(2) \in \mathcal{A}^0_{1/k_2}(S_2) = 0$. □

8.3 Construction of the Stokes Matrices

In the literature, several definitions of Stokes matrices or Stokes multipliers can be found. Some of these definitions seem to depend on choices of bases. Other definitions do not result in matrices that can be interpreted as elements of the differential Galois group of the equation. In this section, we try to give a definition, rather close to the ones in [13, 177, 203, 297], which avoids these problems. The advantage in working with differential modules over the field $\mathbf{C}(\{z\})$ is that the constructions are clearly independent of choices of bases. However, for the readability of the exposition, we have chosen to continue with differential equations in matrix form. As in the earlier part of this chapter, we consider a matrix differential equation $\delta - A$ with A an $n \times n$ matrix with entries in $\mathbf{C}(\{z\})$. The solution space V of this equation is defined as $\ker(\delta - A, \mathrm{UnivR}^n)$, where UnivR is the universal differential ring

$\mathbf{C}((z))[\{e(q)\}, \{z^a\}, l]$. The space V has a decomposition $\oplus V_{q_i}$, where $q_1, \ldots, q_s$ are the eigenvalues of the operator $\delta - A$. Furthermore, the formal monodromy γ acts upon V. The idea is the following. For a direction $d \in \mathbf{R}$, which is not singular with respect to the set $\{q_i - q_j\}$, one uses multisummation in the direction d in order to define a map ψ_d from V to a solution space for $\delta - A$ with entries that are meromorphic functions on a certain sector around d. For a singular direction d, one considers, as before, directions d^+, d^- with $d^- < d < d^+$ and $|d^+ - d^-|$ small. The "difference" $\psi_{d^+}^{-1}\psi_{d^-} \in \mathrm{GL}(V)$ of the two maps will be the Stokes multiplier St_d.

As in the introduction we fix a quasi-split differential equation $\delta - B$ and a formal equivalence $\hat{F}^{-1}(\delta - A)\hat{F} = \delta - B$. By definition, there is a splitting (after taking some m-th-root of z) of $\delta - B$ as a direct sum of equations $\delta - q_i - C_i$, where each C_i is a constant matrix. We note that the matrices C_i are not unique. They can be normalized by requiring that the eigenvalues λ satisfy $0 \leq Re(\lambda) < 1$. Also, $\hat{F}$ is, in general, not unique once one has chosen $\delta - B$. Indeed, any other solution $\hat{G}$ of $\hat{G}^{-1}(\delta - A)\hat{G} = \delta - B$ can be seen to have the form $\hat{G} = \hat{F}C$ with $C \in \mathrm{GL}_n(\mathbf{C})$ such that $C^{-1}BC = B$. The equation $\delta - B$ has a fundamental matrix E with coordinates in the subring $\mathbf{C}(\{z\})[\{e(q)\}, \{z^a\}, l]$ of the universal ring $\mathbf{C}((z))[\{e(q)\}, \{z^a\}, l]$.

Our first concern is to give E an *interpretation* E_S as an invertible matrix of meromorphic functions on a sector S. There is, however, a difficulty. The matrix E has entries involving the symbols $l, z^a, e(q)$. And l, for instance, should have the interpretation as the logarithm of z. To do this correctly, one has to work with sectors T lying on the "Riemann surface of the logarithm of z". This means that one considers the map $\mathbf{C} \to \mathbf{C}^*$, given by $t \mapsto e^{it}$. A sector is then a subset of $\mathbf{C}$, say of the form $\{t \in \mathbf{C} |\ Re(t) \in (a, b) \text{ and } Im(t) > c\}$. The drawback of this formally correct way of stating the constructions and proofs is a rather heavy notation. In the following, we will use sectors T of length $< 2\pi$ on the Riemann surface of $\log z$ and identify T with its projection S on the circle $\mathbf{S}^1$. We keep track of the original sector by specifying for some point of S its original $d \in \mathbf{R}$ lying on T. We will use the complex variable z instead of the above t. Thus we have an interpretation for E_S or E_d as an invertible meromorphic matrix, living above a sector S, actually on the Riemann surface, but with the notation of complex variable z.

Let $\mathcal{M}(S)$ denote the field of the (germs of) meromorphic functions living on the sector S. We note that $\mathcal{M}$ can be seen to be a sheaf on $\mathbf{S}^1$. Then E_d is an invertible matrix with coefficients in $\mathcal{M}(S)$ and is a fundamental matrix for $\delta - B$.

For a suitable sector S we also want to "lift" the matrix $\hat{F}$ to an invertible matrix of meromorphic functions on this sector. We note that $\hat{F}$ is a solution of the differential equation $L(M) := \delta(M) - AM + MB = 0$. The differential operator L acts on $n \times n$ matrices, instead of vectors and thus has order n^2. The expression $\delta(M)$ means that $z\frac{d}{dz}$ is applied to all the entries of M. Using $\hat{F}$ itself, one sees that L is formally equivalent to the quasi-split operator (again acting upon matrices) $\tilde{L} : M \mapsto \delta(M) - BM + MB$. Indeed, $\hat{F}^{-1}L(\hat{F}M)$ is easily calculated to be

$\tilde{L}(M)$. The operator $\tilde{L}$ is quasi-split because $\delta - B$ is quasi-split. Furthermore, the eigenvalues of $\tilde{L}$ are the $\{q_i - q_j\}$. Thus L has the same eigenvalues as $\tilde{L}$ and the singular directions for L are the singular directions for the collection $\{q_i - q_j\}$. For a small enough sector S, there is an asymptotic lift F_S of $\hat{F}$ above S. This means that the entries of F_S lie in $\mathcal{A}(S)$ and have the entries of $\hat{F}$ as asymptotic expansions. Moreover, $L(F_S) = 0$. Since $\hat{F}$ is invertible, we have that F_S is invertible and $F_S^{-1}(\delta - A)F_S = \delta - B$. However, as we know, the lift F_S is, in general, not unique. A remedy for this nonuniqueness is the multisummation process. Let d be a direction that is not singular for the equation L (i.e., nonsingular for the collection $\{q_i - q_j\}$). Then we consider the multisum $S_d(\hat{F})$ in the direction d, which means that the multisummation operator S_d is applied to every entry of $\hat{F}$. The multisum $S_d(\hat{F})$ can be seen as an invertible meromorphic matrix on a certain sector S containing the direction d. Now $S_d(\hat{F})E_d$ is an invertible meromorphic matrix above the sector S and is a fundamental matrix for $\delta - A$. In the following we will use the two differential equations $\delta - A$ and $\delta - B$ simultaneously. Formally, this is done by considering the new matrix differential equation $\delta - \left(\begin{smallmatrix} A & 0 \\ 0 & B \end{smallmatrix}\right)$.

Proposition 8.6 *Let $d \in \mathbf{R}$ be a nonsingular direction for the collection $\{q_i - q_j\}$ and let S be the sector around d defined by the multisummation in the direction d for the differential equation L.*

1. *The $\mathbf{C}(\{z\})$-subalgebra R_2 of the universal ring* UnivR, *i.e., $\mathbf{C}((z))[\{e(q)\}, \{z^a\}, l]$, generated by the entries of E and $\hat{F}$ and the inverses of the determinants of E and $\hat{F}$, is a Picard-Vessiot ring for the combination of the two equations $\delta - A$ and $\delta - B$.*
2. *The $\mathbf{C}(\{z\})$-subalgebra $R_2(S)$ of the field of meromorphic functions $\mathcal{M}(S)$, generated by the entries of E_d and $S_d(\hat{F})$ and the inverses of the determinants of E_d and $S_d(\hat{F})$, is a Picard-Vessiot ring for the combination of the two equations $\delta - A$ and $\delta - B$.*
3. *There is a unique isomorphism of differential rings $\phi_d : R_2 \to R_2(S)$ such that ϕ_d, extended to matrices in the obvious way, has the properties $\phi_d(E) = E_d$ and $\phi_d(\hat{F}) = S_d(\hat{F})$.*
4. *Let R_1 be the $\mathbf{C}(\{z\})$-subalgebra of R_2, generated by the entries of $\hat{F}E$ and let $R_1(S)$ be the $\mathbf{C}(\{z\})$-subalgebra of $R_2(S)$, generated by the entries of $S_d(\hat{F})E_d$. Then R_1 and $R_1(S)$ are Picard-Vessiot rings for $\delta - A$. Moreover, the isomorphism ϕ_d induces an isomorphism $\psi_d : R_1 \to R_1(S)$, which does not depend on the choices for $\delta - B$ and $\hat{F}$.*

Proof. 1 and 2. R_2 is a subring of UnivR. The field of fractions of UnivR has as field of constants $\mathbf{C}$. Thus the same holds for the field of fractions of R_2. Furthermore, R_2 is generated by the entries of the two fundamental matrices and the inverses of their determinants. From the Picard-Vessiot theory (Proposition 1.22), one concludes that R_2 is a Picard-Vessiot ring for the combination of the two equations. The same argument works for the ring $R_2(S)$.

3. The Picard-Vessiot theory tells us that an isomorphism between the differential rings R_2 and $R_2(S)$ exists. The rather *subtle point* is to show that an isomorphism ϕ_d exists, which maps E to E_d and $\hat{F}$ to $S_d(\hat{F})$. The uniqueness of ϕ_d is clear, since the above condition on ϕ_d determines the ϕ_d-images of the generators of R_2. We start by observing that R_2 is the tensor product over $\mathbf{C}(\{z\})$ of the two subalgebras $R_{2,1} := \mathbf{C}(\{z\})[\text{ entries of } \hat{F}, \frac{1}{\det \hat{F}}]$ and $R_{2,2} := \mathbf{C}(\{z\})[\text{ entries of } E, \frac{1}{\det E}]$ of UnivR. Indeed, the map $R_{2,1} \otimes R_{2,2} \to \mathbf{C}((z)) \otimes R_{2,2}$ is injective. Moreover, the obvious map $\mathbf{C}((z)) \otimes R_{2,2} \to$ UnivR is injective, by the very definition of UnivR. We conclude that the natural map $R_{2,1} \otimes R_{2,2} \to$ UnivR is injective. The image of this map is clearly R_2. Now we consider the two $\mathbf{C}(\{z\})$-subalgebras $R_{2,1}(S) := \mathbf{C}(\{z\})[\text{ entries of } S_d(\hat{F}), \frac{1}{\det S_d(\hat{F})}]$ and $R_{2,2}(S) := \mathbf{C}(\{z\})[\text{ entries of } E_d, \frac{1}{\det E_d}]$ of $\mathcal{M}(S)$. The canonical map $J : R_{2,1}(S) \to R_{2,1}$ is an isomorphism, according to Lemma 8.5. The ring $R_{2,2}$ is a localisation of a polynomial ring over the field $\mathbf{C}(\{z\})$ and this implies that there is a unique isomorphism $R_{2,2} \to R_{2,2}(S)$, which, when extended to matrices, sends the matrix E to E_d. Combining this, one finds isomorphisms

$$R_2 \to R_{2,1} \otimes R_{2,2} \to R_{2,1}(S) \otimes R_{2,2}(S).$$

Since $R_{2,1}(S)$ and $R_{2,2}(S)$ are $\mathbf{C}(\{z\})$-subalgebras of $\mathcal{M}(S)$, there is also a canonical morphism $R_{2,1}(S) \otimes R_{2,2}(S) \to \mathcal{M}(S)$. The image of this map is clearly $R_2(S)$. Thus we found a $\mathbf{C}(\{z\})$-linear morphism of differential rings $\phi_d : R_2 \to R_2(S)$, such that $\phi_d(\hat{F}) = S_d(\hat{F})$ and $\phi_d(E) = E_d$. Since R_2 has only trivial differential ideals, ϕ_d is an isomorphism.

4. As in parts 1 and 2, one proves that R_1 and $R_1(S)$ are Picard-Vessiot rings for $\delta - A$. Then clearly ϕ_d must map R_1 bijectively to $R_1(S)$. Finally, we have to see that ψ_d, the restriction of ϕ_d to R_1, does not depend on the choices for $\delta - B$ and $\hat{F}$. Let $\delta - B^*$ be another choice for the quasi-split equation. Then $\delta - B^* = G^{-1}(\delta - B)G$ for some $G \in \mathrm{GL}_n(\mathbf{C}(\{z\}))$. The special form of B and B^* leaves few possibilities for G, but we will not use this fact. Then $(\hat{F}G)^{-1}(\delta - A)(\hat{F}G) = (\delta - B^*)$. All the rings, considered in the proof of part 3, remain unchanged by this change of the pair $(B, \hat{F})$ into $(B^*, \hat{F}G)$. The new fundamental matrices are $\hat{F}G$ and $G^{-1}E$ and their lifts are $S_d(\hat{F}G) = S_d(\hat{F})G$ and $(G^{-1}E)_d = G^{-1}E_d$. The map ϕ_d, extended to matrices, maps again $\hat{F}G$ to $S_d(\hat{F}G)$ and $G^{-1}E$ to $(G^{-1}E)_d$. Thus the ϕ_d for the pair $(B^*, \hat{F}G)$ coincides with the one for the pair $(B, \hat{F})$. The same holds then for ψ_d. The other change of pairs that we can make is $(B, \hat{F}C)$ with $C \in \mathrm{GL}_n(\mathbf{C})$ such that $CB = BC$. In a similar way one shows that ϕ_d and ψ_d do not depend on this change. □

Remark 8.7 *The subtle point of the proof.*
The crucial isomorphism $\phi_d : R_2 \to R_2(S)$ of part 3 of Proposition 8.6, means that every polynomial relation between the entries of the matrices $\hat{F}$ and E over the field $\mathbf{C}(\{z\})$ is also a polynomial relation for the corresponding entries of the matrices $S_d(\hat{F})$ and E_d over $\mathbf{C}(\{z\})$. We have used multisummation to prove this. In general,

it is not true that the same statement holds if the multisum $S_d(\hat{F})$ is replaced by another asymptotic lift F_S of $\hat{F}$ above the sector S (cf. [177]). □

Let $d \in \mathbf{R}$ be a singular direction for the differential equation L. One considers directions d^+, d^- with $d^- < d < d^+$ and $|d^+ - d^-|$ small. Multisummation in the directions d^+ and d^-, yields according to Proposition 8.6, isomorphisms $\psi_{d^+} : R_1 \to R_1(S^+)$ and $\psi_{d^-} : R_1 \to R_1(S^-)$ for suitable sectors S^+, S^- given by the mutisummation process. The intersection $S := S^+ \cap S^-$ is a sector around the direction d. Let $R_1(S) \subset \mathcal{M}(S)$ denote the Picard-Vessiot ring for $\delta - A$ inside the differential field $\mathcal{M}(S)$. The restriction maps $\mathcal{M}(S^+) \to \mathcal{M}(S)$ and $\mathcal{M}(S^-) \to \mathcal{M}(S)$ induce canonical isomorphisms $\text{res}^+ : R_1(S^+) \to R_1(S)$ and $\text{res}^- : R_1(S^-) \to R_1(S)$.

Definition 8.8 The *Stokes map* St_d for the direction d, is defined as $(\text{res}^+\psi_{d^+})^{-1}\text{res}^-\psi_{d^-}$. □

In other words, St_d is defined by the formula $\psi_{d^+} \circ St_d = \text{An} \circ \psi_{d^-}$, in which An denotes the analytical continuation from the sector S^- to the sector S^+. Clearly, St_d is a differential automorphism of the Picard-Vessiot ring R_1. In particular, St_d induces an element of $\text{GL}(V)$. This element is also denoted by St_d and will be called *the Stokes multiplier or the Stokes matrix*. The translation of St_d in matrices can be stated as follows. The symbolic fundamental matrix $\hat{F}E$ of $\delta - A$ is lifted to actual fundamental matrices $S_{d^+}(\hat{F})E_d$ and $S_{d^-}(\hat{F})E_d$, with meromorphic functions as entries. On the intersection S of the sectors S^+ and S^-, one has $S_{d^+}(\hat{F})E_dC = S_{d^-}(\hat{F})E_d$, for some constant matrix $C \in \text{GL}_n(\mathbf{C})$. The columns of $\hat{F}E$ are a basis for V. The columns of $S_{d^+}(\hat{F})E_d$ and $S_{d^-}(\hat{F})E_d$ are the lifts of this basis of V to the sectors S^+ and S^-, obtained by multisummation. The relation between the two lifts is given by C. Thus C is the matrix of St_d with respect to the basis of V defined by the columns of $\hat{F}E$.

From this description of St_d, one sees that if $\delta - A_1$ and $\delta - A_2$ are equivalent equations over K, then, for each direction d, the Stokes maps (as linear maps of V) coincide. This allows us to define the *Stokes maps associated to a differential module M over K* to be the Stokes maps for any associated equation. This allows us to make the following definition.

Definition 8.9 Let M be a differential module over K. We define $\text{Tup}(M)$ to be the tuple $(V, \{V_q\}, \gamma, \{St_d\})$ where $(V, \{V_q\}, \gamma) = \text{Trip}(M)$ is as in Proposition 3.30 and $\{St_d\}$ are the collection of Stokes maps in $\text{GL}(V)$. □

In Chap. 9, we will see that Tup defines a functor that allows us to give a meromorphic classification of differential modules over K.

Theorem 8.10 J.-P. Ramis
The differential Galois group $G \subset \text{GL}(V)$ of the equation $\delta - A$ is generated, as linear algebraic group, by:

1. *The formal differential Galois group, i.e., the differential Galois group over the field* $\mathbf{C}((z))$, *and*
2. *The Stokes matrices, i.e., the collection* $\{St_d\}$, *where d runs in the set of singular directions for the* $\{q_i - q_j\}$.

Moreover, the formal differential Galois group is generated, as a linear algebraic group, by the exponential torus *and the* formal monodromy.

Proof. In Sect. 3.2, we showed that the formal differential Galois group is generated, as a linear algebraic group, by the formal monodromy and the exponential torus (see Proposition 3.40). Let $R_1 \subset R$ denote the Picard-Vessiot ring of $\delta - A$ over $\mathbf{C}(\{z\})$. Its field of fractions $K_1 \subset K$ is the Picard-Vessiot field of $\delta - A$ over $\mathbf{C}(\{z\})$. We have to show that an element $f \in K_1$, which is invariant under the formal monodromy, the exponential torus and the Stokes multipliers belongs to $\mathbf{C}(\{z\})$. Proposition 3.25 states that the invariance under the first two items implies that $f \in \mathbf{C}((z))$. More precisely, from the proof of part 3 of Proposition 8.6 one deduces that f lies in the field of fractions of $\mathbf{C}(\{z\})[$ entries of $\hat{F}, \frac{1}{\det \hat{F}}]$. For any direction d, which is not singular for the collection $\{q_i - q_j\}$, there is a well-defined asymptotic lift on a corresponding sector. Let us write $S_d(f)$ for this lift. For a singular direction d, the two lifts $S_{d^+}(f)$ and $S_{d^-}(f)$ coincide on the sector $S^+ \cap S^-$, since $St_d(f) = f$. In other words, the asymptotic lifts of $f \in \mathbf{C}((z))$ on the sectors at zero glue to an asymptotic lift on the full circle and therefore $f \in \mathbf{C}(\{z\})$. □

Remarks 8.11 1. Theorem 8.10 is stated and a proof is sketched in [239, 240] (a complete proof is presented in [238]). A shorter (and more natural) proof is given in [202].

2. We note that a non-quasi-split equation $\delta - A$ may have the same differential Galois group over $\mathbf{C}((z))$ and $\mathbf{C}(\{z\})$. This occurs when the Stokes matrices already lie in the differential Galois group over $\mathbf{C}((z))$. □

Proposition 8.12 *We use the previous notations.*

1. $\gamma^{-1} St_d \gamma = St_{d+2\pi}$.
2. *Let* $d_1 < \cdots < d_t$ *denote the singular directions (for the collection* $\{q_i - q_j\}$*), then the product* $\gamma St_{d_t} \cdots St_{d_1}$ *is conjugate to the topological monodromy of* $\delta - A$*, considered as an element of* $\mathrm{GL}(V)$.

Proof. 1. We recall the isomorphism $\phi_d : R_2 \to R_2(S)$, constructed in Proposition 8.7. From the construction of ϕ_d one sees that $\phi_{d+2\pi} = \phi_d \circ \gamma$, where γ is the formal monodromy acting on R_1 and V. For the induced isomorphism $\psi_d : R_1 \to R_1(S)$ one also has $\psi_{d+2\pi} = \psi_d \circ \gamma$. Then (omitting the symbol An for analytical continuation), one has $St_{d+2\pi} = \psi^{-1}_{(d+2\pi)^-} \psi_{(d+2\pi)^+}$, which is equal to $\gamma^{-1} St_d \gamma$.

2. The *topological monodromy of* $\delta - A$ is defined as follows. Fix a point p close to the origin. The solution space Sol_p of the equation, locally at p, is a vector space over $\mathbf{C}$ of dimension n. One takes a circle T in the positive direction around 0, starting and ending in p. Analytical continuation of the solutions at p along T produces an invertible map in $\mathrm{GL}(\mathrm{Sol}_p)$. This map is the topological monodromy. After identification of the solution space V with Sol_p, one obtains a topological monodromy map lying in $\mathrm{GL}(V)$. This map is only well defined up to conjugation. If one follows the circle and keeps track of the Stokes multipliers, then one clearly obtains a formula of the type stated in the proposition. By the definition of St_d one has $\psi_{d_1^+} \circ St_{d_1} = \mathrm{An} \circ \psi_{d_1^-}$, where An means analytical continuation from the sector S^- to S^+. Using this formula for all singular directions one finds that

$$\psi_{d_t^+} \circ St_{d_t} \cdots St_{d_1} = \mathrm{An} \circ \psi_{d_1^-}.$$

Moreover $\psi_{d_t^+} = \psi_{(d_1+2\pi)^-} = \psi_{d_1^-} \circ \gamma$ and An is the analytical continuation along a complete circle. This yields $\gamma \circ St_{d_t} \cdots St_{d_1} = \psi_{d_1^-}^{-1} \circ \mathrm{An} \circ \psi_{d_1^-}$, which proves the statement. □

Theorem 8.13 *We use the previous notations. The Stokes multiplier St_d has the form $id + \sum A_{i,j}$, where $A_{i,j}$ denotes a linear map of the form*

$$V \stackrel{\text{projection}}{\rightarrow} V_{q_i} \stackrel{\text{linear}}{\rightarrow} V_{q_j} \stackrel{\text{inclusion}}{\rightarrow} V,$$

and where the sum is taken over all pairs i, j, such that d is a singular direction for $q_i - q_j$.

Proof. The statement of the theorem is quite similar to that of Proposition 8.2. In fact, the theorem can be deduced from that proposition. However, we give a more readable proof, using fundamental matrices for $\delta - A$ and $\delta - B$. The symbolic fundamental matrices for the two equations are $\hat{F}E$ and E. Again, for the readability of the proof we will assume that E is a diagonal matrix with entries $e(q_1), \ldots, e(q_n)$, with distinct elements $q_1, \ldots, q_n \in z^{-1}\mathbf{C}[z^{-1}]$. Thus B is the diagonal matrix with entries $q_1, \ldots, q_n$. The Stokes multiplier St_d is represented by the matrix C satisfying $S_{d^+}(\hat{F})E_dC = S_{d^-}(\hat{F})E_d$. Thus $E_dCE_d^{-1} = S_{d^+}(\hat{F})^{-1}S_{d^-}(\hat{F})$. Let $C = (C_{i,j})$, then the matrix $E_dCE_d^{-1}$ is equal to $M := (e(q_i - q_j)_d C_{i,j})$.

Suppose now, to start with, that each $q_i - q_j$ (with $i \neq j$) has degree k in z^{-1}. The k-summation theorem, Theorem 7.39, implies that $S_{d^+}(\hat{F})^{-1}S_{d^-}(\hat{F}) - 1$ has entries in $\mathcal{A}^0_{1/k}(d - \frac{\pi}{2k}, d + \frac{\pi}{2k})$. The sector has length $\frac{\pi}{k}$ and we conclude that $e(q_i - q_j)_d c_{i,j} = 0$ unless d is a singular direction for $q_i - q_j$. This proves the theorem in this special case.

Suppose now that the degrees with respect to z^{-1} in the collection $\{q_i - q_j \mid i \neq j\}$ are $k_1 < \cdots < k_s$. From the definition of multisummation (and also Proposition 7.58) it follows that the images of the entries of $M - id$ in the sheaf $\mathcal{A}^0_{1/k_1}/\mathcal{A}^0_{1/k_2}$ exist on the interval $(d-\frac{\pi}{2k_1}, d+\frac{\pi}{2k_1})$. Thus for $q_i - q_j$ of degree k_1 one has that $c_{i,j}=0$, unless d is a singular direction for $q_i - q_j$. In the next stage one considers the pairs (q_i, q_j) such that $q_i - q_j$ has degree k_2. Again by the definition of multisummation one has that $c_{i,j}e(q_i - q_j)_d$ must produce a section of $\mathcal{A}^0_{1/k_2}/\mathcal{A}^0_{1/k_3}$ above the sector $(d-\frac{\pi}{2k_2}, d+\frac{\pi}{2k_2})$. This has as a consequence that $c_{i,j}=0$, unless d is a singular direction for $q_i - q_j$. Induction ends the proof.

In the general case, E can, after taking some m-th-root of z, be written as a block matrix, where each block corresponds to a single $e(q)$ and involves some z^as and l. The reasoning above remains valid in this general case. □

Remark 8.14 In Definition 8.9, we associated with any differential module M over K a tuple $\text{Tup}(M) = (V, \{V_q\}, \gamma, \{St_d\})$. This definition, Proposition 8.12, and Theorem 8.13 imply that this tuple has the following properties:

(a) $(V, \{V_q\}, \gamma)$ as an object of Gr_1.

(b) For every $d \in \mathbf{R}$ the element $St_d \in \text{GL}(V)$ has the form $id + \sum A_{i,j}$, where $A_{i,j}$ denotes a linear map of the form $V \overset{\text{projection}}{\rightarrow} V_{q_i} \overset{\text{linear}}{\rightarrow} V_{q_j} \overset{\text{inclusion}}{\rightarrow} V$, and where the sum is taken over all pairs i, j such that d is a singular direction for $q_i - q_j$.

(c) One has $\gamma^{-1} St_d \gamma = St_{V,d+2\pi}$ for all $d \in \mathbf{R}$.

In Chap. 9, we will define a category Gr_2 of such objects and show that Tup defines an equivalence of categories between the category Diff_K of differential modules over K and Gr_2. □

Example 8.15 *The Airy equation.*
The Airy equation $y'' = zy$ has a singular point at $z=\infty$. The translation of the theory developed for the singular point $z=0$ to the point $z=\infty$ is straightforward. The symbolic solution space V at ∞ can be identified with the solutions of the scalar equation in the universal ring at ∞, namely $\mathbf{C}((z^{-1}))[\{e(q)\}, \{z^a\}, l]$. The set that the qs belong to is $\cup_{m\geq 1} z^{1/m}\mathbf{C}[z^{1/m}]$ and z^a and l are again symbols for the functions z^a and $\log(z)$. The two qs of the equation are $q_1 := z^{3/2}$ and $q_2 := -z^{3/2}$. Thus V is the direct sum of two 1-dimensional spaces $V = V_{z^{3/2}} \oplus V_{-z^{3/2}}$. The formal monodromy γ permutes the two 1-dimensional spaces. The differential Galois group of the equation lies in $\text{SL}_2(\mathbf{C})$, since the coefficient of y' in the equation is zero. Therefore, one can give $V_{z^{3/2}}$ and $V_{-z^{3/2}}$ bases such that the matrix of γ with respect to this basis of V reads $\left(\begin{smallmatrix}0 & -1\\ 1 & 0\end{smallmatrix}\right)$. The exponential torus, as subgroup of $\text{SL}(V)$ has, on the same basis, the form $\left(\begin{smallmatrix}t & 0\\ 0 & t^{-1}\end{smallmatrix}\right) | \, t \in \mathbf{C}^*\}$. According to Theorem 8.10, the formal differential Galois group is the infinite Dihedral group $D_\infty \subset \text{SL}_2$ (cf. Exercise 3.33).

The singular directions for $\{q_1 - q_2, q_2 - q_1\}$ are $d = 0, \frac{2\pi}{3}, \frac{4\pi}{3}$ modulo $2\pi\mathbf{Z}$. The topological monodromy is trivial, since there are two independent entire solutions for $y'' = zy$. Using Theorem 8.13, we see that the formal monodromy is not trivial. The three Stokes matrices $St_0, St_{\frac{2\pi}{3}}, St_{\frac{4\pi}{3}}$ have the form $\left(\begin{smallmatrix}1 & *\\ 0 & 1\end{smallmatrix}\right)$, $\left(\begin{smallmatrix}1 & 0\\ * & 1\end{smallmatrix}\right)$, and $\left(\begin{smallmatrix}1 & *\\ 0 & 1\end{smallmatrix}\right)$ with respect to the decomposition $V = V_{z^{3/2}} \oplus V_{-z^{3/2}}$. Their product is γ^{-1} according to Proposition 8.12, and this is only possible if each one is $\neq id$. Theorem 8.10 (and the discussion before Exercise 1.36) implies that the differential Galois group of the Airy equation over $\mathbf{C}(z)$ is SL_2. □

Exercise 8.16 Consider the equation $y'' = ry$ with $r \in \mathbf{C}[z]$ a polynomial of odd degree. Let V denote the symbolic solution space at $z = \infty$. Calculate the qs, γ, the formal differential Galois group, the singular directions, the Stokes matrices, and the differential Galois group. □

Example 8.17 The asymptotic behavior of the following differential equation has been studied by W. Jurkat, D.A. Lutz and A. Peyerimhoff [149, 150] and Martinet and Ramis [201].

$$\delta + A := \delta + z^{-1}\begin{pmatrix} \lambda_1 & 0 \\ 0 & \lambda_2 \end{pmatrix} + \begin{pmatrix} 0 & a \\ b & 0 \end{pmatrix}.$$

We will apply the theory of this chapter to the equation. Let $B = z^{-1}\left(\begin{smallmatrix} \lambda_1 & 0 \\ 0 & \lambda_2 \end{smallmatrix}\right)$. We claim that there is a unique ϕ of the form $1 + \phi_1 z + \phi_2 z^2 + \cdots$ (where the ϕ_i are 2×2-matrices) with $\phi^{-1}(\delta + A)\phi = \delta + B$. This can be proven by solving the equation

$$\delta(\phi) = \left[z^{-1}\begin{pmatrix} \lambda_1 & 0 \\ 0 & \lambda_2 \end{pmatrix} + \begin{pmatrix} 0 & a \\ b & 0 \end{pmatrix}\phi - \phi(z^{-1}\begin{pmatrix} \lambda_1 & 0 \\ 0 & \lambda_2 \end{pmatrix})\right]$$

and the corresponding sequence of equations for the ϕ_n step-wise by "brute force". Explicit formulas for the entries of the ϕ_n can be derived but they are rather complicated. One observes that the expressions for these entries contain truncations of the product formula for the function $\frac{2\sin(\pi\sqrt{ab})}{\sqrt{ab}}$. One defines a transformation ψ by replacing truncations in the entries of all the ϕ_n by the corresponding infinite products. The difference between the two formal transformations ϕ and ψ is a convergent transformation. In particular, one can explicitly calculate the Stokes matrices in this way, but we will find another way to compute them.

The two eigenvalues of $\delta + A$ are $q_1 = -\lambda_1 z^{-1}$ and $q_2 = -\lambda_2 z^{-1}$. There are two singular directions for $\{q_1 - q_2, q_2 - q_1\}$, differing by π. On the given basis for $\delta + A$ and $\delta + B$, the two Stokes matrices have, according to Theorem 8.13, the forms $\left(\begin{smallmatrix} 1 & x_1 \\ 0 & 1 \end{smallmatrix}\right)$ and $\left(\begin{smallmatrix} 1 & 0 \\ x_2 & 1 \end{smallmatrix}\right)$. The formal monodromy of $\delta - A$ is the identity and thus $\left(\begin{smallmatrix} 1+x_1x_2 & x_1 \\ x_2 & 1 \end{smallmatrix}\right)$ is conjugate to the topological monodromy. The topological monodromy can be easily calculated at the point $z = \infty$. For general a, b it has the matrix $exp\left[\,2\pi i\left(\begin{smallmatrix} 0 & -a \\ -b & 0 \end{smallmatrix}\right)\right]$. The trace of the monodromy matrix $e^{2\pi i\sqrt{ab}} + e^{-2\pi i\sqrt{ab}}$ is equal to the other trace $2 + x_1x_2$. Therefore, $x_1x_2 = -(2\sin(\pi\sqrt{ab}))^2$. We consider $x_1 = x_1(a, b)$ and $x_2 = x_2(a, b)$ as functions of (a, b), and we want to find an explicit formula for the map $(a, b) \mapsto (x_1(a, b), x_2(a, b))$. A first observation is that conjugation of all ingredients with the constant matrix $\binom{\lambda\ 0}{0\ 1}$ leads to $(\lambda a, \lambda^{-1}b) \mapsto (\lambda x_1(a,b), \lambda^{-1}x_2(a,b))$. This means that $\frac{x_1(a,b)}{a}$ and $\frac{x_2(a,b)}{b}$ depend only on ab. Thus $(x_1, x_2) = (\alpha(ab)a, \beta(ab)b)$ for certain functions α and β.

The final information that we need comes from transposing the equation and thus interchanging a and b. Let $\hat{F}$ denote the formal fundamental matrix of the equation. A comparison of two asymptotic lifts of $\hat{F}$ produces the values x_1, x_2 as functions of a, b. Put $\hat{G}(z) = (\hat{F}^*)^{-1}(-z)$, where $*$ means the transposed matrix. Then $\hat{G}$ is a fundamental matrix for the equation

$$z^2\frac{d}{dz} + \begin{pmatrix} \lambda_1 & 0 \\ 0 & \lambda_2 \end{pmatrix} + z\begin{pmatrix} 0 & -b \\ -a & 0 \end{pmatrix}.$$

The two Stokes matrices for $\hat{G}$ are obtained from the ones for $\hat{F}$ by taking inverses, transposition, and interchanging their order. This yields the formula $(x_1(-b,-a), x_2(-b,-a)) = (-x_2(a,b), -x_1(a,b))$. One concludes that $\alpha(ab) = \beta(ab) = \frac{2i\sin(\pi\sqrt{ab})}{\sqrt{ab}}$. The formula that we find is then

$$(x_1, x_2) = \frac{2i\sin(\pi\sqrt{ab})}{\sqrt{ab}} \cdot (a, b).$$

We note that we have proven this formula under the mild restrictions that $ab \neq 0$ and the difference of the eigenvalues of the matrix $\binom{0\ a}{b\ 0}$ is not an integer $\neq 0$. It can be verified that the formula holds for all a, b.

The map $\tau : (a, b) \to \frac{2i\sin(\pi\sqrt{ab})}{\sqrt{ab}} \cdot (a, b)$ is easily seen to be a surjective map from $\mathbf{C}^2$ to itself. This demonstrates in this example the third statement made in the introduction about Stokes matrices. This example will also play a role in Chap. 12 where moduli of singular differential equations are studied. □

Remark 8.18 One can calculate the Stokes matrices of linear differential equations when one has explicit formulae for the solutions of these equations. Examples of this are given in [88, 208, 209]. □

9 Stokes Matrices and Meromorphic Classification

9.1 Introduction

We will denote the differential fields $\mathbf{C}(\{z\})$ and $\mathbf{C}((z))$ by K and $\hat{K}$. The classification of differential modules over $\hat{K}$, given in Sect. 3.2, associates with a differential module M a triple $\mathrm{Trip}(M) = (V, \{V_q\}, \gamma)$. More precisely, a tannakian category Gr_1 was defined, which has as objects the above triples. The functor $\mathrm{Trip} : \mathrm{Diff}_{\hat{K}} \to \mathrm{Gr}_1$ from the category of the differential modules over $\hat{K}$ to the category of triples was shown to be an equivalence of tannakian categories.

In Chap. 8, this was extended by associating to a differential module M over K a tuple $\mathrm{Tup}(M) = (V, \{V_q\}, \gamma, \{St_d\})$. We will introduce a tannakian category Gr_2, whose objects are the above tuples. The main goal of this chapter is to show that $\mathrm{Tup} : \mathrm{Diff}_K \to \mathrm{Gr}_2$ is an equivalence of tannakian categories. In other words, the tuples provide the classification of the differential modules over K, i.e., the meromorphic classification. There are natural functors of tannakian categories $\mathrm{Diff}_K \to \mathrm{Diff}_{\hat{K}}$, given by $M \mapsto \hat{K} \otimes_K M$, and the forgetful functor $\mathrm{Gr}_2 \to \mathrm{Gr}_1$, given by $(V, \{V_q\}, \gamma, \{St_d\}) \mapsto (V, \{V_q\}, \gamma)$. The following commutative diagram of functors and categories clarifies and summarizes the main features of the "Stokes theory".

$$\begin{array}{ccc} \mathrm{Diff}_K & \stackrel{\mathrm{Tup}}{\to} & \mathrm{Gr}_2 \\ \downarrow & & \downarrow \\ \mathrm{Diff}_{\hat{K}} & \stackrel{\mathrm{Trip}}{\to} & \mathrm{Gr}_1 \,. \end{array}$$

The description of the differential Galois group of a differential module over $\hat{K}$ (see sect. 3.2) and of a differential module over K (see Theorem 8.10) are easy consequences of this tannakian description. The main difficulty is to prove that every object $(V, \{V_q\}, \gamma, \{St_d\})$ of Gr_2 is isomorphic to $\mathrm{Tup}(M)$ for some differential module M over K. In terms of matrix differential equations this amounts to the following:

> There is a quasi-split differential operator $\delta - B$ that has the triple $(V, \{V_q\}, \gamma)$. One wants to produce a matrix differential operator $\delta - A$ over K and a $\hat{F} \in \mathrm{GL}_n(\hat{K})$ such that $\hat{F}^{-1}(\delta - B)\hat{F} = \delta - B$ and such that the

Stokes maps associated to $\delta - A$ are the prescribed $\{St_d\}$. (See also the introduction of Chap. 8.)

An important tool for the proof is the *Stokes sheaf STS* associated to $\delta - B$. This is a sheaf on the circle of directions $\mathbf{S}^1$, given by: $STS(a, b)$ consists of the invertible holomorphic matrices T, living on the sector (a, b), having the identity matrix as asymptotic expansion and satisfying $T(\delta - B) = (\delta - B)T$. The Stokes sheaf is a sheaf of, in general noncommutative, groups. A theorem of Malgrange and Sibuya states that the cohomology set $H^1(\mathbf{S}^1, STS)$ classifies the equivalence classes of the above pairs $(\delta - A, \hat{F})$. The final step in the proof is a theorem of M. Loday-Richaud, which gives a natural bijection between the set of all Stokes maps $\{St_d\}$ (with $(V, \{V_q\}, \gamma)$ fixed) and the cohomology set $H^1(\mathbf{S}^1, STS)$. Thus paper is rather close to a much earlier construction by Jurkat [148].

We finish this chapter by giving the cohomology set $H^1(\mathbf{S}^1, STS)$ a natural structure of an affine algebraic variety and by showing that this variety is isomorphic with the affine space $\mathbf{A}^N_{\mathbf{C}}$, where N is the irregularity of the differential operator $M \mapsto \delta(M) - BM + MB$, acting upon matrices.

9.2 The Category Gr_2

The objects of Gr_2 are tuples $(V, \{V_q\}, \gamma_V, \{St_{V,d}\})$ with:

(a) $(V, \{V_q\}, \gamma_V)$ as an object of Gr_1.

(b) For every $d \in \mathbf{R}$ the element $St_{V,d} \in \mathrm{GL}(V)$ has the form $id + \sum A_{i,j}$, where $A_{i,j}$ denotes a linear map of the form $V \overset{\text{projection}}{\rightarrow} V_{q_i} \overset{\text{linear}}{\rightarrow} V_{q_j} \overset{\text{inclusion}}{\rightarrow} V$, and where the sum is taken over all pairs i, j such that d is a singular direction for $q_i - q_j$.

(c) One requires that $\gamma_V^{-1} St_{V,d} \gamma_V = St_{V,d+2\pi}$ for all $d \in \mathbf{R}$.

Remarks 9.1 We analyze the data $\{St_{V,d}\}$. Let $q_1, \ldots, q_r$ denote the set of $q \in \mathcal{Q}$, such that $V_q \neq 0$. If d is not a singular direction for any of the $q_i - q_j$, then $St_{V,d} = id$. Using requirement (c), it suffices to consider the $d \in \mathbf{R}$ such that $0 \leq d < 2\pi$ and d is a singular direction for some $q_i - q_j$. Each $A_{i,j}$ is given by a matrix with $\dim V_{q_i} \cdot \dim V_{q_j}$ entries. Thus the data $\{St_{V,d}\}$ (for fixed $(V, \{V_q\}, \gamma_V)$) can be described by a point in an affine space $\mathbf{A}^N_{\mathbf{C}}$. One defines the degree $\deg q$ of an element $q \in \mathcal{Q}$ to be λ if $q = cz^{-\lambda} +$ lower order terms (and of course $c \neq 0$). By counting the number of singular directions in $[0, 2\pi)$ one arrives at the formula $N = \sum_{i,j} \deg(q_i - q_j) \cdot \dim V_{q_i} \cdot \dim V_{q_j}$.

Let M denote the quasi-split differential module over K that has the formal triple $(V, \{V_q\}, \gamma_V)$. Then one easily calculates that the (quasi-split) differential module $\mathrm{Hom}(M, M)$ has irregularity N. Or, in terms of matrices: let $\delta - B$ be the quasi-split matrix differential operator with formal triple $(V, \{V_q\}, \gamma_V)$. Then the differential operator, acting on matrices, $T \mapsto \delta(T) - BT + TB$, has irregularity N. □

We continue the description of the tannakian category Gr_2. A morphism $f : \underline{V} = (V, \{V_q\}, \gamma_V, \{St_{V,d}\}) \to \underline{W} = (W, \{W_q\}, \gamma_W, \{St_{W,d}\})$ is a $\mathbf{C}$-linear map $f : V \to W$ that preserves all data, i.e., $f(V_q) \subset W_q$, $\gamma_W \circ f = f \circ \gamma_V$, $St_{W,d} \circ f = f \circ St_{V,d}$. The set of all morphisms between two objects is obviously a linear space over $\mathbf{C}$. The tensor product of $\underline{V}$ and $\underline{W}$ is the ordinary tensor product $X := V \otimes_{\mathbf{C}} W$ with the data $X_q = \sum_{q_1, q_2,\ q_1+q_2=q} V_{q_1} \otimes W_{q_2}$, $\gamma_X = \gamma_V \otimes \gamma_W$, $St_{X,d} = St_{V,d} \otimes St_{W,d}$. The internal $\mathrm{Hom}(\underline{V}, \underline{W})$ is the linear space $X := \mathrm{Hom}_{\mathbf{C}}(V, W)$ with the additional structure: $X_q = \sum_{q_1, q_2,\ -q_1+q_2=q} \mathrm{Hom}(V_{q_1}, W_{q_2})$, $\gamma_X(h) = \gamma_W \circ h \circ \gamma_V^{-1}$, $St_{X,d}(h) = St_{W,d} \circ h \circ St_{V,d}$ (where h denotes any element of X). The unit element $\mathbf{1}$ is a 1-dimensional vector space V with $V = V_0$, $\gamma_V = id$, $St_{V,d} = id$. The dual $\underline{V}^*$ is defined as $\mathrm{Hom}(\underline{V}, \mathbf{1})$. The fibre functor $\mathrm{Gr}_2 \to \mathrm{Vect}_{\mathbf{C}}$, is given by $(V, \{V_q\}, \gamma_V, \{St_{V,d}\}) \mapsto V$ (where $\mathrm{Vect}_{\mathbf{C}}$ denotes the category of the finite dimensional vector spaces over $\mathbf{C}$). It is easy to verify that the above data define a neutral tannakian category. The following lemma is an exercise (cf. Appendix B).

Lemma 9.2 *Let* $\underline{V} = (V, \{V_q\}, \gamma_V, \{St_{V,d}\})$ *be an object of* Gr_2 *and let* $\{\{\underline{V}\}\}$ *denote the tannakian subcategory generated by* $\underline{V}$, *i.e., the full subcategory of* Gr_2 *generated by all* $\underline{V} \otimes \cdots \otimes \underline{V} \otimes \underline{V}^* \otimes \cdots \otimes \underline{V}^*$. *Then* $\{\{\underline{V}\}\}$ *is again a neutral tannakian category. Let* G *be the smallest algebraic subgroup of* $\mathrm{GL}(V)$ *that contains* γ_V, *the exponential torus and the* $St_{V,d}$. *Then the restriction of the above fibre functor to* $\{\{\underline{V}\}\}$ *yields an identification of this tannakian category with* Repr_G, *i.e., the category of the (algebraic) representations of* G *on finite dimensional vector spaces over* $\mathbf{C}$.

Lemma 9.3 Tup *is a well-defined functor between the tannakian categories* Diff_K *and* Gr_2. *The functor* Tup *is fully faithful.*

Proof. The first statement follows from Remark 9.1, the unicity of the multisummation (for nonsingular directions) and the definitions of the Stokes maps. The second statement means that the $\mathbf{C}$-linear map

$$\mathrm{Hom}_{\mathrm{Diff}_K}(M_1, M_2) \to \mathrm{Hom}_{\mathrm{Gr}_2}(\mathrm{Tup}(M_1), \mathrm{Tup}(M_2))$$

is a bijection. It suffices to prove this statement with $M_1 = \mathbf{1}$ (this is the 1-dimensional trivial differential module over K) and $M_2 = M$ (any differential module over K). Indeed, $\mathrm{Hom}_{\mathrm{Diff}_K}(M_1, M_2)$ is isomorphic to $\mathrm{Hom}_{\mathrm{Diff}_K}(\mathbf{1}, M_1^* \otimes M_2)$.

In considering this situation, one sees that $\mathrm{Hom}_{\mathrm{Diff}_K}(M_1, M_2)$ is equal to $\{m \in M |\ \delta(m) = 0\}$. Let $\mathrm{Tup}(M) = (V, \{V_q\}, \gamma_V, \{St_{V,d}\})$. One has that $\mathrm{Hom}_{\mathrm{Gr}_2}(\mathrm{Tup}(M_1), \mathrm{Tup}(M_2))$ is the set S consisting of the elements $v \in V$ belonging to V_0 and invariant under γ_V and all $St_{V,d}$. The map $\{m \in M |\ \delta(m) = 0\} \to S$ is clearly injective. An element $v \in S$ has its coordinates in $\hat{K}$, since it lies in V_0 and is invariant under the formal monodromy γ_V. The multisums of v in the nonsingular directions glue around $z = 0$ since v is invariant under all the Stokes maps $St_{V,d}$. It follows that the coordinates of v lie in K and thus $v \in M$ and $\delta(v) = 0$. □

Remark 9.4 Let M be a differential module over K and write $\underline{V} := \mathrm{Tup}(M)$. Let $\{\{M\}\}$ denote the tannakian subcategory of Diff_K generated by M. According to Lemma 9.3 the tannakian categories $\{\{M\}\}$ and $\{\{\underline{V}\}\}$ are isomorphic. From Lemma 9.2 one draws the conclusion that the differential Galois group of M is the smallest algebraic subgroup of $\mathrm{GL}(V)$ containing the formal monodromy, the exponential torus and the Stokes maps. Thus the above provides a tannakian proof of Theorem 8.10. □

9.3 The Cohomology Set $H^1(\mathbf{S}^1, STS)$

We start by recalling the definition and some properties of *the cohomology set* $H^1(X, G)$, where X is any topological space and G a sheaf of (not necessarily commutative) groups on X (see [13, 101, 119] for a fuller discussion). For notational convenience we write $G(\emptyset) = \{1\}$. Let $\mathcal{U} = \{U_i\}_{i \in I}$ denote a covering of X by open sets U_i. A 1-cocycle for G and $\mathcal{U}$ is an element $g = \{g_{i,j}\}_{i,j \in I} \in \prod G(U_i \cap U_j)$ satisfying the conditions: $g_{i,i} = 1$, $g_{i,j}g_{j,i} = 1$ and $g_{i,j}g_{j,k}g_{k,i} = 1$ holds on $U_i \cap U_j \cap U_k$ for all i, j, k.

We note that the last condition is empty if $U_i \cap U_j \cap U_k = \emptyset$. Moreover, the second condition follows from the first and the third conditions by considering i, j, k with $k = i$. In some situations it is convenient to fix a total order on I and to define a 1-cocycle g to be an element of $\prod_{i<j} G(U_i \cap U_j)$ satisfying $g_{i,j}g_{j,k} = g_{i,k}$ on $U_i \cap U_j \cap U_k$ whenever $i < j < k$ and $U_i \cap U_j \cap U_k \neq \emptyset$.

Two 1-cocyles g and h are called equivalent if there are elements $l_i \in G(U_i)$ such that $l_i g_{i,j} l_j^{-1} = h_{i,j}$ holds for all i, j. The set of equivalence classes of 1-cocycles (for G and $\mathcal{U}$) is denoted by $\check{H}^1(\mathcal{U}, G)$. This set has a distinguished point, namely the (equivalence class of the) trivial 1-cocycle g with all $g_{i,j} = 1$. For a covering $\mathcal{V}$ that is finer than $\mathcal{U}$, there is a natural map $\check{H}^1(\mathcal{U}, G) \to \check{H}^1(\mathcal{V}, G)$. This map does not depend on the way $\mathcal{V}$ is seen as a refinement of $\mathcal{U}$. Moreover, the map $\check{H}^1(\mathcal{U}, G) \to \check{H}^1(\mathcal{V}, G)$ turns out to be injective. The cohomology set $H^1(X, G)$ is defined as the direct limit (in this case this is a union) of all $\check{H}^1(\mathcal{U}, G)$. The distinguished point of $H^1(X, G)$ will be denoted by 1. The map $\check{H}^1(\mathcal{U}, G) \to H^1(X, G)$ is bijective if $H^1(U_i, G) = 1$ for each $U_i \in \mathcal{U}$. This is Leray's theorem for the case of sheaves of (not necessarily abelian) groups. These properties are stated and proved in [100] for the case of sheaves of abelian groups (see also Appendix C). One easily sees that the proofs extend to the case of sheaves of (not necessarily abelian) groups.

We apply this cohomology for the topological space $\mathbf{S}^1$ and various sheaves of matrices. The first two examples are the sheaves $\mathrm{GL}_n(\mathcal{A})$ and its subsheaf $\mathrm{GL}_n(\mathcal{A})^0$

consisting of the matrices that have the identity as asymptotic expansion. We now present the results of Malgrange and Sibuya (see [13, 177, 188, 191, 201, 263]). The cohomological formulation of the next theorem is due to B. Malgrange.

Theorem 9.5 (G. Birkhoff and Y. Sibuya)
The natural map $H^1(\mathbf{S}^1, \mathrm{GL}_n(\mathcal{A})^0) \to H^1(\mathbf{S}^1, \mathrm{GL}_n(\mathcal{A}))$ *has image* $\{1\}$.

Proof. We only give a sketch of the proof. For detailed proof, we refer to [191] and [13].

As in the proof of Proposition 7.24, one considers the simplest covering $U = (a_1, b_1) \cup (a_2, b_2)$ with $(a_1, b_1) \cap (a_2, b_2) = (a_2, b_1)$, i.e., inequalities $a_1 < a_2 < b_1 < b_2$ for the directions on $\mathbf{S}^1$ and $U \neq \mathbf{S}^1$. A 1-cocycle for this covering and the sheaf $\mathrm{GL}_n(\mathcal{A})^0$ is just an element $M \in \mathrm{GL}_n(\mathcal{A})^0(a_2, b_1)$. We will indicate a proof that the image of this 1-cocycle in $H^1(U, \mathrm{GL}_n(\mathcal{A}))$ is equal to 1. More precisely, we will show that for small enough $\epsilon > 0$ there are invertible matrices M_1, M_2 with coefficients in $\mathcal{A}(a_1, b_1 - \epsilon)$ and $\mathcal{A}(a_2 + \epsilon, b_2)$ such that $M = M_1 M_2$. Let us call this the "multiplicative statement". This statement easily generalizes to a proof that the image of $H^1(\mathbf{S}^1, \mathrm{GL}(n, \mathcal{A})^0)$ in the set $H^1(\mathbf{S}^1, \mathrm{GL}_n(\mathcal{A}))$ is the element 1. The "additive statement for matrices" is the following. Given an $n \times n$ matrix M with coefficients in $\mathcal{A}^0(a_2, b_1)$, then there are matrices M_i, $i = 1, 2$ with coefficients in $\mathcal{A}(a_i, b_i)$ such that $M = M_1 + M_2$. This latter statement follows at once from Proposition 7.24.

The step from this additive statement to the multiplicative statement can be performed in a similar manner as the proof of the classical Cartan's Lemma, (see [120] p. 192–201). A quick (and slightly wrong) description of this method is as follows. Write M as $1 + C$ where C has its entries in $\mathcal{A}^0(a_2, b_1)$. Then $C = A_1 + B_1$ where A_1, B_1 are small and have their entries in $\mathcal{A}(a_1, b_1)$ and $\mathcal{A}(a_2, b_2)$. Since A_1, B_1 are small, $I + A_1$ and $I + B_1$ and we can define a matrix C_1 by the equation $(1 + A_1)(1 + C_1)(1 + B_1) = (1 + C)$. Then C_1 again has entries in $\mathcal{A}^0(a_2, b_1)$ and C_1 is "smaller than" C. The next step is a similar formula $(1 + A_2)(1 + C_1)(1 + B_2) = 1 + C_2$. By induction, one constructs $A_n, B_n C_n$ with equalities $(1 + A_n)(1 + C_n)(1 + B_n) = 1 + C_{n-1}$. Finally, the products $(1 + A_n) \cdots (1 + A_1)$ and $(1 + B_1) \cdots (1 + B_n)$ converge to invertible matrices M_1 and M_2 with entries $\mathcal{A}(a_1, b_1)$ and $\mathcal{A}(a_2, b_2)$ such that $M = M_1 M_2$. We now make this more precise.

As in the proof of Proposition 7.24, we consider a closed path γ_1 consisting of three parts: the line segment from 0 to $(r + \epsilon)e^{i(a_2 + (1-1/2)\epsilon)}$, the circle segment from $(r + \epsilon)e^{i(a_2 + (1-1/2)\epsilon)}$ to $(r + \epsilon)e^{i(b_1 - (1-1/2)\epsilon)}$ and the line segment from $(r + \epsilon)e^{i(b_1 - (1-1/2)\epsilon)}$ to 0. This path is divided into halves γ_1^+ and γ_1^-. As above we are given an element $M = 1 + C$, where the matrix C has entries in $\mathcal{A}^0(a_2, b_1)$. We define the decomposition $C = A_1 + B_1$ by letting A_1 be the integral $\frac{1}{2\pi i} \int_{\gamma_1^+} \frac{C(\zeta)}{\zeta - z} d\zeta$ and B_1 be the integral with the same integrand and with path γ_1^-. We will see below how to select r small enough to ensure that A_1 and B_1 are small and so $1 + A_1$ and $1 + B_1$ are invertible. The matrix C_1 is defined by the equality $(1 + A_1)(1 + C_1)(1 + B_1) = 1 + C$. Clearly the entries of C_1 are sections of the sheaf $\mathcal{A}^0$ and live on a slightly smaller interval. In the next step one has to replace the path γ_1 by a path γ_2 that is slightly smaller. One obtains the path γ_2 by replacing $r + \epsilon$ by $r + \epsilon/2$, replacing $a_2 + (1 - 1/2)\epsilon$ by $a_2 + (1 - 1/4)\epsilon$ and finally replacing $b_1 - (1 - 1/2)\epsilon$ by $b_1 - (1 - 1/4)\epsilon$. The decomposition $C_1 = A_2 + B_2$ is given by integrating $\frac{C_1(\zeta)}{\zeta - z} d\zeta$ over the two halves γ_2^+ and γ_2^- of γ_2. The matrix C_2 is defined by the equality $(1 + A_2)(1 + C_2)(1 + B_2) = 1 + C_1$. By induction one defines sequences of paths γ_k and matrices A_k, B_k, C_k. Now we indicate the estimates that lead to showing that the limit of the products $(1 + A_n) \cdots (1 + A_1)$ and $(1 + B_1) \cdots (1 + B_n)$ converge to invertible matrices M_1 and M_2 with entries $\mathcal{A}(a_1, b_1 - \epsilon)$ and $\mathcal{A}(a_2 + \epsilon, b_2)$. The required equality $M_1 M_2 = M$ follows from the construction.

For a complex matrix $M = (m_{i,j})$, we use the norm $|M| := (\sum |m_{i,j}|^2)^{1/2}$. We recall the useful Lemma 5, page 196 of [120]:

> *There exists an absolute constant P such that for any matrices A and B with $|A|, |B| \leq 1/2$ and C defined by the equality $(1 + A)(1 + C)(1 + B) = (1 + A + B)$ one has $|C| \leq P|A| \cdot |B|$.*

Adapted to our situation this yields $|C_k(z)| \leq P|A_k(z)| \cdot |B_k(z)|$. One chooses r small enough so that one can apply the above inequalities and the supremum of $|A_k(z)|$, $|B_k(z)$, $|C_k(z)|$ on the sets, given by the inequalities $0 < |z| \leq r$ and arguments in $[a_2 + \epsilon, b_2)$, $(a_1, b_1 - \epsilon]$, and $[a_2 + \epsilon, b_1 - \epsilon]$, are bounded by ρ^k for some ρ, $0 < \rho < 1$. For the estimates leading to this, one has in particular to calculate the infimum of $|1 - \frac{\zeta}{z}|$ for ζ on the path of integration and z in the bounded domain under consideration. Details can be copied and adapted from the proof in [120] (for one complex variable and sectors replacing the compact sets K, K', K''). Then the expressions $(1 + A_n) \cdots (1 + A_1)$ and $(1 + B_1) \cdots (1 + B_n)$ converge uniformly to invertible matrices M_1 and M_2. The entries of these matrices are holomorphic on the two sets given by $0 < |z| < r$ and arguments in $(a_2 + \epsilon, b_2)$ and $(a_1, b_1 - \epsilon)$, respectively. To see that the entries of the two matrices are sections of the sheaf $\mathcal{A}$ one has to adapt the estimates given in the proof of Proposition 7.24. □

Remark 9.6 Theorem 9.5 remains valid when GL_n is replaced by any connected linear algebraic group G. The proof is then modified by replacing the expression $M = 1 + C$ by $M = exp(C)$ with C in the Lie algebra of G. One then makes the decomposition $C = A_1 - B + 1$ in the Lie algebra and considers $exp(A_1) \cdot M \cdot exp(-B_1) = M_1$, and so on by induction. □

Let $\{U_i\}$ be a covering of $\mathbf{S}^1$ consisting of proper open subsets. Any $\hat{F} \in \mathrm{GL}_n(\hat{K})$ can be lifted to some element $F_i \in \mathrm{GL}_n(\mathcal{A})(U_i)$ with asymptotic expansion $\hat{F}$. This produces a 1-cocycle $F_i F_j^{-1}$ for the sheaf $\mathrm{GL}_n(\mathcal{A})^0$ and an element

$\xi \in H^1(\mathbf{S}^1, \mathrm{GL}_n(\mathcal{A})^0)$. One sees at once that $\hat{F}$ and $\hat{F}G$, with $G \in \mathrm{GL}_n(K)$, produce the same element ξ in the cohomology set. This leads to the following result.

Corollary 9.7 (B. Malgrange and Y. Sibuya)
The natural map $\mathrm{GL}_n(\hat{K})\backslash\mathrm{GL}_n(K) \to H^1(\mathbf{S}^1, \mathrm{GL}_n(\mathcal{A})^0)$ *is a bijection.*

Proof. Let a 1-cocycle $g = \{g_{i,j}\}$ for the sheaf $\mathrm{GL}_n(\mathcal{A})^0$ and the covering $\{U_i\}$ be given. By Theorem 9.5, there are elements $F_i \in \mathrm{GL}_n(\mathcal{A})(U_i)$ with $g_{i,j} = F_i F_j^{-1}$. The asymptotic expansion of all the F_i is the same $\hat{F} \in \mathrm{GL}_n(\hat{K})$. Thus g is equivalent to a 1-cocycle produced by $\hat{F}$ and the map is surjective. Suppose now that $\hat{F}$ and $\hat{F}\hat{G}$ produce equivalent 1-cocycles. Liftings of $\hat{F}$ and $\hat{G}$ on the sector U_i are denoted by F_i and G_i. We are given that $F_i F_j^{-1} = L_i(F_i G_i G_j^{-1} F_j^{-1})L_j^{-1}$ holds for certain elements $L_i \in \mathrm{GL}_n(\mathcal{A})^0(U_i)$. Then $F_i^{-1} L_i F_i G_i$ is also a lift of $\hat{G}$ on the sector U_i. From $F_i^{-1} L_i F_i G_i = F_j^{-1} L_j F_j G_j$ for all i, j it follows that the lifts glue around $z = 0$ and thus $\hat{G} \in \mathrm{GL}_n(K)$. We conclude that the map is injective. □

We return now to the situation explained in the introduction: A quasi-split differential operator in matrix form $\delta - B$, the associated Stokes sheaf STS, which is the subsheaf of $\mathrm{GL}(n, \mathcal{A})^0$ consisting of the sections satisfying $T(\delta - B) = (\delta - B)T$, and the pairs $(\delta - A, \hat{F})$ satisfying $\hat{F}^{-1}(\delta - A)\hat{F} = \delta - B$ with $\hat{F} \in \mathrm{GL}_n(\hat{K})$ and A has entries in K.

Two pairs $(\delta - A_1, \hat{F}_1)$ and $(\delta - A_2, \hat{F}_2)$ are called *equivalent* or *cohomologous* if there is a $G \in \mathrm{GL}_n(K)$ such that $G(\delta - A_1)G^{-1} = \delta - A_2$ and $\hat{F}_2 = \hat{F}_1 G$. Consider a pair $(\delta - A, \hat{F})$. By the main asymptotic existence theorem (Theorem 7.10), there is an open covering $\{U_i\}$ and lifts F_i of $\hat{F}$ above U_i such that $F_i^{-1}(\delta - A)F_i = \delta - B$. The elements $F_i^{-1}F_j$ are sections of STS above $U_i \cap U_j$. In fact, $\{F_i^{-1}F_j\}$ is a 1-cocycle for STS and its image in $H^1(\mathbf{S}^1, STS)$ depends only on the equivalence class of the pair $(\delta - A, \hat{F})$.

Corollary 9.8 (B. Malgrange and Y. Sibuya)
The map described above is a bijection between the set of equivalence classes of pairs $(\delta - A, \hat{F})$ *and* $H^1(\mathbf{S}^1, STS)$.

Proof. If the pairs $(\delta - A_i, \hat{F}_i)$ for $i = 1, 2$ define the same element in the cohomology set, then they also define the same element in the cohomology set $H^1(\mathbf{S}^1, \mathrm{GL}_n(\mathcal{A})^0)$. According to Corollary 9.7 one has $\hat{F}_2 = \hat{F}_1 G$ for some $G \in \mathrm{GL}_n(K)$ and it follows that the pairs are equivalent. Therefore the map is injective.

Consider a 1-cocycle $\xi = \{\xi_{i,j}\}$ for the cohomology set $H^1(\mathbf{S}^1, STS)$. According to Corollary 9.7 there is an $\hat{F} \in \mathrm{GL}_n(\hat{K})$ and there are lifts F_i of $\hat{F}$ on the U_i such that $\xi_{i,j} = F_i^{-1}F_j$. From $\xi_{i,j}(\delta - B) = (\delta - B)\xi_{i,j}$ it follows that $F_j(\delta - B)F_j^{-1} =$

$F_i(\delta - B)F_i^{-1}$. Thus the $F_i(\delta - B)F_i^{-1}$ glue around $z = 0$ to a $\delta - A$ with entries in K. Moreover, $\hat{F}^{-1}(\delta - A)\hat{F} = \delta - B$ and the F_i are lifts of $\hat{F}$. This proves that the map is also surjective. □

Remark 9.9 Corollary 9.8 and its proof are valid for any differential operator $\delta - B$ over K, i.e., the property "quasi-split" of $\delta - B$ is not used in the proof. □

9.4 Explicit 1-cocycles for $H^1(\mathbf{S}^1, STS)$

This section is a variation on [177]. We will first state the main result. Let $\delta - B$ be quasi-split and let STS denote the associated Stokes sheaf on $\mathbf{S}^1$. The sheaf of the meromorphic solutions of $(\delta - B)y = 0$ can be seen as a locally constant sheaf of n-dimensional vector spaces on the circle $\mathbf{S}^1$. It is more convenient to consider the universal covering $pr : \mathbf{R} \to \mathbf{R}/2\pi\mathbf{Z} = \mathbf{S}^1$ of the circle and the sheaf pr^*STS on $\mathbf{R}$. Let W denote the solution space of $\delta - B$ with its decomposition $W_{q_1} \oplus \cdots \oplus W_{q_r}$. Then W and the W_{q_i} can be seen as constant sheaves on $\mathbf{R}$. Moreover, pr^*STS can be identified with a subsheaf of the constant sheaf $\mathrm{GL}(W)$ on $\mathbf{R}$. In more detail, $pr^*STS(a, b)$ consists of the linear maps of the form $id + \sum A_{i,j}$, where $A_{i,j}$ denotes a linear map of the type $W \stackrel{\text{projection}}{\to} W_{q_i} \stackrel{\text{linear}}{\to} W_{q_j} \stackrel{\text{inclusion}}{\to} W$ and where the sum is taken over all pairs i, j such that the function $e^{\int (q_i - q_j)\frac{dz}{z}}$ has asymptotic expansion 0 on (a, b). For each *singular* direction d we consider the subgroup $pr^*STS_d^*$ of the stalk pr^*STS_d consisting of the elements of the form $id + \sum A_{i,j}$, where $A_{i,j}$ denotes a linear map of the type $W \stackrel{\text{projection}}{\to} W_{q_i} \stackrel{\text{linear}}{\to} W_{q_j} \stackrel{\text{inclusion}}{\to} W$ and where the sum is taken over all pairs i, j such that d is *singular* for $q_i - q_j$.

For a sector $S \subset \mathbf{S}^1$ one chooses a connected component S' of $pr^{-1}(S)$ and one can identify $STS(S)$ with $pr^*STS(S')$. Similarly, one can identify the stalk STS_d for $d \in \mathbf{S}^1$ with $pr^*STS_{d'}$, where d' is a point with $pr(d') = d$. In particular, for a singular direction $d \in \mathbf{S}^1$ the subgroup STS_d^* of the stalk STS_d is well defined. Let $d_0 < \cdots < d_{m-1} < d_0(+2\pi) = d_m$ denote the singular directions for all $q_i - q_j$ (with the obvious periodic notation). Consider the covering $\mathcal{B} = \{B_i\}_{i=0,\ldots,m-1}$, $B_i = (d_{i-1} - \epsilon, d_i + \epsilon)$ with small enough $\epsilon > 0$. The set of 1-cocycles for the covering is clearly $\prod_{i=0,\ldots,m-1} STS(B_i \cap B_{i+1})$ and contains $\prod_{i=0,\ldots,m-1} STS_{d_i}^*$. This allows us to define a map $h : \prod_{i=0,\ldots,m-1} STS_{d_i}^* \to \check{H}^1(\mathcal{B}, STS) \to H^1(\mathbf{S}^1, STS)$. The main result is the following theorem.

Theorem 9.10 (M. Loday-Richaud [177])
The canonical map

$$h : \prod_{i=0,\ldots,m-1} STS_{d_i}^* \to \check{H}^1(\mathcal{B}, STS) \to H^1(\mathbf{S}^1, STS) \text{ is a bijection.}$$

Theorem 9.11 *The functor* Tup : $\mathrm{Diff}_K \rightarrow \mathrm{Gr}_2$ *is an equivalence of tannakian categories.*

Proof. We will deduce this from Theorem 9.10. In fact only the statement that h is injective will be needed, since the surjectivity of h will follow from Corollary 9.8 and the construction of the Stokes matrices.

Let us first give a quick proof of the *surjectivity* of the map h. According to Corollary 9.8 any element ξ of the cohomology set $H^1(\mathbf{S}^1, STS)$ can be represented by a pair $(\delta - A, \hat{F})$. For a direction d that is not singular for the collection $q_i - q_j$, there is a multisum $S_d(\hat{F})$. For $d \in (d_{i-1}, d_i)$ this multisum is independent of d and produces a multisum F_i of $\hat{F}$ above the interval B_i. The element $F_i^{-1}F_{i+1} = S_{d_i^-}(\hat{F})^{-1}S_{d_i^+}(\hat{F}) \in STS(B_i \cap B_{i+1})$ lies in the subgroup $STS^*_{d_i}$ of $STS(B_i \cap B_{i+1})$. Thus $\{F_i^{-1}F_{i+1}\}$ can be seen as an element of $\prod_{i=0,\ldots,m-1} STS^*_{d_i}$ and has, by construction, image ξ under h. In other words, we have found a map $\tilde{h} : H^1(\mathbf{S}^1, STS) \rightarrow \prod STS^*_{d_i}$ with $h \circ \tilde{h}$ is the identity.

Now we start the proof of Theorem 9.11. Using the previous notations, it suffices to produce a pair $(\delta - A, \hat{F})$ with $\hat{F}^{-1}(\delta - A)\hat{F} = \delta - B$ and prescribed Stokes maps at the singular directions $d_0, \ldots, d_{m-1}$. We recall that the Stokes maps St_{d_i} are given in matrix form by $S_{d_i^+}(\hat{F})E_{d_i}St_{d_i} = S_{d_i^-}(\hat{F})E_{d_i}$, where E_* is a fundamental matrix for $\delta - B$ and $S_*(\ )$ denotes multisummation. Therefore, we have to produce a pair $(\delta - A, \hat{F})$ with prescribed $S_{d_i^+}(\hat{F})^{-1}S_{d_i^-}(\hat{F}) \in STS^*_{d_i}$. Assuming that h is injective, one has that h is the inverse of $\tilde{h}$ and the statement is clear. □

Before we give the proof of Theorem 9.10, we introduce some terminology. One defines the *level* or the degree of some $q_i - q_j$ to be λ if $q_i - q_j = *z^{-\lambda}$ + terms of lower order and with $* \neq 0$. If d is a singular direction for $q_i - q_j$ then one attaches to d the level λ. We recall that the differential operator L, acting upon matrices, associated with our problem has the form $L(M) = \delta(M) - BM + MB$. The eigenvalues of L are the $q_i - q_j$ and the singular directions of L are the singular directions for the $\{q_i - q_j\}$. A singular direction d for L can be a singular direction for more than one $q_i - q_j$. In particular, a singular direction can have several levels.

Remark 9.12 *On Theorem 9.10*

1. Suppose that (a, b) is not contained in any interval $(d - \frac{\pi}{2k}, d + \frac{\pi}{2k})$, where d is a singular direction with level k, then $STS(a, b) = 1$. Furthermore, $H^1((a, b), STS) = \{1\}$ if (a, b) does not contain $[d - \frac{\pi}{2k}, d + \frac{\pi}{2k}]$, where d is a singular direction with level k. This follows easily from the similar properties of kernel of the above operator L acting upon $M(n \times n, \mathcal{A}^0)$ (see Corollary 7.21). The link between STS and L is given by $STS(a, b) = 1 + ker(L, M(n \times n, \mathcal{A}^0(a, b)))$.

2. The injectivity of h is not easily deduced from the material that we have at this point. We will give a combinatorial proof of Theorem 9.10 like the one given in [177]

that only uses the structure of the sheaf *STS* and is independent of the nonconstructive result of Malgrange and Sibuya, i.e., Corollary 9.8. The ingredients for this proof are the various levels in the sheaf *STS* and a method to change $\mathcal{B}$ into coverings adapted to those levels.

The given proof of Theorem 9.10 does not appeal to any result on multisummation. In [177], Theorem 9.10 is used to prove that an element $\hat{F} \in GL_n(\hat{K})$ such that $\hat{F}^{-1}(\delta - A)\hat{F} = \delta - B$ for a meromorphic A can be written in an essentially unique way as a product of k_ℓ-summable factors, where the k_ℓ are the levels of the associated $\{q_i - q_j\}$. Hence, this yields, in particular, the multisummability of such an $\hat{F}$.

3. In this setting, the proof will also be valid if one replaces W, W_{q_i} by $R \otimes_{\mathbf{C}} W$, $R \otimes_{\mathbf{C}} W_{q_i}$ for any $\mathbf{C}$-algebra R (commutative and with a unit element). In accordance with this the sheaf *STS* is replaced by the sheaf STS_R that has sections similar to the sheaf *STS*, but where $A_{i,j}$ is built from R-linear maps $R \otimes_{\mathbf{C}} W_{q_i} \to R \otimes_{\mathbf{C}} W_{q_j}$. □

9.4.1 One Level *k*

The assumption is that the collection $\{q_i - q_j\}$ has only one level k, i.e., for $i \neq j$ one has that $q_i - q_j = *z^{-k}+$ terms of lower order and $* \neq 0$. Our first concern is to construct a covering of $\mathbf{S}^1$ adapted to this situation. The covering $\mathcal{B}$ of Theorem 9.10 is such that there are no triple intersections. This is convenient for the purpose of writing 1-cocycles. The inconvenience is that there are many equivalent 1-cocycles. One replaces the covering $\mathcal{B}$ by a covering that does have triple intersections but few possibilities for equivalent 1-cocycles. We will do this in a systematic way.

Definition 9.13 An *m-periodic covering* of $\mathbf{R}$ is defined as a covering by distinct sets $U_i = (a_i, b_i)$, $i \in \mathbf{Z}$ satisfying:

1. $a_i \leq a_{i+1}$, $b_i \leq b_{i+1}$ and $b_i - a_i < 2\pi$ for all i.
2. $a_{i+m} = a_i + 2\pi$ and $b_{i+m} = b_i + 2\pi$ for all i.

The images $\bar{U}_i$ of the U_i under the map $pr : \mathbf{R} \to \mathbf{R}/2\pi\mathbf{Z} = \mathbf{S}^1$, form a covering of $\mathbf{S}^1$ that we will call a *cyclic covering*. For convenience, we will only consider $m > 2$. □

Lemma 9.14 *Let $\mathcal{G}$ be any sheaf of groups on $\mathbf{S}^1$ and let $\mathcal{U} = \{\bar{U}_i\}_{i=0,\dots,m-1}$ be a cyclic covering of $\mathbf{S}^1$. Let C denote the set of 1-cocycles for $\mathcal{G}$ and $\mathcal{U}$. Then the map $r : C \to \prod_{i=0}^{m-1} G(\bar{U}_i \cap \bar{U}_{i+1})$, given by $\{g_{i,j}\} \mapsto \{g_{i,i+1}\}$, is a bijection.*

Proof. One replaces $\mathbf{S}^1$ by its covering $\mathbf{R}$, $\mathcal{G}$ by the sheaf $pr^*\mathcal{G}$ and C by the pr^*C, the set of 1-cocycles for pr^*G and $\{U_i\}$. Suppose that we have shown that the natural map $r^* : pr^*C \to \prod_i pr^*\mathcal{G}(U_i \cap U_{i+1})$ is bijective. Then this bijection induces a bijection between the m-period elements of pr^*C and the m-period elements of $\prod_i pr^*\mathcal{G}(U_i \cap U_{i+1})$. As a consequence r is bijective.

Let elements $g_{i,i+1} \in pr^*\mathcal{G}(U_i \cap U_{i+1})$ be given. It suffices to show that these data extend in a unique way to a 1-cocycle for $pr^*\mathcal{G}$. One observes that for $i < j - 1$ one has $U_i \cap U_j = (U_i \cap U_{i+1}) \cap \cdots \cap (U_{j-1} \cap U_j)$. Now one defines $g_{i,j} := g_{i,i+1} \cdots g_{j-1,j}$. The rule $g_{i,j}g_{j,k} = g_{i,k}$ (for $i < j < k$) is rather obvious. Thus $\{g_{i,j}\}$ is a 1-cocycle and clearly the unique one extending the data $\{g_{i,i+1}\}$. □

Proof of Theorem 9.10 The cyclic covering that we take here is $\mathcal{U} = \{\bar{U}_i\}$ with $U_i := (d_{i-1} - \frac{\pi}{2k}, d_i + \frac{\pi}{2k})$. By Remark 9.12 one has $STS(\bar{U}_i) = 1$, $STS(\bar{U}_i \cap \bar{U}_{i+1}) = STS^*_{d_i}$ and $H^1(\bar{U}_i, STS) = \{1\}$. Thus $\check{H}^1(\mathcal{U}, STS) \to H^1(\mathbf{S}^1, STS)$ is an isomorphism. The map from the 1-cocycles for $\mathcal{U}$ to $\check{H}^1(\mathcal{U}, STS)$ is bijective. By Lemma 9.14 the set of 1-cocycles is $\prod_{i=0,\dots,m-1} STS^*_{d_i}$. Finally, the covering $\mathcal{B}$ of the theorem refines the covering $\mathcal{U}$ and thus the theorem follows. □

In the proof of the induction step for the case of more levels, we will use the following result.

Lemma 9.15 *The elements $\xi, \eta \in \prod_{i=0,\dots,m-1} STS^*_{d_i}$ are seen as 1-cocycles for the covering $\mathcal{B}$. Suppose that there are elements $F_i \in STS(B_i)$ such that $\xi_i = F_i \eta_i F_{i+1}^{-1}$ holds for all i. Then $\xi = \eta$ and all $F_i = 1$.*

Proof. We have just shown that $\xi = \eta$. In proving that all $F_i = 1$ we will work on $\mathbf{R}$ with the sheaf pr^*STS and the m-periodic covering. The first observation is that if $F_{i_0} = 1$ holds for some i_0 then also $F_{i_0+1} = 1$ and $F_{i_0-1} = 1$. Thus all $F_i = 1$. In the sequel we will suppose that all $F_i \neq 1$ and derive a contradiction.

The section F_i has a maximal interval of definition of the form: $(d_{\alpha(i)} - \frac{\pi}{2k}, d_{\beta(i)} + \frac{\pi}{2k})$, because of the special nature of the sheaf STS. If $\alpha(i) < \beta(i)$ it would follow that $F_i = 1$, since the interval then has length $> \frac{\pi}{k}$. Thus $\beta(i) \le \alpha(i)$.

The equality $F_i = \xi_i F_{i+1} \xi_i^{-1}$ implies that F_i also exists above the interval $(d_i - \frac{\pi}{2k}, d_i + \frac{\pi}{2k}) \cap (d_{\alpha(i+1)} - \frac{\pi}{2k}, d_{\beta(i+1)} + \frac{\pi}{2k})$. Therefore $d_{\beta(i)} + \frac{\pi}{2k} \ge \min(d_i + \frac{\pi}{2k}, d_{\beta(i+1)} + \frac{\pi}{2k})$. Thus $\min(i, \beta(i+1)) \le \beta(i)$.

From $F_{i+1} = \xi_i^{-1} F_i \xi_i$ it follows that F_{i+1} is also defined above the interval $(d_i - \frac{\pi}{2k}, d_i + \frac{\pi}{2k}) \cap (d_{\alpha(i)} - \frac{\pi}{2k}, d_{\beta(i)} + \frac{\pi}{2k})$. Thus $d_{\alpha(i+1)} - \frac{\pi}{2k} \le \max(d_i - \frac{\pi}{2k}, d_{\alpha(i)} - \frac{\pi}{2k})$. Therefore, $\alpha(i+1) \le \max(i, \alpha(i))$.

We continue with the inequalities $\min(i, \beta(i+1)) \le \beta(i)$. By m-periodicity, e.g., $\beta(i+m) = \beta(i) + 2\pi$, we conclude that for some i_0 one has $\beta(i_0 + 1) > \beta(i_0)$. Hence $i_0 \le \beta(i_0)$. The inequality $\min(i_0 - 1, \beta(i_0)) \le \beta(i_0 - 1)$ implies $i_0 - 1 \le \beta(i_0 - 1)$. Therefore, $i \le \beta(i)$ holds for all $i \le i_0$ and by m-periodicity this inequality holds for all $i \in \mathbf{Z}$. We then also have that $i \le \alpha(i)$ holds for all i, since $\beta(i) \le \alpha(i)$.

From $\alpha(i+1) \le \max(i, \alpha(i))$ one concludes $\alpha(i+1) \le \alpha(i)$ for all i. Then also $\alpha(i+m) \le \alpha(i)$. But this contradicts $\alpha(i+m) = \alpha(i) + 2\pi$. □

9.4.2 Two Levels $k_1 < k_2$

A choice of the covering $\mathcal{U}$. As always one assumes that $1/2 < k_1$. Let $\mathcal{U} = \{\bar{U}_i\}$ be the cyclic covering of $\mathbf{S}^1$ derived from the m-periodic covering $\{(d_{i-1} - \frac{\pi}{2k_2} -\epsilon(i-1), d_i + \frac{\pi}{2k_2} + \epsilon(i))\}$, where $\epsilon(i) = 0$ if d_i has k_2 as level and $\epsilon(i)$ is positive and small if the only level of d_i is k_1.

One sees that $\bar{U}_i$ does not contain $[d - \frac{\pi}{2k}, d + \frac{\pi}{2k}]$ for any singular point d that has a level k_2. Furthermore, $\bar{U}_i$ can be contained in some $(d - \frac{\pi}{2k_1}, d + \frac{\pi}{2k_1})$ with d singular with level k_1. However, $\bar{U}_i$ cannot be contained in some $(d - \frac{\pi}{2k_2}, d + \frac{\pi}{2k_2})$ with d singular with a level k_2. From Remark 9.12 and the non-abelian version of Theorem C.26, it follows that $\check{H}^1(\mathcal{U}, STS) \to H^1(\mathbf{S}^1, STS)$ is a bijection.

A decomposition of the sheaf STS. For $k \in \{k_1, k_2\}$ one defines the subsheaf of groups $STS(k)$ of STS by $STS(k)$ contains only sections of the type $id + \sum A_{i,j}$ where the level of $q_i - q_j$ is k. Let $i_1 < i_2 < i_3$ be such that $q_{i_1} - q_{i_2}$ and $q_{i_2} - q_{i_3}$ have level k, then $q_{i_1} - q_{i_3}$ has level $\le k$. This shows that $STS(k_1)$ is a subsheaf of groups. Furthermore, $STS(k_2)$ consists of the sections T of $\mathrm{GL}_n(\mathcal{A})^0$ (satisfying $T(\delta - B) = (\delta - B)T$) and such that $T - 1$ has coordinates in $\mathcal{A}^0_{1/k_2}$. This implies that $STS(k_2)$ is a subsheaf of groups and, moreover, $STS(k_2)(a,b)$ is a normal subgroup of $STS(a,b)$. The subgroup $STS(k_1)(a,b)$ maps bijectively to $STS(a,b)/STS(k_2)(a,b)$. We conclude the following lemma.

Lemma 9.16 *$STS(a,b)$ is a semidirect product of the normal subgroup $STS(k_2)(a,b)$ and the subgroup $STS(k_1)(a,b)$.*

Proof of the surjectivity of h.
By Lemma 9.14 the map $h : \prod_{i=0,\dots,m-1} STS^*_{d_i} \to \check{H}^1(\mathcal{B}, STS) \to H^1(\mathbf{S}^1, STS)$ factors over $\check{H}^1(\mathcal{U}, STS)$ and, moreover, $\check{H}^1(\mathcal{U}, STS) \to H^1(\mathbf{S}^1, STS)$ is a bijection. Therefore, it suffices to prove that the map $\prod_{i=0,\dots,m-1} STS^*_{d_i} \to \check{H}^1(\mathcal{U}, STS)$ is bijective. Consider a 1-cocycle $\xi = \{\xi_i\}$ for $\mathcal{U}$ and STS. Each ξ_i can (uniquely) be written as $\xi_i(k_2)\xi_i(k_1)$ with $\xi_i(k_2), \xi_i(k_1)$ sections of the sheaves $STS(k_2)$ and $STS(k_1)$. The collection $\{\xi_i(k_1)\}$ can be considered as a 1-cocycle for $STS(k_1)$ and the covering $\mathcal{U}$. This 1-cocycle does not, in general, satisfy $\xi_i(k_1) \in STS(k_1)^*_{d_i}$. We will replace $\{\xi_i(k_1)\}$ by an equivalent 1-cocycle that has this property.

For the sheaf $STS(k_1)$ we consider the singular directions $e_0 < e_1 < \cdots < e_{s-1}$. These are the elements in $\{d_0, \dots, d_{m-1}\}$ that have a level k_1. Furthermore, we consider the cyclic covering $\mathcal{V}$ of $\mathbf{S}^1$, corresponding with the s-periodic covering $\{(e_{i-1} - \frac{\pi}{2k_1}, e_i + \frac{\pi}{2k_1})\}$ of $\mathbf{R}$. The covering $\mathcal{U}$ is finer than $\mathcal{V}$. For each U_i we choose the inclusion $U_i \subset V_j$, where $e_{j-1} \le d_{i-1} < e_j$. Let $\eta = \{\eta_j\}$ be a 1-cocycle for $STS(k_1)$

and $\mathcal{V}$, satisfying $\eta_j \in STS(k_1)^*_{e_j}$ and which has the same image in $H^1(\mathbf{S}^1, STS(k_1))$ as the 1-cocycle $\{\xi_i(k_1)\}$. The 1-cocycle η is transported to a 1-cocycle $\tilde{\eta}$ for $STS(k_1)$ and $\mathcal{U}$. One sees that $\tilde{\eta}_i \in STS(k_1)^*_{d_i}$ holds for all i. Furthermore, there are elements $F_i \in STS(k_1)(\bar{U}_i)$ such that $F_i\xi_i(k_1)F_{i+1}^{-1} = \tilde{\eta}_i$ for all i.

Consider now the 1-cocycle $\{F_i\xi_iF_{i+1}^{-1}\}$, which is equivalent to ξ. One has $F_i\xi_iF_{i+1}^{-1} = F_i\xi_i(k_2)F_i^{-1}\tilde{\eta}_i$. Now $F_i\xi_i(k_2)F_i^{-1}$ lies in $STS(k_2)(\bar{U}_i \cap \bar{U}_{i+1})$. The only possible singular direction d with a level k_2 such that $\bar{U}_i \cap \bar{U}_{i+1} \subset (d - \frac{\pi}{2k_2}, d + \frac{\pi}{2k_2})$ is $d = d_i$. Hence, $F_i\xi_i(k_2)F_i^{-1} \in STS(k_2)^*_{d_i}$. We conclude that $F_i\xi_iF_{i+1}^{-1} \in STS^*_{d_i}$ and thus the surjectivity has been proven. □

Proof of the injectivity of *h*.
Since the covering $\mathcal{B}$ of the theorem refines $\mathcal{U}$, we need to show that $\prod_{i=0,\dots,m-1} STS^*_{d_i} \to \check{H}^1(\mathcal{U}, STS)$ is injective. As before, an element $\xi = \{\xi_i\}$ of the left-hand side is decomposed as $\xi_i = \xi_i(k_2)\xi_i(k_1)$, where $\xi_i(k_2)$ and $\xi_i(k_1)$ are elements of the groups $STS(k_2)^*_{d_i}$ and $STS(k_1)^*_{d_i}$. For another element η in the set on the left-hand side we use a similar notation. Suppose that ξ and η are equivalent. Then there are elements $F_i \in STS(\bar{U}_i) = STS(k_1)(\bar{U}_i)$ such that $\xi_i(k_2)\xi_i(k_1) = F_i\eta_i(k_2)\eta_i(k_1)F_{i+1}^{-1} = F_i\eta_i(k_2)F_i^{-1}F_i\eta_i(k_1)F_{i+1}^{-1}$. It follows that $\xi_i(k_2) = F_i\eta_i(k_2)F_{i+1}^{-1}$ and $\xi_i(k_1) = F_i\eta_i(k_1)F_{i+1}^{-1}$. From the latter equalities and Lemma 9.15 we conclude that $\xi_i(k_1) = \eta_i(k_1)$ and all $F_i = 1$. □

9.4.3 The General Case

In the general case with levels $k_1 < k_2 < \cdots < k_s$ (and $1/2 < k_1$) the sheaf STS is a semidirect product of the sheaf of normal subgroups $STS(k_s)$, which contains only sections with level k_s, and the sheaf of subgroups $STS(\leq k_{s-1})$, which contains only levels $\leq k_{s-1}$. The cyclic covering $\mathcal{U}$, is associated with the m-periodic covering of $\mathbf{R}$ given by $U_i = (d_{i-1} - \frac{\pi}{2k_s} - \epsilon(i-1), d_i + \frac{\pi}{2k_s} + \epsilon(i))$, where $\epsilon(i) = 0$ if d_i contains a level k_s and otherwise $\epsilon(i) > 0$ and small enough.

The *surjectivity* of the map h (with the covering $\mathcal{B}$ replaced by $\mathcal{U}$) is proved as follows. Decompose a general 1-cocycle $\xi = \{\xi_i\}$ as $\xi_i = \xi_i(k_s)\xi_i(\leq k_{s-1})$. By induction, there are elements $F_i \in STS(\leq k_s)(\bar{U}_i)$ such that all $\eta_i := F_i\xi_i(\leq k_{s-1})F_{i+1}^{-1}$ lie in $STS(\leq k_{s-1})^*_{d_i}$. Then $F_i\xi_iF_{i+1}^{-1} = F_i\xi_i(k_s)F_i^{-1}\eta_i$. If a singular direction d, which has a level k_s, satisfies $\bar{U}_i \cap \bar{U}_{i+1} \subset (d - \frac{\pi}{2k_s}, d + \frac{\pi}{2k_s})$ then $d = d_i$. This implies $F_i\xi_i(k_s)F_i^{-1} \in STS(k_s)^*_{d_i}$ and ends the proof.

The *injectivity* of h is also proved by induction with respect to the number of levels involved. The reasoning is rather involved and we will make the case of three levels $k_1 < k_2 < k_3$ explicit. The arguments for more than three levels are similar. The sheaf STS has subsheaves of normal subgroups $STS(k_3)$ and $STS(\geq k_2)$ (using only sections with level k_3 or with levels k_2 and k_3). There is a subsheaf of groups

$STS(k_1)$ consisting of the sections that only use level k_1. The sheaf $STS(\geq k_2)$ has a subsheaf of groups $STS(k_2)$ of the sections that only use level k_2. Furthermore, STS is a semidirect product of $STS(\geq k_2)$ and $STS(k_1)$. Also, $STS(\geq k_2)$ is a semidirect product of $STS(k_3)$ and $STS(k_2)$. Finally, every section F of STS can uniquely be written as a product $F(k_3)F(k_2)F(k_1)$ of sections for the sheaves $STS(k_i)$.

One considers two elements $\xi, \eta \in \prod_{i=0,\dots,m-1} STS^*_{d_i}$ and sections F_i of the sheaf $STS(\bar{U}_i) = STS(\leq k_2)(\bar{U}_i)$ such that $\xi_i = F_i \eta_i F_{i+1}^{-1}$ holds. Then $\xi_i(k_3)\xi_i(k_2)\xi_i(k_1) = F_i \eta_i(k_3)\eta_i(k_2)\eta_i(k_1)F_{i+1}^{-1}$. Working modulo the normal subgroups $STS(k_3)$ one finds $\xi_i(k_2)\xi_i(k_1) = F_i\eta_i(k_2)\eta_i(k_1)F_{i+1}^{-1}$. This is a situation with two levels and we have proved that then $\xi_i(k_2) = \eta_i(k_2)$, $\xi_i(k_1) = \eta_i(k_1)$. From the equalities $\xi_i(k_2)\xi_i(k_1) = F_i\xi_i(k_2)\xi_i(k_1)F_{i+1}^{-1}$, we want to deduce that all $F_i = 1$. The latter statement would end the proof.

Working modulo the normal subgroups $STS(k_2)$ and using Lemma 9.15 one obtains that all F_i are sections of $STS(k_2)$. The above equalities hold for the covering $\mathcal{U}$ corresponding to the intervals $(d_{i-1} - \frac{\pi}{2k_3} - \epsilon(i-1), d_i + \frac{\pi}{2k_3} + \epsilon(i))$. Since the singular directions d that have only level k_3 play no role here, one may change $\mathcal{U}$ into the cyclic covering corresponding with the periodic covering $(e_{i-1} - \frac{\pi}{2k_3}$ $-\epsilon(i-1), e_i + \frac{\pi}{2k_3} + \epsilon(i))$, where the e_i are the singular directions having a level in $\{k_1, k_2\}$. The above equalities remain the same. Now one has to adapt the proof of Lemma 9.15 for this situation. If some F_{i_0} happens to be 1, then all $F_i = 1$. One considers the possibility that $F_i \neq 1$ for all i. Then F_i has a maximal interval of definition of the form $(e_{\alpha(i)} - \frac{\pi}{2k_2}, e_{\beta(i)} + \frac{\pi}{2k_2})$. Using the above equalities one arrives at a contradiction.

Remark 9.17 In [177], Theorem 9.10 is proved directly from the Main Asymptotic Existence Theorem without appeal to results on multisummation. In that paper this result is used to prove that an element $\hat{F} \in \mathrm{GL}_n(\hat{K})$ with $\hat{F}^{-1}(\delta - B)\hat{F} = B$ for some quasi-split B can be written as the product of k_l-summable factors, where the k_l are the levels of the associated $\{q_i - q_j\}$ and so yields the multisummability of such an $\hat{F}$. These results were achieved before the publication of [198].

Furthermore, combining Theorem 9.10 with Corollary 9.8, one sees that there is a natural bijection b from "$\{(\delta - A, \hat{F})\}$ modulo equivalence" to $\prod_{i=0}^{m-1} STS^*_{d_i}$. This makes the set "$\{(\delta - A, \hat{F})\}$ modulo equivalence" explicit. Using multisummation, one concludes that b associates to $(\delta - A, \hat{F})$ the elements $\{E_{d_i} St_{d_i} E_{d_i}^{-1} | i = 1, \dots, m-1\}$ or equivalently, the set of Stokes matrices $\{St_{d_i} | i = 1, \dots, m-1\}$, since the E_{d_i} are known. □

9.5 $H^1(\mathbf{S}^1, STS)$ as an Algebraic Variety

The idea is to convert this cohomology set into a covariant functor $\mathcal{F}$ from the category of the $\mathbf{C}$-algebras (always commutative and with a unit element) to the category of sets. For a $\mathbf{C}$-algebra R one considers the free R-module $W_R := R \otimes_{\mathbf{C}} W$ and the sheaf of groups STS_R on $\mathbf{S}^1$, defined by its pull back $pr^* STS_R$ on $\mathbf{R}$, which is given by $pr^* STS_R(a, b)$ are the R-linear automorphisms of W_R of the form $id + \sum A_{i,j}$, where $A_{i,j}$ denotes a linear map of the type $W_R \overset{\text{projection}}{\to} (W_i)_R \overset{\text{linear}}{\to} (W_j)_R \overset{\text{inclusion}}{\to} W_R$ and where the sum is taken over all pairs i, j such that $e^{\int (q_i - q_j) \frac{dz}{z}}$ has asymptotic expansion 0 on (a, b). In a similar way one defines the subgroup $(STS_R)^*_d$ of the stalk $(STS_R)_d$. The functor is given by $\mathcal{F}(R) = H^1(\mathbf{S}^1, STS_R)$. Theorem 9.10 and its proof remain valid in this new situation and provides a functorial isomorphism $\prod_{d \text{ singular}} (STS_R)^*_d \to \mathcal{F}(R)$. It follows that this functor is representable (see Definition B.18) and is represented by the affine space $\mathbf{A}^N_{\mathbf{C}}$, which describes all the possible Stokes matrices.

In [13], the following *local moduli problem* is studied:

Fix a quasi-split differential operator $\delta - B$ and consider pairs $(\delta - A, \hat{F})$, where A has entries in K, $\hat{F} \in \mathrm{GL}_n(\hat{K})$ and $\hat{F}^{-1}(\delta - A)\hat{F} = \delta - B$.

Corollary 9.8 states that the set E of equivalence classes of pairs can be identified with the cohomology set $H^1(\mathbf{S}^1, STS)$. We just proved that this cohomology set has a natural structure as the affine space. Also in [13] the cohomology set is given the structure of an algebraic variety over $\mathbf{C}$. It can be seen that the two structures coincide.

The bijection $E \to H^1(\mathbf{S}^1, STS)$ induces an algebraic structure on E of the same type. However, E with this structure is not a fine moduli space for the local moduli problem (see [228]). We will return to the problem of families of differential equations and moduli spaces of differential equations.

10 Universal Picard-Vessiot Rings and Galois Groups

10.1 Introduction

Let K denote any differential field such that its field of constants $C = \{a \in K \mid a' = 0\}$ is algebraically closed, has characteristic 0 and is different from K. The neutral tannakian category Diff_K of differential modules over K is equivalent to the category Repr_H of all finite dimensional representations (over C) of some affine group scheme H over C (see Appendices B.2 and B.3 for the definition and properties). Let $\mathcal{C}$ be a full subcategory of Diff_K that is closed under all operations of linear algebra, i.e., kernels, cokernels, direct sums, and tensor products. Then $\mathcal{C}$ is also a neutral tannakian category and equivalent to Repr_G for some affine group scheme G.

Consider a differential module M over K and let $\mathcal{C}$ denote the full subcategory of Diff_K, whose objects are subquotients of direct sums of modules of the form $M \otimes \cdots \otimes M \otimes M^* \otimes \cdots \otimes M^*$. This category is equivalent to Repr_G, where G is the differential Galois group of M. In this special case there is also a Picard-Vessiot ring R_M and G consists of the K-linear automorphisms of R_M that commute with the differentiation on M (see also Theorem 2.33).

This special case generalizes to arbitrary $\mathcal{C}$ as above. We define *a universal Picard-Vessiot ring* UnivR *for* $\mathcal{C}$ as follows:

1. UnivR is a K-algebra and there is given a differentiation $r \mapsto r'$ that extends the differentiation on K.
2. The only differential ideals of UnivR are $\{0\}$ and UnivR.
3. For every differential equation $y' = Ay$ belonging to $\mathcal{C}$ there is a fundamental matrix F with coefficients in UnivR.
4. R is generated, as K-algebra, by the entries of the fundamental matrices F and $\det(F)^{-1}$ for all equations in $\mathcal{C}$.

It can be shown that UnivR exists and is unique up to K-linear differential isomorphism. Moreover, UnivR has no zero divisors and the constant field of its field of fractions is again C. We shall call this field of fractions the *universal Picard-Vessiot field* of $\mathcal{C}$ and denote it by UnivF. Furthermore, one easily sees that UnivR is the direct limit $\varinjlim R_M$, taken over all differential modules M in $\mathcal{C}$. Finally, the affine group scheme G such that $\mathcal{C}$ is equivalent with Repr_G, can be seen to be the group of the K-linear automorphisms of UnivR that commute with the differentiation of UnivR. We will call UnivG *the universal differential Galois group of* $\mathcal{C}$. The way the group UnivG of automorphism of UnivR is considered as affine group scheme over C will now be made more explicit.

For every commutative C-algebra A one considers the $A \otimes_C K$-algebra $A \otimes_C$ UnivR. The differentiation of UnivR extends to a unique A-linear differentiation on $A \otimes_C$ UnivR. Now one introduces a functor $\mathcal{F}$ from the category of the commutative C-algebras to the category of all groups by defining $\mathcal{F}(A)$ to be the group of the $A \otimes_C K$-linear automorphisms of $A \otimes_C$ UnivR that commute with the differentiation of $A \otimes_C$ UnivR. It can be seen that this functor is representable and according to Appendix B.2, $\mathcal{F}$ defines an affine group scheme. The group UnivG above is this affine group scheme.

The theme of this chapter is to present examples of differential fields K and subcategories $\mathcal{C}$ (with the above conditions) of Diff_K such that both the universal Picard-Vessiot ring and the differential Galois group of $\mathcal{C}$ are explicit. One may compare this with the following problem for ordinary Galois theory: *Produce examples of a field F and a collection $\mathcal{C}$ of finite Galois extensions of F such that the compositum $\tilde{F}$ of all fields in $\mathcal{C}$ and the (infinite) Galois group of $\tilde{F}/F$ are both explicit.* For example, If $F = \mathbf{Q}$ and $\mathcal{C}$ is the collection of all abelian extensions of $\mathbf{Q}$, then the Galois group of $\tilde{F}/F$ is the projective limit of the groups of of the invertible elements $(\mathbf{Z}/n\mathbf{Z})^*$ of $\mathbf{Z}/n\mathbf{Z}$. Other known examples are:

(a) F is a local field and $\tilde{F}$ is the separable algebraic closure of F.
(b) F is a global field and $\mathcal{C}$ is the collection of all abelian extensions of F.
See, for example, [62] and [261].

10.2 Regular Singular Differential Equations

The differential field will be $\widehat{K} = \mathbf{C}((z))$, the field of the formal Laurent series. The category $\mathcal{C}$ will be the full subcategory of $\text{Diff}_{\widehat{K}}$ whose objects are the regular singular differential modules over $\widehat{K}$. We recall from Sect. 3.1.1 that a differential module M is regular singular if there is a $\mathbf{C}[[z]]$-lattice $\Lambda \subset M$ that is invariant under the operator $z \cdot \partial_M$. It has been shown that a regular singular differential module has a basis such that the corresponding matrix differential equation has the form $\frac{d}{dz}y = \frac{B}{z}y$ with B a constant matrix. The symbols $\text{UnivR}_{regsing}$ and $\text{UnivG}_{regsing}$ denote the universal Picard-Vessiot ring and the universal differential Galois group of $\mathcal{C}$.

Proposition 10.1

1. *$\mathcal{C}$ is equivalent to the neutral tannakian category* $\text{Repr}_{\mathbf{Z}}$ *and* $\text{UnivG}_{regsing}$ *is isomorphic to the algebraic hull of* $\mathbf{Z}$.

2. *The universal Picard-Vessiot ring* $\text{UnivR}_{regsing}$ *is equal to* $\widehat{K}[\{z^a\}_{a\in\mathbf{C}}, \ell]$.
3. $\text{UnivG}_{regsing} = \text{Spec}(B)$ *and the Hopf algebra* B *is given by:*
   a) B *equals* $\mathbf{C}[\{s(a)\}_{a\in\mathbf{C}}, t]$ *where the only relations between the generators* $\{s(a)\}_{a\in\mathbf{C}}$, t *are* $s(a+b) = s(a)\cdot s(b)$ *for all* $a, b \in \mathbf{C}$.
   b) *The comultiplication* Δ *on* B *is given by the formulas:* $\Delta(s(a)) = s(a) \otimes s(a)$ *and* $\Delta(t) = (t\otimes 1) + (1\otimes t)$.

Proof. We note that the $\widehat{K}$-algebra $\text{UnivR}_{regsing} := \widehat{K}[\{z^a\}_{a\in\mathbf{C}}, \ell]$ is defined by the relations: $z^{a+b} = z^a \cdot z^b$ for all $a, b \in \mathbf{C}$ and for any $a \in \mathbf{Z}$ the symbol z^a is equal to z^a as element of $\widehat{K}$. The differentiation in $\text{UnivR}_{regsing}$ is given by $\frac{d}{dz}z^a = az^{a-1}$ and $\frac{d}{dz}\ell = z^{-1}$. From the fact that every regular singular differential module can be represented by a matrix differential equation $y' = \frac{B}{z}y$, with B a constant matrix, one easily deduces that $\text{UnivR}_{regsing}$ is indeed the universal Picard-Vessiot ring of $\mathcal{C}$. This proves point 2. The formal monodromy γ is defined as the $\widehat{K}$-linear automorphism of $\text{UnivR}_{regsing}$ given by the formulas $\gamma(z^a) = e^{2\pi i a}z^a$ and $\gamma\ell = \ell + 2\pi i$. Clearly $\gamma \in \text{UnivG}_{regsing}$.

The solution space V_M of a regular singular differential module M is the space $V_M = \ker(\partial_M, \text{UnivR}_{regsing} \otimes_{\widehat{K}} M)$. The action of γ on $R_{regsing}$ induces a $\mathbf{C}$-linear action γ_M on V_M. One associates to M above the pair (V_M, γ_M). The latter is an object of $\text{Repr}_{\mathbf{Z}}$. It is easily verified that one obtains in this way an equivalence $\mathcal{C} \to \text{Repr}_{\mathbf{Z}}$ of tannakian categories. According to part B of the appendix, $\text{UnivG}_{regsing}$ is isomorphic to the algebraic hull of $\mathbf{Z}$.

For the last part of the proposition one considers a commutative $\mathbf{C}$-algebra A and one has to investigate the group $\mathcal{F}(A)$ of the $A\otimes_{\mathbf{C}}\widehat{K}$-automorphisms σ of $A\otimes_{\mathbf{C}}$ $\text{UnivR}_{regsing}$ that commute with the differentiation on $A\otimes_{\mathbf{C}} \text{UnivR}_{regsing}$. For any $a\in\mathbf{C}$ one has $\sigma z^a = h(a)\cdot z^a$ with $h(a) \in A^*$. Furthermore, h is seen to be a group homomorphism $h : \mathbf{C}/\mathbf{Z} \to A^*$. There is a $c \in A$ such that $\sigma\ell = \ell + c$. On the other hand, any choice of a homomorphism h and a $c\in A$ define a unique $\sigma \in \mathcal{F}(A)$. Therefore, one can identify $\mathcal{F}(A)$ with $\text{Hom}_{\mathbf{C}}(B, A)$, the set of the $\mathbf{C}$-algebra homomorphisms from B to A. This set has a group structure induced by Δ. It is obvious that the group structures on $\mathcal{F}(A)$ and $\text{Hom}_{\mathbf{C}}(B, A)$ coincide. □

10.3 Formal Differential Equations

Equations Again $\widehat{K} = \mathbf{C}((z))$. For convenience one considers the differentiation $\delta := z\frac{d}{dz}$ on $\widehat{K}$. Differential equations (or differential modules) over $\widehat{K}$ are called *formal differential equations*.

Theorem 10.2 *Consider the neutral tannakian category* $\mathrm{Diff}_{\widehat{K}}$.

1. *The universal Picard-Vessiot ring is* $\mathrm{UnivR}_{formal} := \widehat{K}[\{z^a\}_{a\in\mathbf{C}}, \ell, \{e(q)\}_{q\in\mathcal{Q}}]$, *(see Sect. 3.2).*
2. *The differential Galois group* UnivG_{formal} *of* $\mathrm{Diff}_{\widehat{K}}$ *has the following structure: There is a split exact sequence of affine group schemes*

$$1 \to \mathrm{Hom}(\mathcal{Q}, \mathbf{C}^*) \to \mathrm{UnivG}_{formal} \to \mathrm{UnivG}_{regsing} \to 1.$$

The affine group scheme $\mathrm{Hom}(\mathcal{Q}, \mathbf{C}^*)$ *is called the* exponential torus. *The formal monodromy* $\gamma \in \mathrm{UnivR}_{regsing}$ *acts on* $\mathcal{Q}$ *in an obvious way. This induces an action of* γ *on the exponential torus. The latter coincides with the action by conjugation of* γ *on the exponential torus. The action, by conjugation, of* $\mathrm{UnivG}_{regsing}$ *on the exponential torus is deduced from the fact that* $\mathrm{UnivG}_{regsing}$ *is the algebraic hull of the group* $\langle\gamma\rangle \cong \mathbf{Z}$.

Proof. The first part has been proved in Sect. 3.2. The morphism $\mathrm{UnivG}_{formal} \to \mathrm{UnivG}_{regsing}$ is derived from the inclusion $\mathrm{UnivR}_{regsing} \subset \mathrm{UnivR}_{formal}$. One associates to the automorphism $\sigma \in \mathrm{UnivG}_{formal}$ its restriction to $\mathrm{UnivR}_{regsing}$. Any automorphism $\tau \in \mathrm{UnivG}_{regsing}$ of $\mathrm{UnivR}_{regsing}$ is extended to the automorphism σ of UnivR_{formal} by putting $\sigma e(q) = e(q)$ for all $q \in \mathcal{Q}$. This provides the morphism $\mathrm{UnivG}_{regsing} \to \mathrm{UnivG}_{formal}$. An element σ in the kernel of $\mathrm{UnivG}_{formal} \to \mathrm{UnivG}_{regsing}$ acts on UnivR_{formal} by fixing each z^a and ℓ and by $\sigma e(q) = h(q)\cdot e(q)$ where $h : \mathcal{Q} \to \mathbf{C}^*$ is a homomorphism. This yields the identification of this kernel with the affine group scheme $\mathrm{Hom}(\mathcal{Q}, \mathbf{C}^*)$. Finally, the algebraic closure of $\widehat{K}$ is contained in $\mathrm{UnivR}_{regsing}$ and in particular γ acts on the algebraic closure of $\widehat{K}$ by sending each z^λ (with $\lambda \in \mathbf{Q}$) to $e^{2\pi i\lambda}z^\lambda$. There is an induced action on $\mathcal{Q}$, considered as a subset of the algebraic closure of $\widehat{K}$. A straightforward calculation proves the rest of the theorem. □

10.4 Meromorphic Differential Equations

The differential field is $K = \mathbf{C}(\{z\})$, the field of the convergent Laurent series over $\mathbf{C}$. On both fields K and $\widehat{K} = \mathbf{C}((z))$ we will use the differentiation $\delta = z\frac{d}{dz}$. In this section we will treat the most interesting example and describe the universal Picard-Vessiot ring UnivR_{conv} and the universal differential Galois group UnivG_{conv} for the category Diff_K of all differential modules over K. Differential modules over K, or their associated matrix differential equations over K, are called *meromorphic differential equations*. In this section we present a complete proof of the description of UnivG_{conv} given in the inspiring paper [203].

Our first claim that there is a more or less explicit expression for the universal Picard-Vessiot ring UnivR_{conv} of Diff_K. For this purpose we define a K-algebra $\mathcal{D}$

with $K \subset \mathcal{D} \subset \widehat{K}$ as follows: $f \in \widehat{K}$ belongs to $\mathcal{D}$ if and only if f satisfies some linear scalar differential equation $f^{(n)} + a_{n-1}f^{(n-1)} + \cdots + a_1 f^{(1)} + a_0 f = 0$ with all coefficients $a_i \in K$. This condition on f can be restated as follows: f belongs to $\mathcal{D}$ if and only the K-linear subspace of $\widehat{K}$ generated by all the derivatives of f is finite dimensional. It follows easily that $\mathcal{D}$ is an algebra over K stable under differentiation. The following example shows that $\mathcal{D}$ *is not a field.*

Example 10.3 The differential equation $y^{(2)} = z^{-3}y$ (here we have used the ordinary differentiation $\frac{d}{dz}$) has a solution $f = \sum_{n \geq 2} a_n z^n \in \widehat{K}$ given by $a_2 = 1$ and $a_{n+1} = n(n-1)a_n$ for $n \geq 2$. Clearly f is a divergent power series and by definition $f \in \mathcal{D}$. Suppose also that $f^{-1} \in \mathcal{D}$. Then also $u := \frac{f'}{f}$ lies in $\mathcal{D}$ and there is a finite dimensional K-vector space W with $K \subset W \subset \hat{K}$ that is invariant under differentiation and contains u. We note that $u' + u^2 = z^{-3}$ and consequently $u^2 \in W$. Suppose that $u^n \in W$. Then $(u^n)' = nu^{n-1}u' = nu^{n-1}(-u^2 + z^{-3}) \in W$ and thus $u^{n+1} \in W$. Since all the powers of u belong to W the element u must be algebraic over K. It is known that K is algebraically closed in $\widehat{K}$ and thus $u \in K$. The element u can be written as $\frac{2}{z} + b_0 + b_1 z + \cdots$ and since $f' = uf$ one finds $f = z^2 \cdot exp(b_0 z + b_1 \frac{z^2}{2} + \cdots)$. The latter is a convergent power series and we have obtained a contradiction. We note that $\mathcal{D}$ can be seen as the linear differential closure of K into $\widehat{K}$. It seems difficult to make the K-algebra $\mathcal{D}$ really explicit. (See Exercise 1.39 for a general approach to functions f such that f and $1/f$ both satisfy linear differential equations.) □

Lemma 10.4 *The universal Picard-Vessiot ring for the category of all meromorphic differential equations is* $\text{UnivR}_{conv} := \mathcal{D}[\{z^a\}_{a \in \mathbf{C}}, \ell, \{e(q)\}_{q \in \mathcal{Q}}]$.

Proof. The algebra UnivR_{formal} contains UnivR_{conv} and UnivR_{conv} is generated, as a K-algebra, by the entries of F and $\det(F)^{-1}$ of all fundamental matrices F of meromorphic equations. The entries of a fundamental matrix are expressions in $z^a, \ell, e(q)$ and formal Laurent series. The formal Laurent series that occur satisfy some linear scalar differential equation over K. From this the lemma follows. □

The universal differential Galois group for Diff_K is denoted by UnivG_{conv}. The inclusion $\text{UnivR}_{conv} \subset \text{UnivR}_{formal}$ induces an injective morphism of affine group schemes $\text{UnivG}_{formal} \to \text{UnivG}_{conv}$. One can also define a morphism $\text{UnivG}_{conv} \to \text{UnivG}_{formal}$ of affine group schemes. In order to do this correctly we replace UnivG_{conv} and UnivG_{formal} by their functors $\mathcal{G}_{conv}$ and $\mathcal{G}_{formal}$ from the category of the commutative $\mathbf{C}$-algebras to the category of groups. Let A be a commutative $\mathbf{C}$-algebra. One defines $\mathcal{G}_{conv}(A) \to \mathcal{G}_{formal}(A)$ by sending any automorphism $\sigma \in \mathcal{G}_{conv}(A)$ to $\tau \in \mathcal{G}_{formal}(A)$ defined by the formula $\tau(g) = \sigma(g)$ for $g = z^a, \ell, e(q)$. The group homomorphism $\mathcal{G}_{formal}(A) \to \mathcal{G}_{conv}(A)$ is defined by sending τ to its restriction σ on the subring $A \otimes_{\mathbf{C}} \text{UnivR}_{conv}$ of $A \otimes_{\mathbf{C}} \text{UnivR}_{formal}$. The functor $\mathcal{N}$ is defined by letting $\mathcal{N}(A)$ be the kernel of the surjective group homomorphism $\mathcal{G}_{conv}(A) \to \mathcal{G}_{formal}(A)$. In other words, $\mathcal{N}(A)$ consists of the automorphisms $\sigma \in \mathcal{G}_{conv}(A)$ satisfying $\sigma(g) = g$ for $g = z^a, \ell, e(q)$. It can be seen that $\mathcal{N}$ is representable and thus defines an affine group scheme N. Thus we have demonstrated the following lemma

Lemma 10.5 *There is a split exact sequence of affine group schemes*

$$1 \to N \to \mathrm{UnivG}_{conv} \to \mathrm{UnivG}_{formal} \to 1.$$

The above lemma reduces the description of the structure of UnivG_{conv} to a description of N and the action of UnivG_{formal} on N. In the following we will study the structure of the Lie algebra Lie(N) of N. We are working with affine group schemes G that are not linear algebraic groups, and consequently we have to be somewhat careful about their Lie algebras $\mathrm{Lie}(G)$.

Definition 10.6 A *pro-Lie algebra* L over $\mathbf{C}$ is the projective limit $\varprojlim L_j$ of finite dimensional Lie algebras. □

Clearly L has the structure of Lie algebra. We have to introduce a topology on L in order to find the "correct" finite dimensional representations of L. This can be done as follows. An ideal $I \subset L$ will be called closed if I contains $\cap_{j\in F}\ker\,(L \to L_j)$ for some finite set of indices F.

Definition 10.7 A *representation* of a pro-Lie algebra L on a finite dimensional vector space W over $\mathbf{C}$ will be a homomorphism of complex Lie algebras $L \to \mathrm{End}(W)$, such that its kernel is a closed ideal. □

For an affine group scheme G, which is the projective limit $\varprojlim G_j$ of linear algebraic groups G_j, one defines $\mathrm{Lie}(G)$ as the pro-Lie algebra $\varprojlim \mathrm{Lie}(G_j)$. Suppose that G is connected, then we *claim* that any finite dimensional complex representation of G yields a finite dimensional representation of $\mathrm{Lie}(G)$. Indeed, this statement is known for linear algebraic groups over $\mathbf{C}$. Thus $\mathrm{Lie}(G_j)$ and G_j have the same finite dimensional complex representations. Since every finite dimensional complex representation of G or of the pro-Lie algebra $\mathrm{Lie}(G)$ factors over some G_j or some $\mathrm{Lie}(G_j)$, the claim follows.

Now we return to the pro-Lie algebra Lie(N). The identification of the affine group scheme N with a group of automorphisms of UnivR_{conv} leads to the identification of Lie(N) with the complex Lie algebra of the K-linear derivations $D : \mathrm{UnivR}_{conv} \to \mathrm{UnivR}_{conv}$, commuting with the differentiation on UnivR_{conv} and satisfying $D(g) = 0$ for $g = z^a, \ell, e(q)$. A derivation $D \in$ Lie(N) is therefore determined by its restriction to $\mathcal{D} \subset \mathrm{UnivR}_{conv}$. One can show that an ideal I in Lie(N) is closed if and only if there are finitely many elements $f_1, \ldots f_s \in \mathcal{D}$ such that $I \supset \{D \in \mathrm{Lie}(N) |\ D(f_1) = \cdots = D(f_s) = 0\}$.

We search now for elements in N and Lie(N). For any direction $d \in \mathbf{R}$ and any meromorphic differential module M one has defined in Sect. 8.3 an element St_d

acting on the solution space V_M of M. In fact, St_d is a K-linear automorphism of the Picard-Vessiot ring R_M of M, commuting with the differentiation on R_M. The functoriality of the multisummation implies that St_d depends functorially on M and induces an automorphism of the direct limit UnivR_{conv} of all Picard-Vessiot rings R_M. By construction, St_d leaves z^a, ℓ, $e(q)$ invariant and therefore St_d lies in N. The action of St_d on any solution space V_M is unipotent. The Picard-Vessiot ring R_M is as a K-algebra generated by the coordinates of the solution space $V_M = \ker(\partial, R_M \otimes M)$ in R_M. It follows that every finite subset of R_M lies in a finite dimensional K-vector space, invariant under St_d and such that the action of St_d is unipotent. The same holds for the action of St_d on UnivR_{conv}. We refer to this property by saying: *St_d acts locally unipotent on R_{conv}.*

The above property of St_d implies that $\Delta_d := \log St_d$ is a well-defined K-linear map $\mathrm{UnivR}_{conv} \to \mathrm{UnivR}_{conv}$. Clearly Δ_d is a derivation on UnivR_{conv}, belongs to $\mathrm{Lie}(N)$ and is *locally nilpotent*. The algebra UnivR_{conv} has a direct sum decomposition $\mathrm{UnivR}_{conv} = \oplus_{q\in\mathcal{Q}} \mathrm{UnivR}_{conv,\,q}$ where $\mathrm{UnivR}_{conv,\,q} := \mathcal{D}[\{z^a\}_{a\in\mathbf{C}}, \ell]e(q)$. This allows us to decompose $\Delta_d : \mathcal{D} \to \mathrm{UnivR}_{conv}$ as a direct sum $\sum_{q\in\mathcal{Q}} \Delta_{d,q}$ by the formula $\Delta_d(f) = \sum_{q\in\mathcal{Q}} \Delta_{d,q}(f)$ and where $\Delta_{d,q}(f) \in \mathrm{UnivR}_{conv,\,q}$ for each $q \in \mathcal{Q}$. We note that $\Delta_{d,q} = 0$ if d is not a singular direction for q. The map $\Delta_{d,q} : \mathcal{D} \to \mathrm{UnivR}_{conv}$ has a unique extension to an element in $\mathrm{Lie}(N)$.

Definition 10.8 The elements $\{\Delta_{d,q} |\ d \text{ singular direction for } q\}$ are called *alien derivations.* □

We note that the above construction and the term *alien derivation* are due to J. Écalle [92]. This concept is the main ingredient for his theory of resurgence.

The group $\mathrm{UnivG}_{formal} \subset \mathrm{UnivG}_{conv}$ acts on $\mathrm{Lie}(N)$ by conjugation. For a homomorphism $h : \mathcal{Q} \to \mathbf{C}^*$ one writes τ_h for the element of this group that is defined by the properties that τ_h leaves z^a, and ℓ invariant and $\tau e(q) = h(q) \cdot e(q)$. Let γ denote, as before, the formal monodromy. According to the structure of UnivG_{formal} described in Chap. 3, it suffices to know the action by conjugation of the τ_h and γ on $\mathrm{Lie}(N)$. For the elements $\Delta_{d,q}$ one has the explicit formulas:

(a) $\gamma\Delta_{d,q}\gamma^{-1} = \Delta_{d-2\pi,\gamma(q)}$.
(b) $\tau_h\Delta_{d,q}\tau_h^{-1} = h(q) \cdot \Delta_{d,q}$.

Consider the set $S := \{\Delta_{d,q} |\ d \in \mathbf{R},\ q \in \mathcal{Q}, d \text{ singular for } q\}$. We would like to state that S generates the Lie algebra $\mathrm{Lie}(N)$ and that these elements are independent. This is close to being correct. The fact that the $\Delta_{d,q}$ act locally nilpotent on UnivR_{conv}, however, complicates the final statement. In order to be more precise we have to go through some general constructions with Lie algebras.

A Construction with Free Lie Algebras We recall some classical constructions, see [144], Chap. V.4. Let S be any set. Let W denote a vector space over $\mathbf{C}$ with basis S. By $W^{\otimes m}$ we denote the m-fold tensor product $W \otimes_{\mathbf{C}} \cdots \otimes_{\mathbf{C}} W$ (note that this is *not* the symmetric tensor product). Then $F\{S\} := \mathbf{C} \oplus \sum_{m\geq 1}^{\oplus} W^{\otimes m}$ is the *free associative algebra on the set S*. It comes equipped with a map $i : S \to F\{S\}$. The universal property of $(i, F\{S\})$ reads:

> For any associative **C**-algebra B and any map $\phi : S \to B$ there is a unique **C**-algebra homomorphism $\phi' : F\{S\} \to B$ with $\phi' \circ i = \phi$.

The algebra $F\{S\}$ is also a Lie algebra with respect to the Lie brackets [,] defined by $[A, B] = AB - BA$. The *free Lie algebra on the set S* is denoted by Lie$\{S\}$ and is defined as the Lie subalgebra of $F\{S\}$ generated by $W \subset F\{S\}$. This Lie algebra is equipped with an obvious map $i : S \to \text{Lie}\{S\}$ and the pair $(i, \text{Lie}\{S\})$ has the following universal property:

> For any complex Lie algebra L and any map $\phi : S \to L$ there is a unique homomorphism $\phi' : \text{Lie}\{S\} \to L$ of complex Lie algebras such that $\phi' \circ i = \phi$.

Furthermore, for any associative complex algebra B and any homomorphism $\psi : \text{Lie}\{S\} \to B$ of complex Lie algebras (where B is given its canonical structure as complex Lie algebra) there is a unique homomorphism $\psi' : F\{S\} \to B$ of complex algebras such that the restriction of ψ' to Lie$\{S\}$ coincides with ψ.

Consider now a finite dimensional complex vector space W and an action of Lie$\{S\}$ on W. This amounts to a homomorphism of complex Lie algebras $\psi : \text{Lie}\{S\} \to \text{End}(W)$ or to a **C**-algebra homomorphism $\psi' : F\{S\} \to \text{End}(W)$. Here, we are only interested in those ψ such that:

(1) $\psi(s) = \psi'(s)$ is nilpotent for all $s \in S$.
(2) there are only finitely many $s \in S$ with $\psi(s) \neq 0$.

For any ψ satisfying (1) and (2) one considers the ideal kerψ in the Lie algebra Lie$\{S\}$ and its quotient Lie algebra Lie$\{S\}$/ker ψ. One now defines a sort of completion $\widehat{\text{Lie}\{S\}}$ of Lie$\{S\}$ as the projective limit of the Lie$\{S\}$/ker ψ, taken over all ψ satisfying (1) and (2).

Lemma 10.9 *Let W be a finite dimensional complex vector space and let $N_1, \ldots, N_s$ denote nilpotent elements of* End(W). *Then the Lie algebra L generated by $N_1, \ldots, N_s$ is algebraic, i.e., it is the Lie algebra of a connected algebraic subgroup of* GL(W).

Proof. Let $N \in \text{End}(W)$ be a nilpotent map and suppose $N \neq 0$. Then the map $t \in \mathbf{G}_{a,\mathbf{C}} \mapsto \exp(tN) \in \text{GL}(W)$ is a morphism of algebraic groups. Its image is an algebraic subgroup H of GL(W), isomorphic to $\mathbf{G}_{a,\mathbf{C}}$. The Lie algebra of H is equal to $\mathbf{C}N$.

Let $G_1, \ldots, G_s$ be the algebraic subgroups of GL(W), each one isomorphic to $\mathbf{G}_{a,\mathbf{C}}$, with Lie algebras $\mathbf{C}N_1, \ldots, \mathbf{C}N_s$. The algebraic group G generated by $G_1, \ldots, G_s$ is equal to $H_1 \cdot H_2 \cdot \dots \cdot H_m$ for some m and some choice for $H_1, \dots, H_m \in \{G_1, \dots, G_s\}$ ([141], Proposition 7.5). Then G is connected and from this representation one concludes that the Lie algebra of G is the Lie algebra generated by $N_1, \dots, N_s$. □

We apply the lemma to the Lie algebra $L_\psi := \mathrm{Lie}\{S\}/\ker \psi$, considered above. By definition this is a Lie algebra in End(W) generated by finitely many nilpotent elements. Let G_ψ denote the connected algebraic group with $\mathrm{Lie}(G_\psi) = L_\psi$. The connected linear algebraic groups G_ψ form a projective system. We will denote the corresponding projective limit by M. The pro-Lie algebra $\widehat{\mathrm{Lie}\{S\}}$ is clearly the pro-Lie algebra of M.

In the sequel S will be the collection of all alien derivations $S := \{\Delta_{d,q} | \ d \in \mathbf{R}, q \in \mathcal{Q}, d$ is singular for $q\}$. The action of UnivG_{formal} on the set of the alien derivations induces an action on $\widehat{\mathrm{Lie}\{S\}}$ and an action on the affine group scheme M. The affine variety $M \times \mathrm{UnivG}_{formal}$ is made into an affine group scheme by the formula $(m_1, g_1) \cdot (m_2, g_2) = (m_1 \cdot g_1 m_2 g_1^{-1}, g_1 g_2)$ for the composition. The precise interpretation of this formula is obtained by replacing M and UnivG_{formal} by their corresponding functors $\mathcal{M}$ and $\mathcal{G}_{formal}$ and define for every commutative $\mathbf{C}$-algebra A the group structure on $\mathcal{M}(A) \times \mathcal{G}_{formal}(A)$ by the above formula, where $g_1 m_2 g_1^{-1}$ stands for the known action of UnivG_{formal} on M. The result is an affine group scheme that is a semidirect product $M \rtimes \mathrm{UnivG}_{formal}$. We can now formulate the description of J. Martinet and J.-P. Ramis for the structure of UnivG_{conv} and Lie(N) in the following theorem.

Theorem 10.10 *The affine group scheme $M \rtimes \mathrm{UnivG}_{formal}$ is canonically isomorphic to* UnivG_{conv}. *In particular, N is isomorphic to M and therefore N is connected. Let S denote again the set of all alien derivations $\{\Delta_{d,q} | \ d \in \mathbf{R}, q \in \mathcal{Q}$, d is singular for $q\}$. Then there exists an isomorphism of complex pro-Lie algebra $\psi : \widehat{\mathrm{Lie}\{S\}} \to \mathrm{Lie}(N)$ that respects the* UnivG_{formal}*-action on both pro-Lie algebras.*

Proof. By definition, the tannakian categories Diff_K and $\mathrm{Repr}_{\mathrm{UnivG}_{conv}}$ are equivalent. According to Sect. 9.2 the tannakian categories Diff_K and Gr_2 are also equivalent. Now we consider the tannakian category $\mathrm{Repr}_{M \rtimes \mathrm{UnivG}_{formal}}$. An object of this category is a finite dimensional complex vector space W provided with an action of $M \rtimes \mathrm{UnivG}_{formal}$. The action of UnivG_{formal} on W gives W the structure of an object of Gr_1, namely a direct sum decomposition $W = \oplus_{q \in \mathcal{Q}} W_q$ and the action of the formal monodromy γ on W has image $\gamma_W \in \mathrm{GL}(W)$ satisfying the required properties. The additional action of M on W translates into an action of its pro-Lie algebra $\widehat{\mathrm{Lie}\{S\}}$ on W. According to the definition of this pro-Lie algebra the latter translates into a set of nilpotent elements $\{\Delta_{W,d,q}\} \subset \mathrm{End}(W)$, where $\Delta_{W,d,q}$ denotes the action of $\Delta_{d,q}$ on W. By definition, there are only finitely many nonzero $\Delta_{W,d,q}$ and every $\Delta_{W,d,q}$ is nilpotent. Using the structure of the semidirect product $M \rtimes \mathrm{UnivG}_{formal}$ and, in particular, the action of UnivG_{formal} on $\widehat{\mathrm{Lie}\{S\}}$ one finds the properties:

(a) $\gamma_W \Delta_{W,d,q} \gamma_W^{-1} = \Delta_{W,d-2\pi,\gamma(q)}$, and
(b) $\Delta_{W,d,q}$ is a $\mathbf{C}$-linear map that maps each summand $W_{q'}$ of W to $W_{q+q'}$.

Define now $\Delta_{W,d} := \oplus_{q \in \mathcal{Q}} \Delta_{W,d,q}$. This is easily seen to be a nilpotent map. Define $St_{W,d} := \exp(\Delta_{W,d})$. Then it is obvious that the resulting tuple $(W, \{W_q\}, \gamma_W, \{St_{W,d}\})$ is an object of Gr_2. The converse, i.e., every object of Gr_2 induces a representation of $M \rtimes \mathrm{UnivG}_{formal}$, is also true. The conclusion is that the tannakian categories $\mathrm{Repr}_{M \rtimes \mathrm{UnivG}_{formal}}$ and Gr_2 are equivalent. Then the tannakian categories $\mathrm{Repr}_{M \rtimes \mathrm{UnivG}_{formal}}$ and $\mathrm{Repr}_{\mathrm{UnivG}_{conv}}$ are equivalent and the affine group schemes $M \rtimes \mathrm{UnivG}_{formal}$ and UnivG_{conv} are isomorphic. If one follows the equivalences between the above tannakian categories then one obtains an isomorphism ϕ of affine group schemes $M \rtimes \mathrm{UnivG}_{formal} \to \mathrm{UnivG}_{conv}$ that induces the identity from UnivG_{formal} to $\mathrm{UnivG}_{conv}/N \cong \mathrm{UnivG}_{formal}$. Therefore, ϕ induces an isomorphism $M \to N$ and the rest of the theorem is then obvious. □

Remarks 10.11
(1) Let W be a finite dimensional complex representation of UnivG_{conv}. Then the image of $N \subset \mathrm{UnivG}_{conv}$ in $\mathrm{GL}(W)$ contains all St_d operating on W. As in the above proof, W can be seen as an object of the category Gr_2. One can build examples such that the smallest algebraic subgroup of $\mathrm{GL}(W)$ containing all St_d is not a normal subgroup of the differential Galois group, i.e., the image of $\mathrm{UnivG}_{conv} \to \mathrm{GL}(W)$. The above theorem implies that the smallest *normal* algebraic subgroup of $\mathrm{GL}(W)$ containing all the St_d is the image of $N \to \mathrm{GL}(W)$.

(2) Theorem 10.10 can be seen as a differential analog of a conjecture of I.R. Shafarevich concerning the Galois group $\mathrm{Gal}(\overline{\mathbf{Q}}/\mathbf{Q})$. Let $\mathbf{Q}(\mu_\infty)$ denote the maximal cyclotomic extension of $\mathbf{Q}$. Furthermore, S denotes an explicitly given countable subset of the Galois group $\mathrm{Gal}(\overline{\mathbf{Q}}/\mathbf{Q}(\mu_\infty))$. The conjecture states that the above Galois group is a profinite completion $\widehat{F}$ of the free non-abelian group F on S. This profinite completion $\widehat{F}$ is defined as the projective limit of the F/H, where H runs in the set of the normal subgroups of finite index of G such that H contains a cofinite subset of S. □

11 Inverse Problems

11.1 Introduction

In this chapter we continue the investigation of Chap. 10 concerning the differential Galois theory for special classes of differential modules. Recall that K is a differential field such that its field of constants $C = \{a \in K|\ a' = 0\}$ has characteristic 0, is algebraically closed and different from K. Furthermore, $\mathcal{C}$ is a full subcategory of the category Diff_K of all differential modules over K, which is closed under all operations of linear algebra, i.e., kernels, cokernels, direct sums, and tensor products. Then $\mathcal{C}$ is a neutral tannakian category and thus isomorphic to Repr_G for some affine group scheme G over C. The *inverse problem of differential Galois theory for the category* $\mathcal{C}$ asks for a description of the linear algebraic groups H that occur as a differential Galois group of some object in $\mathcal{C}$. We note that H occurs as a differential Galois group if and only if there exists a surjective morphism $G \to H$ of affine group schemes over C. The very few examples where an explicit description of G is known are treated in Chap. 10. In the present chapter we investigate the, a priori, easier inverse problem for certain categories $\mathcal{C}$. This is a reworked version of [230].

It is interesting to compare this with Abhyankar's conjecture [1] and its solution. The simplest form of this conjecture concerns the projective line $\mathbf{P}^1_k = \mathbf{A}^1_k \cup \{\infty\}$ over an algebraically closed field k of characteristic $p > 0$. A covering of $\mathbf{P}^1_k$, unramified outside ∞, is a finite morphism $f : X \to \mathbf{P}^1_k$ of projective nonsingular curves such that f is unramified at every point $x \in X$ with $f(x) \neq \infty$. The covering f is called a Galois covering if the group H of the automorphisms $h : X \to X$ with $f \circ h = f$ has the property that X/H is isomorphic to $\mathbf{P}^1_k$. If, moreover, X is irreducible then one calls $f : X \to \mathbf{P}^1_k$ a (connected) *Galois cover*. Abhyankar's conjecture states that a finite group H is the Galois group of a Galois cover of $\mathbf{P}^1_k$ unramified outside ∞ if and only if $H = p(H)$ where $p(H)$ is the subgroup of H generated by its elements of order a power of p. We note in passing that $p(H)$ is also the subgroup of H generated by all its p-Sylow subgroups.

This conjecture has been proved by M. Raynaud [244] (and in greater generality by Harbater [121]). The collection of all coverings of $\mathbf{P}^1_k$, unramified outside ∞, is easily seen to be a Galois category (see Appendix B). In particular, there exists a profinite group G, such that this category is isomorphic to Perm_G. A finite group H is the Galois group of a Galois cover of $\mathbf{P}^1_k$, unramified outside ∞, if and only if there exists a surjective continuous homomorphism of groups $G \to H$. No explicit description of the profinite group G is known although the collection of its finite continuous images is given by Raynaud's theorem. We will return to Abhyankar's conjecture in Sect. 11.6 for a closer look at the analogy with the inverse problem for differential equations.

Examples 11.1 *Some easy cases for the inverse problem.*
1. Let $\mathcal{C}$ denote the category of the regular singular differential modules over the differential field $\mathbf{C}((z))$. Corollary 3.32 states that the Galois group of such a module is the Zariski closure of a subgroup generated by one element. Conversely, given a constant $n \times n$ matrix D, let C be an $n \times n$ constant matrix satisfying $D = e^{2\pi i C}$. Theorem 5.1 implies that the local monodromy of $z\partial_z Y = CY$ is given by D and that this coincides with the formal monodromy. Therefore, we can conclude that a linear algebraic group G is a differential Galois group for an object in $\mathcal{C}$ if and only if G is topologically generated by one element (i.e., there exists a subgroup H of G generated by one element that is dense in G for the Zariski topology). This also follows from the results of Sect. 10.2.

2. Let X be a compact Riemann surface of genus g and $S \subset X$ a finite set of cardinality s. The differential field K is the field of meromorphic functions on X. Let $q \in X$ be a point and t a local parameter at q. Then the field of the locally defined meromorphic functions K_q at q is isomorphic to $\mathbf{C}(\{t\})$. One calls a differential module M over K regular or regular singular at q if $M \otimes_K K_q$ is regular or regular singular (over K_q). Now one defines the full subcategory $\mathcal{C}$ of Diff_K whose objects are the differential modules M over K that are regular for every $q \notin S$ and regular singular (or regular) at the points $q \in S$. The answer to the inverse problem is:

Let $\pi_1(X \setminus S)$ denote the fundamental group of $X \setminus S$. A linear algebraic group G is a differential Galois group for the category $\mathcal{C}$ if and only if there exists a homomorphism $\pi_1(X \setminus S) \to G$ such that its image is dense in G for the Zariski topology. In particular, if $s \geq 1$ then G is a differential group for $\mathcal{C}$ if and only if G is topologically generated by at most $2g+s-1$ elements (i.e., there is a Zariski-dense subgroup H of G generated by at most $2g+s-1$ elements).

The proof goes as follows. The solution of the weak form of the Riemann-Hilbert Problem (Theorem 6.15) extends to the present situation. Thus an object M of $\mathcal{C}$ corresponds to a representation $\rho : \pi_1(X \setminus S) \to \mathrm{GL}(V)$, where V is a finite dimensional vector space over $\mathbf{C}$. The Zariski closure of the image of ρ coincides with the differential Galois group of M. Indeed, Theorem 5.8 and its proof are valid in this more general situation. Finally, the fundamental group of $X \setminus S$ is known to be the free group on $2g + s - 1$ elements, if $s \geq 1$. □

Although the universal differential Galois group has been determined for Diff_K and the differential fields $K = \mathbf{C}((z))$ and $K = \mathbf{C}(\{z\})$, it is not at all evident how to characterize the linear algebraic groups that are factors of this universal differential Galois group. Theorems 11.2 and 11.13 give such a characterization. In this chapter we shall assume a greater familiarity with the theory of linear algebraic groups and Lie algebras. Besides the specific references given below, general references are [45, 141, 280].

11.2 The Inverse Problem for $\mathbf{C}((z))$

Theorem 11.2 *A linear algebraic group G over $\mathbf{C}$ is a differential Galois group of a differential module over $\mathbf{C}((z))$ if and only if G contains a normal subgroup T such that T is a torus and G/T is topologically generated by one element.*

Proof. In Sect. 3.2 we introduced a category Gr_1 of triples that is equivalent to the tannakian category $\mathrm{Diff}_{\mathbf{C}((z))}$. If the differential module M corresponds to the triple $(V, \{V_q\}, \gamma_V)$, then, according to Corollary 3.32, the differential Galois group G of M is the smallest algebraic subgroup of $\mathrm{GL}(V)$ containing the exponential torus T and the formal monodromy γ_V. The exponential torus T is a normal subgroup of G and G/T is topologically generated by the image of γ_V. This proves one direction of the theorem.

For the proof of the other direction, we fix an embedding of G into $\mathrm{GL}(V)$ for some finite dimensional $\mathbf{C}$-vector space (in other words V is a faithful G-module). The action of T on V is given by distinct characters $\chi_1, \ldots, \chi_s$ of T and a decomposition $V = \oplus_{i=1}^{s} V_{\chi_i}$ such that for every $t \in T$, every i and every $v \in V_{\chi_i}$ one has $tv = \chi_i(t) \cdot v$. Let $A \in G$ map to a topological generator of G/T. Then A permutes the vector spaces V_{χ_i}. Indeed, for $t \in T$ and $v \in V_{\chi_i}$ one has $tAv = A(A^{-1}tA)v = \chi_i(A^{-1}tA)Av$. Define the character χ_j by $\chi_j(t) = \chi_i(A^{-1}tA)$. Then $AV_{\chi_i} = V_{\chi_j}$. The proof will be finished if we can find a triple $(V, \{V_q\}, \gamma_V)$ in Gr_1 such that $V = V_{q_1} \oplus \cdots \oplus V_{q_s}$ with $V_{q_i} = V_{\chi_i}$ for all $i = 1, \ldots, s$ and $A = \gamma_V$. This will be shown in the next lemma. □

We will write $A\chi_i = \chi_j$. Recall that $\mathcal{Q} = \cup_{m \geq 1} z^{-1/m}\mathbf{C}[z^{-1/m}]$ carries an action of γ given by $\gamma z^{\lambda} = e^{2\pi i \lambda} z^{\lambda}$.

Lemma 11.3 *There are elements $q_1, \ldots, q_s \in \mathcal{Q}$ such that*

1. *if $A\chi_i = \chi_j$ then $\gamma(q_i) = q_j$.*
2. *if $n_1\chi_1 + \cdots + n_s\chi_s = 0$ (here with additive notation for characters) for some $n_1, \ldots, n_s \in \mathbf{Z}$, then $n_1q_1 + \cdots + n_sq_s = 0$.*

Furthermoremore, if $N \geq 1$ is an integer, then there exists $q_1, \ldots, q_s$ satisfying conditions 1 and 2 and such that for any $i \neq j$ the degrees of $q_i - q_j$ in z^{-1} are $\geq N$.

Proof. Conditions 1 and 2 can be translated into: the **Z**-module $\mathbf{Z}\chi_1 + \cdots + \mathbf{Z}\chi_s$ can be embedded into $\mathcal{Q}$ such that the action of A is compatible with the action of γ on $\mathcal{Q}$. Let $\overline{\mathbf{Q}} \subset \mathbf{C}$ denote the algebraic closure of $\mathbf{Q}$.

Consider $M := \overline{\mathbf{Q}} \otimes (\mathbf{Z}\chi_1 + \cdots + \mathbf{Z}\chi_s) \supset \mathbf{Z}\chi_1 + \cdots + \mathbf{Z}\chi_s$ with the induced A action. Since a power of A acts as the identity on M we may decompose $M = \oplus_{\zeta_j} M_{\zeta_j}$, where ζ_j runs over a finite set of roots of unity and A acts on M_{ζ_j} as multiplication by ζ_j. Furthermore, $\mathcal{Q} = \oplus_{\lambda \in \mathbf{Q},\, \lambda < 0} \mathbf{C} z^{\lambda}$ and γ acts on $\mathbf{C}z^{\lambda}$ as multiplication by $e^{2\pi i \lambda}$. Define $\lambda_j \in \mathbf{Q}, \lambda_j < 0$ by $e^{2\pi i \lambda_j} = \zeta_j$ and λ_j is maximal. Choose for every j an embedding of $\overline{\mathbf{Q}}$-vector spaces $M_{\zeta_j} \subset z^{\lambda_j} z^{-N} \mathbf{C}[z^{-1}]$. Then the resulting embedding $M \subset \mathcal{Q}$ has the required properties. Moreover, any nonzero element of M is mapped to an element of degree $\geq N$ in the variable z^{-1}. □

Remark 11.4 In Proposition 20 of [164], Kovacic characterizes (in a different way) those connected solvable groups that appear as differential Galois groups over $\mathbf{C}((z))$. In this case, one can show that the two characterizations coincide. We note that the differential Galois group for a Picard-Vessiot extension of $\mathbf{C}((z))$ is always solvable, although not always connected. □

11.3 Some Topics on Linear Algebraic Groups

For the formulations of the solution of some inverse problem by J.-P. Ramis, C. Mitschi and M.F. Singer we will need constructions with linear algebraic groups that are not standard. Let C be an algebraically closed field of characteristic zero.

Definition 11.5 *The groups $L(G)$ and $V(G) = G/L(G)$.*
Let G be a linear algebraic group over C. The subgroup $L(G)$ of G is defined as the group generated by all (maximal) tori lying in G. The group $V(G)$ is defined to be $G/L(G)$. □

It is clear that $L(G)$ is a normal subgroup of G, contained in the connected component of the identity G^o of G. According to the Proposition and its Corollary of Chap. 7.5 of [141] (or Proposition 7.5 and Theorem 7.6 of [45]) $L(G)$ is a (connected) algebraic subgroup of G. The factor group $V(G) := G/L(G)$ is again a linear algebraic group.

Let G be a connected linear algebraic group over C. The unipotent radical of G is denoted by R_u. Let $P \subset G$ be a Levi factor, i.e., a closed subgroup such that G is the semidirect product of R_u and P (see [141], p. 184). The group $R_u/(R_u, R_u)$, where (R_u, R_u) is the (closed) commutator subgroup, is a commutative unipotent group and hence isomorphic to $\mathbf{G}_a^n(C) = C^n$. The group P acts on R_u by conjugation and this induces an action on $R_u/(R_u, R_u)$. Therefore, we may write $R_u/(R_u, R_u) = U_1^{n_1} \oplus \cdots \oplus U_s^{n_s}$, where each U_i is an irreducible P-module. For notational convenience, we suppose that U_1 is the trivial 1-dimensional P-module and $n_1 \geq 0$. Since P is reductive, one can write $P = T \cdot H$, where T is a torus and H is a semisimple group. Define $m_i := n_i$ if the action of H on U_i is trivial and $m_i := n_i + 1$ if the action of H on U_i is not trivial. Define $N = 0$ if H is trivial and $N = 1$ otherwise.

Definition 11.6 *The defect and the excess of a linear algebraic group.*
The *defect* $d(G)$ is defined to be n_1 and the *excess* $e(G)$ is defined as $\max(N, m_2, \ldots, m_s)$. □

Since two Levi factors are conjugate, these numbers do not depend on the choice of P. The results of Ramis are stated in terms of the group $V(G)$, whereas the results of Mitschi-Singer are stated in terms of the defect and excess of a connected linear algebraic group. We wish to explain the connection between $V(G)$ and the defect. We start with the following lemma.

Lemma 11.7 *If G is a connected linear algebraic group defined over C, then $d(G)$ $= \dim_C R_u/(G, R_u)$.*

Proof. We note that $R_u/(G, R_u)$ is a commutative unipotent group and so can be identified with a vector space over C. As in the definition of defect, we let P be a Levi factor and write $R_u/(R_u, R_u) = U_1^{n_1} \oplus \ldots \oplus U_s^{n_s}$, where each U_i is an irreducible P-module and U_1 is the trivial one-dimensional P-module. Since $(R_u, R_u) \subset (G, R_u)$ we have a canonical surjective homomorphism $\pi : R_u/(R_u, R_u) \to R_u/(G, R_u)$. We shall show that the kernel of this homomorphism is $U_2^{n_2} \oplus \ldots \oplus U_s^{n_s}$ and so $d(G) = n_1 = \dim R_u/(G, R_u)$.

First note that the kernel of π is $(G, R_u)/(R_u, R_u)$ and that this latter group is $(P, R_u)/(R_u, R_u)$. To see this second statement note that for $g \in G$ we may write $g = pu$, $p \in P, u \in R_u$. For any $w \in R_u$, $gwg^{-1}w^{-1} = puwu^{-1}p^{-1}w^{-1} = p(uwu^{-1})p^{-1}(uw^{-1}u^{-1})(uwu^{-1}w^{-1})$. This has the following consequence $(G, R_u)/(R_u, R_u) \subset (P, R_u)/(R_u, R_u)$. The reverse inclusion is clear.

Next we will show that $(P, R_u)/(R_u, R_u) = U_2^{n_2} \oplus \ldots \oplus U_s^{n_s}$. We write this latter group additively and note that the group $(P, R_u)/(R_u, R_u)$ is the subgroup of $U_1^{n_1} \oplus \ldots \oplus U_s^{n_s}$ generated by the elements $pvp^{-1} - v$ where $p \in P, v \in U_1^{n_1} \oplus \ldots \oplus U_s^{n_s}$. Since the action of P on U_1 via conjugation is trivial, we see that any element $pvp^{-1} - v$ as above must lie in $U_2^{n_2} \oplus \ldots \oplus U_s^{n_s}$. Furthermore, note that for each i, the image of the map $P \times U_i \to U_i$ given by $(p, u) \mapsto pup^{-1} - u$ generates a P-invariant subspace of U_i. For $i \geq 2$, this image is nontrivial hence, since U_i is an irreducible P-module, we have that the image generates all of U_i. This implies that the elements $pvp^{-1} - v$, where $p \in P, v \in U_1^{n_1} \oplus \ldots \oplus U_s^{n_s}$, generate all of $U_2^{n_2} \oplus \ldots \oplus U_s^{n_s}$ and completes the proof. □

We now introduce the normal subgroup (R_u, G^o) of G generated by the commutators $\{aba^{-1}b^{-1} |\ a \in R_u,\ b \in G^o\}$ and the semidirect product $S(G) = R_u/(R_u, G^o)$

$\rtimes G/G^o$ of $R_u/(R_u, G^o)$ and G/G^o, with respect to the action (by conjugation) of G/G^o on $R_u/(R_u, G^o)$ (see [210]).

Proposition 11.8 (Ramis [235])
Let $(V(G)^o, V(G)^o)$ denote the normal subgroup of $V(G)$ generated by the commutators $\{aba^{-1}b^{-1} |\ a, b \in V(G)^o\}$. There is an isomorphism of linear algebraic groups $S(G) \to V(G)/(V(G)^o, V(G)^o)$.

Proof. (1) We start by proving that a reductive group M has the property $L(M) = M^o$. We recall that in characteristic 0, a linear algebraic group is reductive if and only if any finite dimensional representation is completely reducible (see the Appendix of [32]). It follows at once that $N := M/L(M)$ is also reductive. Thus the unipotent radical of N is trivial. By construction, N^o has a trivial maximal torus and is, therefore, unipotent and equal to the unipotent radical of N. Thus $N^o = 1$. Since $L(M)$ is connected, the statement follows.

(2) Let G be a connected linear algebraic group and R_u its unipotent radical. As noted above G is the semidirect product $R_u \rtimes M$, where M is a Levi-factor. We note that a maximal torus of M is also a maximal torus of G. Let $\tau_1 : G = R_u \rtimes M \to V(G) := G/L(G)$ denote the canonical map. Since $L(M) = M$, the kernel of τ_1 is the smallest normal subgroup of G containing M. The kernel of the natural map $\tau_2 : G = R_u \rtimes M \to V(G)/(V(G), V(G))$ is the smallest normal subgroup containing M and (G, G). This is the same as the smallest normal subgroup containing M and (R_u, G), since $G = R_u \cdot M$. Thus the map τ_2 induces an isomorphism $R_u/(R_u, G) \to V(G)/(V(G), V(G))$.

(3) Let G be any linear algebraic group. We consider $V(G) := G/L(G)$ and $S'(G) := V(G)/(V(G)^o, V(G)^o)$. The component of the identity $S'(G)^o$ is easily identified with $S'(G^o)$ and according to (2) isomorphic to the unipotent group $R_u/(R_u, G^o)$. Furthermore, $S'(G)/S'(G)^o$ is canonically isomorphic to G/G^o. Using that $S'(G)^o$ is unipotent, one can construct a left inverse for the surjective homomorphism $S'(G) \to G/G^o$ (see Lemma 11.10.1). Thus $S'(G)$ is isomorphic to the semidirect product of $R_u/(G^o, R_u)$ and G/G^o, given by the action of G/G^o on $R_u/(R_u, G^o)$, defined by conjugation. Therefore, $S(G)$ and $S'(G)$ are isomorphic. □

To finish the proof of Proposition 11.8 (and show the connection between the defect of G and $V(G)$ in Corollary 11.11), we need to prove Lemma 11.10 below. We first prove an auxiliary lemma. Recall that a group G is *divisible* if for any $g \in G$ and $n \in \mathbf{N} - \{0\}$, there is an $h \in G$ such that $h^n = g$.

Lemma 11.9 *Let H be a linear algebraic group such that H^o is abelian, divisible and has a finite number of elements of any given finite order. Then H contains a finite subgroup $\tilde{H}$ such that $H = \tilde{H}H^o$.*

Proof. We follow the presentation of this fact given in [303], Lemma 10.10. Let $t_1, \ldots, t_s$ be coset representatives of H/H^o. For any $g \in H$ we have $t_i g = t_{i\sigma} a_i$ for some $a_i \in H^o$ and permutation σ. Let $\tilde{H} = \{g \in H \mid a_1 a_2 \cdot \ldots \cdot a_s = 1\}$. One can check that $\tilde{H}$ is a group and that the map $\phi : g \mapsto \sigma$ that associates each $g \in \tilde{H}$ to the σ described above maps $\tilde{H}$ homomorphically into the symmetric group $\mathcal{S}_s$. If $g \in \tilde{H}$ and $\phi(g) = id$ then $a_i = g$ for all i and so $g \in H^o$ and of finite order s. Therefore, the kernel of ϕ is a subset of the set of elements of order at most s in H^o. Therefore, the kernel of ϕ is finite and so $\tilde{H}$ is finite.

To finish the proof we will show that $H = \tilde{H}H^o$. The inclusion $\tilde{H}H^o \subset H$ is clear so it is enough to show that for any $g \in H$ there is a $g_o \in H^o$ such that $gg_o^{-1} \in \tilde{H}$. Let $t_i g = t_{i\sigma} a_i$ and let $g_o \in H^o$ such that $g_o^s = a_1 \cdot \ldots \cdot a_s$. Such an element exists because H^o is divisible. We then have that $t_i g g_o^{-1} = t_{i\sigma} a_i g_o^{-1}$ and $a_1 g_o^{-1} a_2 g_o^{-1} \cdot \ldots \cdot a_s g_o^{-1} = a_1 \cdot \ldots \cdot a_s g_o^{-n} = 1$ so $gg_o^{-1} \in \tilde{H}$. □

Lemma 11.10 *Let H be a linear algebraic group such that H^o is unipotent. Then*

1. *H is isomorphic to a semidirect product of H^o and H/H^o.*
2. *H and $H/(H^o, H^o)$ have the same minimal number of topological (for the Zariski topology) generators.*

Proof. 1. Define the closed normal subgroups H_k^o of H by $H_0^o = H^o$ and $H_{k+1}^o = (H^o, H_k^o)$ for $k \geq 0$. Since H^o is a unipotent group $H_k^o = \{1\}$ for large k. Let $l(H)$ denote the smallest integer $m \geq 0$ such that $H_k^o = \{1\}$ for all $m > k$. We shall proceed by induction on $l(H)$. The induction hypothesis implies that H contains a subgroup H' such that H'/H_k^o is finite and $H = H'H^o$. The group H_k^o is abelian, connected, and unipotent. Since the base field is assumed to be of characteristic zero, a unipotent matrix other than the identity cannot be of finite order. Therefore, for any non-negative integer s the map $g \mapsto g^s$ from H_k^o to itself must be an isomorphism. This implies that H_k^o is divisible and has no elements of finite order other than the identity. Lemma 11.9 implies that $H' = \tilde{H}H_k^o$ for some finite group $\tilde{H}$ and so $H = \tilde{H}H^o$. Since H^o is unipotent, we have that $\tilde{H} \cap H^o$ is trivial and the conclusion follows.

2. It suffices to show the following:

> Let $a_1, \ldots, a_n \in H$ be such that their images in $H/(H^o, H^o)$ are topological generators, then $a_1, \ldots, a_n$ are topological generators of H itself.

We will prove the above statement by induction on $l(H)$. The cases $l(H) = 0, 1$ are trivial. Suppose $l(H) = n > 1$ and let M denote the closed subgroup of H generated by $a_1, \ldots, a_n$. The induction hypothesis implies that the natural homomorphism $M \to H/H_n^o$ is surjective. It suffices to show that $M \supset H_n^o$. Take $a \in H^o, b \in H_{n-1}^o$ and consider the element $aba^{-1}b^{-1} \in H_n^o$. One can write $a = m_1 A,\ b = m_2 B$ with $m_1, m_2 \in M$ and $A, B \in H_n^o$. Since H_n^o lies in the center of H^o, one has $aba^{-1}b^{-1} = m_1 A m_2 B A^{-1} m_1^{-1} B^{-1} m_2^{-1} = m_1 m_2 m_1^{-1} m_2^{-1} \in M$. □

Corollary 11.11 *The linear algebraic groups $S(G)$, $V(G) := G/L(G)$ and $V(G)/(V(G)^o, V(G)^o)$ have the same minimal number of topological generators (for the Zariski topology). Moreover, for connected linear algebraic groups, the defect $d(G)$ of Definition 11.6, coincides with the minimal number of topological generators of $G/L(G)$.*

Remark 11.12 Lemma 11.10.1 can be partially generalized to arbitrary linear algebraic groups. The result (due to Platanov; see [303], Lemma 10.10) is: *If C is an algebraically closed field and G is a linear algebraic group defined over C, then $G = HG^o$ for some finite subgroup H of G.*

To prove this, let B be a Borel subgroup of G^o (see Chap. VIII of [141]) and N be the normalizer of B. We claim that $G = NG^o$. Let $g \in G$. Since all Borel subgroups of G^o are conjugate there exists an $h \in G^o$ such that $gBg^{-1} = hBh^{-1}$. Therefore, $h^{-1}g \in N$ and so $G \subset NG^o$. The reverse inclusion is clear.

It is therefore sufficient to prove the theorem for N and so we may assume that G is a group whose identity component is solvable. We therefore have a composition series $G^o = G_m \supset G_{m-1} \supset \ldots \supset G_1 \supset G_o = \{e\}$, where each G_i/G_{i-1} is isomorphic to $\mathbf{G}_a(C)$ or $\mathbf{G}_m(C)$. By induction on m we may assume that there is a subgroup K of G such that K/G_1 is finite and $G = KG^o$. This allows us to assume that G^o is itself isomorphic to $\mathbf{G}_a(C)$ or $\mathbf{G}_m(C)$. One now applies Lemma 11.9.

Platonov's Theorem, combined with Jordan's Theorem, can also be used to prove Proposition 4.18 (see [303], Theorem 3.6 and Corollary 10.11). □

11.4 The Local Theorem

In this section we give a proof of Ramis's solution of the inverse problem for Diff_K with $K = \mathbf{C}(\{z\})$ (cf. [235, 241, 242]).

Theorem 11.13 (J.-P. Ramis) The local theorem.
A linear algebraic group G is a differential Galois group over the field $\mathbf{C}(\{z\})$ if and only if $G/L(G)$ is topologically (for the Zariski topology) generated by one element.

The proof of the "only if" part is more or less obvious. Suppose that G is the differential Galois group of some differential module $\mathcal{M}$ over $\mathbf{C}(\{z\})$. The differential Galois theory implies, see Proposition 1.34 part 2 and Remarks 2.34 (3), that $G/L(G)$ is also the differential Galois group of some differential module $\mathcal{N}$ over $\mathbf{C}(\{z\})$. The differential Galois group of $\mathbf{C}((z)) \otimes \mathcal{N}$, and, in particular, its exponential torus, is a subgroup of $G/L(G)$. Since $G/L(G)$ has a trivial maximal torus, Corollary 3.32 implies that $\mathcal{N}$ is regular singular and so its Galois group of $\mathbf{C}((z)) \otimes \mathcal{N}$ is generated by the formal monodromy. This element corresponds to the topological monodromy in the Galois group of $\mathcal{N}$ so $G/L(G)$ is topologically generated by the topological monodromy of $\mathcal{N}$. The latter is the image of the topological monodromy of $\mathcal{M}$. This proves the "only if" part of Theorem 11.13 and yields the next corollary.

Corollary 11.14 *Suppose that G is a differential Galois group over the field $\mathbf{C}(\{z\})$, then $G/L(G)$ is topologically generated by the image of the topological monodromy.*

The proof of the "if" part of Theorem 11.13 is made more transparent by the introduction of yet another tannakian category Gr_3.

Definition 11.15 *The category* Gr_3.
The objects of the category Gr_3 are tuples $(V, \{V_q\}, \gamma_V, st_{V,d})$ with

1. $(V, \{V_q\}, \gamma_V)$ is an object of Gr_1.
2. For every $d \in \mathbf{R}$ there is given a $st_{V,d} \in \oplus_{q_i,q_j}\mathrm{Hom}(V_{q_i}, V_{q_j})$, where the sum is taken over all pairs i, j with $V_{q_i} \neq 0,\ V_{q_j} \neq 0$ and d is a singular direction for $q_i - q_j$.
3. For every $d \in \mathbf{R}$ one requires that $\gamma_V^{-1} st_{V,d}\gamma_V = st_{V,d+2\pi}$.

The morphisms of Gr_3 are defined as follows. We identify any linear map $A_{i,j} : V_{q_i} \to V_{q_j}$ with the linear map $V \overset{\text{projection}}{\to} V_{q_i} \overset{A_{i,j}}{\to} V_{q_j} \overset{\text{inclusion}}{\to} V$. In this way, $\mathrm{Hom}(V_{q_i}, V_{q_j})$ is identified with a subspace of $\mathrm{End}(V)$. A morphism $f : (V, \{V_q\}, \gamma_V, st_{V,d}) \to (W, \{W_q\}, \gamma_W, st_{W,d})$ is a linear map $V \to W$ satisfying $f(V_q) \subset W_q,\ f \circ \gamma_V = \gamma_W \circ f,\ f \circ st_{V,d} = st_{W,d} \circ f$ for all d. □

The tensor product of two objects $(V, \{V_q\}, \gamma_V, st_{V,d}), (W, \{W_q\}, \gamma_W, st_{W,d})$ is the vector space $V \otimes W$ with the data $(V \otimes W)_q = \oplus_{q_1,q_2;q_1+q_2=q} V_{q_1} \otimes W_{q_2}$, $\gamma_{V\otimes W} = \gamma_V \otimes \gamma_W$ and $st_{V\otimes W,d} = st_{V,d} \otimes id_W + id_V \otimes st_{W,d}$. It is easily seen that Gr_3 is again a neutral tannakian category. In fact, we will show that the tannakian categories Gr_2 and Gr_3 are isomorphic.

Lemma 11.16

1. *The exponential map* $\exp : \mathrm{Gr}_3 \to \mathrm{Gr}_2$ *induces an equivalence of tannakian categories.*
2. *Let the object $(V, \{V_q\}, \gamma_V, st_{V,d})$ be associated to a differential equation over $\mathbf{C}(\{z\})$. Then the differential Galois group $G \subset \mathrm{GL}(V)$ is the smallest algebraic subgroup such that:*
   (a) *The exponential torus and γ_V belong to G.*
   (b) *All $st_{V,d}$ belong to the Lie algebra of G.*

Proof. The exponential map associates to $(V, \{V_q\}, \gamma_V, st_{V,d})$ the object $(V, \{V_q\}, \gamma_V, St_{V,d})$ with $St_{V,d} = exp(st_{V,d})$ for all d. It is easily seen that this results in an equivalence of tannakian categories. The second part of the lemma is a reformulation of Theorem 8.10. □

Let G be given such that $G/L(G)$ is topologically generated by one element. The following lemma will be a guide for the construction of an object in the category Gr_3 having G as differential Galois group. For this we need to consider G as a subgroup of GL(V) for some finite dimensional $\mathbf{C}$-vector space V and describe the action of G on V in some detail. We shall consider linear algebraic groups with the following data:

Let V be a finite dimensional vector space over $\mathbf{C}$ and $G \subset \mathrm{GL}(V)$ an algebraic subgroup. Let T denote a maximal torus and $\underline{g}, \underline{t} \subset \mathrm{End}(V)$ be the Lie algebras of G and T. The action of T on V yields a decomposition $V = \oplus_{i=1}^{s} V_{\chi_i}$, where the χ_i are distinct characters of T and the nontrivial spaces V_{χ_i} are defined as $\{v \in V |\ tv = \chi_i(t)v \text{ for all } t \in T\}$.

For each i, j one identifies $\mathrm{Hom}(V_{\chi_i}, V_{\chi_j})$ with a linear subset of End(V) by identifying $\phi \in \mathrm{Hom}(V_{\chi_i}, V_{\chi_j})$ with $V \stackrel{pr_i}{\to} V_{\chi_i} \stackrel{\phi}{\to} V_{\chi_j} \subset V$, where pr_i denotes the projection onto V_{χ_i}, along $\oplus_{k \neq i} V_{\chi_k}$. The adjoint action of T on $\underline{g}$ yields a decomposition $\underline{g} = \underline{g}_0 \oplus \sum_{\alpha \neq 0} \underline{g}_\alpha$. By definition, the adjoint action of T on $\underline{g}_0$ is the identity and is multiplication by the character $\alpha \neq 0$ on the spaces $\underline{g}_\alpha$. (We note that here the additive notation for characters is used. In particular, $\alpha \neq 0$ means that α is not the trivial character.)

Any $B \in \underline{g}$ can be written as $\sum B_{i,j}$ with $B_{i,j} \in \mathrm{Hom}(V_{\chi_i}, V_{\chi_j})$. The adjoint action of $t \in T$ on B has the form $Ad(t)B = \sum \chi_i^{-1}\chi_j(t)B_{i,j}$. It follows that the $\alpha \neq 0$ with $\underline{g}_\alpha \neq 0$ have the form $\chi_i^{-1}\chi_j$. In particular $\sum_{i,j;\chi_i^{-1}\chi_j=\alpha} B_{i,j} \in \underline{g}_\alpha \subset \underline{g}$. Let $L(G)$ denote, as before, the subgroup of G generated by all the conjugates of T.

Lemma 11.17 (J.-P. Ramis)
The Lie algebra of $L(G)$ is generated by the subspaces $\underline{t}$ and $\{\underline{g}_\alpha\}_{\alpha \neq 0}$.

Proof. Consider some $\alpha \neq 0$ and a nonzero element $\xi \in \underline{g}_\alpha$. From the definition of $\underline{g}_\alpha$ and $\alpha \neq 0$ it follows that there is an ordering, denoted by $V_1, \ldots, V_s$, of the spaces $\{V_{\chi_i}\}$, such that ξ maps each V_i into some V_j with $j > i$. In particular, ξ is nilpotent and $\mathbf{C}\xi$ is an algebraic Lie algebra corresponding to the algebraic subgroup $\{\exp(c\xi) | c \in \mathbf{C}\}$ of G. Let $\underline{h}$ denote the Lie algebra generated by the algebraic Lie algebras $\underline{t}$ and $\mathbf{C}\xi$ for all $\xi \in \underline{g}_\alpha$ with $\alpha \neq 0$. Then $\underline{h}$ is an algebraic Lie algebra. (see [45] Proposition (7.5), p. 190, and Theorem (7.6), p. 192).

Take an element $t \in T$ such that all $\chi_i(t)$ are distinct. Then clearly $t \cdot exp(\xi)$ is semisimple and lies therefore in a conjugate of the maximal torus T. Thus $\exp(\xi) = t^{-1} \cdot (t \cdot \exp(\xi)) \in L(G)$ and ξ lies in the Lie algebra of $L(G)$. This proves that the Lie algebra $\underline{h}$ is a subset of the Lie algebra of $L(G)$.

On the other hand, $\underline{h}$ is easily seen to be an ideal in $\underline{g}$. The connected normal algebraic subgroup $H \subset G^o$ corresponding to $\underline{h}$ contains T and therefore $L(G)$. This proves the other inclusion. □

Continuation of the proof of Theorem 11.13
We add to the above data for G the assumption that $G/L(G)$ is topologically generated by one element a. The aim is to produce an object $(V, \{V_q\}, \gamma_V, st_{V,d})$ of Gr_3 such that the group defined by the conditions (a) and (b) of the second part of Lemma 11.16 is equal to G.

Choose a representative $A \in G$ of $a \in G/L(G)$. Since T is also a maximal torus of $L(G)$, there exists $B \in L(G)$ such that $ATA^{-1} = BTB^{-1}$. After replacing A by $B^{-1}A$ we may suppose that $ATA^{-1} = T$. The element $A \in G$ permutes the spaces V_{χ_i} in the decomposition $V = \oplus_{i=1}^{s} V_{\chi_i}$ with respect to the action of T. As before we will write $\tilde{A}\chi_i = \chi_j$ if $\chi_j(t) = \chi_i(A^{-1}tA)$ for all $t \in T$. Lemma 11.3 produces an object $(V, \{V_q\}, \gamma_V)$ of Gr_1 such that $V_{\chi_i} = V_{q_i}$ for $i = 1, \ldots, s$ and $\gamma_V = A$.

As before, $\underline{g}$ denotes the Lie algebra of G (or G^o). The decomposition of $\underline{g}$ with respect to the adjoint action of the torus T has already been made explicit, namely

$$\underline{g}_{\alpha} = \oplus_{i,j;\chi_i^{-1}\chi_j=\alpha}\underline{g} \cap \mathrm{Hom}(V_{q_i}, V_{q_j}).$$

The above object $(V, \{V_q\}, \gamma_V)$ in Gr_1 is made into an object of Gr_3 by choosing arbitrary elements $st_{V,d} \in \underline{g}_{\alpha}$, where $\alpha = \chi_i^{-1}\chi_j$ and $0 \leq d < 2\pi$ is a singular direction for $q_i - q_j$. The number of singular directions d modulo 2π for $q_i - q_j$ is, by construction, sufficiently large to ensure a choice of the set $\{st_{V,d}\}$ such that these elements generate the vector space $\oplus_{\alpha \neq 0}\underline{g}_{\alpha}$.

Finally, we verify that the algebraic group $\tilde{G}$, associated to the object $(V, \{V_q\}, \gamma_V, st_{V,d})$, is equal to G. By construction, $\tilde{G} \subset G$ and by definition $\tilde{G}$ is the smallest algebraic group with:

(a) The exponential torus and γ_V lie in $\tilde{G}$.
(b) The Lie algebra of $\tilde{G}$ contains all $st_{V,d}$.

By construction, the exponential torus is equal to T and lies in $\tilde{G}$. The Lie algebra $\underline{t}$ lies in the Lie algebra of $\tilde{G}$. Again by construction, each $\underline{g}_{\alpha}$ (with $\alpha \neq 0$) belongs to the Lie algebra of $\tilde{G}$. Lemma 11.17 implies that $L(G)$ is contained in $\tilde{G}$. The choice of $\gamma_V = A$ implies that $\tilde{G} = G$. □

Examples 11.18 *Differential Galois groups in* GL(2) *for* $\mathbf{C}(\{z\})$.
For convenience we consider order two equations over $\mathbf{C}(\{z\})$ with differential Galois group in SL(2). The well-known classification of the algebraic subgroups G of SL(2) (see Sect. 1.4) can be used to determine the Gs such that $G/L(G)$ is topologically generated by one element. The list (of conjugacy classes) that one finds is:

$$\mathrm{SL}_2,\ B,\ \mathbf{G}_a,\ \{\pm 1\} \times \mathbf{G}_a,\ \mathbf{G}_m,\ D_{\infty},\ \text{finite cyclic},$$

where B is the Borel subgroup, $\mathbf{G}_a$ the additive group, $\mathbf{G}_m$ the multiplicative group and D_{∞} is the infinite dihedral group, i.e., the subgroup of SL_2 leaving the union of two lines $L_1 \cup L_2 \subset \mathbf{C}^2$ (through the origin) invariant.

Every group in the above list can be realized by a scalar differential equation

$$y'' + mz^{-1}y' - a_0 y = 0 \text{ where } m \in \{0, 1\} \text{ and } a_0 \in \mathbf{C}[z, z^{-1}].$$

In fact, the choices (m, a_0)= $(0, z)$, $(0, z^2 + 3z + 5/4)$, $(1, 0)$, $(0, -\frac{z^{-2}}{4})$, $(0, z^{-2})$, $(0, -\frac{3}{16}z^{-2} + z^{-1})$, $(0, \frac{1}{4}(-1 + (\frac{t}{n})^2)z^{-2})$ with $\frac{t}{n} \in \mathbf{Q}$ produce the above list of differential Galois groups. □

Remark 11.19 In Proposition 21 of [164], Kovacic gives a characterization of those connected solvable linear algebraic groups that appear as differential Galois groups over $\mathbf{C}(z)$. Once again, this characterization can be shown to be equivalent to the above in this case. □

11.5 The Global Theorem

In this section, X is a compact (connected) Riemann surface of genus g and $S = \{p_1, \ldots, p_s\}$ is a finite subset of X. The differential field K is the field of the meromorphic functions on X. Let $\mathrm{Diff}_{(X,S)}$ denote the full subcategory of Diff_K whose objects are the differential modules that are regular at every point $q \in X \setminus S$.

Proposition 11.20 *Let $\pi_1(X \setminus S)$ denote the fundamental group of $X \setminus S$. Let $\mathcal{M}$ be a differential module in the category* $\mathrm{Diff}_{(X,S)}$ *having differential Galois group G. There is a natural homomorphism $\pi_1(X \setminus S) \to G/L(G)$ that has a dense image with respect to the Zariski topology. In particular, $G/L(G)$ is topologically generated by, at most, $2g + s - 1$ elements if $s > 0$.*

Proof. The tannakian approach implies that there exists a differential module $\mathcal{N}$ in the category $\mathrm{Diff}_{(X,S)}$ with differential Galois group $V(G) = G/L(G)$ (see Remarks 2.34 (3)). Consider a point $q \in X$ with local parameter t, which is singular for $\mathcal{N}$. Let K_q denote the field of meromorphic functions at q. Then $K_q \cong \mathbf{C}(\{t\})$. The differential Galois group of $\mathbf{C}((t)) \otimes \mathcal{N}$ over $\mathbf{C}((t))$ can be embedded as a subgroup of $V(G)$. Since the maximal torus of $V(G)$ is trivial, we conclude that every singular point of $\mathcal{N}$ is regular singular. Now, Example 11.1.2 finishes the proof. □

It is interesting to note that Proposition 11.20 implies a result of O. Gabber, namely that $\pi_1(X \setminus S) \to G/G^o$ is surjective (see [152], 1.2.5).

For nonempty S one defines the full subcategory $\mathcal{C}$ of Diff_K to be the subcategory whose objects are the differential modules $\mathcal{M}$ such that $\mathcal{M}$ is regular for any $q \notin S$ and $\mathcal{M}$ is regular singular at $p_1, \ldots, p_{s-1}$.

Theorem 11.21 (J.-P. Ramis) The global theorem.
A linear algebraic group G is the differential Galois group of a differential module $\mathcal{M}$ in $\mathcal{C}$ if and only if $G/L(G)$ is topologically generated by $2g + s - 1$ elements.

Proof. The "only if" part is proved in Proposition 11.20. Consider a linear algebraic group G such that $G/L(G)$ is topologically generated by, at most, $2g+s-1$ elements.

One chooses small disjoint disks $X_1, \ldots, X_s$ around the points $p_1, \ldots, p_s$. Let $X_i^* = X_i \setminus \{p_i\}$. In X_s^* one chooses a point c. The fundamental group $\pi_1(X_0, c)$, where $X_0 = X \setminus \{p_1, \ldots, p_s\}$, is generated by $a_1, b_1, .., a_g, b_g, \lambda_1, .., \lambda_s$ and has one relation $a_1 b_1 a_1^{-1} b_1^{-1} \cdots \lambda_1 \cdots \lambda_s = 1$. The element λ_s is a loop in X_s^* around p_s and the other λ_i are loops around $p_1, \ldots, p_{s-1}$. The differential module over X is constructed by gluing certain connections $\mathcal{M}_0, \ldots, \mathcal{M}_s$ (with possibly singularities), living above the spaces $X_0, \ldots, X_s$.

Let $pr : G \to G/L(G)$ denote the canonical homomorphism. One chooses a homomorphism $\rho : \pi_1(X_0, c) \to G \subset \mathrm{GL}(V)$, such that the homomorphism $pr\rho$ has Zariski-dense image. Consider the algebraic group $G' = pr^{-1}(<< pr\rho(\lambda_s) >>)$, where $<< a >>$ denotes the algebraic subgroup generated by the element a. The group G' contains $L(G)$ and so $G'/L(G')$ is topologically generated by the image of $\rho(\lambda_s)$. According to the Ramis local theorem, G' is the differential Galois group of a differential equation over the field K_{p_s}. One can extend this very local object to a differential module $\mathcal{M}_s$, living above X_s, with only p_s as singular point. The solution space at the point $c \in X_s$ and the action of G' on this space can be identified with V and $G' \subset \mathrm{GL}(V)$. The topological monodromy corresponding to λ_s can be arranged to be $\rho(\lambda_s) \in G'$.

The usual solution of the Riemann-Hilbert problem (in weak form) provides a differential module $\mathcal{M}_0$ above X_0 such that the monodromy action is equal to ρ. The restrictions of $\mathcal{M}_0$ to X_s^* and $\mathcal{M}_s$ to X_s^* are determined by their local monodromies. These are both equal to $\rho(\lambda_s) \in G \subset \mathrm{GL}(V)$. Thus we have a canonical way to glue $\mathcal{M}_0$ and $\mathcal{M}_s$ over the open subset X_s^*. For each point p_i, $i = 1, \ldots, s-1$ one can consider the restriction of $\mathcal{M}_0$ to X_i^*. This restriction is determined by its local monodromy around the point p_i. Clearly, the restriction of $\mathcal{M}_0$ to X_i^* can be extended to a differential module $\mathcal{M}_i$ above X_i with a regular singular point at p_i.

The modules $\mathcal{M}_0, \mathcal{M}_1, \ldots, \mathcal{M}_s$ (or rather the corresponding analytic vector bundles) are in this way glued to a vector bundle $\mathcal{M}$ above X. The connections can be written as $\nabla : \mathcal{M}_0 \to \Omega_X \otimes \mathcal{M}_0$, $\nabla : \mathcal{M}_i \to \Omega_X(p_i) \otimes \mathcal{M}_i$ for $i = 1, \ldots, s-1$ and $\nabla : \mathcal{M}_s \to \Omega_X(dp_s) \otimes \mathcal{M}_s$ for a suitable integer $d \geq 0$. The connections also glue to a connection $\nabla : \mathcal{M} \to \Omega_X(p_1 + \cdots + p_{s-1} + dp_s) \otimes \mathcal{M}$. Let $\mathcal{M}_*$ denote the set of all meromorphic sections of $\mathcal{M}$. Then $\mathcal{M}_*$ is a vector space over K of dimension n with a connection $\nabla : \mathcal{M}_* \to \Omega_{K/\mathbf{C}} \otimes \mathcal{M}_*$. Thus we have found a differential module over K with the correct singularities. K has a natural embedding in the field K_c. The differential module is trivial over K_c and its Picard-Vessiot field PV over K can be seen as a subfield of K_c. The solution space is $V \subset PV \subset K_c$.

Finally, we have to show that the differential Galois group $H \subset \mathrm{GL}(V)$ is actually G. By construction, $G' \subset H$ and also the image of ρ lies in H. This implies that $G \subset H$. In order to conclude that $G = H$ we can use the Galois correspondence.

Thus we have finally to prove that an element $f \in PV \subset K_c$, which is invariant under G, belongs to K.

Since G' is, by construction, the differential Galois group of $\mathcal{M}_s$ above X_s, we conclude from the invariance of f under G' that f extends to a meromorphic solution of the differential equation above X_s. The invariance of f under the image of ρ implies that f extends to a meromorphic solution of the differential equation above $X_0 \cup X_s$. The points $p_1, \ldots, p_{s-1}$ are regular singular and any solution of the differential equation above X_i^*, $i = 1, \ldots, s-1$ extends meromorphically to X_i. Thus f is meromorphic on X and belongs to K. □

Remark 11.22 For the case of $X = \mathbf{P}^1$ and $p_s = \infty$, it is possible to refine the above reasoning to prove the following statement. □

Corollary 11.23 *Let $G \subset \mathrm{GL}_n(\mathbf{C})$ be an algebraic group such that $G/L(G)$ is topologically generated by $s-1$ elements. Then there are constant matrices $A_1, \ldots, A_{s-1}$ and there is a matrix A_∞ with polynomial coefficients (all matrices of order $n \times n$) such that the matrix differential equation*

$$y' = (\frac{A_1}{z-p_1} + \cdots + \frac{A_{s-1}}{z-p_{s-1}} + A_\infty)y$$

has differential Galois group $G \subset \mathrm{GL}_n(\mathbf{C})$.

Examples 11.24 *Some differential Galois groups for* $\mathrm{Diff}_{(\mathbf{P}^1_{\mathbf{C}},S)}$

1. $(\mathbf{P}^1, \{0, \infty\})$ *and equations of order two.*
The list of possible groups $G \subset \mathrm{SL}(2)$ coincides with the list given in Examples 11.18. This list is, in fact, the theoretical background for the simplification of the Kovacic algorithm for order two differential equations having, at most, two singular points, presented in [232].

2. $(\mathbf{P}^1, \{0, 1, \infty\})$ *and equations of order two.*
Every algebraic subgroup G of GL(2) can be realized for this pair, since $G/L(G)$ is topologically generated by, at most, two elements. More precisely, every algebraic subgroup $G \subset \mathrm{GL}(2)$ is the differential Galois group of an equation

$$y'' + \frac{a_1(z)}{z(z-1)}y' + \frac{a_2(z)}{z^2(z-1)^2}y = 0,$$

where $a_1(z), a_2(z) \in \mathbf{C}[z]$. This equation is regular singular at 0, 1 and has an arbitrary singularity at ∞. □

11.6 More on Abhyankar's Conjecture

The base field k is an algebraically closed field of characteristic $p > 0$, e.g., $\overline{\mathbf{F}}_p$. One considers a curve X/k (irreducible, smooth, projective) of genus g and a finite subset $S \subset X$ with cardinality $s \geq 1$. Abhyankar's conjecture is concerned with the Galois covers of X that are unramified outside S. For any group G we write $p(G)$ for the subgroup of G generated by all elements with order a power of p. The group $p(G)$ is a normal subgroup of G and $G/p(G)$ is the largest quotient of G that has no elements $\neq 1$ of order p. We recall the well-known theorems.

Theorem 11.25 (M. Raynaud [244])
The finite group G is the Galois group of a covering of $\mathbf{P}^1$, unramified outside ∞, if and only if $G = p(G)$.

Theorem 11.26 (D. Harbater [121])

1. *Let the pair (X, S) be as above. The finite group G is a Galois group of a Galois cover of X, unramified outside S, if and only if $G/p(G)$ is generated by $2g+s-1$ elements.*
2. *If G is a Galois group for the pair (X, S), then the natural homomorphism $\pi^{(p)}(X \setminus S, *) \to G/p(G)$ is surjective.*
3. *Suppose that $G/p(G)$ is generated by $2g+s-1$ elements. Then there is a Galois cover of X with Galois group G, wildly ramified in at most one (prescribed) point of S, tamely ramified at the other points of S and unramified outside S.*

Remark 11.27 The group $\pi^{(p)}(X \setminus S, *)$ denotes the prime to p, algebraic fundamental group of $X \setminus S$. This group is known to be isomorphic to the profinite completion of the free group on $2g + s - 1$ generators. □

The following transformation rules seem to link to two subjects:

Differential	**Characteristic** $p > 0$
$X/\mathbf{C}$ curve, finite $S \subset X,\ S \neq \emptyset$	X/k curve, finite $S \subset X,\ S \neq \emptyset$
differential equation	equation with Galois covering of X
singular point (in S)	ramified point (in S)
local differential Galois group	inertia group
regular singular	tamely ramified
irregular singular	wildly ramified
linear algebraic group	finite group
$L(G)$	$p(G)$
$\pi_1(X \setminus S) \to G/L(G)$ Zariski dense	$\pi^{(p)}(X \setminus S) \to G/p(G)$ surjective.

In the work of P. Deligne, N. Katz and G. Laumon on (rigid) differential equations there is also a link, this time more concrete, between differential equations and certain sheaves living in characteristic p (see [153, 156, 157]). Another link between the two theories is provided by recent work of B. H. Matzat and M. van der Put [205] on differential equations in positive characteristic.

11.7 The Constructive Inverse Problem

In this section, the differential field is $C(z)$, where C is an algebraically closed field of characteristic 0. The aim is to give a theoretical algorithm that produces a linear differential equation over $C(z)$ that has a prescribed differential Galois group G. If the group G is not connected, then there is not much hope for an explicit algorithm. Even for a finite group G it is doubtful whether a reasonable algorithm for the construction of a corresponding linear differential equation over $C(z)$ exists. However, for special cases, e.g., finite subgroups of $\mathrm{GL}_n(C)$ with $n = 2, 3, 4$, a reasonable algorithm has been developed in [233]. In this section, we describe the work of C. Mitschi and M. F. Singer [210] (see also [211]) that concerns connected linear algebraic groups. The main result is the following theorem.

Theorem 11.28 (C. Mitschi and M. F. Singer [210])
Let $G \subset \mathrm{GL}(V)$ be a connected linear algebraic group over C with defect $d(G)$ and excess $e(G)$. Let $a_1, \ldots, a_{d(G)}$ denote arbitrary distinct points of C. There is an algorithm, based on the given structure of the group G, which determines matrices $A_1, \ldots, A_{d(G)} \in \mathrm{End}(V)$ and a polynomial matrix $A_\infty \in C[z] \otimes \mathrm{End}(V)$ of degree at most $e(G)$, such that the linear differential equation

$$y' = (\frac{A_1}{z - a_1} + \cdots + \frac{A_{d(G)}}{z - a_{d(G)}} + A_\infty)y,$$

has differential Galois group G. In particular, the points $a_1, \ldots, a_{d(G)}$ are regular singular for this equation and ∞ is possibly an irregular singular point.

Remark 11.29 This result is rather close to Corollary 11.23, since we have seen (Corollary 11.11) that $d(G)$ is the minimal number of topological generators for $G/L(G)$ There is, however, a difference. In Corollary 11.11 there seems to be no bound on the degree of A_∞. In Theorem 11.28 (only for connected G) there is a bound on the degree of A_∞. The proof of the above result is purely algebraic and, moreover, constructive. The special case of the theorem, where the group G is supposed to be connected and reductive, is rather striking. It states that G is the differential Galois group of a matrix differential equation $y' = (A + Bz)y$ with A and B constant matrices. We will present the explicit proof of this statement when G is connected and semisimple. This is at the heart of [210]. The general result is then achieved by following the program given by Kovacic in his papers [164, 165]. We refer to [210] for details. □

Theorem 11.30 (C. Mitschi and M. F. Singer [210])
The field C is supposed to be algebraically closed and of characteristic 0. Every connected semisimple linear algebraic group is the differential Galois group of an equation $y' = (A_0 + A_1 z)y$ over $C(z)$, where A_0, A_1 are constant matrices.

The basic idea of the proof is simple: select matrices A_0, A_1 such that we can guarantee that the Galois group of the equation is firstly a subgroup of G and secondly is not equal to a proper subgroup of G. To guarantee the first property it will be enough to select A_0, A_1 to lie in the Lie algebra of G (see Proposition 1.31). The second step requires more work. We present here the simplified proof presented in [230]. We will begin by reviewing the tannakian approach to differential Galois theory (cf. Appendix C).

For a differential module $\mathcal{M}$ (over $C(z)$) one denotes by $\{\{\mathcal{M}\}\}$ the full tannakian subcategory of $\mathrm{Diff}_{C(z)}$ "generated" by $\mathcal{M}$. By definition, the objects of $\{\{\mathcal{M}\}\}$ are the differential modules isomorphic to subquotients $\mathcal{M}_1/\mathcal{M}_2$ of finite direct sums of tensor products of the form $\mathcal{M} \otimes \cdots \otimes \mathcal{M} \otimes \mathcal{M}^* \otimes \cdots \otimes \mathcal{M}^*$ (i.e., any number of terms $\mathcal{M}$ and its dual $\mathcal{M}^*$). For any linear algebraic group H (over C) one denotes by Repr_H the tannakian category of the finite dimensional representations of H (over C). The differential Galois group of $\mathcal{M}$ is H if there is an equivalence between the tannakian categories $\{\{\mathcal{M}\}\}$ and Repr_H.

Let V be a finite dimensional vector space over C and $G \subset \mathrm{GL}(V)$ be a linear algebraic group. Our first aim is to produce a differential module $\mathcal{M} = (C(z) \otimes V, \partial)$ and a functor of tannakian categories $\mathrm{Repr}_G \to \{\{\mathcal{M}\}\}$.

The Functor $\mathrm{Repr}_G \to \{\{\mathcal{M}\}\}$. The Lie algebra of G will be written as $\underline{g} \subset \mathrm{End}(V)$. One chooses a matrix $A(z) \in C(z) \otimes \underline{g} \subset C(z) \otimes \mathrm{End}(V)$. There corresponds to this choice a differential equation $y' = A(z)y$ and a differential module $\mathcal{M} := (C(z) \otimes V, \partial)$ with ∂ defined by $\partial(v) = -A(z)v$ for all $v \in V$. Let a representation $\rho : G \to \mathrm{GL}(W)$ be given. The induced maps $\underline{g} \to \mathrm{End}(W)$ and $C(z) \otimes \underline{g} \to C(z) \otimes \mathrm{End}(W)$ are also denoted by ρ. One associates to (W, ρ) the differential module $(C(z) \otimes W, \partial)$ with $\partial(w) = -\rho(A(z))w$ for all $w \in W$. The corresponding differential equation is $y' = \rho(A(z))y$. In this way one obtains a functor of tannakian categories $\mathrm{Repr}_G \to \mathrm{Diff}_{C(z)}$. *We claim that every* $(C(z) \otimes W, \partial)$, *as above, lies, in fact, in* $\{\{\mathcal{M}\}\}$.

Indeed, consider $\{\{V\}\}$, the full subcategory of Repr_G generated by V (the definition is similar to the definition of $\{\{\mathcal{M}\}\}$). It is known that $\{\{V\}\} = \mathrm{Repr}_G$ (see [82] or Appendix C.3). The representation $V \otimes \cdots \otimes V \otimes V^* \otimes \cdots \otimes V^*$ is mapped to the differential module $\mathcal{M} \otimes \cdots \otimes \mathcal{M} \otimes \mathcal{M}^* \otimes \cdots \otimes \mathcal{M}^*$. Let $V_2 \subset V_1$ be G-invariant subspaces of $V \otimes \cdots \otimes V \otimes V^* \otimes \cdots \otimes V^*$. Then V_2 and V_1 are invariant under the action of $\underline{g}$. Since $A(z) \in C(z) \otimes \underline{g}$, one has that $C(z) \otimes V_2$ and $C(z) \otimes V_1$ are differential submodules of $\mathcal{M} \otimes \cdots \otimes \mathcal{M} \otimes \mathcal{M}^* \otimes \cdots \otimes \mathcal{M}^*$. It follows that the differential module $(C(z) \otimes V_1/V_2, \partial)$ lies in $\{\{\mathcal{M}\}\}$. This proves the claim.

The next step is to make the differential Galois group H of $\mathcal{M}$ and the equivalence $\{\{\mathcal{M}\}\} \to \mathrm{Repr}_H$ as concrete as we can.

Differential modules over O. Fix some $c \in C$ and let O denote the localization of $C[z]$ at $(z-c)$. A differential module $(\mathcal{N}, \partial)$ over O is a finitely generated O-module equipped with a C-linear map ∂ satisfying $\partial(fn) = f'n + f\partial(n)$ for all $f \in O$ and $n \in \mathcal{N}$. It is an exercise to show that $\mathcal{N}$ has no torsion elements. It follows that $\mathcal{N}$ is a free, finitely generated O-module. Let Diff_O denote the category of the differential modules over O. The functor $\mathcal{N} \mapsto C(z) \otimes_O \mathcal{N}$ induces an equivalence of Diff_O with a full subcategory of $\mathrm{Diff}_{C(z)}$. It is easily seen that this full subcategory has as objects the differential modules over $C(z)$ that are regular at $z = c$.

Let $\widehat{O} = C[[z - c]]$ denote the completion of O. For any differential module $\mathcal{N}$ over O of rank n, one writes $\widehat{\mathcal{N}}$ for $\widehat{O} \otimes \mathcal{N}$. We note that $\widehat{\mathcal{N}}$ is a differential module over $\widehat{O}$ and that $\widehat{\mathcal{N}}$ is, in fact, a trivial differential module. The space $ker(\partial, \widehat{\mathcal{N}})$ is a vector space over C of dimension n, which will be called $\mathrm{Sol}_c(\mathcal{N})$, the solution space of $\mathcal{N}$ over $C[[z - c]]$ (and also over $C((z - c))$). The canonical map $\mathrm{Sol}_c(\mathcal{N}) \to \widehat{\mathcal{N}}/(z - c)\widehat{\mathcal{N}} = \mathcal{N}/(z - c)\mathcal{N}$ is an isomorphism. Let Vect_C denote the tannakian category of the finite dimensional vector spaces over C. The above construction $\mathcal{N} \mapsto \mathcal{N}/(z - c)\mathcal{N}$ is a fibre functor of $\mathrm{Diff}_O \to \mathrm{Vect}_C$. For a fixed object $\mathcal{M}$ of Diff_O one can consider the restriction $\omega : \{\{\mathcal{M}\}\} \to \mathrm{Vect}_C$, which is again a fibre functor. The differential Galois group H of $\mathcal{M}$ is defined in [82] or Appendix C.3 as $\underline{\mathrm{Aut}}^{\otimes}(\omega)$. By definition, H acts on $\omega(\mathcal{N})$ for every object $\mathcal{N}$ of $\{\{\mathcal{M}\}\}$. Thus we find a functor $\{\{\mathcal{M}\}\} \to \mathrm{Repr}_H$, which is an equivalence of tannakian categories. The composition $\{\{\mathcal{M}\}\} \to \mathrm{Repr}_H \to \mathrm{Vect}_C$ (where the last arrow is the forgetful functor) is the same as ω. (We note that the above remains valid if we replace O by any localization of $C[z]$).

Remark 11.31 Let G again be an algebraic subgroup of $\mathrm{GL}(V)$, let $A(z) \in C(z) \otimes \underline{g}$ be chosen. Suppose that the matrix $A(z)$ has no poles at $z = c$. Then we have functors of tannakian categories $\mathrm{Repr}_G \to \{\{\mathcal{M}\}\}$ and $\{\{\mathcal{M}\}\} \to \mathrm{Repr}_H$. The last functor is made by considering $\mathcal{M}$ as a differential module over O. The composition of the two functors maps a representation (W, ρ) of G to a representation of H on $O/(z - c)O \otimes W$, which is canonically isomorphic to W. Therefore, the tannakian approach allows us to conclude directly (without an appeal to Proposition 1.31) that H is an algebraic subgroup of G. □

We now return to the problem of insuring that the differential Galois group of our equation is not a proper subgroup of G. Suppose that H is a proper subgroup of G, then there exists a representation (W, ρ) of G and a line $\tilde{W} \subset W$ such that H stabilizes $\tilde{W}$ and G does not (cf. [141], Chap. 11.2). The differential module $(C(z) \otimes W, \partial)$ has as image in Repr_H the space W with its H-action. The H-invariant subspace $\tilde{W}$ corresponds with a one-dimensional (differential) submodule $C(z)w \subset C(z) \otimes W$. After multiplication of w by an element in $C(z)$, we may suppose that $w \in C[z] \otimes W$ and that the coordinates of w with respect to a basis of W have g.c.d. 1. Let us write $\frac{d}{dz}$ for the differentiation on $C(z) \otimes W$, given by $\frac{d}{dz} fa = f'a$ for $f \in C(z)$ and $a \in W$. Then one finds the equation

$$[\frac{d}{dz} - \rho(A(z))]w = cw \text{ for some } c \in C(z). \tag{11.1}$$

The idea for the rest of the proof is to make a choice for $A(z)$ that contradicts the equation for w above. For a given proper algebraic subgroup H' of G one can produce a suitable $A(z)$ that contradicts the statement that H lies in a conjugate of H'. In general, however, one has to consider infinitely many (conjugacy classes) of proper algebraic subgroups of G. This will probably not lead to a *construction* of the matrix $A(z)$. In the following we will make two restrictions, namely $A(z)$ is a polynomial matrix (i.e., $A(z) \in C[z] \otimes \underline{g}$) and that G is connected and semisimple. As we will see in Lemma 11.32 the first restriction implies that the differential Galois group is a *connected* algebraic subgroup of G. The second restriction implies that G has finitely many conjugacy classes of maximal proper connected subgroups.

Lemma 11.32 *Let W be a finite dimensional C-vector space and let $A_0, \dots, A_m$ be elements of* End(W). *Then the differential Galois group G of the differential equation $y' = (A_0 + A_1 z + \cdots + A_m z^m)y$ over $C(z)$ is connected.*

Proof. Let E denote the Picard-Vessiot ring and let G^o= the component of the identity of G. The field $F = E^{G^o}$ is a finite Galois extension F of $C(z)$ with Galois group G/G^o. The extension $C(z) \subset F$ can be ramified only above the singular points of the differential equation. The only singular point of the differential equation is ∞. It follows that $C(z) = F$ and by the Galois correspondence $G = G^o$. □

As mentioned above, the key to the proof of Theorem 11.30 is the existence of G-modules that allow one to distinguish a connected semisimple group from its connected proper subgroups. These modules are defined below.

Definition 11.33 A faithful representation $\rho : G \to \mathrm{GL}(W)$, in other words a faithful G-module W, will be called a *Chevalley module* if:

(a) G leaves no line in W invariant.
(b) Any proper connected closed subgroup of G has an invariant line. □

We will postpone to the end of this section the proof that a connected semisimple G has a Chevalley module.

Proof of Theorem 11.30: We now return to the construction of a differential equation of the form $y' = (A_0 + A_1 z)y$ having as differential Galois group a given connected semisimple group G. We shall describe the choices for A_0 and A_1 in $\underline{g} \subset \mathrm{End}(V)$.

The connected semisimple group G is given as an algebraic subgroup $G \subset \mathrm{GL}(V)$, where V is a finite dimensional vector space over C. We recall that G is semisimple if and only if its Lie algebra $\underline{g}$ is semisimple. For the construction of the equation we will need *the root space decomposition of* $\underline{g}$. This decomposition reads (see [107] and [143]): $\underline{g} = \underline{h} \oplus (\oplus_\alpha \underline{g}_\alpha)$, where $\underline{h}$ is a Cartan subalgebra and the one-dimensional spaces $\underline{g}_\alpha = CX_\alpha$ are the eigenspaces for the adjoint action of $\underline{h}$ on $\underline{g}$ corresponding to the nonzero roots $\alpha : \underline{h} \to C$. More precisely, the adjoint action of $\underline{h}$ on $\underline{h}$ is zero and for any $\alpha \neq 0$ one has $[h, X_\alpha] = \alpha(h)X_\alpha$ for all $h \in \underline{h}$.

We fix a Chevalley module $\rho : G \to \mathrm{GL}(W)$. The induced (injective) morphism of Lie algebras $\underline{g} \to \mathrm{End}(W)$ is also denoted by ρ. The action of $\underline{h}$ on W gives a decomposition of $W = \oplus W_\beta$ into eigenspaces for a collection of linear maps $\beta : \underline{h} \to C$. The βs are called the *weights* of the representation ρ.

For A_0 one chooses $\sum_{\alpha \neq 0} X_\alpha$. For A_1 one chooses an element in $\underline{h}$ satisfying conditions (a), (b), and (c) below.

(a) The $\alpha(A_1)$ are nonzero and distinct (for the nonzero roots α of $\underline{g}$).
(b) The $\beta(A_1)$ are nonzero and distinct (for the nonzero weights β of the representation ρ.)
(c) If the integer m is an eigenvalue of the operator $\sum_{\alpha \neq 0} \frac{1}{-\alpha(A_1)} \rho(X_{-\alpha})\rho(X_\alpha)$ on W, then $m = 0$.

It is clear that A_1 satisfying (a) and (b) exists. Choose such an A_1. If A_1 does not yet satisfy (c) then a suitable multiple cA_1, with $c \in C^*$, satisfies all three conditions. We now claim:

> *Let $A_0, A_1 \in \underline{g} \subset \mathrm{End}(V)$ be chosen as above, then the action of the differential Galois group of $y' = (A_0 + A_1 z)y$ on the solution space can be identified with $G \subset \mathrm{GL}(V)$.*

The differential Galois group of the proposed equation is a connected algebraic subgroup H of G by Remark 11.31 and Lemma 11.32. If $H \neq G$, then by definition the group H has an invariant line in W. Furthermore, there exists an equation similar to Equation (11.1), that is an equation $[\frac{d}{dz} - (\rho(A_0) + \rho(A_1)z)]w = cw$ with $c \in C(z)$ and a nontrivial solution $w \in C[z] \otimes W$ such that the g.c.d. of the coordinates of w is 1. It follows that $c \in C[z]$ and by comparing degrees one finds that the degree of c is at most 1. More explicitly, one has

$$[\frac{d}{dz} - (\rho(A_0) + \rho(A_1)z)]w = (c_0 + c_1 z)w,$$

with $w = w_m z^m + \cdots + w_1 z + w_0$, all $w_i \in W$, $w_m \neq 0$ and $c_0, c_1 \in C$. Comparing the coefficients of z^{m+1}, z^m, z^{m-1} one obtains three linear equations:

$$\begin{aligned} \rho(A_1)(w_m) &= -c_1 w_m, \\ \rho(A_0)(w_m) + \rho(A_1)(w_{m-1}) &= -c_0 w_m - c_1 w_{m-1}, \\ -m w_m + \rho(A_0)(w_{m-1}) + \rho(A_1)(w_{m-2}) &= -c_0 w_{m-1} - c_1 w_{m-2}. \end{aligned}$$

We let W_β be one of the eigenspaces for the action of $\rho(A_1)$ on W corresponding to the eigenvalue $b := \beta(A_1)$ of $\rho(A_1)$. We will write $W_b = W_\beta$. Any element $\tilde{w} \in W$ is written as $\tilde{w} = \sum_b \tilde{w}_b$, with $\tilde{w}_b \in W_b$ and where b runs over the set $\{\beta(A_1) \mid \beta \text{ is a weight}\}$ of the eigenvalues of $\rho(A_1)$. The relation $[A_1, X_\alpha] = \alpha(A_1)X_\alpha$ implies that $\rho(X_\alpha)(W_d) \subset W_{d+\alpha(A_1)}$ for any eigenvalue d of $\rho(A_1)$. One concludes that $A_0 = \sum_{\alpha \neq 0} X_\alpha$ has the property $\rho(A_0)(W_d) \subset \oplus_{b \neq d} W_b$.

We analyze now the three equations. The first equation can only be solved with $w_m \in W_d, w_m \neq 0$ and $d = -c_1$. The second equation, which can be read as

$c_0 w_m = -\rho(A_0)(w_m) + (-\rho(A_1) - c_1)w_{m-1}$, imposes $c_0 = 0$. Indeed, the two right-hand side terms $-\rho(A_0)(w_m)$ and $(-\rho(A_1) - c_1)w_{m-1}$ have no component in the eigenspace W_d for $\rho(A_1)$ to which w_m belongs. Further

$$w_{m-1} = \sum_{b \neq d} \frac{1}{-b+d}\rho(A_0)(w_m)_b + v_d = \sum_{\alpha \neq 0} \frac{1}{-\alpha(A_1)}\rho(X_\alpha)(w_m) + v_d,$$

for some $v_d \in W_d$. The third equation can be read as

$$-mw_m + \rho(A_0)(w_{m-1}) = (-\rho(A_1) - c_1)w_{m-2}.$$

A necessary condition for this equation to have a solution w_{m-2} is that the left-hand side has 0 as component in W_d. The component in W_d of the left-hand side is easily calculated to be

$$-mw_m + (\rho(A_0)(w_{m-1}))_d = (-m + \sum_{\alpha \neq 0} \frac{1}{-\alpha(A_1)}\rho(X_{-\alpha})\rho(X_\alpha)\ )(w_m).$$

Since this component is zero, the integer m is an eigenvalue of the operator $\sum_{\alpha \neq 0} \frac{1}{-\alpha(A_1)}\rho(X_{-\alpha})\rho(X_\alpha)$. It follows from our assumption on A_1 that $m = 0$. This leaves us with the equation $[\frac{d}{dz} - (\rho(A_0) + \rho(A_1)z)]w = c_1 z w$ and $w \in W$. Since $\frac{d}{dz}w = 0$, one finds that Cw is invariant under $\rho(A_0)$ and $\rho(A_1)$. The Lie algebra $\underline{g}$ is generated by A_0 and A_1 ([49], Chap. 8.2, Exercise 8, p. 221). Thus Cw is invariant under $\underline{g}$ and under G. Our assumptions on the G-module W lead to the contradiction that $w = 0$. The proof of Theorem 11.30 is completed by a proof of the existence of a Chevalley module. □

Lemma 11.34 (C. Mitschi and M. F. Singer [210]).
Every connected semisimple linear algebraic group has a Chevalley module.

Proof. Let the connected semisimple closed subgroup $G \subset \mathrm{GL}(V)$ be given. Chevalley's theorem (see [141], Sect. 11.2) states that for any proper algebraic subgroup H there is a G-module E and a line $L \subset E$ such that H is the stabilizer of that line. Since G is semisimple, E is a direct sum of irreducible modules. The projection of L to one of these irreducible components is again a line. Thus we find that H stabilizes a line in some irreducible G-module E of dimension greater than one. Any subgroup of G, conjugated to H, also stabilizes a line in E. Dynkin's theorem [91] implies that there are only finitely many conjugacy classes of maximal connected proper algebraic subgroups of G. One chooses an irreducible G-module W_i, $i = 1, \dots, m$ for each class and one chooses an irreducible faithful module W_0. Then $W = W_0 \oplus \cdots \oplus W_m$ has the required properties. □

Examples 11.35 *Chevalley modules for* SL_2 *and* SL_3.
1. The standard action of $\mathrm{SL}_2(C)$ on C^2 is a Chevalley module. The elements A_0, A_1 constructed in the proof of Theorem 11.30 are

$$A_0 = \begin{pmatrix} 0 & 1 \\ 1 & 0 \end{pmatrix}, \quad A_1 = \begin{pmatrix} 1 & 0 \\ 0 & -1 \end{pmatrix}.$$

Therefore, the equation

$$y' = \begin{pmatrix} z & 1 \\ 1 & -z \end{pmatrix} y$$

has differential Galois group $\mathrm{SL}_2(C)$.

2. The standard action of $\mathrm{SL}_3(C)$ on C^3 will be called V. The induced representation on

$$W = V \oplus (\Lambda^2 V) \oplus (\mathrm{sym}^2 V) = V \oplus (V \otimes V)$$

is a Chevalley module. Here, $\Lambda^2 V$ is the second exterior power and $\mathrm{sym}^2 V$ is the second symmetric power. Indeed, let H be a maximal proper connected subgroup of SL_3. Then H leaves a line in V invariant or leaves a plane in V invariant or H is conjugated with $\mathrm{PSL}_2 \subset \mathrm{SL}_3$. In the second case H leaves a line in $\Lambda^2 V$ invariant and in the third case H leaves a line in $\mathrm{sym}^2 V$ invariant. Further SL_3 leaves no line in W invariant. □

Remarks 11.36 1. In [210], Theorem 11.28 is used to show:

Let C be an algebraically closed field of characteristic zero, G a connected linear algebraic group defined over C, and k a differential field containing C as its field of constants and of finite transcendence degree over C; then G can be realized as a Galois group of a Picard-Vessiot extension of k.

2. We now give a brief history of work on the inverse problem in differential Galois theory. An early contribution to this problem is due to Bialynicki-Birula [35] who showed that, for any differential field k of characteristic zero with algebraically closed field of constants C, if the transcendence degree of k over C is finite and nonzero then any connected nilpotent group is a Galois group over k. This result was generalized by Kovacic, who showed the same is true for any connected solvable group. In [164, 165] Kovacic introduced powerful machinery to solve the inverse problem. In particular, he developed an inductive technique that gave criteria to lift a solution of the inverse problem for G/R_u to a solution for the full group G. Using this, Kovacic showed that to give a complete solution of the inverse problem, one needed only to solve the problem for reductive groups (note that G/R_u is reductive). He was able to solve the problem for tori and so could give a solution when G/R_u is such a group (i.e., when G is solvable). He also reduced the problem for reductive groups to the problem for powers of simple groups. The work of [212] described in this chapter together with Kovacic's work yields a solution to the inverse problem for connected groups.

When one considers specific fields, more is known. As described above, Kovacic [164, 165] characterized those connected solvable linear algebraic groups that can occur as differential Galois groups over $\mathbf{C}((z))$ and $\mathbf{C}(\{z\})$ and Ramis [235, 241, 242] gave a complete characterization of those linear algebraic groups that occur as differential Galois groups over $\mathbf{C}(\{z\})$. The complete characterization of linear algebraic groups that occur as differential Galois groups over $\mathbf{C}((z))$ is new.

As described in Sect. 5.2, Tretkoff and Tretkoff [283] showed that any linear algebraic group is a Galois group over $C(z)$ when $C = \mathbf{C}$, the field of complex numbers. For arbitrary C, Singer [270] showed that a class of linear algebraic groups (including all connected groups and large classes of nonconnected linear algebraic groups) are Galois groups over $C(z)$. The proof used the result of Tretkoff and Tretkoff and a transfer principle to go from $\mathbf{C}$ to any algebraically closed field of characteristic zero. In the second edition of [183], Magid gives a technique for showing that some classes of connected linear algebraic groups can be realized as differential Galois groups over $C(z)$. As described above, the complete solution of the inverse problem over $\mathbf{C}(z)$ was given by Ramis and, for connected groups over $C(z)$, by Mitschi and Singer.
Added in proof: A solution for solvable by finite groups over $C(z)$ is given by C. Mitschi and M.F. Singer in "Solvable by Finite Groups as Galois Groups", preprint, 2000 and for the general problem over $C(z)$ by J. Hartmann in "On the Inverse Problem in Differential Galois Theory", dissertation, Ruprecht-Karls-Universitaet, Heidelberg, 2002.

Another approach to the inverse problem was given by Goldman and Miller. In [111], Goldman developed the notion of a generic differential equation with group G analogous to what E. Noether did for algebraic equations. He showed that many groups have such an equation. In his thesis [206], Miller developed the notion of a differentially Hilbertian differential field and gave a sufficient condition for the generic equation of a group to specialize over such a field to an equation having this group as Galois group. Regrettably, this condition gave a stronger hypothesis than in the analogous theory of algebraic equations. This condition made it difficult to apply the theory and Miller was unable to apply this to any groups that were not already known to be Galois groups. Another approach using generic differential equations to solve the inverse problem for GL_n is given by Juan in [147].

Finally, many groups have been shown to appear as Galois groups for classical families of linear differential equations. The family of generalized hypergeometric equations has been particularly accessible to computation, either by algebraic methods as in Beukers and Heckmann [33], Katz [154] and Boussel [50], or by mixed analytic and algebraic methods, as in Duval and Mitschi [88] or Mitschi [207, 208, 209]. These equations, in particular, provide classical groups and the exceptional group G_2. Other examples were treated algorithmically, as in Duval and Loday-Richaud [87] or Ulmer and Weil [292] using the Kovacic algorithm for second order equations, or in Singer and Ulmer [274]. Finally, van der Put and Ulmer [233] give a method for constructing linear differential equations with Galois group a finite subgroup of $\mathrm{GL}_n(C)$. □

12 Moduli for Singular Differential Equations

12.1 Introduction

The aim of this chapter is to produce a fine moduli space for irregular singular differential equations over $\mathbf{C}(\{z\})$ with a prescribed formal structure over $\mathbf{C}((z))$. In Sect. 9.5, it is remarked that this local moduli problem, studied in [13], leads to a set E of meromorphic equivalence classes, which can be given the structure of an affine algebraic variety. In fact, E is for this structure isomorphic to $\mathbf{A}^N_{\mathbf{C}}$ for some integer $N \geq 1$. However, it can be shown that there does not exist a universal family of equations parametrized by E (see [228]). This situation is somewhat similar to the construction of moduli spaces for algebraic curves of a given genus $g \geq 1$. In order to obtain a fine moduli space one has to consider curves of genus g with additional finite data, namely a suitable level structure. The corresponding moduli functor is then representable and is represented by a fine moduli space (see Proposition 12.3).

In our context, we apply a result of Birkhoff (see Lemma 12.1) that states that any differential module M over $\mathbf{C}(\{z\})$ is isomorphic to $\mathbf{C}(\{z\}) \otimes_{\mathbf{C}(z)} N$, where N is a differential module over $\mathbf{C}(z)$ having singular points at 0 and ∞. Moreover, the singular point ∞ can be chosen to be a regular singularity. In considering differential modules N over $\mathbf{C}(z)$ with the above type of singularities, the topology of the field $\mathbf{C}$ no longer plays a role. This makes it possible to define a moduli functor $\mathcal{F}$ from the category of $\mathbf{C}$-algebras (i.e., the commutative rings with unit element and containing the field $\mathbf{C}$) to the category of sets. The additional data attached to a differential module (in analogy to the level structure for curves of a given genus) are a prescribed free vector bundle and a fixed isomorphism with a formal differential module over $\mathbf{C}((z))$. The functor $\mathcal{F}$ turns out to be representable by an affine algebraic variety $\mathbf{A}^N_{\mathbf{C}}$. There is a well-defined map from this fine moduli space (which is also isomorphic to $\mathbf{A}^N_{\mathbf{C}}$) to E. This map is analytic, has an open image and its fibres are, in general, discrete infinite subsets of $\mathbf{A}^N_{\mathbf{C}}$. This means that the "level" data that we have added to a differential equation, is not finite. The "level" that we have introduced can be interpreted as prescribing a conjugacy class of a logarithm of the local topological monodromy matrix of the differential equation.

In Sect. 12.2 we introduce the formal data and the moduli functor for the problem. A special case of this moduli functor, where the calculations are very explicit and relatively easy, is presented in Sect. 12.3. The variation of the differential Galois group on the moduli space is studied.

The construction of the moduli space for a general irregular singularity is somewhat technical in nature. First, in Sect. 12.4 the "unramified case" is studied in detail. The more complicated "ramified case" is reduced in Sect. 12.5 to the former one. Finally, some explicit examples are given and the comparison with the "local moduli problem" of [13] is made explicit in examples.

We note that the method presented here can be modified to study fine moduli spaces for differential equations on $P^1_{\mathbf{C}}$ with a number of prescribed singular points and with prescribed formal type at those points.

Lemma 12.1 (G. Birkhoff) *Let M be a differential module over $\mathbf{C}(\{z\})$. There is an algebraic vector bundle $\mathcal{M}$ on $P^1(\mathbf{C})$ and a connection $\nabla : \mathcal{M} \to \Omega(a[0] +[\infty]) \otimes \mathcal{M}$, such that the differential modules $\mathbf{C}(\{z\}) \otimes \mathcal{M}_0$ and M are isomorphic over $\mathbf{C}(\{z\})$ (where $\mathcal{M}_0$ is the stalk at the origin). If the topological local monodromy of M is semisimple then $\mathcal{M}$ can be chosen to be a free vector bundle.*

Proof. The differential module M can be represented by a matrix differential equation $y' = Ay$ such that the entries of the matrix A are meromorphic functions on some neighborhood of 0 having only poles at 0 of order $\geq -a$, for some integer $a \geq 0$. Thus M extends to a connection on some neighborhood $U_1 = \{z \in \mathbf{C} | \, |z| < \epsilon\}$ of 0, having a certain singularity at 0. This connection can be written as $\nabla_1 : \mathcal{M}_1 \to \Omega(a[0]) \otimes \mathcal{M}_1$, where $\mathcal{M}_1$ is an analytic vector bundle on U_1 with rank equal to the dimension of M over $\mathbf{C}(\{z\})$. The restriction of this connection to $U_1^* := U_1 \setminus \{0\}$ has no singularity and is therefore determined by its topological monodromy T. More precisely, let V denote the local solution space of the connection ∇_1 at the point $\epsilon/2 \in U_1$. Then $T : V \to V$ is the map obtained by analytical continuation of solutions along the circle $\{e^{i\phi} \cdot \epsilon/2 | 0 \leq \phi \leq 2\pi\}$. Put $U_2 = P^1(\mathbf{C}) \setminus \{0\}$ and consider the connection $\nabla_2 : \mathcal{M}_2 \to \Omega([\infty]) \otimes \mathcal{M}_2$ above U_2 given by the data:

(a) $\mathcal{M}_2 = O \otimes_{\mathbf{C}} V$, where O is the sheaf of holomorphic functions on U_2.
(b) ∇_2 is determined by the requirement that for $v \in V$ one has $\nabla_2(v) = \frac{dz}{z} \otimes L(v)$, where $L : V \to V$ is a linear map satisfying $e^{2\pi i L} = T$.

The restrictions of the connections $(\mathcal{M}_i, \nabla_i)$, for $i = 1, 2$, to $U_1^* = U_1 \cap U_2$ are isomorphic. After choosing an isomorphism one glues two connections to a connection $(\mathcal{M}, \nabla)$ on $P^1(\mathbf{C})$. This connection has clearly the required properties. We recall from the GAGA principle (see Example 6.6.5), that $(\mathcal{M}, \nabla)$ is the analytification of an algebraic vector bundle provided with an algebraic connection.

When T is semisimple then one can take for L also a diagonal matrix. The eigenvalues of L can be shifted over integers. This suffices to produce a connection such that the vector bundle $\mathcal{M}$ is free. (See Remark 6.23.2.) □

12.2 The Moduli Functor

Let C be an algebraically closed field of characteristic 0. The *data on* $P^1(C)$ for the moduli problem are:

(i) a vector space V of dimension m over C;
(ii) a formal connection ∇_0 on $N_0 := C[[z]] \otimes V$ of the form $\nabla_0 : N_0 \to C[[z]]z^{-k}dz \otimes N_0$ with $k \geq 2$.

We note that $k \leq 1$ corresponds to a regular singular differential equation and these equations are not interesting for our moduli problem. The objects over C that we consider are tuples $(\mathcal{M}, \nabla, \phi)$ consisting of:

(a) a free vector bundle $\mathcal{M}$ on $P^1(C)$ of rank m provided with a connection $\nabla : \mathcal{M} \to \Omega(k[0] + [\infty]) \otimes \mathcal{M}$;
(b) an isomorphism $\phi : C[[z]] \otimes \mathcal{M}_0 \to N_0$ such that $(id \otimes \phi) \circ \nabla = \nabla_0 \circ \phi$ (where $\mathcal{M}_0$ is the stalk of $\mathcal{M}$ at 0).

Two objects over C, $(\mathcal{M}, \nabla, \phi)$ and $(\mathcal{M}', \nabla', \phi')$ are called isomorphic if there exists an isomorphism $f : \mathcal{M} \to \mathcal{M}'$ of the free vector bundles that is compatible with the connections and the prescribed isomorphisms ϕ and ϕ'. For the moduli functor $\mathcal{F}$ from the category of the C-algebras (always commutative and with a unit element) to the category of sets, that we are in the process of defining, we prescribe that $\mathcal{F}(C)$ is the set of equivalence classes of objects over C. In the following remarks we will make $\mathcal{F}(C)$ more explicit and provide the complete definition of the functor $\mathcal{F}$.

Remarks 12.2 1. Let W denote the vector space $H^0(P^1(C), \mathcal{M})$. Then ∇ is determined by its restriction to W. This restriction is a linear map $L : W \to H^0(P^1(C), \Omega(k[0] + [\infty])) \otimes W$. Furthermore, $\phi : C[[z]] \otimes \mathcal{M}_0 = C[[z]] \otimes W \to N_0 = C[[z]] \otimes V$ is determined by its restriction to W. The latter is given by a sequence of linear maps $\phi_n : W \to V$, for $n \geq 0$, such that $\phi(w) = \sum_{n \geq 0} \phi_n(w) z^n$ holds for $w \in W$. The conditions in part (b) are equivalent to the conditions that ϕ_0 is an isomorphism and certain relations hold between the linear map L and the sequence of linear maps $\{\phi_n\}$. These relations can be made explicit if ∇_0 is given explicitly (see Sect. 12.3 for an example). In other words, (a) and (b) are equivalent to giving a vector space W of dimension m and a set of linear maps $L, \{\phi_n\}$ having certain relations.

An object equivalent to the given $(\mathcal{M}, \nabla, \phi)$ is, in terms of vector spaces and linear maps, given by a vector space V' and an isomorphism $V' \to V$ compatible with the other data. If we use the map ϕ_0 to identify W and V, then we have taken a representative in each equivalence class and the elements of $\mathcal{F}(C)$ can be described by pairs (∇, ϕ) with:

(a') $\nabla : \mathcal{M} \to \Omega(k[0]+[\infty])\otimes \mathcal{M}$ is a connection on the free vector bundle $\mathcal{M} := O_{P^1(C)} \otimes V$.
(b') ϕ is an isomorphism $C[[z]]\otimes \mathcal{M}_0 \to N_0$ such that $(id\otimes\phi)\circ\nabla = \nabla_0\circ\phi$ and such that ϕ modulo (z) is the identity from V to itself.

2. Let R be any C-algebra. The elements of $\mathcal{F}(R)$ are given by:

(a') A connection $\nabla : \mathcal{M} \to \Omega(k[0]+[\infty])\otimes \mathcal{M}$ on the free vector bundle $\mathcal{M} := O_{P^1(R)} \otimes V$.
(b') An isomorphism $\phi : R[[z]] \otimes \mathcal{M}_0 \to R[[z]] \otimes N_0$ such that $(id\otimes\phi)\circ\nabla = \nabla_0\circ\phi$ and such that ϕ modulo (z) is the identity from $R\otimes V$ to itself.

As in the first remark, one can translate an object into a set of R-linear maps $L : R\otimes V \to H^0(P^1(R), \Omega(k[0]+[\infty]))\otimes V$ and $\phi_n : R\otimes V \to R\otimes V$ for $n \geq 0$, such that $\phi(v) = \sum_{n\geq 0}\phi_n(v)z^n$ for $v \in R\otimes V$. The conditions are that ϕ_0 is the identity and the relations that translate $(id\otimes\phi)\circ\nabla = \nabla_0\circ\phi$. □

We will show that the translation of $\mathcal{F}(R)$ in terms of maps implies that $\mathcal{F}$ is representable by some affine scheme $\mathrm{Spec}(A)$ over C (see Definitions B.8 and B.18).

Proposition 12.3 *The functor $\mathcal{F}$ described above is representable.*

Proof. Fix a basis of V and consider the basis $\{z^{-s}dz|\ s = 1,\dots,k\}$ of $H^0(P^1(C), \Omega(k[0]+[\infty]))$. The connection ∇ or, what amounts to the same data, the linear map L can be decomposed as $L(v) = \sum_{s=1}^{k} z^{-s}dz\otimes L_s(v)$ where $L_1,\dots,L_k$ are linear maps form V to itself. The entries of the matrices of $L_1,\dots,L_k$ and the ϕ_n for $n \geq 1$ (with respect to the given basis of V) are first seen as a collection of variables $\{X_i\}_{i\in I}$. The condition $(id\otimes\phi)\circ\nabla = \nabla_0\circ\phi$ induces a set of polynomials $\{F_j\}_{j\in J}$ in the ring $C[\{X_i\}_{i\in I}]$ and generates some ideal S. The C-algebra $A := C[\{X_i\}_{i\in I}]/S$ has the property that $\mathrm{Spec}(A)$ represents $\mathcal{F}$. □

$\mathrm{Spec}(A)$ is referred to as a *fine moduli space*. We recall the formalism of representable functors. There is a bijection $\alpha_A : \mathrm{Hom}_k(A, A) \to \mathcal{F}(A)$. Let $\xi = \alpha_A(id_A) \in \mathcal{F}(A)$. This ξ is called the *universal family above* $\mathrm{Spec}(A)$. For any $\eta \in \mathcal{F}(R)$ there exists a unique k-algebra homomorphism $\psi : A \to R$ such that $\psi(\xi) = \eta$. One can make ξ more explicit by writing it as a pair $(\sum_{s=1}^{k} z^{-s}dz\otimes L_s, \phi)$ where the $L_s : V \to A\otimes_C V$ are C-linear, where $\phi \in \mathrm{GL}(A[[z]]\otimes V)$ such that $\phi \equiv id \bmod(z)$ and $\phi(\sum_{s=1}^{k} z^{-s}dz\otimes L_s)\phi^{-1} = \nabla_0$, viewed as a linear map from V to $C[[z]]z^{-k}dz\otimes V$. Then $\psi(\xi)$ is obtained by applying ψ to the coordinates of $L_1,\dots,L_k$ and ϕ. The aim of this chapter is to make A explicit and, in particular, to show that $A \cong C[Y_1,\dots,Y_N]$ for a certain integer $N \geq 1$.

12.3 An Example

12.3.1 Construction of the Moduli Space

The *data* for the moduli functor $\mathcal{F}$ are:

A vector space V of dimension m over C and a linear map $D : V \to V$ having distinct eigenvalues $\lambda_1, \ldots, \lambda_m$. The formal connection at $z = 0$ is given by $\nabla_0 : N \to z^{-2}dz \otimes N_0$, where $N_0 = C[[z]] \otimes V$ and $\nabla_0(v) = z^{-2}dz \otimes D(v)$ for $v \in V$.

The *moduli problem*, stated over C for convenience, asks for a description of the pairs (∇, ϕ) satisfying:

(a) ∇ is a connection $\mathcal{M} \to \Omega(2 \cdot [0] + 1 \cdot [\infty]) \otimes \mathcal{M}$ on the free vector bundle $\mathcal{M} = O_{P^1(C)} \otimes V$.
(b) ϕ is an isomorphism between the formal differential modules $C[[z]] \otimes \mathcal{M}_0$ and N_0 over $C[[z]]$.

Theorem 12.4 *The moduli functor $\mathcal{F}$ is represented by the affine space*

$$\mathbf{A}_C^{m(m-1)} = \mathrm{Spec}(C[\{T_{i,j}\}_{i \neq j}]).$$

For notational convenience we put $T_{i,i} = 0$. The universal family of differential modules is given in matrix form by the operator

$$z^2 \frac{d}{dz} + \begin{pmatrix} \lambda_1 & & & & \\ & \lambda_2 & & & \\ & & \cdot & & \\ & & & \cdot & \\ & & & & \lambda_m \end{pmatrix} + z \cdot (T_{i,j}).$$

Proof. The connection on $\mathcal{M}$ is given by a map ∇ from V to $H^0(P^1(C), \Omega(2 \cdot [0] + 1 \cdot [\infty])) \otimes V$. After replacing the ∇ by $\nabla_{z^2 \frac{d}{dz}}$ one finds a map $C[z] \otimes V \to C[z] \otimes V$ of the form $m \mapsto z^2 \frac{d}{dz} m + A_0(m) + zA_1(m)$ with $A_0, A_1 : V \to V$ linear maps (extended to $C[z]$-linear maps on $C[z] \otimes V$). In the above one has only used condition (a). Condition (b) needs only to be stated for elements in V and it can be written as $(z^2 \frac{d}{dz} + D) \circ \phi = \phi \circ (z^2 \frac{d}{dz} + A_0 + zA_1)$. This translates into

$$\sum n\phi_n z^{n+1} + \sum D\phi_n z^n = \sum \phi_n A_1 z^{n+1} + \sum \phi_n A_0 z^n \text{ and } \phi_0 = 1.$$

Comparing the coefficients of the above formula one finds the relations

$$D = A_0,\ D\phi_1 = A_1 + \phi_1 A_0,\ (n-1)\phi_{n-1} + D\phi_n = \phi_{n-1} A_1 + \phi_n A_0$$

for $n \geq 2$. Or, in more convenient form, $D = A_0$ and

$$D\phi_1 - \phi_1 D = A_1, \ D\phi_n - \phi_n D = \phi_{n-1}A_1 - (n-1)\phi_{n-1} \text{ for } n \geq 2.$$

The map D determines a decomposition of V as a direct sum of m lines V_j. We will call a map $B : V \to V$ *diagonal* if $BV_j \subset V_j$ for all j and *antidiagonal* if $BV_j \subset \oplus_{i \neq j} V_i$ for all j. Every map B is a unique direct sum $B_d + B_a$ of a diagonal map and an antidiagonal map. We now start with the first equality $D\phi_1 - \phi_1 D = A_1$ and conclude that A_1 is antidiagonal. In the following we will show that for any choice of an antidiagonal A_1 there is a unique collection $\{\phi_n\}$ such that all the equalities are satisfied.

The first equation $D\phi_1 - \phi_1 D$ determines uniquely the antidiagonal part of ϕ_1. The second equation $D\phi_2 - \phi_2 D = \phi_1 A_1 - \phi_1$ can only be solved if the right-hand side is antidiagonal. This determines uniquely the diagonal part of ϕ_1. The second equation determines the antidiagonal part of ϕ_2 and the third equation determines the diagonal part of ϕ_2, etc.

It is obvious that the above calculation remains valid if one replaces C by any C-algebra R and prescribes A_1 as an antidiagonal element of $\text{Hom}_R(R \otimes V, R \otimes V)$. We conclude that there is a fine moduli space $\mathbf{A}^{m(m-1)} = \text{Spec}(C[\{T_{i,j}\}_{i \neq j}])$ for the moduli problem considered above. The universal object is thus given by $A_0 = D$ and A_1 is the antidiagonal matrix with entries $T_{i,j}$ outside the diagonal. Furthermore, $\phi_0 = id$ and the coordinates of the ϕ_n are certain expressions in the ring $C[\{T_{i,j}\}_{i \neq j}]$. □

Exercise 12.5 Compute the moduli space and the universal family for the functor $\mathcal{F}$ given by the same data as in Theorem 12.4, but with D replaced by any semisimple (i.e., diagonalizable) linear map from V to itself. Hint: Consider the decomposition $V = V_1 \oplus \cdots \oplus V_s$ of V according to the distinct eigenvalues $\lambda_1, \ldots, \lambda_s$ of D. A linear map L on V will be called *diagonal* if $L(V_i) \subset V_i$ for all i. The map L is called *antidiagonal* if $L(V_i) \subset \oplus_{j \neq i} V_j$ holds for all i. Show that the universal family can be given by $z^2 \frac{d}{dz} + D + zA_1$ where A_1 is the "generic" antidiagonal map. □

12.3.2 Comparison with the Meromorphic Classification

We consider the case $C = \mathbf{C}$ of the example of the last subsection in order to compare the moduli space with the analytic classification of Chap. 9. Let $K = \mathbf{C}(\{z\})$ and $\hat{K} = \mathbf{C}((z))$. As before, an m-dimensional $\mathbf{C}$-vector space V and a linear $D : V \to V$ with distinct eigenvalues $\lambda_1, \ldots, \lambda_m$ are given. Then $N_0 := \mathbf{C}[[z]] \otimes V$ and $\nabla_0 : N_0 \to z^{-2}dz \otimes N_0$ satisfies $\nabla_0(v) = z^{-2}dz \otimes D(v)$ for all $v \in V$. Let N denote the differential module $\hat{K} \otimes N_0$ over $\hat{K}$.

We recall that the analytic classification describes the collection of isomorphism classes E of pairs (M, ψ) such that M is a differential module over $K := \mathbf{C}(\{z\})$

and $\psi : \hat{K} \otimes M \to N$ is an isomorphism of differential modules. In Chap. 9 it is shown that this set of isomorphism classes E is described by the cohomology set $H^1(S^1, STS)$, where S^1 is the circle of directions at $z = 0$ and STS the *Stokes sheaf*. The explicit choice of 1-cocycles for this cohomology set leads to an isomorphism $H^1(S^1, STS) \to \mathbf{C}^{m(m-1)}$. The interpretation of this isomorphism is that one associates to each (isomorphism class) (M, ψ) the Stokes matrices for all singular directions of N.

The moduli space $\mathbf{A}_{\mathbf{C}}^{m(m-1)}$ of Theorem 12.4 (identified with the point set $\mathbf{C}^{m(m-1)}$) has an obvious map to $H^1(S^1, STS)$. This map associates to any $(\mathcal{M}, \nabla, \phi)$ the differential module $M := K \otimes \mathcal{M}_0$ and the isomorphism $\psi : \hat{K} \otimes M \to N$ induced by $\phi : \mathbf{C}[[z]] \otimes \mathcal{M}_0 \to N_0$. In other words, any $\mathbf{C}$-valued point of the moduli space corresponds to a differential operator of the form

$$z^2 \frac{d}{dz} + \begin{pmatrix} \lambda_1 & & & \\ & \lambda_2 & & \\ & & \cdot & \\ & & & \cdot \\ & & & & \lambda_m \end{pmatrix} + z \cdot (t_{i,j}), \text{ with } t_{i,j} \in \mathbf{C} \text{ and } t_{i,i} = 0.$$

The map associates to this differential operator its collection of Stokes matrices (i.e., this explicit 1-cocycle) and the latter is again a point in $\mathbf{C}^{m(m-1)}$. We will show later that this map $\alpha : \mathbf{A}_{\mathbf{C}}^{m(m-1)} \to E = H^1(S^1, STS) = \mathbf{C}^{m(m-1)}$ is a complex analytic map.

The image of α and the fibres of α are of interest. We will briefly discuss these issues. Let a point (M, ψ) of $H^1(S^1, STS)$ be given. Let M_0 denote the $\mathbf{C}\{z\}$-lattice in M such that $\mathbf{C}[[z]] \otimes M_0$ is mapped by the isomorphism ψ to $N_0 \subset N$. We denote the restriction of ψ to $\mathbf{C}[[z]] \otimes M_0$ by ϕ. The differential module M_0 over $\mathbf{C}\{z\}$ extends to some neighborhood of $z = 0$ and has a topological monodromy. According to Birkhoff's Lemma 12.4 one chooses a logarithm of the topological monodromy around the point $z = 0$ and, by gluing, one obtains a vector bundle $\mathcal{M}$ on $P^1(\mathbf{C})$ having all the required data except for the possibility that $\mathcal{M}$ is not free. At the point 0 one cannot change this vector bundle. At ∞ one is allowed any change. When the topological monodromy is semisimple one can make the bundle free. Thus the point (M, ψ) lies in the image of α. In the general case this may not be possible.

It is easily calculated that the jacobian determinant of the map α at the point $0 \in \mathbf{A}_{\mathbf{C}}^{m(m-1)}$ is nonzero. In particular, the image of α contains points (M, ψ) such that the topological monodromy has m distinct eigenvalues. The formula (see Proposition 8.12) that expresses the topological monodromy in Stokes matrices and the formal monodromy implies that the subset of E where the topological monodromy has m distinct eigenvalues is Zariski open (and nonempty) in $E = \mathbf{C}^{m(m-1)}$. The image of α contains this Zariski-open subset.

The surjectivity of the map α is also related to *Birkhoff's Problem* of representing a singular differential module over K by a matrix differential equation involving only polynomials in z^{-1} of a degree restricted by the "irregularity" of the equation at $z = 0$.

We now consider the fibre over a point (M, ψ) in E such that the topological monodromy has m distinct eigenvalues $\mu_1, \dots, \mu_m$. In the above construction of an object $(\mathcal{M}, \nabla, \phi) \in \mathbf{A}_{\mathbf{C}}^{m(m-1)}$ the only freedom is the choice of a logarithm of the topological monodromy. This amounts to making a choice of complex numbers $c_1, \dots, c_m$ such that $e^{2\pi i c_j} = \mu_j,\ j = 1, \dots, m$ such that the corresponding vector bundle $\mathcal{M}$ is free. Let $c_1, \dots, c_m$ be a good choice. Then $c_1 + n_1, \dots, c_m + n_m$ is also a good choice if all $n_j \in \mathbf{Z}$ and $\sum n_j = 0$. Thus the fibre $\alpha^{-1}(M, \psi)$ is countable and discrete in $\mathbf{A}_{\mathbf{C}}^{m(m-1)}$ since α is analytic. In other cases, e.g., the topological monodromy is semisimple and has multiple eigenvalues, the fibre will be a discrete union of varieties of positive dimension.

We now illustrate the above with an explicit formula for α in case $m = 2$.

The universal family is given by the operator in matrix form

$$z^2 \frac{d}{dz} + \begin{pmatrix} \lambda_1 & 0 \\ 0 & \lambda_2 \end{pmatrix} + z \begin{pmatrix} 0 & a \\ b & 0 \end{pmatrix}.$$

The $\lambda_1, \lambda_2 \in \mathbf{C}$ are fixed and distinct. The a, b are variable and $(a, b) \in \mathbf{C}^2$ is a point of the moduli space. In Example 8.17 we showed that the equation has two Stokes matrices of the form $\left(\begin{smallmatrix} 1 & x_1 \\ 0 & 1 \end{smallmatrix}\right)$ and $\left(\begin{smallmatrix} 1 & 0 \\ x_2 & 1 \end{smallmatrix}\right)$. Moreover, (x_1, x_2) is a point of $E \cong \mathbf{C}^2$. Furthermore, the calculation in this example shows the following proposition.

Proposition 12.6 *The map* $\alpha : \mathbf{A}_{\mathbf{C}}^2 \to E = H^1(S^1, STS) = \mathbf{C}^2$ *has the form* $(a, b) \mapsto (x_1, x_2) = f(ab) \cdot (a, b)$ *with* $f(t) := \frac{2i \sin(\pi \sqrt{t})}{\sqrt{t}}$.

We now give some details about the map α. A list of its fibres is:

1. $\alpha^{-1}(0, 0) = \{(a, b) |\ ab \text{ is the square of an integer}\}$.
2. If $x_1 \neq 0$, then $\alpha^{-1}(x_1, 0) = \{(\frac{x_1}{2\pi i}, 0)\}$.
3. If $x_2 \neq 0$, then $\alpha^{-1}(0, x_2) = \{(0, \frac{x_2}{2\pi i})\}$.
4. If $x_1 x_2 \neq 0$, then $\alpha^{-1}(x_1, x_2) = \{\lambda(x_1, x_2) |$ where $\frac{2i \sin(\lambda \pi \sqrt{x_1 x_2})}{\sqrt{x_1 x_2}} = 1\}$. The set of λs satisfying this condition is infinite and discrete.

In particular, α is surjective. For the topological monodromy matrix $\left(\begin{smallmatrix} 1+x_1x_2 & x_1 \\ x_2 & 1 \end{smallmatrix}\right)$ one can distinguish the following cases:

1. $(x_1, x_2) = (0, 0)$ and the monodromy is the identity.
2. $x_1 \neq 0,\ x_2 = 0$ and the monodromy is unipotent.
3. $x_1 = 0,\ x_2 \neq 0$ and the monodromy is unipotent.

4. $x_1x_2 = -4$ and the monodromy has only the eigenvalue -1 and is different from $-id$.

5. $x_1x_2 \neq 0, -4$ and the monodromy has two distinct eigenvalues.

Let $S \subset \mathbf{C}^2$ denote the set of points where the map α is smooth, i.e., is locally an isomorphism. The points of S are the points where the jacobian determinant $-f(ab)(f(ab)+2abf'(ab))$ of α is nonzero. The points where this determinant is 0 are:

1. $f(ab) = 0$. This is equivalent to $ab \neq 0$ is the square of an integer.
2. $f(ab) \neq 0$ and $f(ab) + 2abf'(ab) = 0$. This is equivalent to the condition that $4ab$ is the square of an odd integer.

A point (a, b) where the map α is not smooth corresponds, according to the above calculation, to a point where the eigenvalues of the "candidate" for the monodromy matrix $\begin{pmatrix} 0 & a \\ b & 0 \end{pmatrix}$ has eigenvalues that differ by an integer $\neq 0$. Let $S \subset \mathbf{C}^2$ denote the set where the map α is smooth, i.e, the jacobian determinant is $\neq 0$. The above calculations show that $\alpha(S) = \{(x_1, x_2) |\ x_1x_2 \neq -4\}$. Then $\alpha(S)$ is the Zariski-open subset of $E = \mathbf{C}^2$, where the monodromy has two distinct eigenvalues. The fibre of a point $(x_1, x_2) \in \alpha(S)$ can be identified with the set of conjugacy classes of the 2×2 matrices L with trace 0 and with $\exp(2\pi i L)$ being the topological monodromy of the differential equation corresponding to (x_1, x_2).

Another interesting aspect of the example is that the dependence of the differential Galois group on the parameters a, b can be given. According to a theorem of J. Martinet and J.-P. Ramis (see Theorem 8.10) the differential Galois group is the algebraic subgroup of GL(2) generated by the formal monodromy, the exponential torus and the Stokes matrices. From this, one deduces that the differential equation has a 1-dimensional submodule if and only if $ab = 0$ or $ab \neq 0$ and $\sin(\pi\sqrt{ab}) = 0$. In the first case the differential Galois group is one of the two standard Borel subgroups of GL(2) if $a \neq 0$ or $b \neq 0$. The second case is equivalent to $ab = d^2$ for some integer $d \geq 1$. The two Stokes matrices are both the identity, the equation is over $\mathbf{C}(\{z\})$ equivalent with $z^2\frac{d}{dz} + \begin{pmatrix} \lambda_1 & 0 \\ 0 & \lambda_2 \end{pmatrix}$ and the differential Galois group is the standard torus in GL(2) (assuming λ_1 and λ_2 are linearly independent over the rationals). We return now to the moduli space and the universal family of Theorem 12.4 and investigate the existence of invariant line bundles as a first step in the study of the variation of the differential Galois group on the moduli space.

12.3.3 Invariant Line Bundles

We consider the moduli problem of Exercise 12.5. Let V be a vector space of dimension m and $D : V \to V$ a semisimple linear map. The (distinct) eigenvalues of D are $\lambda_1, \ldots, \lambda_s$ and V_i is the eigenspace corresponding to the eigenvalue λ_i. The dimension of V_i is denoted by m_i. The data for the moduli functor $\mathcal{F}$ is the formal differential module $N_0 = C[[z]] \otimes V$ with connection $\nabla_0 : N_0 \to C[[z]]z^{-2} \otimes N_0$ given by $\nabla_0(v) = z^{-2}dz \otimes D(v)$ for all $v \in V$. The moduli space for this functor is $\mathbf{A}_C^N$ with $N = \sum_{i \neq j} m_i m_j$.

Let $(\mathcal{M}, \nabla, \phi)$ be an object over C corresponding to a (closed) point of this moduli space $\mathbf{A}_C^N$. This object is represented by a differential operator of the form $z^2 \frac{d}{dz} + D$ $+zA_1$ where A_1 is an antidiagonal matrix. The generic fibre $\mathcal{M}_\eta$ is a differential module over $C(z)$. We want to investigate the possibility of a 1-dimensional submodule L of $\mathcal{M}_\eta$. Any L corresponds uniquely to a line bundle $\mathcal{L} \subset \mathcal{M}$ such that $\mathcal{M}/\mathcal{L}$ is a vector bundle of rank $m-1$ and $\nabla : \mathcal{L} \to \Omega(2[0] + [\infty]) \otimes \mathcal{L}$. Let the degree of $\mathcal{L}$ be $-d \leq 0$. Then $\mathcal{L}(d \cdot [\infty]) \subset \mathcal{M}(d \cdot [\infty])$ is free and generated by an element $e = v_0 + v_1 z + \cdots + v_d z^d$ with all $v_i \in V = H^0(P^1(C), \mathcal{M})$ and $v_d \neq 0$. The invariance of $\mathcal{L}$ under ∇ can be formulated as $(z^2 \frac{d}{dz} + D + A_1 z)e = (t_0 + t_1 z)e$, for certain $t_0, t_1 \in C$. The condition that $\mathcal{M}/\mathcal{L}$ is again a vector bundle implies that $v_0 \neq 0$. The equation is equivalent to a sequence of linear equations:

$$
\begin{aligned}
&(D - t_0)v_0 = 0, \\
&(D - t_0)v_1 = (-A_1 + t_1)v_0, \\
&(D - t_0)v_i = (-A_1 - (i-1) + t_1)v_{i-1} \text{ for } i = 2, \ldots, d, \\
&0 = (-A_1 - d + t_1)v_d.
\end{aligned}
$$

The first equation implies that t_0 is an eigenvalue λ_i of D and $v_0 \in V_i$, $v_0 \neq 0$. The second equation can only have a solution if $t_1 = 0$. Moreover, the components of v_1 in V_j for $j \neq i$ are uniquely determined by v_0. The third equation determines the component of v_1 in V_i and the components of v_2 in V_j for $j \neq i$, etc. The last equation determines v_d completely in terms of v_0 and the map A_1. The last equation can be read as a set of homogeneous linear equations for the vector $v_0 \in V_i$. The coordinates of these equations are polynomial expressions in the entries of A_1. The conclusion is the following lemma.

Lemma 12.7 *The condition that there exists an invariant line bundle $\mathcal{L}$ of degree $-s$ with $s \leq d$ determines a Zariski-closed subset of the moduli space $\mathbf{A}_C^N$.*

Example 12.8 $z^2 \frac{d}{dz} + D + z \cdot A_1$, *where* $D = \begin{pmatrix} \lambda_1 & 0 \\ 0 & \lambda_2 \end{pmatrix}$ *and* $A_1 = \begin{pmatrix} 0 & a \\ b & 0 \end{pmatrix}$.
As above, we assume here that $\lambda_1 \neq \lambda_2$. We consider first the case $d = 0$. The line bundle $\mathcal{L}$ is generated by some element $v \in V$, $v \neq 0$ and the condition is $(D + A_1 z)v = (t_0 + t_1 z)v$. Clearly t_0 is one of the two eigenvectors of D and a or b is 0.
Consider now $d \geq 1$ and $ab \neq 0$. Let $e = v_0 + \cdots + v_d z^d$ with $v_0 \neq 0 \neq v_d$ satisfy $(z^2 \frac{d}{dz} + D + z \cdot A_1)e = (t_0 + t_1 z)e$. We make the choice $t_0 = \lambda_1$ and v_0 is the first basis vector. As before $t_1 = 0$. A somewhat lengthy calculation shows that the existence of e above is equivalent with the equation $ab = d^2$. If one starts with the second eigenvalue λ_2 and the second eigenvector, then the same equation $ab = d^2$ is found. We note that the results found here agree completely with the calculations in Sect. 12.3.2. □

12.3.4 The Differential Galois Group

We continue the moduli problem of Exercise 12.5 and Sect. 12.3.3 and keep the same notations. Our aim is to investigate the variation of the differential Galois group on the moduli space $\mathbf{A}^N_C$. The first goal is to define a natural action of the differential Galois group of an object $(\mathcal{M}, \nabla, \phi)$ on the space $V = H^0(P^1(C), \mathcal{M})$. For this we introduce symbols $f_1, \ldots, f_s$ having the properties $z^2\frac{d}{dz}f_i = \lambda_i f_i$, where $\lambda_1, \ldots, \lambda_s$ are the distinct eigenvalues of D. The ring $S = C[[z]][f_1, f_1^{-1}, \ldots, f_s, f_s^{-1}]/I$ where I is the ideal generated by the set of all polynomials $f_1^{m_1} \cdots f_s^{m_s} - 1$ with m_i integers such that $m_1\lambda_1 + \ldots + m_s\lambda_s = 0$. The differentiation $z^2\frac{d}{dz}$ on S is defined by $z^2\frac{d}{dz}z = z^2$ and $z^2\frac{d}{dz}f_i = \lambda_i f_i$. In this way, S is a differential ring. For any $\mathcal{M} := (\mathcal{M}, \nabla, \phi)$, the solution space $Sol(\mathcal{M})$ can be identified with the kernel of the operator $z^2\frac{d}{dz} + D + A_1 z$ on $S \otimes_{C[z]_{(z)}} \mathcal{M}_0 = S \otimes_C V$ (note that our assumption on the formal normal form of the equation implies that there is no formal monodromy and so the equation has a full set of solutions in $S \otimes_C V$). This space has dimension m over C. The ring homomorphism $C[[z]][f_1, f_1^{-1}, \ldots, f_s, f_s^{-1}] \to C$, given by $z \mapsto 0,\ f_1, \ldots, f_s \mapsto 1$, induces a bijection $Sol(\mathcal{M}) \to V$. The smallest ring R with $C[z]_{(z)} \subset R \subset S$, which contains all the coordinates of the elements of $Sol(\mathcal{M})$ with respect to V has the property: R is a differential ring for the operator $z^2\frac{d}{dz}$ and the field of fractions of R is the Picard-Vessiot field of $\mathcal{M}_\eta$ over $C(z)$. The differential Galois group $Gal(\mathcal{M})$, acting upon this field of fractions, leaves R invariant. Thus $Gal(\mathcal{M})$ acts on $Sol(\mathcal{M})$ and on V according to our chosen identification $Sol(\mathcal{M}) \to V$. We note that the formal Galois group at $z = 0$, which is a subgroup of $\mathbf{G}^s_{m,C}$, is a subgroup of $Gal(\mathcal{M})$. We can now formulate our result.

Proposition 12.9 *For any algebraic subgroup $G \subset \mathrm{GL}(V)$, the set of the $\mathcal{M} := (\mathcal{M}, \nabla, \phi) \in \mathbf{A}^N_C$ with $Gal(\mathcal{M}) \subset G$, is a countable union of Zariski-closed subsets.*

Proof. By Chevalley's theorem, there is a vector space W over C obtained from V by a construction of linear algebra and a line $L \subset W$, such that G consists of the elements $g \in \mathrm{GL}(V)$ with $gL \subset L$. This construction of linear algebra can be extended to a construction of an object $(\mathcal{N}, \nabla, \psi)$ from $(\mathcal{M}, \nabla, \phi)$ corresponding to new formal data at $z = 0$ (of the same type that we have been considering here) and regular singularity at $z = \infty$. The invariance of L under the differential Galois group is equivalent to the existence of a line bundle $\mathcal{L} \subset \mathcal{N}$, invariant under ∇, such that $\mathcal{N}/\mathcal{L}$ is again a vector bundle and $\mathcal{L}_0/z\mathcal{L}_0 = L \subset \mathcal{N}_0/z\mathcal{N}_0 = W$. If we bound the degree $-s$ of $\mathcal{L}$ by $s \leq d$ then the existence of $\mathcal{L}$ defines an algebraic subset of the corresponding moduli space, by Lemma 12.7. The proposition now follows. □

Remarks 12.10 1. The occurrence of countable unions of algebraic subsets of the moduli space $\mathbf{A}^N_C$ corresponding to the existence of an invariant line bundle or a condition $Gal(\mathcal{M}) \subset G$, where $G \subset \mathrm{GL}(V)$ is a fixed algebraic subgroup, is due to our choice of not prescribing the regular singularity at ∞. Indeed, let us add to the moduli functor a regular singular module $N_\infty := C[[z^{-1}]] \otimes V$ with some ∇_∞

and an isomorphism $C[[z^{-1}]] \otimes \mathcal{M}_\infty \to N_\infty$ of differential modules. We will show that there is a bound B, depending on the moduli problem, such that the existence of an invariant line bundle implies that its degree $-d$ satisfies $d \leq B$.

To prove this assertion, let $\mathcal{L}$ be an invariant line bundle of degree $-d$. There is given an inclusion $C[[z^{-1}]] \otimes \mathcal{L}_\infty \subset N_\infty$, which induces an inclusion $\mathcal{L}_\infty/(z^{-1}) \subset N_\infty/(z^{-1})$. The operator $\nabla_{z\frac{d}{dz}}$ has on $\mathcal{L}_\infty/(z^{-1})$ an eigenvalue μ, which is one of the, at most, m eigenvalues of the corresponding operator on $N_\infty/(z^{-1})$. Let $e = v_0 + v_1 z + \cdots + v_d z^d$, with $v_0 \neq 0 \neq v_d$ be the generator of $H^0(P^1(C), \mathcal{L}(d \cdot [\infty]))$. As before we have an equation $(z^2\frac{d}{dz} + D + A_1 z)e = (t_0 + t_1 z)e$. From $v_0 \neq 0$ it follows that $t_1 = 0$. This implies that $\nabla_{z\frac{d}{dz}}$ on $\mathcal{L}(d \cdot [\infty])_\infty/(z^{-1})$ has eigenvalue 0. According to the shift that we have made at $z = \infty$ this eigenvalue is also $d + \mu$.

We conclude that after prescribing the regular singularity at $z = \infty$, the set corresponding to the condition $Gal(\mathcal{M}) \subset G$ is an algebraic subspace of the moduli space.

2. The question of how the Galois group varies in a family of differential equations is also considered in [270]. In this paper one fixes integers m and n and considers the set $\mathbf{L}_{n,m}$ of linear differential operators of the form

$$L = \sum_{i=0}^{n}(\sum_{j=0}^{m} a_{i,j} z^j)(\frac{d}{dz})^i$$

of order n with coefficients in $C[z]$ of degree at most m. Such an operator may be identified with the vector $(a_{i,j})$ and so $\mathbf{L}_{n,m}$ may be identified with $C^{(m+1)(n+1)}$. Let $\mathcal{S}$ be a finite subset of $C \cup \mathcal{Q} = \cup_{m \geq 1} z^{-1/m} C[z^{-1/m}]$ and let $\mathbf{L}_{n,m}(\mathcal{S})$ be the set of operators in $\mathbf{L}_{n,m}$ having exponents and eigenvalues (see Definition 3.26) in $\mathcal{S}$ at each singular point. Note that we do not fix the singular points. In [270], it is shown that for many linear algebraic groups G (e.g., G finite, G connected, G^0 unipotent) the set of operators in $\mathbf{L}_{n,m}(\mathcal{S})$ with Galois group G is a constructible subset of $C^{(m+1)(n+1)}$. An example is also given to show that this is not necessarily true for all groups. □

12.4 Unramified Irregular Singularities

A connection (N, ∇) over $\hat{K} := C((z))$ is called *unramified* if its canonical form does not use roots of z. For our formulation of this canonical form we will use the operator $\delta = \nabla_{z\frac{d}{dz}}$ on N. For $q \in z^{-1}C[z^{-1}]$ we write $E(q) = Ke$ for the 1-dimensional connection with $\delta e = qe$. Furthermore, we fix a set of representatives for $C/\mathbf{Z}$. Any regular singular connection over $\hat{K}$ can (uniquely) be written as $\hat{K} \otimes_C V$, where V is a finite dimensional vector space over C and with δ given on V

as a linear map $l : V \to V$ such that all its eigenvalues are in the set of representatives of $C/\mathbf{Z}$ (see Theorem 3.1). The *canonical form for an unramified connection* (N, ∇) *over* $\hat{K}$ is given by:

(a) distinct elements $q_1, \ldots, q_s \in z^{-1}C[z^{-1}]$;
(b) finite dimensional C-vector spaces V_i and linear maps $l_i : V_i \to V_i$ for $i = 1, \ldots, s$ with eigenvalues in the set of representatives of $C/\mathbf{Z}$.

The unramified connection with these data is $N := \oplus_{i=1}^{s} \hat{K} e_i \otimes_C V_i$ with the action of $\delta = \nabla_{z\frac{d}{dz}}$, given by $\delta(e_i \otimes v_i) = q_i e_i \otimes v_i + e_i \otimes l_i(v_i)$. We note that this presentation is unique. We write $N_0 := \oplus_{i=1}^{s} C[[z]]e_i \otimes_C V_i$ and define k_i to be the degree of the q_i in the variable z^{-1}. Put $k = \max\ k_i$. Write $V := \oplus V_i$. One identifies N_0 with $C[[z]] \otimes V$ by $e_i \otimes v_i \mapsto v_i$ for all i and $v_i \in V_i$. The connection on N_0 is denoted by ∇_0.

The moduli problem that we consider is given by the connection (N_0, ∇_0) at $z = 0$ and a nonspecified regular singularity at $z = \infty$. More precisely, we consider (equivalence classes of) tuples $(\mathcal{M}, \nabla, \phi)$ with:

(a) $\mathcal{M}$ is a free vector bundle of rank m on $P^1(C)$ and $\nabla : \mathcal{M} \to \Omega((k+1) \cdot [0] + [\infty]) \otimes \mathcal{M}$ is a connection.
(b) $\phi : C[[z]] \otimes \mathcal{M}_0 \to N_0$ is an isomorphism, compatible with the connections.

Theorem 12.11 *The functor associated to the above moduli problem is represented by the affine space* $\mathbf{A}_C^N$, *where* $N = \sum_{i \neq j} \deg_{z^{-1}}(q_i - q_j) \cdot \dim V_i \cdot \dim V_j$.

The proof of this theorem is rather involved. We start by writing the functor $\mathcal{F}$ from the category of C-algebras to the category of sets in a more convenient form. Let δ_0 denote the differential $(\nabla_0)_{z\frac{d}{dz}} : N_0 = C[[z]] \otimes V \to C((z)) \otimes V$. For any C-algebra R, δ_0 induces a differential $R[[z]] \otimes V \to R[[z]][z^{-1}] \otimes V$, which will also be denoted by δ_0.

For any C-algebra R, one defines $\mathbf{G}(R)$ as the group of the $R[[z]]$-linear automorphisms g of $R[[z]] \otimes_C V$ such that g is the identity modulo z. One can make this more explicit by considering the restriction of g to $R \otimes V$. This map is supposed to have the form $g(w) = \sum_{n \geq 0} g_n(w) z^n$, where each $g_n : R \otimes V \to R \otimes V$ is R-linear. Moreover, g_0 is required to be the identity. The extension of any $g \in \mathbf{G}(R)$ to an automorphism of $R[[z]][z^{-1}] \otimes V$ is also denoted by g.

We now define another functor $\mathcal{G}$ by letting $\mathcal{G}(R)$ be the set of tuples (g, δ) with $g \in \mathbf{G}(R)$ such that the restriction of the differential $g\delta_0 g^{-1} : R[[z]] \otimes V \to R[[z]][z^{-1}] \otimes V$ maps V into $R[z^{-1}] \otimes V$. This restriction is denoted by δ.

Lemma 12.12 *The functors* $\mathcal{F}$ *and* $\mathcal{G}$ *from the category of* C*-algebras to the category of sets are isomorphic.*

Proof. Let R be a C-algebra. An element of $\mathcal{F}(R)$ is the equivalence class of some $(\mathcal{M}, \nabla, \phi)$. A representative for this equivalence class is chosen by taking for $\mathcal{M}$ the trivial vector bundle $O_{P^1(R)} \otimes V$ and requiring that ϕ modulo (z) is the identity. Thus ϕ is an $R[[z]]$-linear automorphism of $R[[z]] \otimes V$ and the identity modulo (z). Furthermore, $\nabla_{z\frac{d}{dz}}$ is equal to $\phi^{-1}\delta_0\phi$. By assumption, $\nabla : R \otimes V \to H^0(P^1(C), \Omega((k+1)\cdot[0]+[\infty])) \otimes_C (R \otimes V)$. This implies that the image of V under $\nabla_{z\frac{d}{dz}}$ lies in $R[z^{-1}] \otimes V$ and therefore $(\phi^{-1}, \delta) \in \mathcal{G}(R)$, where $\delta = \phi^{-1}\delta_0\phi$. In this way, one obtains a map $\mathcal{F}(R) \to \mathcal{G}(R)$ and, in fact, a morphism of functors $\mathcal{F} \to \mathcal{G}$. It is easily seen that the map $\mathcal{F}(R) \to \mathcal{G}(R)$ is bijective for every R. □

Now we proceed by proving that the functor $\mathcal{G}$ is representable.

Lemma 12.13 *Let $(g, \delta) \in \mathcal{G}(R)$. Then g is uniquely determined by δ.*

Proof. Suppose that $(g_1, \delta), (g_2, \delta) \in \mathcal{G}(R)$. Then there exists $h \in \mathbf{G}(R)$ (i.e., h is an $R[[z]]$-linear automorphism h of $R[[z]] \otimes V$, which is the identity modulo (z)) such that $h\delta_0 = \delta_0 h$. It suffices to show that $h = 1$.

We introduce some notations. $R((z))$ will denote $R[[z]][z^{-1}]$. A "linear map" will mean linear with respect to the ring $R((z))$. For a linear map $L : R((z)) \otimes V \to R((z)) \otimes V$ one writes $L = (L_{ji})$ where the $L_{ji} : R((z)) \otimes V_i \to R((z)) \otimes V_j$ are again linear maps. For a linear map L_{ji} one writes L'_{ji} for the linear map with matrix (w.r.t. bases of V_i and V_j) obtained by applying $' = z\frac{d}{dz}$ to all the coefficients of the matrix of L_{ji}. Furthermore, $z\frac{d}{dz} : R((z)) \otimes V \to R((z)) \otimes V$ denotes the obvious derivation, i.e., this derivation is 0 on V. Then clearly $L'_{ji} = z\frac{d}{dz} \circ L_{ji} - L_{ji} \circ z\frac{d}{dz}$. Write the prescribed δ_0 as $z\frac{d}{dz} + L$ where $L = (L_{ji})$ is linear. According to the definition of N_0 one has $L_{ji} = 0$ if $i \neq j$ and $L_{ii} = q_i + l_i$. Write, as above, $h = (h_{ji})$. Then $\delta_0 h - h\delta_0 = 0$ implies that

$$h'_{ji} + h_{ji}l_i - l_j h_{ji} + (q_i - q_j)h_{ji} = 0 \text{ for all } i, j.$$

Suppose that $h_{ji} \neq 0$ for some $i \neq j$. Let n be maximal such that $h_{ji} \equiv 0$ modulo (z^n). One finds the contradiction $(q_i - q_j)h_{ji} \equiv 0$ modulo (z^n). So $h_{ji} = 0$ for $i \neq j$.

For $i = j$ one finds $h'_{ii} + h_{ii}l_i - l_i h_{ii} = 0$. Write $h_{ii} = \sum_{n\geq 0} h_{ii}(n)z^n$, where $h_{ii}(n) : R \otimes V_i \to R \otimes V_i$ are R-linear maps. Then $nh_{ii}(n) + h_{ii}(n)l_i - l_i h_{ii}(n) = 0$ for all $n \geq 0$. The assumption on the eigenvalues of l_i implies that a nonzero difference of eigenvalues cannot be an integer. This implies that the maps $\mathrm{End}(R \otimes V_i) \to \mathrm{End}(R \otimes V_i)$, given by $A \mapsto nA + Al_i - l_i A$, are bijective for all $n > 0$. Hence $h_{ii}(n) = 0$ for $n > 0$. Since h is the identity modulo z we also have that all $h_{ii}(0)$ are the identity. Hence $h = 1$. □

We introduce now the concept of *principal parts*. The principal part $\mathrm{Pr}(f)$ of $f = \sum r_n z^n \in R((z))$ is defined as $\mathrm{Pr}(f) := \sum_{n\leq 0} r_n z^n$. Let $L : R((z)) \otimes V$

$\to R((z)) \otimes V$ be $R((z))$-linear. Choose a basis $\{v_1, \ldots, v_m\}$ of V and consider the matrix of L with respect to this basis given by $Lv_i = \sum_j \alpha_{j,i} v_j$. Then the principal part $\mathrm{Pr}(L)$ of L is the $R((z))$-linear map defined by $\mathrm{Pr}(L)v_i = \sum_j \mathrm{Pr}(\alpha_{j,i})v_j$. It is easily seen that the definition of $\mathrm{Pr}(L)$ does not depend on the choice of this basis. Any derivation δ of $R((z)) \otimes V$ has the form $z\frac{d}{dz} + L$ where L is an $R((z))$-linear map. The principal part $\mathrm{Pr}(\delta)$ of δ is defined as $z\frac{d}{dz} + \mathrm{Pr}(L)$.

Lemma 12.14 *To every $g \in \mathbf{G}(R)$ one associates the derivation $\mathrm{Pr}(g\delta_0 g^{-1})$. Let $\mathbf{H}(R)$ denote the subset of $\mathbf{G}(R)$ consisting of the elements h such that $\mathrm{Pr}(h\delta_0 h^{-1}) = \delta_0$. Then:*

1. *$\mathbf{H}(R)$ is a subgroup of $\mathbf{G}(R)$. Let $d_{i,j}$ denote the degree of $q_i - q_j$ with respect to the variable z^{-1}. Then $g \in \mathbf{G}(R)$ belongs to $\mathbf{H}(R)$ if and only if $g - 1$ maps each V_i into $\oplus_{j=1}^{s} z^{d_{i,j}+1} R[[z]] \otimes V_j$.*
2. *$\mathrm{Pr}(g_1\delta_0 g_1^{-1}) = \mathrm{Pr}(g_2\delta_0 g_2^{-1})$ if and only if $g_1\mathbf{H}(R) = g_2\mathbf{H}(R)$.*
3. *For every differential module $(R((z)) \otimes V, \delta)$ such that $\mathrm{Pr}(\delta) = \delta_0$ there is a unique $h \in \mathbf{H}(R)$ with $h\delta_0 h^{-1} = \delta$.*

Proof. 1. For $g \in \mathbf{G}(R)$ one defines (a "remainder") $\mathrm{Rem}(g\delta_0 g^{-1})$ by the formula $g\delta_0 g^{-1} = \mathrm{Pr}(g\delta_0 g^{-1}) + \mathrm{Rem}(g\delta_0 g^{-1})$. Hence, $\mathrm{Rem}(g\delta_0 g^{-1})$ is linear and maps V into $zR[[z]] \otimes V$. For any $g_1, g_2 \in \mathbf{G}(R)$ we also have that $g_1 \mathrm{Rem}(g_2\delta_0 g_2^{-1}) g_1^{-1}$ maps V into $zR[[z]] \otimes V$ and so $\mathrm{Pr}(g_1(\mathrm{Rem}(g_2\delta_0 g_2^{-1})g_1^{-1}) = 0$. Hence, $\mathrm{Pr}((g_1g_2)\delta_0(g_1g_2)^{-1}) = \mathrm{Pr}(g_1\mathrm{Pr}(g_2\delta_0 g_2^{-1})g_1^{-1})$. This formula easily implies that $\mathbf{H}(R)$ is a subgroup of $\mathbf{G}(R)$.

Let $g \in \mathbf{G}(R)$ and write $g - 1 := (L_{i,j})$, where $L_{i,j}$ is a $R[[z]]$-linear map $R[[z]] \otimes V_j \to R[[z]] \otimes V_i$. The condition $g \in \mathbf{H}(R)$ is equivalent to the condition that $g\delta_0 - \delta_0 g$ maps V into $zR[[z]] \otimes V$. The last condition means that (for all i, j) the map $L_{i,j}\delta_0 - \delta_0 L_{i,j}$ maps V_j into $zR[[z]] \otimes V_i$. This is seen to be equivalent to $(q_j - q_i)L_{i,j}$ maps V_j into $zR[[z]] \otimes V_i$ or equivalently $L_{i,j}V_j \subset z^{d_{i,j}+1}R[[z]] \otimes V_i$.

2. $\mathrm{Pr}(g_1\delta_0 g_1^{-1}) = \mathrm{Pr}(g_2\delta_0 g_2^{-1})$ is equivalent to the condition that $g_1\delta_0 g_1^{-1} - g_2\delta_0 g_2^{-1}$ maps $R[[z]] \otimes V$ into $zR[[z]] \otimes V$. The latter is equivalent to the condition that $g_2^{-1} g_1 \delta_0 g_1^{-1} g_2 - \delta_0$ maps $R[[z]] \otimes V$ into $zR[[z]] \otimes V$. This is again the same as $\mathrm{Pr}(g_2^{-1} g_1 \delta_0 g_1^{-1} g_2) = \delta_0$. The last statement translates into $g_1\mathbf{H}(R) = g_2\mathbf{H}(R)$.

3. Suppose now that $\mathrm{Pr}(\delta) = \delta_0$. Then we try to solve $h\delta_0 h^{-1} = \delta$ with $h \in \mathbf{H}(R)$. From the step-by-step construction that we will give, the uniqueness of h will also follow. We remark that the uniqueness is also a consequence of Lemma 12.13. The problem is equivalent to solving $h\delta_0 h^{-1} - \delta_0 = M$ for any $R[[z]]$-linear map $M : R[[z]] \otimes V \to zR[[z]] \otimes V$. This is again equivalent to solving $h\delta_0 - \delta_0 h = Mh$ modulo z^N for all $N \geq 1$. For $N = 1$, a solution is $h = 1$. Let a solution h_{N-1} modulo z^{N-1} be given. Then $h_{N-1}\delta_0 - \delta_0 h_{N-1} = Mh_{N-1} + z^{N-1}S$ with $S : R[[z]] \otimes V \to R[[z]] \otimes V$. Consider a candidate $h_N = h_{N-1} + z^{N-1}T$ for a solution modulo

z^N with T given in block form $(T_{j,i})$ by maps $T_{j,i} : R[[z]] \otimes V_i \to z^{d_{j,i}} R[[z]] \otimes V_j$. Then we have to solve $T\delta_0 - \delta_0 T - (N-1)T = -S$ modulo z. The linear map $T\delta_0 - \delta_0 T - (N-1)T$ has block form $(-(z\frac{d}{dz})(T_{j,i}) + T_{j,i}l_j - l_i T_{j,i} - (N-1)T_{j,i}$ $+(q_j - q_i)T_{j,i})$. Let the constant map $L_{j,i}$ be equivalent to $z^{-d_{j,i}}T_{j,i}$ modulo z and let $c_{j,i}$ be the leading coefficient of $q_j - q_i$ (for $j \neq i$). Then for $i \neq j$ the block for the pair j, i is modulo z congruent to $c_{j,i}L_{j,i}$. The block for the pair i, i is modulo z equivalent to $L_{i,i}l_i - l_i L_{i,i} - (N-1)L_{i,i}$. Since the nonzero differences of the eigenvalues of l_i are not in $\mathbf{Z}$, the map $A \in \mathrm{End}(V_i) \mapsto (Al_i - l_i A - (N-1)A)$ $\in \mathrm{End}(V_i)$ is bijective. We conclude from this that the required T exists. This shows that there is an element $h \in \mathbf{H}(R)$ with $h\delta_0 h^{-1} = \delta$. □

Corollary 12.15 *1. The functors $R \mapsto \mathbf{G}(R)/\mathbf{H}(R)$ and $\mathcal{G}$ are isomorphic.*
2. The functor $\mathcal{F}$ is representable by the affine space $\mathbf{A}_C^N$, where $N = \sum_{i\neq j} \deg_{z^{-1}}$ $(q_i - q_j) \cdot \dim V_i \cdot \dim V_j$.

Proof. 1. Define the map $\alpha_R : \mathbf{G}(R)/\mathbf{H}(R) \to \mathcal{G}(R)$ by by $g \mapsto (\tilde{g}, \mathrm{Pr}(g\delta_0 g^{-1}))$, where $\tilde{g} = gh$ with $h \in \mathbf{H}(R)$ the unique element with $h\delta_0 h^{-1} = \delta :=$ $g^{-1}\mathrm{Pr}(g\delta_0 g^{-1})g = \delta_0 - R(g\delta_0 g^{-1})$. From Lemma 12.14, α_R is a bijection. Moreover, α_R depends functorially on R.

2. The coset $\mathbf{G}(R)/\mathbf{H}(R)$ has as set of representatives the gs of the form $g = 1 + L$ with $L = (L_{j,i})$, where $L_{i,i} = 0$ and $L_{j,i}$, for $i \neq j$, is an R-linear map $R \otimes V_i$ $\to Rz \otimes V_j \oplus Rz^2 \otimes V_j \oplus \cdots \oplus Rz^{d_{i,j}} \otimes V_j$. Thus the functor $R \mapsto \mathbf{G}(R)/\mathbf{H}(R)$ is represented by the affine space $\oplus_{i\neq j}\mathrm{Hom}(V_i, V_j)^{d_{i,j}}$. □

We note that Theorem 12.4 and Exercise 12.5 are special cases of Corollary 12.15.

12.5 The Ramified Case

Let (N, ∇) be a connection over $\hat{K} = C((z))$. We define $\delta : N \to N$ by $\delta = \nabla_{z\frac{d}{dz}}$. For any integer $e \geq 1$ we write $\hat{K}_e = C((t))$ with $t^e = z$. The ramification index of N is defined as the smallest integer $e \geq 1$ such that $M := \hat{K}_e \otimes N$ is unramified as defined in Sect. 12.4. The idea of the construction of the moduli space for the ramified case given by N (or rather given by some lattice $N_0 \subset N$) is the following. One considers for the unramified case M over $C((t))$ a suitable lattice M_0 on which the Galois group of $C((t))/C((z))$ operates. For the ramified case one chooses for the lattice N_0 the invariants of the lattice M_0 under the action of the Galois group. Then one has two moduli functors, namely $\mathcal{F}$ for N_0 and $\tilde{\mathcal{F}}$ for M_0. The second functor is, according to Sect. 12.4, representable by some $\mathbf{A}_C^N$. Moreover, the Galois group of $C((t))/C((z))$ acts on $\tilde{\mathcal{F}}$ and its moduli space. A canonical isomorphism $\mathcal{F}(R) \to \tilde{\mathcal{F}}(R)^{inv}$, where ^{inv} means the invariants under this Galois group and R is any C-algebra, shows that $\mathcal{F}$ is representable by the $(\mathbf{A}_C^N)^{inv}$. The latter space turns out to be isomorphic with $\mathbf{A}_C^M$ for some integer $M \geq 1$. Although the functors $\mathcal{F}$

and $\tilde{\mathcal{F}}$ are essentially independent of the chosen lattices, a rather delicate choice of the lattices is needed in order to make this proof work.

We will now describe how one makes this choice of lattices and give a fuller description of the functors.

The decomposition $M = \oplus_{i=1}^{s} E(q_i) \otimes M_i$, with distinct $q_1, \ldots, q_s \in t^{-1}C[t^{-1}]$, $E(q_i) = \hat{K}_e e_i$ with $\delta e_i = q_i e_i$ and M_i regular singular, is unique. We fix a set of representatives of $C/(\frac{1}{e}\mathbf{Z})$. Then each M_i can uniquely be written as $\hat{K}_e \otimes_C V_i$, where V_i is a finite dimensional vector space over C and such that $\delta(V_i) \subset V_i$ and the eigenvalues of the restriction of δ to V_i lie in this set of representatives. The uniqueness follows from the description of V_i as the direct sum of the generalized eigenspaces of δ on M_i taken over all the eigenvalues belonging to the chosen set of representatives.

Fix a generator σ of the Galois group of $\hat{K}_e/\hat{K}$ by $\sigma(t) = \zeta t$ and ζ a primitive e-th root of unity. Then σ acts on M in the obvious way and commutes with the δ on M. Furthermore, $\sigma(fm) = \sigma(f)\sigma(m)$ for $f \in \hat{K}_e$, $m \in M$. Thus σ preserves the above decomposition. In particular, if $\sigma(q_i) = q_j$ then $\sigma(E(q_i) \otimes M_i) = E(q_j) \otimes M_j$. We make the convention that σ is the bijection from $E(q_i)$ to $E(q_j)$, which maps e_i to e_j. Using this convention one defines the map $L_{j,i} : M_i \to M_j$ by $\sigma(e_i \otimes m_i) = e_j \otimes L_{i,j}(m_i)$. It is easily seen that $L_{i,j}$ commutes with the δs and $L_{j,i}(fm_i) = \sigma(f)L_{j,i}(m_i)$. From the description of V_i and V_j it follows that $L_{j,i}(V_i) = V_j$.

We note that $L_{j,i}$ need not be the identity if $q_i = q_j$. The reason for this is that $C/(\mathbf{Z})$ and $C/(\frac{1}{e}\mathbf{Z})$ do not have the same set of representatives. In particular, a regular singular differential module N over $\hat{K}$ and a set of representatives of $C/(\mathbf{Z})$ determines an isomorphism $N \cong \hat{K} \otimes W$. The extended module $M = \hat{K}_e \otimes N$ is isomorphic to $\hat{K}_e \otimes W$, but the eigenvalues of δ on W may differ by elements in $\frac{1}{e}\mathbf{Z}$. Thus for the isomorphism $M = \hat{K}_e \otimes V$ corresponding to a set of representatives of $C/(\frac{1}{e}\mathbf{Z})$ one may have that $V \neq W$.

We can summarize the above as follows: The extended differential module $M := \hat{K}_e \otimes N$ is given by the following data:

(a) Distinct elements $q_1, \ldots, q_s \in t^{-1}C[t^{-1}]$.
(b) Finite-dimensional vector spaces $V_1, \ldots, V_s$ and linear maps $l_i : V_i \to V_i$ such that the eigenvalues of l_i lie in a set of representatives of $C/(\frac{1}{e}\mathbf{Z})$.
(c) σ permutes the set $\{q_1, \ldots, q_s\}$ and for every pair i, j with $\sigma q_i = q_j$, there is given a C-linear bijection $\sigma_{j,i} : V_i \to V_j$ such that $\sigma_{j,i} \circ l_i = l_j \circ \sigma_{j,i}$.

The data define a lattice $M_0 = \oplus C[[t]]e_i \otimes V_i$ in the differential module M, with $\delta e_i \otimes v_i = q_i e_i \otimes v_i + e_i \otimes l_i(v_i)$ such that $\delta fm = f\delta m + 1/e \cdot t\frac{df}{dt}m$. Furthermore, the data define an automorphism on M_0, also denoted by σ, which has the properties: $\sigma(fm) = \sigma(f)\sigma(m)$ and if $\sigma(q_i) = q_j$, then $\sigma(e_i \otimes v_i) = e_j \otimes \sigma_{j,i}v_i$.

We consider now the lattice $N_0 = M_0^\sigma$, i.e., the elements invariant under the action of σ, in the differential module N over $\hat{K}$. We will call this the *standard ramified case*.

Again we consider the moduli problem for connections $(\mathcal{N}, \nabla, \psi)$ on $P^1(C)$; $\mathcal{N}$ a free vector bundle; the connection $(\mathcal{N}, \nabla)$ with the two singular points $0, \infty$; the point ∞ regular singular; $\psi : C[[z]] \otimes \mathcal{N}_0 \to N_0$ an isomorphism compatible with the two connections. This defines the functor $\mathcal{F}$ on the category of the C-algebras, that we want to represent by an affine space over C.

Let $X \to P^1(C)$ denote the covering of $P^1(C)$ given by $t^e = z$. We consider above X the moduli problem (of the unramified case): tuples $(\mathcal{M}, \nabla, \phi)$ with a free vector bundle $\mathcal{M}$; a connection $(\mathcal{M}, \nabla)$ with singularities at 0 and ∞; the singularity at ∞ is regular singular; furthermore, an isomorphism $\phi : C[[t]] \otimes \mathcal{M}_0 \to M_0$. This defines a functor $\tilde{\mathcal{F}}$ on the category of the C-algebras. The important observation is that σ acts canonically on $\tilde{\mathcal{F}}(R)$. Indeed, an element $(\mathcal{M}, \nabla, \phi) \in \tilde{\mathcal{F}}$ is given by R-linear maps $\nabla : H^0(X \otimes R, \mathcal{M}) \to H^0(X, \Omega(k \cdot [0] + [\infty])) \otimes H^0(X \otimes R, \mathcal{M})$ and $\phi : H^0(X \otimes R, \mathcal{M}) \to R[[t]] \otimes M_0$ having some compatibility relation. One defines $\sigma(\mathcal{M}, \nabla, \phi) = (\mathcal{M}, \nabla, \sigma \circ \phi)$.

Lemma 12.16 *There is a functorial isomorphism $\mathcal{F}(R) \to \tilde{\mathcal{F}}(R)^\sigma$.*

Proof. We mean by $\tilde{\mathcal{F}}(R)^\sigma$ the set of σ-invariant elements. For convenience we will identify $e_i \otimes V_i$ with V_i. Put $V = \oplus V_i$, then $M_0 = C[[t]] \otimes V$. The map σ on V has eigenvalues $1, \zeta, \ldots, \zeta^{e-1}$. Let $V = \oplus_{i=0}^{e-1} V(i)$ be the decomposition in eigenspaces. Put $W := V(0) \oplus t^{e-1}V(1) \oplus t^{e-2}V(2) \oplus \cdots \oplus tV(e-1)$. Then one has $N_0 = C[[z]] \otimes W$.

The functor $\mathcal{F}$ is "normalized" by identifying $\mathcal{N}$ with $O_{P^1(R)} \otimes W$ and by requiring that ψ_0 is the identity. The same normalization will be made for $\tilde{\mathcal{F}}$. We start now by defining the map $\mathcal{F}(R) \to \tilde{\mathcal{F}}(R)^\sigma$. For notational convenience we will omit the C-algebra R in the notations. An element on the left-hand side is given by $\nabla : W \to H^0(\Omega(k[0] + [\infty])) \otimes W$ and a sequence of linear maps $\psi_n : W \to W$ with $\psi_0 = id$, satisfying some compatibility condition. The isomorphism $\psi : C[[z]] \otimes W \to N_0$ extends to a $C[[t]]$-linear map $C[[t]] \otimes W \to C[[t]] \otimes_{C[[z]]} N_0 \subset M_0$. Call this map also ψ. Then ψ maps W identically into the subset $W \subset N_0 \subset M_0$. The latter W has been written as a direct sum $\oplus_{i=0}^{e-1} t^{e-i} V(i)$. On the left-hand side one can embed $C[[t]] \otimes W$ into $C[[t]] \otimes V$ (with V as above) and extend ψ uniquely to an isomorphism $\phi : C[[t]] \otimes V \to M_0$ such that ϕ_0 is the identity. The $\nabla : W \to H^0(P^1(C), \Omega(k[0] + [\infty])) \otimes W$ extends in a unique way to a $\nabla : V \to H^0(X, \Omega_X(e \cdot k \cdot [0] + [\infty])) \otimes V$ such that the compatibility relations hold. Moreover, one observes that the element in $\tilde{\mathcal{F}}(R)$ that we have defined is invariant under σ.

On the other hand, starting with a σ-invariant element of $\tilde{\mathcal{F}}(R)$ one has a σ-equivariant isomorphism $\phi : C[[t]] \otimes V \to M_0$ with $\phi_0 = id$. After taking invariants one obtains an isomorphism $\psi : C[[z]] \otimes W \to N_0$, with $\psi_0 = id$. The given ∇ induces a $\nabla : W \to H^0(P^1(C), \Omega(k[0] + [\infty])) \otimes W$. In total, one has defined an element of $\mathcal{F}(R)$. The two maps that we have described depend in a functorial way on R and are each other's inverses. □

Corollary 12.17 *There is a fine moduli space for the standard ramified case. This space is the affine space* $\mathbf{A}^N_{\mathbf{C}}$, *with* N *equal to* $\sum_{i \neq j} \deg_{z^{-1}}(q_i - q_j) \cdot \dim V_i \cdot \dim V_j$.

Proof. We keep the above notations. The functor $\tilde{\mathcal{F}}$ is represented by the affine space $\oplus_{i \neq j} \mathrm{Hom}(V_i, V_j)^{d_{i,j}}$, where $d_{i,j}$ is the degree of $q_i - q_j$ with respect to the variable t^{-1}. On this space σ acts in a linear way. The standard ramified case is represented by the σ-invariant elements. From the description of the σ-action on $\oplus V_i$ and the last lemma the statement follows. □

Example 12.18 Take $e = 2$, $t^2 = z$ and M_0 the $C[[t]]$-module generated by e_1, e_2. The derivation δ_0 is given by $\delta_0 e_1 = t^{-1} e_1$ and $\delta_0 e_2 = -t^{-1} e_2$. Let σ be the generator of the Galois group of $C((t))/C((z))$. We let σ act on M_0 by interchanging e_1 and e_2. Thus σ commutes with δ_0. Then $N_0 = M_0^\sigma$ is the $C[[z]]$-module generated by $f_1 = e_1 + e_2,\ f_2 = t(e_1 - e_2)$. The action of δ_0 with respect to this basis is equal to $z\frac{d}{dz} + Ez^{-1} + B$, where E, B are the matrices $\begin{pmatrix} 0 & 0 \\ 1 & 0 \end{pmatrix}$ and $\begin{pmatrix} 0 & 1 \\ 0 & 1/2 \end{pmatrix}$ The universal object for the unramified case is given in matrix form by

$$\delta = z\frac{d}{dz} + t^{-1}\begin{pmatrix} 1 & 0 \\ 0 & -1 \end{pmatrix} + \begin{pmatrix} 0 & a \\ b & 0 \end{pmatrix}.$$

The action of σ on the universal object permutes a and b. Thus the universal σ-invariant object is

$$\delta = z\frac{d}{dz} + t^{-1}\begin{pmatrix} 1 & 0 \\ 0 & -1 \end{pmatrix} + \begin{pmatrix} 0 & a \\ a & 0 \end{pmatrix}.$$

This δ has, with respect to the basis f_1, f_2, the matrix form

$$\delta = z\frac{d}{dz} + z^{-1}\begin{pmatrix} 0 & 0 \\ 1 & 0 \end{pmatrix} + \begin{pmatrix} a & 1 \\ 0 & 1/2 - a \end{pmatrix}.$$

For $a = 0$ one has, of course, the standard module in the ramified case. The above differential operator is the universal family above the moduli space, which is $\mathbf{A}^1_{\mathbf{C}}$. □

12.6 The Meromorphic Classification

Let C be the field of complex numbers $\mathbf{C}$. We consider a moduli functor $\mathcal{F}$ associated to a formal differential module (N_0, ∇_0) as in Sect. 12.4 or 12.5. Its fine moduli space is denoted by $\mathbf{A}^N_{\mathbf{C}}$. The meromorphic classification, attached to $\hat{K} = \mathbf{C}((z))$, is described by the cohomology set $H^1(\mathbf{S}^1, STS)$ or equivalently by the set of Stokes matrices. One identifies, as before, $H^1(\mathbf{S}^1, STS)$ with $\mathbf{C}^N$.

Theorem 12.19 *The canonical map* $\alpha : \mathbf{A}^N_{\mathbf{C}} \to H^1(\mathbf{S}^1, STS) \cong \mathbf{C}^N$ *is complex analytic. The image of* α *contains the Zariski-open subset of* $\mathbf{C}^N$ *consisting of the points* ξ *for which the topological monodromy has* m *distinct eigenvalues. The fibre of a point* $\xi \in \mathbf{C}^N$*, such that its topological monodromy has* m *distinct eigenvalues, is a discrete infinite subset of* $\mathbf{A}^N_{\mathbf{C}}$.

Proof. The map α is defined as in Sect. 12.3.2 and associates to a $\mathbf{C}$-valued point of $\mathbf{A}^N_{\mathbf{C}}$, represented by $(\mathcal{M}, \nabla, \phi)$, the pair (M, ψ), where $M := \mathbf{C}(\{z\}) \otimes H^0(P^1(\mathbf{C}), \mathcal{M})$ with the connection induced by ∇ and where ψ is the isomorphism $\mathbf{C}((z)) \otimes M \to \mathbf{C}((z)) \otimes N_0$ induced by ϕ. Write U for the algebra of regular functions on $\mathbf{A}^N_{\mathbf{C}}$ and write $(g_u, \delta_u) \in \mathcal{G}(U)$ for the universal element. Then $\delta_0 = z\frac{d}{dz} + A_0$ and $\delta_u = z\frac{d}{dz} + A$, where the matrices A_0 and A have coordinates in $\mathbf{C}[z^{-1}]$ and $U[z^{-1}]$. Furthermore, g_u is a formal solution of the differential equation $z\frac{d}{dz}(g_u) + Ag_u - g_u A_0 = 0$. Let d be a nonsingular direction of this differential equation. Multisummation yields a unique lift $S_d(g_u)$ of g_u valid in a fixed sector S around d. Suppose that one knows that $S_d(g_u)$ is an analytic function on $S \times \mathbf{A}^N_{\mathbf{C}}$. Consider now a singular direction d. Then $S_{d^+}(g_u)$ and $S_{d^-}(g_u)$ are both analytic functions on $S \times \mathbf{A}^N_{\mathbf{C}}$ (where S is a suitable sector around the direction d). Then it follows that the Stokes matrix for direction d is an analytic function on $\mathbf{A}^N_{\mathbf{C}}$. One concludes that α is an analytic map. The other statements of the theorem follow from the arguments given in Sect. 12.3.2. □

Thus the theorem is a consequence of the following result in the theory of multisummation.

Proposition 12.20 (B.L.J. Braaksma) *Let* $\underline{x}$ *denote a set of* n *variables. Consider a matrix differential equation*

$$z\frac{d}{dz}y - Ay = h, \text{ where } A \text{ and } h \text{ have coefficients in } \mathbf{C}[z^{-1}, \underline{x}].$$

Let a formal solution $\hat{f}$ *that has coefficients in* $\mathbf{C}[\underline{x}][[z]]$ *be given and assume that* $z\frac{d}{dz} - A$ *is equivalent, via a* $g \in \mathrm{GL}(m, \mathbf{C}[\underline{x}][[z]])$ *such that* g *is the identity modulo* z*, with a (standard) differential equation over* $\mathbf{C}[z^{-1}]$ *(not involving* $\underline{x}$*). Let* d *be a nonsingular direction for* $z\frac{d}{dz} - A$ *and* S *the fixed sector with bisector* d*, given by the multisummation process.*

Then the multisum $S_d(\hat{f})(z, \underline{x})$ *in the direction* d *is holomorphic on* $S \times \mathbf{C}^n$.

Proof. It suffices to prove that $S_d(\hat{f})(z, \underline{x})$ depends locally holomorphically on $\underline{x}$. This means that we must verify that $S_d(\hat{f})(z, \underline{x})$ is holomorphic on $S \times \{a \in \mathbf{C}^n | \ \|a\| < \epsilon\}$ for the required sector and some positive ϵ. The analytic way to produce the multisummation S_d by formal Borel and Laplace integrals (see Example 7.45 and Remarks 7.61) will imply the required result without too much extra effort. Indeed, the various Borel and Laplace transforms of $\hat{f}$ are given by integrals and these integrals depend locally holomorphically on $\underline{x}$. In our more algebraic setting of multisummation, we will have to show that after each step in the construction the result depends locally holomorphically on $\underline{x}$. We only sketch the procedure.

The Main Asymptotic Existence Theorem (Theorem 7.10) has to be adapted to the case of parameters $\underline{x}$. For this, one considers the scalar equation $(\delta - q)\hat{f} = g$ with $q \in z^{-1}\mathbf{C}[z^{-1}]$ and $g = g(z, \underline{x})$ depending holomorphically on $\underline{x}$. A version of the Borel-Ritt Theorem (Theorem 7.3) with parameters that can be applied to $\hat{f}$ and this reduces the problem to the special case where g is flat, uniformly in $\underline{x}$ in some neighborhood of $0 \in \mathbf{C}^n$. One then extends Lemma 7.13 to the case of parameters. A somewhat tedious calculation shows that the estimates of the integrals, involved in the proof of Lemma 7.13, hold uniformly for $\underline{x}$ in a neighborhood of $0 \in \mathbf{C}^n$. A similar verification can be done for the proof of Lemma 7.17. The conclusion is that Theorem 7.10 holds for the case of parameters. We therefore have that $\hat{f}$ has asymptotic lifts f_i with respect to some open cover $\{S_i\}_{i \in I}$ of $\mathbf{S}^1$ and, furthermore, that these lifts depend holomorphically on $\underline{x}$. This induces a 1-cocycle $\xi = \{f_i - f_j\}$ for the sheaf $K_A = \ker(\delta - A, (\mathcal{A}^0_{1/k})^m)$ and the open cover $\{S_i\}$ of $\mathbf{S}^1$ and that this cocycle depends holomorphically on $\underline{x}$ (see Lemma 7.40).

It is given that $\delta - A$ is equivalent, by a transformation $g \in \mathrm{GL}_m(\mathbf{C}[\underline{x}][[z]])$ with $g \equiv 1 \bmod z$, to $\delta - B$ where B is independent of $\underline{x}$. For convenience we suppose that $\delta - A$ has only one positive slope k. One can verify that Lemma 7.41 remains valid for our case of parameters. This means that for a sector $S = (d - \alpha, d + \alpha)$ with d not a singular direction and some $\alpha > \frac{\pi}{2k}$ the sheaf K_A is isomorphic to $K_B \otimes_{\mathbf{C}} O$, where O denotes the ring of holomorphic functions on $\{a \in \mathbf{C}^n \mid \|a\| < \epsilon\}$. Both $H^0(S, K_B)$ and $H^1(S, K_B)$ are zero. Therefore, the restriction of the 1-cocycle ξ to S is the image of a (unique) element $\eta = \{\eta_i\}$ in $\prod_i K_A(S \cap S_i)$ (depending holomorphically on $\underline{x}$). The new choice of lifts $\{f_i - \eta_i\}$ for the cover $\{S \cap S_i\}$ of S glue to together to form the k-sum $S_d(\hat{F})$ on S. Thus $S_d(\hat{f})$ depends holomorphically on $\underline{x}$. The general case, involving more than one positive slope, can be handled in the same way (and with some more effort). □

13 Positive Characteristic

Linear differential equations over differential fields of characteristic $p > 0$ have been studied for a long time ([139, 152, 153, 8],...). Grothendieck's conjecture on p-curvatures is one of the motivations for this. Another motivation is the observation that for the factorization of differential operators over, say, the differential field $\mathbf{Q}(z)$ the reductions modulo prime numbers yield useful information.

In this chapter we first develop the classification of differential modules over differential fields K with $[K : K^p] = p$. It turns out that this classification is rather explicit and easy. It might be compared with Turrittin's classification of differential modules over $\mathbf{C}((z))$. Algorithms are developed to construct and obtain standard forms for differential modules.

From the viewpoint of differential Galois theory, these linear differential equations in characteristic p do not behave well. A completely different class of equations, namely the "linear iterative differential equations", is introduced. These equations have many features in common with linear differential equations in characteristic 0. We will give a survey and explain the connection with p-adic differential equations.

13.1 Classification of Differential Modules

In this chapter, K denotes a field of characteristic $p > 0$ satisfying $[K : K^p] = p$. The universal differential module Ω_K of K has dimension 1 over K. Indeed, choose an element $z \in K \setminus K^p$. Then K has basis $1, z, \dots, z^{p-1}$ over K^p and this implies that $\Omega_K = Kdz$. Let $f \to f' = \frac{df}{dz}$ denote the derivation given by

$$(a_0 + a_1 z + \cdots + a_{p-1} z^{p-1})' = a_1 + 2a_2 z + \cdots + (p-1)a_{p-1} z^{p-2},$$

for any $a_0, \dots, a_{p-1} \in K^p$. Every derivation on K has the form $g\frac{d}{dz}$ for a unique $g \in K$.

Examples of fields K with $[K : K^p] = p$ are $k(z)$ and $k((z))$ with k a perfect field of charactersitic p. We note that any separable algebraic extension $L \supset K$ of a field with $[K : K^p] = p$ again has the property $[L : L^p] = p$.

A connection over the field K is a pair (∇, M) where M is a finite dimensional vector space over K and where $\nabla : M \to \Omega_K \otimes M$ is an additive map satisfying the usual rule $\nabla(fm) = df \otimes m + f\nabla(m)$ for $f \in K$ and $m \in M$. One observes that ∇ is determined by $\partial := \nabla_{\frac{d}{dz}} : M \to M$ given by $\partial = \ell \otimes id \circ \nabla$, where $\ell : \Omega_K \to K$ is the K-linear map defined by $\ell(dz) = 1$. In what follows we will fix z, the derivation $f \mapsto f' = \frac{df}{dz}$ and consider instead of connections differential modules (M, ∂) defined by: M is a finite dimensional vector space over K and $\partial : M \to M$ is an additive map satisfying $\partial(fm) = f'm + f\partial m$ for $f \in K$ and $m \in M$.

Our aim is to show that, under some condition on K, the tannakian category Diff_K of all differential modules over K is equivalent to the tannakian category $\mathrm{Mod}_{K^p[T]}$ of all $K^p[T]$-modules, which are finite dimensional as vector spaces over the field K^p. The objects of the latter category can be given as pairs (N, t_N) consisting of a finite dimensional vector space N over K^p and a K^p-linear map $t_N : N \to N$. The tensor product is defined by $(N_1, t_1) \otimes (N_2, t_2) := (N_1 \otimes_{K^p} N_2, t_3)$, where t_3 is the map given by the formula $t_3(n_1 \otimes n_2) = (t_1 n_1) \otimes n_2 + n_1 \otimes (t_2 n_2)$. We note that this tensor product is not at all the same as $N_1 \otimes_{K^p[T]} N_2$. All further details of the structure of the tannakian category $\mathrm{Mod}_{K^p[T]}$ are easily deduced from the definition of the tensor product.

Exercise 13.1 Let $f \in K^p[T]$ be a *separable* irreducible polynomial and $m \geq 1$. Prove that $K^p[T]/(f^m)$ is isomorphic to $K^p[T]/(f) \otimes_{K^p} K^p[T]/(T^m)$.
Hint: Let e_1, e_2 denote the images of 1 in $K^p[T]/(f)$ and in $K^p[T]/(T^m)$. For any polynomial $Q \in K^p[T]$ one defines a sequence of polynomials $Q_0, Q_1, Q_2, \ldots$ by the formula $Q(T+U) = \sum_{n \geq 0} Q_n(T)U^n$.
Prove that $Q(e_1 \otimes e_2) = \sum_{n=0}^{m-1}(Q_n(T)e_1) \otimes (T^n e_2)$.
Prove that $f^m(e_1 \otimes e_2) = 0$ and that $f^{m-1}(e_1 \otimes e_2) \neq 0$. □

Every object of $\mathrm{Mod}_{K^p[T]}$ is isomorphic to a unique finite direct sum $\oplus(K^p[T]/(f^n))^{m(f,n)}$, where f runs in the set of monic irreducible polynomials in $K^p[T]$ and $n \geq 1$ and $m(f, n) \geq 0$. The classification of the objects in Diff_K follows from the above equivalence of tannakian categories. Let $I(f, n)$ denote the differential module over K corresponding to $K^p[T]/(f^n)$. Then every differential module over K is isomorphic to a finite direct sum $\oplus I(f, n)^{m(f,n)}$ for uniquely determined $m(f, n) \geq 0$. For two differential modules M_1, M_2 over K the set of morphisms $\mathrm{Mor}(M_1, M_2)$ consists of the K-linear maps $\ell : M_1 \to M_2$ satisfying $\ell\partial = \partial\ell$. This is a vector space over K^p and, by the equivalence, equal to $\mathrm{Mor}(N_1, N_2)$ for suitable objects N_1, N_2 of $\mathrm{Mod}_{K^p[T]}$. Using the direct sum decompositions of N_1 and N_2, one easily finds $\mathrm{Mor}(N_1, N_2)$.

The functor $\mathcal{F} : \mathrm{Mod}_{K^p[T]} \to \mathrm{Diff}_K$, which provides the equivalence, will send an object (N, t_N) to an object (M, ∂_M) with $M = K \otimes_{K^p} N$ and ∂_M is chosen such that $\partial_M^p = 1_K \otimes_{K^p} t_N$. At this point there are many details to be explained.

For a differential module (M, ∂_M) one defines its *p-curvature* to be the map $\partial_M^p : M \to M$. The map ∂_M is clearly K^p-linear and thus the same holds for the p-curvature. The p-curvature is, in fact, a K-linear map. An easy way to see this is to consider $\mathcal{D} := K[\partial]$, the ring of differential operators over K. A differential module (M, ∂_M) is the same thing as a left $\mathcal{D}$-module of finite dimension over K. The action of $\partial \in \mathcal{D}$ on M is then ∂_M on M. Now $\partial^p \in \mathcal{D}$ commutes with K and thus ∂_M^p is K-linear. The importance of the p-curvature is already apparent from the next result.

Lemma 13.2 P. Cartier. *Let (M, ∂) be a differential module of dimension d over K. The following statements are equivalent:*
(1) *The p-curvature ∂^p is zero.*
(2) *The differential module (M, ∂) is trivial, i.e., there is a basis $e_1, \ldots, e_d$ of M over K such that $\partial e_j = 0$ for all j.*

Proof. (2)⇒(1) is obvious from $\partial^p e_i = 0$ for all i and ∂^p is K-linear.
Suppose that $\partial^p = 0$. Then the K^p-linear operator ∂ is nilpotent. In particular, there exists a nonzero $e_1 \in M$ with $\partial e_1 = 0$. By induction on the dimension of M one may assume that the differential module M/Ke_1 has a basis $\tilde{e}_2, \ldots, \tilde{e}_d$ with $\partial \tilde{e}_i = 0$ for $i = 2, \ldots, d$. Let $e_2, \ldots, e_d \in M$ be preimages of $\tilde{e}_2, \ldots, \tilde{e}_d$. Then $\partial e_i = f_i e_1$, $i = 2, \ldots, d$ for certain elements $f_i \in K$. Now $\partial^p(e_i) = f_i^{(p-1)} e_1 = 0$. Thus $f_i^{(p-1)} = 0$ and this easily implies that there are elements $g_i \in K$ with $f_i = g_i'$. Now $e_1, e_2 - g_2 e_1, \ldots, e_d - g_d e_1$ is the required basis. □

Much more is true, namely the p-curvature "determines" the differential module M completely. A precise formulation is, in fact, the equivalence of tannakian categories $\mathcal{F} : \mathrm{Mod}_{K^p[T]} \to \mathrm{Diff}_K$ that we want to establish.

Exercise 13.3
(1) Show that a trivial differential module M over K of dimension strictly greater than p has no cyclic vector.
(2) Show that a differential module M of dimension $\leq p$ over K has a cyclic vector. Hint: Try to adapt the proof of Katz to characteristic p. □

It is convenient to regard $Z := K^p[T]$ as a subring of $\mathcal{D}$ by identifying T with $\partial^p \in \mathcal{D}$. An object (N, t_N) is then a left Z-module and $M = K \otimes_{K^p} N$ is a left module over the subring $Z[z] = K \otimes_{K^p} Z$ of $\mathcal{D}$. The aim is to extend this to a structure of left $\mathcal{D}$-module by defining an action of ∂ on M.

Let us consider the simplest case, $N = K^p e$ with $t_N e = ae$ and $a \in K^p$. Then $M = Ke$. For any $b \in K$ one considers Ke as a differential module by $\partial e = be$. Define the element $\tau(b)$ by $\partial^p e = \tau(b)e$. Apply ∂ to both sides of the last equation. This yields $\tau(b)' = 0$ and so $\tau(b) \in K^p$. We are looking for a $b \in K$ such that $\tau(b) = a$. The answer is the following lemma.

Lemma 13.4 (see Lemma 1.4.2 of [227])
(1) $\tau(b) = b^{(p-1)} + b^p$.
(2) *$\tau : K \to K^p$ is additive and has kernel $\{\frac{f'}{f} | f \in K^*\}$.*
(3) *τ is surjective if there are no skew fields of degree p^2 over its center K^p.*

This lemma is the main ingredient for the proof of the following theorem, given in [227].

Theorem 13.5 *Suppose that K has the property that no skew fields of degree p^2 over a center L exist with L a finite extension of K. Then there exists an equivalence $\mathcal{F} : \mathrm{Mod}_{K^p[T]} \to \mathrm{Diff}_K$ of tannakian categories.*

Sketch of the construction of $\mathcal{F}(N)$.
Since $\mathcal{F}$ respects direct sums, it suffices to consider N of the form $K^p[T]/(f^m)$ where f is a monic irreducible polynomial and $m \geq 1$. In the case that f is separable, $K^p[T]/(f^m) \cong K^p[T]/(f) \otimes_{K^p} K^p[T]/(T^m)$ according to Exercise 13.1. Since $\mathcal{F}$ respects tensor products, $\mathcal{F}(K^p[T]/(f^m))$ is isomorphic to $\mathcal{F}(K^p[T]/(f)) \otimes_K \mathcal{F}(K^p[T]/(T^m)$. Now we have to consider the following cases for the $K^p[T]$-module N:

(i) $N = K^p[T]/(T^m)$.
Define the element $c_\infty \in K[[T]]$ by the formula $c_\infty := -z^{-1}\sum_{n\geq 0}(z^pT)^{p^n}$. One easily verifies that $c_\infty^{(p-1)} + c_\infty^p = T$. Consider, for $m \geq 1$ the image c_m of c_∞ in $K[[T]]/(T^m) = K[T]/(T^m)$. Then also $c_m^{(p-1)} + c_m^p = T \mathrm{mod}\ (T^m)$. Thus the differential module $M = K[T]/(T^m)e$ with $\partial e = c_m e$ has the property $\partial^p e = (T\mathrm{mod}(T^m))e$. These explicit formulas justify the definition $\mathcal{F}(Z/(T^m)) := K[T]/(T^m)e$ with $\partial e = c_m e$.

(ii) $N = K^p[T]/(f)$ with f (monic) irreducible and separable.
Write L^p for the field $K^p[T]/(f)$. Then $M = K \otimes_{K^p} N = Le$, where $L = K \otimes_{K^p} L^p = K[T]/(f) = L^p(z)$ is again a field. In this situation $t_N e = ae$, where a is the image of T in L^p, and we search for a $b \in L$ with $\partial_M e = be$ such that $\tau(b) = a$. Lemma 13.4 and the assumption of the theorem guarantee the existence of b. Thus $\mathcal{F}(N) = M$ with $\partial e = be$ for some choice of b. We observe that $\mathcal{F}(N)$ is well defined up to isomorphism. Indeed, according to part (2) of the lemma, any other solution $\tilde{b}$ of the equation $\tau(\tilde{b}) = a$ has the form $\tilde{b} = b + \frac{f'}{f}$. Thus the differential module obtained from the choice $\tilde{b}$ is isomorphic to $\mathcal{F}(N)$ with the choice $\partial e = be$.

(iii) $N = K^p[T]/(f)$ with f monic, irreducible, and inseparable.
Then $K \otimes_{K^p} K^p[T]/(f) = K^p[T]/(f)[X]/(X^p - z^p)$. Since $K^p \subset K^p[T]/(f)$ is inseparable, there exists an element $g \in K^p[T]/(f)$ with $g^p = z^p$. Let s be the image of $X - g$. Then $K \otimes_{K^p} K^p[T]/(f) = K^p[T]/(f)[s]$ with the properties $s^p = 0$ and $s' = 1$. The map $\tau : K^p[T]/(f)[s] \to K^p[T]/(f)$ has the form $\tau(a_0 + a_1 s + \cdots + a_{p-1}s^{p-1}) = -a_{p-1} + a_0^p$. Clearly, τ is surjective. In particular, $\mathcal{F}(N) = K^p[T]/(f)[s]e$ with $\partial e = -as^{p-1}e$, where a is the image of T in $K^p[T]/(f)$. Again $\mathcal{F}(N)$ does not depend, up to isomorphism, on the choice of the solution $\tau(.) = a$.

(iv) $N = K^p[T]/(f^m)$ with f monic, irreducible, inseparable and $m \geq 1$.
For Exercise 13.1 the condition f separable is essential. Therefore, we have to extend (iii) "by hand". Consider $M = K \otimes_{K^p} K^p[T]/(f^m)e$ and let a denote the image of

T in $K^p[T]/(f^m)$. We have to produce an element $b \in K \otimes_{K^p} K^p[T]/(f^m)$ with $\tau(b) = a$. Then M with $\partial(e) = be$ is the definition of $\mathcal{F}(N)$.
Choose a $g \in K^p[T]/(f^m)$ such that $g^p = z^p - fR$ with $R \in K^p[T]/(f^m)$. Put $s = z \otimes 1 - 1 \otimes g$. Then $K \otimes_{K^p} K^p[T]/(f^m) = K^p[T]/(f^m)[s]$, where s has the properties $s^p = fR$ and $s' = 1$. Now $\tau(-bs^{p-1}) = b - b^p(fR)^{p-1}$ for any $b \in K^p[T]/(f^m)$. Thus we have to solve $b - b^p(fR)^{p-1} = a$. Since f is nilpotent in $K^p[T]/(f^m)$, one easily computes a solution b of the equation, by a simple recursion.

Remarks 13.6
(1) For a C_1-field K, see Definition A.52, there are no skew fields of finite dimension over its center K. Furthermore, any finite extension of a C_1-field is again a C_1-field. Therefore, any C_1-field K satisfies the condition of the theorem. Well known examples of C_1-fields are $k(z)$ and $k((z))$ where k is algebraically closed.
(2) If K does not satisfy the condition of the theorem, then there is still a simple classification of the differential modules over K. For every irreducible monic $f \in Z$ and every integer $n \geq 1$ there is an indecomposable differential module $I(f, n)$. There are two possibilities for the central simple algebra $\mathcal{D}/f\mathcal{D}$ with center Z/fZ. It can be a skew field or it is isomorphic to the matrix algebra $\mathrm{Matr}(p, Z/fZ)$. In the first case, $I(f, n) = \mathcal{D}/f^n\mathcal{D}$, and in the second case, $I(f, n) \cong K \otimes_{K^p} Z/f^n Z$, provided with an action of ∂ as explained above. Moreover, any differential module over K is isomorphic with a finite direct sum $\oplus I(f, n)^{m(f,n)}$ with uniquely determined integers $m(f, n) \geq 0$.
(3) The category Diff_K (for a field K satisfying the condition of Theorem 13.5) can be seen to be a neutral tannakian category. However, the Picard-Vessiot theory fails and although differential Galois groups do exist they are rather strange objects. □

Examples 13.7
(1) Suppose $p > 2$ and let $K = \mathbf{F}_p(z)$. The differential modules of dimension 2 over K are:
(i) $I(T^2 + aT + b)$ with $T^2 + aT + b \in \mathbf{F}_p(z^p)[T]$ irreducible.
(ii) $I(T - a) \oplus I(T - b)$ with $a, b \in \mathbf{F}_p(z^p)$.
(iii) $I(T - a, 2)$ with $a \in \mathbf{F}_p(z^p)$.
These differential modules can be made explicit by solving the equation $\tau(B) = A$ in the appropriate field or ring. In case (i), the field is $L = \mathbf{F}_p(z)[T]/(T^2 + aT + b)$ and the equation is $\tau(B) = A$, where A is the image of T in $L^p = \mathbf{F}_p(z^p)[T]/(T^2 + aT + b)$. The differential module is then Le with ∂ given by $\partial e = Be$.
(2) Let $K = \mathbf{F}_2(z)$. The element $(\frac{z}{z^2+z+1})^2$ does not lie in the image of $\tau : K \to K^2$. Indeed, essentially the only element in K that could have the required τ-image is $\frac{a_0+a_1 z}{z^2+z+1}$. This leads to the equation $(a_1 + a_1^2)z^2 + a_1 + a_0 + a_0^2 = z^2$. Thus $a_1 = a_0 + a_0^2$ and $a_0^4 + a_0 = 1$. Thus $a_0 \in \mathbf{F}_{16} \setminus \mathbf{F}_4$. The algebra $F := K[\partial]/K[\partial](T - (\frac{z}{z^2+z+1})^2)$ can be shown to be a central simple algebra over K^2. It has dimension 4 over K^2. There are two possibilities: F is isomorphic to the matrix algebra $\mathrm{M}_2(K^2)$ or F is a skew field. The fact that $\tau(B) = (\frac{z}{z^2+z+1})^2$ has no solution translates into F is

a skew field. The differential module $I(T-(\frac{z}{z^2+z+1})^2)$ is now Fe and has dimension 2 over K. This illustrates (the converse of) part (3) of Lemma 13.4. □

13.2 Algorithmic Aspects

Making Theorem 13.5 effective has two aspects. For a given $K^p[T]$-module N of finite dimension over K^p one has to solve explicitly some equations of the form $b^{(p-1)}+b^p=a$ in order to obtain $\mathcal{F}(N)$.

The other aspect is to construct an algorithm that produces for a given differential module M over K the $K^p[T]$-module N with $\mathcal{F}(N)\cong M$. We introduce some notations. Let V be a finite dimensonal vector space over a field F and let $L:V\to V$ be an F-linear operator. Then $\min_F(L,V)$ and $\mathrm{char}_F(L,V)$ denote the minimal polynomial and the characteristic polynomial of L. Let $\mathcal{F}(N)=M$ and write t_N and t_M for the action of T on N and M. Thus $t_M=\partial^p$ acting upon M. One has the following formulas:

$$\min_{K^p}(t_N,N)=\min_K(t_M,M)\text{ and }\mathrm{char}_{K^p}(t_N,N)=\mathrm{char}_K(t_M,M).$$

For N of the form $K^p[T]/(G)$ with G monic, one has $M=K[T]/(G)$. Thus G is the mimimal polynomial and the characteristic polynomial of both N and M. In the general case, $N=\oplus_{i=1}^r K^p[T]/(G_i)$ and the mimimal polynomial of both t_N and t_M is the least common multiple of $G_1,\dots,G_r$. Furthermore, the product of all G_i is the characteristic polynomial of both t_N and t_M.

Suppose that t_M and its characteristic (or minimal) polynomial F are known. One factors $F\in K^p[Y]$ as $f_1^{m_1}\cdots f_s^{m_s}$, where $f_1,\dots,f_s\in K^p[Y]$ are distinct monic irreducible polynomials. Then $N=\oplus_{i=1}^s(K^p[T]/(f_i^n))^{m(f_i,n)}$ and $M=\oplus_{i=1}^s(K[T]/(f_i^n))^{m(f_i,n)}$ with still unknown multiplicities $m(f_i,n)$. These numbers follow from the Jordan decomposition of the operator t_M and can be read off from the dimensions of the cokernels (or kernels) of $f_i(t_M)^a$ acting on M. More precisely,

$$\dim_K\mathrm{coker}(f_i(t_M)^a:M\to M)=\sum_n m(f_i,n)\cdot\min(a,n)\cdot\deg f_i.$$

We note that the characteristic and the minimal polynomials of t_M and ∂ are linked by the formulas:

$$\min_K(t_M,M)(Y^p)=\min_{K^p}(\partial,M)(Y)\text{ and}$$
$$\mathrm{char}_K(t_M,M)(Y^p)=\mathrm{char}_{K^p}(\partial,M)(Y).$$

For the proof, it suffices to consider the case $M=K\otimes_{K^p}K^p[T]/(f^m)e$ with $f\in K^p[T]$ irreducible and $m\geq 1$. The equalities follow from the observation that the element $z^{p-1}e$ is a cyclic vector for the K^p-linear action of ∂ on M and also for the K-linear action of t_M on M.

Thus, a calculation of the characteristic polynomial or the minimal polynomial of the K^p-linear operator ∂ on M produces the characteristic polynomial or the minimal polynomial of t_M and t_N.

13.2.1 The Equation $b^{(p-1)} + b^p = a$

We discuss here an algorithmic version of Lemma 13.4. We know already that $\tau : K \to K^p$ is surjective for $K = \overline{\mathbf{F}}_p(z)$. This can also be seen by the following formulas for the map

$\tau^{1/p} : K \to K$.
$\tau(az^n)^{1/p} = az^n - a^{1/p}z^{(n-p+1)/p}$ if $n \equiv -1 \bmod p$.
$\tau(az^n)^{1/p} = az^n$ if $n \not\equiv -1 \bmod p$.
$\tau(\frac{a}{(z-b)^n})^{1/p} = \frac{a}{(z-b)^n} + \frac{-a^{1/p}}{(z-b)^{(n+p-1)/p}}$ if $n \equiv 1 \bmod p$.
$\tau(\frac{a}{(z-b)^n})^{1/p} = \frac{a}{(z-b)^n}$ if $n \not\equiv 1 \bmod p$.

After decomposing an $a \in K$ in partial fractions the formulas lead to an explicit solution of $(\tau b)^{1/p} = a$.

For the field $K = \mathbf{F}_p(z)$, the map $\tau : K \to K^p$ is not surjective. Still, we would like to solve $(\tau b)^{1/p} = a$ (whenever there is a solution $b \in K$) without involving the partial fractions decomposition of a. One writes $a = A + \frac{T}{F}$ with $A, T, F \in \mathbf{F}_p[z]$, T and F relatively prime, and $\deg T < \deg F$. Let d be the degree of A. The map $\tau^{1/p}$ induces an $\mathbf{F}_p$-linear bijection on the $\mathbf{F}_p$-vector space $\{B \in \mathbf{F}_p[z] \mid \deg B \leq d\}$. This map is, in fact, unipotent and the solution of $\tau(B)^{1/p} = A$ is easily computed.

Put $\overline{V} := \{\frac{B}{F} \mid B \in \overline{\mathbf{F}}_p[z],\ \deg B < \deg F\}$. There is a solution $w \in \overline{V}$ of $\tau(w)^{1/p} = \frac{T}{F}$. Any other solution has the form $w + \sum_{c \in S} \frac{1}{z-c}$, where S is a finite subset of $\overline{\mathbf{F}}_p$. We conclude that $\tau(C)^{1/p} = \frac{T}{F}$ has a solution in $\mathbf{F}_p(z)$ if and only if there is a solution in $V := \mathbf{F}_p(z) \cap \overline{V}$. Thus trying to solve the $\mathbf{F}_p$-linear equation $\tau(v)^{1/p} = \frac{T}{F}$ with $v \in V$, leads to the answer of our problem.
In the special case where F is irreducible and separable, the kernel of $\tau^{1/p} : V \to V$ is $\mathbf{F}_p\frac{F'}{F}$ and the image of $\tau^{1/p}$ is $\{\frac{B}{F} \mid B \in \mathbf{F}_p[z],\ \deg(B) \leq -2 + \deg F\}$.

Exercise 13.8 Let $K = \overline{\mathbf{F}}_p((z))$. Give an explicit formula for the action of $\tau^{1/p}$ on K. Use this formula to show that $\tau^{1/p}$ is surjective. □

Let $L \supset K = \overline{\mathbf{F}}_p(z)$ be a separable extension of degree d. The map $\tau^{1/p} : L \to L$ is, as we know, surjective. It is not evident how to construct an efficient algorithm for solving $\tau(b)^{1/p} = a$. A possibility is the following. One takes a subring O of the integral closure of $\overline{F}_p[z]$ in L, having basis $b_1, \ldots, b_d$ over $\overline{\mathbf{F}}_p[z]$. Since $L \supset K$ is separable, $b_1^p, \ldots, b_d^p$ is also a basis of L over K. For elements $f_1, \ldots, f_d \in K$

one has the following formula $\tau(\sum_{i=1}^{d} f_f b_i^p)^{1/p} = \sum_{i=1}^{d}(f_i^{(p-1)})^{1/p} b_i + \sum_{i=1}^{d} f_i b_i^p$. Each b_i^p has the form $\sum_{j=1}^{d} c_{j,i}(z) b_j$ with $c_{j,i}(z) \in \overline{\mathbf{F}}_p[z]$. There results a system of semilinear equations for $f_1, \ldots, f_d \in K$, namely

$$(f_i^{(p-1)})^{1/p} + \sum_{j=1}^{d} c_{i,j}(z) f_j = g_i \text{ for } i = 1, \ldots, d,$$

where $g_1, \ldots, g_d \in K$ are given. A decomposition of the $g_1, \ldots, g_d$ in partial fractions leads to equations and solutions for the partial fraction expansions of the $f_1, \ldots, f_d$.

We illustrate this for the case $d = 2$ and $p > 2$. There is a basis $\{1, b\}$ of L over K such that $b^2 = h \in \overline{\mathbf{F}}_p[z]$ has simple zeros. Now

$$\tau(f_0 + f_1 b)^{1/p} = \{(f_0^{(p-1)})^{1/p} + f_0\} + \{(f_1^{(p-1)})^{1/p} + f_1 h^{\frac{p-1}{2}}\} b.$$

The new equation that one has to solve is $T(f) := (f^{(p-1)})^{1/p} + f h^{\frac{p-1}{2}} = g$, where $g \in K$ is given. For $T(f)$ one has the following formulas:
If $f = az^n$, then $T(f) = \epsilon a^{1/p} z^{(n-p+1)/p} + az^n h^{\frac{p-1}{2}}$, where $\epsilon = 1$ if $n \equiv -1 \bmod p$ and $\epsilon = 0$ otherwise.
If $f = a(z-b)^n$, $n < 0$, then $T(f) = \epsilon a^{1/p}(z-b)^{(n-p+1)/p} + a(z-b)^n h^{\frac{p-1}{2}}$, where $\epsilon = 1$ if $n \equiv -1 \bmod p$ and $\epsilon = 0$ otherwise.

The above, combined with the sketch of the proof of Theorem 13.5, presents a solution of the first algorithmic aspect of this theorem. The following subsection continues and ends the second algorithmic aspect of the theorem.

13.2.2 The p-Curvature and Its Minimal Polynomial

For explicity we will suppose that K is either $\mathbf{F}_p(z)$ or $\overline{\mathbf{F}}_p(z)$. In the first case we will restrict ourselves to differential modules over K with dimension strictly less than p. In the second case the field K satisfies the requirements of Theorem 13.5. For a given differential module M over K we will develop several algorithms in order to obtain the decomposition $M \cong \oplus I(f, n)^{m(f,n)}$ that classifies M.

Assume that the differential module M is given as a matrix differential operator $\frac{d}{dz} + A$ acting on $K^m \cong M$. According to Katz [153] the following algorithm produces the p-curvature. Define the sequence of matrices $A(i)$ by $A(1) = A$ and $A(i+1) = A(i)' + A \cdot A(i)$. Then $A(p)$ is the matrix of the p-curvature.

If the differential module M is given in the form $\mathcal{D}/\mathcal{D}L$, with $\mathcal{D} = K[\partial]$ and $L \in \mathcal{D}$ an operator of degree m, then there are cheaper ways to calculate the p-curvature. One possibility is the following:

Let e denote the image of 1 in $\mathcal{D}/\mathcal{D}L$. Then $e, \partial e, \ldots, \partial^{m-1} e$ is a basis of $\mathcal{D}/\mathcal{D}L$ over K. The *p-curvature t of L* is, by definition, the operator ∂^p acting upon $\mathcal{D}/\mathcal{D}L$.

Write $L = \partial^m + \ell_{m-1}\partial^{m-1} + \cdots + \ell_1\partial + \ell_0$.

Define the sequence $R_n \in \mathcal{D}$ by: $\partial^n = *L + R_n$, where $*$ means some element in $\mathcal{D}$ (which we do not want to calculate) and where R_n has degree $< m$. The calculation of the

$R_n = R(n,0) + R(n,1)\partial + \cdots + R(n,m-1)\partial^{m-1}$ is the recursion:

$R_n = \partial^n$ for $n = 0, \ldots, m-1$ and for $n \geq m$: $R_n = *L + \partial R_{n-1}$

$$= \sum_{i=0}^{m-1} R(n-1,i)'\partial^i + \sum_{i=0}^{m-2} R(n-1,i)\partial^{i+1} - \sum_{i=0}^{m-1} R(n-1,m-1)\ell_i\partial^i,$$

and thus $R(n,0) = R(n-1,0)' - R(n-1,m-1)l_0$ and for $0 < i < m$ one has $R(n,i) = R(n-1,i-1) + R(n-1,i)' - R(n-1,m-1)\ell_i$.
The expressions for $R_p, R_{p+1}, \ldots, R_{p+m-1}$ form the columns of the matrix of $t = \partial^p$ acting upon $\mathcal{D}/\mathcal{D}L$. Indeed, for $0 \leq i < m$ the term $\partial^p\partial^i e$ equals

$$R_{p+i}e = R(p+i,0)e + R(p+i,1)\partial e + \cdots + R(p+i,m-1)\partial^{m-1}e.$$

For the calculation of the minimal polynomial of t one has also to compute R_{ip} for $i = 2, \ldots, m$. In the sequel we will write $T = \partial^p$. Thus $T = *L + R_p$ and

$$\begin{aligned} T^2 &= *L + R_p\partial^p \\ &= *L + \sum_{i=0}^{m-1} R(p,i) \sum_{j=0}^{m-1} R(p+i,j)\partial^j. \end{aligned}$$

More generally, the recurrence step is

$$\begin{aligned} T^k &= *L + R_{p(k-1)}\partial^p \\ &= *L + \sum_{i=0}^{m-1} R(p(k-1),i) \sum_{j=0}^{m-1} R(p+i,j)\partial^j. \end{aligned}$$

The minimal monic polynomial $F \in K^p[T]$, satisfied by the p-curvature, is written as $\sum_{i=0}^{m_0} c_i T^i$ with $1 \leq m_0 \leq m$, $c_i \in \mathbf{F}_p(z^p)$ and $c_{m_0} = 1$. We note that $\sum_{i=0}^{m_0} c_i T^i$ (with $1 \leq m_0 \leq m$, $c_i \in \mathbf{F}_p(z^p)$ and $c_{m_0} = 1$) is the minimal polynomial if and only if $(\sum_{i=0} c_i T^i)e = 0$. This translates into $\sum_{i=0}^{m_0} c_i R_{pi} = 0$. The latter expression can be seen as a system of m linear equations over the field $\mathbf{F}_p(z)$ in the unknowns $c_0, \ldots, c_{m_0-1}$. The solution $c_0, \ldots, c_{m_0-1}$ (with minimal m_0) has the property that all $c_j \in \mathbf{F}_p(z^p)$.

For the classification of the module $M := \mathcal{D}/\mathcal{D}L$ one has to factor $F \in K^p[T]$. Let $F = f_1^{m_1} \cdots f_s^{m_s}$ with distinct monic irreducible $f_1, \ldots, f_s$. The dimensions of the K-vector spaces $M/f_i^a M$ (for $i = 1, \ldots, s$ and $a = 1, \ldots, m_i - 1$) determine the multiplicities $m(f_i, n)$. The space $M/f_i^a M$ is the same as $\mathcal{D}/(\mathcal{D}L + \mathcal{D}f_i^a)$. Furthermore, $\mathcal{D}L + \mathcal{D}f_i^a = \mathcal{D}R$, where R is the greatest common right divisor of L and f_i^a.

13.2.3 Example: Operators of Order Two

Consider a differential operator $L = \partial^2 - r$ with $r \in \mathbf{Q}(z)$. For all but finitely many primes p, the obvious reduction $L_p = \partial^2 - r_p$ modulo p makes sense. We will suppose that p is small and not too small, say $p > 3$. The aim is to compute the classification of L_p and to indicate how this information can be used to find liouvillian solutions of $y'' = ry$ such that $\frac{y'}{y}$ is algebraic over $\mathbf{Q}(z)$. We recall from Chap. 4 that these liouvillian solutions correspond to 1-dimensional submodules of $\mathrm{sym}^m \mathcal{D}/\mathcal{D}L$ for some $m \geq 1$ where $\mathcal{D} = \mathbf{Q}(z)[\partial]$. For order two differential equations it is often possible to obtain relevant information without going through a maybe costly computation of the p-curvature.

Let $\mathcal{D}_p := \mathbf{F}_p(z)[\partial]$. There are four possibilities for $\mathcal{D}_p/\mathcal{D}_pL_p$, namely:

(i) $I(T^2 - D)$ with $T^2 - D \in \mathbf{F}_p(z^p)[T]$ irreducible. The second symmetric power is $I(T) \oplus I(T^2 - 4D)$.
(ii) $I(T - w) \oplus I(T + w)$ with $w \in \mathbf{F}_p(z^p)$, $w \neq 0$. The second symmetric power is $I(T) \oplus I(T - 2w) \oplus I(T + 2w)$.
(iii) $I(T^2)$. The second symmetric power is $I(T^3)$.
(iv) $I(T)^2$. The second symmetric power is $I(T)^3$.

The cases (iii) and (iv) are characterized (and detected) by the dimension of the kernel of the $\mathbf{F}_p(z^p)$-linear map $\partial^2 - r_p : \mathbf{F}_p(z) \to \mathbf{F}_p(z)$ being 1 and 2.

For the actual computation we consider the usual basis $e_0 = e, e_1 = \partial e$ of $\mathcal{D}_p/\mathcal{D}_pL_p$ and from the last subsection one can conclude that the matrix of the p-curvature w.r.t. this basis has the form

$$\begin{pmatrix} -f'/2 & rf - f''/2 \\ f & f'/2 \end{pmatrix} \text{ with } f \in \mathbf{F}_p(z) \text{ s.t. } f^{(3)} - 4r_p f^{(1)} - 2r_p' f = 0.$$

This equation is the second symmetric power of L_p. In case (iv) the p-curvature is 0. For the moment we will exclude this case. The 1-dimensional kernel of the $\mathbf{F}_p(z^p)$-linear map on $\mathbf{F}_p(z)$, given by $y \mapsto y^{(3)} - 4r_p y^{(1)} - 2r_p' y$, is easily computed. So $f \neq 0$ is known up to multiplication by a nonzero element of $\mathbf{F}_p(z^p)$ and the same holds for the p-curvature. Now we can give a list of the 1-dimensional submodules of $\mathrm{sym}^m \mathcal{D}_p/\mathcal{D}_pL_p$. We will restrict ourselves to $m = 1, 2$. Put $E := -\frac{1}{4}(\frac{f'}{f})^2 - \frac{1}{2}(\frac{f'}{f})' + r_p$. For the elements in the second symmetric power we will simplify the notation by omitting tensors, e.g., e_0e_1 denotes $e_0 \otimes e_1$.

Case (i). For $m = 1$ there is no 1-dimensional submodule. For $m = 2$, the 1-dimensional submodule has generator

$$e_1^2 - \frac{f'}{f}e_0e_1 + (\frac{1}{2}(\frac{f'}{f})^2 + \frac{1}{2}(\frac{f'}{f})' - r)e_0^2.$$

Case (ii). For $m = 1$ the 1-dimensional submodules have generators $e_1 - u_{\pm}e_0$ with $u_{\pm} := \frac{1}{2}\frac{f'}{f} \pm E^{1/2}$. For $m = 2$ the generators of the 1-dimensional submodules are $(e_1 - u_+e_0)^2$, $(e_1 - u_-e_0)^2$, $(e_1 - u_+e_0)(e_1 - u_-e_0)$.

Case (iii). For $m = 1$ the unique 1-dimensional submodule has generator $e_1 - \frac{1}{2}\frac{f'}{f}e_0$ and for $m = 2$ the unique 1-dimensional submodule has generator $(e_1 - \frac{1}{2}\frac{f'}{f}e_0)^2$.

Consider a possible factorization $L = (\partial + v)(\partial - v)$ with v algebraic of degree 1 or 2 over $\mathbf{Q}(z)$. A candidate for $v \in \mathbf{Q}(z)$ is a lift of $u_\pm$ for case (ii) or $\frac{1}{2}\frac{f'}{f}$ for case (iii). A candidate for the degree two polynomial over $\mathbf{Q}(z)$ for v is a lift of:

$X^2 - \frac{f'}{f}X + (\frac{1}{2}(\frac{f'}{f})^2 + \frac{1}{2}(\frac{f'}{f})' - r)$ for case (i),
$(X - u_+)^2$, $(X - u_-)^2$, $(X - u_+)(X - u_-)$ for case (ii),
$(X - \frac{1}{2}\frac{f'}{f})^2$ for case (iii).

Lifting an element of $\mathbf{F}_p(z)$ to an element $\frac{T}{N} \in \mathbf{Q}(z)$ can be done by prescribing a maximum for the degrees of the polynomials T and N and using LLL to obtain T and N with small coefficients. One can also combine the information in positive characteristic for various primes and even for powers of primes.

In case (iv), the p-curvature is 0 and this leads to a large set of factorizations of L_p. This occurs, in particular, when the differential Galois group of L is finite. One can still try to produce lifts of factorization of L_p with "small terms". A better method seems to refine the notion of p-curvature. One calculates in characteristic 0 (or alternatively modulo p^2) the operator ∂^p acting upon $\mathcal{D}/\mathcal{D}L$. The operator $\partial^{(p)} := \frac{1}{p!}\partial^p$ has a reduction modulo p, which we will also call $\partial^{(p)}$. The space where the latter operates is the 2-dimensional $\mathbf{F}_p(z^p)$-vector space W, given as the kernel of ∂ on $\mathbf{F}_p(z)e_0 + \mathbf{F}_p(z)e_1$. The field $\mathbf{F}_p(z^p)$ is made into a differential field by the formula $(z^p)' = 1$. Then $(W, \partial^{(p)})$ is again a differential module. The p-curvature of $\partial^{(p)}$ is an $\mathbf{F}_p(z^p)$-linear map on W. Lifts of the eigenvectors for this new p-curvature on symmetric powers $\mathrm{sym}^m W$ provide candidates for 1-dimensional submodules of $\mathrm{sym}^m \mathcal{D}/\mathcal{D}L$.

13.3 Iterative Differential Modules

Differential modules in positive characteristic have, as we have seen, some attractive properties. However, the absence of a Picard-Vessiot theory and suitable differential Galois groups is a good reason for considering other theories in positive characteristic. The basic idea, proposed by B.H. Matzat, is to consider "higher differentiations and corresponding higher differential equations". The most elementary setting of this reads as follows:
On the field $K = C(z)$ one considers the higher derivation $\{\partial^{(n)}\}_{n\geq 0}$, which is a sequence of additive maps given by the formulas $\partial^{(n)}z^m = \binom{m}{n}z^{m-n}$. One verifies (at least for f, g powers of z) the rules:
(i) $\partial^{(0)}$ is the identity.
(ii) $\partial^{(n)}(fg) = \sum_{a+b=n} \partial^{(a)}f \cdot \partial^{(b)}g$.
(iii) $\partial^{(n)} \circ \partial^{(m)} = \binom{n+m}{n}\partial^{n+m}$.

If we assume that the $\partial^{(n)}$ are C-linear and that these rules hold for all $f, g \in C(z)$, then the $\partial^{(n)}$ are uniquely defined. We remark further that $\partial^{(1)}$ is the ordinary differentiation and that $\partial^{(n)}$ is a substitute for $\frac{1}{n!}(\partial^{(1)})^n$. Higher differentiations, or here sometimes called iterative differentiations, were invented and studied by H. Hasse and F. K. Schmidt [125]. The definition of an iterative differential equation over say $C(z)$ is most easily formulated in module form. An iterative differential module over K is a finite dimensional vector space M over K equipped with a sequence of additive maps $\partial_M^{(n)} : M \to M$ having the properties:

(a) $\partial_M^{(0)}$ is the identity.

(b) $\partial_M^{(n)}(fm) = \sum_{a+b=n} \partial^{(a)} f \cdot \partial_M^{(b)} m$ for all $f \in K$ and all $m \in M$.

(c) $\partial_M^{(n)} \circ \partial_M^{(m)} = \binom{n+m}{n} \partial_M^{(n+m)}$.

After choosing a basis of M over K one can translate the above into a sequence of matrix equation $\partial^{(n)} y = A_n y$, where each A_n is a matrix with coefficients in K.

The theory of iterative differential equations has recently been developed, see [205]. In this section we will give a survey of the main results. It was a surprise to learn that (linear iterative) differential equations were, in fact, introduced as early as 1963 by H. Okugawa [217]. He proposed a Picard-Vessiot theory along the lines of E.R. Kolchin's work on linear differential equations in characteristic 0. His theory remained incomplete since efficient tools for handling Picard-Vessiot theory and differential Galois groups were not available at that time.

13.3.1 Picard-Vessiot Theory and Some Examples

The field K is assumed to have characteristic $p > 0$ and to be provided with an iterative differentiation $\{\partial^{(n)}\}_{n \geq 0}$ satisfying the rules (i)–(iii) given above. We assume that $\partial^{(1)} \neq 0$ and that the field C of differential constants, i.e., the elements $a \in K$ with $\partial^{(n)} a = 0$ for all $n \geq 1$, is algebraically closed. An iterative differential module M over K is defined by the rules (a)–(c) above. Let M have dimension d over K.

We want to indicate the construction of a Picard-Vessiot field L for M. Like the characteristic zero case, one defines L by:

(PV1) On $L \supset K$ there is given an iterative differentiation, extending the one of K.

(PV2) C is the field of constants of L.

(PV3) $V := \{v \in L \otimes_K M | \partial^{(n)} v = 0 \text{ for all } n \geq 1\}$ is a vector space over C of dimension d.

(PV4) L is minimal, or equivalently L is generated over K by the set of coefficients of all elements of $V \subset L \otimes_K M$ w.r.t. a given basis of M over K.

After choosing a basis of M over K, the iterative differential module is translated into a sequence of matrix equations $\partial^{(n)} y = A_n y$ (for $n \geq 0$). Consider a matrix of indeterminates $(X_{i,j})_{i,j=1}^{d}$ and write D for its determinant. The ring $K[X_{i,j}, \frac{1}{D}]$ is given an iterative differentiation, extending that of K, by putting $(\partial^{(n)} X_{i,j}) = A_n \cdot (X_{i,j})$ for all $n \geq 0$. An iterative differential ideal $J \subset K[X_{i,j}, \frac{1}{D}]$ is an ideal such that $\partial^{(n)} f \in J$ for all $n \geq 0$ and $f \in J$. Let I denote an iterative differential ideal that is maximal among the collection of all iterative differential ideals. Then I can be shown to be a prime ideal. Furthermore, the field of fractions L of $K[X_{i,j}, \frac{1}{D}]/I$ inherits an iterative differentiation. The field of differential constants of L is again C and the matrix $(x_{i,j})$, where the $x_{i,j} \in L$ are the images of the $X_{i,j}$, is a fundamental matrix for the above iterative differential equation. The C-vector space V generated by the columns of $(x_{i,j})$ is the solution space of the iterative differential equation. The differential Galois group G is, as in the characteristic 0 case, defined as the group of the K-linear automorphisms of L commuting with all $\partial^{(n)}$. This group operates on V and is actually a reduced algebraic subgroup of GL(V). In short, the Picard-Vessiot theory can be copied, almost verbatim, from the characteristic zero situation.

Exercise 13.9 Verify that the proofs of Chap. 1 carry over to the case of iterative differential modules.

In contrast to the characteristic zero case, it is not easy to produce interesting examples of iterative differential modules. A first example, which poses no great difficulties, is given by a Galois extension $L \supset K$ of degree $d > 1$. By a theorem of F.K. Schmidt, the higher differential of K extends in a unique way to a higher differential on L. Let the iterative differential module M be this field L and let $\{\partial_M^{(n)}\}_{n \geq 0}$ be the higher differentiaton on L. One observes that $\{m \in M |\ \partial_M^{(n)} m = 0 \text{ for all } n \geq 1\}$ is a 1-dimensional vector space over C. Indeed, since C is algebraically closed, it is also the field of constants of L. We *claim* that L is the Picard-Vessiot field for M. This has the consequence that the differential Galois group of M coincides with the ordinary Galois group of L/K.

A proof of the claim goes as follows. The object $L \otimes_K M = L \otimes_K L$ is also a ring and the formula $\partial^{(n)}(ab) = \sum_{\alpha+\beta=n} \partial^{(\alpha)}(a)\partial^{(\beta)}(b)$ holds for the multiplication in this ring. It is well known that $L \otimes_K L$ is, as ring, isomorphic to $\oplus_{i=1}^{d} Le_i$, the direct sum of d copies of L. Each e_i is idempotent. This easily implies that $\partial^{(n)} e_i =$ for all $n \geq 1$. Then $\{v \in L \otimes_K M |\ \partial^{(n)} v = 0 \text{ for all } n \geq 1\}$ is clearly $\oplus_{i=1}^{d} Ce_i$. Finally, for any field N with $K \subset N \subset L$ one has again that $N \otimes_K M = N \otimes_K L$ is a direct sum of copies of L. If $\{v \in N \otimes_K M |\ \partial^{(n)} v = 0 \text{ for all } n \geq 1\}$ has dimension d, then also $[N : K] = d$ and $N = L$.

As a *small experiment* we will try to make an iterative differential module M of rank one. Let e be a basis of M, then we have to produce a sequence of elements $a_n \in K$ with $\partial_M^{(n)} e = a_n e$ for $n \geq 0$, such that the defining properties are satisfied. First of all, $a_0 = 1$, furthermore, $\partial_M^{(n)} fe = \sum_{s+t=n} \partial^{(s)}(f) a_t e$ must hold for any $f \in K$. This does not pose conditions on the a_ns. The final requirement $\partial_M^{(n)} \circ \partial_M^{(m)} e = \binom{n+m}{n} \partial_M^{(n+m)} e$ translates into quite a number of conditions on the a_ns, namely

$$\sum_{s+t=n} \partial^{(s)}(a_m)a_t = \binom{n+m}{n} a_{n+m} \text{ for all } n, m \geq 0.$$

It seems almost hopeless to give an interesting solution of this set of equations. We have to develop some more theory to produce examples.

Consider an iterative differential module M of dimension one. Thus $M = Ke$. If one applies the formula $\partial_M^{(n)} \circ \partial_M^{(m)} = \binom{n+m}{n}\partial_M^{(n+m)}$ a number of times then one obtains $(\partial_M^{(1)})^p = 0$. This means that the ordinary differential module $(M, \partial_M^{(1)})$ has p-curvature 0. According to Lemma 13.2, there is a basis e_1 of M such that $\partial_M^{(1)} e_1 = 0$. Let K_1 be the subfield of K consisting of the elements $a \in K$ with $\partial^{(1)} a = 0$. Then the kernel M_1 of $\partial_M^{(1)}$ on M is equal to $K_1 e_1$. Since $\partial_M^{(1)}$ commutes with $\partial_M^{(p)}$ one has that $\partial_M^{(p)}$ maps M_1 into itself. The pair $(M_1, \partial_M^{(p)})$ can be seen as an ordinary differential module over the differential field K_1 with differentiation $\partial_M^{(p)}$. Again $(\partial_M^{(p)})^p = 0$ and there is an element e_2 with $K_1 e_1 = K_1 e_2$ and $\partial_M^{(p)} e_2 = 0$. Let K_2 denote the subfield of K_1 consisting of the elements $a \in K_1$ with $\partial^{(p)} a = 0$. The kernel M_2 of $\partial_M^{(p)}$ on M_1 is equal to $K_2 e_2$. One can continue in this way and define by induction:

(a) Subfields K_s of K by $K_{s+1} = \{a \in K_s |\ \partial^{(p^s)} a = 0\}$.

(b) Subsets M_s of M by $M_{s+1} = \{m \in M_s |\ \partial_M^{(p^s)} m = 0\}$.

(c) Elements $e_s \in M$ such that $M_s = K_s e_s$.

We note that the element e_s is unique up to multiplication by an element in K_s^*. We write now $e_s = f_s e$ with $f_s \in K^*$. Then the sequence $\{f_s \bmod K_s^*\}$ is a projective sequence and determines an element ξ in the projective limit $\varprojlim K^*/K_s^*$. On the other hand, for a fixed choice of the basis e of M, any ξ in this projective limit determines a sequences of "subspaces" $M \supset M_1 \supset M_2 \supset \cdots$. From this sequence one can produce a unique iterative differential module structure $\{\partial_M^{(n)}\}_{n \geq 0}$ on M by requiring that $\partial_M^{(p^s)}$ is zero on M_n for $s < n$. If one changes the original basis e of M by fe with $f \in K^*$, then the element ξ in the projective limit is changed into $f\xi$. We conclude that the cokernel $\mathcal{Q}$ of the natural map $K^* \to \varprojlim K^*/K_s^*$ describes the set of isomorphisms classes of the one-dimensional iterative differential modules over K. The group structure of $\mathcal{Q}$ corresponds with the tensor product of iterative differential modules. For some fields K one can make $\mathcal{Q}$ explicit.

Proposition 13.10 *Let C denote an algebraically closed field of characteristic $p > 0$. The field $K = C((z))$ is provided with the higher differentiation $\{\partial^{(n)}\}_{n \geq 0}$ given by the formulas $\partial^{(n)} \sum_m a_m z^m = \sum_m \binom{m}{n} a_m z^{m-n}$. Then the group of isomorphism classes $\mathcal{Q}$ of the one-dimensional iterative differential modules over K is isomorphic with $\mathbf{Z}_p/\mathbf{Z}$.*

Proof. We note that $K_s = C((z^{p^s}))$ for all $s \geq 1$. The group K^* can be written as $z^{\mathbf{Z}} \times C^* \times (1 + zC[[z]])$, where $1 + zC[[z]]$ is seen as a multiplicative group. A similar decomposition holds for K_s^*. Then K^*/K_s^* has a decomposition $z^{\mathbf{Z}/p^s\mathbf{Z}} \times (1 + zC[[z]])/(1 + z^{p^s}C[[z^{p^s}]])$. The projective limit is isomorphic to $z^{\mathbf{Z}_p} \times (1 + C[[z]])$ since the natural map $(1 + zC[[z]]) \to \lim\limits_{\leftarrow} (1 + zC[[z]]) /(1 + z^{p^s}C[[z^{p^s}]])$ is a bijection. Thus the cokernel $\mathcal{Q}$ of $K^* \to \lim\limits_{\leftarrow} K^*/K_s^*$ is equal to $\mathbf{Z}_p/\mathbf{Z}$. □

In particular, there exists for every p-adic integer $\alpha \in \mathbf{Z}_p$ an iterative differential module of dimension one. For every $n \geq 0$ the expression $\binom{\alpha}{n}$ is the "binomial coefficient" $\frac{\alpha(\alpha-1)\cdots(\alpha-n+1)}{n!}$. This expression is an element of $\mathbf{Z}_p$. By $\overline{\binom{\alpha}{n}}$ we mean the image of $\binom{\alpha}{n}$ in $\mathbf{F}_p = \mathbf{Z}_p/p\mathbf{Z}_p$. The iterative differential module $M = Ke$ corresponding to α can now be given by the explicit formula $\partial_M^{(n)} e = \overline{\binom{\alpha}{n}} z^{-n} e$ for all $n \geq 0$. We note that for $\alpha \in \mathbf{Z}$ this module is trivial since $f = z^{-\alpha}e$ satisfies $\partial_M^{(n)} f = 0$ for all $n \geq 1$. The differential Galois group of the module M can easily be computed. If α is not a rational number then this group is the multiplicative group $\mathbf{G}_{m,C}$. If α is rational and has denominator $m \geq 1$, then the differential Galois group is the cyclic group of order m. Indeed, the expression $z^{-\alpha}$ is an algebraic function over K and $f = z^{-\alpha}e$ again satisfies $\partial_M^{(n)} f = 0$ for all $n \geq 1$.

The *local theory* of iterative differential modules concerns those objects over the field $K = C((z))$ provided with the higher derivation, as in Proposition 13.10. The main result, which somewhat resembles the Turrittin classification of ordinary differential modules over $\mathbf{C}((z))$, is the following.

Theorem 13.11 *Let $K = C((z))$ be as above.*
(1) *Every iterative differential module over K is a multiple extension of one-dimensional iterative differential modules.*
(2) *A reduced linear algebraic group G over C is the differential Galois group of an iterative differential module over K if and only if the following conditions are satisfied:*
(a) *G is a solvable group.*
(b) *G/G^o is an extension of a cyclic group with order prime to p by a p-group.*

13.3.2 Global Iterative Differential Equations

Let C be an algebraically closed field of characteristic $p > 0$ and let X be a smooth, irreducible projective curve over C of genus g. The function field K of X is provided with a higher derivation $\{\partial_K^{(n)}\}_{n \geq 0}$ determined by the formulas $\partial_K^{(n)} z^m = \binom{m}{n} z^{m-n}$ for all $n, m \geq 0$. Here, $z \in K$ is chosen such that $C(z) \subset K$ is a finite separable extension. The global theory is concerned with iterative differential modules over K. In order to avoid pathological examples one fixes a finite subset S in X with cardinality $s \geq 1$. One considers those iterative differential modules M over K such that M is regular outside S. This means that for every point $x \notin S$ there exists a local coordinate t at x, a Zariski-open neighborhood U of x and a free $O_X(U)$-module

$N \subset M$ that generates M and such that N is invariant under all $\partial_t^{(n)}$. Here, the $\partial_t^{(n)}$ are adaptations of the $\partial_M^{(n)}$ with respect to the local parameter t.

The main issue is the following conjecture.

Conjecture 13.12 *Let (X, S) be as above. Then a reduced linear algebraic group G over C can be realized as the differential Galois group of an iterative differential module over K that is regular outside S if and only $G/p(G)$ can be realized.*

The group $p(G)$ is defined to be the subgroup of G generated by all the elements of order a power of p (see also Sect. 11.6). It turns out that $p(G)$ is a normal algebraic subgroup of G and thus $H := G/p(G)$ is again a reduced linear algebraic group over C. The group H has no elements of order p. This implies that H has the properties: H^o is either a torus or equal to 1, and H/H^o is a finite group with order prime to p. The question when such H can be realized for the pair (X, S) depends on the nature of the jacobian variety of X. For "general" X and large enough field C the answer is:
H as above can be realized for the pair (X, S) if and only if H contains a Zariski-dense subgroup generated by, at most, $2g + s - 1$ elements.

For a finite group G Conjecture 13.12 is actually equivalent to Abhyankar's conjecture. Indeed, let $L \supset K$ denote the Galois extension corresponding to a Galois cover $Y \to X$ that is unramified outside S. Then L seen as an iterative differential module over K (see above) is regular outside S. Furthermore, the Galois group of L/K is the differential Galois group. On the other hand, suppose that an iterative differential module over K, which is regular outside S, has a Picard-Vessiot extension $L \supset K$ that is a finite Galois extension with group G. Then the corresponding cover $Y \to X$ is unramified outside S. In particular, according to the work of Raynaud and Harbater (see Sect. 11.6), Conjecture 13.12 holds for finite groups. However, we have not found an independent proof.

The conjecture can also be seen as a characteristic p analog of Ramis' Theorem 11.21. One of the main results of [205] is:

Theorem 13.13 *The conjecture holds for (reduced) connected linear algebraic groups over C.*

A special case, which is the guiding example in the proof of Theorem 13.13, is:

The group $\mathrm{SL}_2(C)$ can be realized as a differential Galois group for the affine line over C, i.e., $X = \mathbf{P}^1_C$ and $S = \{\infty\}$.

We note that this is a special case of the Conjecture 13.12. Indeed, one easily verifies that $p(\mathrm{SL}_2(C)) = \mathrm{SL}_2(C)$.

13.3.3 p-Adic Differential Equations

In this subsection we will compare iterative differential equations and p-adic differential equations. In order to avoid technical complications we simplify the setup somewhat. In the examples we allow ourselves more freedom.

Let R_0 be a complete discrete valuation ring of characteristic zero such that pR_0 is the maximal ideal of R_0. The residue field R_0/pR_0 is denoted by C. The field of fractions of R_0 is denoted by L. On $L(z)$ we consider the Gauss norm as valuation. For a polynomial $\sum a_i z^i$ this Gauss norm is defined as $|\sum a_i z^i|_{gauss} = \max |a_i|$. For a rational function $\frac{T}{N} \in L(z)$ one defines $|\frac{T}{N}|_{gauss} = \frac{|T|_{gauss}}{|N|_{gauss}}$. Let F denote the completion of $L(z)$ with respect to the Gauss norm. Then F is again a discretely valued field. Let R denote the valuation ring of F. Then p generates the maximal ideal of R and the residue field R/pR of F is equal to $K = C(z)$.
The differentiation $f \mapsto \frac{df}{dz}$ on $L(z)$ is continuous with respect the Gauss norm and extends uniquely to a differentiation, denoted by ∂_F, on F such that $\partial_F R \subset R$. A small calculation shows that for every $n \geq 1$ one has that $\frac{1}{n!}\partial_F^n(R) \subset R$. Moreover, the reduction of $\frac{1}{n!}\partial_F^n$ modulo pR coincides with the standard higher differentiation $\partial_K^{(n)}$ on K given by the formula $\partial_K^{(n)} z^m = \binom{m}{n} z^{m-n}$.

A *p-adic differential equation* is a differential equation over the field $L(z)$ or over its completion F. These equations have attracted much attention, mainly because of their number theoretical aspects. B. Dwork is one of the initiators of the subject. We will investigate the following question, which is rather important in our context.

Which p-adic differential equations can be reduced modulo p to an iterative differential equation over K?

We need to introduce some terminology in order to formulate an answer to this question. Let M be a finite dimensional vector space over F. A *norm* on M is a map $\| \ \| : M \to \mathbf{R}$ having the properties:

(i) $\|m\| \geq 0$ for all $m \in M$.
(ii) $\|m\| = 0$ if and only $m = 0$.
(iii) $\|m_1 + m_2\| \leq \max(\|m_1\|, \|m_2\|)$.
(iv) $\|fm\| = |f| \cdot \|m\|$ for $f \in F$ and $m \in M$.

Any two norms $\| \ \|$, $\| \ \|^*$ are equivalent, which means that there are positive constants c, C such that $c\|m\| \leq \|m\|^* \leq C\|m\|$ for all $m \in M$. For any additive map $A : M \to M$ one defines $\|A\| = \sup\{\frac{\|A(m)\|}{\|m\|} \mid m \in M, \ m \neq 0\}$. In general, $\|A\|$ can be ∞.
An R-lattice $\Lambda \subset M$ is an R-submodule of M generated by a basis of M over F.

Theorem 13.14 *Let* (M, ∂_M) *be a differential module over* F. *Let* $\partial_M^{(n)}$ *denote the operator* $\frac{1}{n!}\partial_M^n$. *The following properties are equivalent.*

(1) *There is an R-lattice* $\Lambda \subset M$ *that is invariant under all* $\partial_M^{(n)}$.
(2) *Let* $\| \ \|$ *be any norm on* M, *then* $\sup_{n\geq 0} \|\partial_M^{(n)}\| < \infty$.

Both conditions are independent of the chosen norm. Furthermore, condition (2) can be made explicit for a matrix differential equation $y' = Ay$ with coefficients in the field F. By differentiating this equation, one defines a sequence of matrices A_n with coefficients in F satisfying $\frac{1}{n!}(\frac{d}{dz})^n y = A_n y$. Let the norm of a matrix $B = (b_{i,j})$ with coefficients in F be given by $\|B\| = \max |b_{i,j}|$. Then condition (2) is equivalent to $\sup_{n\geq 0} \|A_n\| < \infty$.

Suppose that (M, ∂_M) has the equivalent properties of Theorem 13.14. Then the $\partial_M^{(n)}$ induce maps $\partial^{(n)}$ on $N = \Lambda/p\Lambda$, which make the latter into an iterative differential module over K. The next theorem states that every iterative differential module over K can be obtained in this way.

Theorem 13.15 *Let* N *be an iterative differential module over* K. *Then there exists a differential module* (M, ∂) *over* F *and an R-lattice* $\Lambda \subset M$ *such that* Λ *is invariant under all* $\frac{1}{n!}\partial_M^n$ *and such that the induced iterative differential module* $\Lambda/p\Lambda$ *is isomorphic to* N.

Let (M, ∂) and N be as in Theorem 13.15. The differential module (M, ∂) has a differential Galois group G over the algebraic closure $\overline{L}$ of L. The iterative differential module N has a differential Galois group defined over the algebraic closure $\overline{C}$ of C. The two groups are clearly related. We formulate the following conjecture concerning this relation.

Conjecture 13.16 *There is a group scheme* $\mathcal{G}$ *over the ring of integers* O *of the algebraic closure* $\overline{L}$ *of* L *such that* $\mathcal{G} \otimes_O \overline{L} = G$ *and* $\mathcal{G} \otimes_O \overline{C}$ *contains* H. *Moreover, if* G *is a finite group then, after replacing the differential equation by an equivalent one, the groups* G *and* H *coincide.*

As we have seen above, it is not easy to produce interesting iterative differential modules. The same holds for p-adic differential modules that have the equivalent properties of Theorem 13.14. We will discuss some examples.

Example 13.17 Consider the matrix differential equation $y' = Ay$ with A a constant matrix with coefficients in the algebraic closure of $\mathbf{Q}_p$. Then $A_n = \frac{1}{n!}A^n$. One can verify that $\sup_{n\geq 0} \|A_n\| < \infty$ is equivalent to: every eigenvalue α of A satisfies $|\alpha| \leq p^{-1/(p-1)}$.

Suppose now that the eigenvalues of A have this property. Let $\alpha_1, \dots, \alpha_s$ denote the eigenvalues of A with absolute value equal to $p^{-1/(p-1)}$. One can calculate that the differential Galois group G of the equation $y' = Ay$ over F is equal to the quotient of $(\mathbf{Z}/p\mathbf{Z})^s$ with respect to the subgroup $\{(m_1, \dots, m_s) | \, |m_1\alpha_1 + \cdots + m_s\alpha_s| < p^{-1/(p-1)}\}$.

The more or less obvious reduction of $y' = Ay$ to an iterative differential module over K produces a trivial iterative differential module, i.e., one with differential Galois group $\{1\}$. This situation is not very satisfactory in view of the conjecture. The phenomenon behind this is that the straightforward reduction modulo p of a p-cyclic extension in characteristic 0 is not a p-cyclic extension in characteristic p. There is, however, a "deformation from Artin-Schreier to Kummer" (cf. [98]) which can be applied here. The equation $y' = Ay$ has to be replaced by an equivalent equation $y' = By$. The latter reduces to an iterative differential equation over K such that H is equal to G.

Example 13.18 Consider the equation $y' = Az^{-1}y$, where A is a constant matrix with coefficients in the algebraic closure of $\mathbf{Q}_p$. This equation satisfies the equivalent properties of Theorem 13.14 if and only if A is semisimple and all its eigenvalues are in $\mathbf{Z}_p$.

Assume that A satisfies this property and let $\alpha_1, \ldots, \alpha_s$ denote the distinct eigenvalues of A. The differential Galois group G over F is equal to the subgroup of the elements $t = (t_1, \ldots, t_s)$ in the torus $\mathbf{G}^s_{m,\overline{L}}$ satisfying $t_1^{m_1} \cdots t_s^{m_s} = 1$ for all $(m_1, \ldots, m_s) \in \mathbf{Z}^s$ such that $m_1\alpha_1 + \cdots + m_s\alpha_s \in \mathbf{Z}$.
The differential Galois group H of the corresponding iterative differential equation over K can be computed to be the subgroup of the torus $\mathbf{G}^s_{m,\overline{C}}$ given by the same relations.

Example 13.19 The hypergeometric equation reads

$$z(z-1)y'' + ((a+b+1)z - c)y' + aby = 0.$$

This equation with coefficients $a, b, c \in \mathbf{Z}_p$ has been studied extensively by B. Dwork and others. Using Dwork's ideas (see [89], Theorem 9.2) and with the help of F. Beukers, the following result was found.

Theorem 13.20 *Put $A = -a$, $B = -b$, $C = -c$ and let the p-adic expansions of A, B and C be $\sum A_n p^n$, $\sum B_n p^n$, and $\sum C_n p^n$. The hypergeometric equations with parameters a, b, c satisfies the equivalent properties of Theorem 13.14 if for each i one has $A_i < C_i < B_i$ or $B_i < C_i < A_i$.*

Appendices

A Algebraic Geometry

Affine varieties are ubiquitous in Differential Galois Theory. For many results (e.g., the definition of the differential Galois group and some of its basic properties) it is enough to assume that the varieties are defined over algebraically closed fields and study their properties over these fields. Yet, to understand the finer structure of Picard-Vessiot extensions it is necessary to understand how varieties behave over fields that are not necessarily algebraically closed. In this section we shall develop basic material concerning algebraic varieties taking these needs into account, while at the same time restricting ourselves only to the topics we will use.

Classically, algebraic geometry is the study of solutions of systems of equations $\{f_\alpha(X_1, \dots, X_n) = 0\}$, $f_\alpha \in \mathbf{C}[X_1, \dots, X_n]$, where $\mathbf{C}$ is the field of complex numbers. To give the reader a taste of the contents of this appendix, we give a brief description of the algebraic geometry of $\mathbf{C}^n$. Proofs of these results will be given in this appendix in a more general context.

One says that a set $S \subset \mathbf{C}^n$ is an affine variety if it is precisely the set of zeros of such a system of polynomial equations. For $n = 1$, the affine varieties are finite or all of $\mathbf{C}$ and for $n = 2$, they are the whole space or unions of points and curves (i.e., zeros of a polynomial $f(X_1, X_2)$). The collection of affine varieties is closed under finite intersection and arbitrary unions and so forms the closed sets of a topology, called the Zariski topology. Given a subset $S \subset \mathbf{C}^n$, one can define an ideal $I(S) = \{f \in \mathbf{C}[X_1, \dots, X_n] \mid f(c_1, \dots, c_n) = 0 \text{ for all } (c_1, \dots, c_n) \in \mathbf{C}^n\}$ $\subset \mathbf{C}[X_1, \dots, X_n]$. A fundamental result (the Hilbert Basissatz) states that any ideal of $\mathbf{C}[X_1, \dots, X_n]$ is finitely generated and so any affine variety is determined by a finite set of polynomials. One can show that $I(S)$ is a radial ideal, that is, if $f^m \in I(S)$ for some $m > 0$, then $f \in I(S)$. Given an ideal $I \subset \mathbf{C}[X_1, \dots, X_n]$ one can define a variety $Z(I) = \{(c_1, \dots, c_n) \in \mathbf{C}^n \mid f(c_1, \dots, c_n) = 0 \text{ for all } f \in I\}$ $\subset \mathbf{C}^n$. Another result of Hilbert (the Hilbert Nullstellensatz) states for any *proper* ideal $I \subset \mathbf{C}[X_1, \dots, X_n]$, the set $Z(I)$ is not empty. This allows one to show that maps $V \mapsto I(S)$ and $I \mapsto Z(I)$ define a bijective correspondence between the collection of affine varieties in $\mathbf{C}^n$ and the collection of radical ideals in $\mathbf{C}[X_1, \dots, X_n]$.

Given a variety V, one can consider a polynomial f in $\mathbf{C}[X_1, \dots, X_n]$ as a function $f : V \to \mathbf{C}$. The process of restricting such polynomials to V yields a homomorphism from $\mathbf{C}[X_1, \dots, X_n]$ to $\mathbf{C}[X_1, \dots, X_n]/I(V)$ and allows one to identify

$\mathbf{C}[X_1, \ldots, X_n]/I(V)$ with the collection of polynomial functions on V. This latter ring is called the coordinate ring of V and denoted by $\mathbf{C}[V]$. The ring $\mathbf{C}[V]$ is a finitely generated $\mathbf{C}$-algebra and any finitely generated $\mathbf{C}$-algebra R may be written as $R = \mathbf{C}[X_1, \ldots, X_n]/I$ for some ideal I. I will be the ideal of an affine variety if it is a radical ideal or, equivalently, when R has no nilpotent elements. Therefore, there is a correspondence between affine varieties and finitely generated $\mathbf{C}$-algebras without nilpotents.

More generally, if $V \subset \mathbf{C}^n$ and $W \subset \mathbf{C}^m$ are affine varieties, a map $\phi : V \to W$ is said to be a regular map if it is the restriction of a $\Phi = (\Phi_1, \ldots, \Phi_m) : \mathbf{C}^n \to \mathbf{C}^m$, where each Φ_i is a polynomial in n variables. Given an element $f \in \mathbf{C}[W]$, one sees that $f \circ \phi$ is an element of $\mathbf{C}[V]$. In this way, the regular map ϕ induces a $\mathbf{C}$-algebra homomorphism from $\mathbf{C}[W]$ to $\mathbf{C}[V]$. Conversely, any such $\mathbf{C}$-algebra homomorphism arises in this way. Two affine varieties V and W are said to be isomorphic if there are regular maps $\phi : V \to W$ and $\psi : W \to V$ such that $\psi \circ \phi = id_V$ and $\phi \circ \psi = id_W$. Two affine varieties are isomorphic if and only if their coordinate rings are isomorphic as $\mathbf{C}$-algebras.

We say that an affine variety is irreducible if it is not the union of two proper affine varieties and irreducible if this is not the case. One sees that an affine variety V is irreducible if and only if $I(V)$ is a prime ideal or, equivalently, if and only if its coordinate ring is an integral domain. The Basissatz can be, furthermore, used to show that any affine variety can be written as the finite union of irreducible affine varieties. If one has such a decomposition where no irreducible affine variety is contained in the union of the others, then this decomposition is unique and we refer to the irreducible affine varieties appearing as the components of V. This allows us to frequently restrict our attention to irreducible affine varieties. All of the above concepts are put in a more general context in Sect. A.1.1.

One peculiarity of the Zariski topology is that the Zariski topology of $\mathbf{C}^2 = \mathbf{C} \times \mathbf{C}$ is not the product topology. For example, $V(X_1^2 + X_2^2)$ is not a finite union of sets of the form $\{pt\} \times \{pt\}$, $\{pt\} \times \mathbf{C}$, $\mathbf{C} \times \{pt\}$, or $\mathbf{C} \times \mathbf{C}$. We shall have occasion to deal with products of affine varieties. For example, the Galois theory of differential equations leads one to consider the affine groups G and these are defined as affine varieties where the group law is a regular map from $G \times G \to G$ (as well as insisting that the map taking an element to its inverse is a regular map $G \to G$). To do this efficiently we wish to give an intrinsic definition of the product of two varieties. In Sect. A.1.2, we show that for affine varieties V and W the tensor product $\mathbf{C}[V] \otimes_{\mathbf{C}} \mathbf{C}[W]$ of $\mathbf{C}[V]$ and $\mathbf{C}[W]$ is a $\mathbf{C}$-algebra that has no nilpotent elements. We define the product of V and W to be the affine variety associated with the ring $\mathbf{C}[V] \otimes_{\mathbf{C}} \mathbf{C}[W]$. If $V \subset \mathbf{C}^n$ and $W \subset \mathbf{C}^m$ then we can identify $V \times W$ with point set $V \times W \subset \mathbf{C}^{n+m}$. This set is Zariski closed and has the above coordinate ring.

The Basissatz implies that any decreasing chain of affine varieties $V \supsetneq V_1 \supsetneq \cdots \supsetneq V_t \supsetneq \ldots$ must be finite. One can show that the length of such a chain is uniformly bounded and one can define the dimension of an affine variety V to be the largest number d for which there is a chain of nonempty affine varieties $V \supsetneq V_1 \supsetneq \cdots \supsetneq V_d$. The dimension of an affine variety is the largest dimension of its irreducible components. For an irreducible affine variety V this coincides with the transcendence degree of $\mathbf{C}(V)$ over $\mathbf{C}$, where $\mathbf{C}(V)$ is called the function field of V and is the quotient field of $\mathbf{C}[V]$. These concepts are further discussed in Sect. A.1.3.

Let V be an irreducible variety of dimension d and let $p \in V$. We may write the coordinate ring $\mathbf{C}[V]$ as $\mathbf{C}[X_1, \ldots , X_n]/(f_1, \ldots , f_t)$. One can show that the matrix $(\frac{\partial f_i}{\partial X_j}(p))$ has rank, at most, $n-d$. We say that p is a nonsingular point of V if the rank is exactly $n - d$. This will happen at a Zariski-open set of points on V. The Implicit Function Theorem implies that in a (Euclidean) neighborhood of a nonsingular point, V will be a complex manifold of dimension d. One can define the tangent space of V at a nonsingular point $p = (p_1, \ldots , p_n)$ to be the zero set of the linear equations

$$\sum_{i=1}^{n} \frac{\partial f_j}{\partial X_i}(p)(X_i - p_i) = 0 \text{ for } j = 1, \ldots , t\,.$$

This formulation of the notions of nonsingular point and tangent space appear to depend on the choice of the f_i and are not intrinsic. Furthermore, one would like to define the tangent space at nonsingular points as well. In Sect. A.1.4, we give an intrinsic definition of nonsingularity and tangent space at an arbitrary point of a (not necessarily irreducible) affine variety and show that these concepts are equivalent to the above in the classical case.

A major use of the algebraic geometry that we develop will be to describe linear algebraic groups and sets on which they act. The prototypical example of a linear algebraic group is the group $\mathrm{GL}_n(\mathbf{C})$ of invertible $n \times n$ matrices with entries in $\mathbf{C}$. We can identify this group with an affine variety in $\mathbf{C}^{n^2+1}$ via the map sending $A \in \mathrm{GL}_n(\mathbf{C})$ to $(A, (\det(A))^{-1})$. The ideal in $\mathbf{C}[X_{1,1}, \ldots , X_{n,n}, Z]$ defining this set is generated by $Z \det(X_{i,j}) - 1$. The entries of a product of two matrices A and B are clearly polynomials in the entries of A and B. Cramer's rule implies that the entries of the inverse of a matrix A can be expressed as polynomials in the entries of A and $(\det(A))^{-1}$. In general, a linear algebraic group is defined to be an affine variety G such that the multiplication is a regular map from $G \times G$ to G and the inverse is a regular map from G to G. It can be shown that all such groups can be considered as Zariski-closed subgroups of $\mathrm{GL}_N(\mathbf{C})$ for a suitable N. In Sect. A.2.1, we develop the basic properties of linear algebraic groups ending with a proof of the Lie-Kolchin Theorem that states that a solvable linear algebraic group $G \subset \mathrm{GL}_n$, connected in the Zariski topology, is conjugate to a group of upper triangular matrices. In Sect. A.2.2, we show that the tangent space of a linear algebraic group at the identity has the structure of a Lie algebra and derive some further properties.

In Sect. A.2.3, we examine the action of a linear algebraic group on an affine variety. We say that an affine variety V is a torsor or principal homogeneous space for a linear algebraic group G if there is a regular map $\phi : G \times V \to V$ such that for any $v, w \in V$ there is a unique $g \in G$ such that $\phi(g, v) = w$. In our present context, working over the algebraically closed field $\mathbf{C}$, this concept is not too interesting. Picking a point $p \in V$ one sees that the map $G \to V$ given by $g \mapsto \phi(g, p)$ gives an isomorphism between G and V. A key fact in differential Galois theory is that a Picard-Vessiot extension of a differential field k is isomorphic to the function field of a torsor for the Galois group. The field k need not be algebraically closed and this is a principal reason for developing algebraic geometry over fields that are not algebraically closed. In fact, in Sect. A.2.3 we show that the usual Galois theory of polynomials can be recast in the language of torsors and we end this outline with an example of this.

Example A.1 Consider the affine variety $W = \{\sqrt{-1}, -\sqrt{-1}\} \subset \mathbf{C}^1$ defined by $X^2 + 1 = 0$. The linear algebraic group $G = \{1, -1\} \subset \mathrm{GL}_1(\mathbf{C})$ acts on W by multiplication $(g, w) \mapsto gw$ and this action makes W into a torsor for G. It is easy to see that V and G are isomorphic affine varieties (for example, $f(X) = \sqrt{-1}X$ defines an isomorphism). We say that an affine variety $V \subset \mathbf{C}^n$ is defined over $k \subset \mathbf{C}$ if $I(V) \subset \mathbf{C}[X_1, \dots, X_n]$ has a set of generators in k. We define the k-coordinate ring of V to be $k[V] = k[X_1, \dots, X_n]/(I \cap k[X_1, \dots, X_n])$. It is clear that both W and G are defined over $\mathbf{Q}$ and $\mathbf{Q}[W] = \mathbf{Q}[X]/(X^2 + 1) \simeq \mathbf{Q}(\sqrt{-1})$, $\mathbf{Q}[G] = \mathbf{Q}[X]/(X^2 - 1) \simeq \mathbf{Q} \oplus \mathbf{Q}$. The action of G on W is defined by polynomials with coefficients in $\mathbf{Q}$ as well. On the other hand, there is no isomorphism between G and W defined by polynomials over $\mathbf{Q}$.

In fact, any finite group can be realized (for example, via permutation matrices) as a linear algebraic group defined over $\mathbf{Q}$ and any Galois extension of $\mathbf{Q}$ with Galois group G is the $\mathbf{Q}$-coordinate ring of a torsor for G defined over $\mathbf{Q}$ as well (see Exercise A.50). □

One could develop the theory of varieties defined over an arbitrary field k using the theory of varieties defined over the algebraic closure $\bar{k}$ and carefully keeping track of the "field of definition". In the next sections we have chosen instead to develop the theory directly for fields that are not necessarily algebraically closed. Although we present the following material ab initio, the reader completely unfamiliar with most of the above ideas of algebraic geometry would profit from looking at [75] or the introductory chapters of [124, 214, 262].

A.1 Affine Varieties

A.1.1 Basic Definitions and Results

We will let k denote a field and $\bar{k}$ an algebraic closure of k. *Throughout Appendix A we shall assume, unless otherwise stated, that k has characteristic zero.* We shall occasionally comment on how the results need to be modified for fields of nonzero characteristic. A *k-algebra R* is a commutative ring, having a unit element 1, and containing k as a subring such that $1 \in k$. A homomorphism $\phi : A \to B$ of k-algebras is a ring homomorphism such that ϕ is k-linear (or what is the same, the identity on k). A k-algebra R is called *finitely generated* if there are elements $f_1, \ldots, f_n \in R$ such that every element in R is a (finite) k-linear combination of the elements $f_1^{m_1} \cdots f_n^{m_n}$ with all $m_i \in \mathbf{Z},\ m_i \geq 0$. The $f_1, \ldots, f_n$ are called generators for R over k.

Suppose that the k-algebra R is generated by $f_1, \ldots, f_n$ over k. Define the homomorphism of k-algebras $\phi : k[X_1, \ldots, X_n] \to R$ by $\phi(X_i) = f_i$ for all i. Then clearly ϕ is surjective. The kernel of ϕ is an ideal $I \subset k[X_1, \ldots, X_n]$ and one has $k[X_1, \ldots, X_n]/I \cong R$. Conversely, any k-algebra of the form $k[X_1, \ldots, X_n]/I$ is finitely generated.

A k-algebra R is called *reduced* if $r^n = 0$ (with $r \in R$ and $n \geq 1$) implies that $r = 0$. An ideal I in a (commutative) ring R is called *radical* if $r^n \in I$ (with $n \geq 1$ and $r \in R$) implies that $r \in I$. Thus $k[X_1, \ldots, X_n]/I$ is a reduced finitely generated k-algebra if and only if the ideal I is radical.

The principal definition in this section is the following one.

Definition A.2 An *affine variety over k* is a pair $X := (\max(A), A)$, where A is a finitely generated k-algebra and $\max(A)$ is the set of all maximal ideals of A. This affine variety is called *reduced* if A is reduced. □

Of course, the set $\max(A)$ is completely determined by A and it may seem superfluous to make it part of the definition. Nonetheless, we have included it because $\max(A)$ will be used to state many ring theoretic properties of A in a more geometric way and so is the basic geometric counterpart of the ring A. The set $X := \max(A)$ can be given more structure, namely a topology (see below) and a sheaf O_X of k-algebras (the structure sheaf). Both structures are determined by the k-algebra A. In this way, one obtains a "ringed space" (X, O_X). Since $A = H^0(X, O_X)$, the ringed space determines A. The more usual definition of an affine variety over the field k is: a ringed space that is isomorphic to the ringed space of a finitely generated k-algebra. Thus the above definition of affine variety over k is equivalent to the usual one. We have chosen this definition in order to simplify the exposition. One can reformulate the above definition by saying that the category of the affine varieties over k is the opposite (i.e., the arrows go in the opposite way) of the category of the finitely generated k-algebras.

For an affine variety X, the set $\max(A)$ is provided with a topology, called the *Zariski topology*. To define this topology it is sufficient to describe the closed sets. A subset $S \subset \max(A)$ is called (Zariski-)closed if there are elements $\{f_i\}_{i \in I} \subset A$ such that a maximal ideal $\underline{m}$ of A belongs to S if and only if $\{f_i\}_{i \in I} \subset \underline{m}$. We will use the notation $S = Z(\{f_i\}_{i \in I})$.

The following statements are easily verified:

(1) If $\{G_j\}_{j\in J}$ is a family of closed sets, then $\cap_{j\in J} G_j$ is a closed set.
(2) The union of two (or any finite number of) closed sets is closed.
(3) The empty set and $\max(A)$ are closed.
(4) Every finite set is closed.
(5) Any closed set S is of the form $Z(J)$ for some ideal $J \subset A$.

Statement (5) can be refined using the Hilbert Basissatz. A commutative ring (with 1) R is called *noetherian* if every ideal $I \subset R$ is finitely generated, i.e., there are elements $f_1, \dots, f_s \in I$ such that $I = (f_1, \dots, f_s) := \{g_1 f_1 + \dots + g_s f_s | \, g_1, \dots, g_s \in R\}$.

Hilbert Basissatz: If R is a noetherian ring then $R[x]$ is a noetherian ring. In particular, this implies that $k[X_1, \dots, X_n]$ is noetherian and so any finitely generated k-algebra is noetherian.

We refer to [170], Chap. IV, §4 for a proof of this result. Statement (5) above can now be restated as: Any closed set S is of the form $Z(f_1, \dots, f_m)$ for some finite set $\{f_1, \dots, f_m\} \in A$.

The above definitions are rather formal in nature and we will spend some time on examples in order to convey their meaning and the geometry involved.

Example A.3 *The affine line* $\mathbf{A}_k^1$ *over* k

By definition $\mathbf{A}_k^1 = (\max(k[X]), k[X])$. Every ideal of $k[X]$ is principal, i.e., generated by a single element $F \in k[X]$. The ideal (F) is maximal if and only if F is an irreducible (nonconstant) polynomial. Thus the set $\max(k[X])$ can be identified with the set of monic irreducible polynomials in $k[X]$. The closed subsets of $\max(k[X])$ are the finite sets, the empty set, and $\max(k[X])$ itself. The (Zariski-)open sets are the cofinite sets and the empty subset of $\max(k[X])$.
Suppose now that $k = \overline{k}$. Then every monic irreducible polynomial has the form $X - a$ with $a \in k$. Thus we can identify $\max(k[X])$ with k itself in this case. The closed sets for the (Zariski) topology on k are the finite sets and k itself.

Now we consider the case where $k \neq \overline{k}$. Let F be a monic irreducible element of $k[X]$. Since $\overline{k}$ is algebraically closed, there is a zero $a \in \overline{k}$ of F. Consider the k-algebra homomorphism $\phi : k[X] \to \overline{k}$, given by $\phi(X) = a$. The kernel of ϕ is easily seen to be this maximal ideal (F). This ideal gives rise to a surjective map $\tau : \overline{k} \to \max(k[X])$, defined by $\tau(a)$ is the kernel of the k-algebra homomorphism $k[X] \to \overline{k}$, which sends X to a. The map τ is not bijective, since a monic irreducible polynomial $F \in k[X]$ can have more than one zero in $\overline{k}$. Let us introduce on $\overline{k}$ the equivalence relation $\sim$ by $a \sim b$ if a and b satisfy the same monic minimal polynomial over k. Then $\overline{k}/\sim$ is in bijective correspondence with $\max(k[X])$. □

One can generalize Example A.3 and define the n-dimensional affine space $\mathbf{A}_k^n$ over k to be $\mathbf{A}_k^n = (\max(k[X_1, \ldots, X_n]), k[X_1, \ldots, X_n])$. To describe the structure of the maximal ideals we will need:

Hilbert Nullstellensatz: For every maximal ideal $\underline{m}$ of $k[X_1, \ldots, X_n]$ the field $k[X_1, \ldots, X_n]/\underline{m}$ has a finite dimension over k.

Although this result is well known ([170], Chap. IX, §1), we shall give a proof when the characteristic of k is 0 since the proof uses ideas that we have occasion to use again (see Lemma 1.17). A proof of this result is also outlined in Exercise A.25. We start with the following lemma.

Lemma A.4 *Let F be a field of characteristic zero, R a finitely generated integral domain over F and $x \in R$ such that $S = \{c \in F \mid x - c$ is invertible in $R\}$ is infinite. Then x is algebraic over F.*

Proof. (Rosenlicht) We may write $R = F[x_1, \ldots x_n]$ for some $x_i \in R$ and $x_1 = x$. Assume that x_1 is not algebraic over F and let K be the quotient field of R. Let $x_1, \ldots, x_r$ be a transcendence basis of K over F and let $y \in R$ be a primitive element of K over $F(x_1, \ldots, x_r)$. Let $G \in F[x_1, \ldots, x_r]$ be chosen so that the minimum polynomial of y over $F[x_1, \ldots x_r]$ has leading coefficient dividing G and $x_1, \ldots, x_n \in F[x_1, \ldots x_r, y, G^{-1}]$. Since S is infinite, there exist $c_1, \ldots, c_r \in S$ such that $G(c_1, \ldots, c_r) \neq 0$. One can then define a homomorphism of $F[x_1, \ldots x_r, y, G^{-1}]$ to $\overline{F}$, the algebraic closure of F, such that $x_i \mapsto c_i$ for $i = 1, \ldots, r$. Since $R \subset F[x_1, \ldots x_r, y, G^{-1}]$, this contradicts the fact that $x_1 - c_1$ is invertible in R. □

Note that the hypothesis that F is of characteristic zero is only used when we invoke the Primitive Element Theorem and so the proof remains valid when the characteristic of k is $p \neq 0$ and $F^p = F$. To prove the Hilbert Nullstellensatz, it is enough to show that the image x_i of each X_i in $K = k[X_1, \ldots, X_n]/\underline{m}$ is algebraic over k. Since x_i can equal at most one element of k, there are an infinite number of $c \in k$ such that $x_i - c$ is invertible. Lemma A.4 yields the desired conclusion. A proof in the same spirit as above that holds in all characteristics is given in [204].

Exercise A.5 *Hilbert Nullstellensatz*
1. Show that a set of polynomials $\{f_\alpha\} \subset k[X_1, \ldots, X_n]$ have a common zero in some algebraic extension of k if and only if $1 \notin I$, where I is the ideal generated by $\{f_\alpha\}$.
2. Let $a_1, \ldots, a_n \in k$. Show that the ideal $(X_1 - a_1, \ldots X_n - a_n)$ is a maximal ideal in $k[X_1, \ldots, X_n]$.
3. Assume that k is algebraically closed. Show that the maximal ideals of $k[X_1, \ldots, X_n]$ are of the form $(X_1 - a_1, \ldots X_n - a_n)$ for some $a_i \in k$. Hint: If $\underline{m}$ is maximal, the Hilbert Nullstellensatz says that $k[X_1, \ldots, X_n]/\underline{m}$ is an algebraic extension of k and so equal to k. □

We now turn to a description of $\mathbf{A}_k^n$.

Example A.6 *The n-dimensional affine space* $\mathbf{A}_k^n$ *over k*

By definition $\mathbf{A}_k^n = (\max(k[X_1, \ldots, X_n]), k[X_1, \ldots, X_n])$. The Hilbert Nullstellensatz clarifies the structure of the maximal ideals. Let us first consider the case where k is algebraically closed, i.e., $k = \overline{k}$. From Exercise A.5, we can conclude that any maximal ideal $\underline{m}$ is of the form $(X_1 - a_1, \ldots, X_n - a_n)$ for some $a_i \in k$. Thus we can identify $\max(k[X_1, \ldots, X_n])$ with k^n. We use the terminology "affine space" since the structure of k^n as a linear vector space over k is not included in our definition of $\max(k[X_1, \ldots, X_n])$.

In the general case, where $k \neq \overline{k}$, things are somewhat more complicated. Let $\underline{m}$ be a maximal ideal. The field $K := k[X_1, \ldots, X_n]/\underline{m}$ is a finite extension of k so there is a k-linear embedding of K into $\overline{k}$. For notational convenience, we will suppose that $K \subset \overline{k}$. Thus we have a k-algebra homomorphism $\phi : k[X_1, \ldots, X_n] \to \overline{k}$ with kernel $\underline{m}$. This homomorphism is given by $\phi(X_i) = a_i$ $(i = 1, \ldots, n$ and certain elements $a_i \in \overline{k})$. On the other hand, for any point $a = (a_1, \ldots, a_n) \in \overline{k}^n$, the k-algebra homomorphism ϕ, which sends X_i to a_i, has as kernel a maximal ideal of $k[X_1, \ldots, X_n]$. Thus we find a surjective map $\overline{k}^n \to \max(k[X_1, \ldots, X_n])$. On $\overline{k}^n$ we introduce the equivalence relation $a \sim b$ by, if $F(a) = 0$ for any $F \in k[X_1, \ldots, X_n]$ implies $F(b) = 0$. Then $\overline{k}^n / \sim$ is in bijective correspondence with $\max(k[X_1, \ldots, X_n])$. □

Exercise A.7 *Radical ideals and closed sets*
One considers two sets: $\mathcal{R}$, the set of all radical ideals of $k[X_1, \ldots, X_n]$ and $\mathcal{Z}$, the set of all closed subsets of $\max(k[X_1, \ldots, X_n])$. For any closed subset V we denote by $I(V)$ the ideal consisting of all $F \in k[X_1, \ldots, X_n]$ with $F \in \underline{m}$ for all $\underline{m} \in V$. For any radical ideal I we consider

$$Z(I) := \{\underline{m} \in \max(k[X_1, \ldots, X_n]) | \; I \subset \underline{m}\} .$$

1. Prove that the maps $Z : \mathcal{R} \to \mathcal{Z}$ and $id : \mathcal{Z} \to \mathcal{R}$ are inverses of each other.
Hint: Suppose that I is a radical ideal and that $f \notin I$. To prove that there is a maximal ideal $\underline{m} \supset I$ with $f \notin \underline{m}$, consider the ideal $J = (I, YF - 1)$ in the polynomial ring $k[X_1, \ldots, X_n, Y]$. If $1 \in J$, then

$$\begin{aligned} 1 = & \; g(X_1, \ldots, X_n, Y) \cdot (Yf(X_1, \ldots, X_n) - 1) \\ & + \sum g_\alpha(X_1, \ldots, X_n, Y) f_\alpha(X_1, \ldots, X_n) \end{aligned}$$

with the $f_\alpha \in I$ and $g, g_\alpha \in k[X_1, \ldots, X_n, Y]$. Substituting $Y \mapsto \frac{1}{f}$ and clearing denominators implies that $f^n \in I$ for some positive integer n. Therefore, $1 \notin J$ and so there exists a maximal ideal $\underline{m}' \supset J$. Let $\underline{m} = \underline{m}' \cap k[X_1, \ldots, X_n]$.
2. Assume that k is algebraically closed. Define a subset $\mathcal{X} \subset k^n$ to be closed if $\mathcal{X}$ is the set of common zeros of a collection of polynomials in $k[X_1, \ldots, X_n]$. For

any closed $\mathcal{X} \subset k^n$ let $\mathcal{I}(\mathcal{X})$ be the set of polynomials vanishing on $\mathcal{X}$. For any ideal I define $\mathcal{Z}(I)$ to be the set of common zeros in k^n of the elements of I. Use the Hilbert Nullstellensatz and part 1 to show that the maps $\mathcal{Z}$ and $\mathcal{I}$ define a bijective correspondence between the set of radical ideals of $k[X_1, \dots, X_n]$ and the collection of closed subsets of k^n. □

For an affine variety $X = (\max(A), A)$ one sometimes writes X for the topological space $\max(A)$ and $O(X)$ for A. One calls $O(X)$ or A the ring of *regular functions on* X. Indeed, any $g \in A$ can be seen to be a function on $\max(A)$. The value $g(\underline{m})$ is defined as the image of g under the map $A \to A/\underline{m}$. When $k = \overline{k}$, each $A/\underline{m}$ identifies with k, and so any $g \in A$ can be seen as an ordinary function on $\max(A)$ with values in k. We shall frequently identify $g \in O(X)$ with the map it induces from $\max(A)$ to $A/\underline{m}$. For example, the set $\{x \in X \mid g(x) \neq 0\}$ denotes the set of maximal ideals in A not containing g. Exercise A.7 implies that the intersection of all maximal ideals is $\{0\}$ so the identification of f with the function it induces is injective. One also calls $O(X)$ the *coordinate ring of* X. A *morphism* $X = (\max(A), A) \to Y = (\max(B), B)$ of affine varieties over k, is defined to be a pair (f, ϕ) satisfying:

1. $\phi : B \to A$ is a k-algebra homomorphism.
2. $f : \max(A) \to \max(B)$ is induced by ϕ in the following manner: for any maximal ideal $\underline{m}$ of A, $f(\underline{m}) = \phi^{-1}(\underline{m})$.

We note that since A and B are finitely generated over k, if $\underline{m}$ is a maximal ideal of B and $\phi : B \to A$ is a k-algebra homomorphism, then $\phi^{-1}(\underline{m})$ is always a maximal ideal of A. The Nullstellensatz implies that $B/\underline{m}$ is an algebraic extension of k and so the induced map $\overline{\phi} : A/\phi^{-1}(\underline{m}) \to B/\underline{m}$ maps $A/\phi^{-1}(\underline{m})$ onto a finitely generated k-subalgebra of $B/\underline{m}$. Therefore, $A/\phi^{-1}(\underline{m})$ is again a field and so $\phi^{-1}(\underline{m})$ is again a maximal ideal.

In concrete terms, let $A = k[X_1, \dots, X_n]/I$, $B = k[Y_1, \dots, Y_m]/J$ and let $f_1, \dots, f_m \in k[X_1, \dots, X_n]$ have the property that for any $G(Y_1, \dots, Y_m) \in J$, $G(f_1, \dots, f_m) \in I$. Then the map $\phi : B \to A$ given by $\phi(Y_i) = f_i$ determines a k-homomorphism and yields a morphism from X to Y. Furthermore, any such morphism arises in this way. If $\tilde{f}_1, \dots, \tilde{f}_m \in k[X_1, \dots, X_n]$ also satisfy $G(\tilde{f}_1, \dots, \tilde{f}_m) = 0$ for all $G \in J$ and ψ is defined by $\psi(Y_i) = \tilde{f}_i$, then ϕ and ψ yield the same morphism if and only if $f_i - \tilde{f}_i \in I$ for $i = 1, \dots, m$.

We note that f is a continuous map. One sometimes uses the notations $f = \phi^*$ and $\phi = f^*$. The important thing to note is that only very special continuous maps $\max(A) \to \max(B)$ are of the form ϕ^* for some k-algebra homomorphism ϕ. Moreover, only for reduced affine varieties will the topological map $f : \max(A) \to \max(B)$ determine ϕ.

Exercise A.8 *Continuous maps on* $\max(A)$.
Let $X = (\max(A), A)$ and $Y = (\max(B), B)$ be reduced affine varieties over an

algebraically closed field k. Then $O(X)$ and $O(Y)$ can be considered as rings of functions on the spaces $\max(A)$ and $\max(B)$. Let $f : \max(A) \to \max(B)$ be a continuous map.1. Show that there is a k-algebra homomorphism $\phi : B \to A$ with $f = \phi^*$ if and only for every $b \in B$ the function $\max(A) \xrightarrow{f} \max(B) \xrightarrow{b} k$ belongs to A.2. Suppose that f satisfies the condition of (1). Show that the ϕ with $f = \phi^*$ is unique. □

Let $X = (\max(A), A)$ be a reduced affine variety. A *closed reduced subvariety* Y of X is defined as a pair $(\max(A/I), A/I)$, where I is a radical ideal of A.

Exercises A.9 *Subvarieties*

1. Determine the Zariski-closed subsets of $\mathbf{A}_k^1$.

2. Let V be a reduced closed subvariety of $\mathbf{A}_k^1$. Determine $O(V)$.

3. Let $X := (\max(A), A)$ be a reduced affine variety and consider an $f \in A$ with $f \neq 0$. Define $(W, O(W))$ by $O(W) = A[1/f] = A[T]/(Tf - 1)$ and $W \subset \max(A)$ is the open subset $\{\underline{m} | f \notin \underline{m}\}$ with the induced topology. Prove that $(W, O(W))$ is a reduced affine variety and show that $(W, O(W))$ is isomorphic to the closed reduced subspace $(\max(A[T]/(Tf - 1)), A[T]/(Tf - 1))$ of $(\max(A[T]), A[T])$.

4. Let V be a reduced affine variety. Prove that there is a 1-1 relation between the closed subsets of V and the radical ideals of $O(V)$.

5. Let V be a reduced affine variety. Prove that there is no infinite decreasing set of closed subspaces. Hint: Such a sequence would correspond to an increasing sequence of (radical) ideals. Prove that the ring $O(V)$ is also noetherian and deduce that an infinite increasing sequence of ideals in $O(V)$ cannot exist.

6. Let V be a reduced affine variety and S a subset of V. The *Zariski closure* of S is defined as the smallest closed subset of V containing S. Show that the Zariski closure exists. Show that the Zariski closure corresponds to the radical ideal $I \subset O(V)$ consisting of all regular functions vanishing on S.

7. Determine all the morphisms from $\mathbf{A}_k^1$ to itself.

8. Suppose that the reduced affine varieties X and Y are given as closed subsets of $\mathbf{A}_k^n$ and $\mathbf{A}_k^m$. Prove that every morphism $f : X \to Y$ is the restriction of a morphism $F : \mathbf{A}_k^n \to \mathbf{A}_k^m$ that satisfies $F(X) \subset Y$.

9. Show by example that the image of a morphism $f : X \to \mathbf{A}_k^1$ is, in general, not a closed subset of $\mathbf{A}_k^1$. □

In connection with the last exercise we formulate a useful result about the image $f(X) \subset Y$ of a morphism of affine varieties: *$f(X)$ is a finite union of subsets of Y of* *the form $V \cap W$ with V closed and W open.* We note that the subsets of Y described in the above statement are called *constructible*. For a proof of the statement we refer to [141], p. 33.

In the following, all affine varieties are supposed to be reduced and we will omit the adjective "reduced". An affine variety X is called *reducible* if X can be written as the union of two proper closed subvarieties. For "not reducible" one uses the term *irreducible*.

Lemma A.10
1. The affine variety X is irreducible if and only if $O(X)$ has no zero divisors.
2. Every affine variety X can be written as a finite union $X_1 \cup \cdots \cup X_s$ of irreducible closed subsets.
3. If one supposes that no X_i is contained in X_j for $j \neq i$, then this decomposition is unique up to the order of the X_i and the X_i are called the irreducible components *of X.*

Proof. 1. Suppose that $f, g \in O(X)$ satisfy $f \neq 0 \neq g$ and $fg=0$. Put $X_1 = \{a \in X |\ f(a) = 0\}$ and $X_2 = \{a \in X |\ g(a) = 0\}$. Then $X = X_1 \cup X_2$ and X is reducible. The other implication can be proved in a similar way.

2. If X is reducible, then one can write $X = Y \cup Z$ with the Y, Z proper closed subsets. If both Y and Z are irreducible then we can stop. If, say, Y is reducible then we write $Y = D \cup E$ and find $X = Z \cup D \cup E$, and so on. If this process does not stop, then we find a decreasing sequence of closed subsets, say $F_1 \supset F_2 \supset F_3 \supset \cdots$ of X. By Exercise A.9.5, this cannot happen. Thus X can be written as $X_1 \cup X_2 \cup \cdots \cup X_s$, which each X_i closed and irreducible.

3. Suppose that there is no inclusion between the X_i. Let $Y \subset X$ be a closed irreducible subset. Then $Y = (Y \cap X_1) \cup \cdots \cup (Y \cap X_s)$ and since Y is irreducible one finds that $Y = Y \cap X_i$ for some i. In other words, $Y \subset X_i$ for some i. This easily implies the uniqueness of the decomposition. □

Exercise A.11 *Rational functions on a variety.*
Let $X = (\max(A), A)$ be an affine variety. We define the *ring of rational functions* $k(X)$ *on* X to be the total quotient ring $Qt(A)$ of A. This is the localization of A with respect to the multiplicative set of nonzero divisors of A (see Definition 1.5.1(d)). Note that the definition of localization specializes in this case to: $(r_1, s_1) \sim (r_2, s_2)$ if $r_1 s_2 - r_2 s_1 = 0$. We say that $f \in k(X)$ is *defined at* $\underline{m} \in \max(Z)$ if there exist $g, h \in A$ such that $f = g/h$ and $h \notin \underline{m}$.

1. Show that if X is irreducible, then $k(X)$ is a field.

2. Show that for $f \in k(X)$ there exists an open dense subset $U \subset X$ such that f is defined at all points of X.

3. Let $X = \cup_{i=1}^{t} X_i$ be the decomposition of X into irreducible components. For each i we have the map $g \in O(X) \mapsto g|_{X_i} \in O(X_i)$. This induces a map $k(X) \to k(X_i)$ sending $f \in k(X)$ to $f|_{X_i}$. Show that the map $k(X) \to k(X_1) \times \dots k(X_t)$ defined by $f \mapsto (f|_{X_1}, \dots, f|_{X_t})$ is an isomorphism of k-algebras.

4. Show that, for $f \in k(X)$, $f \in A$ if and only if f is defined at $\underline{m}$ for all $\underline{m} \in \max(A)$. Hint: Let $I \subset A$ be the ideal generated by all $h \in A$ such that there exists an element $g \in A$ with $f = g/h$. If f defined at all $\underline{m} \in \max(A)$, then $I = (1)$. Therefore, there exist $g_1, \dots, g_m, h_1, \dots, h_m, t_1, \dots, t_m \in A$ such that $1 = \sum_{i=1}^{m} t_i h_i$ and, for each i, $f = g_i/h_i$. Show that $f = \sum_{i=1}^{m} t_i g_i \in A$. □

We end this section with the following concept. If $S \subset k[X_1, \dots, X_n]$ is a set of polynomials and $k' \supset k$ is an extension field of k, it is intuitively clear what is meant by a common zero of S in $(k')^n$. We shall need to talk about common zeros of a set of polynomials in any k-algebra R as well as some functorial properties of this notion. We formalize this with the following definition.

Definition A.12 Let k be a field and X an affine variety defined over k. For any k-algebra R, we define the *set of R-points of X*, $X(R)$ to be the set of k-algebra homomorphisms $O(X) \to R$. □

Examples A.13 *R-points*

1. Let $k = \mathbf{Q}$ and let X be the affine variety corresponding to the ring $\mathbf{Q}[X]/(X^2+1)$. In this case $X(\mathbf{Q})$ and $X(\mathbf{R})$ are both empty, while $X(\mathbf{C})$ has two elements.

2. Assume that k is algebraically closed. The Hilbert Nullstellensatz implies that $X(k)$ corresponds to the set of maximal ideals of $O(X)$ (see Example A.6). □

One can show that every k-algebra homomorphism $R_1 \to R_2$ induces the obvious map $X(R_1) \to X(R_2)$. Furthermore, if F is a morphism from X to Y, then F induces a map from $X(R)$ to $Y(R)$. In particular, an element f of $O(X)$ can be considered as a morphism from X to $\mathbf{A}_k^1$ and so gives a map f_R from $X(R)$ to $\mathbf{A}_k^1(R) = R$. In fact, one can show that the map $R \mapsto X(R)$ is a covariant functor from k-algebras to sets. This is an example of a representable functor (see Definition B.18).

Exercises A.14 *$\overline{k}$-points.*

Let $\overline{k}$ be the algebraic closure of k and let X and Y be affine varieties over k.

1. Use the Hilbert Nullstellensatz to show that for any $f \in O(X)$, $f = 0$ if and only if f is identically zero on $X(\overline{k})$. Hint: Let $O(X) = k[X_1, \dots, X_n]/\underline{q}$, $\underline{q}$ a radical ideal. Use Exercise A.7.1 to show that if $f \notin \underline{q}$ then there exists a maximal ideal $\underline{m} \supset \underline{q}$ with $f \notin \underline{m}$. $O(X)/\underline{m}$ is algebraic over k and so embeds in $\overline{k}$.

2. Let $f : X \mapsto Y, g : X \mapsto Y$ be morphisms. Show that $f = g$ if and only if $f = g$ on $X(\bar{k})$.

3. Show that max $O(X)$ is finite if and only if $X(\bar{k})$ is finite.

4. Assume that X is irreducible. Show that $|\max O(X)| < \infty$ if and only if $|\max O(X)| = 1$. Conclude that if $|\max O(X)| < \infty$, then $O(X)$ is a field. Hint: For each nonzero maximal ideal $\underline{m}$ of $O(X)$, let $0 \neq f_{\underline{m}} \in \underline{m}$. Then $g = \prod f_{\underline{m}}$ is zero on $X(\bar{k})$ so $g = 0$ contradicting $O(X)$ being a domain. Therefore, $O(X)$ has no nonzero maximal ideals. □

A.1.2 Products of Affine Varieties over *k*

For the construction of *products of affine varieties* we need another technical tool, namely *tensor products over a field k*. We begin with a review of their important properties.

Let V, W, and Z be vector spaces over a field k. A *bilinear map* $f : V \times W \to Z$ is a map $(v, w) \mapsto f(v, w) \in Z$, which has the properties $f(v_1 + v_2, w) = f(v_1, w) + f(v_2, w)$, $f(v, w_1 + w_2) = f(v, w_1) + f(v_2, w)$ and $f(\lambda v, w) = f(v, \lambda w) = \lambda f(v, w)$ for all $\lambda \in k$. The *tensor product* $V \otimes_k W$ is a new vector space over k together with a bilinear map $u : V \times W \to V \otimes_k W$ such that for any bilinear map $f : V \times W \to Z$ there exists a unique linear map $F : V \otimes_k W \to Z$ such that $f = F \circ u$ (see [170], Chap. 16 for a proof that tensor products exist and are unique as well as for a more complete discussion of the subject). For $v \in V, w \in W$ we denote $u(v, w)$ by $v \otimes w$ and, when this will not lead to confusion, we denote $V \otimes_k W$ by $V \otimes W$. The bilinearity of u then translates as the following three rules:

$$\begin{aligned}(v_1 + v_2) \otimes w &= (v_1 \otimes w) + (v_2 \otimes w),\\ v \otimes (w_1 + w_2) &= (v \otimes w_1) + (v \otimes w_2),\\ \lambda(v \otimes w) &= (\lambda v) \otimes w \;=\; v \otimes (\lambda w) \text{ for all } \lambda \in K.\end{aligned}$$

If $\{v_i\}_{i \in I}$ is a basis of V and $\{w_j\}_{j \in J}$ is a basis of W, then one can show that $\{v_i \otimes w_j\}_{i \in I, j \in J}$ is a basis of $V \otimes W$.

Exercises A.15 *Elementary properties of tensor products*

1. Use the universal property of the map u to show that if $\{v_1, \ldots v_n\}$ are linear independent elements of V then $\sum v_i \otimes w_i = 0$ implies that each $w_i = 0$. Hint: for each $i = 1, \ldots, n$ let $f_i : V \times W \to W$ be a bilinear map such that $f(v_i, w) = w$ and $f(v_j, w) = 0$ if $j \neq i$ for all $w \in W$.

2. Show that if $v_1, v_2 \in V \backslash \{0\}$ and $w_1, w_2 \in W \backslash \{0\}$ then $v_1 \otimes w_1 = v_2 \otimes w_2$ implies that there exists an element $a \in k$ such that $v_1 = av_2$ and $w_1 = \frac{1}{a} w_2$. In particular, if $v \neq 0$ and $w \neq 0$ the $v \otimes w \neq 0$.

3. Show that if $\{v_i\}_{i\in I}$ is a basis of V and $\{w_j\}_{j\in J}$ is a basis of W, then $\{v_i \otimes w_j\}_{i\in I, j\in J}$ is a basis of $V \otimes W$.

4. Let $V_1 \subset V_2$ and W be vector space over k. Prove that there is a canonical isomorphism $(V_2 \otimes W)/(V_1 \otimes W) \cong (V_1/V_2) \otimes W$. □

Let R_1 and R_2 be commutative k-algebras with a unit element. One can define a multiplication on the tensor product $R_1 \otimes_k R_2$ by the formula $(r_1 \otimes r_2)(\tilde{r}_1 \otimes \tilde{r}_2) = (r_1\tilde{r}_1) \otimes (r_2\tilde{r}_2)$ (one uses the universal property of u to show that this is well defined and gives $R_1 \otimes R_2$ the structure of a k-algebra). In the special case $R_1 = k[X_1, \ldots, X_n]$ and $R_2 = k[Y_1, \ldots, Y_m]$ it is easily verified that $R_1 \otimes R_2$ is, in fact, the polynomial ring $k[X_1, \ldots, X_n, Y_1, \ldots, Y_m]$. More generally, let R_1, R_2 be finitely generated K-algebras. Represent R_1 and R_2 as $R_1 = k[X_1, \ldots, X_n]/(f_1, \ldots, f_s)$ and $R_2 = k[Y_1, \ldots, Y_m]/(g_1, \ldots, g_t)$.
Using Exercise A.15.4 one can show that $R_1 \otimes R_2$ is isomorphic to $k[X_1, \ldots, X_n, Y_1, \ldots, Y_m]/(f_1, \ldots f_s, g_1, \ldots, g_t)$.

We wish to study how reduced algebras behave under tensor products. Assume that k has characteristic $p > 0$ and let $a \in k$ be an element such that $b^p = a$ has no solution in k. If we let $R_1 = R_2 = k[X]/(X^p - a)$, then R_1 and R_2 are fields. The tensor product $R_1 \otimes R_2$ is isomorphic to $k[X, Y]/(X^p - a, Y^p - a)$. The element $t = X - Y$ modulo $(X^p - a, Y^p - a)$ has the property $t^p = 0$. Thus $R_1 \otimes_k R_2$ contains nilpotent elements! This is an unpleasant characteristic p-phenomenon that we want to avoid. A field k of characteristic $p > 0$ is called *perfect* if every element is a p-th power. In other words, the map $a \mapsto a^p$ is a bijection on k. One can show that an irreducible polynomial over such a field has no repeated roots and so all algebraic extensions of k are separable. The following technical lemma tells us that the above example is more or less the only case where nilpotents can occur in a tensor product of k-algebras without nilpotents.

Lemma A.16 *Let R_1, R_2 be k-algebras having no nilpotent elements. Suppose that either the characteristic of k is zero or that the characteristic of k is $p > 0$ and k is perfect. Then $R_1 \otimes_k R_2$ has no nilpotent elements.*

Proof. Suppose that $a \in R_1 \otimes R_2$ satisfies $a \neq 0$ and $a^2 = 0$. From this we want to derive a contradiction. It is easily verified that for inclusions of k-algebras $R_1 \subset S_1$ and $R_2 \subset S_2$, one has an inclusion $R_1 \otimes R_2 \to S_1 \otimes S_2$. Thus we may suppose that R_1 and R_2 are finitely generated over k. Take a k-basis $\{e_i\}$ of R_2. The element a can be written as a finite sum $\sum_i a_i \otimes e_i$ with all $a_i \in R_1$. Let $a_j \neq 0$. Because a_j is not nilpotent, there is a maximal ideal $\underline{m}$ of R_1 that does not contain a_j. The residue class field $L := R_1/\underline{m}$ is, according to Hilbert's theorem, a finite extension of k. Since the image of a in $L \otimes R_2$ is not zero, we may assume that R_1 is a finite field extension of k. Likewise we may suppose that R_2 is a finite field extension of k. According to the primitive element theorem [170], one can write $R_2 = k[X]/(F)$, where F is an irreducible and separable polynomial. Then $R_1 \otimes R_2 \cong L[X]/(F)$. The latter ring has no nilpotents since F is a separable polynomial. □

Corollary A.17 *Let k be a field as in Lemma A.16 and let q be a prime ideal in $k[X_1, \dots, X_n]$. If K is an extension of k, then $qK[X_1, \dots, X_n]$ is a radical ideal in $K[X_1, \dots, X_n]$.*

Proof. From Exercise A.15.4, one sees that $K[X_1, \dots, X_n]/qK[X_1, \dots, X_n]$ is isomorphic to $k[X_1, \dots, X_n]/q \otimes_k K$. This latter ring has no nilpotents by Lemma A.16, so $qK[X_1, \dots, X_n]$ is radical. □

We note that one cannot strengthen Corollary A.17 to say that if p is a prime ideal in $k[X_1, \dots, X_n]$ then $pK[X_1, \dots, X_n]$ is a prime ideal in $K[X_1, \dots, X_n]$. For example, $X^2 + 1$ generates a prime ideal in $\mathbf{Q}[X]$ but it generates a nonprime radical ideal in $\mathbf{C}[X]$.

We will assume that the characteristic of k is zero or that the characteristic of k is $p > 0$ and k is perfect. As we have seen there is a bijective translation between reduced affine varieties over k and finitely generated reduced k-algebras. For two reduced affine varieties X_1 and X_2 we want to define a product $X_1 \times X_2$, which should again have the structure of a reduced affine variety over k. Of course, the product $\mathbf{A}_k^n \times \mathbf{A}_k^m$ should be $\mathbf{A}_k^{n+m}$. For reduced affine varieties $V \subset \mathbf{A}_k^n$, $W \subset \mathbf{A}_k^m$ the product should be $V \times W$, seen as a reduced affine subvariety of $\mathbf{A}_k^{n+m}$. This is true, but there is the problem that V and W can be embedded as reduced subvarieties of the affine varieties $\mathbf{A}_k^{m+n}$ in many ways and that we have to prove that the definition of the product is independent of the embedding. This is where the tensor product comes in.

Definition A.18 Let X_1, X_2 be reduced affine varieties over k. The *product* $X_1 \times_k X_2$ is the reduced affine variety corresponding to the tensor product $O(X_1) \otimes_k O(X_2)$. □

We will sometimes use the notation $X_1 \times X_2$ instead of $X_1 \times_k X_2$ when the field k is clear from the context. We have verified that $O(X_1) \otimes_k O(X_2)$ is a finitely generated reduced k-algebra. Thus the definition makes sense. If X_1 and X_2 are presented as reduced subvarieties V of $\mathbf{A}_k^n$ and W of $\mathbf{A}_k^m$ then the rings $O(X_1)$ and $O(X_2)$ are presented as $k[Y_1, \dots, Y_n]/(f_1, \dots f_s)$ and $k[Z_1, \dots, Z_m]/(g_1, \dots, g_t)$. The tensor product can therefore be presented as $k[Y_1, \dots, Y_n, Z_1, \dots, Z_m]/(f_1, \dots, f_s, g_1, \dots, g_t)$. The ideal $(f_1, \dots f_s, g_1, \dots, g_t)$ is a radical ideal, since the tensor product has no nilpotent elements. The zero set of this ideal is easily seen to be $V \times W$. When k is algebraically closed, then one can identify this zero set with the Cartesian product of the set of points of V and the set of points of W.

It will be necessary to "lift" a variety defined over a field k to a larger field $K \supset k$ and this can also be done using tensor products. If $V = (\max(A), A)$ is an affine variety defined over k, we define V_K to be the variety $(\max(A \otimes_k K), A \otimes_k K)$. Note that the k-algebra $A \otimes_k K$ has the structure of a K-algebra where $a \mapsto 1 \otimes a$ defines an embedding of K into $A \otimes_k K$. If we present the ring A as $k[X_1, \dots, X_n]/q$ then Exercise A.15 implies that the ring $A \otimes_k K = K[X_1, \dots, X_n]/qK[X_1, \dots, X_n]$.

In general, if k is not algebraically closed, then the product of irreducible varieties is not necessarily irreducible (see Exercise A.20.3). When k is algebraically closed this phenomenon cannot happen.

Lemma A.19 *Let k be an algebraically closed field and let X, Y be irreducible affine varieties over k. Then $X \times Y$ is irreducible.*

Proof. Since k is algebraically closed, it is enough to show that $X \times Y(k)$ is not the union of two proper, closed subsets. Let $X \times Y = V_1 \cup V_2$ where V_1, V_2 are closed sets. For any $x \in X(k)$, the set $\{x\} \times Y(k)$ is closed and irreducible over k. Therefore, $\{x\} \times Y(k) \subset V_1$ or $\{x\} \times Y(k) \subset V_2$. Let $X_i = \{x \in X \mid \{x\} \times Y(k) \subset V_i\}$. We claim that X_1 is closed. To see this, note that for each $y \in Y(k)$, the set W_y of $x \in X(k)$ such that $x \times y \in V_1$ is closed and $X_1 = \cap_{y \in Y(k)} W_y$. Similarly, X_2 is closed. Therefore $X = X_1$ or $X = X_2$, and therefore either $X \times Y = V_1$ or $X \times Y = V_2$. □

Exercises A.20 *Products*

1. Show that $\mathbf{A}_k^n \times \mathbf{A}_k^m \simeq \mathbf{A}_k^{n+m}$.

2. Show that the Zariski topology on $\mathbf{A}_k^2$ is *not* the same as the product topology on $\mathbf{A}_k^1 \times \mathbf{A}_k^1$.

3. Let k be a field of characteristic zero or a perfect field of characteristic $p > 0$, and let K be an algebraic extension of k with $[K : k] = n$. Show that the ring $K \otimes_k K$ is isomorphic to the sum of n copies of K. □

Let $\overline{k}$ be the algebraic closure of k. The following lemma will give a criterion for an affine variety V over $\overline{k}$ to be of the form $W_{\overline{k}}$ for some affine variety W over k, that is a criterion for V to be *defined over* k. We shall assume that V is a subvariety of $\mathbf{A}_{\overline{k}}^n$, that is, its coordinate ring is of the form $\overline{k}[X_1, \ldots X_n]/q$ for some ideal $q \subset \overline{k}[X_1, \ldots X_n]$. We can then identify the points $V(\overline{k})$ with a subset of $\overline{k}^n$. The Galois group $Aut(\overline{k}/k)$ acts on $\overline{k}^n$ coordinate-wise.

Lemma A.21 *Let $\overline{k}$ be the algebraic closure of k. An affine variety V over $\overline{k}$ is of the form $W_{\overline{k}}$ for some affine variety W over k if and only if $V(\overline{k})$ is stable under the action of $Aut(\overline{k}/k)$.*

Proof. If $V = W_{\overline{k}}$, then $V(\overline{k})$ is precisely the set of common zeros of an ideal $q \subset k[X_1, \ldots, X_n]$. This implies that $V(\overline{k})$ is stable under the above action.

Conversely, assume that $V(\overline{k})$ is stable under the action of $Aut(\overline{k}/k)$ and let $O(V) = \overline{k}[X_1, \ldots, X_n]/q$ for some ideal $q \in \overline{k}[X_1, \ldots, X_n]$. The action of $Aut(\overline{k}/k)$ on $\overline{k}$ extends to an action on $\overline{k}[X_1, \ldots X_n]$. The Nullstellensatz implies that q is stable under this action. We claim that q is generated by $q \cap k[X_1, \ldots, X_n]$. Let S be the $\overline{k}$ vector space generated by $q \cap k[X_1, \ldots, X_n]$. We will show that $S = q$. Assume this is not the case. Let $\{\alpha_i\}_{i\in I}$ be a k-basis of $k[X_1, \dots, X_n]$ such that for some $J \subset I$, $\{\alpha_i\}_{i\in J}$ is a k-basis of $q \cap k[X_1, \dots, X_n]$. Note that $\{\alpha_i\}_{i\in I}$ is also a $\overline{k}$-basis of $\overline{k}[X_1, \dots, X_n]$. Let $f = \sum_{i\in I\setminus J} c_i\alpha_i + \sum_{i\in J} c_i\alpha_i \in q$ and among all such elements select one such that the set of nonzero c_i, $i \in I\setminus J$ is as small as possible. We may assume that one of these nonzero c_i is 1. For any $\sigma \in Aut(\overline{k}/k)$, minimality implies that $f - \sigma(f) \in S$ and therefore that for any $i \in I\setminus J$, $c_i \in k$. Therefore, $\sum_{i\in I\setminus J} c_i\alpha_i = f - \sum_{i\in J} c_i\alpha_i \in q \cap k[X_1, \dots, X_n]$, and so $f \in S$. □

Exercise A.22 *$\overline{k}$-morphisms defined over k*
Let V and W be varieties over k.
1. Let $f \in O(V)\otimes_k\overline{k}$. The group $Aut(\overline{k}/k)$ acts on $O(V)\otimes_k\overline{k}$ via $\sigma(h\otimes g) = h\otimes\sigma(g)$. Show that $f \in O(V) \subset O(V) \otimes_k \overline{k}$ if and only if $\sigma(f) = f$ for all $\sigma \in Aut(\overline{k}/k)$.

2. We say that a morphism $f : V_{\overline{k}} \to W_{\overline{k}}$ is *defined over k* if $f^* : O(W) \otimes_k \overline{k} \to O(V) \otimes_k \overline{k}$ is of the form $g^* \otimes 1$ where g is a morphism from V to W. Show that f is defined over k if and only if $f^*(\sigma(v)) = \sigma(f^*(v))$, for all $v \in V(\overline{k})$ and $\sigma \in Aut(\overline{k}/k)$. □

Remark A.23 Since we are using the action of the Galois group in Lemma A.21 and Exercise A.22 we need to assume that either k is a perfect field (i.e., $k^p = k$) or replace $\overline{k}$ with the separable closure k^{sep} when the characteristic is nonzero. □

A.1.3 Dimension of an Affine Variety

The *dimension* of an affine variety X is defined as the maximal number d for which there exists a sequence $X_0 \supsetneq X_1 \supsetneq \cdots \supsetneq X_d$ of closed irreducible (nonempty) subsets of X. It is, a priori, not clear that d exists (i.e., is finite). It is clear, however, that the dimension of X is the maximum of the dimensions of its irreducible components. Easy examples are the following.

Examples A.24 1. If X is finite, then its dimension is 0.
2. The dimension of $\mathbf{A}_k^1$ is 1.

3. The dimension of $\mathbf{A}_k^n$ is $\geq n$ since one has the sequence of closed irreducible subsets $\{0\} \subset \mathbf{A}_k^1 \subset \mathbf{A}_k^2 \subset \cdots \subset \mathbf{A}_k^n$. □

The dimension of $\mathbf{A}_k^n$ should, of course, be n, but it is not so easy to prove this. One ingredient of the proof is formulated in the next exercises.

Exercises A.25 1. *Integral elements*
If $A \subset B$ are rings, we say that an element $b \in B$ is *integral over A* if it is the root of a polynomial $X^n + a_{n-1}X^{n-1} + \cdots + a_0$ with coefficients $a_i \in A$ and $n \geq 1$, ([170], Chap. VII, §1).
(a) Show that if $b \in B$ is integral over A then b belongs to a subring $B' \supset A$ of B that is finitely generated as an A-module.

(b) Show that if b belongs to a subring $B' \supset A$ of B that is finitely generated as an A-module, then b is integral over A. Hint: Let $b_1, \ldots, b_n$ be generators of B' as an A-module. There exist $a_{i,j} \in A$ such that $bb_i = \sum_{j=1}^{n} a_{i,j} b_j$. Therefore, the determinant

$$d = \det \begin{pmatrix} b - a_{1,1} & a_{1,2} & \ldots & a_{1,n} \\ a_{2,1} & b - a_{2,2} & \ldots & a_{2,n} \\ \vdots & \vdots & \vdots & \vdots \\ a_{n,1} & a_{n,2} & \ldots & b - a_{n,n} \end{pmatrix}$$

must be zero. This gives the desired polynomial.
(c) The ring B is said to be integral over A if each of its elements is integral over A. Show that if B is integral over A and C is integral over B the C is integral over A.
(d) Let B be integral over A and assume that B has no zero divisors. Show that A is a field if and only if B is a field.

2. *Noether Normalization Theorem*
In this exercise, we propose a proof of

> Assume that the field k is infinite and let $R = k[x_1, \ldots, x_n]$ be a finitely generated k-algebra. Then for some $0 \leq m \leq n$, there exist elements $y_1, \ldots, y_m \in R$, algebraically independent over k such that R is integral over $k[y_1, \ldots, y_m]$.

Let $R = k[X_1, \ldots, X_n]/I$ for some ideal I in the polynomial ring $k[X_1, \ldots, X_n]$.
(a) We say that $f \in k[X_1, \ldots, X_n]$ is in "Weierstrass form with respect to X_n", if $f = a_e X_n^e + a_{e-1} X_n^{e-1} + \cdots + a_1 X_n + a_0$ with all $a_i \in k[X_1, \ldots, X_{n-1}]$ and $a_e \in k^*$. Prove that for any element $g \in k[X_1, \ldots, X_n] \backslash k[X_1, \ldots, X_{n-1}]$ there exists an invertible linear transformation of the form $X_i \mapsto X_i + a_i X_n$ with $a_i \in k$ such that after this transformation the element f is in Weierstrass form with respect to X_n. Give a proof of the Noether normalization for the ring $R = k[X_1, \ldots, X_n]/(g)$.
(b) Let $f \in I$, $f \notin k[X_1, \ldots, X_{n-1}]$. Produce a linear change of the variables $X_1, \ldots, X_n$ as in (a) such that after this change of variables, f is in Weierstrass form with respect to X_n. Let $z_i = x_i + a_i x_n$ and show that R is integral over $S = k[z_1, \ldots, z_{n-1}]$. Use induction on n to show that there exist $y_1, \ldots, y_m \in S$, algebraically independent over k such that S is integral over $k[y_1, \ldots, y_m]$. Conclude that R is integral over $k[y_1, \ldots, y_m]$.
Remark: The Noether normalization theorem is valid for finite fields as well. If d is an integer greater than any exponent appearing in the polynomial f in (a), then the transformation $X_i \mapsto X_i + X_n^{d^i}$ will transform f into a polynomial in Weierstrass form and one can proceed as above.

(3) *Hilbert's Nullstellensatz*
Deduce this result from the Noether normalization theorem. Hint: Let $\underline{m}$ be a maximal ideal in $k[X_1, \ldots, X_n]$ and let $R = k[X_1, \ldots, X_n]/\underline{m}$. Assume R is integral over $S = k[y_1, \ldots, y_m]$ with $y_1, \ldots, y_m$ algebraically independent over k and $m \geq 1$.

By part 1.d above, S is a field, yielding a contradiction. Therefore, R is integral over k and so algebraic over k. □

Let X be an affine variety. We say that an injective k-algebra morphism $k[X_1, \ldots, X_d] \to O(X)$ is a Noether normalization if $O(X)$ is integral over the image of $k[X_1, \ldots, X_d]$.

Proposition A.26
1. Let X be an affine variety and let $k[X_1, \ldots, X_d] \to O(X)$ be a Noether normalization. Then the dimension of X is d.

2. Let X be an irreducible affine variety. Then its dimension is equal to the transcendence degree of the fraction field of $O(X)$ over k.

Proof. 1. We need again some results from ring theory, which carry the names "going up" and "lying over" theorems (cf. [10], Corollary 5.9 and Theorem 5.11, or [141]). We refer to the literature for proofs. They can be formulated as follows:

Given are $R_1 \subset R_2$, two finitely generated k-algebras, such that R_2 is integral over R_1. Then for every strictly increasing chain of prime ideals $\underline{p}_1 \subset \cdots \subset \underline{p}_s$ of R_2 the sequence of prime ideals $(\underline{p}_1 \cap R_1) \subset \cdots \subset (\underline{p}_s \cap R_1)$ is strictly increasing. Moreover, for any strictly increasing sequence of prime ideals $\underline{q}_1 \subset \cdots \subset \underline{q}_s$ in R_1 there is a (strictly) increasing sequence of prime ideals $\underline{p}_1 \subset \cdots \subset \underline{p}_s$ of R_2 with $\underline{p}_i \cap R_1 = \underline{q}_i$ for all i.

This statement implies that R_1 and R_2 have the same maximum length for increasing sequences of prime ideals. In the situation of Noether normalization $k[X_1, \ldots, X_d] \subset O(X)$, where X is an affine variety, this implies that the dimensions of X and $\mathbf{A}_k^d$ are equal.

Finally, we will prove by induction that the dimension of $\mathbf{A}_k^n$ is $\leq n$. Let $V \subset \mathbf{A}_k^n$ be a proper closed irreducible subset. Apply the Noether normalization theorem to the ring $O(V) = k[X_1, \ldots, X_n]/I$ with $I \neq 0$. This yields $\dim V \leq n-1$ and thus $\dim \mathbf{A}_k^n \leq n$.

2. Let $k[X_1, \ldots, X_d] \to O(X)$ be a Noether normalization. Then the fraction field of $O(X)$ is a finite extension of the fraction field $k(X_1, \ldots, X_d)$ of $k[X_1, \ldots, X_d]$. Thus the transcendence degree of the fraction field of $O(X)$ is d. By part 1, the dimension of X is also d. □

A.1.4 Tangent Spaces, Smooth Points, and Singular Points

We will again assume that the characteristic of k is either 0 or that k is a perfect field of positive characteristic. Let W be a reduced affine variety over k. For every

$f \in O(W)$, $f \neq 0$ the open subset $U = \{w \in W | \ f(w) \neq 0\}$ of W is again a reduced affine variety. The coordinate ring of U is $O(W)[1/f]$. Let us call U a *special affine subset of* W. The special affine subsets form a basis for the Zariski topology, i.e., every open subset of W is a (finite) union of special affine subsets. Consider a point $P \in W$, that is, an element of $\max(O(W))$. The *dimension of* W *at* P is defined to be the minimum of the dimensions of the special affine neighborhoods of P. The *local ring* $O_{W,P}$ *of the point* P *on* W is defined as the ring of functions f, defined and regular in a neighborhood of P. More precisely, the elements of $O_{W,P}$ are pairs (f, U), with U a special affine neighborhood of P and $f \in O(U)$. Two pairs (f_1, U_1) and (f_2, U_2) are identified if there is a pair (f_3, U_3) with $U_3 \subset U_1 \cap U_2$ and f_3 is the restriction of both f_1 and f_2. Since P is a maximal ideal, the set $S = O(W) \setminus \underline{m}$ is a multiplicative set. Using the definitions of Example 1.5.1(d) one sees that $O_{W,P}$ is, in fact, the localization $S^{-1}O(W)$ of $O(W)$ with respect to S. Some relevant properties of $O_{W,P}$ are formulated in the next exercise.

Exercise A.27 *Local ring of a point.* Show the following

1. $O_{W,P}$ is a noetherian ring.

2. $O_{W,P}$ has a unique maximal ideal, namely $M_P := \{f \in O_{W,P} | \ f(P) = 0\}$, that is, $O_{W,P}$ is a *local ring*. The residue field $k' := O_{W,P}/M_P$ is a finite extension of k. We note that $k' \supset k$ is also separable because k is assumed to be perfect if its characteristic is positive.

3. Let $M_P = (f_1, \ldots, f_s)$ and let M_P^2 denote the ideal generated by all products $f_i f_j$. Then M_P/M_P^2 is a vector space over k' of dimension $\leq s$.

4. Suppose that the above s is minimally chosen. Prove that s is equal to the dimension of M_P/M_P^2. Hint: Use Nakayama's lemma: *Let A be a local ring with maximal ideal m, E a finitely generated A-module and $F \subset E$ a submodule such that $E = F + mE$. Then $E = F$* ([170], Chap. X, §4). □

The *tangent space* $T_{W,P}$ *of* W *at* P is defined to be $(M_P/M_P^2)^*$, i.e., the dual of the vector space M_P/M_P^2. The point P is called *nonsingular or regular* if the dimension of the vector space $T_{W,P}$ coincides with the dimension of W at P. The point P is called *smooth* (over k) if P is regular and the field extension $k \subset O_{W,P}/M_P$ is separable.

Remark A.28 Under our assumption that k has either characteristic 0 or that k is perfect in positive characteristic, any finite extension of k is separable and so the notions smooth (over k) and nonsingular coincide. For nonperfect fields in positive characteristic a point can be nonsingular, but not smooth over k. □

Under our assumptions, a point that is not smooth is called *singular*. We now give some examples.

Examples A.29 Let k be algebraically closed.

1. We will identify $\mathbf{A}_k^n$ with k^n. For $P = (a_1, \ldots, a_n) \in k^n$ one finds that $M_P = (X_1 - a_1, \ldots, X_n - a_n)$ and M_P/M_P^2 has dimension n. Therefore, every point of k^n is smooth.

2. Let $W \subset k^3$ be the reduced affine variety given by the equation $X_1^2 + X_2^2 + X_3^2$ (and suppose that the characteristic of k is not 2). Then $O(W) = k[X_1, X_2, X_3] /(X_1^2 + X_2^2 + X_3^2) = k[x_1, x_2, x_3]$. Consider the point $P = (0, 0, 0) \in W$. The dimension of W at P is two. The ideal $M_P = (x_1, x_2, x_3)$ and the dimension of M_P/M_P^2 is three. Therefore, P is a singular point. □

Exercise A.30 Let K be algebraically closed and let $W \subset k^2$ be the affine reduced curve given by the equation $Y^2 + XY + X^3 = 0$. Calculate the tangent space at each of its points. Show that $(0, 0)$ is the unique singular point. Draw a picture of a neighborhood of that point. □

We shall need the following two results. Their proofs may be found in [141], Theorem 5.2.

Let W be a reduced affine variety.
(a) For every point $P \in W$ the dimension of $T_{W,P}$ is $\geq$ the dimension of W at P.
(b) There are smooth points.

We formulate now the *Jacobian criterion for smoothness*:

Proposition A.31 *Let $W \subset \mathbf{A}_k^n$ be a reduced affine variety and let W have dimension d at $P = 0 \in W$. The coordinate ring $O(W)$ has the form $k[X_1, \ldots, X_n]/ (f_1, \ldots, f_m)$. The* jacobian matrix *is given by $(\frac{\partial f_i}{\partial x_j})_{j=1,\ldots,n}^{i=1,\ldots,m}$. Let $\Delta_1, \ldots, \Delta_s$ denote the set of all the determinants of the square submatrices of size $(n-d) \times (n-d)$ (called the minors of size $n-d$). Then P is smooth if and only if $\Delta_i(0) \neq 0$ for some i.*

Proof. The ideal M_P has the form $(X_1, \ldots, X_n)/(f_1, \ldots, f_m)$ and M_P/M_P^2 equals $(X_1, \ldots, X_n)/(X_1^2, X_1X_2, \ldots, X_n^2, L(f_1), \ldots, L(f_m))$, where for any $f \in (X_1, \ldots, X_n)$ we write $L(f)$ for the linear part of f in its expansion as polynomial in the variables $X_1, \ldots, X_n$. From the above results we know that the dimension of M_P/M_P^2 is at least d. The stated condition on the minors of the jacobian matrix translates into: the rank of the vector space generated by $L(f_1), \ldots, L(f_m)$ is $\geq n-d$. Thus the condition on the minors is equivalent to stating that the dimension of M_P/M_P^2 is $\leq d$. □

The Jacobian criterion implies that the set of smooth points of a reduced affine variety W is open (and not empty by the above results). In the following we will use a handy formulation for the tangent space $T_{W,P}$. Let R be a k-algebra. Recall that $W(R)$ is the set of K-algebra maps $O(W) \to R$ and that every k-algebra homomorphism $R_1 \to R_2$ induces an obvious map $W(R_1) \to W(R_2)$. For the ring R we make a special choice, namely $R = k[\epsilon] = k \cdot 1 + k \cdot \epsilon$ and with multiplication given by $\epsilon^2 = 0$. The k-algebra homomorphism $k[\epsilon] \to k$ induces a map $W(k[\epsilon]) \to W(k)$. We will call the following lemma *the epsilon trick*.

Lemma A.32 *Let $P \in W(k)$ be given. There is a natural bijection between the set $\{q \in W(k[\epsilon])|\ q$ maps to $P\}$ and $T_{W,P}$.*

Proof. To be more precise, the qs that we consider are the k-algebra homomorphisms $O_{W,P} \to k[\epsilon]$ such that $O_{W,P} \xrightarrow{q} k[\epsilon] \to k$ is P. Clearly q maps M_P to $k \cdot \epsilon$ and thus M_P^2 is mapped to zero. The k-algebra $O_{W,P}/M_P^2$ can be written as $k \oplus (M_P/M_P^2)$. The map $\tilde{q} : k \oplus (M_P/M_P^2) \to k[\epsilon]$, induced by q, has the form $\tilde{q}(c+v) = c + l_q(v)\epsilon$, with $c \in k,\ v \in (M_P/M_P^2)$ and $l_q : (M_P/M_P^2) \to k$ a k-linear map. In this way, q is mapped to an element in $l_q \in T_{W,P}$. It is easily seen that the map $q \mapsto l_q$ gives the required bijection. □

A.2 Linear Algebraic Groups

A.2.1 Basic Definitions and Results

We begin with the abstract definition. Throughout this section C will denote an algebraically closed field of characteristic zero and all affine varieties, unless otherwise stated, will be defined over C. Therefore, for any affine variety, we will not have to distinguish between $\max(O(W))$ and $W(C)$.

Definition A.33 A *linear algebraic group* G over C is given by the following data:

(a) A reduced affine variety G over C;
(b) A morphism $m : G \times G \to G$ of affine varieties;
(c) A point $e \in G$;
(d) A morphism of affine varieties $i : G \to G$;

subject to the conditions that G as a set is a group with respect to the composition m, the point e is the unit element and i is the map that sends every element to its inverse. □

Let $O(G)$ denote the coordinate ring of G. The morphisms $m : G \times G \to G$ and $i : G \to G$ correspond to C-algebra homomorphisms $m^* : O(G) \to O(G) \otimes_C O(G)$ and $i^* : O(G) \to O(G)$. Note that $e \in \max(O(G)) = G(C)$ corresponds to a C-algebra homomorphism $e^* : O(G) \to C$.

Examples A.34 *Linear algebraic groups*

1. The *additive group* $\mathbf{G}_a$ (or better, $\mathbf{G}_a(C)$) over C. This is, in fact, the affine line $\mathbf{A}^1_C$ over C with coordinate ring $C[x]$. The composition m is the usual addition. Thus m^* maps x to $x \otimes 1 + 1 \otimes x$ and $i^*(x) = -x$.

2. The *multiplicative group* $\mathbf{G}_m$ (or better, $\mathbf{G}_m(C)$) over C. This is an affine variety $\mathbf{A}^1_C \setminus \{0\}$ with coordinate ring $C[x, x^{-1}]$. The composition is the usual multiplication. Thus m^* sends x to $x \otimes x$ and $i^*(x) = x^{-1}$.

3. A *torus T of dimension n*. This is the direct product (as a group and as an affine variety) of n copies of $\mathbf{G}_m(C)$. The coordinate ring is $O(T) = C[x_1, x_1^{-1}, \dots, x_n, x_n^{-1}]$, The C-algebra homomorphisms m^* and i^* are given by $m^*(x_i) = x_i \otimes x_i$ and $i^*(x_i) = x_i^{-1}$ (for all $i = 1, \dots, n$).

4. The group GL_n of the invertible $n \times n$ matrices over C. The coordinate ring is $C[x_{i,j}, \frac{1}{d}]$, where $x_{i,j}$ are n^2 indeterminates and d denotes the determinant of the matrix of indeterminates $(x_{i,j})$. From the usual formula for the multiplication of matrices one sees that m^* must have the form $m^*(x_{i,j}) = \sum_{k=1}^{n} x_{i,k} \otimes x_{k,j}$. Using Cramer's rule, one can find an explicit expression for $i^*(x_{i,j})$. We do not write this expression down but conclude from its existence that i is really a morphism of affine varieties.

5. Let $G \subset \mathrm{GL}_n(C)$ be a subgroup, which is at the same time a Zariski-closed subset. Let I be the ideal of G. Then the coordinate ring $O(G)$ of G is $C[x_{i,j}, \frac{1}{d}]/I$. It can be seen that the maps m^* and i^* have the property $m^*(I) \subset (C[x_{i,j}, \frac{1}{d}] \otimes I)$ $+(I \otimes C[x_{i,j}, \frac{1}{d}])$ and $i^*(I) \subset I$. Therefore m^* and i^* induce C-algebra homomorphisms $O(G) \to O(G) \otimes O(G)$ and $O(G) \to O(G)$. Thus G is a linear algebraic group. In general, if G is a linear algebraic group over C and $H \subset G(C)$ is a subgroup of the form $V(I)$ for some ideal $I \subset O(G)$ then H is a linear algebraic group whose coordinate ring is $O(G)/I$.

6. Every finite group G can be seen as a linear algebraic group. The coordinate ring $O(G)$ is simply the ring of all functions on G with values in C. The map $m^* : O(G) \to O(G) \otimes O(G) = O(G \times G)$ is defined by specifying that $m^*(f)$ is the function on $G \times G$ given by $m^*(f)(a, b) = f(ab)$. Furthermore, $i^*(f)(a) = f(a^{-1})$. □

Exercise A.35 Show that the linear algebraic groups $\mathbf{G}_a(C), \mathbf{G}_m(C), T$, defined above, can be seen as Zariski-closed subgroups of a suitable $\mathrm{GL}_n(C)$. □

Exercise A.36 *Hopf algebras.*
1. Let $A = O(G)$. Show that the maps m^*, i^*, and e^* satisfy the following commutative diagrams:

$$\text{Coassociative} \quad \begin{array}{ccc} A \otimes_k A \otimes_k A & \overset{m^* \times id_A}{\longleftarrow} & A \otimes_k A \\ {\scriptstyle id_A \times m^*}\uparrow & & \uparrow{\scriptstyle m^*} \\ A \otimes_k A & \underset{m^*}{\longleftarrow} & A \end{array}, \tag{A.1}$$

$$\text{Counit} \quad \begin{array}{ccc} A & \overset{p^* \times id_A}{\longleftarrow} & A \times_k A \\ {\scriptstyle id_A \times p^*}\uparrow & \nwarrow^{id_A} & \uparrow{\scriptstyle m^*} \\ A \otimes_k A & \underset{m^*}{\longleftarrow} & A \end{array}, \tag{A.2}$$

$$\text{Coinverse} \quad \begin{array}{ccc} A & \overset{i^* \times id_A}{\longleftarrow} & A \otimes_k A \\ {\scriptstyle id_A \times i^*}\uparrow & \nwarrow^{p^*} & \uparrow{\scriptstyle m^*} \\ A \otimes_k A & \underset{m^*}{\longleftarrow} & A \end{array}, \tag{A.3}$$

where $p^* : A \to A$ is defined by $p^* = e^* \circ incl$ and $incl$ is the inclusion $k \hookrightarrow A$. A C-algebra A with maps m^*, i^*, and e^* satisfying these conditions is called a *Hopf algebra.*

2. Let A be a finitely generated C-algebra without nilpotents that is a Hopf algebra as well. Show that A is the coordinate ring of a linear algebraic group. (Since we are assuming that C has characteristic zero, the assumption of no nilpotents is not actually needed by a nontrivial result of Cartier, cf. [302], Chap. 11.4.) □

A *morphism* $f : G_1 \to G_2$ *of linear algebraic groups* is a morphism of affine varieties that respects the group structures.

In fact, every linear algebraic group G is isomorphic to a Zariski-closed subgroup of some $\mathrm{GL}_n(C)$ ([141], Theorem 11.2). One can see this property as an analog of the statement: "Every finite group is isomorphic with a subgroup of some S_n". The next proposition gathers together some general facts about linear algebraic groups, subgroups, and morphisms.

Proposition A.37 *Let G be a linear algebraic group.*

1. *The irreducible components of G are disjoint. If $G^o \subset G$ is the irreducible component of G that contains the point $1 \in G$, then G^o is a normal open subgroup of G of finite index.*
2. *If H is a subgroup of G, then the Zariski closure $\overline{H}$ of H is a Zariski-closed subgroup of G.*
3. *Every point of G is smooth.*
4. *If S is a Zariski connected subset of G containing 1, then the subgroup of G generated by S is also connected.*
5. *The commutator subgroup (i.e., the group generated by all commutators $g_1g_2g_1^{-1}g_2^{-1}$, $g_1, g_2 \in G$) of a connected linear algebraic group is connected.*

6. *Let $f : G_1 \to G_2$ be a morphism of linear algebraic groups. Then $f(G_1)$ is again a linear algebraic group.*

Proof. 1. Let $G_1, \ldots, G_s$ be the irreducible components of G. Each of these components contains a point not contained in any other component. For any fixed element $h \in G$, let $L_h : G \to G$ be left translation by h, given by $g \mapsto hg$. The map L_h is a morphism of affine varieties and, given any $g_1, g_2 \in G$ there is a unique $h \in G$ such that $L_h(g_1) = g_2$. From this it follows that any element of G is contained in a unique component of G. Therefore, G contains a unique component G^o containing 1. Since the components of G are disjoint, one sees that each of these is both open and closed in G. For every $h \in G$, the above isomorphism L_h permutes the irreducible components. For every $h \in G^o$ one has that $L_h(G^o) \cap G^o \neq \emptyset$. Therefore $L_h(G^o) = G^o$. The map $i : G \to G$, i.e., $i(g) = g^{-1}$ for all $g \in G$, is also an automorphism of G and permutes the irreducible components of G. It follows that $i(G^o) = G^o$. We conclude that G^o is an open and closed subgroup of G. For any $a \in G$, one considers the automorphism of G, given by $g \mapsto aga^{-1}$. This automorphism permutes the irreducible components of G. In particular, $aG^oa^{-1} = G^o$. This shows that G^o is a normal subgroup. The other irreducible components of G are the left (or right) cosets of G^o. Thus G^o has finite index in G.

2. We claim that $\overline{H}$ is a group. Indeed, inversion on G is an isomorphism and so $\overline{H}^{-1} = \overline{H^{-1}} = \overline{H}$. Moreover, left multiplication L_x on G by an element x is an isomorphism. Thus for $x \in H$ one has $L_x(\overline{H}) = \overline{L_x(H)} = \overline{H}$. Thus $L_x(\overline{H}) \subset \overline{H}$. Furthermore, let $x \in \overline{H}$ and let R_x denote the morphism given by right multiplication. We then have $H \subset \overline{H}$ and as a consequence $R_x(\overline{H}) \subset \overline{H}$. Thus $\overline{H}$ is a group.

3. The results of Sect. A.1.4 imply that the group G contains a smooth point p. Since, for every $h \in G$, the map $L_h : G \to G$ is an isomorphism of affine varieties, the image point $L_h(p) = hp$ is smooth. Thus every point of G is smooth.

4. Note that the set $S \cup S^{-1}$ is a connected set, so we assume that S contains the inverse of each of its elements. Since multiplication is continuous, the sets $S_2 = \{s_1 s_2 \mid s_1, s_2 \in S\} \subset S_3 = \{s_1 s_2 s_3 \mid s_1, s_2, s_3 \in S\} \subset \ldots$ are all connected. Therefore, their union is also connected and this is just the group generated by S.

5. Note that (1) above implies that the notions of connected and irreducible are the same for linear algebraic groups over C. Since G is irreducible, Lemma A.19 implies that $G \times G$ is connected. The map $G \times G \to G$ defined by $(g_1, g_2) \mapsto g_1 g_2 g_1^{-1} g_2^{-1}$ is continuous. Therefore the set of commutators is connected and so generates a connected group.

6. Let $H := f(G_1)$. We have seen that $\overline{H}$ is a group as well. Let $U \subset \overline{H}$ be an open dense subset. Then we claim that $U \cdot U = \overline{H}$. Indeed, take $x \in \overline{H}$. The set xU^{-1} is also an open dense subset of $\overline{H}$ and must meet U. This shows that $xu_1^{-1} = u_2$ holds for certain elements $u_1, u_2 \in U$. Finally, we use that H is a constructible subset (see

the discussion following Exercises A.9). The definition of constructible implies that H contains an open dense subset U of $\overline{H}$. Since H is a group and $U \cdot U = \overline{H}$ we have that $H = \overline{H}$. □

We will need the following technical corollary (cf. [151], Lemma 4.9) in Sect. 1.5.

Corollary A.38 *Let G be an algebraic group and H an algebraic subgroup. Assume that either H has finite index in G or that H is normal and G/H is abelian. If the identity component H^o of H is solvable then the identity component G^o of G is solvable.*

Proof. If H has finite index in G then $H^o = G^o$ so the conclusion is obvious. Now assume that H is normal and that G/H is abelian. In this case, H contains the commutator subgroup of G and so also contains the commutator subgroup K of G^o. By Proposition A.37 this latter commutator subgroup is connected and so is contained in H^o. Since H^o is solvable, we have that K is solvable. Since G^o/K is abelian, we have that G^o is solvable. □

Exercises A.39 1. *Characters of groups*
A *character of a linear algebraic group G* is a morphism of linear algebraic groups $\chi : G \to \mathbf{G}_{m,C}$. By definition, χ is determined by a C-algebra homomorphism $\chi^* : O(\mathbf{G}_m) = C[x, x^{-1}] \to O(G)$. Furthermore, χ^* is determined by an element $\chi^*(x) = a \in O(G)$.
(a) Show that the conditions on a (for χ to be a character) are a is invertible in $O(G)$ and $m^*(a) = a \otimes a$.
(b) Show that $\mathbf{G}_{a,C}$ has only the trivial character, i.e., $\chi(b) = 1$ for all $b \in \mathbf{G}_{a,C}$.
(c) Let T be a torus with $O(T) = C[x_1, x_1^{-1}, \ldots, x_n, x_n^{-1}]$ and $m^*(x_i) = x_i \otimes x_i$ for all $i = 1, \ldots, n$. Show that every character χ of T is given by $\chi^*(x) = x_1^{m_1} \cdots x_n^{m_n}$ with all $m_i \in \mathbf{Z}$. In this way, the group of all characters of T can be identified with the group $\mathbf{Z}^n$.
(d) What are the characters of $\mathrm{GL}_n(C)$? Hint: $SL_n(C)$ equals its commutator subgroup.

2. *Kernels of homomorphisms*
Let $f : G_1 \to G_2$ be a morphism of linear algebraic groups. Prove that the kernel of f is again a linear algebraic group.

3. *Centers of groups*
Show that the center of a linear algebraic group is Zariski closed. □

Remarks A.40 If one thinks of linear algebraic groups as groups with some extra structure, then it is natural to ask what the structure of G/H is for G a linear algebraic group and H a Zariski-closed subgroup of G. The answers are:

(a) G/H has the structure of a variety over C, but, in general, not an affine variety (in fact, G/H is a quasi-projective variety).

(b) If H is a normal (and Zariski-closed) subgroup of G then G/H is again a linear algebraic group and $O(G/H) = O(G)^H$, i.e., the regular functions on G/H are the H-invariant regular functions on G.

Both (a) and (b) have long and complicated proofs for which we refer to [141], Chap. 11.5 and Chap. 12. □

Exercises A.41 *Subgroups*

1. Let $A \in \mathrm{GL}_n(C)$ be a diagonal matrix with diagonal entries $\lambda_1, \dots, \lambda_n$. Then $< A >$ denotes the subgroup of $\mathrm{GL}_n(C)$ generated by A. In general, this subgroup is not Zariski closed. Let $G := \overline{< A >}$ denote the Zariski closure of $< A >$. The proof of Proposition A.37 tells us that G is again a group. Prove that G consists of the diagonal matrices $diag(d_1, \dots, d_n)$ given by the equations: If $(m_1, \dots, m_n) \in \mathbf{Z}^n$ satisfies $\lambda_1^{m_1} \cdots \lambda_n^{m_n} = 1$, then $d_1^{m_1} \cdots d_n^{m_n} = 1$.

2. Let $A \in \mathrm{GL}_2(C)$ be the matrix $\binom{a\ b}{0\ a}$ (with $a \neq 0$). Determine the algebraic group $\overline{< A >}$ for all possibilities of a and b.

3. For two matrices $A, B \in \mathrm{SL}_2(C)$ we denote by $< A, B >$ the subgroup generated by A and B. Furthermore, $\overline{< A, B >}$ denotes the Zariski closure of $< A, B >$. Use the classification of the algebraic subgroups of SL_2 to show that every algebraic subgroup of SL_2 has the form $\overline{< A, B >}$ for suitable A and B (see the remarks before Exercises 1.36). □

Definition A.42 A *representation* of a linear algebraic group G (also called a em G-module) is a C-morphism $\rho : G \to \mathrm{GL}(V)$, where V is a finite dimensional vector space over C. The representation is called *faithful* if ρ is injective. □

We have remarked above that any linear algebraic group is isomorphic to a closed subgroup of some $\mathrm{GL}_n(C)$. In other words, a faithful representation always exists.

Exercise A.43 *Representations*
Let $G = (\max(A), A)$ be a linear algebraic group over C. As before, m^*, i^*, e^* are the maps defining the Hopf algebra structure of A. Consider a pair (V, τ) consisting of finite dimensional C-vector space V and a C-linear map $\tau : V \to A \otimes_C V$, satisfying the following rules:

(i) $(e^* \otimes id_V) \circ \tau : V \to A \otimes_C V \to C \otimes_C V = V$ is the identity map.

(ii) The maps $(m^* \otimes id_V) \circ \tau$ and $(id_A \otimes \tau) \circ \tau$ from V to $A \otimes_C A \otimes_C V$ coincide.

Show that there is a natural bijection between the pairs (V, τ) and the homomorphism $\rho : G \to \mathrm{GL}(V)$ of algebraic groups. Hint: For convenience we use a basis $\{v_i\}$ of V over C. We note that the data for ρ is equivalent to a C-algebra homomorphism $\rho^* : C[\{X_{i,j}\}, \frac{1}{det}] \to A$ and thus to an invertible matrix $(\rho^*(X_{i,j}))$ with coefficients

in A (having certain properties). One associates to ρ the C-linear map τ given by $\tau v_i = \sum \rho^*(X_{i,j}) \otimes v_j$.
On the other hand, one associates to a given τ with $\tau v_i = \sum a_{i,j} \otimes v_j$ the ρ with $\rho^*(X_{i,j}) = a_{i,j}$.
In Appendix B2 we will return to this exercise. □

Exercise A.44 *Representations of* $\mathbf{G}_m$ *and* $(\mathbf{G}_m)^r$
1. For any representation $\rho : \mathbf{G}_m \to \mathrm{GL}(V)$ there is a basis $v_1, \ldots, v_n$ of V such that $\rho(x)$ is a diagonal matrix w.r.t. this basis and such that the diagonal entries are integral powers of $x \in \mathbf{G}_m(C)$. Hint: Any commutative group of matrices can be conjugated to a group of upper triangular matrices. An upper triangular matrix of finite order is diagonal. The elements of finite order are dense in G_m. Finally, use Exercise A.39.3.

2. Generalize this to show that for any representation $\rho : (\mathbf{G}_m)^r \to \mathrm{GL}(V)$ there is a basis $v_1, \ldots, v_n$ of V such that $\rho(x)$ is a diagonal matrix w.r.t. this basis. □

We close this section with a proof of the Lie-Kolchin Theorem. Before we do this we need to characterize Zariski-closed subgroups of a torus. This is done in the second part of the following lemma.

Lemma A.45 *Let G be a proper Zariski-closed subgroup of $T \subset \mathrm{GL}_n$. Then*

1. *There exists a nonempty subset $\mathcal{S} \subset \mathbf{Z}^n$ such that $I(G) \subset C[x_1, x_1^{-1}, \ldots, x_n, x_n^{-1}]$ is generated by $\{x_1^{\nu_1} x_2^{\nu_2} \cdots x_n^{\nu_n} - 1 \mid (\nu_1, \ldots, \nu_n) \in \mathcal{S}\}$, and*
2. *G is isomorphic to a direct product $\mathbf{G}_m^r \times H$ where $0 \leq r < n$ and H is the direct product of $n - r$ cyclic groups of finite order.*
3. *The points of finite order are dense in G.*

Proof. (see [250])
1. Let

$$F(x_1, \ldots, x_n) = \sum c_{\nu_1,\ldots,\nu_n} x_1^{\nu_1} \cdots x_n^{\nu_n} \in C[x_1, x_1^{-1}, .., x_n, x_n^{-1}], \tag{A.4}$$

where each $c_{\nu_1,\ldots,\nu_n} \in C\backslash\{0\}$ and $(\nu_1, \ldots, \nu_n) \in \mathbf{Z}^n$. We say that F is G-homogeneous if for any $(a_1, \ldots a_n) \in G$ all the terms $a_1^{\nu_1} \cdots a_n^{\nu_n}$ are equal.

We claim that any $F(x) \in C[x_1, x_1^{-1}, \ldots x_n, x_n^{-1}]$ vanishing on G is the sum of G-homogeneous elements of $C[x_1, x_1^{-1}, \ldots, x_n, x_n^{-1}]$, each of which also vanishes on G. If $F(x)$ is not homogeneous then there exist elements $a = (a_1, \ldots, a_n) \in G$ such that a linear combination of $F(x)$ and $F(ax)$ is nonzero, contains only terms appearing in F and has fewer nonzero terms than F. Note that $F(ax)$ also vanishes on G. Making two judicious choices of a, we see that F can be written as the sum of two polynomials, each vanishing on G and each having fewer terms than F. Therefore, induction on the number of nonzero terms of F yields the claim.

Let $F \in C[x_1, x_1^{-1}, \dots, x_n, x_n^{-1}]$ as in (A.4) be G-homogeneous and vanish on G. Dividing by a monomial if necessary we may assume that one of the terms appearing in G is 1. Since $F(1, \dots, 1) = 0$ we have that $\sum c_{\nu_1,\dots,\nu_n} = 0$. Furthermore, G-homogeneity implies that $a_1^{\nu_1} \cdots, a_n^{\nu_n} = 1$ for all $(a_1, \dots, a_n) \in G$ and all terms $x_1^{\nu_1} \cdots x_n^{\nu_n}$ in F. Therefore,

$$\begin{aligned} F(x) &= \sum c_{\nu_1,\dots,\nu_n} x_1^{\nu_1} \cdots x_n^{\nu_n} \\ &= \sum c_{\nu_1,\dots,\nu_n} (x_1^{\nu_1} \cdots x_n^{\nu_n} - 1). \end{aligned}$$

The totality of all such $x_1^{\nu_1} \cdots x_n^{\nu_n} - 1$ generate $I(G)$.

2. The set of $(\nu_1, \dots, \nu_n)$ such that $x_1^{\nu_1} \cdots x_n^{\nu_n} - 1$ vanishes on G forms an additive subgroup S of $\mathbf{Z}^n$. The theory of finitely generated modules over a principal ideal domain (Theorem 7.8 in Chap. III, §7 of [170]) implies that there exists a free set of generators $\{a_i = (a_{1,i}, \dots, a_{n,i})\}_{i=1,\dots,n}$ for $\mathbf{Z}^n$ and integers $d_1, \dots, d_n \geq 0$ such that S is generated by $\{d_i a_i\}_{i=1,\dots,n}$. The map $(x_1, \dots, x_n) \mapsto (x_1^{a_{1,1}}, \dots x_n^{a_{n,1}}, \dots, x_1^{a_{1,n}}, \dots, x_n^{a_{n,n}})$ is an automorphism of T and sends G onto the subgroup defined by the equations $\{x_i^{d_i} - 1 = 0\}_{i=1,\dots,n}$.

3. Using part 2, we see it is enough to show than the points of finite order are dense in G_m and this is obvious. □

Theorem A.46 (Lie-Kolchin) *Let G be a solvable connected subgroup of* GL_n. *Then G is conjugate to a subgroup of upper triangular matrices.*

Proof. We follow the proof given in [250]. Recall that a group is solvable if the descending chain of commutator subgroups ends in the trivial group. Lemma A.37(6) implies that each of the elements of this chain is connected. Since this chain is left invariant by conjugation by elements of G, each element in the chain is normal in G. Furthermore, the penultimate element is commutative. Therefore, either G is commutative or its commutator subgroup contains a connected commutative subgroup $H \neq \{1\}$. We identify GL_n with $GL(V)$ where V is an n-dimensional vector space over C and proceed by induction on n.

If G is commutative, then it is well known that G is conjugate to a subgroup of upper triangular matrices (even without the assumption of connectivity). If V has a nontrivial G-invariant subspace W then the images of G in $\mathrm{GL}(W)$ and $\mathrm{GL}(V/W)$ are connected and solvable and we can proceed by induction using appropriate bases of W and V/W to construct a basis of V in which G is upper triangular. Therefore, we can assume that G is not commutative and leaves no nontrivial proper subspace of V invariant.

Since H is commutative, there exists a $v \in V$ that is a joint eigenvector of the elements of H, that is, there is a character χ on H such that $hv = \chi(h)v$ for all $h \in H$. For any $g \in G$, $hgv = g(g^{-1}hgv) = \chi(g^{-1}hg)gv$ so gv is again a joint eigenvector of H. Therefore the space spanned by joint eigenvectors of H is G-invariant. Our assumptions imply that V has a basis of joint eigenvectors of H and so we may assume that the elements of H are diagonal. The Zariski closure $\overline{H}$ of H is again diagonal and since H is normal in G, we have that $\overline{H}$ is also normal in G. The group $\overline{H}$ is a torus and so, by Lemma A.45(2), we see that the set of points of any given finite order N is finite. The group G acts on $\overline{H}$ by conjugation, leaving these sets invariant. Since G is connected, it must leave each element of order N fixed. Therefore G commutes with the points of finite order in $\overline{H}$. Lemma A.45 again implies that the points of finite order are dense in $\overline{H}$ and so that H is in the center of G.

Let χ be a character of H such that $V_\chi = \{v \in V \mid hv = \chi(h)v \text{ for all } h \in H\}$ has a nonzero element. As noted above, such a character exists. For any $g \in G$, a calculation similar to that in the preceding paragraph shows that $gV_\chi = V_\chi$. Therefore, we must have $V_\chi = V$ and H must consist of constant matrices. Since H is a subgroup of the commutator subgroup of G, we have that the determinant of any element of H is 1. Therefore H is a finite group and so must be trivial since it is connected. This contradiction proves the theorem. □

Finally, the above proof is valid without the restriction that C has characteristic 0. We note that the Lie-Kolchin Theorem is not true if we do not assume that G is connected. To see this, let $G \subset \mathrm{GL}_n$ be any finite, noncommutative, solvable group. If G were a subgroup of the group of upper triangular matrices, then since each element of G has finite order, each element must be diagonal (recall that the characteristic of C is 0). This would imply that G is commutative.

A.2.2 The Lie Algebra of a Linear Algebraic Group

The Lie algebra $\mathfrak{g}$ of a linear algebraic group G is defined as the tangent space $T_{G,1}$ of G at $1 \in G$. It is clear that G and G^o have the same tangent space and that its dimension is equal to the dimension of G, which we denote by r. The Lie algebra structure on $\mathfrak{g}$ has still to be defined. For convenience we suppose that G is given as a closed subgroup of some $\mathrm{GL}_n(C)$. We apply the "epsilon trick" of Lemma A.32 first to $\mathrm{GL}_n(C)$ itself. The tangent space $\mathfrak{g}$ of G at the point 1 is then identified with the matrices $A \in M_n(C)$ such that $1 + \epsilon A \in G(C[\epsilon])$. We first note that the smoothness of the point $1 \in G$ allows us to use Proposition A.31 and the formal implicit function theorem to produce a formal power series $F(z_1, \dots, z_r) = 1 + A_1 z_1 + \dots + A_r z_r +$ higher order terms with the $A_i \in \mathrm{M}_n(C)$ and such that $F \in G(C[[z_1, \dots, z_r]])$ and such that the A_i are linearly independent over C. Substituting $z_i = \epsilon$, $z_j = 0$ for $j \neq i$ allows us to conclude that each $A_i \in \mathfrak{g}$. For any $A = c_1 A_1 + \cdots + c_r A_r$, the substitution $z_i = c_i t$ for $i = 1, \dots r$ gives an element $f = I + At + \dots$ in the power series ring $C[[t]]$ with $f \in G(C[[t]])$ (see Exercise A.48, for another way of finding such an f).

In order to show that $\mathfrak{g}$ is, in fact, a Lie subalgebra of $M_n(C)$, we extend the epsilon trick and consider the ring $C[\alpha]$ with $\alpha^3 = 0$. From the previous discussion, one can lift $1+\epsilon A \in G(C[\epsilon])$ to a point $1+At+A_1t^2+\cdots \in G(C[[t]])$. Mapping t to $\alpha \in C[\alpha]$, yields an element $1+\alpha A+\alpha^2 A_1 \in G(C[\alpha])$. Thus for $A, B \in \mathfrak{g}$ we find two points $a = 1+\alpha A+\alpha^2 A_1$, $b = 1+\alpha B+\alpha^2 B_1 \in G(C[\alpha])$. The commutator $aba^{-1}b^{-1}$ is equal to $1+\alpha^2(AB-BA)$. A calculation shows that this implies that $1+\epsilon(AB-BA) \in G(C[\epsilon])$. Thus $[A, B] = AB-BA \in \mathfrak{g}$. An important feature is the action of G on $\mathfrak{g}$, which is called *the adjoint action Ad of G on* $\mathfrak{g}$. The definition is quite simple, for $g \in G$ and $A \in \mathfrak{g}$ one defines $Ad(g)A = gAg^{-1}$. The only thing that one has to verify is $gAg^{-1} \in \mathfrak{g}$. This follows from the formula $g(1+\epsilon A)g^{-1} = 1+\epsilon(gAg^{-1})$, which is valid in $G(C[\epsilon])$.

We note that the Lie algebra $M_n(C)$ has many Lie subalgebras, a minority of them are the Lie algebras of algebraic subgroups of $\mathrm{GL}_n(C)$. Those that do come from algebraic subgroups are called *algebraic Lie subalgebras of* $M_n(C)$.

Exercises A.47 *Lie algebras*

1. Let T denote the group of the diagonal matrices in $\mathrm{GL}_n(C)$. The Lie algebra of T is denoted by $\mathfrak{t}$. Prove that the Lie algebra $\mathfrak{t}$ is "commutative", i.e., $[a, b] = 0$ for all $a, b \in \mathfrak{t}$. determine with the help of Lemma A.45 the algebraic Lie subalgebras of $\mathfrak{t}$.

2. Consider $A = \left(\begin{smallmatrix} a & b \\ 0 & a \end{smallmatrix}\right) \in \mathrm{GL}_2(C)$ and the linear algebraic group $\overline{< A >} \subset \mathrm{GL}_2(C)$. Calculate the Lie algebra of this group (for all possible cases). Hint: See Exercise A.41. □

Exercise A.48 *Lie algebras and exponentials*
Let $G \subset \mathrm{GL}_n(C)$ be a linear algebraic group with Lie algebra $\mathfrak{g} \subset \mathrm{M}_n(C)$. For any $A \in \mathrm{M}_n(C)$, define

$$exp(tA) = 1 + At + \frac{A^2}{2!}t^2 + \frac{A^3}{3!}t^3 + \dots \quad \in \mathrm{M}_n(C[[t]]),$$

where t is an indeterminate. The aim of this exercise is to show that $A \in \mathfrak{g}(C)$ if and only if $exp(tA) \in G(C[[t]])$, cf. [67], Théorème 7, Chap. II.12.

1. Show that if $exp(tA) \in G(C[[t]])$, then $A \in \mathfrak{g}$. Hint: Consider the homomorphism $\phi : C[[t]] \to C[\epsilon]$ given by $t \mapsto \epsilon$.

2. Let I be the ideal defining G in $C[X_{1,1}, \dots, X_{n,n}, \frac{1}{\det}]$ and let $P \in I$. Show that if $A \in \mathfrak{g}(C)$ then $\sum \frac{\partial P}{\partial X_{i,j}}(AX)_{i,j} \in I$, where $X = (X_{i,j})$. Hint: Since $1+\epsilon A \in G(C[\epsilon])$, we have $P(X(1+\epsilon A)) \in I \cdot C[\epsilon]$. Furthermore, $P(X+\epsilon XA) = P(X) + \epsilon \sum \frac{\partial P}{\partial X_{i,j}}(AX)_{i,j}$.

3. Assume $A \in \mathfrak{g}(C)$. Let $J \subset C[[t]]$ be the ideal generated by $\{P(exp(tA)) \mid P \in A\}$. Show that J is left invariant by $\frac{d}{dt}$ and that $J \subset tC[[t]]$. Hint: Use part 2 for the first part and note that $P(1) = 0$ for all $P \in I$ for the second part.

4. Let J be as in part 3. Show that $J = (0)$ and therefore that $exp(tA) \in G(C[[t]])$. Hint: If not, $J = (t^m)$ for some integer $m \geq 0$. From part 3, we have that $m \geq 1$ and that $t^{m-1} \in J$. □

A.2.3 Torsors

Let G be a linear algebraic group over the algebraically closed field C of characteristic 0. Recall from Sect. A.1.2 that if $k \supset C$, G_k is defined to be the variety associated to the ring $O(G) \otimes_C k$.

Definition A.49 A *G-torsor Z over a field $k \supset C$* is an affine variety over k with a G-action, i.e., a morphism $G_k \times_k Z \to Z$ denoted by $(g, z) \mapsto zg$, such that:

1. For all $x \in Z(\bar{k})$, $g_1, g_2 \in G(\bar{k})$, we have $z1 = z$; $z(g_1g_2) = (zg_1)g_2$.
2. The morphism $G_k \times_k Z \to Z \times_k Z$, given by $(g, z) \mapsto (zg, z)$, is an isomorphism.

□

The last condition can be restated as: for any $v, w \in Z(\bar{k})$ there exists a unique $g \in G(\bar{k})$ such that $v = wg$. A torsor is often referred to as a *principal homogeneous space over G*.

Exercise A.50 *Galois extensions and torsors of finite groups*
Let k be a field of characteristic zero and let G be a finite group of order n. We consider G as an affine algebraic group as in Example A.34.6. Note that the k-points of the variety G correspond to the elements of the group G. Let Z be a G-torsor over k and assume that Z is irreducible.

1. Show that $Z(\bar{k})$ is finite and so $K = O(Z)$ is a field. Hint: Use Exercise A.14 (4).

2. For each $g \in G$, the map $z \mapsto zg$ is an isomorphism of Z to itself and so gives a k-automorphism σ_g of $O(Z)$. Show that $g \mapsto \sigma_g$ is an injective homomorphism of G to $Aut(K/k)$. Hint: If $\sigma_g = id$, then $g = id$ on $Z(\bar{k})$.

3. Show that K is a Galois extension of k with Galois group G. Hint: Let $[K : k] = m$. Comparing dimensions, show that $m = n$. Since $n = |G| \leq |Aut(K/k)| \leq n$, Galois theory gives the conclusion.

4. Conversely, let K be a Galois extension of k with Galois group G. For $g \in G$ let $\sigma_g \in Aut(K/k)$ be the corresponding automorphism. Consider the map $K \otimes_k K \to O(G) \otimes_k K$ given by

$$f \otimes 1 \mapsto \sum_{g \in G} \chi_g \otimes \sigma_g(f),$$

$$1 \otimes h \mapsto \sum_{g \in G} \chi_g \otimes h,$$

where $\chi_g \in O(G)$ is the function that is 1 on g and 0 on the rest of G. Show that this is an isomorphism. Conclude that $K = O(X)$ for some connected G-torsor. Hint:

Since the two spaces have the same k-dimension, it suffices to show that the map is injective. Let $u = \sum_i f_i \otimes h_i$ be an element that maps to zero. Using properties of the tensor product and noting that $[K : k] = n$, we can assume that the f_i are linearly independent over k. The image of u is $\sum_{g\in G} \chi_g \otimes (\sum_i \sigma_g(f_i)h_i)$. Therefore, for each $g \in G$, $\sum_i \sigma_g(f_i)h_i = 0$. Since $\det(\sigma_g(f_i)) \neq 0$ (cf. [170], Chap. VI, §5, Corollary 5.4), each $h_i = 0$. □

The *trivial G-torsor over k* is defined by $Z = G_k := G \otimes_C k$ and $G_k \times_k G_k \to G_k$ is the multiplication map $(g, z) \mapsto z \cdot g$. Two torsors Z_1, Z_2 over k are defined to be isomorphic over k if there exist a k-isomorphism $f : Z_1 \to Z_2$ such that $f(zg) = f(z)g$ for all $z \in Z_1,\ g \in G$. Any G-torsor over k, isomorphic to the trivial one, is called trivial.

Suppose that Z has a k-rational point b, i.e., $b \in Z(k)$. The map $G_k \to Z$, given by $g \mapsto bg$, is an isomorphism. It follows that Z is a trivial G-torsor over k. Thus the torsor Z is trivial if and only if Z has a k-rational point. In particular, if k is algebraically closed, every G-torsor is trivial.

Let Z be any G-torsor over k. Choose a point $b \in Z(\bar{k})$, where $\bar{k}$ is the algebraic closure of k. Then $Z(\bar{k}) = bG(\bar{k})$. For any $\sigma \in \mathrm{Aut}(\bar{k}/k)$, the Galois group of $\bar{k}$ over k, one has $\sigma(b) = bc(\sigma)$ with $c(\sigma) \in G(\bar{k})$. The map $\sigma \mapsto c(\sigma)$ from $\mathrm{Aut}(\bar{k}/k)$ to $G(\bar{k})$ satisfies the relation

$$c(\sigma_1) \cdot \sigma_1(c(\sigma_2)) = c(\sigma_1\sigma_2).$$

A map $c : \mathrm{Aut}(\bar{k}/k) \to G(\bar{k})$ with this property is called a *1-cocycle for* $\mathrm{Aut}(\bar{k}/k)$ *acting on* $G(\bar{k})$. Two 1-cocycles c_1, c_2 are called *equivalent* if there is an element $a \in G(\bar{k})$ such that

$$c_2(\sigma) = a^{-1} \cdot c_1(\sigma) \cdot \sigma(a) \text{ for all } \sigma \in \mathrm{Aut}(\bar{k}/k).$$

The set of all equivalence classes of 1-cocycles is, by definition, *the cohomology set* $H^1(\mathrm{Aut}(\bar{k}/k), G(\bar{k}))$. This set has a special point 1, namely the image of the trivial 1-cocycle.

Take another point $\tilde{b} \in Z(\bar{k})$. This defines a 1-cocycle $\tilde{c}$. Write $\tilde{b} = ba$ with $a \in G(\bar{k})$. Then one finds that $\tilde{c}(\sigma) = a^{-1} \cdot c(\sigma) \cdot \sigma(a)$ for all $\sigma \in \mathrm{Aut}(\bar{k}/k)$. Thus $\tilde{c}$ is equivalent to c and the torsor Z defines a unique element c_Z of $H^1(\mathrm{Aut}(\bar{k}/k), G(\bar{k}))$. For the next lemma we shall need the fact that $H^1(\mathrm{Aut}(\bar{k}/k), \mathrm{GL}_n(\bar{k})) = \{1\}$ ([170], Chap. VII, Exercise 31; [260], p. 159).

Lemma A.51 *The map $Z \to c_Z$ induces a bijection between the set of isomorphism classes of G-torsors over k and $H^1(\mathrm{Aut}(\bar{k}/k), G(\bar{k}))$.*

Proof. The map $Z \to c_Z$ is injective. Indeed, let Z_1 and Z_2 be torsors, $b_1 \in Z_1(\bar{k})$ and $b_2 \in Z_2(\bar{k})$ two points defining equivalent 1-cocycles. After changing the point

b_2 we may suppose that the two 1-cocycles are identical. One defines $f : Z_1(\bar{k}) \to Z_2(\bar{k})$ by $f(b_1 g) = b_2 g$ for all $g \in G(\bar{k})$. f defines an isomorphism $(Z_1)_{\bar{k}} \to (Z_2)_{\bar{k}}$. By construction f is invariant under the action of $\mathrm{Aut}(\bar{k}/k)$. Therefore Exercise A.22 implies that f is induced by an isomorphism $\tilde{f} : Z_1 \to Z_2$ of G-torsors.

Let an element of $H^1(\mathrm{Aut}(\bar{k}/k), G(\bar{k}))$ be represented by a 1-cocycle c. The group G is an algebraic subgroup of $\mathrm{GL}_n(C)$. Since $H^1(\mathrm{Aut}(\bar{k}/k), \mathrm{GL}_n(\bar{k})) = \{1\}$, there is a $B \in \mathrm{GL}_n(\bar{k})$ with $c(\sigma) = B^{-1}\sigma(B)$ for all $\sigma \in \mathrm{Aut}(\bar{k}/k)$. The subset $BG(\bar{k}) \in \mathrm{GL}_n(\bar{k})$ is Zariski closed and defines an algebraic variety $Z \subset \mathrm{GL}_n(\bar{k})$. For $\sigma \in \mathrm{Aut}(\bar{k}/k)$ one has $\sigma(BG(\bar{k})) = \sigma(B)G(\bar{k}) = Bc(\sigma)G(\bar{k}) = BG(\bar{k})$. Thus, Lemma A.21 implies that Z is defined over k. It is clear that Z is a G-torsor over k. Furthermore, $B \in Z(\bar{k})$ defines the 1-cocycle c. This shows the map $Z \mapsto c_Z$ is also surjective. □

We have already noted that $H^1(\mathrm{Aut}(\bar{k}/k), \mathrm{GL}_n(\bar{k})) = \{1\}$ for any field k. Hilbert's Theorem 90 implies that $H^1(\mathrm{Aut}(\bar{k}/k), \mathbf{G_m}(\bar{k})) = \{1\}$ and $H^1(\mathrm{Aut}(\bar{k}/k), \mathbf{G_a}(\bar{k})) = \{1\}$, [170], Chap. VI, §10. Furthermore, the triviality of H^1 for these latter two groups can be used to show that $H^1(\mathrm{Aut}(\bar{k}/k), G(\bar{k})) = \{1\}$ when G is a connected solvable group, [260]. We will discuss another situation when $H^1(\mathrm{Aut}(\bar{k}/k), G(\bar{k})) = \{1\}$. For this we need the following definition.

Definition A.52 A field F is called a *C_1-field* if every homogeneous polynomial $f \in F[X_1, \ldots, X_n]$ of degree less than n has a nontrivial zero in F^n. □

It is known that the fields $C(z)$, $C((z))$, $\mathbf{C}(\{z\})$ are C_1-fields if C is algebraically closed, [169]. The field $C(z, e^z)$, with C algebraically closed, is not a C_1-field.

Theorem A.53 (T. A. Springer, [260] p. 150)
Let G be a connected linear algebraic group over the field k of characteristic 0. *Suppose that k is a C_1-field. Then* $H^1(\mathrm{Aut}(\bar{k}/k), G(\bar{k})) = \{1\}$.

B Tannakian Categories

In this appendix we examine the question: *when is a category the category of representations of a group G and how do we recover G from such a category?* When G is a compact Lie group, Tannaka showed that G can be recovered from its category of finite dimensional representations and Krein characterized those categories that are equivalent to the category of finite dimensional representations of such a group (see [52] and [182]). In this appendix, we shall first discuss this question when G is a finite (or profinite) group. The question here is answered via the theory of Galois categories (introduced in [118]). We will then consider the situation when G is an affine (or proaffine) algebraic group. In this case, the theory of tannakian categories furnishes an answer. Original sources for the theory of tannakian categories are [251, 81, 82] (see also [52]). The very definition of tannakian category is rather long and its terminology has undergone some changes. In the following we will both expand and abbreviate a part of the paper [82] and our terminology is more or less that of [82]. For the basic definitions from category theory we refer to [170], Chap. I, §11.

B.1 Galois Categories

We wish to characterize those categories that are equivalent to the category of finite sets on which a fixed profinite group acts. We begin by giving the definition of a profinite group (cf. [304]).

Definition B.1 (1) Let $(I, \leq)$ be a partially ordered set such that for every two elements $i_1, i_2 \in I$ there exists an $i_3 \in I$ with $i_1 \leq i_3$ and $i_2 \leq i_3$. Assume, furthermore, that for each $i \in I$, we are given a finite group G_i and for every pair $i_1 \leq i_2$ a homomorphism $m(i_2, i_1) : G_{i_2} \to G_{i_1}$. Furthermore, assume that the $m(i_2, i_1)$ verify the rules: $m(i, i) = id$ and $m(i_2, i_1) \circ m(i_3, i_2) = m(i_3, i_1)$ if $i_1 \leq i_2 \leq i_3$. The above data are called an *inverse system of abelian groups*.

The *projective limit* of this system will be denoted by $\lim\limits_{\leftarrow} B_i$ and is defined as follows: Let $G = \prod G_i$ be the product of the family. Let each G_i have the discrete topology and let G have the product topology. Then $\lim\limits_{\leftarrow} B_i$ is the subset of G consisting of those elements (g_i), $g_i \in G_i$ such that for all i and $j \geq i$, one has $m(j, i)(g_j) = g_i$.

We consider $\varprojlim B_i$ a topological group with the induced topology. Such a group is called a *profinite group*. □

Example B.2 Let p be a prime number, $I = \{0, 1, 2, \dots\}$ and let $G_n = \mathbf{Z}/p^{n+1}\mathbf{Z}$. For $i \geq j$ let $m(j, i) : \mathbf{Z}/p^{j+1}\mathbf{Z} \to \mathbf{Z}/p^{i+1}\mathbf{Z}$ be the canonical homomorphism. The projective limit is called the *p-adic integers* $\mathbf{Z}_p$. □

Remarks B.3 1. The projective limit is also known as the *inverse limit*.

2. There are several characterizations of profinite groups (cf. [304] p.19). For example, a topological group is profinite if and only if it is compact and totally disconnected. Also, a topological group is profinite if and only if it is isomorphic (as a topological group) to a closed subgroup of a product of finite groups. □

The theory of Galois categories concerns characterizing those categories equivalent to the category of finite sets on which a finite (or profinite) group acts.

Definition B.4 Let G be a finite group. The category Perm_G is defined as follows. An object (F, ρ) is a finite set F with a G-action on it. More explicitly, a homomorphism of groups $\rho : G \to \mathrm{Perm}(F)$ is given, where $\mathrm{Perm}(F)$ denotes the group of all permutations of F. A morphism $m : (F_1, \rho_1) \to (F_2, \rho_2)$ is a map $m : F_1 \to F_2$ with $m \circ \rho_1 - \rho_2 \circ m$. One calls (F, ρ) also a finite G-set and the action of G on F will also be denoted by $g \cdot f := \rho(g)(f)$ for $g \in G$ and $f \in F$.

We extend this definition to the case when G is a profinite group. An object of Perm_G is now a pair (F, ρ), with F a finite set and $\rho : G \to \mathrm{Perm}(F)$ a homomorphism such that thc kcrncl is an open subgroup of G. Morphisms are defined as above. □

We want to recognize when a category is equivalent to Perm_G for some group G. In order to do so, we have to investigate the structure of Perm_G. For two finite G sets X_1, X_2 one can form the disjoint union $X_1 \coprod X_2$, provided with the obvious G-action. This is, in fact, the categorical sum of X_1 and X_2, which means:

1. There are given morphisms $a_i : X_i \to X_1 \coprod X_2$ for $i = 1, 2$.
2. For any pair of morphism $b_i : X_i \to Y$, there is a unique morphism $c : X_1 \coprod X_2 \to Y$ such that $b_i = c \circ a_i$ for $i = 1, 2$.

Let Fsets denote the category of the finite sets. There is an obvious functor $\omega : \mathrm{Perm}_G \to \mathrm{Fsets}$ given by $\omega((F, \rho)) = F$. This functor is called a forgetful functor since it forgets the G-action on F. An *automorphism* σ *of* ω is defined by giving, for each element X of Perm_G, an element $\sigma(X) \in \mathrm{Perm}(\omega(X))$ such that: For every morphism $f : X \to Y$ one has $\sigma(Y) \circ \omega(f) = \omega(f) \circ \sigma(X)$. One says that the automorphism σ *respects* $\coprod$ if the action of $\sigma(X_1 \coprod X_2)$ on $\omega(X_1 \coprod X_2) = \omega(X_1) \coprod \omega(X_2)$ is the sum of the actions of $\sigma(X_i)$ on the sets $\omega(X_i)$.

The key to the characterization of G from the category Perm_G is the following simple lemma.

Lemma B.5 *Let* $\mathrm{Aut}^{\coprod}(\omega)$ *denote the group of the automorphisms of* ω *that respect* $\coprod$. *The natural map* $G \to \mathrm{Aut}^{\coprod}(\omega)$ *is an isomorphism of profinite groups.*

Proof. The definition of $G' := \mathrm{Aut}^{\coprod}(\omega)$ yields a map $G' \to \prod_X \mathrm{Perm}(X)$ (the product taken over all isomorphism classes of objects X) that identifies G' with a closed subgroup of $\prod_X \mathrm{Perm}(X)$. Thus G' is also a profinite group. Fix any element $g \in G$ and consider σ_g defined by $\sigma_g(X)e = g \cdot e$ for every object X and point $e \in X$. Thus $g \mapsto \sigma_g$ is a homomorphism from G to G'. This homomorphism is clearly injective. We want to show that it is also surjective. Consider $\sigma \in G'$ and for every open normal subgroup $N \subset G$ the G-set $X_N = G/N$. There is an element $g_N \in G$ such that $\sigma(X_N)N = g_N N$. Multiplication on the right $aN \mapsto aNg$ by an element $g \in G$ is a morphism of the G-set X_N and commutes, therefore, with $\sigma(X_N)$. Then $\sigma(X)gN = \sigma(X)Ng = (\sigma(X)N)g = g_N Ng = g_N gN$. Thus $\sigma(X_N)$ coincides with the action of g_N on X_N. For two open normal subgroups $N_1 \subset N_2$, the map $gN_1 \mapsto gN_2$ is a morphism $X_{N_1} \to X_{N_2}$. It follows that $g_{N_1}N_2 = g_{N_2}N_2$. Thus σ determines an element in the projective limit $\lim_{\leftarrow} G/N$, taken over all open normal subgroups N of G. This projective limit is equal to G and so σ determines an element $g \in G$. The action of $\sigma(X)$ and g coincide for all X of the form G/N with N an open normal subgroup. The same holds then for X of the form G/H where H is an open subgroup. Finally, every G-set is the disjoint union of orbits, each orbit is isomorphic to some G/H with H an open subgroup. Since σ respects disjoint unions, i.e., $\coprod$, one finds that $\sigma(X)$ and g coincide for every G-set X. □

The next step is to produce a set of requirements on a category $\mathcal{C}$ that will imply that $\mathcal{C}$ is equivalent to Perm_G for a suitable profinite group G. There is, of course, no unique answer here. We will give the answer of [118], where a *Galois category* $\mathcal{C}$ is defined by the following rules:

(G1) There is a final object $\mathbf{1}$, i.e., for every object X, the set $\mathrm{Mor}(X, \mathbf{1})$ consists of one element. Moreover, all fibre products $X_1 \times_{X_3} X_2$ exist.

(G2) Finite sums exist as well as the quotient of any object of $\mathcal{C}$ by a finite group of automorphisms.

(G3) Every morphism $f : X \to Y$ can be written as a composition $X \xrightarrow{f_1} Y' \xrightarrow{f_2} Y$ with f_1 a strict epimorphism and f_2 a monomorphism that is an isomorphism onto a direct summand.

(G4) There exists a covariant functor $\omega : \mathcal{C} \to \mathrm{Fsets}$ (called the *fibre functor*) that commutes with fibre products and transforms right units into right units.

(G5) ω commutes with finite direct sums, transforms strict epimorphisms to strict epimorphisms and commutes with forming the quotient by a finite group of automorphisms.

(G6) Let m be a morphism in $\mathcal{C}$. Then m is an isomorphism if $\omega(m)$ is bijective.

One easily checks that any category Perm_G and the forgetful functor ω satisfy the above rules.

One defines an *automorphism σ of ω* in exactly the same way as in the case of the category of G-sets and uses the same definition for the notion that σ preserves $\coprod$. As before, we denote by $\mathrm{Aut}^{\coprod}(\omega)$ the group of the automorphisms of ω that respect $\coprod$. This definition allows us to identify $G = \mathrm{Aut}^{\coprod}(\omega)$ with a closed subgroup of $\prod_X \mathrm{Perm}(\omega(X))$ and so makes G into a profinite group.

Proposition B.6 *Let $\mathcal{C}$ be a Galois category and let G denote the profinite group* $\mathrm{Aut}^{\coprod}(\omega)$. *Then $\mathcal{C}$ is equivalent to the category* Perm_G.

Proof. We only sketch part of the rather long proof. For a complete proof we refer to ([118], p. 119–126). By definition, G acts on each $\omega(X)$. Thus we find a functor $\tau : \mathcal{C} \to \mathrm{Perm}_G$, which associates with each object the finite G-set $\omega(X)$. Now one has to prove two things:

(a) $\mathrm{Mor}(X, Y) \to \mathrm{Mor}(\tau(X), \tau(Y))$ is a bijection.

(b) For every finite G-set F there is an object X such that F is isomorphic to the G-set $\omega(X)$.

As an exercise we will show that the map in (a) is injective. Let two elements f_1, f_2 in the first set of (a) satisfy $\omega(f_1) = \omega(f_2)$. Define $g_i : X \to Z := X \times Y$ as $g := id_X \times f_i$. The fibre product $X \times_Z X$ is defined by the two morphisms g_1, g_2 and consider the morphism $X \times_Z X \stackrel{pr_1}{\to} X$. By (G4), the functor ω commutes with the constructions and $\omega(pr_1)$ is an isomorphism since $\omega(f_1) = \omega(f_2)$. From (G6) it follows that pr_1 is an isomorphism. This implies $f_1 = f_2$. □

Examples B.7 1. Let k be a field. Let k^{sep} denote a separable algebraic closure of k. The category $\mathcal{C}$ will be the dual of the category of the finite dimensional separable k-algebras. Thus the objects are the separable k-algebras of finite dimension and a morphism $R_1 \to R_2$ is a k-algebra homomorphism $R_2 \to R_1$. In this category, the sum $R_1 \coprod R_2$ of two k-algebras is the direct product $R_1 \times R_2$. The fibre functor ω associates with R the set of the maximal ideals of $R \otimes_k k^{sep}$. The profinite group $G = \mathrm{Aut}^{\coprod}(\omega)$ is isomorphic to the Galois group of k^{sep}/k.

2. Finite (topological) coverings of a connected, locally simply connected, topological space X. The objects of this category are the finite topological coverings $Y \to X$. A morphism m between two coverings $u_i : Y_i \to X$ is a continuous map $m : Y_1 \to Y_2$ with $u_2 \circ m = u_1$. Fix a point $x \in X$. A fibre functor ω is then defined by: ω associates with a finite covering $f : Y \to X$ the fibre $f^{-1}(x)$. This category is isomorphic to Perm_G where G is the profinite completion of the fundamental group $\pi(X, x)$.

3. Étale coverings of an algebraic variety [118]. □

B.2 Affine Group Schemes

In Sect. B.1, we studied categories of finite sets on which a finite group acts. This led us naturally to profinite groups, i.e., projective limits of finite groups. In the next section we wish to study categories of finite dimensional representations of a linear algebraic group G over a field k. We recall that G is defined by its coordinate ring $O(G)$ that is a finitely generated k-algebra. Again, projective limits, this time of linear algebraic groups, are needed. These projective limits correspond to direct limits of the coordinate rings of these linear algebraic groups. A direct limit of this sort is, in general, no longer a finitely generated k-algebra. Although one could proceed in an ad hoc manner working with these limits, the natural (and usual) way to proceed is to introduce the notion of an *affine group scheme*. We shall briefly introduce affine schemes (over a field). Then specialize to affine group schemes and commutative Hopf algebras (over a field). In addition, we shall show that an affine group scheme is a projective limit of linear algebraic groups. In the application to differential Galois theory (see Chap. 10), affine groups schemes arise naturally as representable functors (from the category of k-algebras to the category of groups). We shall define this latter notion below and show how these objects can be used to define affine group schemes.

For a k-algebra homomorphism $\phi : B \to A$ between finitely generated k-algebras, one has that for every maximal ideal $\underline{m}$ of A the ideal $\phi^{-1}(\underline{m})$ is also maximal. This fact makes the geometric object $\max(A)$, introduced in Sect. A.1.1, meaningful for a finitely generated k-algebra A. In the sequel, we will work with k-algebras that are not finitely generated. For these algebras $\max(A)$ is not the correct geometric object. Here is an example:
Let $B = k[T]$ and $A = k(T)$ and let $\phi : B \to A$ be the inclusion. Then (0) is the (only) maximal ideal of A and $\phi^{-1}((0))$ is not a maximal ideal of B.
The correct geometric object is given in the following definition.

Definition B.8 Let A be any commutative unitary ring. The set of prime ideals of A is called the *spectrum* of A and is denoted by $\mathrm{Spec}(A)$. An *affine scheme* X is a pair $X := (\mathrm{Spec}(A), A)$. □

Remark B.9 When A is an algebra over the field k, one calls $(\mathrm{Spec}(A), A)$ an *affine scheme over* k. In the extensive literature on schemes (e.g., [94, 124, 262]), the definition of the affine scheme of a commutative ring A with 1 is more involved. It is, in fact, $\mathrm{Spec}(A)$, provided with a topology and a sheaf of unitary commutative rings, called the structure sheaf. These additional structures are determined by A and also determine A. The main observation is that a morphism of affine schemes (with the additional structures) $f : \mathrm{Spec}(A) \to \mathrm{Spec}(B)$ is derived from a unique ring homomorphism $\phi : B \to A$ and, moreover, for any prime ideal $\underline{p} \in \mathrm{Spec}(A)$ the image $f(\underline{p})$ is the prime ideal $\phi^{-1}(\underline{p}) \in \mathrm{Spec}(B)$. In other words, the category of affine schemes is the opposite of the category of the commutative rings with 1. Since we will only need some geometric language and not the full knowledge of these additional

structures we may define the affine scheme of A as above. Furthermore, a *morphism of affine schemes* $(\mathrm{Spec}(A), A) \to (\mathrm{Spec}(B), B)$ is a ring homomorphism $\phi : B \to A$ and the corresponding map $f : \mathrm{Spec}(A) \to \mathrm{Spec}(B)$ defined by $f(\underline{p}) = \phi^{-1}(\underline{p})$. A *morphism of affine k-schemes* $\Phi : X = (\mathrm{Spec}(A), A) \to Y = (\mathrm{Spec}(B), B)$ is a pair $\Phi = (f, \phi)$ satisfying:
1. $\phi : B \to A$ is a k-algebra homomorphism.
2. $f : \mathrm{Spec}(A) \to \mathrm{Spec}(B)$ is induced by ϕ in the following manner: for any prime ideal $\underline{p}$ of A, $f(\underline{p}) = \phi^{-1}(\underline{p})$.
This will suffice for our purposes. We note that the same method is applied in Appendix A w.r.t. the definition of affine varieties. □

Examples B.10 *Affine Schemes*

1. Let k be algebraically closed and let $A = k[X, Y]$. The affine k-variety $X = (\max(A), A)$ of Appendix A has as points the maximal ideals $(X - a, Y - b)$ with $(a, b) \in k^2$. The affine k-scheme $(\mathrm{Spec}(A), A)$ has more points. Namely, the prime ideal (0) and the prime ideals $(p(X, Y))$ with $p(X, Y) \in k[X, Y]$ an irreducible polynomial. Geometrically, the points of $\mathrm{Spec}(A)$ correspond to the whole space k^2, irreducible curves in k^2 and ordinary points of k^2.

2. If $k \subset K$ are fields and K is not a finite algebraic extension of k, then $(\mathrm{Spec}(K), K)$ is an affine scheme that does not correspond to an affine variety.

3. Let n be a positive integer and let $A = k[x]/(x^n - 1)$. We define the affine scheme $\mu_{n,k} = (\mathrm{Spec}(A), A)$. Note that if k has characteristic $p > 0$ and $p|n$, then A has nilpotent elements. □

The topology on $\mathrm{Spec}(A)$ is called the *Zariski topology*. By definition, a subset $S \subset \mathrm{Spec}(A)$ is called (Zariski-)closed if there are elements $\{f_i\}_{i \in I} \subset A$ such that a prime ideal $\underline{p} \in S$ if and only if $\{f_i\}_{i \in I} \subset \underline{p}$. For any subset $\{f_i\}_{i \in I} \subset A$, we define $V(\{f_i\}_{i \in I}) = \{\underline{p} \in \mathrm{Spec}(A) \mid \{f_i\}_{i \in I} \subset \underline{p}\}$. For any $f \in A$, we define X_f to be the open set $U(f) := \mathrm{Spec}(A) \setminus V(f)$. The family $\{U(f)\}$ is a basis for the Zariski topology. For completeness we give the definition of the *structure sheaf* O_X on $\mathrm{Spec}(A)$. For any $f \in A$, one defines $O_X(U(f)) = A_f = A[T]/(Tf - 1)$, the localization of A w.r.t. the element f. A general open U in $\mathrm{Spec}(A)$ is written as a union $\cup U(f_i)$. The sheaf property of O_X determines $O_X(U)$.

Let $X = (\mathrm{Spec}(A), A)$ and $Y = (\mathrm{Spec}(B), B)$ be affine schemes over k. The *product* of X and Y is the affine scheme $X \times_k Y = (\mathrm{Spec}(A \otimes_k B), A \otimes_k B)$. This is, of course, analogous to the definition of the product of affine varieties over k.

The definition, given below, of an affine group scheme G over k is again analogous to the definition of a linear algebraic group (Definition A.33). The only change that one has to make is to replace the affine k-varieties by affine k-schemes.

Definition B.11 An *affine group scheme over* k is an affine k-scheme $G = (\mathrm{Spec}(A), A)$ together with morphisms $m : G \times_k G \to G$, $i : G \to G$ and $e : (\mathrm{Spec}(k), k) \to G$, such that the following diagrams are commutative.

$$\text{Associative} \quad \begin{array}{ccc} G \times_k G \times_k G & \overset{m \times id_G}{\longrightarrow} & G \times_k G \\ {\scriptstyle id_G \times m}\downarrow & & \downarrow{\scriptstyle m} \\ G \times_k G & \underset{m}{\longrightarrow} & G \end{array}, \tag{B.1}$$

$$\text{Unit} \quad \begin{array}{ccc} G & \overset{p \times id_G}{\longrightarrow} & G \times_k G \\ {\scriptstyle id_G \times p}\downarrow & \searrow^{id_G} & \downarrow{\scriptstyle m} \\ G \times_k G & \underset{m}{\longrightarrow} & G \end{array}, \tag{B.2}$$

$$\text{Inverse} \quad \begin{array}{ccc} G & \overset{i \times id_G}{\longrightarrow} & G \times_k G \\ {\scriptstyle id_G \times i}\downarrow & \searrow^{p} & \downarrow{\scriptstyle m} \\ G \times_k G & \underset{m}{\longrightarrow} & G \end{array}, \tag{B.3}$$

where $p : G \to G$ is defined by $p = e \circ \kappa$ and $\kappa : G \to (\mathrm{Spec}(k), k)$ is the morphism induced by the algebra inclusion $k \hookrightarrow A$. □

By definition, the m, i, e in this definition correspond to k-algebra homomorphisms $\Delta : A \to A \otimes_k A$, $\iota : A \to A$, $\epsilon : A \to k$ satisfying conditions dual to (B.1), (B.2), and (B.3). According to the next definition, one can reformulate the data defining the affine group scheme $G = (\mathrm{Spec}(A), A)$ over k by: A is a *commutative Hopf algebra* over k.

Definition B.12 A *commutative Hopf algebra over* k is a k-algebra A equipped with k-algebra homomorphisms $\Delta : A \to A \otimes_k A$ (the comultiplication), $\iota : A \to A$ (the antipode or coinverse) and $\epsilon : A \to k$ (the counit) making the following diagrams commutative.

$$\text{Coassociative} \quad \begin{array}{ccc} A \otimes_k A \otimes_k A & \overset{\Delta \times id_A}{\longleftarrow} & A \otimes_k A \\ {\scriptstyle id_A \times \Delta}\uparrow & & \uparrow{\scriptstyle \Delta} \\ A \otimes_k A & \underset{\Delta}{\longleftarrow} & A \end{array}, \tag{B.4}$$

$$\text{Counit} \quad \begin{array}{ccc} A & \overset{p^* \times id_A}{\longleftarrow} & A \times_k A \\ {\scriptstyle id_A \times p^*}\uparrow & \nwarrow^{id_A} & \uparrow{\scriptstyle \Delta} \\ A \otimes_k A & \underset{\Delta}{\longleftarrow} & A \end{array}, \tag{B.5}$$

$$\text{Coinverse} \quad \begin{array}{ccc} A & \overset{\iota \times id_A}{\longleftarrow} & A \otimes_k A \\ {\scriptstyle id_A \times \iota}\uparrow & \nwarrow^{p^*} & \uparrow{\scriptstyle \Delta} \\ A \otimes_k A & \underset{\Delta}{\longleftarrow} & A \end{array}, \tag{B.6}$$

where $p^* : A \to A$ is defined by $p^* = \epsilon \circ incl$ and $incl$ is the inclusion $k \hookrightarrow A$. □

Examples B.13 *Affine group schemes.*

1. Let $A = k[x_1, x_2, \dots]$ be the polynomial ring in an infinite number of indeterminates x_i. Let $\Delta(x_i) = x_i \otimes 1 + 1 \otimes x_i$, $\iota(x_i) = -x_i$ and $\epsilon(x_i) = 0$. This defines an affine group scheme. Note that A is the direct limit of finitely generated Hopf algebras $A_n = k[x_1, \dots, x_n]$ and that each of these is the coordinate ring of a linear algebraic group $\mathbf{G}_a^n$. Therefore, the affine group scheme $(\mathrm{Spec}(A), A)$ is the inverse limit of affine group schemes coming from linear algebraic groups. We shall show below that this is the case, in general.

2. Let $A = k[x]/(x^n - 1)$ and let $\boldsymbol{\mu}_{n,k} = (\mathrm{Spec}(A), A)$. The homomorphisms defined by $\Delta(x) = x \otimes x$, $\iota(x) = x^{n-1}$ and $\epsilon(x) = 1$ define a commutative Hopf algebra. We observe that for a field k of characteristic $p > 0$, the algebra A is not reduced if $p|n$. A result of Cartier ([302], Chap. 11.4) implies that in characteristic zero, any commutative Hopf algebra over k is reduced. In other words, affine group schemes over a field of characteristic 0 are reduced. □

We will define properties of affine group schemes in terms of the associated ring A. An important concept for linear algebraic groups is that of a *representation* (Definition A.42). We give the analogous definition for affine group schemes here.

Definition B.14 A pair (V, τ) is called a *representation of* $G = (\mathrm{Spec}(A), A)$ (also called a G-module) is k-vector space V and a k-linear map $\tau : V \to A \otimes_k V$ such that

(i) $\tau : V \to A \otimes_k V$ is k-linear.

(ii) $(\epsilon \otimes id) \circ \tau : V \to A \otimes V \to k \otimes V = V$ is the identity.

(iii) The maps $(\Delta \otimes id_V) \circ \tau$ and $(id_A \otimes \tau) \circ \tau$ from V to $A \otimes A \otimes V$ coincide.

A *morphism* $f : (V_1, \tau_1) \to (V_2, \tau_2)$ between representations is a k-linear map satisfying $\tau_2 \circ f = f \circ \tau_1$. □

For the special case that G is a linear algebraic group over k and V is finite dimensional, one recovers from Exercise A.43 the earlier definition of a representation of G, namely a homomorphism $\rho : G \to \mathrm{GL}(V)$ of linear algebraic groups over k. The general situation of the above definition is obtained by taking (direct and projective) limits. We note further that the set $\mathrm{Mor}((V_1, \rho_1), (V_2, \rho_2))$ of all homomorphisms between two representations is a vector space over k. The trivial representation, i.e., a one-dimensional vector space over k on which all elements of G act as the identity, is denoted by $\mathbf{1}$.

Definition B.15 Let G be an affine group scheme over k. The category of all *finite dimensional* representations of G is denoted by Repr_G. □

Lemma B.16 *Let $\tau : V \to A \otimes V$ be any representation of the affine group scheme G over k and let $S \subset V$ be a finite set. Then there exists a finite dimensional $W \subset V$ with $S \subset W$ and $\tau(W) \subset A \otimes W$. In particular, V is the union of finite dimensional representations.*

Proof. It suffices to consider the case where S consists of a single element v. Choose a basis $\{a_i\}$ of A over k and write $\tau(v) = \sum a_i \otimes v_i$. Let W be the finite dimensional subspace of V generated by all v_i. From $(\epsilon \otimes id) \circ \tau(v) = v$ it follows that $v = \sum \epsilon(a_i)v_i$ belongs to W. The equality $(\Delta \otimes id_V) \circ \tau(v) = (id_A \otimes \tau) \circ \tau(v)$ yields $\tau(v_j) \in A \otimes_k W$ for all j. Indeed, one writes $\Delta(a_i) = \sum \delta_{i,j,k} a_j \otimes a_k$. The left-hand side of the equality reads $\sum_{i,j,k} \delta_{i,j,k} a_j \otimes a_k \otimes v_i$ and the right-hand side reads $\sum_j a_j \otimes \tau(v_j)$. So $\tau(v_j) = \sum_{i,k} \delta_{i,j,k} a_k \otimes v_i$. Thus $\tau(W) \subset A \otimes_k W$. □

Corollary B.17 *Let $G = (\mathrm{Spec}(A), A)$ be a group scheme over k. Then A is the union of finitely generated subalgebras B, each of which is invariant under Δ, ϵ, and ι. Each such B defines a linear algebraic group G_B over k. Furthermore, G is the projective limit of the G_B.*

Proof. The map $\Delta : A \to A \otimes A$ makes A into a representation of G. Let $S \subset A$ be a finite set and let $V \subset A$ be the finite dimensional vector space of Lemma B.16 with $S \subset V$ and $\Delta(V) \subset A \otimes V$. Take a basis $\{v_i\}$ of V and define elements $a_{i,j} \in A$ by $\Delta(v_i) = \sum a_{i,j} \otimes v_j$. Then $B := k[v_i, a_{i,j}, \iota(v_i), \iota(a_{i,j})] \subset A$ can be seen to satisfy $\Delta(B) \subset B \otimes_k B$ and $\iota(B) = B$. Thus B defines a linear algebraic group G_B. Now A is the direct limit of subalgebras B, finitely generated over k and invariant under Δ, ϵ, and ι. This translates, by definition, into G is the projective limit of the projective system $\{G_B\}$.

We note that this projective system of linear algebraic groups has an additional property, namely $f : G_{B_1} \to G_{B_2}$ is surjective for $B_2 \subset B_1$. Indeed, it is known that the image of f is a Zariski-closed subgroup H of G_{B_2}. Let $I \subset B_2$ be the ideal of H. Then I is the kernel of $B_2 \to B_1$. Thus $I = 0$ and $H = G_{B_2}$. □

In general, an affine group scheme H over a general field k (or even a linear algebraic group over k) is not determined by its group of k-rational points $H(k)$. We now define an object that is equivalent to a group scheme.

Let $G = (\mathrm{Spec}(A), A)$ be an affine group scheme over k. One associates to G a functor, called FG, from the category of the k-algebras to the category of groups (as usual, by a k-algebra we will mean a commutative algebra over k having a unit element). For a k-algebra R we put $FG(R) = G(R)$, i.e., the set of k-algebra homomorphisms $A \to R$. For two elements $\phi, \psi \in G(R)$ one defines the product $\phi \cdot \psi$ as the k-algebra homomorphism $A \stackrel{\Delta}{\to} A \otimes A \stackrel{\phi\otimes\psi}{\to} R \otimes R \stackrel{prod}{\to} R$, where the last map is just the product in R, i.e., $prod(r_1 \otimes r_2) = r_1 r_2$. One can show that the obvious map from $\mathrm{Mor}(G_1, G_2)$, the set of morphisms of affine group schemes over k, to $\mathrm{Mor}(FG_1, FG_2)$, the set of morphisms between the two functors FG_1 and FG_2, is a bijection.

We note that only rather special functors F from the category of the k-algebras to the category of groups are of the form FG for some affine group scheme G over k. The condition is that F, seen as a functor from k-algebras to the category of sets is *representable*. To define this we need the notion of a morphism of functors, cf. [170], Chap. I, §11.

Definition B.18 *Morphism of functors, representable functors.*
(1) Let F_1, F_2 denote two covariant functors from the category of all k-algebras (or any other category) to the category of sets. A morphism $\alpha : F_1 \to F_2$ (or natural transformation) consists of a family of maps $\alpha_R : F_1(R) \to F_2(R)$ (for every k-algebra R) such that for each morphism $f : R_1 \to R_2$ of k-algebras the relation $F_2(f) \circ \alpha_{R_1} = \alpha_{R_2} \circ F_1(f)$ holds. The morphism α is called an isomorphism if every α_R is a bijection. We will write $\mathrm{Mor}(F_1, F_2)$ for the set of morphisms from the functor F_1 to the functor F_2.

(2) For a k-algebra A, one defines the covariant functor F_A from the category of k-algebras to the category of sets by the formula $F_A(R)$ is the set $\mathrm{Hom}_k(A, R)$ of the k-algebra homomorphisms of A to R. Furthermore, for any morphism $f : R_1 \to R_2$ of k-algebras, $F_A(f) : \mathrm{Hom}_k(A, R_1) \to \mathrm{Hom}_k(A, R_2)$ is the map $h \mapsto f \circ h$.
The *Yoneda Lemma*, which we will admit without proof, concerns two functors F_{A_1}, F_{A_2} as above. The statement is that the obvious map $\mathrm{Hom}_k(A_1, A_2) \to \mathrm{Mor}(F_{A_2}, F_{A_1})$ is a bijection. In particular, the functor F_A determines A.

(3) A functor F from the category of k-algebras (or any other category) to the category of sets is called *representable* if there exists a k-algebra A (or an object A) and an isomorphism $\alpha : F \to F_A$. The Yoneda Lemma implies that A is unique up to a unique isomorphism. □

Suppose that F is a functor from k-algebras to the category of groups and suppose that F is representable as a functor from k-algebras to sets. Let A be the k-algebra representing F (i.e., F is isomorphic to F_A). Then we claim that A has the structure of a commutative Hopf algebra. In other words, $G = (\mathrm{Spec}(A), A)$ is an affine group scheme over k and F is isomorphic to FG, defined above. We will sketch the proof. For more details, see [302], Chaps. 1.3, 1.4.

Consider the functor $F \times F$, defined by $(F \times F)(R) = F(R) \times F(R)$. This functor is represented by $A \otimes_k A$. There is a morphism of functors $\alpha : F \times F \to F$, given by $\alpha_R : F(R) \times F(R) \to F(R)$ is the multiplication in the group $F(R)$. According to Yoneda's Lemma, α defines a morphism of k-algebras $A \to A \otimes_k A$. This morphism will be called Δ. The morphism of functors $\beta : F \to F$, given by $\beta_R : F(R) \to F(R)$ is taking the inverse in the group $F(R)$, induces a k-algebra homomorphism $\iota : A \to A$. Finally, consider the functor E from k-algebras to groups, given by $E(R) = \{1\}$ for every R. This functor is represented by the k-algebra k itself. The morphism of functors $\gamma : F \to E$, given by the only possible map $\gamma_R : F(R) \to E(R) = \{1\}$ for every R, induces a k-algebra homomorphism

$A \to k$, which we will call ϵ. A straightforward verification shows that A equipped with Δ, ι, and ϵ is a commutative Hopf algebra.

Examples B.19 *Representable functors.*

1. Let H be an abelian group, written additively. We associate with H the functor defined by $F(R) = \mathrm{Hom}(H, R^*)$ where $R^* =$ the group of units of R. The group algebra of H over k can be written as $A = \oplus_{h \in H} k t_h$ where the multiplication is given by $t_0 = 1$ and $t_{h_1} \cdot t_{h_2} = t_{h_1+h_2}$. The functor F is clearly represented by A. Thus $G = (\mathrm{Spec}(A), A)$ must be an affine group scheme over k. In particular $\Delta : A \to A \otimes A$ must exist. One easily shows that the formula for Δ must be $\Delta(t_h) = t_h \otimes t_h$ for all $h \in H$. For the group $H = \mathbf{Z}$ one observes that $A = k[t_1, t_1^{-1}]$ and $G = \mathbf{G}_{m,k}$. If H is the cyclic group of order n, then the corresponding G is equal to $\boldsymbol{\mu}_{n,k} = (\mathrm{Spec}(k[T]/(T^n - 1)), k[T]/(T^n - 1))$. In general, for a finitely generated H the group A is the coordinate ring of a commutative linear algebraic group and, moreover, an extension of a torus by a finite group. For $H = \mathbf{Q}$, or more generally a vector space over $\mathbf{Q}$, the affine group scheme G is rather large and no longer a linear algebraic group. In the classification of differential modules over $\mathbf{C}((z))$ an affine group scheme occurs, namely the *exponential torus*. We recall that one considers a complex vector space $\mathcal{Q} = \cup_{m \geq 1} z^{-1/m}\mathbf{C}[z^{-1/m}]$. The complex-valued points of the exponential torus were defined as $\mathrm{Hom}(\mathcal{Q}, \mathbf{C}^*)$. Let G be the affine group scheme corresponding to $\mathcal{Q}$, then the above group is $G(\mathbf{C})$.

2. Let H be any group. Let $\mathcal{C}$ denote the category of the representations of H on finite dimensional vector spaces over k. We will see in the following that $\mathcal{C}$ is a "neutral tannakian category over k", which means that $\mathcal{C}$ is, in a natural way, equivalent to Repr_G for some affine group scheme G. In other terms, this affine group scheme has the same set of "algebraic" representations as the ordinary representations of H on finite dimensional k-vector spaces. The group G can be seen as a sort of "algebraic hull" of H. Even for a simple group like $\mathbf{Z}$ this algebraic hull is rather large and difficult to describe. Again this situation occurs naturally in the classification of differential equations over, say, $\mathbf{C}(z)$ (see Chaps. 10 and 12). □

B.3 Tannakian Categories

One wants to recognize when a category is equivalent to Repr_G for some affine group scheme G over k. We start by recovering G from the category Repr_G. We will now formulate and prove Tannaka's Theorem. In [280], Theorem 2.5.3, this theorem is formulated and proved for a linear algebraic group over an algebraically closed field. We will give an exposition of the general situation.

The main ingredients are the tensor product and the fibre functor $\omega : \mathrm{Repr}_G \to \mathrm{Vect}_k$. The last category is that of the finite dimensional vector spaces over k. The functor ω is again the forgetful functor that associates to the representation (V, ρ)

the finite dimensional k-vector space V (and forgets ρ). In analogy with Galois categories, we will show that we can recover an affine group scheme from the group of automorphisms of the fibre functor (with respect to tensor products). If we naively follow this analogy, we would define an automorphism of ω to be a functorial choice $\sigma(X) \in \mathrm{GL}(\omega(X))$ for each object $X \in \mathrm{Repr}_G$ such that $\sigma(X_1 \otimes X_2) = \sigma(X_1) \otimes \sigma(X_2)$. This approach is a little too naive. Instead, we will define $G' := Aut^{\otimes}(\omega)$ to be a functor from the category of k-algebras to the category of groups and then show that this functor is isomorphic to the functor FG.

Let R be a k-algebra. An element σ of $G'(R)$ is given by a collection of elements $\{\sigma(X)\}_X$, where X runs over the collection of all objects in Repr_G. Each $\sigma(X)$ is an R-linear automorphism of $R \otimes_k \omega(X)$ such that the following hold:

(i) $\sigma(\mathbf{1})$ is the identity on $R \otimes \omega(\mathbf{1}) = R$.

(ii) For every morphism $f : X \to Y$ one has an R-linear map $id_R \otimes \omega(f) : R \otimes \omega(X) \to R \otimes \omega(Y)$. Then $(id_R \otimes \omega(f)) \circ \sigma(X) = \sigma(Y) \circ (id_R \otimes \omega(f))$.

(iii) The R-linear automorphism $\sigma(X \otimes Y)$ on $R \otimes \omega(X \otimes Y) = R \otimes_k \omega(X) \otimes_k \omega(Y) = (R \otimes \omega(X)) \otimes_R (R \otimes \omega(Y))$ is obtained as the tensor product of the two R-linear maps $\sigma(X)$ and $\sigma(Y)$.

It is easy to see that $G'(R)$ is a group and that $R \mapsto G'(R)$ is a functor from k-algebras to groups.

Theorem B.20 (Tannaka's Theorem) *Let G be an affine group scheme over k and let $\omega : \mathrm{Repr}_G \to \mathrm{Vect}_k$ be the forgetful functor. There is an isomorphism of functors $FG \to \mathrm{Aut}^{\otimes}(\omega)$.*

Proof. We write, as above, G' for the functor $\mathrm{Aut}^{\otimes}(\omega)$. First, we have to define, for any k-algebra R, a map $\xi \in G(R) \mapsto \sigma_\xi \in G'(R)$. The element ξ is a k-homomorphism $A \to R$. Let $X = (V, \tau)$ be a representation of G. Then one defines $\sigma_\xi(X)$ as the extension to an R-linear map $R \otimes V \to R \otimes V$ of the k-linear map $V \stackrel{\tau}{\to} A \otimes V \stackrel{\xi \otimes id_V}{\to} R \otimes V$. The verification that this definition leads to a morphism of functors $FG \to G'$ is straightforward. We have to show that $FG(R) \to G'(R)$ is bijective for every R.

Take some element $\sigma \in G'(R)$. Let (V, τ) be *any* G-module. Lemma B.16 writes V as the union of finite dimensional subspaces W with $\tau(W) \subset A \otimes W$. The R-linear automorphism $\sigma(W)$ of $R \otimes W$ glue to an R-linear automorphism $\sigma(V)$ of $R \otimes V$. Thus we have extended σ to the wider category of all G-modules. This extension has again the properties (i), (ii), and (iii). Now consider the G-module (A, τ) with $\tau = \Delta$. We want to find an element $\xi \in G(R)$, i.e., a k-algebra homomorphism $\xi : A \to R$, such that $\sigma = \sigma_\xi$. The restriction of $\sigma_\xi(A, \tau)$ to $A \subset R \otimes A$ was defined by $A \stackrel{\Delta}{\to} A \otimes A \stackrel{\xi \otimes id_A}{\to} R \otimes A$. If we follow this map with $R \otimes A \stackrel{id_R \otimes \epsilon}{\to} R \otimes k = R$

then the result is $\xi : A \to R$. Since we require that $\sigma_\xi(A, \tau) = \sigma(A, \tau)$ the k-algebra homomorphism

$$A \subset R \otimes A \stackrel{\sigma(A,\tau)}{\to} R \otimes A \stackrel{id_R \otimes \epsilon}{\to} R \otimes k = R$$

must be chosen as ξ. In order to see that $\sigma = \sigma_\xi$ one may replace σ by $\sigma_\xi^{-1}\sigma$ and prove that the latter is 1. In other words, we may suppose that $R \otimes A \stackrel{\sigma(A,\tau)}{\to} R \otimes A \stackrel{id_R \otimes \epsilon}{\to} R \otimes k = R$ is equal to $R \otimes A \stackrel{id_R \otimes \epsilon}{\to} R \otimes k = R$ and we have to prove that $\sigma = 1$.

One also has to consider the G-module $(A \otimes A, \mu)$ with $\mu = \Delta \otimes id_A$. Let $\{a_i\}$ be a k-basis of A, then the G-module $(A \otimes A, \mu)$ is the direct sum of the G-modules $A \otimes a_i$. Each of those modules is isomorphic to (A, τ) and therefore $\sigma(A \otimes A, \mu) = \sigma(A, \tau) \otimes id_A$.

The law for the comultiplication shows that $\Delta : A \to A \otimes A$ is a morphism between the G-modules (A, τ) and $(A \otimes A, \mu)$. Now we must relate the various arrows in the following diagrams to the morphisms they represent.

$$\begin{array}{ccccc} R \otimes A & \stackrel{\sigma(A,\tau)}{\longrightarrow} & R \otimes A & & \\ \downarrow & & \downarrow & & \\ R \otimes A \otimes A & \stackrel{\sigma(A \otimes A,\mu)}{\longrightarrow} & R \otimes A \otimes A & \stackrel{id_R \otimes \epsilon \otimes id_A}{\longrightarrow} & R \otimes A \end{array} .$$

Let us write $\Delta_R : R \otimes A \to R \otimes A \otimes A$ for the R-linear extension of Δ. The two "down arrows" in the diagram are Δ_R. The diagram is commutative since $\Delta : (A, \tau) \to (A \otimes A, \mu)$ is a morphism of G-modules.

We want to show that the upper path from $R \otimes A$, in the upper left-hand corner, to $R \otimes A$, in the lower right-hand corner, produces the map $\sigma(A, \tau)$ and that the lower path produces the identity on $R \otimes A$. This would prove $\sigma(A, \tau) = id$.

The rule $A \stackrel{\Delta}{\to} A \otimes A \stackrel{\epsilon \otimes id_A}{\to} A = id_A$ for affine group scheme A implies that $(id_R \otimes \epsilon \otimes id_A) \circ \Delta_R$ is the identity on $R \otimes A$. This proves the statement on the first path. We recall that our assumption on σ is $R \otimes A \stackrel{\sigma(A,\tau)}{\to} R \otimes A \stackrel{id_R \otimes \epsilon}{\to} R \otimes k = R$ is equal to the map $id_R \otimes \epsilon$. Furthermore, $\sigma(A \otimes A, \mu) = \sigma(A, \tau) \otimes id_A$. The composition of the two arrows in the lower row is therefore $id_R \otimes \epsilon \otimes id_A$. The rule $A \stackrel{\Delta}{\to} A \otimes A \stackrel{\epsilon \otimes id_A}{\to} A = id_A$ implies now that the other path yields the identity map on $R \otimes A$.

We conclude that $\sigma(A, \tau) = id$. Consider a G-module (V, μ) of some dimension $d < \infty$. We have to show that $\sigma(V, \mu) = id$. Consider any k-linear map $u : V \to k$ and the composed map $\phi : V \stackrel{\mu}{\to} A \otimes V \stackrel{id_A \otimes u}{\to} A \otimes k = A$. One easily verifies that ϕ is a morphism between the G-modules (V, μ) and (A, τ). By taking a basis of d

elements of the dual of V, one obtains an embedding of the G-module (V, μ) in the G-module $(A, \tau)\oplus\cdots\oplus(A, \tau)$. From $\sigma(A, \tau) = id$ one concludes that $\sigma(V, \mu) = id$. Thus $\sigma = 1$. This shows that the functor gives a bijection $FG(R) \to G'(R)$. □

The next step is to consider a category $\mathcal{C}$ with a "fibre functor" $\omega : \mathcal{C} \to \mathrm{Vect}_k$ and to produce a reasonable set of properties of $\mathcal{C}$ and ω that ensure that $\mathcal{C}$ is equivalent to Repr_G for a suitable affine group scheme G over k. In this equivalence we require that ω is compatible with the forgetful functor $\mathrm{Repr}_G \to \mathrm{Vect}_k$. In particular, the G in this statement must be the affine group scheme over k that represents the functor $\mathrm{Aut}^{\otimes}(\omega)$ from the category of the k-algebras to the category of groups. This leads to the following definition, copied from [82], Definition 2.19, of a *neutral tannakian category* $\mathcal{C}$ *over* k:

(1) The category $\mathcal{C}$ has a *tensor product*, i.e., for every pair of objects X, Y a new object $X \otimes Y$. The tensor product $X \otimes Y$ depends functorially on both X and Y. The tensor product is associative and commutative and there is a unit object, denoted by $\mathbf{1}$. The latter means that $X \otimes \mathbf{1}$ is isomorphic to X for every object X. In the above statements one has to keep track of the isomorphisms, everything must be functorial and one requires many commutative diagrams in order to avoid "fake tensor products".

(2) $\mathcal{C}$ has *internal Homs*. This means the following. Let X, Y denote two objects of $\mathcal{C}$. The internal Hom, denoted by $\underline{\mathrm{Hom}}(X, Y)$, is a new object such that the two functors $T \mapsto \mathrm{Hom}(T \otimes X, Y)$ and $T \mapsto \mathrm{Hom}(T, \underline{\mathrm{Hom}}(X, Y))$ are isomorphic. Let us denote $\underline{\mathrm{Hom}}(X, \mathbf{1})$ by X^*. One requires that the canonical morphism $X \to (X^*)^*$ is an isomorphism. Moreover, one requires that the canonical morphism $\underline{\mathrm{Hom}}(X_1, Y_1) \otimes \underline{\mathrm{Hom}}(X_2, Y_2) \to \underline{\mathrm{Hom}}(X_1 \otimes X_2, Y_1 \otimes Y_2)$ is an isomorphism.

(3) $\mathcal{C}$ is an *abelian category* (see [170], Chap. III, §3). We do not want to recall the definition of an abelian category but note that the statement is equivalent to: $\mathcal{C}$ is isomorphic to a category of left modules over some ring A that is closed under taking kernels, cokernels, and finite direct sums.

(4) An isomorphism between $End(\mathbf{1})$ and k is given.

(5) There is a *fibre functor* $\omega : \mathcal{C} \to \mathrm{Vect}_k$, which means that ω is k-linear, faithful, exact, and commutes with tensor products. We note that (3) and (4) imply that each $\mathrm{Hom}(X, Y)$ is a vector space over k. The k-linearity of ω means that the map $\mathrm{Hom}(X, Y) \to \mathrm{Hom}(\omega(X), \omega(Y))$ is k-linear. Faithful is defined by: $\omega(X) = 0$ implies $X = 0$. Exact means that ω transforms exact sequences into exact sequences.

Remark B.21 One sees that Repr_G is a neutral tannakian category. The definition of a *tannakian category* is a little weaker than that of a neutral tannakian category. The fibre functor in (5) is replaced by a fibre functor $\mathcal{C} \to \mathrm{Vect}_K$ where (say) K is a field extension of k. The problem studied by Saavedra and finally solved by Deligne in [81] was to find a classification of tannakian categories analogous to

Theorem B.22 below. We note that the above condition (2) seems to be replaced in [81] be an apparently weaker condition, namely the existence of a functor $X \mapsto X^*$ having suitable properties. □

Theorem B.22 *A neutral tannakian category* $\mathcal{C}$ *over* k *with fibre functor* $\omega : \mathcal{C} \to \mathrm{Vect}_k$ *is canonically isomorphic to* Repr_G *where* G *represents the functor* $\mathrm{Aut}^{\otimes}(\omega)$.

Proof. We will only explain the beginning of the proof. We write G for the functor $\mathrm{Aut}^{\otimes}(\omega)$. Our first concern is to show that G is an affine group scheme. Let $\{X_i\}_{i\in I}$ denote the collection of all (isomorphism classes of) objects of $\mathcal{C}$. We give I some total order. For each X_i the functor $R \mapsto \mathrm{GL}_R(R \otimes \omega(X_i))$ is the functor associated with the linear algebraic group $\mathrm{GL}(\omega(X_i))$. Let us write B_i for the affine algebra of $\mathrm{GL}(\omega(X_i))$. For any finite subset $F = \{i_1 < \cdots < i_n\} \subset I$ one considers the functor $R \mapsto \prod_{j=1}^n \mathrm{GL}_R(R \otimes \omega(X_{i,j}))$ that is associated with the linear algebraic group $\prod_{j=1}^n \mathrm{GL}(\omega(X_{i,j}))$. The affine algebra B_F of this group is $B_{i_1} \otimes \cdots \otimes B_{i_n}$. For inclusions of finite subsets $F_1 \subset F_2$ of I one has obvious inclusions of k-algebras $B_{F_1} \subset B_{F_2}$. We define B as the direct limit of the B_F, where F runs over the collection of the finite subsets of I. It is rather obvious that B defines an affine group scheme H over k and that $H(R) = \prod_{i\in I} \mathrm{GL}_R(R \otimes \omega(X_i))$ for every k-algebras R. By definition, $G(R)$ is a subgroup of the group $H(R)$. This subgroup is defined by a relation for each morphism $f : X_i \to X_j$ and by a relation for each isomorphism $X_i \otimes X_j \cong X_k$. Each condition imposed on $\sigma = \{\sigma(X_i)\}_{i\in I} \in G(R)$ can be written as a set of polynomial equations with coefficients in k for the entries of the matrices $\sigma(X_i)$ (w.r.t. k-bases for the spaces $\omega(X_i)$). The totality of all those equations generates an ideal $J \subset B$. Put $A := B/J$ then $G(R) = \mathrm{Hom}(A, R) \subset \mathrm{Hom}(B, R)$. In other words, G is the affine group scheme associated with A. For a fixed object X of $\mathcal{C}$ and each k-algebra R one has (by construction) an action of $G(R)$ on $R \otimes \omega(X)$. This makes each $\omega(X)$ into a G-module. The assignment $X \mapsto \omega(X)$ with its G-action, is easily seen to define a functor $\tau : \mathcal{C} \to \mathrm{Repr}_G$. The latter should be the equivalence between the two categories. One has to prove:

(a) $\mathrm{Hom}(X, Y) \to \mathrm{Hom}_G(\omega(X), \omega(Y))$ is a bijection.

(b) For every G-module V of finite dimension over k, there is an object X such that the G-module $\omega(X)$ is isomorphic to V.

The injectivity in (a) follows at once from ω being exact and faithful. We will not go into the technical details of the remaining part of the proof. Complete proofs are in [82] and [81]. Another sketch of the proof can be found in [52], pp. 344–348. □

Example B.23 *Differential modules.*
K denotes a differential field with a field of constants C. Let Diff_K denote the category of the differential modules over K. It is evident that this category has all the properties of a neutral tannakian category over C with the possible exception of a fibre functor $\omega : \mathrm{Diff}_K \to \mathrm{Vect}_C$. There is, however, a "fibre functor" $\tau : \mathrm{Diff}_K$

$\to \mathrm{Vect}_K$ that is the forgetful functor and associates to a differential module (M, ∂) the K-vector space M. When C is algebraically closed and has characteristic 0, this suffices to show that a fibre functor with values in Vect_C exists. This is proved in the work of Deligne [81]. The proof is remarkably complicated. From the existence of this fibre functor Deligne is able to deduce the Picard-Vessiot theory.

On the other hand, if one assumes the Picard-Vessiot theory, then one can build a universal Picard-Vessiot extension $\mathrm{UnivF} \supset K$, which is obtained as the direct limit of the Picard-Vessiot extensions of all differential modules (M, ∂) over K. Let G denote the group of the differential automorphisms of UnivF/K. By restricting the action of G to ordinary Picard-Vessiot fields L with $K \subset L \subset \mathrm{UnivF}$, one finds that G is the projective limit of linear algebraic groups over C. In other words, G is an affine group scheme over C. The equivalence $\mathrm{Diff}_K \to \mathrm{Repr}_G$ is made explicit by associating to a differential module (M, ∂) over K the finite dimensional C-vector space $\ker(\partial, \mathrm{UnivF} \otimes_K M)$ equipped with the induced action of G. Indeed, G acts on UnivF and therefore on $\mathrm{UnivF} \otimes M$. This action commutes with the ∂ on $\mathrm{UnivF} \otimes M$ and thus G acts on $\ker(\partial, \mathrm{UnivF} \otimes_K M)$.

For a fixed differential module M over K, one considers the full subcategory $\{\{M\}\}$ of Diff_K, defined in Sect. 2.4. This is again a neutral tannakian category and by Theorem 2.33 equivalent with Repr_G where G is the differential Galois group of M over K.

For a general differential field K, these equivalences are useful for understanding the structure of differential modules and the relation with the solution spaces of such modules. In a few cases, the universal Picard-Vessiot field UnivF and the group G are known explicitly. An important case is the differential field $K = \mathbf{C}((z))$ with differentiation $\frac{d}{dz}$. See Chap. 10 for a discussion of this and other fields. □

Example B.24 *Connections.*
1. Let X be a connected Riemann surface. A connection (M, ∇) on X is a vector bundle M on X provided with a morphism $\nabla : M \to \Omega_X \otimes M$ having the usual properties (see Sect. 6.2). Let Conn_X denote the category of all connections on X. Choose a point $x \in X$ with local parameter t. Define the functor $\omega : \mathrm{Conn}_X \to \mathrm{Vect}_{\mathbf{C}}$ by $\omega(M, \nabla) = M_x/tM_x$. The only non-trivial part of the verification that $\mathcal{C}$ is a neutral tannakian category over $\mathbf{C}$, is showing that $\mathcal{C}$ is an abelian category. We note that in the category of all vector bundles on X cokernels need not exist. However, for a morphism $f : (M, \nabla_1) \to (N, \nabla_2)$ of connections, the image $f(M) \subset N$ is locally a direct summand, due to the regularity of the connection. Conn_X is equivalent with Repr_G for a suitable affine group scheme G over $\mathbf{C}$. Let π denote the fundamental group $\pi(X, x)$ and let $\mathcal{C}$ denoted the category of the representations of π on finite dimensional complex vector spaces. As in Sects. 5.3 and 6.4, the weak form of the Riemann-Hilbert theorem is valid. This theorem can be formulated as:

The monodromy representation induces an equivalence of categories $\mathcal{M} : \mathrm{Conn}_X \to \mathcal{C}$.

The conclusion is that the affine group scheme G is the "algebraic hull" of the group π, as defined in Example B.19.

2. Let X be again a connected Riemann surface and let S be a finite subset of X. A regular singular connection (M, ∇) for (X, S) consists of a vector bundle and a connection $\nabla : M \to \Omega_X(S) \otimes M$ with the usual rules (see Definition 6.8). $\omega_X(S)$ is the sheaf of differential forms with poles at S of order ≤ 1. If S is not empty, then the category of the regular singular connections is not abelian since cokernels do not always exist.

3. C denotes an algebraically closed field of characteristic 0. Let X be an irreducible, smooth curve over C. The category $\mathrm{AlgConn}_X$ of all connections on X is again a neutral tannakian category over C. In general (even if C is the field of complex numbers), it seems that there is no description of the corresponding affine group scheme. The first explicit example $C = \mathbf{C}$ and $X = \mathbf{P}^1_{\mathbf{C}} \setminus \{0\}$ is rather interesting. We will discuss the results in this special case.

Let K denote the differential field $\mathbf{C}(\{z\})$. One defines a functor α from the category $\mathrm{AlgConn}_X$ to the category Diff_K by

$$(M, \nabla) \mapsto (K \otimes_{\mathbf{C}[z^{-1}]} H^0(X, M), \partial), \text{ where } \partial \text{ is the extension of}$$

$$\nabla_{\frac{d}{dz}} : H^0(X, M) \to z^{-2}\mathbf{C}[z^{-1}] \otimes_{\mathbf{C}[z^{-1}]} H^0(X, M) = z^{-2} H^0(X, M).$$

Explicitly, let $e_1, \ldots, e_n$ be a free basis of the $\mathbf{C}[z^{-1}]$-module $H^0(X, M)$. Then ∇ is determined by the matrix B w.r.t. $\{e_1, \ldots, e_n\}$, having entries in $\mathbf{C}[z^{-1}]$, of the map $\nabla_{\frac{d}{d(z^{-1})}}$. Thus ∇ is represented by the matrix differential equation $\frac{d}{d(z^{-1})} + B$. Rewriting this in the variable z, one obtains the matrix differential equation $\frac{d}{dz} + A = \frac{d}{dz} - z^{-2}B$ over the field K. We note that the coefficients of A lie in $z^{-2}\mathbf{C}[z^{-1}]$.

It is rather clear that α is a morphism of neutral tannakian categories. We start by proving that $\mathrm{Hom}(M_1, M_2) \to \mathrm{Hom}(\alpha(M_1), \alpha(M_2))$ is bijective. It suffices (use internal Hom) to prove this for $M_1 = \mathbf{1}$ and $M_2 = M$ is any object. One can identify $\mathrm{Hom}(\mathbf{1}, M)$ with $\ker(\nabla, H^0(X, M))$ and $\mathrm{Hom}(\mathbf{1}, \alpha(M))$ with $\ker(\partial, K \otimes H^0(X, M))$. The injectivity of the map under consideration is clear. Let $f \in \ker(\partial, K \otimes H^0(X, M))$. Then f is a meromorphic solution of the differential equation in some neighborhood of 0. This solution has a well-defined extension to a meromorphic solution F on all of $\mathbf{P}^1_{\mathbf{C}}$, since the differential equation is regular outside 0 and X is simply connected. Thus F is a rational solution with, at most, a singularity at 0. Therefore, $F \in \ker(\nabla, H^0(X, M))$ and has image f.

The next question is whether each object of Diff_K is isomorphic to some $\alpha(M)$. Apparently this is not the case since the topological monodromy of any $\alpha(M)$ is trivial. This is the only constraint. Indeed, suppose that N is a differential module over K that has trivial topological monodromy. We apply Birkhoff's method (see Lemma 12.1). N extends to a connection on $\{z \in \mathbf{C} |\ |z| < \epsilon\}$ for some positive epsilon and with a singularity at $z = 0$. The restriction of the connection is trivial on $\{z \in \mathbf{C} | \ 0 < |z| < \epsilon\}$, since the topological monodromy is trivial. This trivial connection extends to a trivial connection on $\{z \in \mathbf{P}^1_{\mathbf{C}} | \ 0 < |z| \ \}$. By gluing we find a complex analytic connection, with a singularity at $z = 0$, on all of $\mathbf{P}^1_{\mathbf{C}}$. By GAGA this produces an "algebraic" connection on $\mathbf{P}^1_{\mathbf{C}}$. The restriction (M, ∇) of the latter to X satisfies $\alpha(M, \nabla) \cong N$. Summarizing, we have shown that AlgConn_X is equivalent to the full subcategory of Diff_K whose objects are the differential modules with trivial topological monodromy.

The work of J.-P. Ramis on the differential Galois theory for differential modules over $K = \mathbf{C}(\{z\})$ can be interpreted as a description of the affine group scheme G corresponding to the neutral tannakian category Diff_K. This is fully discussed in Sect. 12.6. The topological monodromy can be interpreted as an element of G (or better $G(\mathbf{C})$). The affine group scheme corresponding to AlgConn_X is the quotient of G by the closed normal subgroup generated by the topological monodromy. □

C Sheaves and Cohomology

C.1 Sheaves: Definition and Examples

The language of sheaves and their cohomology is a tool to understand and formulate the differences between local properties and global ones. We will apply this language especially for the asymptotics properties of formal solutions of differential equations. Other applications that concern us are the formulation and constructions for the Riemann-Hilbert problem and moduli of singularities of linear differential equations.

The aim of this text is to present the ideas and to develop a small amount of technical material; just enough for the applications we have in mind. Proofs will sometimes be rather sketchy or not presented at all. The advantages and the disadvantages of this presentation are obvious. For more information we refer the reader to [100, 120, 124].

The topological spaces that we will use are very simple ones, say subsets of $\mathbf{R}^n$ or $\mathbf{C}^n$ and sometimes algebraic varieties provided with the Zariski topology. We will avoid "pathological" spaces.

Definition C.1 Let X be a topological space. A *sheaf F on X* is given by

1. For every open set $A \subset X$ a set $F(A)$,
2. For every pair of open sets $A \subset B$ a map $\rho_A^B : F(B) \to F(A)$,

and these data should satisfy a list of properties:

1. ρ_A^A is the identity on $F(A)$.
2. For open sets $A \subset B \subset C$ one has $\rho_A^C = \rho_A^B \rho_B^C$.
3. Let an open set A, an open covering $\{A_i\}_{i \in I}$ of A and elements $a_i \in F(A_i)$ for every $i \in I$ be given such that for every pair i, j the following holds

   $$\rho_{A_i \cap A_j}^{A_i} a_i = \rho_{A_i \cap A_j}^{A_j} a_j.$$

   Then there is a unique element $a \in F(A)$ with $\rho_{A_i}^A a = a_i$ for every $i \in I$.

□

If F satisfies all above properties, with the possible exception of the last one, then F is called a *presheaf*. We illustrate the concept "sheaf" with some examples and postpone a fuller discussion of presheaves to Sect. C.1.3.

Examples C.2
1. X is any topological space. One defines F by:
(i) For open $A \subset X$, $F(A)$ is the set of the continuous maps form A to $\mathbf{R}$.
(ii) For any pair of open sets $A \subset B \subset X$ the map ρ_A^B is the restriction map, i.e., $\rho_A^B f$ is the restriction of the continuous map $f : B \to \mathbf{R}$ to a map from A to $\mathbf{R}$.

2. $X = \mathbf{R}^n$ and F is given by:
(i) For open $A \subset \mathbf{R}^n$, $F(A)$ is the set of the C^∞-functions from A to $\mathbf{R}$.
(ii) For every pair of open sets $A \subset B$, the map ρ_A^B is again the restriction map.

3. $X = \mathbf{C}$ *and* O_X, *the sheaf of holomorphic functions* is given by:
(i) For open $A \subset X$, $O_X(A)$ consists of the holomorphic functions $f : A \to \mathbf{C}$.
(ii) ρ_A^B, for open sets $A \subset B$, is again the restriction map.
We recall that a function f is holomorphic on A, if for every point $a \in A$ there is a convergent power series $\sum_{n \geq 0} a_n (z-a)^n$ that is equal to f on some neighborhood of a.

4. $X = \mathbf{C}$ *and* $\mathcal{M}$, *the sheaf of meromorphic functions*, is given by:
(i) For open $A \subset \mathbf{C}$, $\mathcal{M}(A)$ is the set of the meromorphic functions on A.
(ii) ρ_A^B is again the restriction map.
We recall that a "function" f on A is meromorphic if for every point $a \in A$ there is a convergent Laurent series $\sum_{n \geq N} a_n (z-a)^n$ that is equal to f on a neighborhood of a. Another equivalent definition would be that for every point $a \in A$, there is a disk around a in A and holomorphic functions C, D on this disk, D not identical zero, such that the fraction $\frac{C}{D}$ is equal to f on this disk. We remark that D may have zeros and thus f has poles. The set of poles of f is a discrete subset of A.

5. X is any topological space and D is a nonempty set. The *constant sheaf on* X *with values in* D is the sheaf F given by: $F(A)$ consists of the functions $f : A \to D$ such that there exists for every point $a \in A$ a neighborhood U with f constant on U. (In other words, $f(U)$ is one point of D.) ρ_A^B is again the restriction map. The elements of $F(A)$ are sometimes called the locally constant functions on A with values in D.

6. *Direct sum* Let F_1 and F_2 be two sheaves on a topological space X. The presheaf $U \mapsto F_1(U) \times F_2(U)$ is actually a sheaf, called the *direct sum of* F_1 *and* F_2. The notation $F_1 \oplus F_2$ for the direct sum will also be used. □

Exercise C.3 X is a topological space, D a nonempty set and F is the constant sheaf on X with values in D.
(a) Suppose that the open set A is connected. Prove that $F(A)$ consists of the constant functions of A with values in D.

(b) Suppose that the open set A is the disjoint union of open connected subsets A_i, $i \in I$. (The A_i are called the connected components of A.) Prove that $F(A)$ consists of the functions $f : A \to D$ that are constant on each A_i. □

Remark: For most sheaves it is clear what the maps ρ are. In the sequel we will omit the notation ρ and replace $\rho_A^B f$ by $f|_A$, or even omit the ρ_A^B completely.

C.1.1 Germs and Stalks

F denotes a sheaf (or presheaf) on a topological space X. Let x be a point of X. We consider pairs (U, f) with $f \in F(U)$ and U a neighborhood of x. Two pairs (U_1, f_1), (U_2, f_2) are called equivalent if there is a third pair (U_3, f_3) with $U_3 \subset U_1 \cap U_2$ and $f_3 = f_1|_{U_3} = f_2|_{U_3}$. The equivalence class $[U, f]$ of a pair (U, f) is called a *germ* of F at x. The collection of all germs of F at x is called the *stalk* of F at x and is denoted by F_x.

Examples C.4 1. *The sheaf of the real C^∞-functions on* **R** will be denoted by C^∞. The stalk C_0^∞ of this sheaf at 0, is a rather complicated object. It is, in fact, a ring, because one can add and multiply C^∞-functions. One can associate to a germ $[U, f]$ its Taylor series at 0, i.e., $\sum_{n\geq 0} \frac{f^{(n)}(0)}{n!} x^n$. This Taylor series is a formal power series. The collection of all formal power series (in the variable x and with coefficients in **R**) is usually denoted by $\mathbf{R}[[x]]$. The map $C_0^\infty \to \mathbf{R}[[x]]$, which associates to each germ its Taylor series is a homomorphism of rings. A non trivial result is that this map is actually surjective (see Theorem 7.3). The kernel of the map is an ideal, the ideal of the flat germs at 0. A germ $[U, f]$ is called *flat at 0* if $f^{(n)}(0) = 0$ for all $n \geq 0$.

2. *The sheaf of the holomorphic functions on* **C** will be denoted by $O_{\mathbf{C}}$ or simply O. One associates to every germ $[U, f]$ of O at 0 the power series $\sum_{n\geq 0} \frac{f^{(n)}(0)}{n!} z^n$. This power series (in the complex variable z and with coefficients in **C**) is convergent (either by definition or as a consequence of a different definition of holomorphic function). The collection of all convergent power series (in the variable z and with complex coefficients) is denoted by $\mathbf{C}\{z\}$. We now have an isomorphism $O_0 \to \mathbf{C}\{z\}$.

3. The ring $\mathbf{C}\{z\}$ is a rather simple one. The invertible elements are the power series $\sum_{n\geq 0} c_n z^n$ with $c_0 \neq 0$. Every element $f \neq 0$ can uniquely be written as $z^n E$ with $n \geq 0$ and E a unit. One defines the order of $f = z^n E$ at 0 as the above n and one writes this in formula as $ord_0(f) = n$. This is completed by defining $ord_0(0) = +\infty$. The ring $\mathbf{C}\{z\}$ has no zero divisors. Its field of fractions is written as $\mathbf{C}(\{z\})$. The elements of this field can be written as expressions $\sum_{n\geq a} c_n z^n$ ($a \in \mathbf{Z}$ and the $c_n \in \mathbf{C}$ such that there are constants $C, R > 0$ with $|c_n| \leq CR^n$ for all $n \geq a$). The elements of $\mathbf{C}(\{z\})$ are called *convergent Laurent series*. Every convergent Laurent series

$f = \sum f_n z^n \neq 0$ has uniquely the form $z^m E$ with $m \in \mathbf{Z}$ and E a unit of $\mathbf{C}\{z\}$. One defines $ord_0(f) = m$. In this way we have constructed a map

$$ord_0 : \mathbf{C}(\{z\}) \to \mathbf{Z} \cup \{\infty\}$$

with the properties

1. $ord_0(fg) = ord_0(f) + ord_0(g)$.
2. $ord_0(f) = \infty$ if and only if $f = 0$.
3. $ord_0(f + g) \geq \min(ord_0(f), ord_0(g))$.

Every convergent Laurent series can be seen as the germ of a meromorphic function at 0. Let $\mathcal{M}$ denote again the sheaf of the meromorphic functions on $\mathbf{C}$. We conclude that the stalk $\mathcal{M}_0$ is isomorphic to the field $\mathbf{C}(\{z\})$. For any other point $u \in \mathbf{C}$ one makes similar identifications $O_u = \mathbf{C}\{z - u\}$ and $\mathcal{M}_u = \mathbf{C}(\{z - u\})$.

4. *Skyscraper sheaves*
Let X be a topological space where points are closed, $p \in X$ and G an abelian group. We define a sheaf $i_p(G)$ by setting $i_p(G)(U) = G$ if $p \in U$ and $i_p(G)(U) = 0$ if $p \notin U$. The stalk at point q is G if $q = p$ and 0 otherwise. This sheaf is called a *skyscraper sheaf (at p)*. If $p_1, \ldots, p_n$ are distinct points the sheaf $\oplus i_p(G)$ is called the skyscraper sheaf (at $p_1, \ldots p_n$) □

C.1.2 Sheaves of Groups and Rings

A sheaf F on a topological space X is called a *sheaf of groups* if every $F(A)$ is a group and every map ρ_A^B is a homomorphism of groups. In a similar way, one defines sheaves of abelian groups, sheaves of commutative rings, vector spaces, etc. If D is a group, then the constant sheaf on X with values in D is obviously a sheaf of groups. Usually, this sheaf is denoted by D_X, or also by D itself. The sheaves C^∞, O, M are sheaves of commutative rings. The sheaf $\mathrm{GL}_n(O)$ on $\mathbf{C}$ is given by $A \mapsto \mathrm{GL}_n(O)(A)$, which consists of the invertible $n \times n$ matrices with coefficients in $O(A)$, or otherwise stated $\mathrm{GL}_n(O)(A) = \mathrm{GL}_n(O(A))$. It is a sheaf of groups on $\mathbf{C}$. For $n = 1$ it is a sheaf of commutative groups and for $n > 1$ it is a sheaf of noncommutative groups. The restriction of a sheaf F on X to an open subset U is written as $F|_U$. Its definition is more or less obvious, namely $F|_U(A) = F(A)$ for every open subset $A \subset U$.

Definition C.5 A *morphism* $f : F \to G$ between two sheaves of groups, rings, etc., is defined by

1. For every open A a map $f(A) : F(A) \to G(A)$.
2. f commutes with the restriction maps, i.e., for open $A \subset B$ the formula $\rho_A^B f(B) = f(A)\rho_A^B$ holds.

3. Every $f(A)$ is a homomorphism of groups, rings, etc.

□

We make a small excursion in order to demonstrate that sheaves can be used to define global objects. A *ringed space* is a pair (X, O_X) with X a topological space and O_X a sheaf of unitary commutative rings on X. A morphism of ringed spaces is a pair $(f, g) : (X, O_X) \to (Y, O_Y)$ with $f : X \to Y$ a continuous map and g a family $\{g(A)\}_{A \text{ open in } Y}$ of homomorphisms of unitary rings $g(A) : O_Y(A) \to O_X(f^{-1}A)$, compatible with restrictions. The latter means: For open $A_1 \subset A_2 \subset Y$ and $h \in O_Y(A_2)$ one has $g(A_1)(h|_{A_1}) = (g(A_2)(h))|_{f^{-1}(A_1)}$.

Using this terminology one can define various "global objects". We now give two examples.

Examples C.6 1. A C^∞- *variety of dimension n* is a ringed space (M, F) such that M is a Hausdorff topological space and has an open covering $\{M_i\}$ with the property that, for each i, the ringed space (M_i, F_i) (where $F_i = F|_{M_i}$) is isomorphic to the ringed space (B_n, C^∞). The latter is defined by B_n being the open ball with radius 1 in $\mathbf{R}^n$ and C^∞ being the sheaf of the C^∞-functions on B_n. The "global object" is (M, F) and the "local object" is (B_n, C^∞). Our definition of C^∞-variety M can be rephrased by saying that M is obtained by gluing copies of B_n. The sheaf F on M prescribes the way one has to glue.

2. A *Riemann surface* is a ringed space (X, O_X) such that X is a connected Hausdorff space and (X, O_X) is locally isomorphic to (D, O_D). Here "(D, O_D)" means: $D = \{z \in \mathbf{C} |\ |z| < 1\}$ and O_D is the sheaf of the holomorphic functions on D. Further "(X, O_X) locally isomorphic to (D, O_D)" means that X has an open covering $\{X_i\}$ such that each $(X_i, O_X|_{X_i})$ is isomorphic to (D, O_D), as ringed spaces. □

C.1.3 From Presheaf to Sheaf

Let F be a presheaf on some topological space X. The purpose is to construct a sheaf $\hat{F}$ on X that is as close to F as possible. The precise formulation of this is:

1. $\hat{F}$ is a sheaf.
2. There is given a morphism $\tau : F \to \hat{F}$ of presheaves.
3. For any morphism of presheaves $f : F \to G$, with G actually a sheaf, there is a unique morphism of sheaves $\hat{f} : \hat{F} \to G$ such that $\hat{f} \circ \tau = f$.

We note that this definition is formulated in such a way that, once $\hat{F}$ and τ exist they are unique up to (canonical) isomorphism. One calls $\hat{F}$ *the sheaf associated to the presheaf* F. The construction is somewhat formal and uses the stalks F_x of the presheaf F. Define, for any open $A \subset X$ the set $\hat{F}(A)$ as the subset of $\prod_{x \in A} F_x$, given by:

An element $(a_x)_{x\in A}$ belongs to $\hat{F}(A)$ if for every point $y \in A$ there is an open neighborhood U of y and an element $f \in F(U)$ such that for any $u \in U$ the element $a_u \in F_u$ coincides with the image of f in the stalk F_u.

The morphism $\tau : F \to \hat{F}$ is given by maps $\tau(A) : F(A) \to \hat{F}(A)$ for all A (and should be compatible with the restriction maps). The definition of $\tau(A)$ is rather straightforward, namely $f \in F(A)$ is mapped to $(a_x)_{x\in A} \in \hat{F}(A)$ where each $a_x \in F_x$ is the image of f in the stalk F_x.

The verification that $\hat{F}$ and τ as defined above, have the required properties is easy and uninteresting. We note that F and $\hat{F}$ have the same stalks at every point of X.

We will give an example to show the use of "the associated sheaf". Let B be a sheaf of abelian groups on X and let A be an abelian subsheaf of B. This means that $A(U)$ is a subgroup of $B(U)$ for each open set U and that for any pair of open sets $U \subset V$ the restriction map $B(V) \to B(U)$ maps $A(V)$ to $A(U)$. Our purpose is to define a *quotient sheaf of abelian groups B/A on X*. Naively, this should be the sheaf that associates to any open U the group $B(U)/A(U)$. However, this defines only a presheaf P on X. The quotient sheaf B/A is defined as the sheaf associated to the presheaf P. We note that the stalk $(B/A)_x$ is isomorphic to B_x/A_x. This follows from the assertion that the presheaf and its associated sheaf have the same stalks.

Example C.7 Let O denote the sheaf of the holomorphic functions on $\mathbf{C}$. Let $\mathbf{Z}$ be the constant sheaf on $\mathbf{C}$. One can see $\mathbf{Z}$ as an abelian subsheaf of O. Let $O/\mathbf{Z}$ denote the quotient *sheaf*. Then, for general open $U \subset \mathbf{C}$, the map $O(U)/\mathbf{Z}(U) \to (O/\mathbf{Z})(U)$ is not surjective. Indeed, take $U = \mathbf{C}^* \subset \mathbf{C}$ and consider the cover of U by $U_1 = \mathbf{C} \setminus \mathbf{R}_{\geq 0}$ and $U_2 = \mathbf{C} \setminus \mathbf{R}_{\leq 0}$. On each of the two sets there is a determination of the logarithm. Thus $f_1(z) = \frac{1}{2\pi i}\log(z)$ on U_1 and $f_2(z) = \frac{1}{2\pi i}\log(z)$ are well-defined elements of $O(U_1)$ and $O(U_2)$. The f_1, f_2 do not glue to an element of $O(U)$. However, their images g_j in $O(U_j)/\mathbf{Z}$, for $j = 1, 2$, and a fortiori their images h_j in $(O/\mathbf{Z})(U_j)$ do glue to an element $h \in (O/\mathbf{Z})(U)$. This element h is not the image of some element in $O(U)$. This proves the statement. Compare also with Example C.16 and Example C.18. □

Let A and B again be abelian sheaves on X and let $f : A \to B$ be a morphism. Then one would like to define a *kernel of f* as a sheaf of abelian groups on X. The naive approach would be $\ker f(U) := \ker(f(U) : A(U) \to B(U))$. This defines an abelian subsheaf of A. In this case one does not have to make the step from presheaf to sheaf. Moreover, the stalk $(\ker f)_x$ is equal to the kernel of $A_x \to B_x$. The *cokernel of f* is the sheaf associated to the presheaf $U \mapsto B(U)/(\operatorname{im} f(U) : A(U) \to B(U))$. In this case the step from presheaf to sheaf is necessary. The *image of f* is the sheaf associated by the presheaf $U \mapsto \operatorname{im}(f(U) : A(U) \to B(U))$. Again the step from presheaf to sheaf is, in general, needed.

C.1.4 Moving Sheaves

Let $f : X \to Y$ be a continuous map between topological spaces. We want to use f to move sheaves on X to sheaves on Y and vice versa. The definitions are as follows.

Definition C.8 *Direct image.*
Let F be a sheaf on X. The *direct image of* G, f_*F is the sheaf on Y, defined by the formula $f_*F(V) = F(f^{-1}V)$ for any open $V \subset Y$.

It is an exercise to show that the formula really defines a sheaf on Y. It is, in general, difficult, if not impossible, to express the stalk $(f_*F)_y$ in terms of F and $f^{-1}(y)$.

Example C.9 Let $\mathbf{Z}$ be the constant sheaf on $\mathbf{R} \setminus \{0\}$ and let $f : \mathbf{R} \setminus \{0\} \to \mathbf{R}$ be the inclusion map. One then has that the stalk of f_*X at 0 is $\mathbf{Z} \oplus \mathbf{Z}$ since $f_*\mathbf{Z}(-\epsilon, \epsilon) = \mathbf{Z}((-\epsilon, 0) \cup (0, \epsilon)) = \mathbf{Z} \oplus \mathbf{Z}$ for any $\epsilon > 0$. □

Let G be a sheaf on Y, then we would like to define a sheaf f^*G on X by the formula $f^*G(U) = G(fU)$ for any open set $U \subset X$. This is, however, not possible because fU is, in general, not an open set. So we have to make a more careful definition. Let us start by defining a presheaf P on X. For any open set $U \subset X$, let $P(U)$ be the direct limit of $G(V)$, taken over all open $V \supset fU$. As the definition of direct limit occurs a little later in this text, we will say this more explicitly. One considers pairs (V, g) with $V \supset fU$, V open and $g \in G(V)$. Two pairs (V_1, g_1) and (V_2, g_2) are called equivalent if there is a third pair (V_3, g_3) with $V_3 \subset V_1 \cap V_2$ and $g_3 = g_1|_{V_3} = g_2|_{V_3}$. The equivalence classes of pairs (V, g) could be called germs of G for the set fU. Thus we define $P(U)$ as the set of germs of G for the set fU. It turns out that P is, in general, a presheaf and not a sheaf. Thus we end up with the following definition.

Definition C.10 *The inverse image of* G, f^*G is the sheaf associated to the presheaf P.

One rather obvious property of f^*G is that the stalk $(f^*G)_x$ is equal to the stalk $G_{f(x)}$.

A rather special situation is: X is a closed subset of Y. Formally one writes $i : X \to Y$ for the inclusion map. Let F be an abelian sheaf on X. The sheaf i_*F is easily seen to have the stalks $(i_*F)_y = 0$ if $y \notin X$ and $(i_*F)_x = F_x$ for $x \in X$. One calls i_*F the *extension with 0 of* F *to* Y. For a sheaf G on Y, the sheaf i^*G on X is called the *restriction of* G *to* X. The stalk $(i^*G)_x$ is equal to G_x. One can extend i^*G with 0 to Y, i.e., i_*i^*G. There is a natural homomorphism of abelian sheaves $G \to i_*i^*G$ on the space Y. We will return to this situation later.

Exercise C.11 1. Let X be a topological space whose points are closed. Take a point $p \in X$ and let $i : \{p\} \to X$ be the inclusion map. Let $\mathcal{G}$ be the constant sheaf on $\{p\}$ with group G. Show that the skyscraper sheaf $i_p(G)$ is the same as $i_*(\mathcal{G})$.

2. Let X be a closed subset of Y, F a sheaf of abelian groups on X and U an open subset of Y. Show that $i_* i^* F(U) = F(U \cap X)$ if $U \cap X$ is nonempty and is 0 otherwise. □

C.1.5 Complexes and Exact Sequences

We begin by giving some definitions concerning abelian groups:

Definition C.12 *Complexes.*

1. Let $f : A \to B$ be a homomorphism of abelian groups. We define the *kernel of f*, $ker(f) = \{a \in A |\ f(a) = 0\}$, the *image of f*, $im(f) = \{f(a) |\ a \in A\}$ and the *cokernel of f*, $coker(f) = B/im(f)$.
2. A sequence of abelian groups and homomorphisms
$$\cdots A^{i-1} \xrightarrow{f^{i-1}} A^i \xrightarrow{f^i} A^{i+1} \xrightarrow{f^{i+1}} A^{i+2} \cdots$$
is called a *(co)complex* if for every j one has $f^j f^{j-1} = 0$ (Under the assumption that both f^j and f^{j-1} are present. The 0 indicates the 0-map from A^{j-1} to A^{j+1}).
3. A sequence of abelian groups and homomorphisms
$$\cdots A^{i-1} \xrightarrow{f^{i-1}} A^i \xrightarrow{f^i} A^{i+1} \xrightarrow{f^{i+1}} A^{i+2} \cdots$$
is called *exact* if for every j (f^j and f^{j-1} are supposed to be present) one has $im(f^{j-1}) = ker(f^j)$.

□

This last notion needs some explanation and some examples. We remark first that an exact sequence is also a complex, because $im(f^{j-1}) = ker(f^j)$ implies $f^j f^{j-1} = 0$.

Examples C.13 1. $0 \to A \xrightarrow{f} B$ *is exact if and only if f is injective.* Here, the 0 indicates the abelian group 0. The first arrow is not given a name because there is only one homomorphism $0 \to A$, namely the 0-map. The exactness of the sequence translates into: "the image of the 0-map, i.e., $0 \subset A$, is the kernel of f". In other words: $ker(f) = 0$, or f is injective.

2. $A \xrightarrow{f} B \to 0$ *is exact if and only if f is surjective.* The last arrow is not given a name because there is only one homomorphism from B to 0, namely the 0-map. The exactness translates into: "the kernel of the 0-map, this is B itself, is equal to the image of f". Equivalently, $im(f) = B$, or f is surjective.

3. $0 \to A \xrightarrow{f} B \to 0$ *is exact if and only if f is an isomorphism.*

4. $0 \to A \xrightarrow{f} B \xrightarrow{g} C \to 0$ *is exact if and only if* f *is injective and* C *is, via* g*, isomorphic to the cokernel of* f. Indeed, "f is injective, g is surjective and $ker(g) = im(f)$" is the translation of exactness. From $ker(g) = im(f)$ one deduces, using a well-known isomorphy theorem, an isomorphism $B/im(A) \to C$. A sequence as above is called *a short exact sequence*. □

Exercises C.14 *Complexes.*
1. Construct maps for the arrows in the following exact sequence

$$0 \to \mathbf{Z} \to \mathbf{C} \to \mathbf{C}^* \to 0.$$

We note that the operation in an abelian group is usually denoted by $+$. The above sequence is an exception to that, because $\mathbf{C}^* = \mathbf{C} \setminus \{0\}$ is considered as a group for the multiplication.

2. Construct maps for the arrows in the following exact sequence

$$0 \to \mathbf{Z}^2 \to \mathbf{Z}^2 \to \mathbf{Z}/5\mathbf{Z} \to 0.$$

3. Give a complex that is not exact.

4. Let F be a presheaf of abelian groups on a topological space X. For every open $A \subset X$ and open covering $\{A_i\}_{i \in I}$ and (in order to simplify) a chosen total order on the index set I, one considers the sequence of abelian groups and homomorphisms

$$0 \to F(A) \xrightarrow{\epsilon} \prod_i F(A_i) \xrightarrow{d^0} \prod_{i<j} F(A_i \cap A_j)\ ,$$

where

1. $\epsilon(f) := (f|_{A_i})_{i \in I}$.
2. $d^0((f_i)_i) = (f_i|_{A_i \cap A_j} - f_j|_{A_i \cap A_j})_{i<j}$.

(a) Prove that the above sequence is a complex.
(b) Prove that F is a sheaf if and only if the above sequence (for all choices of A and $\{A_i\}_{i \in I}$) is exact. □

Let a complex of (abelian groups) $\cdots \to A^{i-1} \xrightarrow{d^{i-1}} A^i \xrightarrow{d^i} A^{i+1} \cdots$ be given. By definition $d^j d^{j-1} = 0$ holds for all j such that d^j and d^{j-1} are present. This condition is equivalent with $im(d^{j-1}) \subset ker(d^j)$ for all j. The complex is an exact sequence if and only if $im(d^{j-1}) = ker(d^j)$ for all j. One can "measure" the nonexactness of a complex by a calculation of the abelian groups $ker(d^j)/im(d^{j-1})$. This leads to the following definition.

Definition C.15 *The* j*-th cohomology group* H^j *of a complex is the group* $ker(d^j)/im(d^{j-1})$.

Examples C.16 *Cohomology groups.*
1. Consider the complex

$$0 \to A^0 \xrightarrow{d^0} A^1 \xrightarrow{d^1} A^2 \to 0 \text{ with } A^0 = \mathbf{Z},\ A^1 = \mathbf{Z}/8\mathbf{Z},\ A^2 = \mathbf{Z}/2\mathbf{Z}$$
$$\text{and } d^0(n) = 4n \bmod 8,\ d^1(n \bmod 8) = n \bmod 2.$$

The other maps in the complex are 0. One sees that

$$H^0 \cong \mathbf{Z},\ H^1 \cong \mathbf{Z}/2\mathbf{Z},\ H^2 = 0.$$

2. Consider the complex $0 \to O(X) \xrightarrow{d^0} O(X)^* \to 0$, in which X is an open subset of $\mathbf{C}$, $O(X)$, $O(X)^*$ are the groups of the holomorphic and the invertible holomorphic functions on X. This means $O(X)^* = \{f \in O(X) |\ f \text{ has no zeros on } X\}$ and the group operation on $O(X)^*$ is multiplication. The map d^0 is given by $d^0(f) = e^{2\pi i f}$.

H^0 consists of the holomorphic functions $f \in O(X)$ with values in $\mathbf{Z}$. These functions are precisely the locally constant functions with values in $\mathbf{Z}$ and thus $H^0 = \mathbf{Z}(X)$. (and $= \mathbf{Z}$ if X is connected).

The term H^1 measures whether the invertible functions, i.e., $f \in O(X)^*$, are the exponentials of holomorphic functions. This depends on X. We consider some cases:

(a) X is an open disk, say $\{z \in \mathbf{C} |\ |z| < 1\}$. Choose $f \in O(X)^*$. We are looking for a $g \in O(X)$ with $e^{2\pi i g} = f$. This g satisfies the differential equation $g' = \frac{f'}{2\pi i f}$. The function $\frac{f'}{2\pi i f}$ lies in $O(X)$ and is equal to a power series $\sum_{n\geq 0} a_n z^n$ with radius of convergence ≥ 1. One can take for g the expression $b + \sum_{n\geq 0} \frac{a_n}{n+1} z^{n+1}$. The radius of convergence is again ≥ 1 and thus $g \in O(X)$. The constant b is chosen such that $e^{2\pi i b} = f(0)$. The function $e^{2\pi i g} f^{-1}$ has derivative 0 and is equal to 1 in the point $z = 0$. Therefore $e^{2\pi i g} f^{-1}$ is equal to 1 on X and $f = e^{2\pi i g}$.

(b) Let X be an annulus, say $X = \{z \in \mathbf{C} |\ r_1 < |z| < r_2\}$ with $0 \leq r_1 < r_2 \leq \infty$. We admit that every element $f \in O(X)$ can be represented as a convergent Laurent series $\sum_{n\in\mathbf{Z}} a_n z^n$ (with the condition on the absolute values of the a_n expressed by $\sum_{n\in\mathbf{Z}} |a_n| r^n$ converges for every real r with $r_1 < r < r_2$). We are looking for a g with $e^{2\pi i g} = f$. Such a g has to satisfy the differential equation $g' = \frac{f'}{2\pi i f}$. Write $\frac{f'}{2\pi i f} = \sum_n a_n z^n$. Then g exists if and only if $a_{-1} = 0$. The term a_{-1} is not always 0, e.g., for $f = z^k$ one has $a_{-1} = \frac{k}{2\pi i}$. We conclude that $H^1 \neq 0$. Assuming a result from classical complex function theory, namely that $\frac{1}{2\pi i} \int \frac{f'(z)dz}{f(z)}$ is an integer (see [40]), one can easily show that $H^1 \cong \mathbf{Z}$. □

C.2 Cohomology of Sheaves

C.2.1 The Idea and the Formalism

In this section, X is a topological space and F is a sheaf of abelian groups on X. The stalk F_x, for $x \in X$, is, in an obvious way, also an abelian group. A morphism of abelian sheaves $f : F \to G$ induces for every $x \in X$ a homomorphism of groups $f_x : F_x \to G_x$. We will use this to give a definition of an exact sequence of sheaves.

Definition C.17 A sequence of abelian sheaves and morphisms

$$\cdots F^{i-1} \stackrel{f^{i-1}}{\to} F^i \stackrel{f^i}{\to} F^{i+1} \to \cdots$$

on X is called *exact* if for every point $x \in X$ the induced sequence of abelian groups

$$\cdots F_x^{i-1} \stackrel{f_x^{i-1}}{\to} F_x^i \stackrel{f_x^i}{\to} F_x^{i+1} \to \cdots$$

is exact. □

We remark that the literature often uses another equivalent definition of exact sequence of abelian sheaves.
For a given exact sequence of sheaves, as above, and for an open set $A \subset X$ one finds a complex

$$\cdots F^{i-1}(A) \stackrel{f^{i-1}(A)}{\to} F^i(A) \stackrel{f^i(A)}{\to} F^{i+1}(A) \to \cdots .$$

The important observation is that this complex is, in general, not exact!

Examples C.18
1. $X = \mathbf{C}$ and $\mathbf{Z}$, O, O^* are the sheaves on X of the constant functions with values in $\mathbf{Z}$, the holomorphic functions and the invertible holomorphic functions (with multiplication). The exact sequence

$0 \to \mathbf{Z} \to O \to O^* \to 0$ of abelian sheaves on X is given by:

$\mathbf{Z} \to O$ is the inclusion map $f \in \mathbf{Z}(A) \mapsto f \in O(A)$ (i.e., a locally constant function with values in $\mathbf{Z}$ is considered as a holomorphic function).

$O \to O^*$ is defined by $f \in O(A) \mapsto e^{2\pi i f} \in O(A)^*$.

In proving that the sequence is exact we have to show for every point $x \in X$ the exactness of the sequence of stalks. For convenience we take $x = 0$. The sequence of stalks is

$$0 \to \mathbf{Z} \to \mathbf{C}\{z\} \to \mathbf{C}\{z\}^* \to 0.$$

An element $f \in \mathbf{C}\{z\}^*$ has the form $f = a_0(1 + a_1 z + a_2 z^2 + \cdots)$ with $a_0 \neq 0$. Choose b_0 with $e^{2\pi i b_0} = a_0$ and define g as $g = b_0 + \frac{1}{2\pi i}\log(1 + a_1 z + a_2 z^2 + \cdots)$. In this we use for log the formula $\log(1+u) = \sum_{n>0} \frac{(-1)^{n-1}}{n} u^n$.
It is clear that $g \in \mathbf{C}\{z\}$. It is also easy to see that any solution h of $e^{2\pi i h} = f$ has the form $g + n$ with $n \in \mathbf{Z}$. Thus we have proved that the sequence of stalks is exact.

Consider an annulus $A = \{z \in \mathbf{C} |\ r_1 < |z| < r_2\}$ with $0 \leq r_1 < r_2 \leq \infty$. Then

$$0 \to \mathbf{Z}(A) \to O(A) \to O(A)^*$$

is exact, but the last map is not surjective, as we have seen in Example C.16.

2. The circle $\mathbf{S}^1$ can be seen as a 1-dimensional C^∞-variety. We consider three sheaves on it:

- $\mathbf{R}$, the constant sheaf with values in $\mathbf{R}$.
- C^∞, the sheaf of the C^∞-functions.
- Ω, the sheaf of the C^∞-1-forms. The sections of $\Omega(A)$ are expressions $\sum f_i dg_i$ (finite sums, $f_i, g_i \in C^\infty(A)$) obeying the rules $d(g_1+g_2) = dg_1+dg_2, d(g_1g_2) = g_1dg_2 + g_2dg_1$.

Let A be chosen such that there exists a C^∞ isomorphism $t : A \to (0, 1)$. Then $\Omega(A) = C^\infty(A)dt$, in other words, every 1-form is equal to fdt for a unique $f \in C^\infty(A)$. This brings us to an exact sequence

$$0 \to \mathbf{R} \to C^\infty \to \Omega \to 0,$$

in which the first nontrivial arrow is the inclusion and the second nontrivial arrow is the map $f \mapsto df = f'dt$.

We will quickly verify that the sequence is exact. Let a 1-form ω be given in a neighborhood A of a point. As above we will use the function t. Then $\omega = fdt$ and f can be written as $g \circ t$, where g is a C^∞-function on $(0, 1)$. Let G be a primitive function of the function g. Then $G \circ t \in C^\infty(A)$ and $d(G \circ t) = (g \circ t)dt = fdt$. The functions G and $G \circ t$ are unique up to a constant. This proves the exactness. The sequence

$$0 \to \mathbf{R} \to C^\infty(\mathbf{S}^1) \to \Omega(\mathbf{S}^1)$$

is also exact, as one easily sees. The map $C^\infty(\mathbf{S}^1) \to \Omega(\mathbf{S}^1)$ is, however, not surjective. An easy way to see this is obtained by identifying $\mathbf{S}^1$ with $\mathbf{R}/\mathbf{Z}$. The C^∞-functions on $\mathbf{S}^1$ are then the 1-periodic functions on $\mathbf{R}$. The 1-forms on $\mathbf{S}^1$ are the 1-periodic 1-forms on $\mathbf{R}$. Such a 1-periodic 1-form is equal to $h(t)dt$, where h is a C^∞-function on $\mathbf{R}$ having the property $h(t+1) = h(t)$. Let $\omega = h(t)dt$ be given. We are looking for a C^∞-function $f(t)$ with $f'(t) = h(t)$ and $f(t+1) = f(t)$. The first condition yields $f(t) = c + \int_0^t h(s)ds$ with c any constant. The second condition is satisfied if and only if $\int_0^1 h(s)ds = 0$. In general, the latter does not hold. We conclude that the map is not surjective. In fact, the above reasoning proves that the cokernel of the map is isomorphic with $\mathbf{R}$. □

We now give *the formalism of cohomology of sheaves*. Let F be an abelian sheaf on a topological space X. Then there is a sequence of abelian groups, denoted as $H^i(X, F)$, $i = 0, 1, 2, \ldots$. These groups are called the cohomology groups of the sheaf F on X. This collection depends in a "functorial way" on F, which means that for a morphism of abelian sheaves $f : F \to G$ a collection of homomorphisms $H^i(f) : H^i(X, F) \to H^i(X, G)$ is given. All this should satisfy the rules: $H^i(id) = id$, $H^i(f \circ g) = H^i(f) \circ H^i(g)$. Furthermore, the term $H^0(X, F)$ is, by definition, equal to $F(X)$ and the term $H^0(f)$ is, by definition, equal to $f(X) : F(X) \to G(X)$. A definition of the higher $H^i(X, F)$ is rather complicated and will be given later. We continue first with the formalism.

The most important property of the cohomology groups is: For every short exact sequence of (abelian) sheaves

$$0 \to F_1 \to F_2 \to F_3 \to 0$$

there is a long exact sequence of cohomology groups

$$\begin{aligned} 0 &\to H^0(X, F_1) \to H^0(X, F_2) \to H^0(X, F_3) \\ &\to H^1(X, F_1) \to H^1(X, F_2) \to H^1(X, F_3) \\ &\to H^2(X, F_1) \to H^2(X, F_2) \to H^2(X, F_3) \\ &\ldots \\ &\to H^n(X, F_1) \to H^n(X, F_2) \to H^n(X, F_3) \\ &\ldots . \end{aligned}$$

This long exact sequence of cohomology depends "functorially" on the short exact sequence of sheaves. This means that a morphism between two short exact sequences of sheaves induces a morphism between the two long exact sequences of cohomology. Furthermore, the latter is compatible with composition of morphisms and the identity induces the identity. We finally remark that for an open subset $A \subset X$ the groups $H^i(A, F)$ (etc.) are defined by taking the restrictions to A. In particular, $H^i(A, F) = H^i(A, F|_A)$.

The definition of cohomology groups is not only complicated, it also gives no easy way to calculate the groups. We demonstrate the value of the cohomology groups by some results.

Examples C.19 1. Consider again the exact sequence

$$0 \to \mathbf{Z} \to O \to O^* \to 0$$

on $X = \mathbf{C}$. It can be shown that for every open open subset $A \subset \mathbf{C}$ one has $H^i(A, O) = 0$ and $H^i(A, O^*) = 0$ for all $i \geq 1$ (see Theorem C.26). The long exact sequence of cohomology implies then $H^i(A, \mathbf{Z}) = 0$ for $i \geq 2$ and the interesting part of this sequence is

$$0 \to \mathbf{Z}(A) \to O(A) \to O^*(A) \to H^1(A, \mathbf{Z}) \to 0.$$

The cohomology group $H^1(A, \mathbf{Z})$ "measures" the nonsurjectivity of the map $O(A) \to O^*(A)$. One can show that for a connected open subset with g holes the group $H^1(A, \mathbf{Z})$ is isomorphic to $\mathbf{Z}^g$. For $A = \mathbf{C}$ one has $g = 0$ and $H^1(A, \mathbf{Z}) = 0$. For a ring domain A one has $g = 1$ and $H^1(A, \mathbf{Z}) \cong \mathbf{Z}$. This is in conformity with the explicit calculations of Example C.18.

2. Consider the exact sequence of sheaves

$$0 \to \mathbf{R} \to C^\infty \to \Omega \to 0$$

on $\mathbf{S}^1$. One can show that the cohomology group H^i with $i > 1$ is zero for every sheaf on $\mathbf{S}^1$. Moreover, the two sheaves C^∞ and Ω satisfy H^1 is zero. The long exact sequence of cohomology is now rather short, namely

$$0 \to \mathbf{R} \to C^\infty(\mathbf{S}^1) \to \Omega(\mathbf{S}^1) \to H^1(\mathbf{S}^1, \mathbf{R}) \to 0.$$

Moreover, one can show that $H^1(\mathbf{S}^1, A) = A$ for every constant sheaf of abelian A groups on $\mathbf{S}^1$ (see Examples C.22 and C.26). This confirms our earlier explicit calculation. □

C.2.2 Construction of the Cohomology Groups

Given are a sheaf (of abelian groups) F on a topological space X and an open covering $\mathcal{U} = \{U_i\}_{i \in I}$ of X. We choose a total ordering on the index set I, in order to simplify the definition somewhat. The Čech complex for these data is:

$$0 \to C^0(\mathcal{U}, F) \xrightarrow{d^0} C^1(\mathcal{U}, F) \xrightarrow{d^1} C^2(\mathcal{U}, F) \xrightarrow{d^2} C^3 \ldots,$$

given by

1. We write $U_{i_0, i_1, \ldots, i_n}$ for the intersection $U_{i_0} \cap U_{i_1} \cap \cdots \cap U_{i_n}$.
2. $C^0(\mathcal{U}, F) = \prod_{i_0} F(U_{i_0})$.
3. $C^1(\mathcal{U}, F) = \prod_{i_0 < i_1} F(U_{i_0, i_1})$.
4. And, in general: $C^n(\mathcal{U}, F) = \prod_{i_0 < i_1 < \cdots < i_n} F(U_{i_0, \ldots, i_n})$.
5. $d^0((f_i)_i) = (f_j - f_i)_{i<j}$. We have omitted in the formula the symbols for the restriction maps.

6. $d^1((f_{i,j})_{i<j}) = (f_{i,j} - f_{i,k} + f_{j,k})_{i<j<k}$. Again we have omitted the symbols for the restriction maps.
7. And, in general: $d^n((f_{i_0,\dots,i_n})) = (A_{i_0,\dots,i_{n+1}})_{i_0<\cdots<i_{n+1}}$, where

$$A_{i_0,\dots,i_{n+1}} = \sum_{0\le j\le n+1} (-1)^j f_{i_0,\dots,i_{j-1},i_{j+1},\dots,i_{n+1}}.$$

Or, in words, the alternating sum (i.e., provided with a sign) of the terms f_*, where $*$ is obtained from the sequence $i_0, \dots, i_{n+1}$ by omitting one item.

A simple calculation shows that $d^n \circ d^{n-1} = 0$ for all $n \ge 1$. Thus the above sequence is a (co)complex.

Remark C.20 The usual alternating Čech complex

$$0 \to \tilde{C}^0(\mathcal{U}, F) \to \tilde{C}^1(\mathcal{U}, F) \to \cdots$$

is defined by $\tilde{C}^n(\mathcal{U}, F)$ consists of the elements $(f_{i_0,\dots,i_n})$ in $\prod F(U_{i_0,\dots,i_n})$ satisfying $f_{\pi(i_0),\dots,\pi(i_n)} = \text{sign}(\pi) f_{i_0,\dots,i_n}$ for all permutations $\pi \in S_{n+1}$ and $f_{i_0,\dots,i_n} = 0$ if $i_s = i_t$ for some $s \ne t$.

After choosing a total order on I, one identifies $\tilde{C}^n(\mathcal{U}, F)$ with $C^n(\mathcal{U}, F)$. In particular, the Čech cohomology groups defined by means of $0 \to C^0(\mathcal{U}, F) \to C^1(\mathcal{U}, F) \to \cdots$ do not depend on the total order on I.

Definition C.21 The *Čech cohomology groups* of this complex are again defined as $ker(d^n)/im(d^{n-1})$. The notation for the n-th cohomology group is $\check{H}^n(\mathcal{U}, F)$.

For $n = 0$ one adopts the convention that $d_{-1} = 0$ and thus $\check{H}^0(\mathcal{U}, F) = ker(d^0)$. According to Exercise C.14 this group is equal to $F(X)$.

Consider now $n = 1$. The $ker(d^1)$ consists of the elements $(f_{i,j})$ satisfying the relation:

$$f_{i_1,i_2} - f_{i_0,i_2} + f_{i_0,i_1} = 0.$$

This relation is called the *1-cocycle relation*. The elements satisfying this rule are called *1-cocycles*. Thus $ker(d^1)$ is the group of the 1-cocycles. The elements of $im(d^0)$ are called *1-coboundaries*. The first cohomology group is, therefore, the quotient of the group of the 1-cocycles by the subgroup of the 1-coboundaries. We illustrate this with a simple example.

Example C.22 Let X be the circle $\mathbf{S}^1$ and F be the constant sheaf with group A on $\mathbf{S}^1$. The open covering $\{U_1, U_2\}$ of X is given by $U_i = \mathbf{S}^1 \setminus \{p_i\}$, where p_1, p_2 are two distinct points of $\mathbf{S}^1$. The Čech complex is

$$0 \to F(U_1) \times F(U_2) \to F(U_{1,2}) \to 0.$$

Since $U_{1,2}$ has two connected components and the U_i are connected, this complex identifies with

$$0 \to A \times A \overset{d^0}{\to} A \times A \to 0,$$

with $d^0((a_1, a_2)) = a_2 - a_1$. One easily sees that the cohomology groups $\check{H}^n(\mathcal{U}, F)$ of this complex are $A, A, 0, 0, \ldots$ for $n = 0, 1, 2, 3, \ldots$. □

Exercises C.23 *Cohomology groups for a covering.*
1. $X = [0, 1]$, F is the constant sheaf with group A and $\mathcal{U} = \{U_1, U_2, U_3\}$ with $U_1 = [0, 1/2)$, $U_2 = (1/4, 3/4)$, $U_3 = (1/2, 1]$. Calculate the groups $\check{H}^n(\mathcal{U}, F)$.

2. $X = S^2$= the two-dimensional sphere, F is the constant sheaf on X with group A and $\mathcal{U} = \{U_1, U_2, U_3, U_4\}$ is given by:
Choose a "north pole" N and a "south pole" Z on S^2. Choose two distinct half circles L_1, L_2 from N to Z. Define $U_i = S^2 \setminus L_i$ for $i = 1, 2$. Furthermore, U_3 is a small disk around N and U_4 is a small disk around Z. Calculate the groups $\check{H}^n(\mathcal{U}, F)$. □

This gives some impression about the meaning of the group $\check{H}(\mathcal{U}, F)$ for a sheaf F on a topological space X with an open covering $\mathcal{U}$. The Čech cohomology groups depend heavily on the chosen open covering $\mathcal{U}$ and we want, in fact, for a fixed sheaf F, to consider all the open coverings at the same time. We need for this, again, another construction.

Let $\mathcal{U} = \{U_i\}_{i \in I}$ and $\mathcal{V} = \{V_j\}_{j \in J}$ be two open coverings of X. One calls $\mathcal{V}$ *finer* than $\mathcal{U}$ (or *a refinement of* $\mathcal{U}$) if there is a map $\phi : J \to I$ such that for every $j \in J$ there is an inclusion $V_j \subset U_{\phi(j)}$. From a given ϕ one deduces a homomorphism of the complex $C^*(\mathcal{U}, F)$ to the complex $C^*(\mathcal{V}, F)$. This induces morphisms

$$m(\mathcal{U}, \mathcal{V}, n) : \check{H}^n(\mathcal{U}, F) \to \check{H}^n(\mathcal{V}, F)$$

for every $n \geq 0$. The morphisms do not depend on the choice of ϕ. For the definition of the groups $\check{H}^n(X, F)$ we need still another notion, namely the *direct limit*:

Definition C.24 *Direct limit.*
1. Let $(H, \leq)$ be a partially ordered set such that for every two elements $h_1, h_2 \in H$ there is a third $h_3 \in H$ with $h_1 \leq h_3$ and $h_2 \leq h_3$. Assume, furthermore, that for each $h \in H$, we are given an abelian group B_h and for every pair $h_1 \leq h_2$ a homomorphism $m(h_1, h_2) : B_{h_1} \to B_{h_2}$. Furthermore, assume that $m(h_1, h_2)$ verify the rules: $m(h, h) = id$ and $m(h_2, h_3) \circ m(h_1, h_2) = m(h_1, h_3)$ if $h_1 \leq h_2 \leq h_3$. The above data are called a *direct system of abelian groups.*

2. The *direct limit* of this system will be denoted by $B := \varinjlim B_h$ and is defined as follows: Let $\cup_{h \in H} B_h$ be the disjoint union and let $\sim$ be the equivalence relation: $d \sim e$ if $d \in B_{h_1}$, $e \in B_{h_2}$ and there is an h_3 with $h_1 \leq h_3$, $h_2 \leq h_3$, and $m(h_1, h_3)d = m(h_2, h_3)e$. We define B to be the set of equivalence classes $B = (\cup_{h \in H} B_h)/\sim$. □

We have already seen an example of a direct limit. Indeed, for a sheaf F and a point $x \in X$, the stalk F_x is the direct limit of the $F(U)$, where U runs in the set of the open neighborhoods of x. That is, $F_x = \varinjlim F(U)$.

Finally, the collection $\{\check{H}^n(\mathcal{U}, F)\}$ forms a direct system of abelian groups. Every one of these groups is indexed by a $\mathcal{U}$ and the index set consists of the collection of all open coverings of X. The partial ordering on the index set is given by $\mathcal{U} \leq \mathcal{V}$ if $\mathcal{V}$ is finer than $\mathcal{U}$. We define now

$$\check{H}^n(X, F) = \varinjlim \check{H}^n(\mathcal{U}, F).$$

For good spaces, for example paracompact, Hausdorff spaces, the Čech cohomology groups $\check{H}^n(X, F)$ describe the "correct" cohomology and we write them as $H^n(X, F)$. We recall the definition and some properties of paracompact spaces.

Definition C.25 A topological space X is called *paracompact* if every open covering of X can be refined to a covering $\{U_i\}_{i \in I}$ by open sets which is locally finite, i.e., for every point $x \in X$ there is an open neighborhood V such that $V \cap U_i \neq \emptyset$ holds for, at most, finitely many $i \in I$. □

Some properties of paracompact spaces are:

1. A paracompact Hausdorff space is normal, that is, for any two closed subsets X_1, X_2 of X with $X_1 \cap X_2 = \emptyset$ there exist open sets $U_1 \supset X_1$ and $U_2 \supset X_1$ such that $U_1 \cap U_2 = \emptyset$.
2. A closed subset of a paracompact space is also paracompact.
3. A metric space is paracompact.
4. A compact space is paracompact.

One can show that for paracompact, Hausdorff spaces X, $H^*(X, F)$ satisfy the formalism of cohomology.

It will be clear to the reader that we have skipped a large body of proofs. Moreover, the definition of cohomology is too complicated to allow a direct computation of the groups $H^n(X, F)$.

The following theorem of Leray ([120], p. 189) gives some possibilities for explicit calculations.

Theorem C.26 *Let X be a paracompact, Hausdorff space. Suppose that the open covering $\mathcal{U} = \{U_i\}_{i \in I}$ has the property that for all $i_0, \dots, i_m \in I$ and every $n > 0$ the group $H^n(U_{i_0,\dots,i_m}, F)$ is 0. Then the natural map $\check{H}^n(\mathcal{U}, F) \to H^n(X, F)$ is an isomorphism for every $n \geq 0$.*

This means that, in some cases, one needs only to calculate the cohomology groups with respect to a fixed open covering.

C.2.3 More Results and Examples

A topological space X is called *pathwise connected* if any two points of X can be connected by a path. A path wise connected space X is called *simply connected* if any two paths f, g from $a \in X$ to $b \in X$ are homotopic. The latter notion is defined by the existence of a continuous $H : [0, 1] \times [0, 1] \to X$ such that : $H(0, t) = a$ for all t; $H(1, t) = b$ for all t; $H(s, 0) = f(s)$ for all s; $H(s, 1) = g(s)$ for all s. The map H is called a homotopy from f to g. Naively, H is a continuous deformation of the path f to the path g, which leaves the end points fixed.

Further useful results are given in ([110], Chap. 5.12, [51], Chap. II.15):

Theorem C.27 *Let X be an open simply connected subspace of $\mathbf{R}^n$ and A a constant sheaf of abelian groups on X. Then $H^n(X, A) = 0$ for all $i > 0$.*

We note that this result is easily seen to be true for intervals on $\mathbf{R}$ of the form $[a, b]$, $[a, b)$, and $(a, b]$. Indeed, any open covering can be refined to an open covering by intervals such that each interval intersects only its neighbors.

Theorem C.28 *Let X be a "good" topological space of topological dimension n and F any abelian sheaf on X. Then $H^i(X, F) = 0$ for $i > n$.*

A possible definition of "topological dimension" would be: $\dim X \leq n$ if every open covering of X can be refined to an open covering for which the intersection of any $n + 2$ members is empty. From this definition, the theorem follows at once. It is not difficult to prove that the topological dimension of any subset of $\mathbf{R}^n$ is $\leq n$. It is a little more complicated to show that the topological dimension of $\mathbf{R}^n$ is precisely n.

Exercises C.29 Using the formalism of cohomology and the above results, calculate the groups $H^*(X, A)$ for a constant abelian sheaf A and the space X given as:

(a) S^1, S^2, a ring domain.
(b) $\mathbf{R}^2 \setminus D_1 \cup D_2$, where D_1, D_2 are two disjoint closed disks.
(c) $\mathbf{C}^*$, a topological torus.
(d) An n-dimensional topological torus, i.e., $\mathbf{R}^n/\mathbf{Z}^n$. □

D Partial Differential Equations

The Picard-Vessiot theory of linear ordinary differential equations generalizes to certain systems of linear partial differential equations. In the first section of this appendix we characterize these systems in terms of $k[\partial_1, \dots, \partial_r]$-modules, systems of homogeneous linear differential polynomials, integrable systems of matrix equations and integrable connections. In the final section, we sketch the Picard-Vessiot theory for this setting and give some indications concerning other aspects of integrable connections.

D.1 The Ring of Partial Differential Operators

A *$\boldsymbol{\Delta}$-ring* R is a commutative ring with a unit equipped with a set of commuting derivations $\boldsymbol{\Delta} = \{\partial_1, \dots, \partial_r\}$. A $\boldsymbol{\Delta}$-ideal $I \subset R$ is an ideal of R such that $\partial_i I \subset I$ for all $i = 1, \dots, r$. A $\boldsymbol{\Delta}$-field k is a field that is a $\boldsymbol{\Delta}$-ring. If R is a $\boldsymbol{\Delta}$-ring, the set $\{c \in R \mid \partial_i(c) = 0 \text{ for all } i = 1, \dots, r\}$ is called the *constants* of R. This can be seen to be a ring and, if R is a field, then this set will be a field as well. Throughout this appendix we will assume that for any $\boldsymbol{\Delta}$-ring, $\mathbf{Q} \subset R$ and that its ring of constants is an algebraically closed field.

Examples D.1 *$\boldsymbol{\Delta}$-fields*
1. Let C be an algebraically closed field and $t_1, \dots, t_r$ indeterminates. The field $C(t_1, \dots, t_r)$ with derivations $\partial_i, i = 1, \dots, r$ defined by $\partial_i(c) = 0$ for all $c \in C$ and $\partial_i(t_j) = 1$ if $i = j$ and 0 otherwise is a $\boldsymbol{\Delta}$-field.

2. The field of fractions $C((t_1, \dots, t_r))$ of the ring of formal power series in r variables is a $\boldsymbol{\Delta}$-field with the derivations defined as above.

3. For $C = \mathbf{C}$, the complex numbers, the field of fractions $\mathbf{C}(\{t_1, \dots, t_r\})$ of the ring of convergent power series in r variables with $\boldsymbol{\Delta}$ defined as above is again a $\boldsymbol{\Delta}$-field. □

Definition D.2 Let k be a $\boldsymbol{\Delta}$-field. The *ring of (partial) differential operators $k[\partial_1, \dots, \partial_r]$ with coefficients in* k is the noncommutative polynomial ring in the variables ∂_i, where the ∂_i satisfy $\partial_i\partial_j = \partial_j\partial_i$ for all i, j and $\partial_i a = a\partial_i + \partial_i(a)$ for all $a \in k$. □

When $r = 1$, we shall refer to this ring as the ring of *ordinary differential operators* and this is precisely the ring studied in Chap. 2. In the ordinary case, any left ideal in this ring is generated by a single element. This is no longer true for $k[\partial_1, \dots, \partial_r]$ when $r > 1$. For example, the left ideal generated by ∂_1, ∂_2 in $k[\partial_1, \partial_2]$ cannot be generated by a single element.

Definition D.3 A $k[\partial_1, \dots, \partial_r]$*-module* $\mathcal{M}$ is a *finite dimensional* k-vector space that is a left module for the ring $k[\partial_1, \dots, \partial_r]$. □

In the ordinary case, if $I \subset k[\partial]$, $I \neq (0)$, then the quotient $k[\partial]/I$ is finite dimensional k-vector space. This is not necessarily true in the partial case. For example, the left ideal generated by ∂_1 in $k[\partial_1, \partial_2]$ does not give a finite dimensional quotient. We therefore state the following definition.

Definition D.4 The *rank* of a left ideal $I \subset k[\partial_1, \dots, \partial_r]$ is the dimension of the k-vector space $k[\partial_1, \dots, \partial_r]/I$. We say that the ideal I is *zero-dimensional* if its rank is finite. □

The following is an analog of Proposition 2.9 that allows us to deduce the equivalence of $k[\partial_1, \dots, \partial_r]$-modules and zero-dimensional left ideals of $k[\partial_1, \dots, \partial_r]$.

Lemma D.5 *If k contains a nonconstant, then for any $k[\partial_1, \dots, \partial_r]$-module $\mathcal{M}$ there is a zero-dimensional left ideal $I \subset k[\partial_1, \dots, \partial_r]$ such that $\mathcal{M} \simeq k[\partial_1, \dots, \partial_r]/I$ as $k[\partial_1, \dots, \partial_r]$-modules.*

Proof. There is an element $z \in k$ such that $\partial_1(z) \neq 0$. By Lemma 2.11 there is an element $w \in \mathcal{M}$ such that $k[\partial_1]w = \mathcal{M}$. Hence, the $k[\partial_1, \dots, \partial_r]$ homomorphism $k[\partial_1, \dots, \partial_r] \to \mathcal{M}$, given by $L \mapsto Lw$, is surjective. □

Remarks D.6 1. Given a finite set of elements $L_1, \dots, L_s \in k[\partial_1, \dots, \partial_r]$, the Gröbner bases techniques allow one to decide if the left ideal $I \subset k[\partial_1, \dots, \partial_r]$ generated by these elements has finite rank and, if so, to calculate this rank (see [68, 69, 254]).

2. Given a $\boldsymbol{\Delta}$-field k, one can form the ring of differential polynomials $k\{y_1, \dots, y_n\}$ in n variables over k as follows (this should not be confused with the ring of convergent power series $\mathbf{C}\{z_1, \dots, z_n\}$). Let $\boldsymbol{\Theta}$ be the free commutative multiplicative semigroup generated by the elements of $\boldsymbol{\Delta}$ and let $\{\theta y_i\}_{\theta \in \boldsymbol{\Theta}, i \in \{1, \dots, n\}}$ be a set of indeterminates. One defines $k\{y_1, \dots, y_n\}$ to be the ring $k[\theta y_i]_{\theta \in \boldsymbol{\Theta}, i \in \{1, \dots, n\}}$. This ring has a structure of a $\boldsymbol{\Delta}$-ring defined by $\partial_j(\theta y_i) = \partial_j \theta y_i$. We denote the set of homogeneous linear elements of $k\{y_1, \dots, y_n\}$ by $k\{y_1, \dots, y_n\}_1$. Kolchin ([162], Chap. IV.5) defines a $\boldsymbol{\Delta}$-ideal I to be *linear* if I is generated (as a $\boldsymbol{\Delta}$-ideal) by a set $\Lambda \subset k\{y_1, \dots, y_n\}_1$. He further shows that if this is the case then

$$I \cap k\{y_1, \dots, y_n\}_1 = \text{the } k\text{-span of } \{\theta L\}_{\theta \in \boldsymbol{\Theta}, L \in \Lambda} \, . \tag{D.1}$$

The codimension of $I \cap k\{y_1, \ldots, y_n\}_1$ in $k\{y_1, \ldots, y_n\}_1$ is called the *linear dimension of* I (which need not be finite). Let $\mathcal{U}$ be a universal field over k with constants $\mathcal{C}$, that is a $\mathbf{\Delta}$-field that contains a copy of every finitely generated differential extension field of k. Kolchin ([162], Chap. IV.5, Corollary 1) shows that the mapping that sends any finite dimensional $\mathcal{C}$-subspace $\mathcal{V}$ of $\mathcal{U}^n$ to the $\mathbf{\Delta}$-ideal $I(\mathcal{V})$ of elements of $k\{y_1, \ldots, y_n\}$ that vanish on $\mathcal{V}$ is a bijective mapping onto the set of linear $\mathbf{\Delta}$-ideals of finite linear dimension. Furthermore, if $\dim_{\mathcal{C}} \mathcal{V} = m$ then $I(\mathcal{V})$ has linear dimension m. Therefore, one can say that the linear $\mathbf{\Delta}$-ideals of finite linear dimension correspond to systems of homogeneous linear partial differential equations whose solution spaces are finite dimensional.

Let us now consider the case of $n = 1$, that is the $\mathbf{\Delta}$-ring $k\{y\}$. The map $\theta \mapsto \theta y$ induces a k-linear bijection ψ between $k[\partial_1, \ldots, \partial_r]$ and $k\{y\}_1$. If I is a left ideal of $k[\partial_1, \ldots, \partial_r]$, then $\psi(I)$ will generate a linear $\mathbf{\Delta}$-ideal J in $k\{y\}$. Equation (D.1) implies that this yields a bijection between the sets of such ideals. Furthermore, I has finite rank m if and only if J has finite linear dimension m. Therefore, the left ideals I in $k[\partial_1, \ldots, \partial_r]$ of finite rank correspond to systems of homogeneous linear differential equations in one indeterminate having finite dimensional solution spaces in $\mathcal{U}$.

3. One can also study the ring of differential operators with coefficients in a ring. For example, the ring $D = \mathbf{C}[z_1, \ldots, z_r, \partial_1, \ldots, \partial_r]$, where $z_i z_j = z_j z_i$ and $\partial_i \partial_j = \partial_j \partial_i$ for all i, j, and $\partial_i x_j = x_j \partial_i$ if $i \neq j$ and $\partial_i x_i = x_i \partial_i + 1$ is referred to as the *Weyl algebra* and leads to the study of D-modules. We refer to [38] and [74] and the references therein for an exposition of this subject as well as [68, 69, 184, 254] for additional information concerning questions of effectivity. Given a left ideal J in D, one can consider the ideal $I = Jk[\partial_1, \ldots, \partial_r]$ with $k = \mathbf{C}(z_1, \ldots, z_n)$. The holonomic rank of J (see the above references for a definition of this quantity) is the same as the rank of I (see Chap. 1.4 of [254]). □

We now make the connection between $k[\partial_1, \ldots, \partial_r]$-modules and systems of equations of the form

$$\partial_i u = A_i u \quad i = 1, \ldots, r, \tag{D.2}$$

where $u \in k^m$ and each A_i is an $m \times m$ matrix with entries in k. Let $\mathcal{M}$ be a $k[\partial_1, \ldots, \partial_r]$-module and let $e_1, \ldots, e_m$ be a k-basis of $\mathcal{M}$. For each $\ell = 1, \ldots r$, we may write

$$\partial_\ell e_i = -\sum_j a_{j,i,\ell} e_j, \tag{D.3}$$

where $A_\ell = (a_{i,j,\ell})$ is an $m \times m$ matrix with entries in k. If $u = \sum_i u_i e_i \in \mathcal{M}$, then $\partial_\ell u = \sum_i (\partial_\ell(u_i) - \sum_j a_{i,j,\ell} u_j) e_i$ (note that $\partial_\ell u$ denotes the action of ∂_ℓ on $\mathcal{M}$, while $\partial_\ell(u)$ denotes the application of the derivation to an element of the field). Therefore, once a basis of $\mathcal{M}$ has been selected and the identification $\mathcal{M} \simeq k^n$ has been made, we have that the action of ∂_i on k^n is given by $u \mapsto \partial_i(u) - A_i u$, where $\partial_i(u)$ denotes the vector obtained by applying ∂_i to each entry of u. In particular, for $u \in k^n$, u is mapped to zero by the action of ∂_i if and only if u satisfies $\partial_i(u) = A_i u$. Since $\mathcal{M}$ is a $k[\partial_1, \dots, \partial_r]$-module, the actions of ∂_i and ∂_j commute for any i, j and so $(\partial_i - A_i)(\partial_j - A_j) = (\partial_j - A_j)(\partial_i - A_i)$. This is equivalent to

$$\partial_i(A_j) + A_i A_j = \partial_j(A_i) + A_j A_i \text{ for all } i, j. \tag{D.4}$$

These latter equations are called the *integrability conditions* for the operators $\partial_i - A_i$.

Definition D.7 For $i = 1, \dots, r$, let A_i be an $m \times m$ matrix with coefficients in k. We say that the system of linear equations $\{\partial_i u = A_i u\}$ is an *integrable system* if any pair of matrices A_i, A_j satisfy the integrability conditions (D.4). □

We have shown in the discussion preceding the above definition that selecting a k-basis for a $k[\partial_1, \dots, \partial_r]$-module leads to an integrable system. Conversely, given an integrable system, one can define a $k[\partial_1, \dots, \partial_r]$-module structure on k^m via (D.3), where the e_i are the standard basis of k^m. The integrability conditions insure that the actions of any ∂_i and ∂_j commute.

We end this section with a description of the terminology of integrable connections. In the ordinary case, we have encountered this in Sect. 6.1 and this setting most readily generalizes to give a coordinate free way of presenting linear differential equations on manifolds.

In Sect. 6.1 we defined a universal differential module but noted that for many applications this object is too large, and restricted ourselves to smaller modules. All of these fit into the following definition:

Definition D.8 Let $C \subset k$ be fields of characteristic zero with C algebraically closed. A *special differential* (Ω, d) is a finite dimensional k-vector space Ω together with a map $d : k \to \Omega$ such that

1. The map d is C-linear and $d(fg) = fd(g) + gd(f)$ for all $f, g \in k$.
2. The kernel of d is C.
3. Ω is generated as a k-vector space by $d(k)$.
4. The k-linear vector space $\text{der}(k) := \{l \circ d | \ l \in \text{Hom}_k(M, k)\}$, consisting of C-linear derivations on k, is closed under Lie brackets [,] (i.e., for $D_1, D_2 \in \text{der}(k)$ one has that $[D_1, D_2] := D_1 D_2 - D_2 D_1 \in \text{der}(k)$).

□

Consider a special differential $d : k \to \Omega$. Choose elements $t_1, \dots, t_r \in k$ such that $\{dt_1, \dots, dt_r\}$ is a basis of Ω. Let $l_i : \Omega \to k$ be the k-linear map given by $l_i(dt_j) = 1$ for $j = i$ and 0 for $i \neq j$. Put $\partial_i = l_i \circ d$ for $i = 1, \dots, r$. Then $\{\partial_i\}_i$

is a k-basis for the Lie algebra der(k). Consider $D := [\partial_i, \partial_j]$. One verifies that $D(t_s) = 0$ for $s = 1, \ldots, r$. Since $D \in$ der(k), it follows that $D = 0$. Thus a special differential (Ω, d) gives rise to the $\boldsymbol{\Delta}$-field structure on k given by $\{\partial_1, \ldots, \partial_r\}$. This structure is special in the sense that there are elements $t_1, \ldots, t_r \in k$ such that $\partial_i t_j = 1$ for $j = i$ and 0 otherwise. Moreover, the intersection of kernels of the ∂_i is the prescribed field C. On the other hand, a $\boldsymbol{\Delta}$-field with $\boldsymbol{\Delta} = \{\partial_i\}_{i=1}^r$ for which elements $t_1, \ldots, t_r \in k$ exist with the above properties and with field of constants C, induces the special differential $d : k \to \Omega := kdt_1 \oplus \cdots \oplus kdt_r$ with $d(f) := \sum_i \partial_i(f)dt_i$.
We conclude that the concept of a $\boldsymbol{\Delta}$-field k is slightly more general than that of a special differential.

Examples D.9
1. k is an algebraic extension of a purely transcendental extension $C(t_1, \ldots, t_r)$ of C and Ω is the universal differential module $\Omega_{k/C}$ (see Sect. 6.1). In this case, Ω is a k-vector space of dimension r with basis $dt_1, \ldots, dt_r$ and $d : k \to \Omega$ is given by $d(f) = \partial_1(f)dt_1 + \cdots + \partial_r(f)dt_r$, where ∂_i is the unique extension of the derivation $\frac{\partial}{\partial t_i}$ on $C(t_1, \ldots, t_r)$.

2. k is an algebraic extension of $C((t_1, \ldots, t_r))$, where this latter field is defined as in Examples D.1.2. Here, one can take Ω to be the k-vector space of dimension r with basis $dt_1, \ldots, dt_r$ and d is defined as above. Note that Ω is not the universal differential since there are derivations on $C((t_1, \ldots, t_r))$ that are not k-linear combinations of the ∂_i.

3. One can replace in part 2 the field C with $\mathbf{C}$, the complex numbers, and $C((t_1, \ldots, t_r))$ with $\mathbf{C}(\{t_1, \ldots, t_r\})$, the field of fractions of the ring of convergent power series (see Examples D.1.3) and construct Ω in a similar manner. □

Definition D.10 Fix a special differential $d : k \to \Omega$. Let M denote a finite dimensional vector space over k. A *connection* ∇ on M is a map $\nabla : M \to \Omega \otimes_k M$ satisfying:

1. ∇ is a C-linear.
2. $\nabla(\lambda v) = d(\lambda) \otimes v + \lambda\nabla(v)$ for all $\lambda \in k$ and $v \in M$.

For any k-linear map $l : \Omega \to k$ and corresponding derivation $l \circ d : k \to k$ one defines $\nabla_{l \circ d} := M \stackrel{\nabla}{\to} \Omega \otimes M \stackrel{l \otimes id_M}{\to} M$. A connection is *integrable* if for any two k-linear maps $l_1, l_2 : \Omega \to k$ one has

$$[\nabla_{l_1 \circ d}, \nabla_{l_2 \circ d}] = \nabla_{[l_1 \circ d, l_2 \circ d]} \ .$$ □

We now show that the concept of an integrable connection is equivalent to an integrable system of linear partial differential equations. Let (Ω, d) be a special differential for k. We will use the notations following Definition D.8. Let (∇, M)

be a connection. One considers the partial differential operators $\nabla_{\partial_i} : M \to M$ for $i = 1, \ldots, r$. If the connection is integrable then the ∇_{∂_i} commute, since the ∂_i commute. The converse is also easily verified.
As a consequence, all connections are integrable for $r = 1$.

D.2 Picard-Vessiot Theory and Some Remarks

In this section we shall re-examine the material in the first part of this book and discuss to what extent the theory developed there generalizes to partial differential equations.

Our main object of study will be an integrable system $\{\partial_i u = A_i u\}$ where the A_i are $m \times m$ matrices with coefficients in some $\boldsymbol{\Delta}$-field k (see Definition D.7). We shall denote such a system with the notation $\partial u = Au$.

One begins this study as in Sects. 1.1 and 1.2 with the study of $\boldsymbol{\Delta}$-rings and $\boldsymbol{\Delta}$-fields. As we have shown in the previous section, there is a correspondence between integrable systems and zero-dimensional right ideals in $k[\partial_1, \ldots, \partial_r]$, which is analogous to the correspondence between differential equations $Y' = AY$ and operators $L \in k[\partial]$. The results of Sect. 1.2 carry over to the case of integrable systems. A small difference is that one does not have a wronskian matrix. Nonetheless, there is a result corresponding to Lemma 1.12 that is useful in transferring the results of Chap. 1 to the case of partial differential equations. In Remarks D.6.2, we defined $\boldsymbol{\Theta}$ to be the free commutative multiplicative semigroup generated by the elements of $\boldsymbol{\Delta}$. We denote by $\boldsymbol{\Theta}(s)$ the set of elements of $\boldsymbol{\Theta}$ of order less than or equal to s.

Lemma D.11 *Let k be a $\boldsymbol{\Delta}$-field with field of constants C and let $y_1, \ldots, y_n$ be elements of k. If these elements are linearly dependent over C then*

$$\det(\theta_i y_j)_{1 \le i \le n, 1 \le j \le n} = 0 \tag{D.5}$$

for all choices of $\theta_1, \ldots \theta_n \in \boldsymbol{\Theta}$. Conversely, if Equation (D.5) holds for all choices of $\theta_1, \ldots, \theta_n$ with $\theta_i \in \boldsymbol{\Theta}(i-1)$, $1 \le i \le n$, then $y_1, \ldots, y_n$ are linearly dependent over C.

Proof. If $\sum_{i=1}^n c_i y_i = 0$ for $c_1, \ldots, c_n \in C$ not all zero, then $\sum_{i=1}^n c_i \theta y_i = 0$ for all $\theta \in \boldsymbol{\Theta}$ so Equation (D.5) holds.

We prove the converse by induction on n. We may assume that $n > 1$ and that there exist $\theta_i' \in \boldsymbol{\Theta}(i-1)$, $1 \le i \le n-1$, such that $\det N \neq 0$ where

$$N = (\theta_i y_j)_{1 \le i \le n-1, 1 \le j \le n-1} \, .$$

Under our assumption, this implies that the matrix

$$M = (\theta y_j)_{\theta \in \boldsymbol{\Theta}(n-1), 1 \le j \le n-1}$$

has rank precisely $n-1$. Therefore, the space V of vectors $(d_1, \dots, d_n) \in k^n$ such that $\sum_{j=1}^n d_j \theta y_j = 0$ for all $\theta \in \boldsymbol{\Theta}(n-1)$ has dimension 1. Let $(c_1, \dots, c_n)$ be an element of V with some $c_j = 1$. For any $\partial \in \boldsymbol{\Delta}$ we have that

$$0 = \partial(\sum_{j=1}^n c_j \theta y_j) = \sum_{j=1}^n \partial(c_j)\theta y_j + \sum_{j=1}^n c_j \partial\theta y_j \,.$$

However, if $\theta \in \boldsymbol{\Theta}(n-2)$ then $\partial\theta \in \boldsymbol{\Theta}(n-1)$ and so $\sum_{j=1}^n \partial(c_j)\theta y_j = 0$. In particular, this holds for $\theta = \theta_i'$, $1 \le i \le n-1$, as defined above. Since any row of M is a k-linear combination of the rows $(\theta_i' y_1, \dots, \theta_i' y_n)$, $1 \le i \le n-1$, we have that $\sum_{j=1}^n \partial(c_j)\theta y_j = 0$ for any $\theta \in \boldsymbol{\Theta}(n-1)$. Therefore, $(\partial(c_1), \dots, \partial(c_n)) \in V$ and we can conclude that $(\partial(c_1), \dots, \partial(c_n))$ is a k-multiple of $(c_1, \dots, c_n)$. Since $c_j = 1$ and $\partial(c_j) = 0$, we have that each $\partial(c_i) = 0$. This holds for all $\partial \in \boldsymbol{\Delta}$ so each $c_i \in C$. □

Remark D.12 1. In [162], Kolchin proves a result (Chap. II, Theorem 1) that gives criteria similar to Lemma D.11 for a set of n elements in k^t to be linearly dependent over C. The above result gives these criteria for $t = 1$ and the proof is the same as Kolchin's but specialized to this situation. Lemma D.11 is sufficient for the Galois theory of partial differential equations. For example, Corollary 1.40 can be stated and proven for partial differential Picard-Vessiot extensions. In this case, the use of the wronksian matrix $W(y_1, \dots, y_n)$ and reference to Lemma 1.12 are replaced by a nonsingular matrix of the form $(\theta_i y_j)_{1 \le i \le n, 1 \le j \le n}$ for some $\theta_i \in \boldsymbol{\Theta}(i-1)$, $1 \le i \le n$. □

Exercise D.13 Let $d : k \to \Omega$ be a special differential and $\nabla : M \to \Omega \otimes_k M$ an integrable connection. Adapt the proof of Lemma 1.7 in order to show that the dimension of the C-vector space $\{m \in M | \ \nabla m = 0\}$ is less than or equal to the dimension of M over k. □

We now turn to the Picard-Vessiot theory of integrable systems. The field of constants C of the $\boldsymbol{\Delta}$-field is supposed to be algebraically closed of characteristic 0. Analogous to the ordinary case (Definition 1.15), we state the following definition.

Definition D.14 A *Picard-Vessiot ring* over k for the system $\partial u = Au$, is a $\boldsymbol{\Delta}$-ring over k satisfying:

1. R is a simple $\boldsymbol{\Delta}$-ring, i.e., the only $\boldsymbol{\Delta}$-ideals are R and (0).
2. There exists a fundamental matrix B for $\partial u - A$ with coefficients in R, i.e., a matrix $B \in \mathrm{GL}_m(R)$ satisfying $\partial_i B = A_i B$ for $i = 1, \dots, r$.

3. R is generated as a ring by k, the entries of a fundamental matrix B and the inverse of the determinant of B.

□

All of the results of Chap. 1 remain true for integrable systems and the proofs in this context are easy modifications of the proofs given there.

1. Picard-Vessiot extensions exist and are unique up to k-isomorphism.
2. If R is a Picard-Vessiot ring then the set of $\boldsymbol{\Delta}$-k-algebra automorphisms (k-algebra automorphisms σ of R such that $\partial_i(\sigma(f)) = \sigma(\partial_i(f))$ for all $f \in R$) has a natural structure of a linear algebraic group. This group is called the *differential Galois group* of the $\partial - A$ and is denoted by $\mathrm{Aut}^{\boldsymbol{\Delta}}(R/k)$. In particular, if $V = \{v \in R^m \mid \partial_i v = A_i v \text{ for all } i = 1, \dots r\}$ then V is left invariant by $\mathrm{Aut}^{\boldsymbol{\Delta}}(R/k)$ and the image of $\mathrm{Aut}^{\boldsymbol{\Delta}}(R/k)$ is a Zariski-closed subgroup of $\mathrm{GL}(V)$.
3. We define a *Picard-Vessiot field* for the integrable system $\partial - A$ to be the field of fractions of a Picard-Vessiot ring for this equation. As in Proposition 1.22, one has that a $\boldsymbol{\Delta}$-field $L \supset k$ is a Picard-Vessiot field for $\partial - A$ if and only if the field of constants of of L is C, there exists a fundamental matrix $B \in \mathrm{GL}_m(L)$ for this equation and L is generated over k by the entries of B.
4. There is a Galois correspondence precisely as described in Proposition 1.34.
5. As in Theorem 1.28, one can show that a Picard-Vessiot ring over a field k is the coordinate ring of a G-torsor, where G is the Galois group of the equation.
6. Given $\boldsymbol{\Delta}$-fields $k \subset K$, one can define $t \in K$ to be an *integral (of an element of k)* if $\partial t \in k$ for all $\partial \in \boldsymbol{\Delta}$. Similarly, one can define an element $t \in K$ to be an *exponential (of an integral of an element of k)* if $\partial t/t \in k$ for all $\partial \in \boldsymbol{\Delta}$. With these definitions, the results of Sect. 1.5 can be generalized to $\boldsymbol{\Delta}$-fields. In particular, Theorem 1.43 holds.

In [162], Kolchin develops a Galois theory for linear differential ideals in $k\{y_1, \dots, y_r\}$ of finite linear dimension and shows in Chap. VI.6 that Picard-Vessiot extensions for such an ideal are always generated by a fundamental system of zeroes of such an ideal where r can be taken to be equal to 1. Therefore, the theory outlined here is equivalent to Kolchin's theory.

We finish this appendix with some remarks concerning other aspects of connections. For integrable connections, or more generally for $k[\partial_1, \dots, \partial_r]$-modules, one can clearly define the notions of homomorphism, direct sums, tensor product, etc. The tannakian equivalence of Theorem 2.33 remains valid for an integrable connection (∇, M) and its differential Galois group G. In particular, the results in Chap. 2 relating the behavior of differential modules and that of the differential Galois group (e.g., reducibility, complete reducibility) remain valid in the context of integrable connections.

The formal theory concerns the field $k = C((t_1, \ldots, t_r))$ and the special differential $d : k \to \Omega = kdt_1 \oplus \cdots \oplus kdt_r$ (see Examples D.9). No general version of Theorem 3.1 for (formal) integrable connections is known. However, there are classifications of integrable systems of a very special form in [95, 65] and there are some more general ideas in [185] (see also [186]).

Very little has been written concerning algorithms similar to those in Chap. 4 for integrable systems. Algorithms to find rational solutions of integrable systems appear in [174, 215], (see also [63, 64]). An algorithm to find solutions of integrable systems all of whose logarithmic derivatives are rational can be found in [174]. Algorithmic questions concerning the reducibility of an integrable system are dealt with in [287] (see also [286]).

As in the one-dimensional case, the analytic theory of integrable connections, without singularities, on a complex analytic manifold is related to local systems and representations of the fundamental group. Regular singular integrable connections were first presented in [80]. In this book, Deligne investigates the basic properties of connections with regular singularities and gives a solution of the weak Riemann-Hilbert Problem. This theory is further developed in [192]. Meromorphic integrable connections with irregular singularities and their asymptotic properties are treated in [194, 196, 252, 253, 186].

Bibliography

1. S. Abhyankar. Coverings of algebraic curves. *Am. J. Math.*, 79:825–856, 1957.
2. S. A. Abramov. Rational solutions of linear differential and difference equations with polynomial coefficients (in Russian). *J. Computational Math. Math. Physics*, 29(11):1611–1620, 1989.
3. S. A. Abramov and M. Bronstein. On Solutions of Linear Functional Systems. In B. Mourrain, editor, *Proceedings of the 2001 International Symposium on Symbolic and Algebraic Computation (ISSAC'2001)*, pp. 1–6. ACM Press, 2001.
4. S. A. Abramov, M. Bronstein, and M. Petkovšek. On polynomial solutions of linear operator equations. In A. H. M. Levelt, editor, *Proceedings of the 1995 International Symposium on Symbolic and Algebraic Computation (ISSAC'95)*, pp. 290–296. ACM Press, 1995.
5. S. A. Abramov and K. Yu. Kvansenko. Fast algorithms for rational solutions of linear differential equations with polynomial coefficients. In S. M. Watt, editor, *Proceedings of the 1991 International Symposium on Symbolic and Algebraic Computation (ISSAC'91)*, pp. 267–270. ACM Press, 1991.
6. K. Adzamagbo. Sur l'effectivité du lemme du vecteur cyclique. *Comptes Rendues de l'Académie des Sciences*, 306:543–546, 1988.
7. L. Ahlfors. *Complex Analysis*. 2nd edn., McGraw Hill, New York, 1966.
8. Y. André. Sur la conjecture des p-courbures de Grothendieck-Katz. *Institut de Mathématiques de Jussieu, Prépublication 134*, 1997.
9. D. V. Anosov and A. A. Bolibruch. *The Riemann-Hilbert Problem*. Vieweg, Braunschweig, Wiesbaden, 1994.
10. M. F. Atiyah and I. MacDonald. *Introduction to Commutative Algebra*. Addison-Wesley, Reading, Mass., 1969.
11. M. Audin. *Les Systèmes Hamiltoniens et leur Intégrabilité*, Vol. 8 of *Cours Spécialisés*. Société Mathématique de France, Paris, 2001.
12. D. G. Babbitt and V. S. Varadarajan. Formal reduction of meromorphic differential equations: a group theoretic view. *Pacific J. Math.*, 109(1):1–80, 1983.
13. D. G. Babbitt and V. S. Varadarajan. Local moduli for meromorphic differential equations. *Astérisque*, 169-170:1–217, 1989.
14. F. Baldassarri and B. Dwork. On second order linear differential equations with algebraic solutions. *Am. J. Math.*, 101:42–76, 1979.
15. W. Balser. *From Divergent Power Series to Analytic Functions*, Vol. 1582 of *Lecture Notes in Mathematics*. Springer-Verlag, Heidelberg, 1994.
16. W. Balser, B. L. J. Braaksma, J.-P. Ramis, and Y. Sibuya. Multisummability of formal power series solutions of linear ordinary differential equations. *Asymptotic Anal.*, 5:27–45, 1991.

17. W. Balser, W. B. Jurkat, and D. A. Lutz. A general theory of invariants for meromorphic differential equations. Part I: Formal invariants; Part II: Proper Invariants. *Funcialaj Ekvacioj*, 22:197–221; 257–283, 1979.
18. M. Barkatou. On the equivalence problem of linear differential systems and its application for factoring completely reducible systems. In O. Gloor, editor, *Proceedings of the 1998 International Symposium on Symbolic and Algebraic Computation (ISSAC'98)*. ACM Press, 1998.
19. M. A. Barkatou. Rational Newton algorithm for computing formal solutions of linear differential equations. In P. Gianni, editor, *Symbolic and Algebraic Computation - ISSAC'88*, pp. 183–195. ACM Press, 1988.
20. M. A. Barkatou. A rational version of Moser's algorithm. In A. H. M. Levelt, editor, *Proceedings of the 1995 International Symposium on Symbolic and Algebraic Computation (ISSAC'95)*, pp. 290–296. ACM Press, 1995.
21. M. A. Barkatou. An algorithm to compute the exponential part of a formal fundamental matrix solution of a linear differential system. *Applicable Algebra in Engineering, Communication and Computing*, 8(9):1–24, 1997.
22. M. A. Barkatou. On rational solutions of systems of linear differential equations. *J. Symb. Comput.*, 28(4/5):547–568, 1999.
23. M. A. Barkatou and F. Jung. Formal solutions of linear differential and difference equations. *Programmirovanie*, 1, 1997.
24. M. A. Barkatou and E. Pflügel. An algorithm computing the regular formal solutions of a system of linear differential equations. Technical report, IMAC-LMC, Université de Grenoble I, 1997. RR 988.
25. Cassidy Bass, Buium, editor. *Selected Works of Ellis Kolchin with Commentary*. American Mathematical Society, 1999.
26. A. Beauville. Monodromie des systèmes différentiels à pôles simples sur la sphère de Riemann (d'après A. Bolibruch). *Astérisque*, 216:103–119, 1993. Séminaire Bourbaki, No. 765.
27. B. Beckermann and G. Labahn. A uniform approach for the fast computation of matrix-type Padé-approximants. *SIAM J. Matrix Anal. Applic.*, pp. 804–823, 1994.
28. E. Beke. Die Irreducibilität der homogenen Differentialgleichungen. *Math. Annal.*, 45:278–294, 1894.
29. M. Berkenbosch. Field extensions for algebraic Riccati solutions. manuscript, University of Groningen, 2000.
30. D. Bertrand. Constructions effectives de vecteurs cycliques pour un d-module. *Publications du Groupe d'Etude d'Analyse Ultramétrique*, 11, 1984/1985.
31. D. Bertrand. Un analogue différentiel de la théorie de Kummer. In P. Philippon, editor, *Approximations Diophantiennes et Nombres Transcendentes, Luminy 1990*, pp. 39–49. Walter de Gruyter and Co., Berlin, 1992.
32. F. Beukers, D. Brownawell, and G. Heckman. Siegel normality. *Annal. Math.*, 127:279–308, 1988.
33. F. Beukers and G. Heckman. Monodromy for the hypergeometric function ${}_nF_{n-1}$. *Inventiones Mathematicae*, 95:325–354, 1989.
34. A. Bialynicki-Birula. On the Galois theory of fields with operators. *Am. J. Math.*, 84:89–109, 1962.
35. A. Bialynicki-Birula. On the inverse problem of Galois theory of differential fields. *Bull. Am. Math. Soc.*, 69:960–964, 1963.
36. A. Bialynicki-Birula, G. Hochschild, and G. D. Mostow. Extensions of Representations of Algebraic Groups. *Am. J. Math.*, 85:131–144, 1963.

37. G. Birkhoff. *Collected Mathematical Papers*, Vol. 1. Dover Publications, New York, 1968.
38. J. E. Bjork. *Rings of Differential Operators*. North Holland, Amsterdam, 1979.
39. H. F. Blichtfeldt. *Finite Collineation Groups*. University of Chicago Press, Chicago, 1917.
40. R. P. Boas. *Invitation to Complex Variables*. Random House, New York, 1987.
41. A. A. Bolibruch. The Riemann-Hilbert problem. *Russian Math. Surveys*, 45(2):1–47, 1990.
42. A. A. Bolibruch. On sufficient conditions for the positive solvability of the Riemann-Hilbert problem. *Math. Notes Acad. Sci. USSR*, 51(2):110–117, 1992.
43. A. A. Bolibruch. The 21st Hilbert Problem for Linear Fuchsian Systems. *Proceedings of the Steklov Institute of Mathematics*, 206(5):1 – 145, 1995.
44. A. A. Bolibruch. Holomorphic bundles associated with linear differential equations and the Riemann-Hilbert problem. In Braaksma et al., editor, *The Stokes Phenomenon and Hilbert's* 16th *Problem*, pp. 51–70. World Scientific, 1996.
45. A. Borel. *Linear Algebraic Groups, Second Enlarged Edition*. Springer-Verlag, New York, 1991.
46. M. Bouffet. Un lemme de Hensel pour les opérateurs différentiels. *C.R. Acad. Sci. Paris*, 331(4):277–280, 2000.
47. M. Bouffet. Factorisation d'opérateurs différentielles à coefficients dans une extension liouvillienne d'un corps valué. Technical report, Université Paul Sabatier, Toulouse, 2001.
48. A. Boulanger. Contribution à l'étude des équations linéaires homogènes intégrables algébriquement. *J. l'École Polytechnique, Paris*, 4:1 – 122, 1898.
49. N. Bourbaki. *Groupes et Algèbres de Lie,* Chaps 7 and 8. Masson, Paris, 1990.
50. K. Boussel. Groupes de Galois des équations hypergéométriques. *Comptes Rendus Acad. Sci., Paris*, 309:587–589, 1989.
51. G. E. Bredon. *Sheaf Theory*. Graduate Texts in Mathematics. Springer-Verlag, New York, 1997.
52. L. Breen. Tannakian categories. In U. Jannsen et al, editors, *Motives*, Vol. 55 of *Proceedings of Symposia in Pure Mathematics*, pp. 337–376. American Mathematical Society, 1994.
53. A. Brill. Über die Zerfällung einer Ternärform in Linearfactoren. *Math. Ann.*, 50:157–182, 1898.
54. M. Bronstein. The Risch differential equation on an algebraic curve. In P. Gianni, editor, *Symbolic and Algebraic Computation - ISSAC'88*, pp. 64–72. ACM Press, 1988.
55. M. Bronstein. Integration of Elementary Functions. *J. Symb. Comput.*, 9(3):117–174, 1990.
56. M. Bronstein. Linear ordinary differential equations: breaking through the order 2 barrier. In P. Wang, editor, *Proceedings of the International Symposium on Symbolic and Algebraic Computation- ISSAC'92*, pp. 42–48. ACM Press, 1992.
57. M. Bronstein. On Solutions of Linear Ordinary Differential Equations in their Coefficient Field. *J. Symb. Comput.*, 13(4):413 – 440, 1992.
58. M. Bronstein. An improved algorithm for factoring linear ordinary differential operators. In J. von zur Gathen, editor, *Proceedings of the 1994 International Symposium on Symbolic and Algebraic Computation (ISSAC'94)*, pp. 336–347. ACM Press, 1994.
59. M. Bronstein, T. Mulders, and J.-A. Weil. On symmetric powers of differential operators. In W. Küchlin, editor, *Proceedings of the 1997 International Symposium on Symbolic and Algebraic Computation (ISSAC'97)*, pp. 156–163. ACM Press, 1997.

60. M. Bronstein and M. Petkovsek. An introduction to pseudo-linear algebra. *Theoret. Comput. Sci.*, 157:3–33, 1996.
61. J. Calmet and F. Ulmer. On liouvillian solutions of homogeneous linear differential equations. In M. Nagata and S. Watanabe, editors, *Proceedings of the International Symposium on Symbolic and Algebraic Computation- ISSAC'90*, pp. 236–243. ACM Press, 1990.
62. J. W. S. Cassels and J. Fröhlich. *Algebraic Number Theory*. Academic Press, London, 1986.
63. E. Cattani, C. D'Andrea, and A. Dickenstein. The A-hypergeometric system associated with a monomial curve. *Duke Math. J.*, 99:179–207, 1999.
64. E. Cattani, A. Dickenstein, and B. Sturmfels. Rational hypergeometric functions. Technical report, MSRI Publication 1999-051, 1999.
65. H. Charrière and R. Gérard. Formal reduction of irregular integrable connections having a certain kind of irregular singularities. *Analysis*, 1:85–115, 1981.
66. G. Chen. An algorithm for computing the formal solutions of differential systems in the neighborhood of an irregular singular point. In M. Watanabe, editor, *Proceedings of the 1990 International Symposium on Symbolic and Algebraic Computation (ISSAC'90)*, pp. 231–235. ACM Press, 1990.
67. C. Chevalley. *Théorie des Groupes de Lie*, Vol. II, Groupes Algébriques. Hermann, Paris, 1951.
68. F. Chyzak. *Fonctions holonomes en calcul formel.* Thèse universitaire, École polytechnique, 1998. INRIA, TU 0531. 227 pages. Available at `http://www-rocq.inria.fr/algo/chyzak`.
69. F. Chyzak. Groebner bases, symbolic summation and symbolic integration. In B. Buchberger and F. Winkler, editors, *Groebner Bases and Applications (Proc. of the Conference 33 Years of Gröbner Bases)*, Vol. 251 of *London Mathematical Society Lecture Notes Series*, pp. 32–60. Cambridge University Press, 1998.
70. T. Cluzeau and E. Hubert. Resolvent representation for regular differential ideals. Technical Report RR-4200, INRIA Sophia Antipolis, `http://www.inria.fr/rrrt/rr-4200.html`, 2001.
71. E. Compoint and M. F. Singer. Calculating Galois groups of completely reducible linear operators. *J. Symb. Comput.*, 28(4/5):473–494, 1999.
72. F. Cope. Formal solutions of irregular linear differential equations, I. *Am. J. Math.*, 56:411–437, 1934.
73. F. Cope. Formal solutions of irregular linear differential equations, II. *Am. J. Math.*, 58:130–140, 1936.
74. S. C. Coutinho. *A Primer of Algebraic D-Modules*, Vol. 33 of *London Mathematical Society Lecture Notes Series*. Cambridge University Press, Cambridge, 1995.
75. D. Cox, J. Little, and D. O'Shea. *Ideals, Varieties and Algorithms*. Springer, New York, 1991.
76. C. W. Curtis and I. Reiner. *Representation Theory of Finite Groups and Associative Algebras*. John Wiley and Sons, New York, 1962.
77. J. Davenport. *On the Integration of Algebraic Functions*, Vol. 102 of *Lecture Notes in Computer Science*. Springer-Verlag, Heidelberg, 1981.
78. J. Davenport and M. F. Singer. Elementary and liouvillian solutions of linear differential equations. *J. Symb. Comput.*, 2(3):237–260, 1986.
79. W. Dekkers. The matrix of a connection having regular singularities on a vector bundle of rank 2 on $P^1(\mathbf{C})$. In *Équations différentielles et systèmes de Pfaff dans le champ complexe (Sem., Inst. Rech. Math. Avancée, Strasbourg, 1975)*, Vol. 712 of *Lecture Notes in Mathematics*, pp. 33–43. Springer, New York, 1979.

80. P. Deligne. *Equations Différentielles à Points Singuliers Réguliers*, Vol. 163 of *Lecture Notes in Mathematics*. Springer-Verlag, Heidelberg, 1970.
81. P. Deligne. Catégories tannakiennes. In P. Cartier et al., editor, *Grothendieck Festschrift, Vol. 2*, pp. 111–195. Birkhäuser, 1990. Progress in Mathematics, Vol. 87.
82. P. Deligne and J. Milne. Tannakian categories. In P. Deligne et al., editor, *Hodge Cycles, Motives and Shimura Varieties*, pp. 101–228, 1982. Lecture Notes in Mathematics, Vol. 900.
83. J. Della Dora, C. di Crescenzo, and E. Tournier. An algorithm to obtain formal solutions of a linear differential equation at an irregular singular point. In J. Calmet, editor, *Computer Algebra - EUROCAM '82* (Lecture Notes in Computer Science, 144, pp. 273–280, 1982.
84. J. Dieudonné. Notes sur les travaux de C. Jordan relatifs a la théorie des groupes finis. In *Oeuvres de Camille Jordan*, pp. XVII–XLII. Gauthier-Villars, Paris, 1961.
85. J. D. Dixon. *The Structure of Linear Groups*. Van Nostrand Reinhold Company, New York, 1971.
86. L. Dornhoff. *Group Representation Theory*. Marcel Dekker, Inc, New York, 1971.
87. A. Duval and M. Loday-Richaud. Kovacic's algorithm and its application to some families of special functions. *Applicable Algebra in Engineering, Communication and Computing*, 3(3):211–246, 1992.
88. A. Duval and C. Mitschi. Groupe de Galois des équations hypergéométriques confluentes généralisées. *C. R. Acad. Sci. Paris*, 309(1):217–220, 1989.
89. B. Dwork. *Lectures on p-adic differential equations*. Grundlehren der mathematische Wissenschaften 253, Springer Verlag, New York, 1982.
90. B. Dwork, G. Gerotto, and F. J. Sullivan. *An Introduction to G-Functions*, Vol. 133 of *Annals of Mathematics Studies*. Princeton University Press, Princeton, 1994.
91. E. Dynkin. Maximal subgroups of the classical groups. *Trudy Moskov. Mat. Obschetsva*, 1:39–160, 1957. Am. Math. Soc. Transl., Ser. 2, 6, 1957, pp. 245–378.
92. J. Écalle. *Les fonctions résurgentes, Tome I, II, III*. Publications Mathématiques d'Orsay, Orsay, 1981-1985.
93. J. Écalle. *Introduction aux fonctions analysables et preuve constructive de la conjecture de Dulac*. Hermann, Paris, 1992.
94. D. Eisenbud and J. Harris. *The Geometry of Schemes*, Vol. 197 of *Graduate Texts in Mathematics*. Springer, New York, 2000.
95. A. R. P. van den Essen and A. H. M. Levelt. Irregular singularities in several variables. *Mem. Am. Math. Soc.*, 40(270), 1982.
96. H. Bateman et al. *Higher Transcendental Functions*, Vol. 1. McGraw Hill, New York, 1953.
97. G. Ewald. *Combinatorial Convexity and Algebraic Geometry*. Graduate Texts in Mathematics. Springer-Verlag, New York, 1996.
98. T. Sekiguchi F. Oort and F. Suwa. On the deformation of Artin-Schreier to Kummer. *Ann. Sci. Éc. Norm. Sup, 4ème série*, 22:345–375, 1989.
99. E. Fabry. *Sur les intégrales des équations différentielles linéaires à coefficients rationnels*. PhD thesis, Paris, 1885.
100. O. Forster. *Lectures on Riemann Surfaces*. Number 81 in Graduate Texts in Mathematics. Springer-Verlag, New York, 1981.
101. J. Frenkel. Cohomologie non abélienne et espaces fibrés. *Bull. Soc. Math. France*, 85:135–220, 1957.
102. A. Fröhlich and J. C. Shepherdson. Effective procedures in field theory. *Philos. Trans. Royal Soc. London*, 248:407–432, 1955-1956.

103. L. Fuchs. Zur Theorie der linearen Differentialgleichungen mit verändichen coefficienten. *J. Reine Angewandte Math.*, 66:121–160, 1866.
104. L. Fuchs. Ergänzungen zu der in 66-sten Bande dieses Journal enthalten Abhandlung. *J. Reine Angewandte Math.*, 68:354–385, 1868.
105. L. Fuchs. Über die linearen Differentialgleichungen zweiter Ordnung, welche algebraische Integrale besitzen, und eine neue Anwendung der Invariantentheorie. *J. Reine Angewandte Math.*, 81:97–147, 1875.
106. L. Fuchs. Über die linearen Differentialgleichungen zweiter Ordnung, welche algebraische Integrale besitzen, zweite Abhandlung. *J. Reine Angewandte Math.*, 85:1–25, 1878.
107. W. Fulton and J. Harris. *Representation theory. A first course*, Vol. 129 of *Graduate Texts in Mathematics*. Springer, New York, 1991.
108. I. M. Gelfand, M. M. Kapranov, and A. V. Zelevinsky. *Discriminants, Resultants and Multidimensional Determinants*. Birkhäuser, Boston, Basel, Stuttgart, 1994.
109. R. Gérard and A. H. M. Levelt. Invariants mesurant l'irrégularité en un point singulier des systèmes d'équations différientielles linéaires. *Ann. Inst. Fourier*, 23, 1973.
110. R. Godement. *Topologie Algébrique et Théorie des Faisceaux*. Publ. de l'Inst. de Math. de l'Univ. de Strasbourg. Hermann, Paris, 1973.
111. L. Goldman. Specialization and Picard-Vessiot theory. *Trans. Am. Math. Soc.*, 85:327–356, 1957.
112. J. J. Gray. Fuchs and the theory of differential equations. *Bull. Am. Math. Soc.*, 10(1):1–26, 1984.
113. J. J. Gray. *Linear Differential Equations and Group Theory from Riemann to Poincaré*. Birkhäuser, Boston, Basel, Stuttgart, 2nd edn., 2000.
114. P. Griffiths and J. Harris. *Principles of Algebraic Geometry*. John Wiley and Sons, New York, 1978.
115. D. Yu. Grigoriev. Complexity for irreducibility testing for a system of linear ordinary differential equations. In M. Nagata and S. Watanabe, editors, *Proceedings of the International Symposium on Symbolic and Algebraic Computation- ISSAC'90*, pp. 225–230. ACM Press, 1990.
116. D. Yu. Grigoriev. Complexity of factoring and calculating the gcd of linear ordinary differential operators. *J. Symb. Comput.*, 10(1):7–38, 1990.
117. A. Grothendieck. Sur la classification des fibrés holomorphes sur la sphère de Riemann. *Am. J. Math.*, 79:121–138, 1957.
118. A. Grothendieck et al. *Revêtements Etales et Groupes Fondamentaux (SGA 1)*, Vol. 224 of *Lecture Notes in Mathematics*. Springer Verlag, Berlin, 1971.
119. R. C. Gunning. *Lectures on Vector Bundles on Riemann Surfaces*. Princeton University Press, Princeton, NJ, 1967.
120. R. C. Gunning and H. Rossi. *Analytic Functions of Several Complex Variables*. Prentice Hall, Englewood Cliffs, NJ, 1965.
121. D. Harbater. Abhyankar conjecture on Galois groups over curves. *Inventiones Math.*, 117:1–25, 1994.
122. J. Harris. *Algebraic Geometry, A First Course*. Springer-Verlag, New York, 1994.
123. W. A. Harris and Y. Sibuya. The reciprocals of solutions of linear ordinary differential equations. *Adv. Math.*, 58:119–132, 1985.
124. R. Hartshorne. *Algebraic Geometry*. Number 52 in Graduate Texts in Mathematics. Springer-Verlag, New York, 1977.
125. H. Hasse and F. K. Schmidt. Noch eine Begründung der Theorie der höhere Differentialquotienten in einem algebraischen Funktionenkörper einer Unbestimmten. *J. Reine Angewandte Math.*, 117:215–237, 1937.

126. P. A. Hendriks and M. van der Put. Galois action on solutions of a differential equation. *J. Symb. Comput.*, 19(6):559–576, 1995.
127. S. Hessinger. *Computing Galois groups of fourth order linear differential equations*. PhD thesis, North Carolina State University, 1997.
128. A. Hilali. On the algebraic and differential Newton-Puiseux polygons. *J. Symb. Comput.*, 4(3):335–349, 1987.
129. A. Hilali and A. Wazner. Un algorithme de calcul de l'invariant de Katz d'un système différentiel linéaire. *Ann. Inst. Fourier*, 36(3):67–81, 1986.
130. A. Hilali and A. Wazner. Calcul des invariants de Malgrange et de Gérard-Levelt d'un système différentiel linéaire en un point singulier irrégulier. *J. Diff. Eq.*, 69, 1987.
131. A. Hilali and A. Wazner. Formes super-irréductibles des systèmes différentiels linéaires. *Num. Math.*, 50:429–449, 1987.
132. E. Hille. *Ordinary Differential Equations in the Complex Domain*. John Wiley and Sons, New York, 1976.
133. W. V. D. Hodge and D. Pedoe. *Methods of Algebraic Geometry*, Vol. 1. Cambridge University Press, Cambridge, 1947.
134. M. van Hoeij, J.-F. Ragot, F. Ulmer, and J.-A. Weil. Liouvillian solutions of linear differential equations of order three and higher. *J. Symb. Comput.*, 28(4/5):589–610, 1999.
135. M. van Hoeij and J.-A. Weil. An algorithm for computing invariants of differential Galois groups. *J. Pure Appl. Algebra*, 117/118:353–379, 1996.
136. M. van Hoeij. Rational solutions of the mixed differential equation and its application to factorization of differential operators. In *Proceedings of the 1996 International Symposium on Symbolic and Algebraic Computation*, pp. 219–225. ACM Press, 1996.
137. M. van Hoeij. Factorization of differential operators with rational function coefficients. *J. Symb. Comput.*, 24:5:537–561, 1997.
138. M. van Hoeij. Formal solutions and factorization of differential operators with power series coefficients. *J. Symb. Comp.*, 24:1–30, 1997.
139. T. Honda. Algebraic differential equations. *Symp. Math.*, 24:169–204, 1981.
140. E. Hrushovski. Calculating the Galois group of a linear differential equation. *Preprint*, 2001.
141. J. Humphreys. *Linear Algebraic Groups*. Graduate Texts in Mathematics. Springer-Verlag, New York, 1975.
142. A. Hurwitz. Über einige besondere homogene lineare Differentialgleichungen. *Math. Annal.*, 26:117–126, 1896.
143. N. Jacobson. Pseudo-linear transformations. *Annal. Math.*, 38:484–507, 1937.
144. N. Jacobson. *Lie Algebras*. Dover Publications, Inc., New York, 1962.
145. C. Jordan. Mémoire sur les équations différentielles linéaires à intégrale algébrique. *J. Reine Angewandte Math.*, 84:89–215, 1878.
146. C. Jordan. Sur la détermination des groupes d'ordre fini contenus dans le groupe linéaire. *Atti Accad. Napoli*, 8(11):177–218, 1879.
147. L. Juan. On a generic inverse differential galois problem for gl_n. Technical report, MSRI, 2000. MSRI Preprint 2000-033.
148. W. Jurkat. *Meromorphe Differentialgleichungen*. Lecture Notes in Mathematics No 637, Springer Verlag, New York, 1978.
149. W. Jurkat, D. Lutz, and A. Peyerimhoff. Birkhoff invariants and effective calculations for meromorphic linear differential equations I. *J. Math. Anal. Appl.*, 53:438–470, 1976.
150. W. Jurkat, D. Lutz, and A. Peyerimhoff. Birkhoff invariants and effective calculations for meromorphic linear differential equations II. *Houston J. Math*, 2:207–238, 1976.

151. I. Kaplansky. *An Introduction to Differential Algebra.* Hermann, Paris, 2nd edition, 1976.
152. N. Katz. Algebraic solutions of differential equations; p-curvature and the Hodge filtration. *Inv. Math.*, 18:1–118, 1972.
153. N. Katz. A conjecture in the arithmetic theory of differential equations. *Bull. Soc. Math. France*, 110,111:203–239, 347–348, 1982.
154. N. Katz. On the calculation of some differential Galois groups. *Inventiones Mathematicae*, 87:13–61, 1987.
155. N. Katz. A simple algorithm for cyclic vectors. *Am. J. Math.*, 109:65–70, 1987.
156. N. Katz. *Exponential Sums and Differential Equations*, Vol. 124 of *Annals of Mathematics Studies.* Princeton University Press, Princeton, 1990.
157. N. Katz. *Rigid Local Systems*, Vol. 139 of *Annals of Mathematics Studies.* Princeton University Press, Princeton, 1996.
158. F. Klein. Über lineare Differentialgleichungen, I. *Math. Annal.*, 11:115–118, 1877.
159. F. Klein. Über lineare Differentialgleichungen, II. *Math. Annal.*, 12:167–179, 1878.
160. M. Kohno. *Global Analysis in Linear Differential Equations*, Vol. 471 of *Mathematics and its Applications.* Kluwer Academic Publishers, Boston, Dordrecht, 1999.
161. E. R. Kolchin. Algebraic matric groups and the Picard-Vessiot theory of homogeneous linear ordinary differential equations. *Annal. Math.*, 49:1–42, 1948.
162. E. R. Kolchin. *Differential Algebra and Algebraic Groups.* Academic Press, New York, 1976.
163. V. P. Kostov. Fuchsian systems on CP^1 and the Rieman-Hilbert problem. *C.R. Acad. Sci. Paris*, 315:143–148, 1992.
164. J. Kovacic. The inverse problem in the Galois theory of differential fields. *Annal. Math.*, 89:583–608, 1969.
165. J. Kovacic. On the inverse problem in the Galois theory of differential fields. *Annal. Math.*, 93:269–284, 1971.
166. J. Kovacic. An Eisenstein criterion for noncommutative polynomials. *Proceedings of the American Mathematical Society*, 34(1):25–29, 1972.
167. J. Kovacic. An algorithm for solving second order linear homogeneous differential equations. *J. Symb. Comput.*, 2:3–43, 1986.
168. J. Kovacic. Cyclic vectors and Picard-Vessiot theory. Technical report, Prolifics, Inc., 1996.
169. S. Lang. On quasi-algebraic closure. *Annal. Math.*, 15:373–290, 1952.
170. S. Lang. *Algebra.* Addison Wesley, New York, 3rd edn., 1993.
171. I. Lappo-Danilevskii. *Mémoires sur la théorie des systèmes des équations différentielles linéaires.* Chelsea, New York, 1953.
172. A. H. M. Levelt. Jordan decomposition for a class of singular differential operators. *Archiv Math.*, 13(1):1–27, 1975.
173. A. H. M. Levelt. Differential Galois theory and tensor products. *Indagationes Mathematicae, New Series*, 1(4):439 – 450, 1990.
174. Z. Li and F. Schwarz. Rational solutions of Riccati-like partial differential equations. *J. Symb. Comput.*, 31(6):691–716, 2001.
175. M. Loday-Richaud. Calcul des invariants de Birkhoff des systèmes d'ordre deux. *Funkcialaj Ekvacioj*, 33:161–225, 1990.
176. M. Loday-Richaud. Introduction à la multisommabilité. *Gaz. Math., SMF*, 44:41–63, 1990.
177. M. Loday-Richaud. Stokes phenomenon, multisummability and differential Galois groups. *Annal. Inst. Fourier*, 44(3):849–906, 1994.

178. M. Loday-Richaud. Solutions formelles des systèmes différentiels linéaires méromorphes et sommation. *Expositiones Mathematicae*, 13:116–162, 1995.
179. M. Loday-Richaud. Rank reduction, normal forms and Stokes matrices. Technical report, Université D'Angers, 2001.
180. M. Loday-Richaud and G. Pourcin. On index theorems for linear differential operators. *Annal. Inst. Fourier*, 47(5):1379–1424, 1997.
181. A. Magid. Finite generation of class groups of rings of invariants. *Proc. Am. Math. Soc.*, 60:45–48, 1976.
182. A. Magid. *Module Categories of Analytic Groups*, Vol. 81 of *Cambridge Tracts in Mathematics*. Cambridge University Press, Cambridge, 1982.
183. A. Magid. *Lectures on Differential Galois Theory*. University Lecture Series. American Mathematical Society, 1994. 2nd edn.
184. P. Maisonobe. $\mathcal{D}$-modules: an overview towards effectivity. In E. Tournier, editor, *Computer Algebra and Differential Equations*, Vol. 193 of *London Mathematical Society Lecture Notes Series*, pp. 21–56. Cambridge University Press, 1994.
185. H. Majima. On the representation of solutions of completely integrable Pfaffian systems with irregular singular points. *Proc. Sem. at RIMS (Kokyuroku), Kyoto Univ.*, 438, 1981. (in Japanese).
186. H. Majima. *Asymptotic Analysis for Integrable Connections with Irregular Singular Points*, Vol. 1075 of *Lecture Notes in Mathematics*. Springer-Verlag, Heidelberg, 1984.
187. B. Malgrange. Sur les points singuliers des équations différentielles. *Ens. Math.*, 20:149–176, 1974.
188. B. Malgrange. Remarques sur les équations différentielles à points singuliers irréguliers. In R. Gérard and J. P. Ramis, editors, *Équations différentielles et systèmes de Pfaff dans le champ complexe*, pages 77–86, 1979. Lecture Notes in Mathematics, Vol. 712.
189. B. Malgrange. Sur la réduction formelle des équations différentielles à singularités irrégulières. Technical report, Grenoble, 1979.
190. B. Malgrange. Sur la réduction formelle des équations différentielles à singularités irrégulières, manuscript, 1979.
191. B. Malgrange. La classification des connexions irrégulières à une variable. In *Mathématiques et Physique*, pp. 381–390. Birkhäuser, 1983. Progress in Mathematics, Vol. 37.
192. B. Malgrange. Regular connections, after Deligne. In *Algebraic D-Modules,* Borel et al., chapter IV, pp 151–172. Academic Press, 1987.
193. B. Malgrange. *Équations différentielles à coefficients polynomiaux*. Birkhäuser, Boston, 1991.
194. B. Malgrange. Connexions méromorphes. In J.-P. Brasselet, editor, *Singularities*, Vol. 201 of *London Mathematical Society Lecture Note Series*, pp. 251–261. Cambridge University Press, 1994.
195. B. Malgrange. Sommation des séries divergentes. *Exposition. Math.*, 13(2-3):163–222, 1995.
196. B. Malgrange. Connexions méromorphes, II: le réseau canonique. *Inv. Math.*, 124:367–387, 1996.
197. B. Malgrange. Le groupoïde de Galois d'une feuilletage. *Monographie de l'Enseignement Mathématique*, 38:465–501, 2001.
198. B. Malgrange and J.-P. Ramis. Fonctions multisommables. *Ann. Inst. Fourier, Grenoble*, 42(1-2):353–368, 1992.
199. D. Marker and A. Pillay. Differential Galois theory, III: Some inverse problems. *Illinois J. Math.*, 41(43):453–461, 1997.

200. M. F. Marotte. Les équations différentielles linéaires et la théorie des groupes. *Annales de la Faculté des Sciences de Toulouse*, 12(1):H1 – H92, 1887.
201. J. Martinet and J.-P. Ramis. Problèmes de modules pour les équations différentielles non linéaire du premier ordre. *Publications Mathématiques de l'IHES*, 55:63–164, 1982.
202. J. Martinet and J.-P. Ramis. Théorie de Galois différentielle et resommation. In E. Tournier, editor, *Computer Algebra and Differential Equations*, pp. 115–214. Academic Press, 1989.
203. J. Martinet and J.-P. Ramis. Elementary acceleration and multisummability. *Annales de l'Institut Henri Poincaré, Physique Théorique*, 54(4):331–401, 1991.
204. H. Matsumura. *Commutative Ring Theory*. Cambridge University Press, Cambridge, 1986.
205. B. H. Matzat and M. van der Put. Iterative differential equations and the Abhyankar conjecture. Preprint, 2001.
206. J. Miller. *On Differentially Hilbertian Differential Fields*. PhD thesis, Columbia University, 1970.
207. C. Mitschi. Matrices de Stokes et groupe de Galois des équations hypergéométriques confluentes généralisées. *Pacific J. Math.*, 138(1):25–56, 1989.
208. C. Mitschi. Differential Galois groups and G-functions. In M. F. Singer, editor, *Computer Algebra and Differential Equations*, pp. 149–180. Academic Press, 1991.
209. C. Mitschi. Differential Galois groups of confluent generalized hypergeometric equations: An approach using Stokes multipliers. *Pacific J. Math.*, 176(2):365–405, 1996.
210. C. Mitschi and M. F. Singer. Connected linear groups as differential Galois groups. *J. Algebra*, 184:333–361, 1996.
211. C. Mitschi and M. F. Singer. The inverse problem in differential Galois theory. In B. L. J. Braaksma et al., editor, *The Stokes Phenomenon and Hilbert's 16th Problem*, pp. 185–197. World Scientific, River Edge, NJ, 1996.
212. C. Mitschi and M. F. Singer. On Ramis' solution of the local inverse problem of differential Galois theory. *J. Pure Appl. Algebra*, 110:185–194, 1996.
213. J. J. Morales-Ruiz. *Differential Galois Theory and Non-Integrability of Hamiltonian Systems*. Birkhäuser, Basel, 1999.
214. D. Mumford. *Algebraic Geometry I: Complex Projective Varieties*. Springer Verlag, Berlin, 1976.
215. T. Oaku, N. Takayama, and H. Tsai. Polynomial and rational solutions of holonomic systems. Technical report, to appear in *J. Pure Appl. Algebra*, 2000.
216. C. Okonek, M. Schneider, and H. Spindler. *Vector Bundles on Complex Projective Space*. Progress in Mathematics, 3. Birhäuser, Boston, 1980.
217. K. Okugawa. Basic properties of differential fields of an arbitrary characteristic and the Picard-Vessiot theory. *J. Math. Kyoto Univ.*, 2-3:295–322, 1963.
218. P. Painlevé. Sur les équations différentielles linéaires. *Comptes Rendus de l'Académie des Sciences, Paris*, 105:165–168, 1887.
219. P. Th. Pépin. Méthodes pour obtenir les intégrales algébriques des équations différentielles linéaires du second ordre. *Atti dell'Accad. Pont. de Nuovi Lincei*, 36:243–388, 1881.
220. E. Pflügel. An algorithm for computing exponential solutions of first order linear differential systems. In W. Küchlin, editor, *Proceedings of the 1997 International Symposium on Symbolic and Algebraic Computation (ISSAC'97)*, pp. 164–171. ACM Press, 1997.
221. E. Pflügel. Effective formal reduction of linear differential systems. Technical report, LMC-IMAG, Grenoble, 1999. To appear in Applied Algebra in Engineering, Communication, and Computing.

222. A. Pillay. Differential Galois theory, II. *Ann. Pure Appl. Logic*, 88(4):181–191, 1997.
223. A. Pillay. Differential Galois theory, I. *Illinois J. Math.*, 42(4):678–699, 1998.
224. J. Plemelj. *Problems in the sense of Riemann and Klein*. Interscience, New York, 1964.
225. E. G. C. Poole. *Introduction to the Theory of Linear Differential Equations*. Dover Publications, New York, 1960.
226. M. van der Put. Grothendieck's conjecture for the Risch equation $y' = ay + b$. *Indagationes Mathematicae*, 2001.
227. M. van der Put. Differential equations in characteristic p. *Compositio Mathematica*, 97:227–251, 1995.
228. M. van der Put. Singular complex differential equations: An introduction. *Nieuw Archief voor Wiskunde, vierde serie*, 13(3):451–470, 1995.
229. M. van der Put. Reduction modulo p of differential equations. *Indagationes Mathematicae*, 7:367–387, 1996.
230. M. van der Put. Recent work in differential Galois theory. In *Séminaire Bourbaki: Vol. 1997/1998*, Astérisque. Société Mathématique de France, Paris, 1998.
231. M. van der Put. Galois theory of Differential Equations, Algebraic groups and Lie Algebras. *J. Symb. Comput.*, 28:441–472, 1999.
232. M. van der Put. Symbolic analysis of differential equations. In A. M. Cohen, H. Cuypers, and H. Sterk, editors, *Some Tapas of Computer Algebra*, pp. 208–236. Springer-Verlag, Berlin, 1999.
233. M. van der Put and F. Ulmer. Differential equations and finite groups. *J. Algebra*, 226:920–966, 2000.
234. M. Rabin. Computable algebra, general theory and theory of computable fields. *Trans. Am. Math. Soc.*, 95:341–360, 1960.
235. J.-P. Ramis. About the Inverse Problem in Differential Galois Theory: The Differential Abhyankar Conjecture. Unpublished manuscript; version 16-1-1995.
236. J.-P. Ramis. Dévissage gevrey. *Astérisque*, 59/60:173–204, 1978.
237. J.-P. Ramis. *Théorèmes d'indices Gevrey pour les équations différentielles ordinaires*, Vol. 296 of *Memoirs of the American Mathematical Society*. American Mathematical Society, 1984.
238. J.-P. Ramis. Filtration Gevrey sur le groupe de Picard-Vessiot d'une équation différentielle irrégulière. Informes de Matematica, Preprint IMPA, Series A-045/85, 1985.
239. J.-P. Ramis. Phénomène de Stokes et filtration Gevrey sur le groupe de Picard-Vessiot. *Comptes Rendus Acad. Sci., Paris*, 301:165–167, 1985.
240. J.-P. Ramis. Phénomène de Stokes et resommation. *Comptes Rendus Acad. Sci., Paris*, 301:99–102, 1985.
241. J.-P. Ramis. About the inverse problem in differential Galois theory: Solutions of the local inverse problem and of the differential Abhyankar conjecture. Technical report, Université Paul Sabatier, Toulouse, 1996.
242. J.-P. Ramis. About the inverse problem in differential Galois theory: The differential Abhyankar conjecture. In Braaksma et al., editor, *The Stokes Phenomenon and Hilbert's* 16th *Problem*, pp. 261–278. World Scientific, 1996.
243. J.-P. Ramis and Y. Sibuya. Hukuhara's domains and fundamental existence and uniqueness theorems for asymptotic solutions of Gevrey type. *Asymptotic Anal.*, 2:39–94, 1989.
244. M. Raynaud. Revêtements de la droite affine en characteristique $p > 0$ et conjecture d'Abhyankar. *Inventiones Math.*, 116:425–462, 1994.
245. B. Riemann. Beiträge zur Theorie der durch die Gauss'sche Reihe $F(\alpha, \beta, \gamma, x)$ darstellbaren Funktionen. *Abh. Kon. Ges. Wiss. Göttingen*, VII Math. Classe:A–22, 1857.

246. R. H. Risch. The solution of the problem of integration in finite terms. *Bull. Am. Math. Soc.*, 76:605–608, 1970.
247. J. F. Ritt. *Differential Algebra*, Vol. 33 of *American Mathematical Society Colloquium Publication*. American Mathematical Society, New York, 1950.
248. P. Robba. Lemmes de Hensel pour les opérateurs différentiels, Applications à la réduction formelle des équations différentielles. *L'Enseignement Math.*, 26:279–311, 1980.
249. M. Rosenlicht. Toroidal algebraic groups. *Proc. Am. Math. Soc.*, 12:984–988, 1961.
250. M. Rosenlicht. Initial results in the theory of linear algebraic groups. In A. Seidenberg, editor, *Studies in Algebraic Geometry*, Vol. 20 of *MAA Studies in Mathematics*, pp. 1–18. Mathematical Association of America, Washington, D.C., 1980.
251. N. Saavedra Rivano. *Catégories Tannakiennes*, Vol. 265 of *Lecture Notes in Mathematics*. Springer Verlag, Berlin, 1972.
252. C. Sabbah. Stokes phenomenon in dimension two. In B. L. J. Braaksma et al., editors, *The Stokes Phenomenon and Hilbert's 16th Problem*, pp. 279–293. World Scientific, River Edge, NJ, 1996.
253. C. Sabbah. Équations différentielles à points singuliers irréguliers et phénomène de Stokes en dimension 2. *Astérisque*, 263, 2000.
254. M. Sato, B. Sturmfels, and N. Takayama. *Gröbner Deformations of Hypergeometric Differential Equations*, Vol. 6 of *Algorithms and Computation in Mathematics*. Springer-Verlag, Berlin, 2000.
255. L. Schlesinger. *Handbuch der Theorie der Linearen Differentialgleichungen*. Teubner, Leipzig, 1887.
256. I. Schur. Über Gruppen periodischer Substitutionen. *Sitzber. Preuss. Akad. Wiss.*, pp. 619–627, 1911.
257. F. Schwartz. A factorization algorithm for linear ordinary differential equations. In G. Gonnet, editor, *Proceedings of the 1989 International Symposium on Symbolic and Algebraic Computation (ISSAC 89)*, pp. 17–25. ACM Press, 1989.
258. H. A. Schwarz. Über diejenigen Fälle, in welchen die Gaussische hypergeometrische Reihe eine algebraische Funktion ihres vierten Elements darstellt. *J. Reine Angewandte Math.*, 75:292–335, 1872.
259. J.-P. Serre. Géométrie algébrique et géométrie analytique. *Ann. Inst. Fourier*, 6:1–42, 1956.
260. J.-P. Serre. *Cohomologie Galoisienne*. Number 5 in Lecture Notes in Mathematics. Springer-Verlag, New York, 1964.
261. J.-P. Serre. *Local Fields*, Vol. 67 of *Graduate Texts in Mathematics*. Springer-Verlag, New York-Berlin, 1979.
262. I. Shafarevich. *Basic Algebraic Geometry*. Springer, New York, 2nd edn., 1994.
263. Y. Sibuya. *Linear differential equations in the complex domain: problems of analytic continuation*, Vol. 82 of *Translations of Mathematical Monographs*. American Mathematical Society, Providence, 1990.
264. M. F. Singer. Algebraic solutions of nth order linear differential equations. In *Proceedings of the 1979 Queen's Conference on Number Theory*, pp. 379–420. Queen's Papers in Pure and Applied Mathematics, **59**, 1979.
265. M. F. Singer. Liouvillian solutions of nth order homogeneous linear differential equations. *Am. J. Math.*, 103:661–681, 1981.
266. M. F. Singer. Solving homogeneous linear differential equations in terms of second order linear differential equations. *Am. J. Math.*, 107:663–696, 1985.
267. M. F. Singer. Algebraic relations among solutions of linear differential equations. *Trans. Am. Math. Soc.*, 295:753–763, 1986.

268. M. F. Singer. Algebraic relations among solutions of linear differential equations: Fano's theorem. *Am. J. Math.*, 110:115–144, 1988.
269. M. F. Singer. Liouvillian solutions of linear differential equations with liouvillian coefficients. *J. Symb. Comput.*, 11:251–273, 1991.
270. M. F. Singer. Moduli of linear differential equations on the Riemann sphere with fixed Galois group. *Pacific J. Math.*, 106(2):343–395, 1993.
271. M. F. Singer. Testing reducibility of linear differential operators: a group theoretic perspective. *Appl. Algebra Eng., Commun. Comput.*, 7:77–104, 1996.
272. M. F. Singer and F. Ulmer. Galois groups of second and third order linear differential equations. *J. Symb. Comput.*, 16(3):9–36, 1993.
273. M. F. Singer and F. Ulmer. Liouvillian and algebraic solutions of second and third order linear differential equations. *J. Symb. Comput.*, 16(3):37–73, 1993.
274. M. F. Singer and F. Ulmer. Necessary conditions for liouvillian solutions of (third order) linear differential equations. *Appl. Algebra Eng., Commun., Comput.*, 6(1):1–22, 1995.
275. M. F. Singer and F. Ulmer. Linear differential equations and products of linear forms. *J. Pure Appl. Algebra*, 117:549–564, 1997.
276. M. F. Singer and F. Ulmer. Linear differential equations and products of linear forms, II. Technical report, North Carolina State University, 2000.
277. M. F. Singer. Direct and inverse problems in differential Galois theory. In Cassidy Bass, Buium, editor, *Selected Works of Ellis Kolchin with Commentary*, pp. 527–554. American Mathematical Society, 1999.
278. R. Sommeling. *Characteristic classes for irregular singularities.* Thesis, University of Nijmegen, Nijmegen, The Netherlands, 1993.
279. S. Sperber. On solutions of differential equations which satisfy certain algebraic relations. *Pacific J. Math.*, 124:249–256, 1986.
280. T. A. Springer. *Linear Algebraic Groups, Second Edition*, Vol. 9 of *Progress in Mathematics*. Birkhäuser, Boston, 1998.
281. E. Tournier. Solutions formelles d'équations différentielles. Thèse, Faculté des Sciences de Grenoble, 1987.
282. B. M. Trager. *On the Integration of Algebraic Functions.* PhD thesis, MIT, 1984.
283. C. Tretkoff and M. Tretkoff. Solution of the inverse problem in differential Galois theory in the classical case. *Am. J. Math.*, 101:1327–1332, 1979.
284. S. P. Tsarev. Problems that appear during factorization of ordinary linear differential operators. *Program. Comput. Softw.*, 20(1):27–29, 1994.
285. S. P. Tsarev. An algorithm for complete enumeration of all factorizations of a linear ordinary differential operator. In Lakshman Y. N., editor, *Proceedings of the 1996 International Symposium on Symbolic and Algebraic Computation*, pp. 226–231. ACM Press, 1996.
286. S. P. Tsarev. Factorization of linear partial differential operators and Darboux integrability of nonlinear PDE's. *SIGSAM Bull.*, 32(4):21–28, 1998.
287. S. P. Tsarev. Factorzation of overdetermined systems of linear PDE's with finite dimensional solution space. Technical report, Krasnoyarsk State Pedagogical University, 2000.
288. H. Turrittin. Convergent solutions of ordinary differential equations in the neighborhood of an irregular singular point. *Acta Math.*, 93:27–66, 1955.
289. H. Turrittin. Reduction of ordinary differential equations to the Birkhoff Canonical form. *Trans. A.M.S.*, 107:485–507, 1963.
290. F. Ulmer. On liouvillian solutions of differential equations. *Appl. Algebra Eng., Commun. Comput.*, 2:171–193, 1992.

291. F. Ulmer. Irreducible linear differential equations of prime order. *J. Symb. Comput.*, 18(4):385–401, 1994.
292. F. Ulmer and J.-A. Weil. A note on Kovacic's algorithm. *J. Symb. Comput.*, 22(2):179–200, 1996.
293. H. Umemura. Birational automorphism groups and differential equations. *Nagoya Math. J.*, 119:1–80, 1990.
294. H. Umemura. Differential Galois theory of infinite dimension. *Nagoya Math. J.*, 144:59–135, 1996.
295. H. Umemura. Galois theory of algebraic and differential equations. *Nagoya Math. J.*, 144:1–58, 1996.
296. H. Umemura. Lie-Drach-Vessiot Theory. In *CR-Geometry and Overdetermined Systems*, Vol. 25 of *Advanced Studies in Pure Mathematics*, pp. 364–385. North-Holland, 1999.
297. V. S. Varadarajan. Meromorphic differential equations. *Expositiones Mathematicae*, 9(2):97–188, 1991.
298. V. S. Varadarajan. Linear meromorphic differential equations: A modern point of view. *Bull. (New Series) Am. Math. Soc.*, 33(1):1–42, 1996.
299. E. Volcheck. *Resolving Singularities and Computing in the Jacobian of a Plane Algebraic Curve*. PhD thesis, UCLA, 1994.
300. E. Volcheck. Testing torsion divisors for symbolic integration. In Wolfgang W. Küchlin, editor, *The ISSAC'96 Poster Session Abstracts*, Zurich, Switzerland, July 24–26 1996, ETH.
301. W. Wasow. *Asymptotic Expansions for Ordinary Differential Equations*. Interscience Publications, Inc, New York, 1965.
302. W. C. Waterhouse. *Introduction to Affine Group Schemes*, Vol. 66 of *Graduate Texts in Mathematics*. Springer-Verlag, New York, 1979.
303. B. A. F. Wehrfritz. *Infinite Linear Groups*. Ergebnisse der Mathematik. Springer-Verlag, Berlin, 1973.
304. J. S. Wilson. *Profinite Groups*, Vol. 19 of *London Mathematical Society Monographs, New Series*. Clarendon Press, Oxford, 1998.
305. J. Wolf. *Spaces of Constant Curvature, Third Edition*. Publish or Perish, Inc., Boston, 1974.
306. A. Zharkov. Coefficient fields of solutions in Kovacic's algorithm. *J. Symb. Comput.*, 19(5):403–408, 1995.

List of Notation

Index

《国外数学名著系列》(影印版)

(按初版出版时间排序)

1. 拓扑学 I：总论　S. P. Novikov (Ed.)　2006.1
2. 代数学基础　Igor R. Shafarevich　2006.1
3. 现代数论导引 (第二版)　Yu. I. Manin　A. A. Panchishkin　2006.1
4. 现代概率论基础 (第二版)　Olav Kallenberg　2006.1
5. 数值数学　Alfio Quarteroni　Riccardo Sacco　Fausto Saleri　2006.1
6. 数值最优化　Jorge Nocedal　Stephen J. Wright　2006.1
7. 动力系统　Jürgen Jost　2006.1
8. 复杂性理论　Ingo Wegener　2006.1
9. 计算流体力学原理　Pieter Wesseling　2006.1
10. 计算统计学基础　James E. Gentle　2006.1
11. 非线性时间序列　Jianqing Fan　Qiwei Yao　2006.1
12. 函数型数据分析 (第二版)　J. O. Ramsay　B. W. Silverman　2006.1
13. 矩阵迭代分析 (第二版)　Richard S. Varga　2006.1
14. 偏微分方程的并行算法　Petter Bjørstad　Mitchell Luskin(Eds.)　2006.1
15. 非线性问题的牛顿法　Peter Deuflhard　2006.1
16. 区域分解算法：算法与理论　A. Toselli　O. Widlund　2006.1
17. 常微分方程的解法 I：非刚性问题 (第二版)　E. Hairer　S. P. Nørsett　G. Wanner　2006.1
18. 常微分方程的解法 II：刚性与微分代数问题 (第二版)　E. Hairer　G. Wanner　2006.1
19. 偏微分方程与数值方法　Stig Larsson　Vidar Thomée　2006.1
20. 椭圆型微分方程的理论与数值处理　W. Hackbusch　2006.1
21. 几何拓扑：局部性、周期性和伽罗瓦对称性　Dennis P. Sullivan　2006.1
22. 图论编程：分类树算法　Victor N. Kasyanov　Vladimir A. Evstigneev　2006.1
23. 经济、生态与环境科学中的数学模型　Natali Hritonenko　Yuri Yatsenko　2006.1
24. 代数数论　Jürgen Neukirch　2007.1
25. 代数复杂性理论　Peter Bürgisser　Michael Clausen　M. Amin Shokrollahi　2007.1
26. 一致双曲性之外的动力学：一种整体的几何学的与概率论的观点　Christian Bonatti　Lorenzo J. Díaz　Marcelo Viana　2007.1
27. 算子代数理论 I　Masamichi Takesaki　2007.1
28. 离散几何中的研究问题　Peter Brass　William Moser　János Pach　2007.1
29. 数论中未解决的问题 (第三版)　Richard K. Guy　2007.1
30. 黎曼几何 (第二版)　Peter Petersen　2007.1
31. 递归可枚举集和图灵度：可计算函数与可计算生成集研究　Robert I. Soare　2007.1
32. 模型论引论　David Marker　2007.1

33. 线性微分方程的伽罗瓦理论　Marius van der Put　Michael F. Singer　2007.1
34. 代数几何 II：代数簇的上同调，代数曲面　I. R. Shafarevich (Ed.)　2007.1
35. 伯克利数学问题集 (第三版)　Paulo Ney de Souza　Jorge-Nuno Silva　2007.1
36. 陶伯理论：百年进展　Jacob Korevaar　2007.1

37. 同调代数方法 (第二版)　Sergei I. Gelfand　Yuri I. Manin　2009.1
38. 图像处理与分析：变分，PDE，小波及随机方法　Tony F. Chan　Jianhong Shen　2009.1
39. 稀疏线性系统的迭代方法　Yousef Saad　2009.1
40. 模型参数估计的反问题理论与方法　Albert Tarantola　2009.1
41. 常微分方程和微分代数方程的计算机方法　Uri M. Ascher　Linda R. Petzold　2009.1
42. 无约束最优化与非线性方程的数值方法　J. E. Dennis Jr.　Robert B. Schnabel　2009.1
43. 代数几何 I：代数曲线，代数流形与概型　I. R. Shafarevich (Ed.)　2009.1
44. 代数几何 III：复代数簇，代数曲线及雅可比行列式　A. N. Parshin　I. R. Shafarevich (Eds.) 2009.1
45. 代数几何 IV：线性代数群，不变量理论　A. N. Parshin　I. R. Shafarevich (Eds.)　2009.1
46. 代数几何 V：Fano 簇　A. N. Parshin　I. R. Shafarevich (Eds.)　2009.1
47. 交换调和分析 I：总论，古典问题　V. P. Khavin　N. K. Nikol'skij (Eds.)　2009.1
48. 复分析 I：整函数与亚纯函数，多解析函数及其广义性　A. A. Gonchar　V. P. Havin　N. K. Nikolski (Eds.)　2009.1
49. 计算不变量理论　Harm Derksen　Gregor Kemper　2009.1
50. 动力系统 V：分歧理论和突变理论　V. I. Arnol'd (Ed.)　2009.1
51. 动力系统 VII：可积系统，不完整动力系统　V. I. Arnol'd, S. P. Novikov (Eds.)　2009.1
52. 动力系统 VIII：奇异系统 II：应用　V. I. Arnol'd (Ed.)　2009.1
53. 动力系统 IX：带有双曲性的动力系统　D. V. Anosov (Ed.)　2009.1
54. 动力系统 X：旋涡的一般理论　V. V. Kozlov　2009.1
55. 几何 I：微分几何基本思想与概念　R. V. Gamkrelidze (Ed.)　2009.1
56. 几何 II：常曲率空间　E. B. Vinberg (Ed.)　2009.1
57. 几何 III：曲面理论　Yu. D. Burago　V. A. Zalgaller (Eds.)　2009.1
58. 几何 IV：非正规黎曼几何　Yu. G. Reshetnyak (Ed.)　2009.1
59. 几何 V：最小曲面　R. Osserman (Ed.)　2009.1
60. 几何 VI：黎曼几何　M. M. Postnikov　2009.1
61. 李群与李代数 I：李理论基础，李交换群　A. L. Onishchik (Ed.)　2009.1
62. 李群与李代数 II:李群的离散子群,李群与李代数的上同调　A. L. Onishchik　E. B. Vinberg (Eds.)　2009.1
63. 李群与李代数 III：李群与李代数的结构　A. L. Onishchik　E. B. Vinberg (Eds.)　2009.1
64. 经典力学与天体力学中的数学问题　Vladimir I. Arnold　Valery V. Kozlov　Anatoly I. Neishtadt　2009.1
65. 数论 IV：超越数　A. N. Parshin　I. R. Shafarevich (Eds.)　2009.1
66. 偏微分方程 IV：微局部分析和双曲型方程　Yu. V. Egorov　M. A. Shubin (Eds.)　2009.1
67. 拓扑学 II：同伦与同调，经典流形　S. P. Novikov　V. A. Rokhlin (Eds.)　2009.1